Fundamentals of World Regional Geography

3e

Joseph J. Hobbs
University of Missouri, Columbia

Cartography by Andrew Dolan

BROOKS/COLE
CENGAGE Learning

Australia • Brazil • Japan • Korea • Mexico • Singapore • Spain • United Kingdom • United States

Fundamentals of World Regional Geography, Third Edition
Joseph J. Hobbs

Sr. Acquisitions Sponsoring Editor: Aileen Berg

Developmental Editor: Liana Sarkisian

Editorial Assistant: Margaux Cameron

Media Editor: Alexandria Brady

Marketing Manager: Jack B. Cooney

Marketing Program Manager: Darlene Macanan

Marketing Coordinator: Lianne Ramsauer

Content Project Manager: Hal Humphrey

Art Director: Pam Galbreath

Manufacturing Planner: Rebecca Cross

Production Service: Teresa Christie, MPS North America

Rights Acquisitions Specialist: Roberta Broyer

Photo Researcher: Jeremy Glover, Bill Smith Group

Text Researcher: Sue C. Howard

Cartographer: Andrew Dolan

Cover Designer: William Stanton

Cover Image: Mont-Saint-Michel, France; © John Harper/Corbis

For product information and technology assistance, contact us at **Cengage Learning Customer & Sales Support, 1-800-354-9706.**

For permission to use material from this text or product, submit all requests online at **www.cengage.com/permissions.** Further permissions questions can be e-mailed to **permissionrequest@cengage.com.**

Library of Congress Control Number: 2012937193

ISBN-13: 978-1-133-11378-2

ISBN-10: 1-133-11378-8

Brooks/Cole
20 Davis Drive
Belmont, CA 94002-3098
USA

Cengage Learning is a leading provider of customized learning solutions with office locations around the globe, including Singapore, the United Kingdom, Australia, Mexico, Brazil, and Japan. Locate your local office at **www.cengage.com/global.**

Cengage Learning products are represented in Canada by Nelson Education, Ltd.

To learn more about Brooks/Cole, visit **www.cengage.com/brookscole.**

Purchase any of our products at your local college store or at our preferred online store **www.CengageBrain.com.**

Printed in Canada
2 3 4 5 6 7 15 14 13

For Mom
Who has been with me for every step
On every journey

With love

Brief Contents

Contents

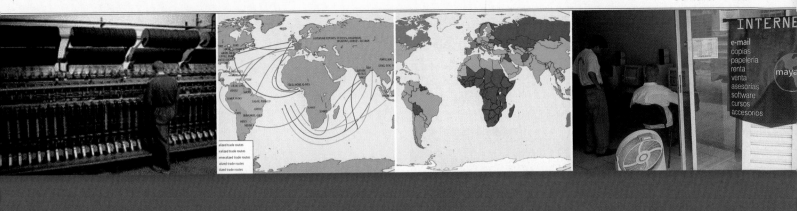

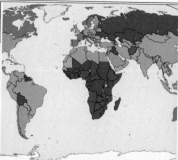

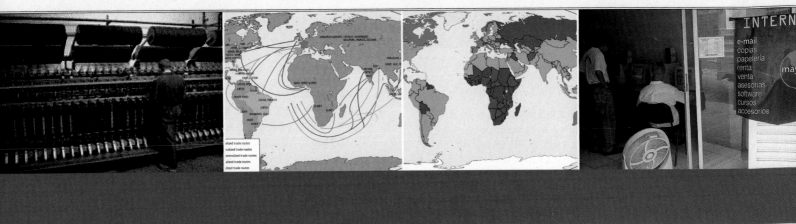

Maps

Preface

To appreciate how the world works today, you will benefit from a solid grounding in the environmental, cultural, historic, economic, and geopolitical contexts of the world's regions and nations. *Fundamentals of World Regional Geography* establishes that foundation for you and offers you an opportunity to explore in more detail the events, issues, and landscapes of your world.

My aim in writing the book is to make the world feel *accessible* to you. I hope that no place or issue will seem too far away or strange. I encourage you to think and talk more about what you read here, including through social media. I've started a Twitter site for us all to communicate with one another about what's happening in the world: follow me on Twitter: *@worldprof*

Chapters 1 through 3 provide the basic concepts, tools, and vocabulary for world regional geography. In the first chapter, geography's uniquely spatial approach to the world is introduced, along with some of the discipline's milestone concepts and its considerable career possibilities—especially those growing from the "geospatial revolution." The second chapter covers the essential characteristics of the world's physical processes and how some of them have been altered by human activity. Climate change and the treaties to control it have a prominent role in that chapter. Chapter 3 traces the modification of landscapes by human actions, describes trends and projections of population growth, and considers efforts to slow destructive trends in resource use.

Then come eight chapters exploring the world's regions through a consistent, thematic approach focusing in turn on five elements: "Area and Population," "Physical Geography and Human Adaptations," "Cultural and Historical Geographies," "Economic Geography," and "Geopolitical Issues." The final section of each chapter, entitled "Regional Issues and Landscapes," contains a selection of short studies of critical problems in global affairs—for example, the Palestinian-Israeli conflict—and exemplary or important problems in human or physical geography, such as ethnicity and petroleum in Nigeria, and the sources and impacts of deforestation in the Amazon Basin. The book is built for a one- or two-semester course. Instructors and students have the freedom to use some or all of content in a given chapter. If time is limited, you can concentrate on the five thematic elements of the chapter and limit the case studies selected from the Regional Issues and Landscapes section. It does not matter what order you read the book in; after the three introductory chapters, no chapter presumes that you have read any other chapter.

New to the Third Edition

Both longtime and first-time users of *Fundamentals of World Regional Geography* should be pleased with this text. I have tried to make the writing more *accessible* than ever. I use first person fairly often so that students can recognize the geographer behind the writing.

- I have written some new *Perspectives from the Field* essays, describing a meeting with the leader of a "Cargo Cult" in Vanuatu and observing landscape changes in Southeast Asia over a period of decades, for example.

- The visual layout is appealing. Cartographer Andrew Dolan builds on his reputation as the best cartographer of the world regional geography texts. For this edition's world maps he uses the Winkel-Trippel projection for the first time, to excellent effect. Also for the first time, I have dipped into my extensive collection of digital photographs. Since the second edition, I have researched, travelled, and photographed across Oceania, South and East Asia, and Latin America.

- There is more *geospatial* content throughout, with extensive use of remotely-sensed imagery and practical applications of GIS.

- For the first time I will correspond regularly with students and professors, especially to track issues related to the book. With the twitter handle *@worldprof*, I will share events and insights with readers.

- The book has, of course, been thoroughly updated. Since the second edition was published, the world has experienced the Great Recession. The economic fallout is a common thread throughout the book. The continued ascendance of China is another pervasive theme.

- I have made it easier to understand and use the important "margin cross-reference" feature which stitches common strands together all across the world.

- A new feature called "Try It" encourages students to use their own critical thinking and problem-solving skills to address a wide range of issues. The student may do these individually or in group exercises assigned by the instructor.

- More attention has been given to solutions, jumping from the principles of sustainable development to specific applications, especially in the developing world.

- I work purposefully to be objective, always trying to present pros and cons of controversial issues (e.g., of large dams) and multiple perspectives of complex political problems (e.g., of the Arab/Israeli conflict).

- Major metropolitan area populations are in tables rather than in the text, where they would interrupt the flow of prose.

- The Arctic Ocean appears to be opening up to trade and minerals exploration as seasonal and permanent sea ice coverage shrinks. In several chapters, we follow the race for Arctic routes and resources.

- I continue using the pattern of five geographic themes for each region: Area and Population; Physical Geography and Human Adaptations; Cultural and Historical Geographies; Economic Geography; and Geopolitical Issues. The book's signature emphasis on geopolitics is as strong as ever. This edition has an enhanced focus on Economic Geography, introducing students to such blocs as the "PIGS" of Europe and the emerging market "BRICS." Conventional and alternative energy sources receive considerable attention in each region.

Chapter-Specific Changes

- Chapter 1, "Objectives and Tools of World Regional Geography," introduces "the Geospatial Revolution," with discussion of GIS and Remote Sensing. Students are encouraged to use Google Earth habitually. The AAG's exciting forecasts about job growth in Geography are shared.

- Chapter 2, "Physical Processes and World Regions," introduces the concept of the Anthropocene and includes a useful new map on the human "footprint." The term *global warming* is largely replaced by the more nuanced and variable *climate change*. Measures for adaptation to and mitigation of climate change are reviewed. There is more information on marine resources. I have removed the detailed information on weather fronts; profs can introduce these with other materials if they wish.

- Chapter 3, "Human Processes and World Regions," digs more deeply into the paradoxes of population, including how "underpopulation" as well as overpopulation can threaten economic well-being.

- Chapter 4 on Europe helps students make sense of the "Euromess:" the continent's sovereign debt crisis, problems with the Euro, and threats to the integrity of the European Union itself. Europe's growing immigration problems, especially in the south, are examined in detail. How can Europe restrict movement while widening Schengenland?

- Chapter 5, "Russia and the Near Abroad," sees Russia deepening its dependence on petroleum while Vladimir Putin consolidates his hold on power. The two are closely tied in what is described as "Putinomics." Russia continues to use oil and natural gas as bludgeons in foreign policy, through a geographic cat's cradle of pipelines.

- Chapter 6 is about the Middle East and North Africa. The major innovation here is a thorough analysis of the Arab Spring, including both the broad underlying themes and fallout, and an individual country survey showing how the revolution played out differently. We examine the near-death and resurrection of Iraq's southern marshes. Iran's prospective nuclear weapons development and the potential for preemptive military action are described. By popular demand, Afghanistan has been moved out of the Middle East and into Asia.

- Chapter Seven, the book's longest (because it covers the most populous region), is now "South and East Asia" rather than "Monsoon Asia." Villagers still outnumber urbanites here, but in anticipation of a rapid change in that equation there is attention to Asia's megacities. I discuss the phenomenon of "language death" due partly to computerized communication, and also the potential of computers to save endangered languages.

- Chapter 8 takes us across Oceania and Antarctica. The "Lucky Country" of Australia continues on it lucky streak on the back of ever-growing commodities demand in China. The Alliance of Small Island States is more alarmed than ever over rising sea levels.

- Chapter 9 on Sub-Saharan Africa portrays the continent at its most hopeful state is decades. Conflicts have subsided. Ironically, some of the greatest threats come from having too much wealth in oil and other natural resources: this is the so-called "resource curse." Technologies are growing quickly in Africa. Mobile phones are providing a shortcut to the information superhighway, and African software developers have contributed an important crowdsourcing tool, Ushahidi.

- Chapter 10 on Latin America describes the unspeakable tragedy of Haiti's 2010 earthquake and the world's other seismically prone cities like Port-au-Prince that are "rubbles in waiting." There are new discussions of religious movements, including those of Santeria and Maria Lionza. Brazil's impressive recent economic growth, based in part on newly discovered offshore oil reserves, is described.

- Chapter 11's disccusion of United States and Canada depicts two developed countries that still have their doors open to immigrants. In the United States, we had a 2010 census data form for 63 distinct ethnic/racial groups. The recent booms and busts of the Great Recession are depicted. We look at the phenomenon of "birth tourism" and the slowdown in illegal immigration due to economic recession. We see the Arid West becoming more arid, with some authorities saying this is the old and the new normal. Meteorological natural hazards of many kinds seem to be on the rise. Canada's tar sands and natural gas released by "fracking" have given North America more energy independence than they have enjoyed in decades. The *Deepwater Horizon* accident and its impacts on the Gulf are here. Outsourcing has begun to give way to insourcing (onshoring) and new life in American manufacturing.

Acknowledgments

I am grateful to everyone who encouraged and helped me with the book, especially Cindy, Katie, Lily, and Mom. Andy Dolan is the best cartographer there is. Andy wrote much of glossary and, as always, provided a lot of guidance on what he thought should be added to and removed from the text. The Cengage Learning team was headed up by Acquisitions Editor Aileen Berg and included Development Editor Liana Sarkisian, Project Manager Teresa Christie of MPS North America, Editorial Assistant Margaux Cameron, Media Editor Alexandria Brady, Marketing Manager Jack B. Cooney, Marketing Coordinator Lianne Ramsauer, Marketing Communications Manager Darlene Macanan, Rights Acquisitions Specialist Roberta Broyer, Photo Researcher Jeremy Glover of Bill Smith Group, Text Permissions Researcher Sue C. Howard, and Art Director Pam Galbreath.

Ancillaries

This text is accompanied by a number of ancillary publications to assist instructors and enhance student learning.

Instructor Resources

PowerLecture with JoinIn and ExamView

This one-stop digital library and presentation tool includes preassembled Microsoft® PowerPoint® lecture slides. In addition to a full Instructor's Manual and Test Bank, PowerLecture also includes an image library, animations, videos, blank maps, Google Earth activities, JoinIn™ Student Response System, and ExamView® testing software with all the test items from the printed Test Bank in electronic format, enabling you to create customized tests in print or online.

Online Instructor's Manual and Test Bank

By Chad Garick of Jones County Junior College. Streamline and maximize the effectiveness of your course preparation using such resources as *chapter summaries and outlines, learning objectives, lecture suggestions, key terms, and answers to in-text questions.* This time-saving resource also includes a Test Bank that offers over 70 questions. Available for download on the instructor's companion Web site.

ExamView Computerized Testing Software

Featuring automatic grading, ExamView® allows you to create, deliver, and customize tests and study guides (both print and online) in minutes.

WebTutor Toolbox™

WebTutor™ Toolbox for WebCT™ or Blackboard® provides access to all the content of this text's rich Book Companion Web site from within your course management system. Robust communication tools—such as course calendar, asynchronous discussion, real-time chat, a whiteboard, and an integrated e-mail system—make it easy for your students to stay connected to the course.

Student Resources

Geography CourseMate

The more you study, the better the results. Make the most of your study time by accessing everything you need to succeed in one place. Geography CourseMate includes:

- An interactive eBook with highlighting, note taking, and an interactive glossary
- Interactive learning tools, including:
  - Quizzes
  - Flashcards
  - Videos
  - Animations
  - … and more!

Joe Hobbs received his B.A. at the University of California Santa Cruz and his M.A. and Ph.D. at the University of Texas Austin. He is a professor and chairman of the Department of Geography at the University of Missouri. He is a geographer of the Middle East with many years of field research on Bedouin peoples and natural environments in the deserts of Egypt. Joe's interest in the region grew from his boyhood in Saudi Arabia. His profession in Geography has roots in all the travels and time abroad with his mom and dad. His research in Egypt has been supported by grants from Fulbright, the American Council of Learned Societies, the American Research Center in Egypt, and the National Geographic Society Committee for Research and Exploration. In the 1990s, he served as the team leader of the Bedouin Support Program, a component of the St. Katherine National Park project in Egypt's Sinai Peninsula. In 2007, he led an effort to establish a national plan for environmental management in the United Arab Emirates. From 2009–2012 he worked with a team funded by the Norwegian Research Council, studying the interactions between nomadic pastoralists and acacia trees in Egypt and the Sudan.

Katie Hobbs

His current research interests include indigenous peoples' participation in national parks and other protected areas, and educational and diplomatic relations between Vietnam and the United States.

Joe is the author of the books *Bedouin Life in the Egyptian Wilderness* and *Mount Sinai* (both University of Texas Press), co-author of *The Birds of Egypt* (Oxford University Press), and co-editor of *Dangerous Harvest: Drug Plants and the Transformation of Indigenous Landscapes* (Oxford University Press).

Joe has taught graduate and undergraduate courses in world regional geography, environmental geography, the geography of the Middle East, the geography of caves, and the geography of global current events, the geographies of drugs and terrorism, and a field course on the ancient Maya geography of Belize. He has received the University of Missouri's highest teaching award, the Kemper Fellowship, as well as MU awards for leadership in international education. He has led adventure tours to remote areas in Latin America, Africa, the Indian Ocean, Asia, Europe, and the High Arctic. Joe lives in Missouri with his wife Cindy, daughters Katie and Lily, and lots of animals.

"If a man takes no thought about what is distant, he will find sorrow near at home."
—CONFUCIUS

The Earth at night. In general, the prosperous countries are lit up, while poor countries are darker. The differences between prosperous and poor countries are discussed often in the book.

C. Mayhew & R. Simmon (NASA/GSFC), NOAA/ NGDC, DMSP Digital Archive

Joe Hobbs

1

Objectives and Tools of World Regional Geography

Before you begin reading, there is an important feature of the book that you need to know about: *cross-referencing*. The book is written with global interconnections in mind. Globalization is understandable as a concept, but how exactly does it work? The page and figure numbers in the book's margins serve to tie the diverse strands of global issues together. For example, when you read in Chapter 4 that Europeans generally dislike genetically modified foods, page numbers in the margin lead you to discussions of other countries' policies on GM foods and how genetic modification can affect ecosystems. Likewise, the discussion of how genetic modification of food crops can affect ecosystems has a page number in the margin, leading you back to the discussion of Europeans and GM foods. As you read about China's economic growth and appetite for raw materials, margin numbers will take you to places all around the world where these forces come into play. I hope you will use this feature often and learn a lot from it.

chapter outline

chapter objectives

This chapter should enable you to

- Appreciate the richness of geography
- Identify the six essential elements and the five themes of geography
- Learn some key concepts in geography
- Appreciate the book's overall objectives
- Learn the basic language of maps
- Understand the "geospatial revolution," geographic information systems (GIS), and remote sensing
- See how geographic knowledge is put to work in the job market

1.1 Welcome to World Regional Geography

About every five years, the National Geographic Society does a survey to see how much Americans between the ages of 18 and 24 know about the world. The findings can be discouraging, especially when it comes to the big issues; take knowledge of Afghanistan, for example. From a base inside Afghanistan, al-Qa'ida planned its attacks against the United States on September 11, 2001. The United States responded by opening a war in Afghanistan that would become the longest in American history. This conflict often dominated the news, and almost every American knew or had seen someone deployed to Afghanistan. But where is Afghanistan? In National Geographic's most recent survey, 90 percent of the Americans polled could not locate Afghanistan on a map.

So what? Does it matter if you can't find Afghanistan on a map? Does it matter if you can't even find the United States on a map? Years ago, geography earned a reputation for forcing students to memorize lists of places and facts (•**Figure 1.1**). And the truth is, by themselves, those pieces of knowledge probably mean little. But geography is all about context and connections. Understanding *where* things are makes it much easier to appreciate and answer the *who, what, when, why,* and *how* questions in life, at every scale—from your daily activities to world affairs. Geography always starts with the "where" question, but it is far more interesting and important than its old reputation for memorizing places suggests. Helping you to understand contexts and relationships, geography can help you make better informed judgments and decisions. See the Try It feature on page 3.

Geographic knowledge has the power to transform our lives and contribute to the welfare of our communities and our countries. By the end of this chapter, you will know what geography is, recognize the benefits you might gain from learning world regional geography, understand the organization and objectives of this book, and learn some of the key concepts and tools of geography.

What Is Geography?

Geography, a term first used by the Greek scholar Eratosthenes in the third century B.C.E.,[1] literally means "description of the Earth" but is probably best characterized as "the study of the Earth as the home of humankind." Focusing on interactions between people and the environments in which we live, the modern academic discipline of geography has its roots in the Greek and Roman civilizations and the Scientific Revolution in Europe.

• **Figure 1.1** Geography used to be associated with memorizing mind-numbing facts. Not any more!

Geography has an important niche in modern science. Its role is well summarized by the National Geographic Society's **"six essential elements of geography."** They are as follows:

1. *The World in Spatial Terms.* Geography studies the relationships among people, places, and environments by mapping information about them into a spatial context (*spatial* means "of or relating to space").

2. *Places and Regions.* The identities and lives of individuals and peoples are rooted in particular places and in human constructs called regions.

3. *Physical Systems.* Physical processes shape the Earth's surface and interact with plant and animal life to create, sustain, and modify ecosystems.

4. *Human Systems.* People are central to geography; human activities, settlements, and structures help shape the Earth's surface, and humans compete for control of the Earth's surface.

5. *Environment and Society.* The physical environment is influenced by the ways in which human societies value and use the Earth's physical features and processes.

6. *Uses of Geography.* Knowledge of geography enables people to develop an understanding of the relationships among people, places, and environments over time—that is, of the Earth as it was, is, and might be.

Source: Geography for Life: National Geography Standards 1994, National Geographic Research and Exploration, Washington, DC, pp. 34–35.

Another summary of what geography is all about is known as the **Five Themes of Geography**. The National Council for Geographic Education and the Association of American Geographers developed this list. Because of its effective simplicity, some geographers prefer using it:

1. Location
2. Place
3. Human–Environment Interaction
4. Movement
5. Region

Although brief, these six elements and five themes cover a lot of ground. Geography is probably the most all-encompassing of the social sciences (a point of pride for geographers). Broadly, the discipline has two major branches, **physical geography** and

Think about U.S. Involvement in Afghanistan

Try It

Why has it been difficult for the United States to be involved in Afghanistan? Suggest two or three reasons.

In trying to answer this important question (important partly because of the sacrifices, costs, and risks that came with U.S. involvement), it may be difficult for you *not* to use a map. You will probably consider Afghanistan in its regional context, looking at its neighbors and how it is positioned relative to major oceans and mountain ranges. You may think about some of the connections and interactions between places, cultures, beliefs, natural resources, politics, and other factors related to Afghanistan.

If it is a tough question to answer now, be patient; this book will help you answer it (on pages 240–244)—and many other questions that ask you to think critically about the world today.

After you have considered this question, you may have more empathy for a man who had to make tough choices about U.S. engagement in Afghanistan. Here is what the former U.S. Secretary of Defense Robert Gates told cadets at the U.S. Military Academy at West Point in 2011:

> Any future defense secretary who advises the president to again send a big American land army into Asia or into the Middle East or Africa should "have his head examined," as General [Douglas] MacArthur so delicately put it . . . Just think about the range of security challenges we face right now beyond Iraq and Afghanistan: terrorism and terrorists in search of weapons of mass destruction, Iran, North Korea, military modernization programs in Russia and China, failed and failing states, revolution in the Middle East, cyber-piracy . . . , natural and man-made disasters, and more. And I must tell you, when it comes to predicting the nature and location of our next military engagements, since Vietnam, our record has been perfect. We have never once gotten it right.

You may have to read those last two sentences more than once to appreciate Gates' purposeful irony. He might have wished that American presidents were advised by geographers! Geography is all about "getting it right" when it comes to understanding how the world works. Geographic knowledge of the *where, who, what, when, why,* and *how* can help guide informed decision making at all scales, from how you might get from point A to point B in town, to whether the United States should commit troops to a ground war in Asia.

human geography, each of which has roots and relationships with other disciplines in the social and physical sciences (•Figure 1.2). Geographers often bridge the social and natural sciences in their research, publication, and teaching (another point of pride for them). As you can see in the center of Figure 1.2, where all the components of the discipline converge, geography is almost always concerned with the theme of **human–environment interaction**. This concern has put geographers at the cutting edge of science and policy in the twenty-first century, because so many of the Earth's most pressing problems—climate change, population growth, and hunger, for example—involve the coupling of human and environmental systems.

Geographers' interests in human–environment interaction, and especially in the ways in which people are changing the face of the Earth, go way back. The great German geographer Alexander von Humboldt (1769–1859) began geography's modern era in a series of classic studies on this theme. From field observations in Venezuela, he concluded, "Felling the trees which cover the sides of the mountains provokes in every

Try It   The Geography of Anyplace

Try using the "five themes of geography" to characterize any place. Staying with the U.S.–Afghanistan connection for now, here is an example to work from, using Ground Zero in Manhattan.

Geographic Characteristics of Ground Zero

Location: Lower Manhattan, New York City (with an exact location of north latitude: 40 degrees, 42 minutes, 43 seconds, and west longitude: 74 degrees 00 minutes, 49 seconds (later in the chapter, we will look at latitude and longitude).

Place: Formerly, office buildings and firms at the heart of one of the world's great financial centers (a reason it was targeted for destruction); now, a place of historical significance and collective grief for people of the United States.

Human–Environment Interaction: Lower Manhattan occupies low-lying ground that once was marshy swampland. Construction of the twin towers of the World Trade Center, as well as the buildings erected after the 9/11 attacks, required special foundations to keep the Hudson River's water from pouring in.

Movement: Before 9/11, the daily comings and goings of office workers in the World Trade Center; on 9/11, the diversion of airplanes to target the buildings; after 9/11, the flow of mourners, tourists, and construction crews to the site.

Region: Situated in region of the United States known as the Northeast, in a humid continental climate region (in the next chapter we will look at such physical regions).

You can use these themes to appreciate any place geographically, from the Great Pyramids of Egypt to where you are now. Try it.

• **Figure 1.2** Selected subfields of geography. These are the main subject areas in human geography and physical geography and their links with the most closely related disciplines in the social and natural sciences.

© Cengage Learning 2013

two disasters for future generations: a want of fuel and a scarcity of water."[2] A century and a half later, we are von Humboldt's future generations. Look at some of the most pressing global environmental issues that concern us today: they include deforestation and shortages of fresh water.

In Humboldt's wake, other geographers in Europe and the United States wrote about environmental changes due to deforestation and the expansion of agriculture and industry. The American geographer Carl Sauer (1889–1975) wrote, "We have accustomed ourselves to think of ever expanding productive capacity, of ever fresh spaces of the world to be filled with people, of ever new discoveries of kinds and sources of raw materials, of continuous technical progress operating indefinitely to solve problems of supply. Yet our modern expansion has been affected in large measure at the cost of an actual and permanent impoverishment of the world."[3] These words have a modern ring to them, but Sauer, a geographer at the University of California–Berkeley, wrote them in 1938. Sauer is credited with founding the **landscape perspective** in American geography, based on the method of studying the transformation through time of a **natural landscape** to a **cultural landscape.** Essentially, Sauer challenged us to think of what the world would look like without people and then understand what people have done to reshape the world through time. Geographers are interested in how the forces of nature and culture have been at work in the creation of the **landscape**—the collection of physical and human geographic features on the Earth's surface—and in particular the roles that human ideas, activities, and cultures play in modifying the landscape. **Culture**—the system of values, beliefs, and attitudes that shapes and influences perception and behavior[4]—underlies many of our decisions about how to use and modify the landscape. That is why geographers are so concerned with cultural features such as ethnicity, language, and religion, and why you will learn much about these issues in this book.

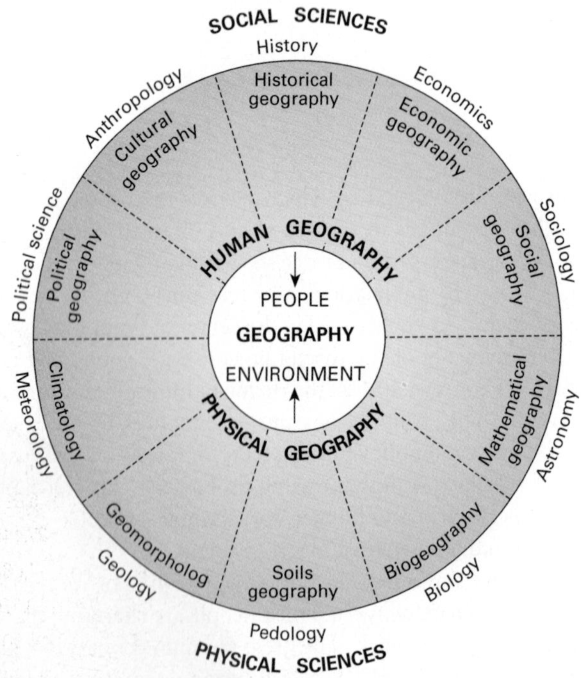

The World Regional Approach to Geography

The world regional approach to geography ranges across the human and physical subfields of geography, synthesizing, simplifying, and characterizing the human experiences of Earth as home. It is impossible to deal with something as large and diverse as our planet without an organizing framework. World regional geography simplifies the task by dividing the world into **regions** (•Figure 1.3 and •Table 1.1). These subdivisions of space are human constructs, not "facts on the

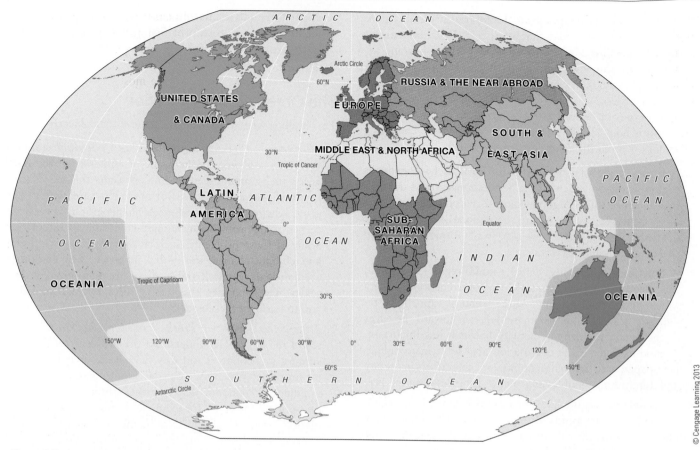

© Cengage Learning 2013

• **Figure 1.3** World regions as identified and used in this book.

Table 1.1   The Major World Regions: Basic Data

Region	Area (thousands; sq mi)	Area (thousands; sq km)	Estimated Population (millions)	Estimated Population Density (sq mi)	Estimated Population Density (sq km)	Annual Rate of Natural Increase (%)	Human Development Index	Urban Population (%)	GNI PPP Per Capita ($US)
Europe	1,959.3	5,072.0	532.3	272	105	0.0	0.873	72	31,070
Russia and the Near Abroad	8,533.2	22,100.8	282.0	33	13	0.2	0.726	64	12,300
Middle East and North Africa	5,416.1	14,027.8	503.4	93	36	1.8	0.650	62	9,460
Monsoon Asia	8,265.3	21,407.1	3,829.2	463	179	1.0	0.621	42	6,080
Oceania	3,306.8	8,564.5	37.3	11	4	1.1	0.822	67	27,440
Sub-Saharan Africa	8,655.2	22,417.2	846.5	98	38	2.5	0.423	32	2,080
Latin America	7,946.2	20,580.7	595.8	75	29	1.2	0.724	79	10,560
North America	8,403.8	21,765.8	346.3	41	16	0.5	0.909	79	44,800
World Regional Totals	**52,485.9**	**135,938.5**	**6,972.8**	**133**	**51**	**1.2**	**0.647**	**51**	**10,240**
United States (for comparison)	3,717.8	9,629.1	311.7	84	32	0.5	0.910	79	45,640

Sources: World Population Data Sheet, Population Reference Bureau, 2011; U.N. Human Development Report, United Nations, 2010; World Factbook, CIA, 2011.

ground." People create and draw boundaries around regions that share relatively similar characteristics. A region is simply a convenience and a generalization, helping us become acquainted with the world and preparing us for more detailed insights. This book recognizes eight world regions; others have more or less.

Three types of regions are recognized by geographers and used in this book. Each is helpful in its own way in conveying information about different parts of the world:

- A **formal region** (also called a **uniform** or **homogeneous region**) is one in which all the population shares a defining trait or set of traits. A good example is a political unit such as a county or a state, where the regional boundaries are defined on a map. Figure 4.1 on page 66 is a formal region map showing the countries of Europe.

- A **functional region** (also called a **nodal region**) is a spatial unit characterized by a central focus on some kind activity (often an economic activity). At the center of a functional region, the activity is most intense, whereas toward the edges of the region the defining activity becomes less important. A good example is the distribution area for a metropolitan newspaper, with the highest numbers of subscribers in the city and diminishing numbers at growing distances from the city.

- A **vernacular region** (or **perceptual region**) is a region that exists in the minds of many people. This region may play an important role in cultural identity, but does not necessarily have official or clear-cut borders. Good examples are the South, the Bible Belt, and the Rust Belt in the United States (•**Figure 1.4**). These regional terms have economic and cultural connotations, but 10 people might have 10 different definitions of the qualities and boundaries of these regions. Vernacular or perceptual regions,

created by individuals and cultures, represent the regional identities that help us organize, simplify, and make sense of the world around us.

The Objectives of This Book

This book is written to help you achieve four objectives:

1. *To understand Earth's problems and potential solutions to these problems.* As the home of humankind, the Earth needs to be maintained. It has problems that need attention and that can be solved or made less threatening. Like geography in general, world regional geography is very concerned with problems in human–environment interaction. Some of these problems, such as overpopulation, poverty, and climate change, are global in scope, whereas others are national, regional, and local, or are manifested at these scales. The next two chapters of the book will introduce you to many of these issues.

2. *To develop a habit of synthesizing information to understand the world.* To understand the problems of Afghanistan (or any country) and why these issues may be relevant to your life, you must consider many factors: resources, population, economic development, ethnicity, history, and geopolitical interests, for example. Is that too much information for you to take in? Not if you have the right tool for filtering and synthesizing the information. World regional geography is that tool. Again, geography is all about context and connections. Using geography's holistic and integrative approach in a regional framework, you will look at relationships between places and people. You will be using information, techniques, and perspectives from both the natural sciences and the social sciences. You will become familiar with issues in political science, history, economics, anthropology, sociology, geology, atmospheric science, and other areas. In pulling these issues and perspectives together and finding the links among them, you will be doing geography.

This approach will be rewarding for you in both tangible and intangible ways. Your overall university experience should become richer as you become better able to link the various courses you are taking. In the future, you will be more likely to interact with the world with new wisdom that can be rewarding both professionally and personally. More complete knowledge of the world—good geography—is also good business. In the competitive environment of the global economy, better understanding of cultures and environments throughout the world helps boost the "bottom line." You may be surprised how much your geographic knowledge will help you in your career, whatever it turns out to be.

3. *To understand current events.* This book is carefully written to set the stage of world events for you. With this book and your professor's guidance, you should become able to read or view news with a much better understanding of the issues underlying world events. Incidents like violence in

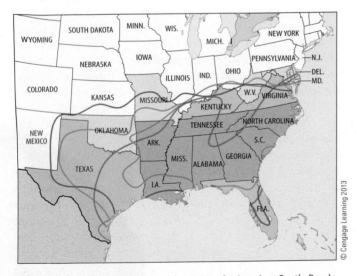

• **Figure 1.4** Definitions of a vernacular region, the American South. Purple shading represents three state-based delineations; colored lines delimit various religious, linguistic, and cultural "Souths." These are just a few of the many different interpretations of the region.

Joe Hobbs

• **Figure 1.5** Where is this place?[5] What clues on the landscape or in the man's appearance might tell you where you are?

the Middle East, earthquakes and tsunamis in the western Pacific, and disease epidemics are not random, unpredictable events. They are rooted in consistent, recognizable problems that have geographic dimensions. You should find it satisfying to feel so well informed about a problem that "suddenly" appears in the news.

4. *To develop the ability to interpret places and "read" landscapes.* As you saw in the exercise on Ground Zero, in doing geography you are concerned both with **space** (the exact placement of locations on the face of the Earth) and with **place** (the imprecise but important physical and cultural contexts of a location). Place is much more subjective than space because, like a vernacular region, it often is defined by the meanings of a particular location. For example, your perceptions of New York City may be very different from those of your friend and may be shaped by personal experience in the "Big Apple" or by photographs or movies you have seen. In this book, there is much discussion of the "sense of place" that individuals and groups have about locations and regions. Perception of place can have a very strong influence on how we make decisions and interact with other people. Perception of place can even have a strong impact on world events. For example, in **Chapter 6** on the Middle East, you will see how Jewish and Muslim perceptions of sacred places located within a few meters of each other in Jerusalem play crucial roles in conflict and peacemaking in the Middle East and beyond.

As an example of how you can define and identify place, study the photograph in •**Figure 1.5,** and try to name the place—or at least the region—where this photograph was taken. What clues in the physical and human geographies of this place help

you locate it? As you use this book and progress through your course, you should be increasingly skilled in looking at images of a place with insight into many of the features that make up its identity. These include the climate, vegetation, and landforms of the physical environment and the language, religion, history, and livelihoods of the people who live in that environment. The skill of interpreting place will also serve you well when you travel. For this place, see note 5 on page 19.

1.2 The Language of Maps

How skilled are you at reading maps? What about imagining or drawing them? (See the Insights feature on page 8.) Did you ever wish you were better at these skills? This section should help you—and it will be useful as you navigate this book.

We turn now to the most important tool that geographers use to explore and explain relationships on our planet: the map. As geographers study people, places, and environments, they usually (but not always) collect and depict information that can be mapped. In other words, they are interested in the **spatial** context of things. As noted in the first essential element of geography, *spatial* means "pertaining to space."

A **map** is a representation of various phenomena over all or a part of the Earth's surface, usually rendered on a flat surface such as paper or a computer monitor. The science of making maps is called **cartography.** There are two basic types of maps: reference maps and thematic maps. **Reference maps** are concerned mainly with depicting the locations of various features, both natural and human-made, on the Earth's surface (road atlas maps are a good example, as are the opening maps for each regional chapter, such as Europe in **Figure 4.1**). **Thematic maps** show the spatial distribution of one or more attributes

INSIGHTS | **Mental Maps**

When someone asks you, "Could you draw me a map of how to get there?" you might quickly draw some lines, write down some street names, talk about some familiar landmarks, and apologize for how crude your map is. Your map would probably end up looking very different from that of another person asked the same question. Our understanding of location is not completely objective. Each of us has a personal sense of space and place and associations with them.

A **mental map**, like a vernacular region, is a collection of personal geographic information that each of us uses to spatially organize the images and facts we have about places, both local and distant. We constantly draw upon that geographic information to make our way through daily life, and are always revising and updating that information as we succeed or fail on our way. Sometimes we use that

information to create actual maps. These maps are not accurate, precise, or scientific, but they portray useful information and tell us much

about the individuals and cultures that create them (• Figure 1.A, B).

A   B

• **Figure 1.A, B** The man in Figure 1.A is Saleh Ali, a Bedouin tribesman from the Eastern Desert of Egypt. During his 75 years of living, he has never used a map, but has drawn on a wealth of personal experience and oral traditions—his mental maps—to find his way across one of the world's most rugged natural environments, the Sahara. On this day, he used his "mental maps" to take the author to a valley called Wadi Abu Haadh, which, he said, was "a place of many vipers." You can see him probing the gravel under an acacia tree—using his mental map of that habitat—looking for one of these poisonous snakes. In Figure 1.B he has found one of the creatures. "Don't step there," he warned me. Can you see the head of this deadly animal?[6] See note 6 on page 19.

---

a given area. Thematic maps can be divided into two categories: quantitative and qualitative. Quantitative thematic maps show the spatial distribution of numerical information (such as population density or income levels, as in Figure 3.14 on page 55), whereas qualitative thematic maps display nonnumeric data (such as the distribution of climates or languages, as in Figure 2.4 on page 24.

As maps are an essential tool in the study of world regional geography, it is important that you know how to read them. The main elements of the "language of maps" are *scale, coordinate systems, projections,* and *symbolization.*

Scale

A map is a reducer; it shrinks an area to the manageable size of a chart, piece of paper, or computer monitor. The amount of reduction appears on the map's **scale**, which shows the actual distance on Earth as represented by a given linear unit on the map. A common way to depict scale is with a fraction or ratio, such as 1:10,000 or 1:10,000,000. In the fraction, one linear unit on the map (for example, 1 inch or 1 centimeter) represents 10,000 or 10,000,000 such real-world units on the ground. A **large-scale map** is one with a relatively large representative fraction (for example, 1:10,000 or even 1:100) that portrays a relatively small area in more detail. A **small-scale map** has a relatively small representative fraction (such as 1:1,000,000 or 1:10,000,000) that portrays a relatively large area in more

generalized terms. Compare the two maps in •Figure 1.6. Figure 1.6a is a small-scale map showing San Francisco and surrounding parts of the Bay Area. Figure 1.6b is a large-scale map that zooms in on part of San Francisco. Remember this inverse relationship: A small-scale map shows a large area, and a large-scale map shows a small area.

Coordinate Systems

Maps cannot convey the subjective meanings associated with place, but they are very effective in conveying information about space and location. In this book, you will be concerned with two kinds of **location**: relative location and absolute location.

Relative location defines a place in relationship to other places and is information you can derive from many maps. It is one of the most basic reference tools of everyday life; you might say you live south of the city, just west of the shopping mall, or next door to a good friend. Relative location will become part of your basic geographic knowledge and your critical thinking about geography. You might look at Figure 10.1. on page 351 to see, for example, how tantalizingly close the landlocked country of Bolivia is to the Pacific Ocean, and wonder, "Why on earth is Bolivia landlocked?!" Understanding the implications of relative location will be quite useful for you in following world affairs; keep your eye on Bolivia's efforts to recover in international courts the sea access it lost to Chile through warfare in 1879.

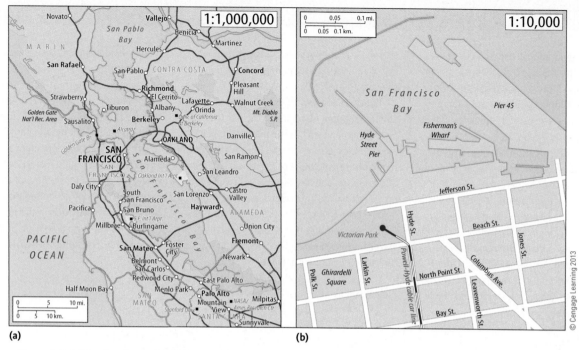

Figure 1.6 **(a)** Small-scale and **(b)** large-scale maps of San Francisco and environs.

Absolute location, also known as **mathematical location,** is crucial for reference maps but is not always necessary for thematic maps. **Coordinate systems** are used to determine absolute location. These coordinate systems use grids consisting of horizontal and vertical lines covering the entire globe. The intersections of these lines create addresses in a global coordinate system, giving each location a specific, unique, and mathematical placement (as appears, for example, as a waypoint in the common GPS device).

The most common coordinate system uses **parallels** of **latitude** and **meridians** of **longitude.** The term *latitude* denotes position with respect to the equator and the poles (see **•Figure 1.7**). Latitude and longitude are measured in **degrees** (°), **minutes** ('), and **seconds** ("). Each degree of latitude, which is made up of 60 minutes, is about 69 miles (111 km) apart; these distances vary a little because Earth is a slightly flattened (oblate) sphere or ellipsoid. Each minute of latitude, which is made up of 60 seconds, is therefore roughly a mile apart. The **equator,** which circles the globe east and west midway between the poles, has a latitude of 0°. All other latitudinal lines are parallel to the equator and to each other, which explains why they are called *parallels.*

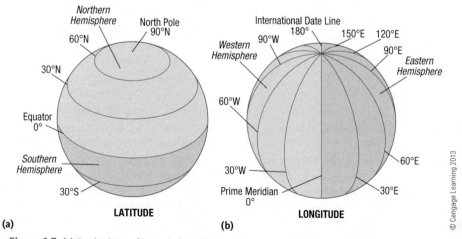

• Figure 1.7 **(a)** Earth's lines of latitude (parallels) in increments of 30 degrees, from the equator (0 degrees) to the North Pole (90 degrees north latitude). **(b)** Earth's lines of longitude (meridians) in increments of 30 degrees.

Among the geographical concepts used throughout this book is that of core location and peripheral location (the region of Europe is subdivided along these lines, for example). Some locales have greater importance in local, regional, or world affairs because they have a central, or core, location relative to others. Other locales are less important because they are situated far from "where the action is." A comparison of two countries, the United Kingdom (U.K.) and New Zealand, provides a good example (• **Figure 1.C**). Both are island countries. Their climates are remarkably similar, although they are in opposite **hemispheres** and are about as far apart as two places on Earth can be. Westerly winds blow

off the surrounding seas, bringing abundant rain and moderate temperatures throughout the year to both.

But there are important differences. The United Kingdom is located in the Northern Hemisphere, which has the bulk of the world's land (it is sometimes described as Earth's **land hemisphere**) and most of its principal centers of population and industry; New Zealand is on the other side of the equator, surrounded by the vast expanses of water in the Southern Hemisphere (sometimes known as Earth's **water hemisphere**) and off the beaten track of the globe's economic activity.

As Figure 1.C illustrates well, the United Kingdom is located near the center of the

world's landmasses. Only a narrow channel separates it from the densely populated industrial areas of western continental Europe. Many major oceanic commercial routes converge on this western seaboard area of Europe. For centuries, the United Kingdom has played a major role in the economic and political development of northwestern Europe. New Zealand, meanwhile, has been a far outpost of that development and history. The United Kingdom has a central or **core location** in the modern framework of human activity on Earth, whereas New Zealand has a **peripheral location**.

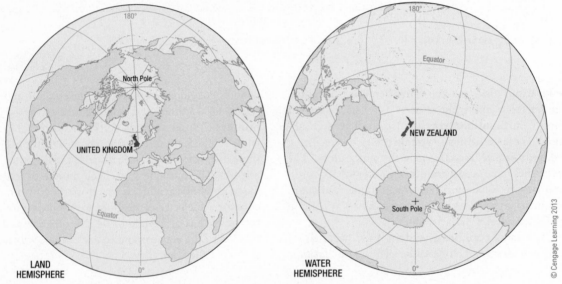

• **Figure 1.C**   In the left map, note how the major landmasses are grouped around the margins of the Atlantic and Arctic Oceans. The British Isles and the northwestern coast of Europe lie in the center of the "land hemisphere," which constitutes 80 percent of the world's total land area and has about 90 percent of the world's population. In the map on the right, New Zealand lies near the center of the opposite hemisphere, or "water hemisphere," which has only 20 percent of the land and about 10 percent of the population.

Every point on a parallel has the same latitude (for example, places on the Equator in both South America and Africa are located at 0° latitude). Places north of the equator are in **north latitude.** Places south of the equator are in **south latitude.** The highest latitude a place can have is 90°N (the **North Pole**) or 90°S (the **South Pole**). Places located between the **Arctic Circle** at 65.56°N and the North Pole, and between the **Antarctic Circle** at 65.56°S and the South Pole, form the most commonly recognized boundaries of the **high latitudes.** Places located between the **Tropic of Cancer**

and the **Tropic of Capricorn**, at 23.44°N and 23.44°S, respectively, are said to be in **low latitudes**. Places occupying an intermediate position with respect to the poles and the equator are said to be in the **middle latitudes.** Incidentally, there are no universally accepted definitions for the boundaries of the high, middle, and low latitudes. The northern half of the Earth between the equator and the North Pole is called the **Northern Hemisphere,** and the southern half between the equator and the South Pole is the **Southern Hemisphere.**

Meridians of longitude are straight lines connecting the poles (see Figure 1.7b). Every meridian runs due north–south. All the meridians converge at the poles and are farthest apart at the equator. Lines of longitude are not the same distance from one another across the globe, so their values vary. At the equator, the distance between lines of longitude is about 69 miles (111 km), whereas at the Arctic Circle it is only about 28 miles (45 km). The meridian most commonly used as a starting point is the one running through the Royal Astronomical Observatory in Greenwich (pronounced "Grenich"), England. Known as the Greenwich Meridian, or **prime meridian,** it has a longitude of 0°. Places east of the prime meridian are in **east longitude;** places west of it are in **west longitude.**

The meridian of 180°, exactly halfway around the world from the prime meridian, is the other dividing line between places east and west of Greenwich. All of the Earth's surface eastward from the prime meridian to 180° is in the **Eastern Hemisphere,** and all of the Earth's surface westward from the prime meridian to 180° is in the **Western Hemisphere.** This meridian of 180° has another purpose in addition to complementing the prime meridian. It serves as the **International Date Line,** where the beginning of one day and the end of another day meet. The Earth has 24 time zones, and there must be a line where the Earth's clock begins and ends. The line has a few zigzags in it for political and practical reasons (especially to fit a country or part of a country in a single time zone; see Figure 8.1, page 285). The date west of the line is one day ahead of the date east of the line. The person traveling west across the International Date Line gains a day (crossing from Monday to Tuesday, for example); and traveling east, loses a day (crossing from Monday to Sunday).

You now have the ability to create and interpret the absolute location of any spot on Earth. Try It—see the feature on this page.

Projections

Because the Earth is spherical, any map created on a flat surface will inevitably have some distortion. **Map projections** are mathematical attempts to minimize this distortion. There are thousands of projections, but there is no single "correct" or "perfect" projection for any particular map. All flat maps have varying amounts of distortion among the five basic properties of a globe: area, shape, scale, distance, and direction. On large-scale maps (such as a map of a city) the distortions are small enough that they may be disregarded for most purposes, but on maps of smaller scales (countries, continents, the entire world) distortion is an increasing problem as scale becomes smaller.

Latitude and Longitude

Here is a useful exercise to ensure that you understand how latitude, longitude, and absolute location work.

On the map in • **Figure 1.D**, the latitude of Madrid, Spain, is approximately 41 degrees north latitude, 4 degrees west longitude (41°N, 4°W).

What are the approximate latitude and longitude coordinates of Oslo, Norway? In which hemispheres (north/south; east/west) is Oslo located[7]?

Understanding absolute location is a simple and indispensable part of the language of maps.

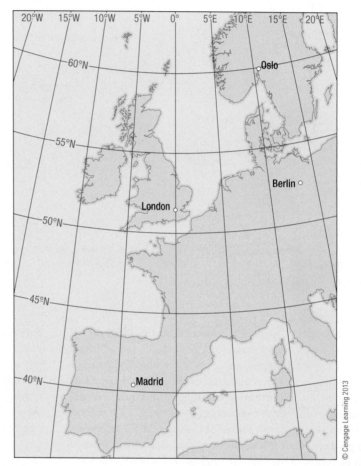

© Cengage Learning 2013

• **Figure 1.D** What are the approximate latitude and longitude coordinates of Oslo, Norway? The answer is in footnote 7 on page 19.

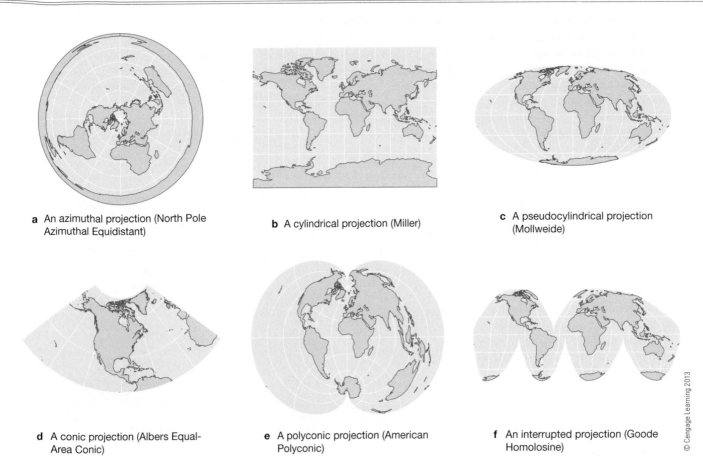

a An azimuthal projection (North Pole Azimuthal Equidistant)

b A cylindrical projection (Miller)

c A pseudocylindrical projection (Mollweide)

d A conic projection (Albers Equal-Area Conic)

e A polyconic projection (American Polyconic)

f An interrupted projection (Goode Homolosine)

© Cengage Learning 2013

• **Figure 1.8** Examples of different map projections.

Most projections work by transferring the features of the spherical Earth onto a "developable surface" (a geometric surface that can be flattened into a plane without tearing or stretching) such as a plane, cylinder, or cone. These projections are referred to as **azimuthal**, **cylindrical**, and **conic**, respectively (there are also notable subsets of these categories, such as pseudocylindrical and polyconic.) See •**Figure 1.8** for examples of these projections. A few projections are interrupted, instead of contiguous; the most famous of these is Goode's Homolosine, which has a "peeled orange" look.

Projections are also classified by which metric property they preserve (or distort the least). **Conformal projections**, which include the famous Mercator Projection discussed in the feature on page 13, preserves shapes well. **Equal-area projections**, as the name suggests, preserve area. But no equal-area projection can preserve shapes, and no conformal projection can preserve areas. **Equidistant projections** preserve distance from a specific point to all other points. Map projections that do not preserve any one metric, or try to distort all properties about equally for aesthetic purposes (making the map "look right"), are called **compromise projections**. The Winkel Tripel projection used for the world maps in this book is a compromise projection.

Another important property of a map is its **orientation**, the relationship between the direction on the map and the corresponding compass directions in reality. Almost all maps place north at the top, but there are sometimes reasons to orient a map differently, and it is possible to present a different perception of geographical space by changing a map's orientation. There are some radical orientations that literally turn the world upside down—see A Different Viewpoint feature on page 14.

Symbolization

Maps allow us to get information, to see patterns of distribution, and to compare these patterns with one another. But no map can be a complete record of a given area. In a process called *cartographic abstraction*, important details are chosen by the cartographer to convey the map's information, while less relevant details are often not shown. Understanding the abstraction process is an important skill to have when using maps: the user knows that no map can be "complete," and many details must be simplified or omitted to keep a map legible. Ironically, all maps must "lie" to some degree to inform their readers. There is a fine book about the use and abuse of maps entitled *How to Lie with Maps*.

The geographic features shown on maps are represented by symbols, such as lines, fills, shapes, colors, and type. For example, reference maps can show different types of roads by using lines of differing width, and make different countries easier to tell apart by using different fill colors for each.

Thematic maps often (but not always) use just one type of symbol to display their data, and they can be classified by which symbol they use. **Choropleth maps**, the most common type of thematic map in this book, display their data by filling in political units with differing colors. A good example of a choropleth map is, again, the map of Europe in Figure 4.1 on page 66. **Isarithmic maps** do not use political units, but instead use lines or bands of color to join points of equal value across the mapped area. A map showing high temperatures or a topographic map showing contour lines are both examples of isarithmic maps, as is the map of vernacular map of the South in Figure 1.4. **Graduated symbol maps** use simple symbols such as circles or bar graphs, scaled proportionally to the quantity of the attribute being mapped. **Dot maps** use dots to represent a stated amount of some phenomenon within a political unit (for example, if one dot equaled 1,000 people, 12 dots in an area would indicate 12,000 people). **Flow maps** use arrows of various widths to indicate the movement of people or goods from one area to another (see the map of people on the move in Figure 11.3 on page 394).

1.3 Geographic Technologies and Careers

To close this chapter, we turn to some of the most innovative tools and breakthroughs in geography and consider how you (or someone you know) might become part of the action in this growing field. This section should also open your eyes to some of the newest tools that will help you have an easier and yet deeper understanding of the world's regional geographies. The opportunities that may open to you by studying geography are abundant and exciting. "We are living in the era of the geographer," Stanford ecologist Hal Mooney wrote.[8]

The Geospatial Revolution

In recent decades, technological advances in the field of geography and related sciences have accelerated so quickly that they can truly be called revolutionary. These advances are not confined to the laboratory or the library. Like information on the Web, they have been democratized: you can use the power of geographic tools on your own computer, tablet, or smartphone.

A relatively new term relates to most of these technological advances: **geospatial**. This adjective means "pertaining to the geographic location and characteristics of natural or constructed features and boundaries on, above, or below the Earth's surface, especially referring to data that is geographic and spatial in nature."[9] Even with simple paper maps, geographers have always worked with geospatial information.

The Mercator Projection

Arguably the most well-known map projection is the one developed by Gerardus Mercator in 1569. Mercator (1512–1594) developed his cylindrical, conformal projection mainly for navigational use. The **Mercator Projection**, which you can see in • **Figure 1.E**, was designed to show lines of constant compass bearing (known as rhumb lines) as straight lines, which was very important to assist sailing vessels in charting their courses. This was a significant improvement over previous projections, and Mercator projections are still used for water navigation today. However, the projection's usefulness at sea makes it largely unsuitable for other purposes, including reference world maps. In order for the rhumb lines to be shown straight, the projection must continually increase the spacing between the parallels away from the Equator. This results in enormous distortions of size approaching the polar areas (in fact, the poles themselves cannot be shown on a Mercator map as they lie at infinity). On a Mercator map, Greenland and Africa appear similarly sized, whereas in reality Africa is about 14 times larger than Greenland!

Despite the objections of cartographers for many decades, the Mercator projection is still a common sight on world maps in school classrooms, TV newscasts, and online mapping services such as Google's. Its straight lines and convenient rectangular shape make it attractive for such uses. The Mercator Projection also makes it possible to show the United States or Europe, despite their northern locations, as being essentially at the "center of the world." This is done by removing most of Antarctica but keeping much of the Arctic in the frame. Indeed, the Mercator projection's exaggeration of mid-northern latitudes made it very popular in the West during the age of European colonialism. The Mercator Projection was gradually accepted as being "authoritative," and some people are still slow to abandon it for less distorted projections. Most atlases and textbooks no longer use Mercator. But it is still prevalent enough in other media that it is important to recognize this projection's drawbacks and to think critically about how a map projection can influence our perceptions of the world.

45

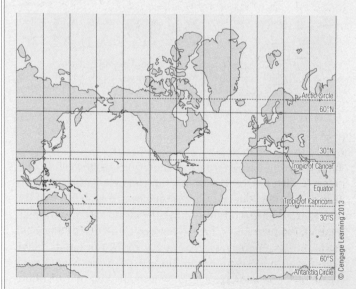

• **Figure 1.E** The Mercator Projection.

A DIFFERENT **viewpoint**

"What's Up: South!"

Now you know how world maps can look different, depending on what projection is used, and how they can be used to produce certain impressions (such as Europe or North America being at the "center of the world"). But imagine *not* being from North America or Europe and getting very tired of how your part of the world is depicted on world maps. Some Australians from "Down Under" have produced a map that challenges the basic premise that "Up Over" people have about the world: that north is at the top. These Aussie cartographers have produced a map called "What's Up: South!" with the world turned upside down (See •**Figure 1.F**). They argue that, like any ball, the Earth has no definitive top and bottom. The conventional choice to put north at the top is just that, a convention. The "South Up" map causes us to see the world in a different way. For one thing, our eyes see a world that is more watery. And we see that Australia and the other "southern" continents now dominate the world! Incidentally, there has been a long history of alternative map orientations. Classical Arab cartographers, producing the world's finest maps of their time, routinely placed south at the map's top (a good example is al-Idrisi's twelfth-century world map). For our purposes in this book, however, unless otherwise noted north is always at the top. This is true of most maps.

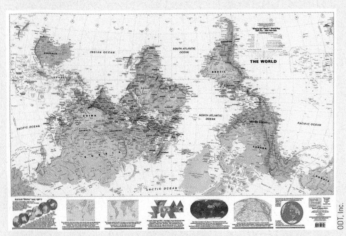

•**Figure 1.F** Who says north has to be at the top of the map? Here is an Australian projection that is literally meant to make you see the world differently.
Source: Copyright © 2006, ODT Maps. All Right Reserved.

information, but underlying them are enormous and sophisticated technologies. Two of the most important geospatial technologies are geographic information systems (GIS) and remote sensing.

GIS—the familiar acronym for **geographic information systems**—represents the cutting edge of geography today. This is one of geography's most powerful tools, and is the field's leading area of growth and employment because of its value in addressing so many real-world problems. GIS is quite complex, but in essence it is a computer-based tool that allows people to create, view, manipulate, analyze, and store geospatial data. These data can be far more than just the latitude–longitude coordinates of a feature on the Earth's surface—a mouse-click on a road, for example, will bring up a box displaying the road's name; its length and width; when it was last repaved; its speed limit; whether it is owned by the city, county, or state; the amount of traffic it receives in a certain time frame; and so forth. These different types of information, called attributes, do not magically appear, but must be created by people who spend an enormous amount of time and energy to compile them from a variety of sources.

GIS data are created and displayed in "layers"—databases storing the locations and attributes of features belonging to a single theme (for example, with layers showing potential customers, the streets on which those customers live, their land ownership, the land's elevation, and land uses in their community, as in •**Figure 1.9**). A GIS can have dozens of layers displaying all kinds of geospatial data for a given area. GIS users can select and highlight certain attributes in various layers that relate to their interests or analysis, while turning off those attributes that are not needed. The ability to selectively

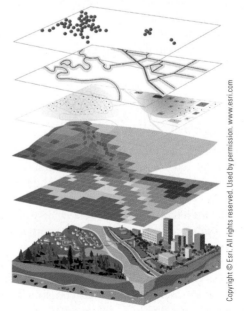

•**Figure 1.9** Geographic information systems (GIS) create and use layers of spatial data. GIS data, images, and models have an enormous range of applications.

But the word geospatial has connotations of powerful computer hardware, software, and information-gathering tools including satellites. These technologies are evolving rapidly. The geospatial revolution is still underway, and it will touch your life in many ways. Have you seen or used a GPS (global positioning system) device, in a car or in the hand? Have you looked up directions on Google Maps or MapQuest? These devices and applications provide you with simple

GIS makes it possible to see geographic patterns, problems, and connections easily and efficiently—a capability that Esri, the world's largest GIS software maker, calls "The Geographic Advantage."[10] One of the best ways to see what GIS can do is to visit Esri's Web site, http://www.gis.com, where you will see an amazing number of ways that GIS is being used all around you and all around the world. Here are a few examples of the geographic advantage:

- *Natural Resource Management.* In forestry, caring for existing and future trees ensures a steady supply of wood for the world's building needs. GIS provides tools to help determine where to cut today and where to seed tomorrow while minimizing negative impacts.

- *Business.* Every day, businesses deliver goods and services to clients all around a city. Each truck driver needs a route of how to most efficiently visit each client. GIS provides tools to create efficient routes that save time and money and reduce pollution.

- *Defense.* In the military, leaders need to understand terrain to make decisions about how and where to deploy their troops, equipment, and expertise. They need to know which areas to avoid and which are safe. GIS provides tools to help get personnel and materials to the place where they can best do their job.

- *Emergency Preparedness and Response.* During floods and hurricanes, emergency response teams save lives and property. GIS provides tools to help locate shelters, distribute food and medicine, and evacuate those in need.

- *Communications and Media.* In telecommunications, when phone service is out, it means part of the network may be disconnected. GIS provides tools to help find out what part of the network is affected and brings that information to the field so workers can get everyone talking again.

- *Planning.* Planners of all kinds—business analysts, city planners, environmental planners, and strategists from all organizations—use GIS to lay out a framework so that growth can occur in a managed way. As you will see in Chapter 7, GIS can even help us plan for a warmer world in which rising sea levels threaten to flood coastal cities and farms.

260

view certain types of geospatial data, as well as to add other data (such as from a handheld GPS device, from a digitized paper map, or from an image taken by an orbiting satellite), allows users to see spatial relationships with extraordinary ease and clarity. The knowledge they gain from understanding geographic relationships is extremely valuable for decision making and explains why GIS is known as a "critical thinking technology" (see Insights, "GIS in Action" on this page). GIS, in other words, is a high-tech example of what geography is all about.

Remote sensing, also known as *Earth observation*, is the science of acquiring information about the Earth's surface without being in direct contact with it. Most remote sensing data is obtained by aerial photography (cameras mounted on airplanes taking pictures of the ground) or by sensors on Earth-orbiting satellites. Remote sensing is not limited to cameras that capture visible light; much important information about processes and features on the surface or in the atmosphere is gleaned from satellite sensors that can "see" other parts of the electromagnetic spectrum such as infrared and microwave wavelengths. Radar (which measures the reflection of satellite-emitted radio waves bouncing off of ground features) and light detection and ranging ([LIDAR], which uses light in the same way) are also examples of remote sensing.

Remote sensing is an exceptionally good tool for helping geographers understand how people and natural processes modify the Earth. This book uses remote sensing to introduce you to many places, patterns, and problems. "A picture is worth a thousand words," the adage says, and a remote sensing image often goes far in describing a landscape. For example, •Figure 1.G (a remotely sensed image taken on April 4, 2010, by a satellite 425 miles above the Earth), is a fine depiction of the cultural landscape of Sendai, in northeast Japan. Just inland from the concrete wave breakers and scalloped shoreline, people have built homes between a greenbelt of trees and an irrigation canal. The canal is a straight and uniform feature—a good indication that it is the handiwork of people rather than of nature. Bridges and streets also stand out as built features. Continuing westward (left) across the image, you see many more homes, including some nestled against a patch of trees (the Japanese prize their forests, and go to great lengths to protect them). Finally, at the far left of the image the long brown rows of empty land are fields for wet rice, which would become green when planted and irrigated from the nearby canal. This satellite image provides an excellent snapshot of how the people of Japan modify nature to suit their needs and desires.

Illustrating dramatically how remote sensing allows us to appreciate landscape change, the next image (Figure 1.H) shows how nature reclaimed Sendai. On March 11, 2011, a magnitude 9.0 earthquake thrust a section of the Earth's crust upward from the seafloor about 80 miles from Sendai. The displaced water roared ashore as devastating

Try It   Reading the Landscape with Remote Sensing

Remote sensing images have a way of bringing out the detective in all of us. Compare Figures G and H, examining and describing the details of this landscape's transformation. What changes has this natural disaster wrought on Sendai?

G

GeoEye/EyeQ

H

GeoEye/EyeQ

• **Figure 1.G, H** The shoreline of Sendai, Japan, before and after the gigantic earthquake and tsunami on March 11, 2011.

tsunami waves as high as 40 feet. The results are breathtaking: almost every feature on the landscape, both natural and cultural, has been destroyed or transformed. See the Try It feature on this page, which invites you to interpret these remote sensing images, and A Closer Look feature on page 17.

254

Careers in Geography

This striking poster in •**Figure 1.10** was produced by the Association of American Geographers (AAG). The employment trends in favor of geography since 2000 have been remarkable. The U.S. Department of Labor has identified geospatial technology as one of the most important emerging and evolving fields in the technology industry. In 2010, this agency projected "much faster than average" growth—more than 20 percent—in jobs for geographers, geoscientists, cartographers, urban and regional planners, and other geographic professionals, through the year 2018.[11]

Much of this job growth will occur in professions in which people feel like they can have a positive impact. The Association of American Geographers had these observations of global trends that are contributing "to a renaissance of geography and its potential for making a difference in society and the world:"

These include globalization at an increasing pace and scale, phenomena that compel greater understanding of the world, places, people, and natural systems that affect us as a planet and as global citizens and consumers. It includes a recent proliferation of geographic technologies, once fairly obscure and now pervasive in our daily lives, such as GPS in cell phones and cars, online mapping at your fingertips, cable news reports using spatial visualizations, and many more applications in modern business and government services that underlie operations, planning, and progress in all sectors everywhere we live and work. It also includes an academic trend toward greater interdisciplinarity, especially a renewed focus on big questions that matter but that require a breadth of knowledge and multiple fields to tackle. Geography's long-standing intellectual traditions in crossing those usual disciplinary boundaries are now better understood, increasingly seen as relevant and more widely respected in scholarly circles. These trends have produced unprecedented growth in the field.[12]

More and more students are graduating with geography degrees—bachelors, masters, and doctoral—and more of them are finding jobs. Here are just a few examples of starting jobs of recent graduates from the University of Texas-Austin Department of Geography. B.A. students became a Nature Conservancy preserve manager, a transportation planner, a natural science teacher, and a geospatial engineer. Masters students became a U.S. Geological Survey water specialist, the chief executive officer of an insurance company in the West Indies, and a vice president for market research in real

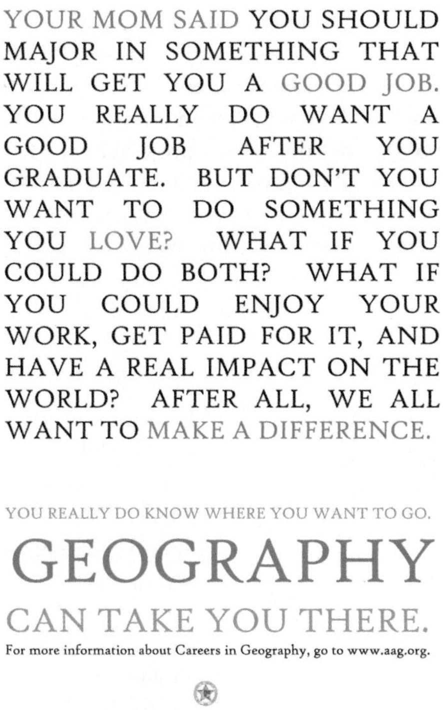

YOUR MOM SAID YOU SHOULD MAJOR IN SOMETHING THAT WILL GET YOU A GOOD JOB. YOU REALLY DO WANT A GOOD JOB AFTER YOU GRADUATE. BUT DON'T YOU WANT TO DO SOMETHING YOU LOVE? WHAT IF YOU COULD DO BOTH? WHAT IF YOU COULD ENJOY YOUR WORK, GET PAID FOR IT, AND HAVE A REAL IMPACT ON THE WORLD? AFTER ALL, WE ALL WANT TO MAKE A DIFFERENCE.

YOU REALLY DO KNOW WHERE YOU WANT TO GO.

GEOGRAPHY
CAN TAKE YOU THERE.
For more information about Careers in Geography, go to www.aag.org.

• **Figure 1.10** Geography is awesome.

Source: Dr. Patricia Solís, Association of American Geographers, copyright registered 2004. Reprint permission granted for educational and dissemination purposes only; please do not reprint, translate, or otherwise alter without express written permission by the author.

estate.[13] Many Ph.D. graduates became geography professors (ask your prof which of the 60 specialty groups of the AAG she or he belongs to), whereas others took executive positions in GIS departments and companies or became decision makers in foundations and research institutions all around the world. But even if you are not a geography major or you have not yet seen an advertisement proclaiming "Geographer Wanted," chances are that many professional opportunities will arise if you use geography's tools to grow knowledge of the world in which you live. This book will help you do just that.

A **closer** LOOK

Google Earth

I would like to recommend two tools that can greatly enhance your understanding and enjoyment of world regional geography. One is a globe. As we saw earlier, flat maps distort representations of the Earth, but a globe puts a representation of the true world in your hands. If you keep a globe handy, and use it as you work your way through the book, it will help you learn more about the Earth and remember what you learn.

The other tool is a virtual globe, map, and geographic information software program like Google Earth or NASA's World Wind. If you have not already done so (and if you have permission to do so), on the PC or Mac, navigate to http://www.earth.google.com (apps are also available for most smartphones and tablets). Download the free Google Earth application, open it, bookmark it, and use it often—*PLEASE!* With Google Earth you are in control of an easy-to-use but incredibly powerful geospatial toolkit. It is so powerful, in fact, that it enables you to become a rather effective spy: you can see not only your home or apartment, but top secret North Korean and Iranian nuclear facilities. A Russian intelligence official grumbled of Google Earth, "Terrorists don't need to reconnoiter their target. Now an American company is working for them!"[14]

By the way, when I photograph a bridge, power plant, or dam (as I do often, thinking I might use the photo in one of my classes, or for this book), I am sometimes detained and questioned by security guards. That is ironic, considering what Google Earth can do.

Google Earth's base map is composed of many thousands of remotely sensed images, from both aerial photography and orbiting satellites, put together as a giant mosaic. And although Google Earth is not a full-fledged GIS, its "Layers" sidebar works in a similar fashion: you can turn on and off certain features like country boundaries and road networks. With the "Weather" layer, you can get real-time views of storm systems and temperatures. How hot or cold is it in Arabia or Siberia right now? It's easy to find out. You might also be interested in the "Global Awareness" layer that lets you explore some of the world's environmental problems. And then there is the "Street View" layer for many locations, infamous for news stories like this: "A woman, checking out a female friend's house on Google Maps, was surprised to see her husband's Range Rover out front. A divorce is underway."[15]

summary

- A recent study showed that Americans generally have poor knowledge of world geography.

- There are six essential elements of the national geography standards: the world in spatial terms, places and regions, physical systems, human systems, environment and society, and the uses of geography.

- Geography means "description of the Earth" and is also defined as "the study of the Earth as the home of humankind."

- The four main objectives of the text are for readers to understand Earth's problems and potential solutions for these problems; develop a habit of synthesizing information to understand the world; understand current events; and develop the ability to interpret places and "read" landscapes.

- Maps are the geographers' most basic tools. The basic language of maps includes the concepts and terms of scale, coordinate systems, projection, and symbolization. Maps can depict spatial data in a variety of ways.

• Individuals and cultures generate their own unique "mental maps." Regions are in effect mental maps that help us make sense of a complex world. Modern geographic thought derives from a long legacy of interest in how people interact with the environment. The dominant approach has been to understand how people have changed the landscape or the face of the Earth.

• The discipline of geography may be divided into regional and systematic specialties, with the systematic fields having the most followers. Their concerns overlap many disciplines in the natural and social sciences. Geographers are employed in many private and public capacities. The strongest growth area with the most jobs is in geographic information systems (GIS). The "geospatial revolution" employs many geographic tools and touches many parts of our daily lives.

Key Terms + Concepts

absolute location (p. 9)
Antarctic Circle (p. 10)
Arctic Circle (p. 10)
cartography (p. 7)
choropleth map (p. 13)
coordinate systems (p. 9)
core location (p. 10)
cultural landscape (p. 4)
culture (p. 4)
degrees (p. 9)
dot map (p. 13)
equator (p. 9)
five themes of geography (p. 3)
flow map (p. 13)
formal region (p. 6)
functional region (p. 6)
geographic information
    systems (GIS) (p. 14)
geography (p. 2)
geospatial (p. 13)
graduated symbol map (p. 13)
hemisphere (p. 10)
    Eastern Hemisphere (p. 11)
    land hemisphere (p. 10)

Northern Hemisphere
    (p. 10)
Southern Hemisphere
    (p. 10)
water hemisphere (p. 10)
Western Hemisphere
    (p. 11)
homogeneous region (p. 6)
human–environment
    interaction (p. 3)
human geography (p. 3)
International Date
    Line (p. 11)
isarithmic map (p. 13)
landscape (p. 4)
landscape perspective
    (p. 4)
large-scale map (p. 8)
latitude (p. 9)
    high (p. 10)
    low (p. 10)
    middle, (p. 10)
    north (p. 10)
    south (p. 10)

location (p. 8)
    absolute location (p. 9)
    relative location (p. 8)
longitude (p. 9)
    east (p. 11)
    west (p. 11)
map (p. 7)
map projection (p. 11)
    azimuthal (p. 12)
    compromise (p. 12)
    conformal (p. 12)
    conic (p. 12)
    cylindrical (p. 12)
    equal-area (p. 12)
    equidistant (p. 12)
    Mercator (p. 13)
mathematical location (p. 9)
mental map (p. 8)
meridian (p. 9)
    prime meridian (p. 11)
minutes (p. 9)
natural landscape (p. 4)
nodal region (p. 6)
North Pole (p. 10)

orientation (p. 12)
parallel (p. 9)
perceptual region (p. 6)
peripheral location (p. 10)
physical geography (p. 3)
place (p. 7)
reference map (p. 7)
region (p. 4)
remote sensing (p. 15)
scale (p. 8)
seconds (p. 9)
six essential elements of
    geography (p. 3)
small-scale map (p. 8)
South Pole (p. 10)
space (p. 7)
spatial (p. 7)
symbolization (p. 12)
thematic map (p. 7)
Tropic of Cancer (p. 10)
Tropic of Capricorn (p. 10)
uniform region (p. 6)
vernacular region (p. 6)

Review Questions

1. What is geography? What are some of its characteristic approaches?

2. What are the six essential elements of geography, as defined by the National Geographic Society? What does each element indicate about geography's concern with space, place, or the environment? What are the five themes of geography? Are you comfortable applying them to your knowledge of places?

3. What does *spatial* mean, and how does geography's interest in space differentiate it from other disciplines?

4. What geographic features make the United Kingdom and New Zealand different?

5. What are the major terms and concepts associated with scale, coordinate systems, projections, and symbolization?

6. Why is a map made with the Mercator projection more suitable for navigation than a map made with other projections?

7. What is the difference between a dot map and a choropleth map?

8. What is a mental map?

9. What is GIS, and what typically makes it different from old-fashioned manual cartography? What are some applications of remote sensing? How can modern geospatial tools help us with longstanding concerns in geography?

10. What do geographers do for a living?

Notes

1. The abbreviation B.C.E. stands for "before the Common Era," which is a reference to the dating system invented by European Christians that sets the birth of Jesus Christ as year 1. In Christian cultures, dates before that year are expressed as B.C., meaning "before Christ," and later years are identified as A.D., which stands for *anno Domini* (Latin, "in the year of our Lord"). Religion-neutral dating systems such as the one used in this book employ B.C.E. (before the Common Era) and C.E. (Common Era), respectively, but the years are numbered the same in the two systems.

2. Quoted in Geoffrey J. Martin and Preston E. James, *All Possible Worlds: A History of Geographical Ideas* (New York: Wiley, 1993), p. 150.

3. Quoted in Andrew Goudie, *The Human Impact on the Natural Environment* (Oxford: Blackwell, 1986), p. 6.

4. The anthropologist Kathleen A. Dahl authored this definition of culture.

5. You are in Monument Valley, northeastern Arizona near the Utah border, in the southwestern region of the United States. Place clues include an arid landscape, sandstone buttes that you have probably seen in American Western movies, and a Native American of the Navajo group, wearing typical Navajo turquoise and silver jewelry.

6. The viper's triangular head is just left of the center of the photo. This snake has buried himself in the shallow sand and gravel, a "virtual land mine" ready to explode on its prey or foe: a mouse, a Bedouin, or a geographer.

7. Using this map, you should determine Oslo's location as about 60 degrees north, 11 degrees east (60°N, 11°E). Oslo is therefore in the Northern Hemisphere, the Eastern Hemisphere, and the land hemisphere.

8. Committee on Strategic Directions for the Geographical Sciences in the Next Decade, National Research Council, 2010, Understanding the Changing Planet: Strategic Directions for the Geographical Sciences, Washington, DC: The National Academies Press.

9. http://www.dictionary.com.

10. Esri, at http://www.gisday.com/cd2009/fliers/what_is_gis.pdf.

11. http://www.aag.org/cs/careers

12. http://www.aag.org/cs/jobs_and_careers/geography_can_take_you_there

13. Information courtesy of Bill Doolittle, University of Texas at Austin.

14. Lieutenant General Leonid Sazhin, Federal Security Service, quoted in Katie Hafner and Saritha Rai, "Google Offers a Bird's-Eye View, and Some Governments Tremble," *New York Times,* December 20, 2005, p. 1.

15. http://gawker.com/5191459/cheating-husband-said-caught-via-google-street-view.

Online Resources

CourseMate: Make the most of your study time by accessing everything you need to succeed in one place. Read your textbook, take notes, review flashcards, watch videos, complete activities, take practice quizzes, and more—online with CourseMate. Log in at **www.cengagebrain.com.**

"The acquisition of wisdom is above that of pearls. The topaz of Ethiopia cannot equal it, Nor can it be valued in pure gold. Where then does wisdom come from? And where is the place of understanding?"

—JOB 28

Joe Hobbs

Joe Hobbs

Russia's Arctic wilderness of Franz Josef Land, not far from the North Pole. The dark patches of ground are permafrost, or "permanently" frozen ground. But as the Arctic warms, permafrost is melting and adding greenhouse gases to the Earth's atmosphere. The Arctic is the "canary in the coal mine" of climate change.

2

Physical Processes and World Regions

Many issues in world regional geography relate to the interaction of people and the natural environment, and how these relationships have changed the face of the Earth. Physical processes are the main focus in this chapter, and human processes the focus in **Chapter 3**. But these processes are so intertwined that human forces receive much attention in this chapter, and natural forces much attention in the following chapter.

This discussion of physical geography introduces you to the four "spheres" that make up the Earth's habitable environment: the **lithosphere**, its outer "rind" of rock; the **hydrosphere**, made up of all the world's water features; the **atmosphere**, the layer of gases surrounding the Earth; and the **biosphere** (also known as the ecosphere), which is the global ecological system including all the relationships played out among the lithosphere, hydrosphere, and atmosphere. Earth's lithosphere is a work in progress, and you will see how its changing surfaces provide both opportunities and threats to people. We will consider the climate and vegetation types that play such large roles in human activities, and appreciate the rich diversity of wild plant and animal species. We will look briefly at the planet's often-overlooked oceans and the resources they hold. Finally, we will examine how the climate may be changing, what these changes may mean for life on Earth, and what people can do to prevent or lessen some of the most dangerous climatic changes.

chapter objectives

This chapter should enable you to

- Understand the tectonic forces behind some of the world's major landforms and natural hazards

- Recognize consistent global patterns in the distribution of vegetation types and climates

- Identify the natural areas most threatened by human activity and explain how natural habitat loss may endanger human welfare

- Appreciate the important roles of the world's oceans

- Describe the potential impacts of global climate change and international efforts to prevent them

2.1 Geologic Processes

The Earth's surface is in motion. Those of you living on the West Coast of the United States have probably had some personal experience of the unforgettable sensation of the ground moving beneath you in a mild "temblor" or even a large earthquake. These events reflect processes that are constantly changing the face of the Earth. Here we explore just one set of geological processes—arguably the most important of them all because of its role in shaping global landforms—plate tectonics.

16, 254, 261

Plate Tectonics

About a century ago, a German geologist named Alfred Wegener came up with what seemed to many like an outlandish theory known as **continental drift.** Pointing to such things as the jigsaw puzzle–like geometry of Africa's west coast and South America's east coast, he proposed that the continents were once joined in a supercontinent (which he named Pangaea) but that they drifted apart over time. However, Wegener was not able to explain the forces behind these movements, and he was derided by a scientific community that seemed unwilling to accept anything other than a *terra firma* that was truly firm. Peers pronounced his conclusions "utter damned rot" and "mere geopoetry."[1] But later discoveries in deep-sea science established Wegener's basic proposition as factual, and today a good deal is known about how continental drift occurs.

The Earth's lithosphere, which varies from 50 to 125 miles in thickness, is made up of about a dozen giant and several smaller sections of rock called **plates.** These plates move in various directions in processes known collectively as **plate tectonics** (•Figure 2.1). On the ocean floors in places such as the Mid-Atlantic Ridge and the East Pacific Rise, new lithosphere is "born" as molten material rises from the Earth's mantle and cools into solid rock (•Figure 2.2). Plate tectonics are often explained by the useful analogy of a conveyor belt (the convection cell in Figure 2.2) in constant motion. On either side of the long, roughly continuous ridges, the two young plates move away from one another, carrying islands with them; this process is called **seafloor spreading.**

Seafloor spreading has few impacts on people, but when the Earth's plates collide, there is cause for great concern: such **tectonic forces** are among the planet's greatest natural hazards. The most dangerous tectonic forces are related to **seismic activity** (*seismic* refers to Earth vibrations, mainly earthquakes) that causes earthquakes and **tsunamis,** and the **volcanism** (movement of molten earth material, especially in volcanoes) often associated with plate movements. The globe's most active and deadly realm of tectonic activity is the so-called "Ring of Fire" on the rim of the Pacific Ocean (see Figure 2.1 and page 358).

Plates collide and converge in different ways and with different consequences; among the most notable consequences are the building of Earth's great mountain ranges. In some parts of the world—off the east coast of Japan, as we saw in Chapter 1, and off the coast of the U.S. Pacific Northwest, for example—one plate "dives" below another in a process known as **subduction** (see Figure 2.2). The descending lithosphere is melted again as it dives into the Earth's superheated mantle along a deep linear feature known as a **trench** (for example, the Mariana Trench off Japan). Subduction is another stage along the "conveyor belt" process that will eventually see this material recycled as new-born lithospheric crust.

16, 254, 358, 396

This subduction process releases enormous amounts of energy. Periodically, the great stress of one plate pushing beneath another is released in the form of an earthquake. The world's largest recorded earthquakes—registering 9.5 (Chile, 1960), 9.2 (United States, 1964), 9.1 (Indonesia, 2004), and 9.0 (Japan, 2011), respectively, on the **Richter scale,** which measures the strength of the earthquake at its source—have struck along these subduction zones. This sudden displacement of a section of oceanic lithosphere is also what triggers a tsunami, one of nature's most powerful and destructive processes. See pages 17, 254, and 288 for more insight into tsunamis.

214, 358, 396

In other places where they meet, lithospheric plates grind and slide along one another, as in California and Turkey. The processes of rock crowding together or pulling apart along these fracture lines is known as **faulting,** with movement along various kinds of **faults** causing earthquakes, the emergence of new landforms, and other consequences.

Volcanism generally takes place along and near **subduction zones** (see Figure 2.2) and also in the world's several dozen geologic hot spots where molten material has broken through the crust as a "plume" (as in Yellowstone National Park and in Hawaii—see page 288). Despite posing a host of natural hazards

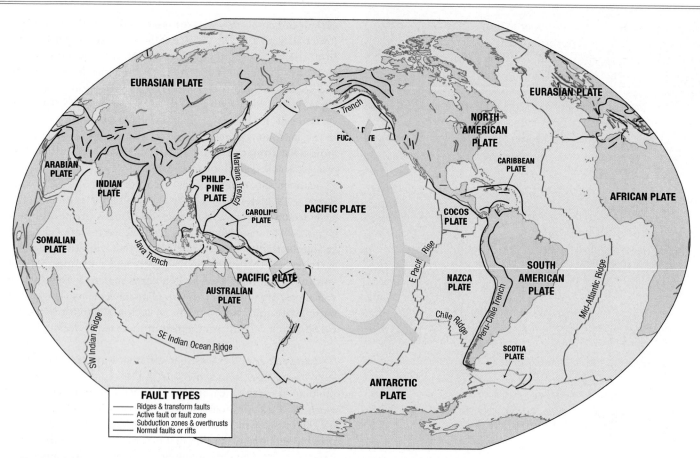

• **Figure 2.1** Major tectonic plates and their general direction of movement. Earthquakes, volcanoes, and other geologic events are concentrated where plates separate, collide, or slide past one another. Where they separate, rifting produces very low land elevations (well below sea level at the Dead Sea of Israel and Jordan, for example) or the emergence of new crust on the ocean floor (in the middle of the Atlantic Ocean, for example). The yellow ring points to the rough outline of the "Ring of Fire."

Source: Adapted from NASA, "Global Tectonic Activity Map of the Earth," DTAM-1, 2002.

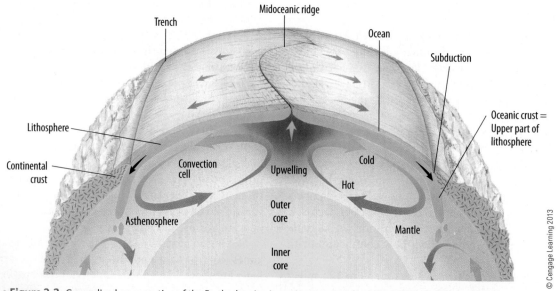

• **Figure 2.2** Generalized cross section of the Earth, showing its main concentric layers and the process by which its lithosphere is recycled.

© Cengage Learning 2013

to people living on their slopes or downwind—including pyroclastic flows (fast-moving currents of rock fragments, hot gases, and ash), lava flows, ashfalls, and other dangers—volcanoes have beneficial qualities. Volcanic rock usually breaks down to form fertile soils (as in Ethiopia and the Nile Valley), and the range of climate and vegetation types on their slopes creates opportunities for growing crops and raising livestock.

181, 311

We turn now to global patterns of climate and vegetation.

2.2 Patterns of Climate and Vegetation

"Everybody talks about the weather, but nobody does anything about it," Mark Twain reportedly said. We will see in this chapter that we have actually done a lot about the weather. First, we look at natural patterns in vegetation and climate that present different opportunities for people around the world. Once you recognize these patterns, you will be more prepared to explore all the world's regions.

As you experience a warm, dry, cloudless summer day or a cold, wet, overcast winter day, you are encountering the **weather**—the atmospheric conditions occurring at a given time and place. **Climate** is the average weather of a place over a long time period. Along with surface conditions such as elevation

and soil type, climatic patterns have a strong correlation with patterns of natural vegetation and in turn with human opportunities and activities on the landscape.

Precipitation and temperature are the key variables in weather and climate. Warm air holds more moisture than cool air, and **precipitation**—rain, snow, sleet, and hail—results from processes that cool the air to release moisture. Water is, of course, essential for life on Earth. In this book, you will see many examples of how the precious resource of fresh water is increasingly imperiled and contested. Some geographers would tell you that the map of the world's precipitation—•**Figure 2.3**—is the most important of all maps in understanding life on Earth.

180, 250, 259, 265, 436

Combinations of precipitation, temperature, latitude, and elevation produce a great variety of local climates. Geographers group these local climates into a number of major climate types, each of which occurs in more than one part of the world (•**Figure 2.4**) and is associated closely with other types of natural features, especially vegetation. Geographers recognize 10 to 20 major types of terrestrial ecosystems, called **biomes,** which are categorized by dominant types of natural vegetation (•**Figures 2.5** and **2.6**). When you are using maps of biomes, it is important to keep in mind that these depict vegetation that would dominate in the absence of human activity. For example, Figure 2.5 depicts a vast, unbroken area of temperate mixed forest in the eastern United States, but of course much of this region has been

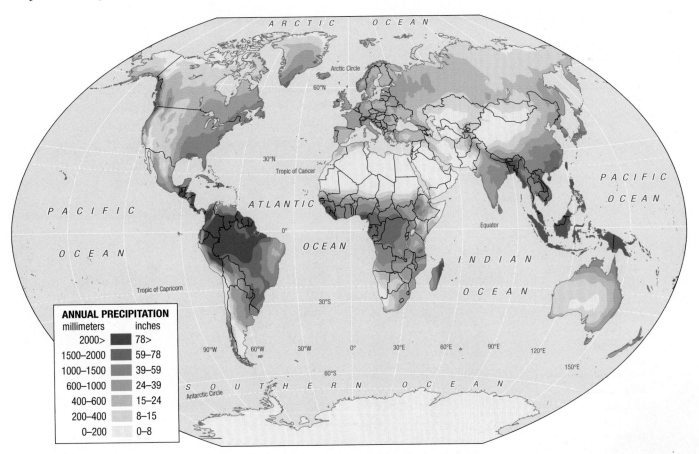

• **Figure 2.3** World precipitation map. Is this, as some authorities say, the most important of all world maps?

Source: Adapted with permission from UNEP/GRID and UEA/CRU, "Mean Annual Precipitation," GRID - Geneva Data Sets - Atmosphere, GNV174.

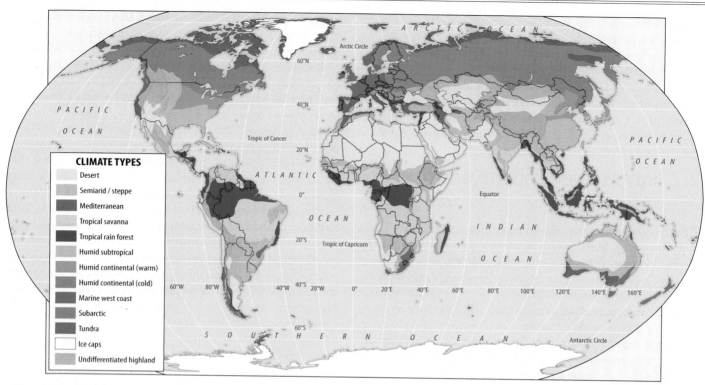

CLIMATE TYPES
- Desert
- Semiarid / steppe
- Mediterranean
- Tropical savanna
- Tropical rain forest
- Humid subtropical
- Humid continental (warm)
- Humid continental (cold)
- Marine west coast
- Subarctic
- Tundra
- Ice caps
- Undifferentiated highland

• **Figure 2.4** World climates.

Source: Based on Rand McNally's Classroom Atlas, 2003.

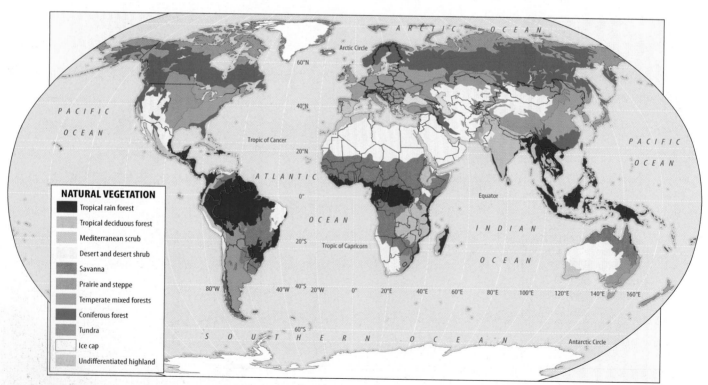

NATURAL VEGETATION
- Tropical rain forest
- Tropical deciduous forest
- Mediterranean scrub
- Desert and desert shrub
- Savanna
- Prairie and steppe
- Temperate mixed forests
- Coniferous forest
- Tundra
- Ice cap
- Undifferentiated highland

• **Figure 2.5** World biomes (natural vegetation) map.

Source: Based on World Wildlife Fund ecoregions data, 1999.

transformed by agriculture and urbanization. The map of the "human footprint" on page 42 in the next chapter shows what people have done to change the face of the Earth.

see page 42

Climate plays the main role in determining the distribution of biomes, but differing soils and landforms may promote different types of vegetation where the climatic pattern is essentially the same. Vegetation and climate types are sufficiently related that many climate types take their names from vegetation types—for example, the tropical rain forest climate and the tundra climate. You may easily see the geographic links between climate and vegetation by comparing the maps in Figures 2.4 and 2.5. The spatial distributions of climate and vegetation types do not overlap perfectly, but there is a high degree of correlation (see Try It on this page).

In the **ice cap, tundra,** and **subarctic** climates, the dominant feature is a long, severely cold winter, making agriculture difficult or even impossible. Summers are short and cool. Perpetual ice in the form of glaciers may be found at very high elevations in the lower latitudes (even in equatorial regions) and in the polar realms. The ice cap biome (•**Figure 2.6a**) has no plant life, except in those very few spots where enough ice or snow melts in the summer to allow tundra vegetation to grow. Tundra vegetation (•**Figure 2.6b**) is made up of mosses, lichens, shrubs, dwarfed trees, and some grasses. Needleleaf evergreen **coniferous trees** can stand long periods when the ground is frozen, depriving them of moisture. **Coniferous forests** (often called **boreal forests** or *taiga,* their Russian name; •**Figure 2.6c**) thrive in areas of subarctic climate, including large parts of Russia's Siberia region.

In **desert** and in **semiarid** or **steppe** climates, the dominant feature is aridity (extreme dryness) or semiaridity (mainly dry conditions, with sparse precipitation). Deserts and steppes occur in both low and middle latitudes. Agriculture in these areas usually requires irrigation. The Earth's

largest dry region extends in a broad band across northern Africa and southwestern and central Asia. The deserts of the middle and low latitudes are generally too dry for either trees or grasslands. They have **desert shrub** vegetation (•**Figure 2.6d**), often only in dry riverbeds, and in many places have no vegetation at all. Grasslands dominate in the moister steppe climate, which represents a transitional zone between very arid

• **Figure 2.6 (b)** Tundra, northern Norway.

• **Figure 2.6 (a)** Ice cap biome, Franz Josef Land, Russian Arctic region.

• **Figure 2.6 (c)** Coniferous forest.

• **Figure 2.6 (d)** Desert shrub, southern Sinai Peninsula, Egypt.

• **Figure 2.6 (f)** Tropical rain forest.

deserts and more humid areas. The biome composed mainly of short grasses is also called the steppe, or **temperate grassland** (•**Figure 2.6e**). The temperate grassland region of the United States and Canada originally supported both tall-grass and short-grass vegetation types, known in those countries as **prairies.**

The **tropical rain forest** and **tropical savanna** climates are rainy, low-latitude climates. The key difference between them is that the tropical savanna type has one or more prolonged dry seasons, whereas the tropical rain forest climate has a steadier rainfall regime. Heat and moisture dominate in the tropical rain forest biome (•**Figure 2.6f**), where broadleaf evergreen trees (which lose their leaves, but not seasonally or in tandem) dominate the vegetation. In dry tropical areas that have enough moisture to support trees, **tropical deciduous forest** replaces the rain forest. Here the broadleaf trees are not green throughout the year; they lose their leaves and are dormant during the dry season and then add foliage and resume their growth during the wet season. The tropical deciduous forest approaches the luxuriance of the tropical rain forest in wetter areas but thins

out to low, sparse scrub and thorn forest (•**Figure 2.6g**) in drier areas. **Savanna** vegetation (•**Figure 2.6h**) has taller grasses than the steppe and usually includes some trees, if found in areas of greater overall rainfall and more pronounced wet and dry seasons. You have seen much savanna vegetation in documentaries about the great animal herds of East Africa.

In the **marine west coast** climate, occupying the western sides of continents in the higher middle latitudes (the Pacific Northwest region of the United States, for example), warm ocean currents moderate the winter temperatures, and summers tend to be cool. Coniferous forest dominates in some cool, wet areas of marine west coast climate; a good example is the redwood and sequoia forest of northern California (•**Figure 2.6i**). These humid middle-latitude regions have mild to hot summers, and winters ranging from mild to cold, with several other types of climate interspersed.

The **Mediterranean** climate (named after its most prevalent area of distribution, the lands around the Mediterranean Sea) is typically found between a marine west coast climate and a

• **Figure 2.6 (e)** Steppe, eastern Turkey.

• **Figure 2.6 (g)** Scrub and thorn forest, northern Zimbabwe.

• **Figure 2.6 (h)** Savanna, southern Kenya.

• **Figure 2.6 (j)** Mediterranean scrub forest (dark area in left foreground) east of Los Angeles. The mountains produce a "rain shadow" effect: the light area in the right and background is desert.

76 lower-latitude steppe or desert climate. In the summer high-sun period, it lies under high atmospheric pressure and is typically rainless. In the winter low-sun period, it usually receives precipitation from low-pressure systems carried by westerly winds. **Mediterranean scrub forest** (•**Figure 2.6j**), known locally by such names as *maquis* and *chaparral*, characterizes Mediterranean climate areas. Many of these shrubs contain oils that are highly aromatic, repelling wild grazing animals but attracting human culinary interest: sage and thyme

are good examples. Much of California has Mediterranean climate and vegetation.

The **humid subtropical** climate occupies the eastern portion of continents between approximately 20 and 40 degrees of latitude and is characterized by hot summers, mild to cool winters, and ample precipitation for agriculture. The **humid continental** climate lies poleward of the humid subtropical type; it has cold winters, warm to hot summers, and enough rainfall for agriculture, with the greater part of the precipitation falling in the summer. This climate type is often subdivided into warm and cold subtypes, indicating the greater severity of winter in the zones closer to the poles. In middle-latitude areas with these humid subtropical and humid continental climate types, a **temperate mixed forest** with mostly broadleaf but also coniferous trees (as in the U.S. Midwest and Northeast; •**Figure 2.6k**) is found. As cold winter temperatures freeze the water within reach of plant roots, broadleaf trees shed their colorful leaves and cease to grow, thus reducing water loss. Then they produce new foliage and grow vigorously during the hot, wet summer.

• **Figure 2.6 (i)** Marine west coast forest type; giant sequoias in King's Canyon National Park, central California.

• **Figure 2.6 (k)** Temperate mixed forest, central Missouri, United States.

• **Figure 2.6 (l)** Undifferentiated highland vegetation, San Juan Mountains, Colorado, United States. Note the "timber line" above which trees do not grow. Here, biomes change with elevation, and so the vegetation types have to be characterized as "undifferentiated" on small scale maps.

Coniferous forests can thrive in some hot and moist locations where porous, sandy soil allows water to escape downward, giving conifers (which can withstand drier soil conditions) an advantage over broadleaf trees. Pine forests on the coastal plains of the southern United States are an example.

Undifferentiated highland climates have a range of conditions according to elevation and exposure to wind and sun. The vegetation in these climates (•**Figure 2.6l**) differs greatly, depending on elevation, degree and direction of slope, and other factors. These climates are "undifferentiated" in the context of world regional geography in that a small mountainous area may contain numerous biomes, and it would be impossible to map them on a small scale. The world's mountain regions have a complex array of natural conditions and opportunities for human use. With increasing elevation, temperatures fall by a predictable **lapse rate**, on average about 3.6°F (2.0°C) for each increase of 1,000 feet (305 m) in elevation. In climbing from sea level to the summit of a high mountain peak near the equator (for example, in western Ecuador), a person would experience many of the major climate and biome types found in a sea-level walk from the equator to the North Pole!

2.3 Biodiversity

All of Earth's natural regions have remarkable life forms. But scientists recognize the exceptional importance of some biomes because of their **biological diversity** (or **biodiversity**)—the number of plant and animal species present and the variety of genetic materials these organisms contain.

The most diverse biome is the tropical rain forest. From a single tree in Peru's Amazon region, the entomologist Terry Erwin recovered about 10,000 insect species. From another tree several yards away, he counted another 10,000, many of which differed from those of the first tree. Alwyn Gentry of the Missouri Botanical Garden recorded 300 tree species in a 1-hectare (2.47-acre) plot of the Peruvian rain forest. Less than 50 years ago, scientists calculated that there were 4 million to 5 million species of plants and animals on Earth. Now, however, their estimates are much higher, in the range of 40 million to 80 million. This startling revision is based on research, still in its infancy, on species inhabiting the rain forest. Many of these are small insects. Incidentally, the number of animal species far exceeds that of plant species.

The Importance of Biodiversity

Such diversity is important in its own right, but it also has vital implications for nature's ongoing evolution and for people's lives on Earth. Humankind now relies on a handful of crops as staple foods. In our agricultural systems, the trend in recent decades has been to develop high-yield varieties of grains and to plant them as vast **monocultures** (single-crop plantings). This trend, which is the cornerstone of the so-called Green Revolution, is controversial. On the one hand, it puts more food on the global table. But on the other, it may render agriculture more vulnerable to pests and diseases and thus pose long-term risks of famine. In evolutionary terms, the Green Revolution has reduced the natural diversity of crop varieties that allows nature and farmer to turn to alternatives when adversity strikes. At the same time, while we remove tropical rain forests and other natural ecosystems to provide ourselves with timber, agriculture, or living space, we may be eliminating the foods, medicines, and raw materials of tomorrow, even before we have collected them and assigned them scientific names. "We are causing the death of birth," lamented biologist Norman Myers.[2]

Regions where human activities are rapidly depleting the rich variety of plant and animal life are known as **biodiversity hot spots,** which scientists believe deserve immediate attention for study and conservation. Thirty-four priority regions identified by Conservation International are depicted in •**Figure 2.7**. By referring to the map of the Earth's biomes in Figure 2.5, you will see that most of these hot spots are in tropical rain forest areas. Many are islands that tend to have high biodiversity because species on them have evolved in isolation to fulfill special roles in these ecosystems and because human pressures on island ecosystems are particularly intense. Efforts are under way in most of these hot spots to establish national parks and other protected areas. Conservationists are also turning their attention increasingly to the state of the world's oceans, which play critical roles in the Earth's physical processes and sustain many people.

2.4 The World's Oceans

Life on Earth would be impossible without the roles and resources of the hydrosphere, which includes both oceans and freshwater sources such as lakes and rivers. This book deals often with issues of freshwater, which is appropriate considering the projection by the World Commission on Water that by 2025, half of the world's population will live under conditions of severe water stress.[3] In this section, the focus is on the world's oceans.

356
249
290, 344
290, 344

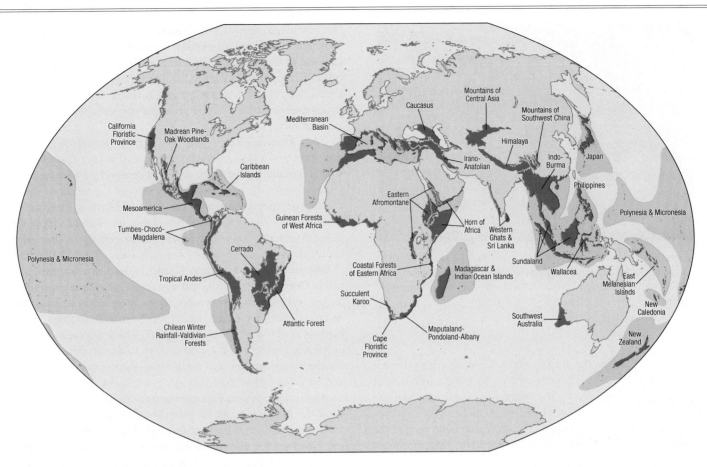

• **Figure 2.7** World biodiversity hot spots.

Source: Adapted from "Threatened Ecoregions" map, National Geographic Family Reference Atlas of the World, p. 40. Copyright © 2002 National Geographic Society. Source data provided by Conservation International.

Why Should We Care about Oceans?

I live in Missouri, where a day at the beach or a plate of fresh seafood is just a dream. Even those of you living close to the sea may have very little interaction with it. Why bother thinking about the world's oceans, much less worrying about them?

It's a watery world. About 70 percent of the world's surface is comprised of water. That is a huge proportion, and it is very important to consider how water shapes life on the planet. Oceans have the largest role in the **hydrologic cycle,** as the sun evaporates seawater into vapor that is then released as freshwater precipitation. Without the seas, the Earth's usable freshwater resources would be extremely limited.

The oceans feed us. About a billion people, or 15 percent of the world's population, rely mainly on fish for their protein (•**Figure 2.8**). Growing human populations, increased buying power among the middle classes in developing countries, and technological improvements in the fishing industry are putting these resources under unprecedented pressure.[4] Since 1980, there has been a spectacular 40 percent growth in worldwide demand for seafood. Some scientists have issued a stern warning on the future of the world's fisheries: Unless major steps are taken quickly, there will be a "global collapse" of all species currently fished by 2050.[5]

Defining "collapse" as a population less than 10 percent of its previous levels, these authorities noted that 30 percent of the fish species currently exploited have already collapsed. The declines are most extreme among stocks of large predatory fish like sharks, tuna, and swordfish. An estimated 75 million sharks are killed each year (mainly by Spanish and Indonesian fishermen) to provide shark fin soup (which can cost up

• **Figure 2.8** Seattle's Pike Street Market. Humanity's demand for fish and seafood is growing, but what about supplies? Stocks are falling, and in many species plummeting.

• **Figure 2.9** Shark fins at a Hong Kong market. There are several streets in this section of the city filled with shops selling shark fins and edible birds' nests. Both are specialty items in high demand among ethnic Chinese people throughout the world. Note the small Daoist prayer niche.

to $100 per bowl) to the world's ethnic Chinese, who believe it bestows health and virility (•**Figure 2.9**). An uncertain but large percentage of these sharks are "finned," a cruel procedure in which the fins are cut off the living fish, which are then thrown back into the sea to die. Public relations campaigns against the consumption of shark fin soup have begun to yield successes. Sales of shark fins in Hong Kong, the main center for the shark fin trade, have declined by about a third in recent years. (In one television ad, retired basketball star Yao Ming says, "Remember, when the buying stops, the killing can too.") Populations of the bluefin tuna—a magnificent, huge fish built for speed and endurance—are under pressure because their flesh is prized for *sushi* and *sashimi*. Not surprisingly, Japan has consistently blocked efforts to put the bluefin on a United Nations (UN) list of protected species.

Some good news is that these grim realities and scenarios can be changed by action on several fronts, including reducing the number of unwanted fish caught in nets and cracking down on overfishing where it is recognized to be a problem. Governments will probably need to intervene. Legislation like the United Nations Convention on the Law of the Sea (see page 272) can encourage marine conservation and include mechanisms to enforce the law. Like the climate change treaties discussed later, it is easy for a country to sign an international agreement. But then this legislation needs to be ratified, and most importantly it needs to be enforced.

More good news is that **aquaculture** (the cultivation of aquatic organisms for food, including **fish farming**) has the potential to increasingly substitute for wild-caught fish stocks. Fish farms already supply half of the seafood consumed around the world. There has been particularly explosive growth in aquaculture in China (which produces 70 percent of the world's farmed fish). Some of this aquaculture is what may be described as "sustainable seafood," including the unlikely harvest of shrimp raised in well water deep in the Arizona desert.[6]

But much commercial fish farming is controversial because of inputs of antibiotics, pesticides, fertilizers, and hormones, and because of the prospect that genetically modified salmon and other fish (dubbed "Frankenfish" by detractors) would escape their confines and alter the genetic heritage of wild fish. These issues mirror many of those raised in conventional land-based agriculture.

They provide energy and other raw materials for human use. As terrestrial sources of conventional energy dwindle, our fossil fuel–addicted economies create demand for new supplies. Deep seas are the final frontier for energy exploration. Under the U.S. territorial waters of the Gulf of Mexico, there are as many as 40 billion barrels of petroleum (enough to supply the U.S. oil demand for about 5 years). But getting to that oil is problematic: much of it lies beneath more than 10,000 feet (3,050 m) of water and then 5 miles (8 km) of rock, salt, and sand. Deep-water drilling presents special engineering challenges and environmental risks. Human error in operating the *Deepwater Horizon* well, which was pumping oil from an unbelievable depth of about 35,000 feet (10,000 m) in the Gulf of Mexico, led to a historic oil spill and cleanup effort in 2010. 417 Similar opportunities and perils lie in the depths of the Arctic and Atlantic Oceans (see pages 427 and 382). Meanwhile, there is enormous potential to capture unconventional energy supplies from the sea, especially by using the power of surface winds and waves and of rising and falling tides to generate electricity. Prospects are increasing for the deep-sea mining of other minerals, including gold, silver, and the copper, cobalt, nickel, and "rare earth" minerals (many of which are used in high-tech devices) held within manganese nodules strewn across much of the world's seafloor. 273

They play important roles in trade and commerce. The seafaring days are far from over. A remarkable 90 percent of global trade is seaborne. You know all those Chinese-made goods that are practically everywhere you look in the United States? About 98 percent of them travel across the Pacific on cargo ships.[7] There is an economic rationale for this. Airfreight can cost 20 times as much as sea freight. As we will see in many cases, protecting the world's sea-lanes from enemies has long been one of the world's most important geopolitical concerns. These concerns are greater now than at any time in history, particularly where oil is involved (see, for example, pages 177 and 240).

All of these features of the world's oceans help to explain a remarkable fact of modern life on Earth: Over half of the world's people live within 60 miles (100 km) of a coastline. Ten percent of the world's people live within 6 miles (10 km) of the coast. You can see these spatial characteristics in the map of the world's population on page 55.

2.5 Global Environmental Change

In the atmosphere—the layer of oxygen, nitrogen, and other gases extending from the Earth's surface to about 60 miles (100 km) above it—changes are occurring that will almost certainly have

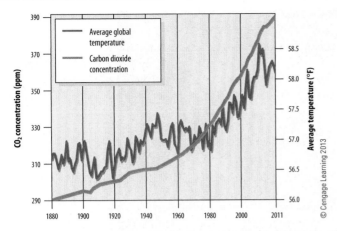

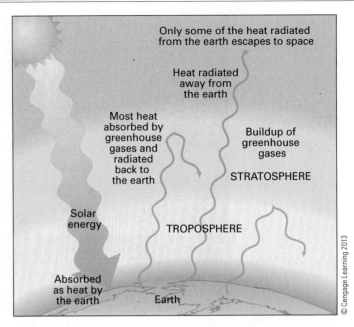

• **Figure 2.10** Industrialization, the burning of tropical forests and other factors have produced a steady increase in carbon dioxide emissions. Many scientists believe that these increased emissions explain the corresponding steady increase in the global mean temperature.

• **Figure 2.11** The greenhouse effect. Some of the solar energy radiated as heat (infrared radiation) from the earth's surface escapes into space, while greenhouse gases trap the rest. Naturally occurring greenhouse gases make the earth habitable, but carbon dioxide and other greenhouse gases emitted by human activities accentuate the greenhouse effect, making the planet unnaturally warmer.

profound effects on natural systems and on human uses of the Earth. Many of these changes are attributable to human activities, and the final section of this chapter is a prelude to the following chapter on human processes affecting the planet.

Until about 2000, there was considerable uncertainty in the science of climate change. Most scientists at that time insisted that human activities were responsible for a documented warming of the Earth's surface (by 1.4° Fahrenheit, or about 0.8° Celsius, since the late nineteenth century), but a significant minority either insisted that warming was not occurring or, if it was, that a natural climatic cycle was responsible. Today, there is far less scientific uncertainty about the human role in climate change.

The roughly 2,500 atmospheric scientists from more than 130 countries who make up the United Nations–sponsored **Intergovernmental Panel on Climate Change (IPCC)** have concluded that global warming is "unequivocal." Human production of **greenhouse gases** such as **carbon dioxide (CO₂)** is, according to the IPCC, "very likely" responsible for this warming (•**Figure 2.10**). In the scientists' jargon, "very likely" means a more than 90 percent probability.[8]

The Greenhouse Effect

In 1827, French mathematician Jean-Baptiste Fourier established the concept of the **greenhouse effect,** noting that the Earth's atmosphere acts like the transparent glass cover of a greenhouse (•**Figure 2.11**). You can also think of the atmosphere as being like a car's windshield. Visible sunlight passes through the glass to strike the planet's surface. Oceans and land, like the floor of the greenhouse or the car's upholstery, reflect the incoming solar energy back as heat (invisible infrared radiation). Acting like the greenhouse glass or car windshield, the Earth's atmosphere traps some of that heat.

The greenhouse effect is not a bad thing. In fact, if not for naturally occurring greenhouse gases such as carbon dioxide

and water vapor, the Earth would be too cold for most forms of life (with average temperatures well below freezing). What has so many people worried is what happens when human activities add newer and greater quantities of greenhouse gases into the atmosphere, trapping unnatural amounts of heat; these are the so-called "anthropogenic" causes of climate change. Carbon dioxide released into the atmosphere from the burning of coal, oil, and natural gas is the greatest source of concern, but **methane** (from rice paddies and the guts of ruminating animals like cattle), **nitrous oxide** (from the breakdown of nitrogen fertilizers), and **chlorofluorocarbons (CFCs),** used as coolants and refrigerants) are also greenhouse gases resulting from human activities. CFCs also destroy stratospheric ozone, a gas that has the important effect of preventing much of the sun's harmful ultraviolet radiation from reaching the Earth's surface. In the car-as-Earth metaphor, continued production of these greenhouse gases has the effect of rolling up the car windows on a sunny day, with the result of increased temperatures and physical overheating of the occupants.

The Effects of Global Warming

The trend toward a warmer world is remarkable. Although formal records of meteorological observations began only about a century ago, past climates left evidence in the form of marine fossils, corals, glacial ice, fossilized pollen, and annual growth rings in trees. These indirect sources, along with formal records dating back to 1861, indicate that the twentieth century was the warmest of the past six centuries and that the seven

• **Figure 2.12** Some of the world's climate changes anticipated with a global mean increase of 7°F (4°C). http://www.egov.vic.gov.au/government-initiatives/environment/climate-change-interactive-map.html

warmest years in the past 600 years were clustered recently; in descending order they were 2010, 1998, 2002, 2003, 2005, and 2007.

And what is the forecast? The IPCC uses a number of different "emission scenarios" that consider such variables as rates of population, economic and industrial growth, as well as efforts that might be made to curb greenhouse gas emissions. The "best estimate range" of the IPCC is that the mean global temperatures will warm from 3°F to 7°F (1.8°C to 4°C) by 2100. Briefly, these are the climate changes that are anticipated in this warming world (see •**Figure 2.12**).

A warmer climate overall, but . . . . Among our first questions in the study of the world regional geography of climate change is *where* the anticipated changes will occur and what effects these changes will have. Computer models suggest that there will not be a uniform temperature increase across the entire globe. Instead, the increases will vary spatially and seasonally. Some places will actually become cooler. Different computer models sometimes produce contradictory results about the timing and impacts of warming. For example, models predicting warmer summers generally forecast that crop productivity will decline, whereas models predicting warmer winters generally forecast an increase in crop yields.

More precipitation overall, but also more pronounced drought. Higher temperatures will mean increased evaporation from the world's oceans. This will result in more precipitation, but it will be distributed irregularly. The IPCC forecasts more precipitation in the higher latitudes and less precipitation, with intense and longer droughts, in the lower latitudes. What the IPCC calls "heavy precipitation events" will be more common. Even unusually large snowfall trends

are expected, and the magnitude of hurricanes (whose strength increases with warmer ocean surface waters) will increase.

Pronounced warming in the polar regions. Geographically, the impacts of climate change are expected to be greatest at the higher latitudes. Permafrost, the "permanently" frozen ground typical of the polar regions, has been melting at an alarming rate. Among other things, this has been forcing Arctic peoples to relocate from traditional settlements. [20] Sailing on a Russian nuclear-powered icebreaker, I saw open water at the North Pole in August 1996. The Russian Ice Master on board said that in 30 years of working on icebreakers in the Arctic, he had never seen anything like this: vast expanses of the Arctic Ocean completely ice-free or just dotted with sea ice and icebergs. Ever since that summer, there have been reports of unprecedented change at sea and on land in the Arctic. The IPCC report estimates that the average coverage of Arctic sea ice has shrunk about 3 percent per decade since 1978, with summertime ice decreasing 7 percent per decade. The trend is self-perpetuating: the darker ocean waters that replace the white ice cover absorb ever more solar radiation (think of cars here again: How hot would the surface of a dark blue car feel to your touch on a summer day compared with the surface of a white car?). That warming melts even more ice, creating more dark surface water and more warming (•**Figure 2.13**). In some parts of the world, this is seen as good news. A further retreat of polar ice would bode well for maritime shipping through the long-icebound Northwest Passage across Arctic Canada, and through the Northern Sea Route across the top of Russia, which in the past has been navigable only in summer and only with the aid of icebreakers.

260, 437

148, 427

Joe Hobbs

• **Figure 2.13** A polar bear feeding on a seal in the Arctic Ocean. The bear's scientific name, *Ursus maritimus*—meaning "marine bear"—aptly suggests that this great predator spends much of its time at sea. But it must come up on the sea ice to feed, sleep, hibernate, and bear young. As sea ice retreats, there are fewer places for the bear to eat and rest, and there are increasing accounts of bears swimming to their death in open water. The sea ice seen here deflects the sun's rays, but the dark blue water absorbs them, contributing to warmer water and more ice melt. See page 32.

Shifting biomes, with species extinction and agricultural changes. There is general agreement that in a warmer world, the distribution of climatic conditions typical of biomes will shift poleward around the world, and upward in mountainous regions. There is particular concern about the Arctic and subarctic biomes of Canada, Scandinavia, and Russia, where studies suggest that climate change could alter 70 percent of these natural habitats' areal coverage, and cause the extinction of 20 percent of the habitats' species.[9]

Many animal species will be able to migrate immediately to keep pace with changing temperatures, but plants, being stationary, will not. Conservationists who have struggled to maintain islands of habitat as protected areas are particularly concerned about rapidly changing climatic conditions. Agriculture would also be affected as growing zones move poleward. And although the right temperature and rainfall combinations for productive agriculture might emerge in new areas, the right soil conditions would have to exist as well (the hard rock of the Canadian Shield will never make good farmland, for example).

Rising sea levels. There is also general agreement that if global temperatures rise, so will sea levels as glacial ice melts, and as seawater warms and occupies more volume (just as mercury in a thermometer does, causing the mercury to climb the glass tube). Sea levels rose 6 to 9 inches (15 to 23 cm) in the twentieth century. The IPCC forecasts a further rise of 7 to 24 inches (18 to 61 cm) rise by 2100, but other scientists call these figures far too conservative because they do not account for any melting of the vast Greenland and Antarctic land ice sheets. Projections of sea level rise based on the melting of these ice sheets are breathtaking.

Decision makers in coastal cities throughout the world, in island countries, and in nations with important urban or agricultural lowland areas adjacent to the sea are worried about the implications of rising sea levels. The low-lying coastal countries of the Netherlands and Bangladesh have for many years been outspoken advocates of reduced greenhouse gases. Their natural allies are the roughly 40 vulnerable island countries around the world who make up the advocacy group known as the Alliance of Small Island States. The president of alliance member country Maldives, an Indian Ocean country composed entirely of islands barely above sea level, pleaded "We are an endangered nation!" 250

Geopolitical instability. There is growing consensus that some of the poorest and most overcrowded nations will be hit hardest by climate change and that political and economic instability will result. The IPCC forecasts that rising sea levels will flood tens of millions of poor people out of their homes each year and that by 2080, possibly 200 million to 600 million people will face starvation because of failing crops. By then, one billion to three billion people in developing countries will face shortages of freshwater related to growing aridity. Epidemics of malaria and other killer diseases will be more widespread.

Such environmental crises will likely lead to large movements of "climate change refugees" and to conflicts over resources, upsetting the fragile balance of power between countries. In a published report, U.S. military thinkers concluded that climate change warming "presents significant national security challenges to the United States," adding, "Projected climate change will seriously exacerbate already marginal living standards in many Asian, African and Middle Eastern nations, causing widespread political instability and the likelihood of failed states. The chaos that results can be an incubator of civil strife, genocide and the growth of terrorism. The U.S. may be drawn more frequently into these situations."[10]

There is some scientific discussion of climate changes that cannot be avoided. From emissions already in the Earth's atmosphere, temperatures will continue to climb because of the oceans' "lag time": they warm up slowly and will take a long time to respond to greenhouse gases already in the atmosphere. Some scientists have begun speaking of a **tipping point** in climate change, a point at which feedback effects amplify temperature changes, producing some changes that are irreversible. The tipping point would bring a rapid acceleration in temperature and of related effects like more violent storms, dramatic crop losses, and spreading deserts.

What Can Be Done about Global Climate Change?

Fortunately, there is growing international political will to take the strong and sometimes costly measures that may be necessary to prevent even the direst scenarios. Scientists and policy-makers are advocating actions to limit future warming of the Earth's atmosphere to no more than 2°C (3.6°F)—well below the IPCC forecast of up to 4°C (7°F) by 2100.

Efforts to reduce the production of greenhouse gases focus on the world's wealthiest nations, but there is a good reason for the growing pressure on China to join the fight. By far the greatest producers of carbon dioxide emissions, China and the United States together account for roughly 40 percent of the entire global emissions output. As recently as 1994, China was predicted to catch up with the United States' CO_2 emissions levels by 2019. The fact that China achieved this distinction in 2007 instead is testimony to the dizzying pace of that country's development—a theme that echoes throughout this book.

There are essentially two approaches to confronting climate change: mitigation and adaptation. **Mitigation** measures aim to avoid the adverse impacts of climate change in the long term; these include steps like switching from coal to cleaner fossil fuels to produce electricity, replacing fossil fuels with other energy resources, reducing energy consumption, and removing greenhouse gases from the atmosphere by boosting photosynthesis (for example, by planting more trees, which absorb carbon dioxide in the photosynthetic process of producing plant tissue and oxygen). **Adaptation** measures are designed to cope with and reduce the unavoidable impacts of climate change in the short and medium terms; these include building sea walls and dikes to prevent flooding related to rising sea levels, relocating people from flood-prone areas to higher ground, and developing crop varieties that are more suited to expected changes in precipitation and temperature. Generally, mitigation is seen as a problem that the world's richer countries plus China should deal with, because they are the main producers of greenhouse gases. Adaptation is typically seen as a problem for the poorer countries, which are expected to suffer the most from climate change. Many of the world's largest population clusters are in low-lying river deltas in poor countries: Vietnam's Mekong, Bangladesh's Brahmaputra/Ganges, and Egypt's Nile, for example. These are the broad associations of adaptation and mitigation. In reality, most poor and wealthy countries alike will have to take measures to both adapt to and mitigate climate change.

Following are some of the main approaches to mitigating climate change, and thereby avoiding the kinds of almost-unthinkable adaptations that would be needed in vulnerable countries like Vietnam (see page 260).

Negotiate and implement international treaties to reduce greenhouse gas emissions. Most scientists and policymakers agree that the best way to confront global climate change is to implement international treaties to reduce emissions (see Geography of Energy, page 35, and •Figure 2.14). Countries have proven they can unite in effective action against greenhouse gases. Thanks to the **Montreal Protocol** and its amendments, signed by 37 countries in the late 1980s, the production of CFCs worldwide was reduced in phases to zero by 2010. As a result there should a marked reduction in the size of stratospheric ozone "holes" that have been observed seasonally over the southern and northern polar regions since about 1985. CFC molecules have very long life spans, but scientists predict that the ozone layer should recover to pre-1980 levels by 2050. This recovery will also depend on the successful phase-out of other ozone-destroying refrigerants known as hydrochlorofluorocarbons (HCFCs).

Cut emissions through market-based incentives. Compared with richer nations, the less developed countries (LDCs), other than China and India, release relatively little CO_2 into the atmosphere through industrial activities; most of their emissions come from deforestation. Because the less developed countries produce relatively little carbon dioxide, policymakers are coming up with innovative ways to keep them from doing so, and to encourage them to develop clean industrial technologies. One approach is the system of **cap-and-trade** or **emissions trading.** In these schemes, already being used by the United Nations, the European Union, and the unregulated "voluntary market," managers require the richer nations to achieve a net reduction in carbon emissions. It is very expensive for the already highly energy-efficient, richer country to cut its greenhouse gas emissions, but relatively inexpensive for the rich country to pay a poorer, energy-inefficient country to cut its emissions. These schemes therefore allow the richer countries either to make the cuts at home or to offset their emissions by buying credits from poorer countries that actually cut carbon emissions. The poorer country selling the credits is obliged to use the income to invest in energy-efficient and nonpolluting technologies. In some of the systems, the LDCs can also sell carbon credits for pollution they are not actually creating; with relatively little manufacturing, their emissions may be lower than what they are allowed to produce in the trading scheme.

China is making big profits from emissions trading, using European money to reduce emissions of HFC-23, a potent greenhouse gas produced in many Chinese refrigerator factories. Russia is another big advocate of emission-trading schemes like these. Its industries are producing well below their benchmark 1990 carbon dioxide emission levels (they declined 36% between 1990 and 2009), so the country has much to gain by selling its unused emission rights. One buyer is Denmark, which is paying to convert coal-fired electricity plants to cleaner-burning gas in Russia's Siberia region. In such a trading scheme, both countries apparently benefit, as does the global atmosphere. However, this practice is still young, and critics say it ultimately lets the big polluters in the industrialized countries avoid the hard work of cutting emissions further at home.

Increase carbon sequestration. **Carbon sequestration** refers mainly to the natural capture and long-term storage of carbon in forests, farmlands, or in the oceans, so that the build-up of carbon dioxide in the atmosphere will reduce or slow (there are also artificial means of carbon sequestration, for example, pumping carbon dioxide deep into underground repositories).[11] In the process of photosynthesis, plants absorb carbon dioxide. Collectively, the world's forests, farmlands, and seas (which contain carbon-absorbing phytoplankton) thus serve as giant **carbon sinks.** In international climate-change negotiations, forest- and farm-rich countries want to receive credit for having these natural buffers against climate change.

Climate change can also be mitigated by reducing deforestation, especially in the poorer countries. People in these countries rely heavily on trees and other vegetation for fuel. When burned, trees release the CO_2 that they had stored in the process of photosynthesis. At the same time, in being burned they cease to exist as organisms that, in the process of photosynthesis, remove CO_2 from the atmosphere. Deforestation for agricultural expansion, conversion to pastureland, infrastructure

GEOGRAPHY OF ENERGY   The Kyoto Protocol and Beyond

In 1997, at a world conference on climate change, 84 countries (of the 160 countries represented at the conference) signed the **Kyoto Protocol**, a landmark international treaty for climate change. The agreement required 38 more developed countries (MDCs, also known as the Annex I countries) to reduce carbon dioxide emissions to 5 percent or more *below their 1990 levels* by the year 2012. The United States pledged to cut its emissions to 7 percent below 1990 levels by that date. The European Union made a promise of 8 percent below 1990 levels, and Japan promised a 6 percent reduction below 1990 levels.

Meeting these pledges would require substantial legislative, economic, and behavioral changes in these more affluent countries. Transportation and other technologies that use fossil fuels would have to become more energy efficient, making these technologies at least temporarily more expensive. Advocates of the protocol argued that the initial sacrifices would soon be rewarded by a more efficient and competitive economy powered by cleaner and cheaper sources of energy derived from the sun. Opponents, however, felt that higher fuel prices would be too costly for the United States and other industrialized economies to bear. Their position made it difficult for some signatories of the Kyoto Protocol to ratify the treaty. But in 2005, after the European Union countries, Japan, and finally Russia ratified the Kyoto Protocol, it went into effect for the countries that ratified it.

The United States signed the Kyoto Protocol during the Clinton Administration. Even then, the political will for ratification did not exist. When President George W. Bush assumed office in 2001, he rejected the Kyoto Protocol outright. The Bush administration had two objections: the potentially high economic

• **Figure 2.14** Protestors calling for action on negotiating the climate change treaties in Copenhagen in 2009.

cost of implementing the treaty and the fact that China, along with all the world's less developed countries, was not required by the Kyoto Protocol to take any steps to reduce greenhouse gas emissions. China argued then, as it still does now, as the world's leading greenhouse gas producer, that it is a developing country whose growth should not be hampered by pressure from countries that became rich by burning fossil fuels. China's economy continues to grow by as much as 10 percent annually, and so do its carbon dioxide emissions, up 206 percent between 1990 and 2009. In the same period, U.S. emissions rose by 7 percent. European Union (EU) member country achievements have been remarkable: Britain's emissions were down 15 percent for that period, and Germany's down 21 percent.

The EU is particularly upset that the United States has not taken leadership on climate change. The 27 countries of the European Union see themselves as the industrialized, wealthy peers of the United States, with the same challenges in investing in leaner and greener manufacturing technologies. But although the Europeans kept "upping the ante"—proposing to cut emissions to at least 20 percent and as much as 40 percent *below* 1990 levels by 2020, and taking actions to keep up with original Kyoto Protocol reduction pledges—the United States talked about cutting emissions. So did China. The American and Chinese political systems could not be more different, but each of these two leading contributors to greenhouse gas emissions has lacked the political will to confront climate change. Much-anticipated climate change talks in Copenhagen in 2009 failed to make progress, and it was not clear what kind of agreement if any would succeed the Kyoto Protocol when it expired in 2012.

98

---

development, logging, and other purposes accounts for about 20 percent of global greenhouse gas emissions (more than the global transportation sector and second only to the energy sector.[12] Since 2008, the United Nations has been developing a program called **Reducing Emissions from Deforestation and Forest Degradation (REDD).** This is an effort to create a financial value for the carbon stored in forests, offering incentives for developing countries to reduce carbon emissions from forested lands and invest in low-carbon development. REDD has other potential benefits, like alleviating poverty, enhancing the status of indigenous peoples (who often have their own systems of protecting environments), and protecting biodiversity. REDD is emerging as part of the carbon-trading scheme described above, allowing richer countries to purchase carbon credits from poorer countries. But in the long run, the richer countries may also be expected to donate funds to REDD in the interest of the planet's welfare.

The next chapter discusses many ways in which the world's developed and less developed countries must confront problems like global climate change differently. Like this chapter, it will conclude with some specific ideas about how to deal with some of the most pressing issues of our time.

253, 384

summary

- The Earth's three layers of habitable space are the hydrosphere, atmosphere, and lithosphere. The lithosphere is made up of separate plates that are in motion, a process known as plate tectonics. These movements result in mountain building, volcanic activity, earthquakes, and other consequences.

- Weather refers to atmospheric conditions prevailing at one time and place. Climate is a typical pattern recognizable in the weather of a region over a long period of time. Climatic patterns have a strong correlation with patterns of vegetation and in turn with human opportunities and activities in the environment.

- Geographers group local climates into major climate types, each of which occurs in more than one part of the world and is associated with other natural features, particularly vegetation. Geographers recognize 10 to 20 major types of ecosystems or biomes, which are categorized by the type of natural vegetation. Vegetation and climate types are so sufficiently related that many climate types are named for the vegetation types.

- Some biomes are particularly important because of their biological diversity—the number of plant and animal species and the variety of genetic materials these organisms contain. Regions where human activities are rapidly depleting a rich variety of plant and animal life are known as biodiversity hot spots, places scientists believe deserve immediate attention for study and conservation.

- Oceans cover about 71 percent of the Earth's surface. They play the key role in the hydrologic cycle, sustain large numbers of people through the protein in fish and seafood, and contain valuable mineral resources.

- Most of the sun's visible short-wave energy that reaches the Earth is absorbed, but some of it returns to the atmosphere in the form of infrared long-wave radiation, which generates heat and helps warm the atmosphere. This is the Earth's natural greenhouse effect.

- The scientific community is convinced with 90 percent or greater certainty that human activities, particularly the production of carbon dioxide and other greenhouse gases, are responsible for global warming. Computer-based climate change models use the scenario of a doubling of atmospheric carbon dioxide and indicate that the mean global temperature might warm up by an additional 11 degrees Fahrenheit (6.1 degrees Celsius) by the year 2100. The general scientific consensus is that with global warming, the distribution of climatic conditions typical of biomes will shift poleward and upward in elevation, sea levels will rise, and mean global precipitation will increase (but with drought intensified in some areas).

- There are two approaches to confronting climate change: Mitigation measures aim to avoid the adverse impacts of climate change in the long term, and are usually associated with more-developed countries (MDCs). Adaptation measures are designed to cope with and reduce the unavoidable impacts of climate change in the short and medium terms, and are usually associated with less-developed countries (LDCs).

- In the policy arena, the European Union countries tend to exert the strongest leadership in fighting climate change. The United States has often been slow or unwilling to take strong, specific actions. China has tried to be exempted from mandatory steps.

- The Kyoto Protocol was an international agreement requiring the industrialized countries that ratified it to make substantial cuts in their carbon dioxide emissions to reduce global warming. The United States dropped its support for the treaty. It went into effect for the ratifying countries after Russia ratified it in 2004 and expired in 2012. Efforts continued to reduce carbon dioxide emissions, especially in the MDCs and China, and to maximize ways to reduce deforestation, especially in the LDCs.

Key Terms + Concepts

adaptation (p. 34)
aquaculture (p. 30)
atmosphere (p. 20)
biodiversity (p. 28)
biodiversity hot spots (p. 28)
biological diversity (p. 28)
biomes (p. 23)
   boreal forest (p. 25)
   coniferous forest (p. 25)
   desert shrub (p. 25)
   Mediterranean scrub forest (p. 27)
   prairie (p. 26)
   savanna (p. 26)

steppe (p. 25)
taiga (p. 26)
temperate grassland (p. 26)
temperate mixed forest (p. 27)
tropical deciduous forest (p. 26)
biosphere (p. 20)
cap-and-trade system (p. 34)
carbon sequestration (p. 34)
carbon sink (p. 34)
climate (p. 23)
   desert (p. 25)
   humid continental (p. 27)

humid subtropical (p. 27)
ice cap (p. 25)
marine west coast (p. 26)
Mediterranean (p. 26)
semiarid (p. 25)
subarctic (p. 25)
tropical rain forest (p. 26)
tropical savanna (p. 26)
tundra (p. 25)
undifferentiated highland (p. 28)
coniferous trees (p. 26)
continental drift (p. 21)
emission-trading system (p. 34)

fault (p. 21)
faulting (p. 21)
fish farming (p. 30)
greenhouse effect (p. 31)
greenhouse gases (p. 31)
   carbon dioxide (CO_2) (p. 31)
   chlorofluorocarbons (CFCs) (p. 31)
   methane (p. 31)
   nitrous oxide (p. 31)
hydrologic cycle (p. 29)
hydrosphere (p. 20)

Intergovernmental Panel on
Climate Change (IPCC)
(p. 31)
Kyoto Protocol (p. 35)
lapse rate (p. 28)
lithosphere (p. 20)
mitigation (p. 34)

monoculture (p. 28)
Montreal Protocol (p. 34)
plates (p. 21)
plate tectonics (p. 21)
precipitation (p. 23)
Reducing Emissions from
Deforestation and Forest

Degradation (REDD)
(p. 35)
Richter scale (p. 21)
seafloor spreading (p. 21)
seismic activity (p. 21)
subduction (p. 21)
subduction zone (p. 21)

tectonic forces (p. 21)
tipping point (p. 33)
trench (p. 21)
tsunami (p. 21)
volcanism (p. 21)
weather (p. 23)

Review Questions

1. What are the three "spheres" of habitable life on Earth?

2. What is plate tectonics, and what are some of the main consequences of tectonic activity?

3. What is the difference between weather and climate? What are the main forces that produce precipitation and aridity?

4. What are the major climate types and their associated biomes? Where do they tend to occur on Earth?

5. Where are the biodiversity hot spots? In what kinds of locations and biomes do many of them occur?

6. What important roles do the world's oceans play, and what are their major resources? In what ways are these resources threatened?

7. What apparent long-term effects on the Earth's atmosphere can be attributed to modern technology? What predictions are there for future changes, and what steps are being taken or considered to avert these changes through mitigation and through adaptation?

Notes

1. Wolf Roder, "Three Near Misses: Are These Science or Pseudoscience?" *Cincinnati Skeptics Newsletter*, June 1996. Accessed November 14, 2006, from http://www.cincinnatiskeptics.org/newsletter/vol5/n5/misses.html.

2. Norman Myers, *Primary Source: Tropical Forests and Our Future* (New York: Norton, 1984), p. x.

3. World Bank, "On World Water Day, World Bank Calls for Investments in Water Infrastructure and Better Governance," press release, March 22, 2007.

4. Review the United Nations' *Annual Report of the State of the World's Fisheries and Aquaculture* at www.fao.org/sof/sofia/index_en.htm.

5. Boris Worm et al., "Impacts of Biodiversity Loss on Ocean Ecosystem Services," *Science*, November 3, 2006, pp. 787–790.

6. See the Monterey Bay Aquarium's guide to sustainable seafood, "Seafood Watch," at www.mbayaq.org/cr/seafoodwatch.asp.

7. Amy Roach Partridge, "Global Trucking Woes," *Global Logistics*, November 2006, www.inboundlogistics.com/articles/global/global1106.shtml. Accessed September 27, 2007.

8. Intergovernment Panel on Climate Change, "Climate Change 2007," www.ipcc.ch. Accessed July 4, 2007.

9. Sarah Lyall, "Global Warming Report Predicts Doom for Many Species," *New York Times*, September 1, 2000.

10. Brad Knickerbocker, "Could Warming Cause War?" *Christian Science Monitor*, April 19, 2007, p. 2.

11. Source on carbon sequestration: www.carbonventures.com/policy/article.php?list=The%20Kyoto%20Protocol&id=4467&link=Carbon%20Sequestration/

12. Source on deforestation and carbon: www.un-redd.org/AboutREDD/tabid/582/Default.aspx/

Online Resources

CourseMate: Make the most of your study time by accessing everything you need to succeed in one place. Read your textbook, take notes, review flashcards, watch videos, complete activities, take practice quizzes, and more—online with CourseMate. Log in at **www.cengagebrain.com**.

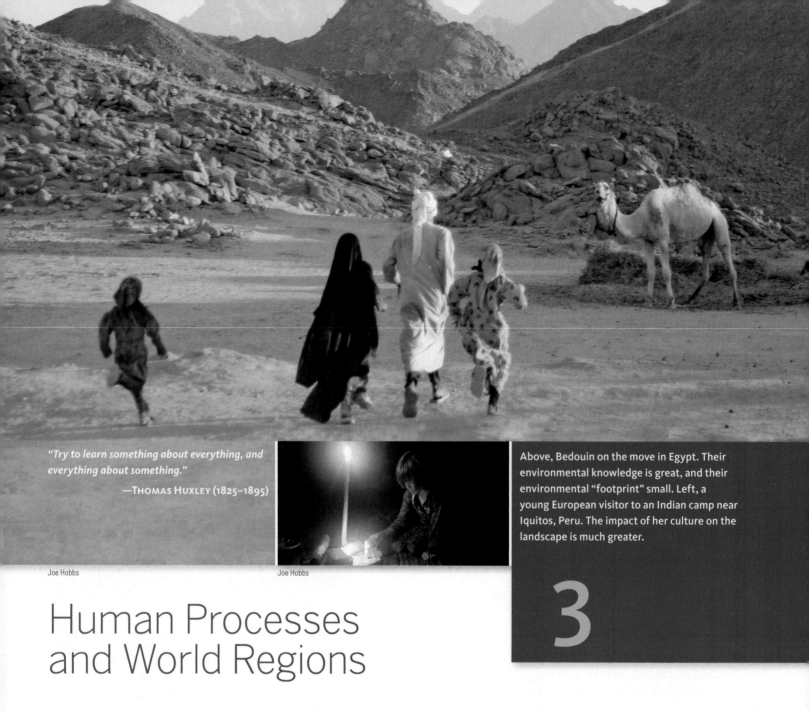

"Try to learn something about everything, and everything about something."

—THOMAS HUXLEY (1825–1895)

Above, Bedouin on the move in Egypt. Their environmental knowledge is great, and their environmental "footprint" small. Left, a young European visitor to an Indian camp near Iquitos, Peru. The impact of her culture on the landscape is much greater.

Joe Hobbs

Joe Hobbs

3

Human Processes and World Regions

This chapter continues your introduction to geography's basic vocabulary, focusing on how people have interacted with the environment to change the face of the Earth, with an emphasis on the human roles. Modern trends in human–environment interactions are seen as the products of revolutionary changes in the past: the arrival of agriculture and of industrialization. The chapter uses some of the concepts of economic geography to explain where rich and poor countries are located on the Earth's surface and account for some of these patterns of prosperity and poverty. It considers where and why populations are increasing and what the implications of that growth are. Finally, it shares ideas about how to solve some of the most important global problems of our time.

chapter objectives

This chapter should enable you to

- Gain a historical perspective on the capacity of human societies to transform environments and landscapes.
- Understand why some countries are rich and others poor and recognize the geographic distribution of wealth and poverty.
- Explain the simultaneous trends of falling population growth in the richer countries and rapid population growth elsewhere.
- Explore the principles of sustainable development.

• **Figure 3.1** Until the relatively recent past—just a few thousand years ago—people were exclusively hunters and gatherers. This rock art in Egypt's Sinai Peninsula was created over a long time span, as it depicts Neolithic-period hunting of ibex, later uses of camels and horses, and writing from the Nabatean period (first century C.E.).

3.1 Two Revolutions That Have Changed the Earth

The geographer's approach to understanding a current landscape—in almost all cases, a cultural landscape that has been fashioned by human activity—is often historical, involving study of its development from the prehuman or early human natural landscape. The current spatial patterns of our relationship with the Earth may be seen as products of two "revolutions": the Agricultural Revolution that began in the Middle East about 10,000 years ago and the Industrial Revolution that began in eighteenth-century Europe. Each of these revolutions transformed humanity's relationship with the natural environment. Each increased substantially our capacity to consume resources, modify landscapes, grow in number, and spread in distribution. Why should we consider these historic processes? The answer lies in the name the Geological Society of America gave to its annual meeting in 2011: "The past is the key to the future."

Hunting and Gathering

Until about 10,000 years ago, our ancestors lived by **hunting and gathering** (also known as **foraging**). We were quite good at it—it served us well for more than 100,000 years, until we began experimenting with the revolutionary technologies of agriculture.

Foraging was quite different from the farming and industrial ways of living that succeeded it. Joined in small bands of extended family members, hunters and gatherers were nomads with no villages, homes, or other fixed dwellings. They moved to take advantage of changing opportunities on the landscape. These foragers scouted large areas to locate foods such as seeds, tubers, foliage, fish, and game animals (•Figure 3.1). Moving their small group from place to place, they had a relatively limited impact on the natural environment, especially compared with the impacts left by agricultural and industrial societies.

Hunters and gatherers may have been the **"original affluent society."**[1] Many scholars have praised these preagricultural people for the apparent harmony they maintained with the natural world in both their economies and their spiritual beliefs. After short periods of work to collect the foods they needed, they enjoyed long stretches of leisure time. Studies of those few hunter-gatherer cultures that lingered into modern times, such as the San (Bushmen) of southern Africa and several Amerindian groups of South America, suggest that although their life expectancy was low, they suffered little from the debilitating social and psychological problems found in industrialized societies.

Hunters and gatherers did modify their landscapes. These foragers were not always at peace with one another or with the natural world. With upright posture, stereoscopic vision, opposable thumbs, an especially large brain, and no mating season, *Homo sapiens* (Latin for "wise man") became, after its emergence in southern Africa about 200,000 years ago, an **ecologically dominant species**—defined as one that competes more successfully than other organisms for nutrition and other essentials of life, or that exerts a greater influence than other species on the environment. Using fire to flush out or create new pastures for the game animals they hunted, preagricultural people shaped the face of the land on a vast scale relative to their small numbers. Many of the world's prairies, savannas, and steppes where grasses now prevail developed as hunters and gatherers repeatedly set fires.

These people also overhunted and in some cases eliminated animal species. The controversial **Pleistocene overkill hypothesis** states that rather than being at harmony with nature, hunters and gatherers of the late Pleistocene Epoch (a geological period from about 2 million to 10,000 years ago) hunted many species to extinction, including the elephant-like mastodon of North America.

Farming: Welcome to the Anthropocene

Despite these excesses, the environmental changes that hunters and gatherers could cause were limited. Humans' power to modify landscapes took a giant step with **domestication**, the controlled breeding and cultivation of plants and animals. Domestication brought about the **Agricultural Revolution**, also known as the **Neolithic Revolution** or the **Food-Producing Revolution**.

This was the dawn of a new age, the likes of which the world had never experienced. In conventional geologic time scales, the Agricultural Revolution that began about 10,000 years ago coincided with a great retreat of glacial ice sheets and the onset of the Holocene Epoch in which we now live (Holocene comes from Greek words meaning "entirely new"). But there is a fascinating scientific effort underway to rewrite our understanding of the age in which we live. The goal is to rename the Holocene as the **Anthropocene Epoch** or simply "the Anthropocene," meaning "the age of humans." Peoples' impacts on the atmosphere, lithosphere, and hydrosphere have been so monumental in the last 10,000 years—and especially within the last 250 years—that they rank alongside the great natural forces that defined past ages.

It all began with agriculture. Why people started to produce rather than continue to hunt and gather plant and animal foods—first in the Middle East and later in Asia, Europe, Africa, and the Americas—is uncertain. Although there are many ideas about this process, surprisingly little hard evidence exists to explain it confidently. Two theories are most often put forward. One is that the climate changed. Increasing drought and reduced plant cover may have forced people and wild plants and animals into smaller areas, where people began to tame wild herbivores and sow wild seeds to produce a more dependable food supply.

A more widely accepted theory is that their own growing populations in areas originally rich in wild foods compelled people to find new food sources, so they began sowing cereal grains and breeding animals. The latter process may have begun about 8000 B.C.E. in the Zagros Mountains of what is now Iran. The culture of domestication spread outward from there (in the geographic process known as **diffusion**), but also developed independently in several world regions.

Now that humans were producing as well as consuming foods, their landscape uses, cultures, social organizations, and other characteristics changed dramatically. Among other things, in choosing to breed plants and animals, people settled down. Gradually abandoning the nomadic life and **extensive land use** of hunting and gathering, they came to favor the **intensive land use** of agriculture and animal husbandry. They could sow and harvest crops in specific places year after year. With less need to move around, they began living in fixed dwellings, at least on a seasonal basis. These developed into villages, small settlements with fewer than 5,000 inhabitants.

• **Figure 3.2** The Tigris River in southeastern Turkey. The brown areas are rainless in the long, hot summers and are capable of producing only one crop a year through dry (unirrigated) farming. The green areas along the river are irrigated and can produce two or three crops a year. Irrigation was thus a revolutionary technology that greatly increased the number of people the land could support.

People in these settlements raised larger and more reliable stocks of food, making it possible to support their growing numbers. Through **dry farming**, which involved planting and harvesting according to the seasonal rainfall cycle, population densities could be 10 to 20 times higher than they were in the hunting and gathering mode. By about 4000 B.C.E., people along the Tigris, Euphrates, and Nile Rivers began **irrigation** of crops—bringing water to the land artificially by using levers, channels, and other technologies—an innovation that allowed them to grow crops year-round, independent of seasonal rainfall or river flooding (•**Figure 3.2**). Irrigation allowed even more people to make a living off the land; irrigated farming yields five to six times more food per unit area than dry farming. In ecological terms, the expanding food surpluses of the Agricultural Revolution raised the Earth's **carrying capacity**, the size of a species' population (in this case, humans) that an ecosystem can support. When domestication of plants and animals began about 10,000 years ago, there were probably fewer than 5 million people on Earth. Today, there are more than 7 billion.

Culture became more complex, and society became more stratified. The steep increase in food production freed more people from the actual work of producing food, and they undertook a wide range of activities unrelated to subsistence needs. Irrigation and the dependable food supplies it provided thus set the stage for the development of **civilization**, the complex culture of urban life characterized by the appearance of writing, economic specialization, social stratification, and high population concentrations. By 3500 B.C.E., for example, 50,000 people lived in the southern Mesopotamian city of Uruk, in what is now Iraq. Other **culture hearths**—regions where civilization followed the domestication of plants and animals—emerged between 8000 and 2500 B.C.E. in China, Southeast Asia, the Indus River Valley, Egypt, West Africa, Mesoamerica, and the Andes.

Human impacts on the natural environment increased. The agriculture-based urban way of life that spread from these culture hearths had larger and more lasting impacts on the natural environment than either hunting and gathering or early agriculture. Acting as agents of humankind, domesticated plants and animals proliferated at the expense of the wild species that people came to regard as pests and competitors. For example, *Bos primigenius,* a wild bovine that was the ancestor of most of the world's domesticated cattle, was hunted to extinction by 1627. Farmers who resented the animals' raids on their crops were probably among the most ardent hunters. Agriculture's permanent and site-specific nature magnified the human imprint on the land, while the pace and distribution of that impact increased with growing numbers of people.

The Industrial Revolution

The human capacity to transform natural landscapes took another giant leap with the **Industrial Revolution**, which began in Europe around 1750 c.e. This new pattern of human-land relations was based on breakthroughs in technology that several factors made possible (•**Figure 3.3**):

- Western Europe had the economic capital necessary for experimentation, innovation, and risk. Much of this money came from the lucrative trade in gold, undertaken initially in the Spanish and Portuguese empires after 1400.

- Significant improvements in agricultural productivity took place in Europe by about 500. New tools such as the heavy plow and more intensive and sustainable use of farmland led to increased crop yields. Human populations grew correspondingly.

- Population growth itself was a factor. More people freed from work in the fields could devote their talents and labor to trying new things, including tinkering with technologies that would improve crop yields. As agricultural innovations spread and industrial productivity improved, more and more of the growing European population was freed from farming, and for the first time in history, a region emerged in which city dwellers outnumbered rural folk. The process

of industrialization, having now spread to nearly all parts of the world, continues to promote urbanization today.

Most geographers see population growth today as a drain on resources, but the Industrial Revolution illustrates that given the right conditions, more people do create more resources. Innovations such as the steam engine tapped the vast energy of fossil fuels—initially coal and later oil and natural gas. Fossil fuels, which are the stored-up carbon products created by photosynthesis in ancient ecosystems, allowed the Earth's carrying capacity for humankind to be raised again, this time into the billions.

Industrialization, Colonization, and Environmental Change

As they began to deplete their local supplies of resources needed for industrial production, Europeans started to look for these materials abroad. As early as their **Age of Discovery**, also known as the **Age of Exploration**, which began in the fifteenth century, Europeans probed ecosystems across the globe to feed a growing appetite for innovation, economic growth, and political power. The process of European **colonization**—the extension of European countries' political and economic control over foreign areas—was thus linked directly to the Industrial Revolution. Mines and plantations from such faraway places as central Africa and India supplied the copper and cotton that fueled economic growth in colonizing countries such as Belgium and England.

No longer dependent on the foods and raw materials they could procure within their own political and ecosystem boundaries, Europeans of the Industrial Revolution had an impact on the natural environment that was far more extensive and permanent than that of any other people in history. The world today is heir to their legacy: we tend to think of countries and economies being "industrialized" or "industrializing," either possessing that enormous power or on the pathway to acquiring it. It is difficult to summarize the impacts of agriculture and industrialization during the Anthropocene; much of this book tells that story. But these are some good examples:[2]

- The vast majority of Earth's ecosystems reflect the presence of people.

- Today, fully 40 percent of the Earth's land-based photosynthetic output is dedicated to human uses, especially in agriculture and forestry.

- Since 1750, the total forested area on Earth has declined by more than 20 percent. Today, there are more trees on farms than in wild forests.

- Since 1750, total cropland has grown by nearly 500 percent, with more expansion in the period from 1950 to today than in the century from 1750 to 1850. Much of this expansion has been made possible by the use of fertilizers produced from fossil carbon.

- Since 1750 people have released quantities of fossil carbon that the planet took hundreds of millions of years to store away. This has given people a commanding role in the planet's carbon cycle.

• **Figure 3.3** L. S. Lowry's *Industrial Landscape* has all the essential elements of industrializing Britain: cotton mills and factories, coal mines, and working families' cottages.

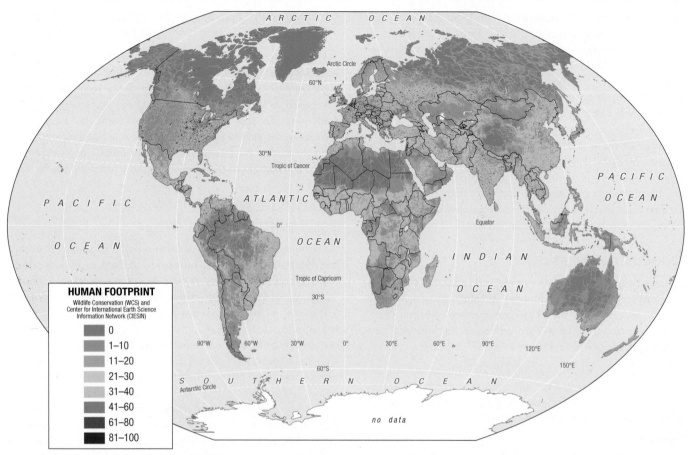

• **Figure 3.4** The biome map on page 24 depicts what the world's vegetation would look like without human activity. Here we see how strongly people have changed the natural environment. This isarithmic map of the human footprint is a quantitative analysis of human impacts on the Earth's biomes. A score of 1 indicates the least human influence in a given biome. A higher number means greater impact. However, because each biome has its own independent scale, a score of 1 in a tropical rainforest might reflect a different level of human activity than in a broadleaf forest.

Source: Wildlife Conservation Society (WCS) and Center for International Earth Science Information Network (CIESIN).

The global impacts of human activities are best summarized with the map of the "human footprint" depicted in •Figure 3.4. It shows uneven distributions of human impacts across the globe. These are the results of economic, population, and other processes we will now explore.

3.2 The Geography of Development

One of the most striking characteristics of human life on Earth is the large disparity between wealthy and poor people, both within and between countries. At a high level of generalization, the world's countries can be divided into "haves" and "have-nots" (•Figure 3.5 and •Table 3.1). Writers refer to these distinctions variously as "developed" and "underdeveloped," "developed" and "developing," "more developed" and "less developed," "industrialized" and "industrializing" or "industrialized" and "nonindustrialized," and "North" and "South," based on the concentration of wealthier countries in the middle latitudes of the Northern Hemisphere and the abundance of

poorer nations in the Southern Hemisphere. This text uses the terms **more developed countries (MDCs)** and **less developed countries (LDCs)**. This framework is an introductory tool and cannot account for the tremendous variations and continuous changes in economic and social welfare that characterize the world today. Some countries, including those known as the "Asian Tigers," are best described as **newly industrializing countries (NICs)** because they do not fit the MDC or LDC idealized types. The relevant regional chapters describe these cases.

Measuring Development

There is no single, universally acceptable standard for measuring wealth and poverty on the global scale. But you are likely to encounter these kinds of indexes and figures as you read about development.

Annual per capita gross domestic product. **Gross domestic product (GDP)** is the total output of goods and services that a country produces for home use in a year. Divided by the country's population, the resulting figure of **per capita GDP** is one

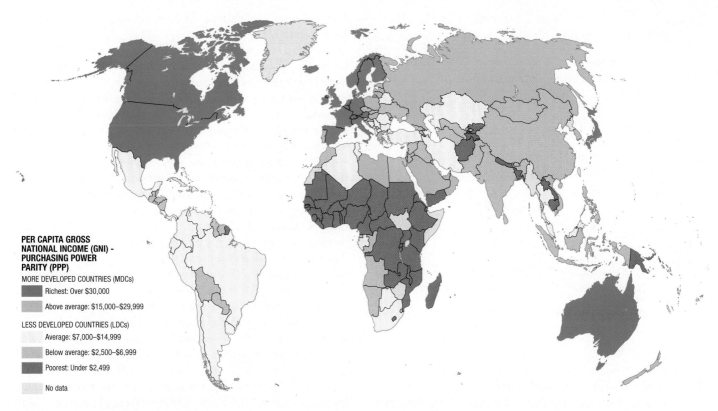

**PER CAPITA GROSS
NATIONAL INCOME (GNI) -
PURCHASING POWER
PARITY (PPP)**

MORE DEVELOPED COUNTRIES (MDCs)

Richest: Over $30,000

Above average: $15,000–$29,999

LESS DEVELOPED COUNTRIES (LDCs)

Average: $7,000–$14,999

Below average: $2,500–$6,999

Poorest: Under $2,499

No data

• **Figure 3.5** Wealth and poverty by country. Note the concentration of wealth in the middle latitudes of the Northern Hemisphere, and Africa's great poverty.

Source: Data from World Factbook, CIA, 2004.

Table 3.1	Characteristics of More Developed Countries (MDCs) and Less Developed Countries (LDCs)	
Characteristic	**MDC**	**LDC**
Per capita GDP and income	High	Low
Percentage of population in the middle class	High	Low
Percentage of population involved in manufacturing	High	Low
Energy use	High	Low
Percentage of population living in cities	High	Low
Percentage of population living in rural areas	Low	High
Birth rate	Low	High
Death rate	Low	High[a]
Population growth rate	Low	High
Percentage of population under age 15	Low	High
Percentage of population that is literate	High	Low
Amount of leisure time available	High	Low
Life expectancy	High	Low

[a]Although death rates are high in the LDCs relative to the MDCs, in most countries they are quite low compared to what they once were in the LDCs.

of the most commonly used measures of economic well-being. A closely related measure is per capita **gross national product (GNP)**, which includes foreign output by domestically owned producers.

Annual per capita gross national income purchasing power parity. Although quite a mouthful, **per capita gross national income purchasing power parity (per capita GNI PPP)** is a useful figure in the study of world regional geography. **Gross national income (GNI)** includes gross domestic production plus income from abroad from sources such as rents, profits, and labor. **Purchasing power parity (PPP)** conversion factors consider differences in the relative prices of goods and services, providing a better overall way of comparing the real value of output between different countries' economies. GNI PPP is measured in current "international dollars," which indicate the amount of goods and services one could buy in the United States with a given amount of money. Definitions vary, but in this book, an MDC is a county with an annual per capita GNI PPP of $15,000 or more; all others are considered LDCs. By the way, economics is known sarcastically as "the dismal science," and you should probably not be too concerned with how the figures were calculated; instead, use them for simple comparisons.

The gulf between the world's richest and poorest countries is startling (•**Table 3.2**). The average per capita GNI PPP in the MDCs is nearly 6 times greater than in the LDCs. With a per capita GNI PPP of $59,950, Luxembourg is the world's richest country, whereas Liberia is the poorest with only $290. Forty-eight percent of the world's people live on less than $2 per day. These raw numbers suggest that economic productivity and income alone characterize **development**, which, according to a common definition, is a process

Table 3.2   Top and Bottom Countries in Per Capita GNI-PPP (U.S. Dollars, 2011)*

Top Countries	GNI-PPP Per Capita	Bottom Countries	GNI-PPP Per Capita
Luxembourg	59,590	Afghanistan	860
Norway	55,420	Togo	850
Singapore	49,780	Sierra Leone	790
Switzerland	47,100	Malawi	780
United States	45,640	Central African Rep.	750
Netherlands	39,740	Niger	680
Denmark	38,780	Eritrea	580
Australia	38,510	Burundi	390
Austria	38,410	Congo, Dem. Rep.	300
Sweden	38,050	Liberia	290

*This table excludes city-states, territories, colonies, and dependencies. Only countries with available data are listed.

Source: 2011 World Population Sheet, Population Reference Bureau.

of improvement in the material conditions of people through diffusion of knowledge and technology.

The Human Development Index. Definitions like that, and statistics like per capita GDP and per capita GNI PPP, reveal little about measures of well-being such as income distribution, gender equality, literacy, and life expectancy. Recognizing the shortcomings of strictly economic definitions, the United Nations Development Programme created the **Human Development Index (HDI)**, a scale that considers attributes of quality of life. This book uses HDI in the Basic Data tables of the chapter that introduce each region (for example, Table 4.1 on page 67). On the HDI scale, a measure of 1.0 means "perfect." According to this index, with a rating of 0.938, Norway is "the world's best place to live," although it ranks second in per capita GNI PPP. In Chapter 4, we will see why the Norwegians and other Scandinavians are living so well. Following Norway, in descending order, are Australia, New Zealand, the United States, and Ireland. In HDI terms, with a measure of just 0.140, Zimbabwe is the world's worst place to live; the Democratic Republic of the Congo, Niger, Burundi, and Mozambique fare only slightly better. Note that all five of these low-rated countries are in Africa, for reasons discussed in Chapter 9.

On the basis of per capita GNI PPP, 1.2 billion, or 18 percent, of the world's people, inhabit the MDCs. Most citizens of these countries, such as the United States, Canada, Japan, South Korea, Australia, New Zealand, and the nations of Europe, enjoy an affluent lifestyle with freedom from hunger. Employed in industries or services, most of the people live in cities rather than in rural areas. Disposable income, the money that people can spend on goods beyond their subsistence needs, is generally high. There is a large middle class. Population growth is low as a result of low birth rates and low death rates (these demographic terms are explained in

Section 3.3 on population). Life expectancy is long, and the literacy rate is high.

Life for a very large number of the planet's other 82 percent, or about 5.7 billion people, is quite different. In the LDCs, including many countries in Latin America, Africa, and Asia, poverty and often hunger prevail. The leading occupation is subsistence agriculture, and the industrial base is small. Most people live in rural areas, but that statistic will change soon; 46 percent of the LDCs' population is urban. The middle class tends to be small but growing, with an enormous gulf between the vast majority of poor and a very small wealthy elite, who own most of the private landholdings. With high birth rates and falling death rates, population growth is high. Life expectancy is short, and the literacy rate is low.

LDCs are not predestined to be in poverty forever. Effective development policies and practices have changed the outlook for many countries. Doomsayers predicted for decades that the colossal country of India would one day collapse under the weight of overpopulation and poverty. As we will see in Chapter 7, India is instead emerging as a global economic power to be reckoned with, for example, by using its human capital to generate wealth from software and outsourced services (see Insights feature on page 235). A few decades ago, China was a desperately poor country, but today it has the world's second largest economy. Still, because of their large populations, both countries face many challenges—including big gaps between their haves and have-nots—and neither ranks as a more developed country.

With four-fifths of the world's people living in the poorer, less developed countries, it is important to understand the root causes of underdevelopment and to appreciate how wealth and poverty affect the global environment in very different but equally profound ways. With this knowledge, it becomes possible to envision solutions.

Why Are Some Countries Rich and Others Poor?

Many theories attempt to explain the disparities between MDCs and LDCs. It is important to recognize that there is no single, widely accepted explanation about development in general or in a particular region or country. Probably the best thing you can do with your critical thinking skills is to weigh the various explanations and see which one, or which combination of them, seems to fit the situation of a given country or region. Here are the main explanations you are likely to come across.

Embraced most strongly in the LDCs, **dependency theory** argues that the worldwide economic pattern established by the Industrial Revolution and by the forces of colonialism that accompanied it persists today. In his book *Ecological Imperialism*, the historian and geographer Alfred Crosby explains how dependency led to the rich–poor divide, depicting the two very different ways in which European powers used foreign lands during the Industrial Revolution.[3] In the pattern of **settler colonization**,

Europeans sought to create new Europes, or **neo-Europes**, in lands much like their own: temperate middle-latitude zones with moderate rainfall and rich soils where they could raise wheat and cattle. And so, between 1630 and 1930, more than 50 million Europeans emigrated from their homelands to create European-style settlements in what are now Canada, the United States, Argentina, Uruguay, Brazil, South Africa, Australia, and New Zealand. These lands were destined to become some of the world's wealthier regions and countries.

In contrast to their preference to settle familiar mid-latitude environments, Europeans viewed the world's tropical lands mainly as sources of raw materials and markets for their manufactured goods. The environment was too different from home to make settlement attractive. In establishing a pattern of **mercantile colonialism**, Crosby explains, Europeans were less inhabitants than conquering occupiers of the colonies, overseeing indigenous peoples and resettled slaves in the production of primary or unfinished products: sugar in the Caribbean; rubber in Latin America, West Africa, and Southeast Asia; and gold and copper in southern Africa, for example. Colonialism required huge migrations of people to extract the Earth's resources, including 30 million slaves and contracted workers from Africa, India, and China to work mines and plantations around the globe.

294, 322, 365, 409

In the mercantile system, the colony provided raw materials to the ruling country in return for finished goods; this meant that people in India would purchase clothing made in England from the raw cotton they themselves had harvested. The relationship was most advantageous to the colonizer. England, for example, would not allow its colony India to purchase finished goods from any country but England. It prohibited India from producing any raw materials the empire already had in abundance, such as salt (India's Mohandas Gandhi defiantly violated this prohibition in his famous "March to the Sea"). Finished products are **value-added products**, meaning they are worth much more than the raw materials they are made from, so manufacturing in the ruling country concentrated wealth there while limiting industrial and economic development in the colony.

The colony was obliged to contribute to, but was prohibited from competing with, the economy of the ruling country—a relationship that dependency theorists insist continues today. Dependency theory asserts that to participate in the world economy, the former colonies, now independent countries, continue to depend on exports of raw materials to and purchases of finished goods from their former colonizers and other MDCs, and this disadvantageous position keeps them poor. Dependency theorists call this relationship **neocolonialism**. With independence, the former colonies needed revenue. To earn that money, they continued to produce the goods for which markets already existed—generally the same unprocessed primary products they supplied in colonial times. Dependency theorists argue that when the former colonies try to break their dependency by becoming exporters of manufactured goods, the MDCs impose trade barriers and quotas to block that development (see Insights, page 420). Dependency theory and neocolonialism also envision the world in terms of core and periphery—a concept of duality you have seen already in Chapter 1 and will encounter often in the book—with wealth flowing from the periphery of the world's poor countries to its core of wealthy countries.

Whether because of neocolonialism or a more complex array of variables, many developing countries continue to rely heavily on income from the export of a handful of raw materials. This makes them vulnerable to the whims of nature and the world economy. The economy of a country heavily dependent on rubber exports, for example, may suffer if an insect pest wipes out the crop or if a foreign laboratory develops a synthetic substitute. When demand for rubber rises, that country may actually harm itself trying to increase its market share by producing more rubber, because in doing so, it drives down the price. (Consider oil: members of the Organization of Petroleum Exporting Countries [OPEC] cartel of oil-producing countries drove oil prices to all-time lows in the mid-1980s when they overproduced oil in a bid to earn more revenue.) If the rubber-producing country withholds production to shore up rubber's price, it provides consuming countries with an incentive to look for substitutes and alternative sources. (Again, look at oil: after OPEC embargoed shipments of oil to the United States in the 1970s, the United States began developing domestic oil supplies and becoming more energy-efficient, thus temporarily reducing oil prices and OPEC's revenues.) The developing country is in a dependent and disadvantaged position—even if it produces something as vital as oil.

Many geographers view dependency theory and neocolonialism as simplistic and politically biased explanations of development. They consider a wider and more complex set of factors, including culture, location, and natural environment, to explain why some countries are wealthy and others are poor. These include the following factors:

Advantageous and disadvantageous location. Does a country's location play a role in how rich or poor it is? In some cases, yes. Location can influence a country's economic fortunes. For example, as discussed in Chapter 1, because it is situated close to a great mainland with which to trade, the island of Great Britain enjoys a core location favorable for economic development. Japan has a similar location relative to the Asian landmass. In contrast, landlocked nations such as Bolivia in South America and numerous nations in Africa have locations unfavorable for trade and economic development, and they have not overcome this disadvantage. But it is important to recognize that geographic location is never the sole decisive factor in development. Like Japan and Britain, Madagascar and Sri Lanka are island nations situated close to large mainlands, but for a variety of (mainly political) reasons discussed in the relevant chapters, neither has experienced prosperity.

Resource wealth or poverty. Having or lacking a diversity or abundance or **natural resources** (naturally occurring materials that people can use) plays a significant role in development.

One of the most remarkable international trends of recent decades has been **globalization**—the spread of free trade, free markets, investments, and ideas across borders and the political and cultural adjustments that accompany this diffusion. There has been much debate and sometimes-violent conflict over the pros and cons of globalization. For good or ill, globalization is a reality. *The World Is Flat*, journalist Thomas Friedman entitled his book on the subject. He observed, ironically, that in the search for India, Christopher Columbus proved the world was round; but in visiting India, Friedman himself found the world to be flat: the "playing field" of the international economy had been leveled: "It is now possible for more people than ever to collaborate and compete in real time with more people on more different kinds of work from more different corners of the planet and on a more equal footing than at any previous time in the history of the world—using computers, e-mail, fiber-optic networks, teleconferencing and dynamic new software."[4]

Supporters of globalization argue that the newly emerging global economy will bring increased prosperity to the entire world. Innovations in one country will be transferred instantly to another country, productivity will increase, and standards of living will improve. They propose that one obvious solution to the problem of the LDCs' inability to compete in the world economy, and therefore escape their dependency, is to reduce the trade barriers that MDCs have erected against them (some of these barriers are discussed on **page 420** in **Chapter 11**). With free trade, free enterprise will prosper, pumping additional capital into national economies and raising incomes for all. Many of the inputs into globalization come from **multinational companies** (also called *transnational companies*, meaning companies with operations outside their home countries) as they increase their investments abroad. Most of the companies are based in the MDCs,

but multinational corporations have also grown in LDCs such as China, India, and Mexico.

Opponents of globalization argue that the process will actually increase the gap between rich and poor countries, and even within countries; a selected few developing nations (or people within a country) will prosper from increased foreign investment and resulting industrialization, but the hoped-for global wealth will bypass other countries (or people within a country) altogether. And the increasing interdependence of the world economies will make all the players more vulnerable to economic and political instability. The multinational companies will recognize huge profits at the expense of poor wage laborers. In addition, environments will be harmed if environmental regulations are reduced to a lowest common denominator (for example, the high standards of air quality demanded by the U.S. Clean Air Act would be deemed noncompliant with World Trade Organization rules because they make it harder for countries with "dirtier" technologies to compete in the marketplace). Such concerns often lead to massive protests against "corporate-led globalization," especially at meetings of the World Trade Organization, the International Monetary Fund, and the World Bank. Many more confrontations like these can be anticipated. Typical protesters' demands are that working conditions be improved in the foreign "sweatshops," where textiles and other goods are produced at low cost for U.S. corporations, and that corporations like Starbucks should sell only "fair trade" coffee beans bought at a price giving peasant coffee growers a living wage rather than at the "exploitive" price typically paid. Corporations like Starbucks have often been quick to respond to the demands of the environmentally and socially conscious consumer.

It is often assumed that the rapid growth and spread of computer and wireless technologies are benefiting everyone, but in fact there is still a **digital divide** to overcome. The divide

is between the handful of countries that are the leaders in **information technology (IT)** and the majority of nations that have little ability to create, purchase, or use new technologies. The fear in the technology-lagging countries is that the growth of e-commerce will concentrate the wealth generated by that commerce in the technology leadership countries. This would enable the leaders to make even further advances in technology and realize even bigger economic gains from a so-called **knowledge economy** based on innovation and services, while the technology laggards fail to catch up and simply become poorer (•**Figure 3.A**). There are many efforts underway to empower the world's poor through access to information technology; the "One Laptop per Child" program seeks to provide the world's children with rugged, low-cost, low-power, Internet-connected laptops (see http://.onelaptop.org). Rapid developments and falling prices in smartphone technology could also help level the playing fields of globalization. 331

• **Figure 3.A** An Internet café in Mérida, Yucatán, Mexico. Information technology—especially in the form of computers, the Internet, and cellular telephones—is spreading rapidly around the world. Some people believe that there is now an unfolding Information Revolution comparable to the Agricultural and Industrial Revolutions in its capacity to change humanity's relationship with the earth. 196

Joe Hobbs

Superabundance of an especially valuable resource (for example, oil in the Persian/Arabian Gulf countries) or a diversity of natural resources has helped some countries become more developed than others. The former Soviet Union and the United States achieved superpower status in the twentieth century in large part by using the enormous natural resources of both countries.

Cultural and historical factors. In some cases, human industriousness has helped compensate for resource limitations and helped to promote development. For example, Japan has a rather small territory with few natural resources (including almost no petroleum). Yet in the second half of the twentieth century, it became an industrial powerhouse largely because the 270

Japanese people united in a common purpose to rebuild from wartime devastation, placing priorities on education, technical training, and seaborne trade from their advantageous island location. Conversely, cultural or political problems like corruption and ethnic factionalism can hinder development in a resource-rich nation, as in the mineral-wealthy Democratic Republic of Congo.

Incidentally, many students have told me they enjoyed reading the geographer Jared Diamond's book *Guns, Germs, and Steel* because he writes so engagingly about the forces that propelled countries forward or held them back. Diamond's book has come to be known as "A Short History of Everybody for the Last 13,000 Years"!

Environmental Impacts of Underdevelopment

Geographers are very interested in the environmental impacts of relations between MDCs and LDCs, especially on the poorer countries. LDCs generally lack the financial resources needed to build roads, dams, energy grids, and other infrastructure critical to development. Then they turn to the World Bank, International Monetary Fund (IMF), and other institutions of the MDCs to borrow funds for these projects. Many borrowers are unable to pay even the interest on these loans, which is sometimes huge; debtor nations have been known to spend as much as 40 percent of annual revenues on interest payments to the lenders, or more than they spend on education and health combined.

This debt cycle can be very harmful to natural environments. When lender institutions threaten to cut off assistance, borrowing countries often try to raise cash quickly to avoid this prospect, generally by one or two methods, or both. One method is to dedicate more land to the production of **cash crops** (also known as **commercial crops**). These are items such as coffee, tea, sugar, coconuts, and bananas exported to the MDCs, where they are luxuries or staples (illegal drug crops are also cash crops, of course, but fall outside most statistics and many of the general associations made here). Governments or foreign corporations often buy out, force out, or otherwise displace subsistence food farmers in the search for new lands on which to grow these commercial crops. In this process, known as **marginalization**, poor subsistence farmers are pushed onto fragile, inferior, or marginal lands that cannot support crops for long and end up depleted by cultivation. In Brazil's Amazon Basin, for example, peasant migrants arrive from Atlantic coastal regions, where government and wealthy private landowners cultivate the best soils for sugarcane, soybeans, and other cash crops. The newcomers to Amazonia slash and burn the rain forest to grow rice and other crops that exhaust the soil's limited fertility in a few years. Then they move on to cultivate new lands, and in their wake come cattle ranchers, whose land use further degrades the soil. This is a very common, important problem in the LDCs, and you will see many examples of it in the following chapters.

The second way for the LDC debtor countries to raise cash quickly is to simply sell off their natural assets. Who is buying

them? The wealthier countries, and increasingly and most strikingly, China—an issue that this book covers often. LDCs often face difficult choices between using natural resources for immediate gain or for longer-term, more sustainable growth. In most cases, they feel compelled to take short-term profits, by cash cropping and other strategies, a choice that can have devastating environmental consequences. Most LDCs have resource-based economies that rely not on industry, but on stocks of productive soils, forests, and fisheries. The countries' long-term economic health could benefit through protection and careful use of these assets. But to pay off international debts and meet other needs, the LDCs often draw down their ecological capital faster than nature can replace it. In ecosystem terms, they exceed **sustainable yield** (also known as the **natural replacement rate**), the highest rate at which a **renewable resource** can be used without decreasing its potential for renewal. One figure that comes up often in the literature is that in African countries as a whole, 28 trees are cut down for every tree that is planted.

A country where 28 trees are cut for every one planted is approaching or is in a state of **ecological bankruptcy**, the exhaustion of environmental capital. This environmental poverty plays itself out in several damaging ways. First, there are negative effects on the health and well-being of the people living in such countries. People in the LDCs feel the impacts of environmental degradation more directly than people in the MDCs (•**Figure 3.6**). Many of the world's poor drink directly from untreated water supplies; an estimated 1 billion people lack access to clean water. They tend to cook their meals with fuelwood rather than with fossil fuels. They are more dependent on nature's abilities to replenish and cleanse itself, and so they suffer more when those abilities are diminished (see Insights, page 48).

In addition, many political and social crises can emerge from the ecological bankruptcy process. Revolutions, wars, and refugee migrations in developing nations often have underlying

• **Figure 3.6** The Thu Bon River in Vietnam. In the LDCs, it is very common for people to rely on polluted water sources for drinking, bathing, cooking, and washing their clothes and eating utensils. Health effects can be severe; one of the most tragic is infant diarrhea, typically a waterborne illness that kills an estimated 2 million youngsters worldwide every year.

Deforestation—the removal of tree cover—has many detrimental effects in the LDCs. You can see how these relationships play out in •**Figure 3.B**. As people remove trees to use as fuel, for construction, or to make room for crops, their existing crop fields lose protection against the erosive force of wind. Less water is available for crops because in the absence of tree roots to funnel water downward into the soil, it runs off quickly. Increased salinity (salt content) generally accompanies increased runoff, causing the quality of irrigation and drinking water downstream to decline. Eroded topsoil resulting from reduced plant cover can choke irrigation channels, reduce water delivery to crops, raise floodplain levels, and increase the chances that floods will destroy fields and settlements. As reservoirs fill with silt, hydroelectric generation, and therefore industrial production, is diminished. Upstream, where the problem began, fewer trees are available to use as fuel.

As they deplete nearby fuelwood supplies from trees and other vegetation, rural people make their own lives more difficult and have to change their activity patterns. Women and children, in almost every society the main fuelwood collectors, must walk farther, and then farther again, to gather fuel. Using animal dung and crop residues for fuel can solve one problem but create another: it deprives the soil of the fertilizers these poor people often use when farming. Reduced food output is the result. A family may eventually give up one cooked meal a day or put up with colder temperatures in their homes because fuel is lacking—steps that have a negative impact on the family's health. These impacts of fuelwood depletion are known collectively as the **fuelwood crisis**.

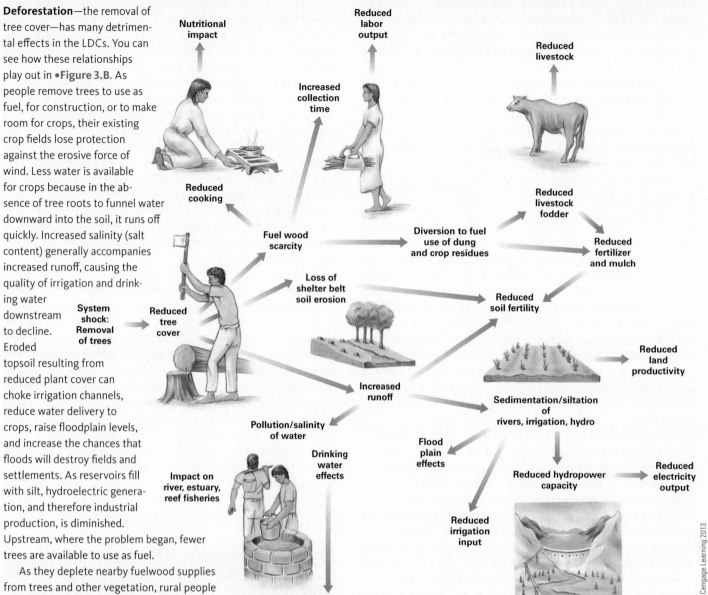

• **Figure 3.B** Impacts of deforestation in the LDCs.

© Cengage Learning 2013

The fuelwood crisis is a problem that governments and societies can confront successfully. A combination of new technologies, government policies, and sound economic growth can halt and even reverse deforestation. The biogas digester, a simple household device that can be used in warm climates, ferments animal manure for the production of methane cooking gas and thus spares fuelwood. Governments can protect forests while also allowing people to earn their livelihoods in them from **nontimber forest products**

(NTFPs), such as handicrafts made from bamboo (see www.ntfp.org). The process of economic development can itself lead to less pressure on forest resources. The world's wealthier countries have more forest now than they did in 1990, and forest cover has actually increased in about half of the world's 50 most forested countries since that time. A key challenge for the wealthier forested countries is to slow their demand for timber from the LDCs, and thus help break one of the cycles of deforestation, debt, and poverty.

environmental causes. Such problems, as related in the earlier discussion of climate change refugees in Chapter 2, are central to the national interests of the United States and other countries far beyond the affected nations. Deforestation in Mexico or Haiti, for example, is among the root causes why rural people from those countries try to move to the United States, legally or otherwise. When their home environments are so degraded that they simply cannot support farming as they once did, people will move.

3.3 The Geography of Population

Population may be the most critical issue in geography. It certainly is one of the most important issues for human life on Earth. The welfare of humanity and of the planet's other species and natural habitats is tied closely to two related issues: the number of people there are and the rates at which we consume resources. In some quarters, there are fears that the human **population explosion** the world has experienced since 1800 will lead to a precipitous crisis: a massive disease outbreak, famine, or some other kind of catastrophe triggered by too many people living too close together without sufficient food and other resources to sustain them. On the other hand, there is optimism that the human population growth rate is slowing, and our numbers are expected to stabilize—ideally at a population that can be sustained with minimal risk of famine or other suffering.

Meanwhile, large numbers of people move across and within political borders, usually by choice but in some cases by force. Migrants bring new cultures, ideas, and opportunities with them, but their reception is not always warm: tension and violence have characterized relations between majority and migrant minority groups in recent years in Europe and Australia, for example, and in the United States there is a heated debate about "immigration reform" directed at illegal migrants, especially from Latin America. Migration is one of the major themes discussed throughout this book.

Geography's interests in population are shared by many other fields, including biology, sociology, anthropology, and political science. The study of population is known as **demography**. The field of demography is most concerned with patterns of birth, death, marriage, and related issues in themselves, with less attention to issues of migration and population distributions. What most distinguishes population geography is its focus on spatial variations. This section of the book examines how many people have lived on Earth and how many we may be in the future, mindful of who these people are and why they have been so few or so many. It pays especially close attention to the fundamentally geographic issue of where they are. This discussion will be a very important reference for you as you work your way forward through the world regions. I hope you will find it interesting, and recommend that you explore the best Internet resource on world population (where you can also download a free world population data table): the Population Reference Bureau site at prb.org.

How Many People Have Ever Lived on Earth?

Around 100,000 years ago, our *Homo sapiens* ancestors came out of Africa across the land bridge of Suez and began to populate Eurasia. By around 10,000 years ago, about the time plants and animals began to be domesticated, there were probably about 5.3 million humans in the world—roughly the current number of residents of the Atlanta metropolitan area or the country of Finland. By 1 c.e., humans numbered between 250 and 300 million, about the population of the United States today. The first billion was reached around 1800. Then a staggering population explosion occurred in the wake of the Industrial Revolution. The second billion came in 1930, the fourth in 1975, the sixth in 1999, and the seventh in 2011 (•Figure 3.7). *Homo sapiens* is now by far the most populous large mammal on Earth and has succeeded where no other animal has in extending its range to the world's farthest corners. Using a 24-hour time period to represent the 200,000 years of our species' history, we reached our first billion within just the last three minutes. The people alive today represent about 6 percent of all the people who have ever lived on Earth!

Figure 3.7 illustrates very dramatically that right after about 1800, the human population surged. What happened after tens of thousands of years that made our numbers suddenly skyrocket? In addition to the factor, addressed earlier, of greater food surpluses that made it possible to feed more people, we answer that question by considering some of the basic vocabulary of population geography.

How Can We Measure Population Changes?

Excluding the issue of migration for now, two main variables determine population change in a given village, city, or country or the entire planet: birth rate and death rate. The **birth rate** is

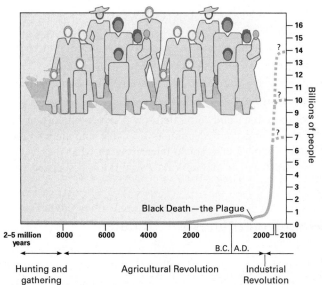

• **Figure 3.7** The human population has exploded since the Industrial Revolution, in a classic J-shaped curve of exponential growth.

© Cengage Learning 2013

the annual number of live births per 1,000 people in a population. The **death rate** is the annual number of deaths in that same sample population of 1,000. The **population change rate**—the figure that is often called the "population growth rate" but that may represent either growth or loss—is the birth rate minus the death rate in that population.

We can put these measures to work to appreciate the Earth's population now. Suppose there were a perfect sample of 1,000 people representing the world's population in 2011. Their birth rate was 20 per 1,000 and their death rate was 8 per 1,000. This means that by the end of that year, among the 1,000 people, 20 babies had been born and 8 people had died, resulting in a net growth of 12. That figure, 12 per 1,000, or 1.2 percent, represents the 2011 population change rate for the world.

We may now look at some of different combinations of birth rates, death rates, and population change rates and the various forces behind them.

What Determines Family Size?

Many factors affect birth rates, which tend to be much higher in the less developed countries of the world and among the poorer residents of more affluent countries. Some of the motivations and circumstances may seem unfamiliar at first, but you will see there are good reasons for them:

- Better-educated and wealthier people have fewer children. The parents of most of you reading this book probably considered the economic cost of raising you and sending you to college and made decisions about family size in part because of their education and yours.

- Conversely, less educated and poorer people generally want and have more children. One reason for this is economic: poorer parents are often convinced that "extra" children will help bring more income to the family by working in fields or factories and will help care for them in their old age.

- People in cities tend to have fewer children than those in rural areas.

- Those who marry earlier generally have more children (their reproductive life span is longer).

- Couples with access to and understanding of contraception generally have fewer children.

- Value systems and cultural norms play very important roles. Even where contraception is available and understood, a couple may decide not to interfere with what they perceive as God's will or may seek the social status associated with a larger family.

What Determines Death Rates?

Death rates are correlated mainly with health factors, particularly the level of nutrition and level of medical care available. Improvements in food production and distribution help reduce death rates. Better sanitation, better hygiene, and cleaner drinking water eliminate fatal diseases such as infant diarrhea, a common cause of infant mortality in the less developed countries. The availability of antibiotics, immunizations, insecticides, and other improvements in medical and public health technologies also correlate with death rates.

Human death rates overall have been on a steady trend of decline for decades. But death rates sometimes rise, of course, especially with the outbreak of epidemics such as HIV/AIDS. 316 Large proportions of the world's population have in fact been killed in disease epidemics and natural disasters. The "Black Death" (caused by bubonic plague) in Europe and Asia killed about 15 percent of the world's population between 1334 and 1349; at least 40 percent of Europe's population died. An influenza (flu) epidemic in 1918–19 killed as many as 100 million people. We cannot assume that such devastation will never happen again. Earlier this century, there were fears that the "bird flu" might kill as many as 150 million people if it mutated to a human strain.

Closely related to the measure of death rates is that of *life expectancy,* the number of years a person may expect to live in a given environment (typically defined as a country—see •Figure 3.8—and differentiated between women, who usually live longer, and men). As death rates fall, life expectancy increases, and the reverse is also true. In the United States in 2011, life expectancy for women was 80 years, and for men, 75 years. In Zimbabwe, a southern African country hit hard by the HIV/AIDS epidemic, a woman could expect to live 45 years and a man 46 years. But as we will see in Chapter 9, that is an improvement for Zimbabwe as it and other sub-Saharan African countries begin to make headway against the scourge of AIDS. 316

What Determines the Population Change Rate?

Throughout history, natural disasters, diseases, and wars have taken huge bites out of our numbers. Overall, however, with birth rates higher than death rates, the trend line has been one of growth—and since 1800, of spectacular growth. In 1968, the rate of population growth hit an all-time high of 2.0 percent. To appreciate how rapid that growth rate was, you can calculate the **doubling time**. Applying the often-used "Rule of 70," in which 70 is divided by the growth rate, the doubling time is the number of years required for the human population to double (assuming that the rate of growth would be unchanged over the entire period). In 1968, the human population was growing at a rate that would, if unchanged, have doubled in 35 years. That rate did slow down, however, illustrating the limited usefulness that doubling time has in projecting future growth. At the 2011 population change rate of 1.2 percent, our numbers would double in 58 years.

Doubling time may be only approximate, but it is a good tool for comparing among countries and also illustrates that human populations have the potential for exponential growth; that is, not an incremental or arithmetic increase from 1 to 2 to 3 to 4, and so on, but geometric growth from 2 to 4 to 8 to 16, for example. (Exponential growth in the context of the famous Malthusian scenario is discussed on page 58.) The annual rate of population change worldwide is depicted by country in •Figure 3.9.

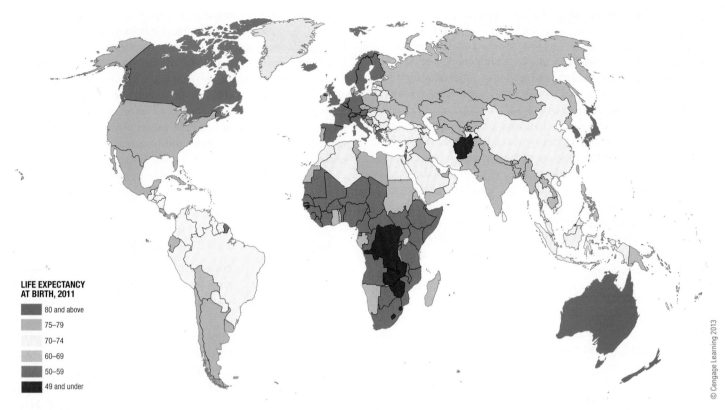

LIFE EXPECTANCY AT BIRTH, 2011

- 80 and above
- 75–79
- 70–74
- 60–69
- 50–59
- 49 and under

• **Figure 3.8** Life expectancy is closely tied to economic well-being; people live longer where they can afford the medicines and other amenities and technologies that prolong life.

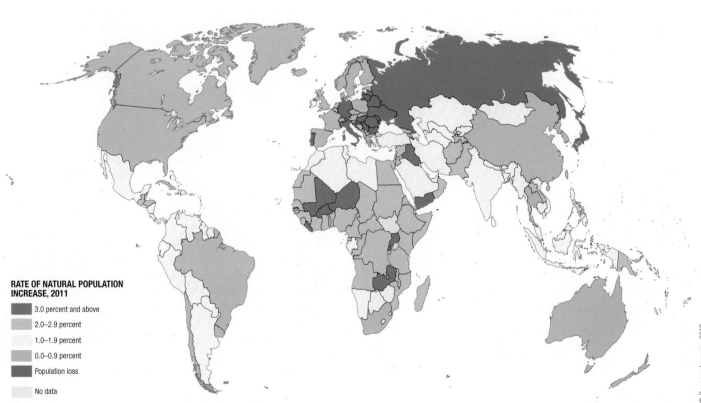

RATE OF NATURAL POPULATION INCREASE, 2011

- 3.0 percent and above
- 2.0–2.9 percent
- 1.0–1.9 percent
- 0.0–0.9 percent
- Population loss
- No data

• **Figure 3.9** Population change rates are highest in the countries of Africa and other regions of the developing world and lowest in the more affluent countries.

Why Has the Human Population "Exploded?"

If both birth rates and death rates are high (as they were when our ancestors were hunters and gatherers and as recently as the dawn of the Industrial Revolution), population growth is minimal: many people are born to a given population in a given period, but many also die in that same period, so they "cancel each other out." Population growth is also negligible if both birth and death rates are low (as they are today in countries like Japan and Italy). But when the birth rate is high and death rates are low, population surges, as seen in •Figure 3.10. This scenario of high birth rates, plunging death rates, and surging growth is exactly what played out for our species beginning around 1800 in western Europe and after about 1950 in the LDCs.

It is vital to appreciate the fact that the explosive growth in world population since the beginning of the Industrial Revolution is the result not of a rise in birth rates, but of a dramatic decline in death rates, particularly in the less developed countries. The death rate has fallen as improvements in agricultural and medical technologies have diffused from the richer to the poor countries. Until recently, however, there were no strong incentives for people in the less developed countries to have fewer children. With birth rates remaining high and death rates falling quickly, the population has grown sharply; the LDCs are generally in or emerging from stage 2 of an important model that demographers call the **demographic transition** (Figure 3.10).

In its entirety, from stages 1 to 4, as seen in Figure 3.10, the demographic transition model depicts the change from high birth rates and high death rates to low birth rates and low death rates that accompanied economic growth in the more developed countries (for example, western European nations, Japan, and the United States). The first two are the same two stages that humankind as a whole experienced from our earliest days until the present. The latter two have generally been experienced only by people in the wealthier countries. Note how birth rates, death rates, population change rates, and economic development correspond in this model:

- In the **first**, or **preindustrial, stage** (from the earliest humans to about 1800 C.E.), birth rates and death rates were high, and population growth was negligible.

- In the **second**, or **transitional, stage**, birth rates remained high, but death rates dropped sharply after about 1800 due to medical and other innovations of the Industrial Revolution.

- In the **third**, or **industrial, stage**, beginning around 1875, birth rates began to fall as affluence spread.

- Finally, after about 1975, some of the industrialized countries entered the **fourth**, or **postindustrial stage**, with both low birth rates and low death rates and therefore (once again, as in stage 1), low population growth.

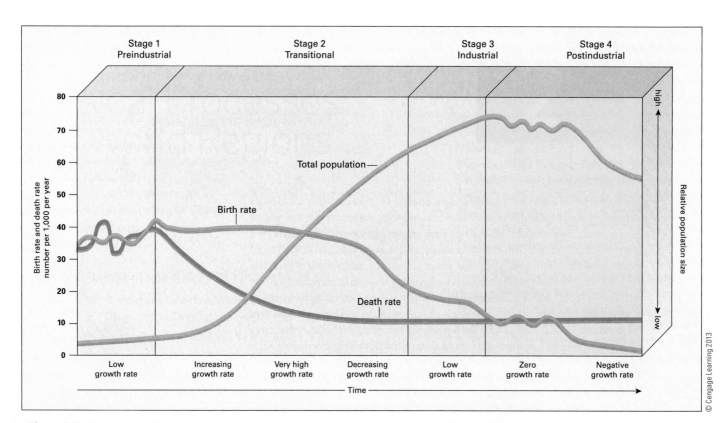

• **Figure 3.10** The demographic transition models population change in the world's wealthier countries. Note how the population surged in the wake of the Industrial Revolution as death rates fell while birth rates remained high, but then leveled out and began to decline as economic development advanced.

In this model, the United States, with population growth of about 0.5 percent per year, is in the early years of the postindustrial stage. Other industrialized, affluent countries have made their way well into that stage. Some wealthier countries, including Austria, Portugal, and Greece, have in recent years officially registered **zero population growth (ZPG)**, and many other European countries are close to ZPG. Because of the growing desire of women in some countries—including Japan and Germany—to pursue their own careers and postpone marriage, birth rates in those countries have fallen below death rates, meaning that these countries are actually losing population or even experiencing a **population implosion**. In other words, the fertility rate of Japan and some other postindustrial countries is below the **population replacement level**, the number of new births required to keep the population steady (generally calculated as 2.1 children per woman in the MDCs). Other countries—notably Russia and Hungary—that qualify as MDCs on the basis of per capita GNI PPP are experiencing population losses, unfortunately due as much to rising death rates as to falling birth rates. Trends of this sort can be most easily appreciated by looking at a very useful and informative device, the age structure diagram.

The Age Structure Diagram

About 9 of every 10 babies born in the world today are in the poorer countries. Both current and projected rates of population growth are distributed quite unevenly between the poorer and richer nations. This phenomenon is apparent in the age structure diagrams typical of these countries. An **age structure diagram** (often called a **population pyramid**) classifies a population by gender and by five-year age increments (•**Figure 3.11**). One important index these profiles show is the percentage of a population under age 15. A country like Niger in Figure 3.11 is typically poor and faces the prospect of increasing poverty because so many new jobs, food, and other resources will have to be created to meet the demands of those children as they mature and have their own children. The bottom-heavy age structure diagram also suggests a continued surge in population as those children grow to enter their reproductive years. A large, youthful population in which competition for jobs, education, and land is intense is a social environment ripe for discord. A study by Population Action International found that 80 percent of the civil conflicts of recent decades took place in countries where at least 60 percent of the population was younger than 30.[5] Disaffection among huge populations of impoverished, unentitled young people was one of the main forces that brought down several authoritarian governments in the "Arab Spring" of 2011.

The bottom-heavy, pyramid-shaped age structure diagram of Niger contrasts markedly with the more chimney-shaped structures of the United States and Germany in Figure 3.12. These wealthier countries have a much more even distribution of population through age groups, with a modest share under age 15. Such profiles suggest that, not considering migration, their population growth will be low in the near future.

Collectively, about 29 percent of the population of the poorer countries is under age 15, whereas the corresponding figure for

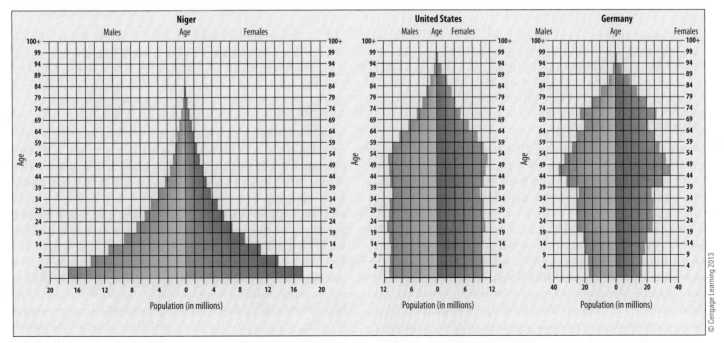

• **Figure 3.11** Population by age and gender in representative countries. The pyramid-shaped age structure diagram for Niger contrasts remarkably with those of the far more affluent United States and postindustrial Germany, with their chimney-like shapes. A poor country, Niger has a relatively high birth rate, with about 49 percent of the population under age 15. The U.S. population is growing slowly, while Germany and some other industrialized nations are losing populations.

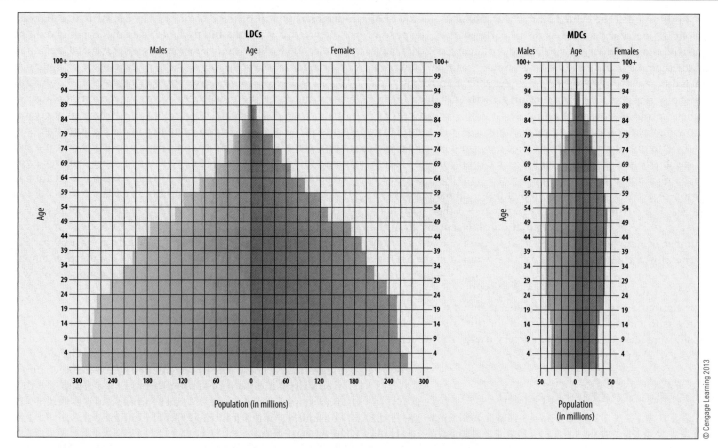

• **Figure 3.12** Population by age and gender worldwide, in terms of economic development. This diagram summarizes some of the most important facts about the human population: the lion's share lives in the poorer countries, and the lion's share of that population is young.

the wealthier countries is 16 percent (•Figure 3.12). As a whole, then, the developing world faces the critical challenge of providing for a surging population in the future, even while struggling to meet the needs of the people alive today.

Where Do We Live?

Where do all these people live? The world population carto-gram in •Figure 3.13 shows clearly that China and India are the most populous countries, with about 1.4 and 1.3 billion people, respectively. Think about it this way: about one-third of all people living on Earth today are either Chinese or Indian!

Why are so many people in just two countries? There are several reasons. Both countries are large. China is 20 percent larger than the contiguous 48 U.S. states, and India is about one-third the size of China. Both have been populated since very ancient times, and successful intensive agriculture has been practiced in both for more than 4,000 years. Both have large areas of productive soils and high rainfall, promoting successful farming. Both are developing countries in which birth rates remained high while death rates fell. Their population explosions prompted China's government to adopt an aggressive population control policy several decades ago, and

India followed suit with less forceful measures. India's higher growth rate—1.5 percent, compared with China's 0.5 percent, means India will likely overtake China in population by 2030.

Although there are many variables to consider in explaining why large numbers of people are clustered in particular countries, including cultural factors and family planning policies, the natural setting is by far the most important factor. The population densities shown in •Figure 3.14 correlate generally with agricultural and other environmental conditions. The deepest reds showing the highest population densities are in the more humid and fertile regions of both China and India. Conversely, in the very western part of China, including the high Tibetan Plateau, very dry conditions limit agriculture to a few favored areas. The world's highest mountains, the Himalayas, rise just north of the deep red area of high population density in northern India. The moisture-laden winds that bring so much productive rainfall to India cannot cross that mountain barrier, which has been a divide between densely and sparsely populated regions for thousands of years. Looking at the lightly populated areas of the world, you can recognize similar environmental factors at work: in the Sahara of northern Africa, the Arctic of northern Canada, and the Amazon Basin of South America, conditions are too dry, too cold, or too wet and infertile to support large numbers of people.

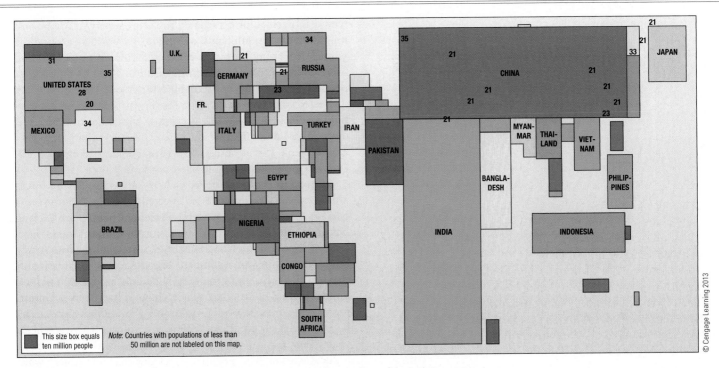

© Cengage Learning 2013

• **Figure 3.13** The demographic heavyweights of China and India stand out in the world population cartogram. The United States and Indonesia, the world's third and fourth most populous countries, are prominent too.

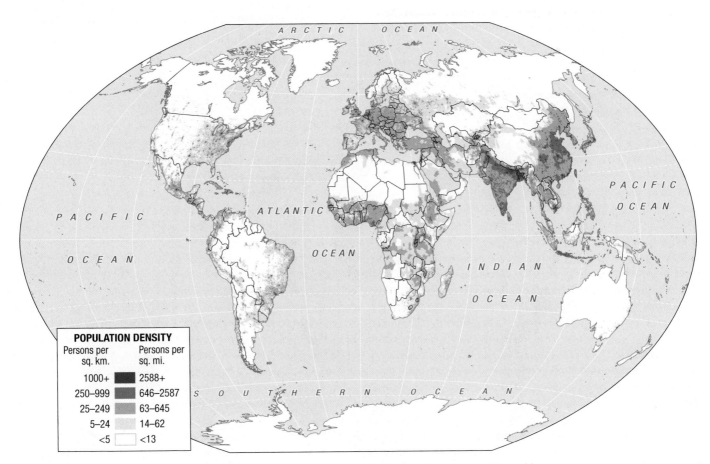

• **Figure 3.14** This isarithmic map of population density shows the approximate distribution of people around the world.

Source: Based on data from Population Reference Bureau, 2004.

The Geography of Migration

Migration refers to the movement of people from one location to another in any setting, whether within a community, within a country, or between countries. Migration is one of the most dynamic and most problematic human processes on Earth (•Figure 3.15). A migrant is always both an **emigrant** (one who moves from a place) and an **immigrant** (one who moves to a place). Migration is usually associated with either **push factors**, as when hunger or lack of land "pushes" peasants out of rural areas into cities or warfare pushes people from one place to another, or **pull factors**, when, for example, an educated villager is "pulled" by a job opportunity in the city or another country. People responding to push factors are often referred to as **nonselective migrants**, and people reacting to pull factors are called **selective migrants**.

Both push and pull forces are behind the **rural-to-urban migration** pattern that is characteristic of most countries. The growth of cities, known as **urbanization**, is due in part to the natural growth rate of people already living in urban areas, but it is particularly strong in the LDCs because of this internal migration of both selective and nonselective migrants.

Within and between countries today, there are considerable movements of **refugees**, the victims of such severe push factors as persecution, political repression, and war. They may be on the move either as illegal immigrants or as people granted **asylum**, or given permission to immigrate on the grounds that they would be harmed or persecuted in their country of origin. Immigration and asylum laws and quotas vary widely around the world, depending on host countries' political, economic, and social systems. Among the most disadvantaged of the world's peoples are the **internally displaced persons (IDPs)** who are dislodged and impoverished by strife in their home country but have little prospect of emigrating. Sub-Saharan Africa has more IDPs (about 12 million in 20 countries) than any other world region. In this book, you will read many accounts of others moved by forces beyond their control.

Migration is almost never clearly "good" or "bad" but is almost always a controversial mixed blessing. Migration is bad for the country whose most talented people are leaving but good for the country they go to; the Indian doctor benefits your community in the United States but may be sorely missed in India, for example. The doctor is illustrative of the **brain drain**, the emigration of educated and talented people from a place that needs them. The Mexican immigrant who does low-wage labor in the United States may be perceived as an **illegal alien** threatening to overwhelm social services and take jobs away from local people, or as a **guest worker** who performs services that few others want to do but that are critical to the country's economic well-being. Some host-country peoples are more welcoming of new cultures that enrich their ethnic mosaic (Americans are generally perceived as among the most accommodating cultures in the world), whereas others fear losing their ethnic majorities and privileges.

226

71,
393

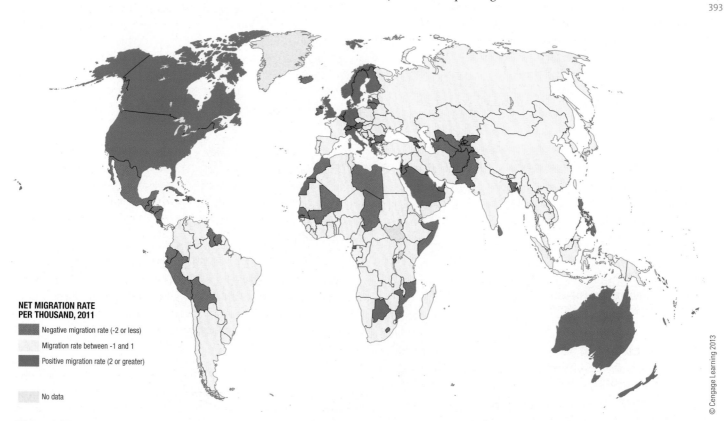

NET MIGRATION RATE PER THOUSAND, 2011

- Negative migration rate (-2 or less)
- Migration rate between -1 and 1
- Positive migration rate (2 or greater)
- No data

• **Figure 3.15** The global picture of people on the move. The major trends are of migrants in search of work in more affluent countries and of refugees driven by warfare or environmental adversity.

© Cengage Learning 2013

How Many People Will Live on Earth?

Population geographers are fairly confident in their calculations of how many people have lived on Earth at various times in the past. Projecting future numbers is another matter. There are many uncertainties. Will birth rates fall faster than expected in the developing world? Will death rates surge because of HIV/AIDS or some other epidemic? These are some of the wild cards in the population deck.

In asking how many people will live on Earth in the future, we are essentially asking, "Will the poorer countries of the world go through the demographic transition?" Will the current, relatively high birth rates in the less developed countries continue their present slow decline? In 1970, Kenya had a birth rate of 51 and Bangladesh had a birth rate of 45; by 2011, Kenya's birth rate had dropped to 37 and Bangladesh's to 22. Considering what might have been, given the country's huge population, China's decline may be the most impressive of all, from a rate of 38 in 1965 to 12 in 2011. The Chinese government's generally harsh **one-child policy**, which uses a combination of incentives and punishments, has been the main reason for the decline. But the demographic transition model also suggests to us that many increasingly affluent Chinese couples would probably have make the free choice to limit their children, if given that opportunity. In Kenya and Bangladesh, growing literacy and the slow but steady economic progress of women have brought down the birth rate.

216

Family planning policies and levels of education and economic well-being have played various roles in the poorer countries, but have collectively combined to bring birth rates down since 1968 (•**Figure 3.16**). If the processes of increasing economic and social development continue in the LDCs as a whole, they should make steady progress through the demographic transition, and the Earth's population should cease to grow and should stabilize.

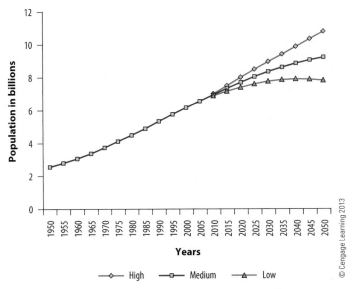

• **Figure 3.17** United Nations projections for population growth.

But will the poorer countries complete the transition successfully? An unsavory but possible scenario would see death rates rise dramatically (due to disease or famine, for example) in at least some of the countries, in effect pushing them back to stage 1 of the demographic transition, where both birth and death rates are high. Others could remain in stage 2, with high birth rates and falling death rates, long enough to bring unexpectedly high numbers of people into the world. With such different scenarios in mind, the United Nations prefers to use a widely respected model with three projections: high, moderate, and low growth (•**Figure 3.17**).

In view of declining birth rates worldwide, the United Nations has revised its projection for modest population growth downward. The agency predicts 9.3 billion by 2050 (half a billion fewer than it had estimated in earlier projections) and stabilization—the maximum number of people that will ever live on Earth at one time—at 10 billion in 2100.

This downward revision has prompted a rash of popular and academic articles proclaiming that "the population explosion is over" and even some essays arguing there would soon be too few people on Earth, particularly in the richer countries. Other experts have cautioned that it is too soon to declare the population bomb defused. "World population growth turned a little slower," a Population Institute report argues. "The difference, however, is comparable to a tidal wave surging toward one of our coastal cities. Whether the tidal wave is 80 or 100 feet high, the impact will be similar."[6] (To put that in some perspective, while population growth rates have been declining since 1968, the population base is so large that more people than ever before—about 83 million—are added to the world's population each year.) The United Nations has pledged to do its best to help stabilize the global population at a lower number than now projected. The organization's plan is to focus on enhancing the education and employment of women in the less developed countries as a means of bringing down birth rates.

• **Figure 3.16** National family planning programs can have a pronounced impact on birth rates if the actions go beyond slogans. This sign on the famous Tahrir Square in Cairo urges parents to have no more than two children "for the sake of a better life." Egypt's birth rate fell from 40 in 1970 to 25 in 2011.

The Malthusian Scenario

Even while the birth rate falls, the population increases. Already large and growing numbers pose fundamental questions: Can the Earth sustain 10 billion or more people? Will we exceed the planet's carrying capacity, and what will happen if we do? Early in the Industrial Revolution, an English clergyman named Thomas Malthus (1766–1834) postulated that human populations, growing geometrically or exponentially, would exceed food supplies, which grow only arithmetically or linearly. He predicted a catastrophic human die-off as a result of this irreconcilable equation (•Figure 3.18). He could not have foreseen that the exploitation of new lands and resources, including tapping the energy of fossil fuels, would permit food production to keep pace with or even outpace population growth for at least the next two centuries.

This **Malthusian scenario** of the lost race between food supplies and mouths to feed remains a source of constant and important debate today. On one side of the debate are optimists, the so-called **technocentrists** or **cornucopians**, who argue that human history provides insight into the future. Thanks to their technological ingenuity, people have always been able to conquer food shortages and other problems and therefore always will (•Figure 3.19). Julian Simon, a University of Maryland economist, argued that far from being a drain on resources, additional people create additional resources. Technocentrists therefore insist that people can raise the Earth's carrying capacity indefinitely and that the die-off that Malthus predicted will always be averted. Our more numerous descendants will instead enjoy more prosperity than we do.

In contrast, the **neo-Malthusians** (heirs of the reasoning of Thomas Malthus) argue that although we have been successful so far, we cannot increase the Earth's carrying capacity indefinitely. There is an upper limit beyond which growth cannot occur, with calculations ranging from 8 to 40 billion people. Neo-Malthusians insist that the poorer countries cannot remain indefinitely in the second stage of the demographic revolution. Either they must intentionally bring birth rates down further and make it successfully through the demographic transition, or they must unwillingly suffer nature's solution, a catastrophic increase in death rates. Thus by either the **birth rate solution** or the **death rate solution** to the problem, the neo-Malthusians argue, the LDCs must confront their population crisis (see pages 59–61).

Whereas technocentrists view the equation between people and resources passively, insisting that no corrective action is needed, neo-Malthusians tend to be activists who describe terrible scenarios of a death rate solution in order to motivate people to adopt the birth rate solution. The biologist Paul Ehrlich thinks that we are increasingly vulnerable to a Malthusian catastrophe, particularly as viruses diffuse around the globe with unprecedented speed. "The only big question that remains," Ehrlich wrote, "is whether civilization will end with the bang of an all-out nuclear war, or the whimper of famine, pestilence and ecological collapse."[7] Such dire warnings have earned many neo-Malthusians the reputation of being "gloom-and-doom pessimists." The neo-Malthusians insist that there are simply too many people already for the world to support, especially in the LDCs, where there are not enough resources to support them. But what does "too many people" mean? How many are too many, and based on what criteria?

What Is "Overpopulation"?

It is probably most useful to think about two distinct types of overpopulation, one characteristic of the poor countries and the other of the rich (•Figure 3.20). **People overpopulation** is an apparent problem in the poorer countries. The environmental problems characteristic of LDCs are intensified by the relatively high rates of human population growth in those countries. More people cut more trees, hunt more wildlife, and otherwise use more resources. Many persons, each using a small quantity of natural resources daily to sustain life, have a great collective impact on the environment and may add up to too many people for the local environment to support. Common consequences are malnutrition and even the famine emergencies in which richer countries are called on to provide relief.

Consumption overpopulation is characteristic of the MDCs. In the wealthier countries, there are fewer persons, but each uses a large quantity of natural resources from ecosystems around the world. Their collective impacts also degrade the environment, and even their smaller numbers may be "too

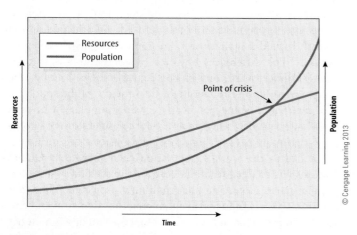

• **Figure 3.18** Malthus envisioned a race between people and resources, in which people lost.

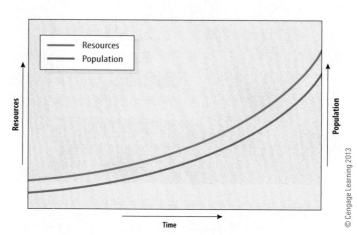

• **Figure 3.19** The technocentrists reason that production of food and other resources will always stay ahead of population growth.

Developing Countries

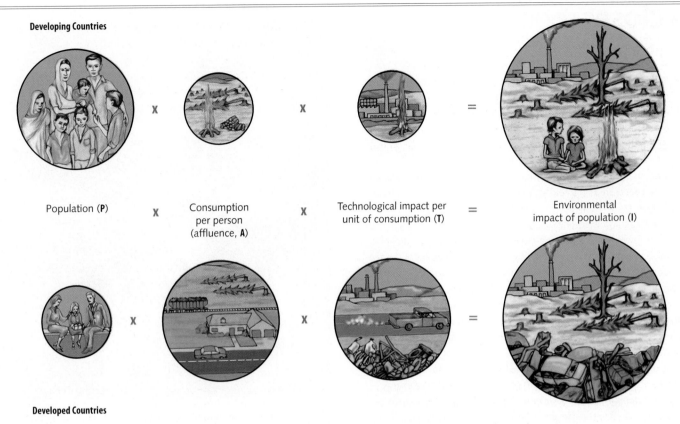

Population (**P**)   ×   Consumption per person (affluence, **A**)   ×   Technological impact per unit of consumption (**T**)   =   Environmental impact of population (**I**)

Developed Countries

• **Figure 3.20** Two types of overpopulation, calculated according to this formula: *number of people × number of units of resources used per person × environmental degradation and pollution per unit of resource used = environmental impact*. Circle size shows the relative importance of each factor. People overpopulation is caused mostly by growing numbers of people and is typical of LDCs. Consumption overpopulation is caused mostly by growing affluence and is typical of MDCs.
Source: From MILLER, Environmental Science, 4E. © 1993 Cengage Learning.

many" at such unsustainably high levels of consumption, particularly with their habit of feeding so high on the food chain; much more energy is required to produce meat than to produce grains. With less than 5 percent of the world's population, the United States may be regarded as the world's leading overconsumer. According to a useful measure known as the **ecological footprint**—the amount of biologically productive land needed to sustain a person's consumption and absorb his or her wastes—for every acre needed to support an average Ethiopian, 69 acres are needed to support an average American. The disparity suggests that if Ethiopia is "underdeveloped," the United States is "overdeveloped" (see A Closer Look feature on page 60).

If the vast majority of the world's population were to consume resources at the rate that U.S. citizens do, the environmental results might be ruinous. Even if consumption levels in the LDCs do not rise substantially, the sheer increase in numbers of people in those countries suggests that degradation of the environment will accelerate in the coming decades.

3.4 An Action Plan for Global Problems

The cornucopian view is comforting: we have nothing to worry about. By keeping up the good, ingenious work as we have in the past, our futures will be secure. If, however, one diverges modestly

from this premise or even accepts the most dire neo-Malthusian view (see Try It, page 61), it is appropriate to ask what can be done to prevent or solve some of the world's critical problems involving natural resources and human population numbers.

The Death Rate Solution and Lifeboat Ethics

Confronted with the Malthusian scenario, one option is the death rate solution, to "let nature take its course" and allow people imperiled by famine or other catastrophe to perish. The neo-Malthusian ecologist Garrett Hardin introduced **lifeboat ethics**—the question of whether or not the wealthy should rescue the "drowning" poor—in a distressing and challenging essay about the Malthusian scenario. "People turn to me," wrote Hardin, "and say, 'My children are starving. It's up to you to keep them alive.' And I say, 'The hell it is. I didn't have those children.'"[9] He described the world not as a single "spaceship *Earth*" or "global village" with a single carrying capacity, as many environmentalists do, but as a number of distinct "lifeboats," each occupied by the citizens of single countries and each having its own carrying capacity. Each rich nation is a lifeboat comfortably seating a few people. The world's poor are in lifeboats so overcrowded that many fall overboard. They swim to the rich lifeboats and beg to be brought aboard. What should the passengers of the rich lifeboat do? The choices pose an ethical dilemma for you to consider (•**Figure 3.21**).

A **closer** LOOK

Population and Energy

This book discusses energy a lot, and for good reasons: energy supply and demand issues impact our lives at many scales. But energy does not refer just to fossil fuels. Food energy, which is made available to ecosystems through the process of photosynthesis, and how that energy flows through ecosystems, are central to the question "How many people can the Earth support?"

Food chains (for example, in which a deer eats grass and a mountain lion eats a deer) are short, seldom consisting of more than four trophic (feeding) levels. The collective weight (known as **biomass**) and absolute number of organisms decline substantially at each successive trophic level (there are fewer leopards than antelopes). The reason is a fundamental rule of nature known as the **second law of thermodynamics** or the second energy law, which states that the amount of high-quality usable energy diminishes dramatically as the energy passes through an ecosystem. In living and dying, organisms use and lose the high-quality, concentrated energy that green plants produce and that is passed up the food chain. As an organism is consumed in any given link in the food chain, about 90 percent of that organism's energy is lost, in the form of heat and feces, to the environment. The amount of animal biomass that can be supported at each successive level thus declines geometrically, and little energy remains to support the top carnivores. So, for example, there are many more deer than there are mountain lions in a natural system.

The second law of thermodynamics has important consequences and implications for the human use of resources. The higher we feed on the food chain (the more meat we eat, for example), the more energy we use. The question of how many people the Earth's resources can support thus depends very much on what we eat. The affluent consumer of meat demands a huge expenditure of food energy because that energy flows from grain (producer) to livestock animal (primary consumer, with a 90 percent energy loss) and then from livestock animal to human consumer (secondary consumer, with another 90 percent energy loss). Each year, the average U.S. citizen eats 100 pounds of beef, 50 pounds of pork, and 45 pounds of poultry. By the time it is slaughtered, a cow has eaten about 10 pounds of grain per pound of its body weight; a pig, 5 pounds; and a chicken, 3 pounds. Thus in a year, a single American consumes the energy captured by two-thirds of a ton of grain (1,330 pounds) in meat products alone. By skipping the meat and eating just the corn that fattens these animals, many people could live on the energy required to sustain just one meat-eater. The point is not to feel guilty about eating meat but to realize that there is no answer to the question "How many people can the world support?" The question needs to be rephrased: "How many people can the world support based on people consuming _____?" (with the blank filled in with a specific number of calories per day or with a general lifestyle).

Development means a more affluent lifestyle, including eating more meat (feeding higher on the food chain). Some analysts fear the burden that the changing dietary habits accompanying development will pose to the planet's food energy supply. Lester Brown of the WorldWatch Institute wrote a provocative essay titled "Who Will Feed China?"[8] Keeping in mind the second law of thermodynamics, he argued that if China continues on its present course of growing affluence, with a move away from reliance on rice as a dietary staple to a diet rich in meats and beer, the entire planet will shudder. How could more than a billion and half people be sustained so high on the food chain?

Joe Hobbs

• **Figure 3.21** A dangerously overcrowded ferry in Tanzania. In Hardin's metaphor, what if this boat capsizes while a less crowded boat floats nearby?

Hardin set out the following scenario. There are 50 rich passengers in a boat with a capacity of 60. Around them are 100 poor swimmers who want to come aboard. The rich boaters have three choices. First, they could take in all the swimmers, capsizing the boat with "complete justice, complete catastrophe." Second, as they enjoy an unused excess capacity of 10, they could admit just 10 from the water. But which 10? And what about the margin of comfort that excess capacity allows them? Finally, the rich could prevent any of the doomed from coming aboard, ensuring their own safety, comfort, and survival—and ensuring the others would die.

Hardin's choice was the third: drowning. To preserve their own standard of living and ensure the planet's safety, the wealthy countries must cease to extend food and other aid to the poor and must close their doors to immigrants from poor countries. "Every life saved this year in a poor country diminishes the quality of life for subsequent generations," Hardin concluded. "For the foreseeable future, survival demands that we govern our actions by the ethics of a lifeboat."[10] Incidentally, Hardin jumped off his lifeboat, committing suicide in 2003.

The Birth Rate Solution and Sustainable Development

In recent decades, new concepts and tools for managing the Earth and its resources in an effective, long-term way have emerged. Known collectively as **sustainable development** (or **ecodevelopment**), these ideas and techniques consider what both MDCs and LDCs can do to avert the possible Malthusian dilemma and improve life on the planet. By promoting the birth rate solution and other concrete actions, sustainable development offers an activist agenda without the peril of doom depicted by the neo-Malthusians.

The World Conservation Union defines sustainable development as "improving the quality of human life while living within the carrying capacity of supporting ecosystems."[11] Sustainable development refutes what its proponents perceive as the current pattern of unsustainable development, whereby economic growth is based in large part on excessive resource use. Advocates of sustainable development point out that a country that depletes its resource base for short-term profits gained through deforestation increases its gross domestic product (GDP) and appears to be more "developed" than a country that protects its forests for a long-term harvest of sustainable yield. Deforestation appears to be beneficial to a country because it raises GDP through the production of pulp, paper, furniture, and charcoal. However, GDP growth does not measure the negative impacts of deforestation, such as erosion, flooding, siltation, and malnutrition. These consequences are known as **external costs**, or **externalities**, and they are not taken into account in the prices of goods and services. Advocates of sustainable development argue that these externalities should be added or "internalized" before a good or a service is marketed because the true high costs of producing these goods and services would be recognized. The lower true costs of less destructive practices would then be evident, providing stronger incentives for individuals, companies, and nations to invest in sustainable practices and technologies.

Sustainable development is a complex assortment of theories and activities, but its proponents call for eight essential changes in the way people perceive and use their environments:

1. People must change their worldviews and value systems, recognizing the finiteness of resources and reducing their expectations to a level more in keeping with the Earth's environmental capabilities. Proponents of sustainable development argue that this change in perspective is needed especially in the MDCs, where instead of trying to "keep up with the Joneses," people should try to enjoy life through more social rather than material pursuits.

2. People should recognize that development and environmental protection are compatible. Rather than viewing environmental conservation as a drain on economies, we should see it as the best guarantor of future economic well-being. This is especially important in the LDCs, with their resource-based economies.

3. People all over the world should consider the needs of future generations more than we do now. Much of the wealth we generate is in effect borrowed or stolen from our descendants. Our economic system values current environmental benefits and costs far more than future benefits and costs, and so we try to improve our standard of living today without regard to tomorrow.

4. Communities and countries should strive for self-reliance, particularly through the use of appropriate technologies. For example, remote villages could rely increasingly on solar power for electricity rather than be linked into national grids of coal-burning plants.

5. LDCs need to limit population growth as a means of avoiding the destructive impacts of people overpopulation. Advances in the status of women, improvements in education and social services, and effective family planning technologies can help limit population growth.

6. Governments need to practice land reform, particularly in the LDCs. Poverty is often not the result of too many people on too little total land area, but of a small, wealthy minority holding a disproportionately high share of quality land. To avoid the environmental and economic consequences of marginalization (as discussed on page 47), a more equitable distribution of land is needed.

7. Economic growth in the MDCs should be slowed to reduce the effects of consumption overpopulation. If economic growth, understood as the result of consumption of natural resources, continues at its present rate in excess of sustainable yield, the Earth's "environmental capital" will continue to diminish rapidly.

8. Wealth should be redistributed between the MDCs and LDCs. Because poverty is such a fundamental cause of environmental degradation, the spread of a reasonable level of prosperity and security to the LDCs is essential. Proponents argue that this does not mean that rich countries should give cash outright to poor countries. Instead, the lending institutions of MDCs can forgive some existing debts owed by LDCs or use such innovations as **debt-for-nature swaps**, in which a certain portion of debt is forgiven in return for the borrower's pledge to invest that amount in national parks or other conservation programs. Reducing or eliminating **trade barriers** that MDCs impose against products from LDCs would also help redistribute global wealth.

Some geographers and other scientists believe that sustainable development will bring about the "**Third Revolution**," a shift in human ways of interacting with the Earth so dramatic that it will be compared with the origins of agriculture and industry. The formidable changes called for in sustainable development are attracting increasing attention, perhaps because there are no other comprehensive strategies for dealing with some of the most critical issues of our time.

Weighing Lifeboat Ethics

Think about these options, with some concrete examples in mind. As an occupant of the rich lifeboat *United States*, for example, what would you do for the drowning refugees from lifeboat *Haiti* or lifeboat *Somalia*? Help bail out everyone in need? Assist just a few? Ignore them altogether? Why did you make this choice?

(This can be a difficult ethical question. I do this as an exercise in my class, with everyone making their choices anonymously on a piece of paper. I tally up the results of "save" and "drown," and we discuss as much as the students want to.)

summary

- A useful way to begin to appreciate the current spatial patterns of our relationship with the Earth is to view these patterns as products of two "revolutions" in the relatively recent past: the Agricultural or Neolithic (New Stone Age) Revolution and the Industrial Revolution. Each transformed humanity's relationship with the natural environment. After the Industrial Revolution, the human impact on Earth grew dramatically.

- The world is markedly divided between the haves and the have-nots, characterized at the largest scale by contrasts between more developed and less developed countries (MDCs and LDCs). Explanations for these disparities include dependency theory, cultural factors, geographic location, and natural resource base.

- LDC economies often rely on the export of a few raw materials or commercial crops and tend to draw down their natural "capital" quickly, with profound impacts on the environments.

- There is disagreement on whether globalization is beneficial or detrimental to poorer countries and to the poorer people who live in those countries.

- Population growth is measured by birth rates, death rates, and migration. Growth rates tend to be higher in the LDCs, due to relatively high but falling birth rates and much lower and falling death rates. The demographic transition, a model of what happened to populations in the MDCs, shows the wealthier countries passing from high birth rates and death rates to low birth rates and death rates.

- The explosive growth in world population since the beginning of the Industrial Revolution is the result not of a rise in birth rates, but of a dramatic decline in death rates, particularly in the LDCs. High rates of human population growth intensify environmental problems characteristic of the LDCs. More people use more resources, a phenomenon that suggests there is a problem of people overpopulation in the poorer countries. Consumption overpopulation is more characteristic of the MDCs, where fewer people use large quantities of natural resources from around the world.

- Human demands for more food and other photosynthetic products are growing, and it is questionable whether supplies can keep pace. The Malthusian scenario, which has not been realized so far, insists that a catastrophic die-off of people will occur when their numbers exceed food supplies.

- New concepts and tools for managing the Earth and its resources in an effective, long-term way have emerged in the past two decades. Known collectively as sustainable development or ecodevelopment, these ideas and techniques consider what humanity can do to avert the Malthusian dilemma and improve life on the planet.

Key Terms + Concepts

Age of Discovery (p. 41)
Age of Exploration (p. 41)
age structure diagram (p. 53)
Agricultural Revolution
  (p. 40)
Anthropocene Epoch (p. 40)
asylum (p. 56)
biomass (p. 60)
birth rate (p. 49)
birth rate solution (p. 58)
brain drain (p. 56)
carrying capacity (p. 40)
cash crops (p. 47)
civilization (p. 40)
colonization (p. 41)
commercial crops (p. 47)
consumption overpopulation
  (p. 58)
cornucopians (p. 58)
culture hearth (p. 40)
death rate (p. 50)
death rate solution (p. 58)
debt-for-nature swap (p. 61)

deforestation (p. 48)
demographic transition (p. 52)
  first (preindustrial) stage
  (p. 52)
  second (transitional) stage
  (p. 52)
  third (industrial) stage
  (p. 52)
  fourth (postindustrial) stage
  (p. 52)
demography (p. 49)
dependency theory (p. 44)
development (p. 43)
diffusion (p. 40)
digital divide (p. 46)
domestication (p. 40)
doubling time (p. 50)
dry farming (p. 40)
ecodevelopment (p. 60)
ecological bankruptcy (p. 47)
ecological footprint (p. 59)
ecologically dominant species
  (p. 39)

emigrant (p. 56)
extensive land use (p. 40)
external costs (p. 61)
externalities (p. 61)
food chain (p. 60)
Food-Producing Revolution
  (p. 40)
foraging (p. 39)
fuelwood crisis (p. 48)
globalization (p. 46)
gross domestic product (GDP)
  (p. 42)
gross national income (GNI)
  (p. 43)
gross national product (GNP)
  (p. 43)
guest worker (p. 56)
Human Development Index
  (HDI) (p. 44)
hunting and gathering (p. 39)
illegal alien (p. 56)
immigrant (p. 56)
Industrial Revolution (p. 41)

information technology (IT)
  (p. 46)
intensive land use (p. 40)
internally displaced persons
  (IDPs) (p. 56)
irrigation (p. 40)
knowledge economy (p. 46)
less developed countries
  (LDCs) (p. 42)
lifeboat ethics (p. 59)
Malthusian scenario (p. 58)
marginalization (p. 47)
mercantile colonialism (p. 45)
migration (p. 56)
more developed countries
  (MDCs) (p. 42)
multinational companies
  (p. 46)
natural replacement rate
  (p. 47)
natural resource (p. 45)
neocolonialism (p. 45)
neo-Europes (p. 45)

Neolithic Revolution (p. 40)
neo-Malthusians (p. 58)
newly industrializing countries (NICs) (p. 42)
nontimber forest products (NTFPs) (p. 48)
nonselective migrants (p. 56)
"original affluent society" (p. 39)
one-child policy (p. 57)
people overpopulation (p. 58)
per capita GDP (p. 42)

per capita gross national income (p. 43)
    purchasing power parity (p. 43)
    (per capita GNI PPP) (p. 43)
Pleistocene overkill hypothesis (p. 39)
population change rate (p. 50)
population explosion (p. 49)
population implosion (p. 53)
population pyramid (p. 53)

population replacement level (p. 53)
pull factors (p. 56)
purchasing power parity (PPP) (p. 43)
push factors (p. 56)
refugees (p. 56)
renewable resource (p. 47)
rural-to-urban migration (p. 56)
second law of thermodynamics (p. 60)

selective migrants (p. 56)
settler colonization (p. 44)
sustainable development (p. 60)
sustainable yield (p. 47)
technocentrists (p. 58)
"Third Revolution" (p. 61)
trade barriers (p. 61)
urbanization (p. 56)
value-added products (p. 45)
zero population growth (ZPG) (p. 53)

Review Questions

1. What were the Agricultural and Industrial Revolutions? In what ways did they initiate important changes in human–Earth relationships?

2. What are the typical differences between MDCs and LDCs?

3. According to dependency theory, what are the causes of disparities between MDCs and LDCs? What other factors may explain global wealth and poverty?

4. What is the fuelwood crisis? What are some of the effects of deforestation on human lives and economies in the LDCs?

5. What are the two types of overpopulation, and how do they differ?

6. What has been the main cause of the world's explosive population growth since the beginning of the Industrial Revolution?

7. What variables distinguish the four stages of the demographic transition, and what explains them?

8. Which world regions have the most and the fewest people, and what factors might account for these differences?

9. How do technocentrists and neo-Malthusians view the balance between people and resources?

10. What do Hardin's lifeboats represent—that is, what is his metaphor?

11. What are the goals and methods of sustainable development?

Notes

1. Marshall B. Sahlins, *Stone Age Economics* (Chicago: Aldine-Atherton, 1972).

2. Many of these points come from this special edition: "The Geology of the Planet: Welcome to the Anthropocene," *The Economist*, May 26, 2011.

3. Alfred W. Crosby, *Ecological Imperialism: The Biological Expansion of Europe, 900–1900* (New York: Cambridge University Press, 1986).

4. Friedman, Thomas. 2005. The World is Flat: A Brief History of the Twenty-First Century. New York: Picador.

5. Celia W. Dugger, "Very Young Populations Contribute to Strife, Study Concludes." *New York Times*, April 4, 2007, p. A6.

6. Cited in Steven A. Holmes, "Global Crisis in Population Still Serious, Group Warns." *New York Times*, December 31, 1997, p. A7.

7. Paul R. Ehrlich, "Populations of People and Other Living Things," in *Earth '88: Changing Geographic Perspectives*, ed. Harm De Blij (Washington, DC: National Geographic Society, 1988), p. 309.

8. Lester R. Brown, "Who Will Feed China?" *WorldWatch Magazine*, September–October 1994, p. 10.

9. Garrett Hardin, "The Tragedy of the Commons." *Science*, 162, 1968, pp. 1243–1248.

10. Garrett Hardin, "Lifeboat Ethics: The Case against Helping the Poor," *Psychology Today*, September 1974, p. 6; and "Living on a Lifeboat," *BioScience*, 24, 1974, p. 568.

11. World Wildlife Fund, *Sustainable Use of Natural Resources: Concepts, Issues and Criteria* (Gland, Switzerland: World Wildlife Fund, 1995), p. 5.

Online Resources

 CourseMate: Make the most of your study time by accessing everything you need to succeed in one place. Read your textbook, take notes, review flashcards, watch videos, complete activities, take practice quizzes, and more—online with CourseMate. Log in at **www.cengagebrain.com**.

"Europe is so well gardened that it resembles a work of art, a scientific theory, a neat metaphysical system. Man has re-created Europe in his own image."

—ALDOUS HUXLEY
(1894-1963)

Above: Europe at night. Left: Built in the 13th Century, Sweden's Kalmar Castle is one of Europe's best-preserved castles. For its strategic location on the country's southeast coast, Kalmar was long known as "The Lock and Key of Sweden."

C. Mayhew & R. Simmon (NASA/GSFC), NOAA/NGDC, DMSP Digital Archive

Joe Hobbs

Europe

4

Many culture hearths have exerted influences on peoples and landscapes around the world, but none more impressively than Europe. For more than 500 years, European nation building, scientific and technological achievements, and colonization reshaped cultures and natural systems around the world. Europe is not as strong as it once was, but the region still has enormous economic and political influence on the rest of the world. This chapter introduces the human and physical geographies of this influential world region and explores some of its characteristics and unique problems.

chapter objectives

This chapter should enable you to

- Recognize Europe as a postindustrial region with a wealthy, declining population.
- Become familiar with Europe's immigration issues.
- Learn the region's distinguishing geographic characteristics, landforms, and climates.
- Get acquainted with Europe's major ethnic groups, languages, and religions.
- Understand how Europe rose to global political and economic dominance and then declined in the twentieth century, setting the stage for its "Eurocrisis."
- Trace Europe's emergence from wartime divisions to supranational unity within the European Union (EU) and know why the European Union is important.
- Appreciate important distinctions between Europeans and Americans.
- Identify the concurrence of economic and other forces that have shaped Europe's core region.
- Recognize the spatial relations between heavy industrial resources and cities in Europe and the factors that have caused the deindustrialization of Europe.
- Relate ways in which the core portion of Europe represents one of the most heavily modified landscapes in the world—for example, through deforestation and land reclamation.
- Contrast the primacy of Paris with the decentralized urban pattern of Germany and other European countries.
- Become acquainted with the roots and issues of some of Europe's persistent ethnic and political struggles, including the Troubles of Northern Ireland and heightened tensions affecting Muslims.
- Appreciate the significance of a reunified Germany, especially the economic costs of reintegration for the more prosperous west, a process that is significant globally because of its relevance to the two Koreas and other pairs of nations with prospects for reunification.
- Recognize some geographic, economic, and political factors that have kept some subregions secondary to the European Core in power and influence.
- Appreciate the benefits and drawbacks that people realize from different political and economic systems, including the welfare state and collectivization.
- Recognize the impacts of the wars and their aftermath on the political geographies of the region.
- Observe the trend toward greater unity under the European Union and the North Atlantic Treaty Organization (NATO) and the concurrent trend of devolution of power from central governments to provinces.
- Relate the often troubled histories of ethnic minorities in Europe and how aims of ethnic peoples have led to war, terrorism, and the redrawing of national boundaries.

4.1 Area and Population

Europe is often classified as one of the world's seven continents. Look at Europe on the globe or on the world maps inside the book's cover, however, and you will see that Europe does not necessarily "deserve" to be a continent. Europe is not a distinct landmass like Australia or Africa. It is instead an appendage or a subcontinent of the world's greatest landmass, Eurasia (a term combining the names *Europe* and *Asia*). Nevertheless, the popular and scholarly designation of Europe as a region distinct from the rest of Eurasia is a time-honored and useful organizing device, and this book honors that tradition.

Europe is a great peninsula, fringed by smaller peninsulas and islands and bordered on its seaward sides by the Arctic and Atlantic Oceans, the Mediterranean Sea, and the Black Sea (•Figure 4.1). In this text, Europe is a political region made up of the countries of Eurasia lying west of Turkey, Russia, and three of the former republics of the Soviet Union: Belarus, Ukraine, and Moldova. (Incidentally, the traditional *physical* dividing line between Europe and Asia is drawn from the Ural Mountains down to the Caucasus, technically placing a small part of Turkey, a large part of Russia, and all three of those former republics—now independent countries—within Europe.)

Europe's Subregions

This text classifies Europe into four subregions, made up of a core and three peripheral areas (•Table 4.1).

The **European Core** spreads across northwestern and north central Europe. It consists of the United Kingdom, Ireland, France, the **"Benelux"** countries (an acronym for Belgium, the Netherlands, and Luxembourg), Switzerland, Austria, and Germany. These countries have the largest populations and the most important economic and political roles in Europe. Within the European Core are also three tiny political entities, known as **microstates:** these are Andorra, Monaco, and Liechtenstein.

Following are Europe's three peripheral areas:

Northern Europe is made up of Denmark, Iceland, Norway, Sweden, and Finland.

Southern Europe includes Portugal, Spain, Italy, Greece, Malta, and Cyprus.

Eastern Europe (in many sources classified as the countries of Central and Eastern Europe, going by the acronym CEE) includes Estonia, Latvia, Lithuania, Poland, the Czech Republic, Slovakia, Hungary, Romania, Bulgaria, Albania, Serbia, Kosovo, Montenegro, Bosnia and Herzegovina, Croatia, Macedonia, and Slovenia.

Small but Powerful Europe

Europe's influence in the world has been disproportionately large, given its size. Europe is only about half as large as the United States' "Lower 48" (the contiguous United States; that comparison is easily seen in Figure 4.3). The average European country is only 50,000 square miles (130,000 sq km) in area, or about the size of Arkansas.

• **Figure 4.1** Reference map of the political and physical geographies of Europe. In nearly all European countries, the political capital is also the largest and most important city.

Within Europe, some countries far outsize the others in population and influence (•**Figure 4.2b**). Four countries—Germany, the United Kingdom, France, and Italy—are very large. Their populations range from about 82 million in Germany to about 63 million each in France and in the United Kingdom, and 61 million in Italy. These four countries together make up about half of Europe's population (and for comparison are equal to about 84 percent of the population of the United States).

Europe has one of the world's great clusters of human population (**Figure 4.2; see also Figure 3.14 and Table 4.1**). You can clearly see this in the chapter-opening image of Earth from space, at night. Europe's population of 532 million is about 1.7 times that of the United States. One of every

Table 4.1   Europe: Basic Data

Political Unit	Area (thousands; sq mi)	Area (thousands; sq km)	Estimated Population (millions)	Estimated Population Density (sq mi)	Estimated Population Density (sq km)	Annual Rate of Natural Increase (%)	Human Development Index	Urban Population (%)	Per Capita GNI PPP ($US)
European Core	**549.5**	**1,422.6**	**253.1**	**461**	**178**	**0.1**	**0.900**	**77**	**36,840**
Andorra	0.2	0.5	0.1	500	193	0.7	0.838	91	N/A
Austria	32.4	83.8	8.4	259	100	0.0	0.885	67	38,410
Belgium	11.8	30.5	11.0	932	360	0.2	0.886	99	36,610
France	212.9	551.2	63.3	297	115	0.4	0.884	77	33,950
Germany	137.8	356.7	81.8	594	229	−0.2	0.905	73	36,850
Ireland	27.1	70.1	4.6	170	66	1.0	0.908	60	33,040
Liechtenstein	1.0	2.6	0.04	40	15	0.5	0.905	15	N/A
Luxembourg	0.1	0.1	0.5	8,333	3218	0.4	0.867	83	59,590
Monaco	0.001	0.002	0.03	30,000	11583	0.0	N/A	100	N/A
Netherlands	15.8	40.9	16.4	1038	401	0.3	0.910	66	39,740
Switzerland	15.9	41.1	7.5	472	182	0.2	0.903	74	47,100
United Kingdom	94.5	244.6	62.7	663	256	0.4	0.863	80	35,860
Northern Europe	**485.8**	**1,257.7**	**25.7**	**53**	**20**	**0.2**	**0.905**	**77**	**40,950**
Denmark	16.6	43.0	5.6	337	130	0.2	0.895	72	38,780
Finland	130.6	338.1	5.4	41	16	0.2	0.882	68	35,280
Iceland	39.8	103.0	0.3	8	3	0.9	0.898	93	32,840
Norway	125.1	323.8	5.0	40	15	0.4	0.943	79	55,420
Sweden	173.7	449.7	9.4	54	21	0.3	0.904	84	38,050
Southern Europe	**401.9**	**1,040.5**	**128.0**	**318**	**123**	**0.0**	**0.885**	**71**	**31,390**
Cyprus	3.6	9.3	1.1	306	118	0.6	0.840	64	30,290
Greece	51.0	132.0	11.3	222	86	0.1	0.861	73	28,800
Italy	116.3	301.1	60.8	523	202	−0.1	0.874	68	31,870
Malta	0.1	0.2	0.4	4,000	1,544	0.2	0.832	100	23,170
Portugal	35.5	91.9	10.7	301	116	−0.1	0.809	38	24,080
San Marino	0.02	0.05	0.03	1,500	579	0.4	N/A	84	N/A
Spain	195.4	505.9	46.2	236	91	0.2	0.878	77	31,490
Vatican City	0.1	0.04	0.001	10	4	0.0	N/A	100	N/A
Eastern Europe	**522.1**	**1,351.2**	**125.5**	**240**	**93**	**−0.1**	**0.802**	**61**	**17,100**
Albania	11.1	28.7	3.2	288	111	0.6	0.739	50	8,640
Bosnia and Herzegovina	19.7	51.0	3.8	193	74	0.0	0.733	46	8,770
Bulgaria	42.8	110.8	7.5	175	68	−0.5	0.771	73	13,260
Croatia	21.8	56.4	4.4	202	78	−0.2	0.796	56	19,200
Czech Republic	30.4	78.7	10.5	345	133	0.1	0.865	74	23,940
Estonia	17.4	45.0	1.3	75	29	0.0	0.835	68	19,120
Hungary	35.9	92.9	10.0	279	108	−0.4	0.816	68	19,090
Kosovo	4.2	10.8	2.3	548	211	1.4	N/A	N/A	N/A
Latvia	24.9	64.4	2.2	88	34	−0.5	0.805	68	17,610
Lithuania	25.2	65.2	3.2	127	49	−0.2	0.810	67	17,310

(continued)

Table 4.1 Europe: Basic Data (*continued*)

Political Unit	Area (thousands; sq mi)	Area (thousands; sq km)	Estimated Population (millions)	Estimated Population Density (sq mi)	Estimated Population Density (sq km)	Annual Rate of Natural Increase (%)	Human Development Index	Urban Population (%)	Per Capita GNI PPP ($US)
Macedonia	9.9	25.6	2.1	212	82	0.3	0.728	65	10,880
Montenegro	5.4	14.0	0.6	111	43	0.4	0.771	64	13,110
Poland	124.8	323.1	38.2	306	118	0.1	0.813	61	18,290
Romania	92.0	238.1	21.4	233	90	−0.2	0.781	55	14,540
Serbia	29.9	77.4	7.3	244	94	−0.5	0.766	58	11,700
Slovakia	18.9	48.9	5.4	286	110	0.1	0.834	55	22,110
Slovenia	7.8	20.2	2.1	269	104	0.2	0.884	50	26,470
Summary Total	**1,959.3**	**5,072.0**	**532.3**	**272**	**105**	**0.0**	**0.873**	**72**	**31,070**

Sources: World Population Data Sheet, Population Reference Bureau, 2011; Human Development Report, United Nations, 2011; World Factbook, CIA, 2011.

Table 4.2 Europe: Metropolitan Populations Data

1	London, U.K.	13.4
2	Paris, France	11.9
3	Cologne, Germany	11.8
4	Amsterdam, Netherlands	6.6
5	Madrid, Spain	6.4
6	Manchester, U.K.	5.3
7	Berlin, Germany	5.1
8	Barcelona, Spain	5
9	Milan, Italy	4.4
10	Frankfurt, Germany	4.1
11	Rome, Italy	4
12	Athens, Greece	3.9
13	Naples, Italy	3.9
14	Birmingham, U.K.	3.8
15	Hamburg, Germany	3.3
16	Lisbon, Portugal	2.6
17	Budapest, Hungary	2.6
18	Mannheim, Germany	2.5
19	Katowice, Poland	2.5
20	Copenhagen, Denmark	2.4

Population in millions.

© Cengage Learning 2013

13 people in the world is a European, and they live in a space half the size of the United States. Within this great cluster of people, there are some remarkable variations; some European countries are lightly populated, and others have large, dense concentrations of people (Table 4.2).

Spatial Patterns of Energy, Industries, and Cities in Europe

Most of Europe's largest and most densely populated modern cities are located where they are for a reason: their geographical *situation* favored them (see Insights on page 70). Cities like Manchester in England, Essen in Germany, Prague in the Czech Republic, and Lille in France are situated near historical sources of coal and hydroelectric power. These energy resources were critical in the location and rapid development of many European cities during and in the wake of the Industrial Revolution (see Figure 4.22 on page 91, and the Industrial Revolution on page 41). Europe has two major belts of industrialization and urbanization. One belt runs generally north–south from the United Kingdom to Italy. The second belt of dense population runs in a more east–west direction from the United Kingdom to Poland (and eastward out of the region into Ukraine).

These two urban-industrial belts cover only a small portion of Europe but produce more economic goods and services than the rest of Europe combined. There are only three other areas on Earth that resemble Europe's urban-industrial belts; these are in eastern North America, Japan, and China.

267, 270, 393

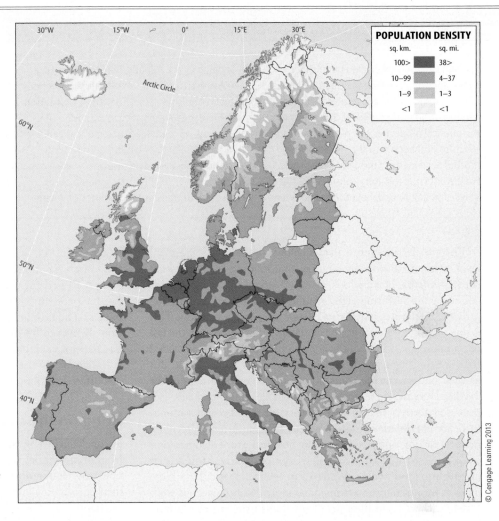

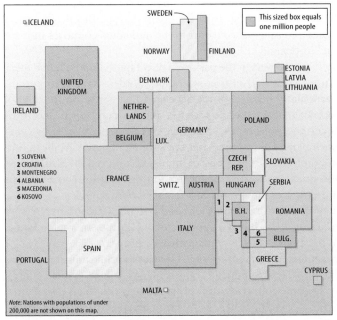

• **Figure 4.2a** Population distribution (right) and **4.2b** population cartogram (below) of Europe. The continent's demographic heavyweights are Germany, the United Kingdom, France, and Italy.

Source: After Hoffman, 1990.

Why Is Europe's Population Declining?

There are several outstanding characteristics of Europe's population: it is urban, aging, and—until the critical feature of migration is added in—shrinking.

Europeans are one of the most urbanized people on Earth. This large urban population is shrinking, as the region has entered the final stage of the demographic transition. It is useful to think of Europe as a real-life illustration of the demographic transition. This region traced a trajectory from preindustrial high birth rates and high death rates to postindustrial low birth rates and low death rates (see Figure 3.10, page 52). Over a period of decades, Europe recovered from the demographic setbacks of two world wars (for example, France lost more than 3 percent of its prewar population in World War I, and Poland lost 19 percent in World War II). The population peaked in 1997 and then began a slow, steady decline to what is widely acknowledged as a **"birth dearth"**. Today, birth rates are low in this region of relative affluence and high urbanization, where increasing numbers of employed and educated women are saying, "There is no time for motherhood, and children are too expensive anyway."[1] Note in the summary total column of Table 4.1 on page 68 that Europe's population growth rate is exactly 0.0, or zero population growth.

Site and situation are related but distinct dimensions of place and space and are crucial to understanding the geography of Europe and the world's other regions. **Site** refers to the physical properties of a piece of land on which something is or will be located: Rome is often described as "the city of seven hills." For a contrast within Italy, consider the site of Venice, on flat ground, penetrated and often inundated by the Adriatic Sea. Known as *La Serenissima*, "the most serene [city]," Venice is in fact often in peril; see Figure 4.35). **Situation** is the larger geographical context of the site; the saying "all roads lead to Rome" indicates that the Eternal City was situated as a transportation hub. In studying Europe, you will see many examples of cities that grew up at the crossroads of transportation routes or where coal, hydropower, and other resources could be obtained nearby; these are examples of how situation plays an important role in urban and economic geography. Sometimes situation and site correspond in the most advantageous ways for human settlement. In other cases, however, the perfect situation is an imperfect site: see New Orleans on page 438.

Europe's fertility rate is in fact below population replacement level, with no European country maintaining its population through births. Have a look at Table 4.1: you can see that most countries are hovering right around or below zero population growth. Birth rates in Eastern Europe are particularly low. This is due mainly to the economic challenges that came with the fall of communism. Many of the communist governments provided free housing, education, and child care to their citizens. But as they underwent the transition to more democratic and free market systems, these governments cut the state subsidies that provided a safety net for their citizens. It then became rational to delay childbirth or decide not to have children. Perceiving better working opportunities elsewhere in Europe, women and couples from the east also began moving westward in search of education and work; Western Europe is a particularly large magnet for immigration from many more distant places as well.

More than 15 million people of non-European origin now live in Europe. The major groups are Turks, Kurds, Arabs, Indians, Pakistanis, Sri Lankans, and a variety of peoples from Sub-Saharan Africa, the Caribbean, and Latin America.

During periods of solid economic growth (especially 1997–2007), immigrants often provided needed labor and were generally welcomed in Europe. Guest workers were encouraged to come to industrial and agricultural centers that were not able to meet their labor needs with domestic workers. West Indians came to Britain to help with labor shortages in the 1950s, for example. Germany invited Turks as guest workers during its economic boom years from 1961–1997, and 2 million Turks responded to that opportunity. Spain welcomed Latin Americans, Moroccans, Romanians, and others when it was enjoying its property and construction boom in the 1990s (with the benefit of hindsight, we can now see that was a classic real estate bubble, similar to the one that built up in the United States during those same years).

But recent events, including the shockwaves of the Great Recession that began in 2007, growing domestic unemployment, terrorist attacks, and tax burdens, have changed native attitudes toward immigrants considerably. Immigration has become a divisive and even dangerous issue in Europe, especially where the migrants are Muslims (see Regional Perspective on page 71, and the introduction to Islam on page 173).

Europe's shrinking population is aging faster than that of any other world region. The median age in Italy is projected to rise from 44 in 2011 to 53 by 2050. The populations of the Czech Republic and Spain are expected to contract by about 20 percent by 2050, and the Netherlands may shrink by 10 percent. How is it possible to get more European babies? The Italian government has tried to reverse its population shrinkage by offering parents a **"baby bounty"** of about $1,200 for each child after their first—but only if the mother is Italian or carries a **European Union (EU)** passport. France has offered similar cash incentives for third children.

Bring on the Immigrants?

Europe's shrinking, aging population could use a youthful boost from immigration, which is already the main source of population growth in most European countries. If birth rates remain at their current low level, the European Union will have a shortfall of 20 million workers by 2030. The EU countries would need an annual inflow of about 3 million migrants a year to prevent the future support ratio from dropping sharply. But that prospect raises a number of concerns in the individual countries and for the European Union itself.

Governments in the past were reluctant to impose harsh measures that would restrict migration. Borders have been porous, particularly because of the Schengen Agreement on freedom of movement (described on page 89). But Europe's open door attitudes and policies are changing, especially because of growing unemployment at home and the surging numbers of illegal immigrants. European views of immigrants shifted markedly when the global economic crisis struck Europe in 2007. Most Europeans now see immigrants as a financial burden on society and as competing with them for jobs and benefits; immigrants threaten to unravel the social safety net of the European welfare state, and they live outside of mainstream European society instead of becoming integrated within it. Germany's prime minister went as far as to declare multiculturalism a "total failure" in her country.

In some countries, notably Denmark and the Netherlands, generous welfare benefits were a magnet for immigration. But Denmark has pushed back, passing tough anti-immigration laws and reducing welfare benefits for new arrivals. In the Netherlands, increasing numbers of Dutch nationals are expressing weariness with paying the tax burden to support literacy, technical training, and health care costs for

REGIONAL PERSPECTIVE | Islamophobia in Europe

Many Europeans are worried about the immigration of Muslims into their countries and the growing cultural presence of their religion, Islam. Some fears relate to terrorism: the London subway bombings of 2005 and the London and Glasgow car bombing attempts of 2007 were attributed to Muslim dissidents of Pakistani and Middle Eastern origin, for example. But more than anything, deep-seated cultural issues—what some have described as a "clash of civilizations"—have led to violence and recrimination. In 2004, a Dutch citizen of Moroccan descent murdered Theo van Gogh, a Dutch filmmaker (and distant relative of the artist Vincent van Gogh) who had criticized the status of women in Muslim societies. In 2005, after a Danish newspaper published satirical cartoons of the Prophet Muhammad (among other things, wearing a bomb in his turban), a wave of sometimes violent demonstrations against Denmark erupted across the Muslim world. For all Muslims except some Shiite groups, it is heresy to depict the Prophet in any form, so such negative and stereotypical caricatures touched a raw nerve. Many European media fueled the fire by publishing the cartoons in their own outlets, and the cycle of taunting and reaction intensified.

Islamophobia—a fear of Islam and the Muslims who practice it—has grown throughout Europe. Muslims face growing prejudice and suspicion. Many native Europeans believe that Islam cannot be accommodated within their value systems and societies. Right-wing political parties with anti-Islamic platforms have been on the rise in countries including Sweden, the Netherlands, France, Belgium, Italy, Switzerland, and Austria. Germany's Interior Ministry was quoted saying that Islam is not a part of the German way of life.[2] Switzerland outlawed the construction of Islamic minarets. (●**Figure 4.A**) French and Belgian governments prohibited Muslim women and girls from wearing scarves (*hijab*) in schools and government offices, and wearing face-covering veils (*niqab* and *burqa*) in public places (see ●**Figure 4.A**). Some governments cited both security concerns (it is important to see a person's face in numerous circumstances,

such as in banking and retail selling or wherever presenting identification is required) and longstanding bans on any religious symbols in secular society. France's secular political system insists on strict separation of church and state. It would be unthinkable for a French politician to take the oath of office using a Bible, for example. Christian students in France are prohibited from wearing crosses, and Jewish students from wearing skullcaps. The French government has banned any "conspicuous signs of religious affiliations" in public schools. But Muslims feel that the restrictions are in reality a form of discrimination aimed particularly at them.

Writers and editorialists on the right have depicted the transformation of the cultural landscape from Europe to **Eurabia**. Islam is "patiently conquering Europe's cities, street by street," wrote the British columnist Christopher Caldwell.[3] He forecasts a Europe swamped by the demographic tide of Muslims, whose birthrates far exceed those of native Europeans and whose radical interpretations

of Islam threaten the continent's security. Polls and statistics, however, reveal a less ominous picture of European Muslims, and a Muslim majority in any Western European country appears unlikely by any reasonable demographic projection. Like most native Europeans, Muslims in Europe are far more preoccupied with their economic well-being than with militant politics, and their birth rates are falling. "If you drew up a composite profile of the average French European Muslim today," wrote journalist Simon Kuper, "it would look something like this: She has two or three children, who attend non-religious schools. She is relatively poor but generally content, although angry about discrimination. She feels more religious than a decade ago, but doesn't wear a headscarf, although she has friends who do. She opposes terrorism, although she probably knows terrorist sympathizers. She votes socialist, and worries about economic issues more than about anything in the Middle East."[4]

● **Figure 4.A** Switzerland implemented a ban against the construction of minarets—an architectural element of Muslim mosques—in 2010. This poster in favor of the ban depicts minarets as missiles puncturing the Swiss flag and presumably threatening the destruction of Swiss and other Western societies. Muslim women wearing the *burqa* are associated with these symbols of evil. The sign reads "Yes to Banning Minarets" Europeans are not of one voice in answering questions about their ethnic and religious diversity, and anti-immigration sentiment has been on the rise.

immigrants, who now make up 10 percent of the country's population.

The numbers of immigrants into Europe have been large. Between 1989 and 2009, 26 million came to the 15 Western European countries making up the core of the European Union (**prior to its "big bang" expansion; see page 87**). About 1.8 million people enter the European Union legally each year, and perhaps 500,000 come illegally. Germany is home to an estimated 1 million "illegals," as Europeans call their **undocumented workers,** and France has 300,000. Most of the illegal aliens cross land or sea borders unlawfully, and about a third are visitors who overstay tourist or student visas.

Italy's Mediterranean island of Lampedusa, which is actually closer to the North African country of Tunisia than it is to mainland Italy, is a favorite transit site for illegal immigrants, mainly from North and Sub-Saharan Africa (**see Lampedusa in Figure 4.1 on page 66**). Some were reportedly forced to flee there by Libyan authorities after Libya's leader, Muammar Gaddafi, vowed to flood Europe with African migrants and "turn Europe black" if European militaries continued to attack Gaddafi's regime.

When they no longer wanted immigrants, European governments, particularly those of Italy and Spain, were able to use generous foreign aid to effectively pay Libya, Tunisia, and Morocco to keep would-be immigrants to Europe bottled up within those countries, or to repatriate them to their home countries. As North African regimes toppled in the Arab Spring, those guarantees disappeared, and tens of thousands of immigrants—mainly Somalis, Eritreans, Senegalese, and Nigerians—began making their way to mainland Europe through Lampedusa. Italian authorities held them for 2 days in Lampedusa, then classified them in mainland Italy as either refugees seeking political asylum, or "economic immigrants" seeking work. And then, much to the chagrin of France, Germany, and other destination countries, Italy issued them national residence permits allowing them to move on elsewhere in Europe. Italy hoped the migrants would exploit the passport-free Schengen area to travel on to countries where many have family members. (Under European law, the country where migrants first arrive is responsible for deciding their status as immigrants, refugees, or asylum seekers. A protocol signed by the European countries gives everyone the right to seek asylum in Europe and gives refugees the right not to be sent to a place where they may be tortured or persecuted). French authorities pointed out that Schengen rules grant freedom of movement only to those with proper passports and the means to support themselves.

When migrants reached their destination countries—Spain is one case—most found themselves destitute and living in squalid camps in the countryside. Spain had prided itself on being immigrant-friendly, especially during its economic boom (bubble) years of 1997–2007. These immigrants contributed to Spain's economic success. But in the depths of the financial crisis after 2007, Spain struggled to halt the surge of immigrants.

4.2 Physical Geography and Human Adaptations

Europe has diverse and impressive physical features. Among its most distinctive physical characteristics are its irregular shape, its jagged coastal outline, its high latitude, and its temperate climate. The main peninsula of Europe is fringed by a number of smaller peninsulas, notably the Scandinavian, Jutland, Iberian, Italian, and Balkan peninsulas. It is useful to think of Europe as being "a peninsula of peninsulas" (**see Figure 4.1**). Offshore, there are numerous islands, including Great Britain, Ireland, Iceland, Sicily, Sardinia, Corsica, and Crete. The island of Cyprus lies far in the eastern Mediterranean. Around the indented shores of Europe, arms of the sea penetrate the land in the form of **estuaries** (the tidal mouths of rivers), and harbors offer protection for shipping. This complex mingling of land and water provides many opportunities for maritime activity, and much of Europe's history has focused on seaborne trade, sea fisheries, and sea power.

On the map in •**Figure 4.3**, you can see that much of Europe lies north of the latitudes of the 48 conterminous United States. The British Isles are at the same latitude as Hudson Bay in Canada. Athens, Greece, is only slightly farther south of the latitude of Saint Louis, Missouri. You may expect Europe to be cold, considering those facts. But generally it is rather warm.

Why Is Europe So Warm?

Europe's mild climates, especially in winter, are particularly surprising given its high latitudes. London, for example, has about the same average temperature in January as Richmond, Virginia, which is 950 miles (c. 1,500 km) farther south. Relatively warm

• **Figure 4.3** Europe in terms of latitude and area compared with the United States and Canada.

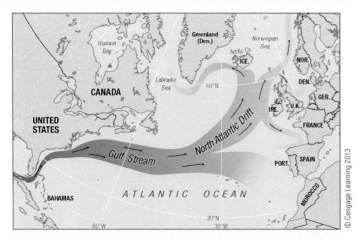

• **Figure 4.4** The Gulf Stream and North Atlantic Drift bathe the western and northwestern region of the Eurasian landmass with relatively warm waters swept up from the Caribbean and the Gulf of Mexico. The warm water makes the air temperature warmer than it otherwise would be, giving the region milder winters than would be expected at such a high latitude.

currents of water account for Europe's warmth. The **Gulf Stream** and its appendage the **North Atlantic Drift** (•**Figure 4.4**), which originate in tropical western parts of the Atlantic Ocean, flow to the north and east and cause the waters around Europe to be much warmer in winter than the latitude would warrant. With winds blowing mainly from the west (from sea to land) along the Atlantic coast of Europe, the moving air in winter absorbs heat from the ocean and transports it to the land. This makes winter temperatures abnormally mild for this latitude, especially along the coast. In the summer, the climatic roles of water and land are reversed: instead of being warmer than the land, the ocean is now cooler, so the air brought to the land by **westerly winds** (*westerlies*) in the summer has a cooling effect. These influences are carried as far north as the Scandinavian Peninsula and into the Barents Sea, making Norwegian ports generally ice-free in winter and giving the Russians a warm-water port at Murmansk, even though it lies above the Arctic Circle.

Northern Europeans—the Norwegians, Swedes, Finns, and Icelanders—enjoy their long summer days. If you are an American visiting this part of the world, your body clock will be surprised, and you will have to accustom yourself to the local rhythms. At midnight, the sky is still bright, and you will have to force yourself to go to sleep. The locals enjoy this "midnight sun" and celebrate the summer solstice with great fervor at "midsummer" festivals. They revel because they know the price they have to pay in winter with its long, dark nights.

The same winds that bring warmth in winter and coolness in summer also bring abundant moisture. Most of this falls as rain, although the higher mountains and more northerly areas have considerable snow. Abundant, well-distributed, and relatively dependable moisture has always been one of Europe's major assets (see the world precipitation map, Figure 2.3 on page 23). The average precipitation in most places in the European lowlands is 20 to 40 inches (c. 50 to 100 cm) per year. This is enough for a wide range of crops, because mild temperatures and high atmospheric humidity reduce the rate at which plants lose their moisture.

Human Settlement on Europe's Varied Landscapes

With plains, plateaus, hills, mountains, and water bodies, Europe's topographic features are very diverse. Enriched by the human associations of an eventful history, Europe offers its residents and visitors the chance to enjoy distinctive and often highly scenic landscapes.

One of the most prominent surface features of Europe is an undulating or rolling plain that extends without a break from near the French-Spanish border, across western and northern France, central and northern Belgium, the Netherlands, Denmark, northern Germany, Poland, and far into Russia. Known as the **North European Plain** (see Figure 4.1), it has outliers in Great Britain, the southern part of the Scandinavian Peninsula, and southern Finland.

Geographic characteristics of the North European Plain have had important impacts on human settlement. The plain contains the largest part of Europe's farmland, much of it on soils formed from deposits of **loess** (a windblown soil of generally high fertility). It is underlain in some places by deposits of coal, iron ore, potash (used in the production of soaps, glass, and some compounds), and other minerals that were important in the region's industrial development. This broad lowland provides an important natural transportation route skirting the highlands to the south. Thanks to these geographic characteristics, many of the largest European cities, including London, Paris, and Berlin, developed on the plain. From northeastern France eastward, there is a band of especially dense population extending along the southern edge of the plain; you can see this "plainly" in Figure 4.2a.

Europe's highlands are very important in the character of European landscapes and peoples. South of the North European Plain, Europe is mainly mountainous or hilly, with scattered plains, valleys, and plateaus. The hills are geologically old, and their elevations have been worn down by erosive forces over long periods of time. But many mountains in Southern Europe are geologically young and often high and ruggedly spectacular, with jagged peaks and snowcapped summits. The highest is Mount Blanc (15,771 ft [4,807 m]) on the French-Italian border, but the most iconic is Switzerland's Matterhorn, at 14,690 ft (4,478 m) (•**Figure 4.5**). Another prominent mountain chain, the Pyrenees, forms a wall up to 10,695 feet (3,404 m) high along the border of Spain and France. The famous Carthaginian general Hannibal crossed both the Pyrenees and the Alps with his formidable army—which included elephants used for transport—in an assault on Rome in the third century B.C.E. The Transylvanian Alps and the Carpathian Mountains, legendary chains of mountains evoking tales of vampires and mysterious woodlands, meet in Romania.

Glaciation—a geologic and climatologic process in which great ice sheets formed in the Arctic and Antarctic and advanced toward the equator—played a major role in shaping the landscapes of Europe. Continental ice sheets formed over the continent, beginning on the Scandinavian Peninsula and Scotland (•**Figure 4.6**), during the **Ice Age** of the Pleistocene (c. 2 million to 10,000 years ago). Evidence suggests that there were four

Joe Hobbs

• **Figure 4.5** The iconic Matterhorn rising to 14,690 feet (4478 m) above the Swiss resort of Zermatt. This village is carefully managed as a pedestrian-only jewel of international tourism.

major periods of widespread glacial coverage. Today, landscapes of **glacial scouring**—the erosive action of ice masses in motion—characterize most of Norway and Finland, much of Sweden, parts of the British Isles, and Iceland. These changes created many favorable sites for hydroelectric installations. **Glacial deposition**—the process of offloading rock and soil in glacial retreat or lateral movement—had a major effect on the

Joe Hobbs

• **Figure 4.7** Norway's scenic fjords were carved by glaciers and then filled by rising sea levels. This is the head of Geirangerfjord on the country's west coast. Note the village of Geiranger where the fjord's waters terminate.

present landscape. Glacial deposits were left behind on most of the North European Plain and are productively farmed today. The weight of the glaciers was so great, and their retreat so recent in geologic terms, that parts of the Scandinavian Peninsular are still rising—rather like a sponge from which a rock has been removed. This process, called **isostatic rebound,** has the curious effect of raising some of Scandinavia's harbors and forcing people to relocate port facilities. These are perhaps some of the few places on Earth where rising sea levels due to climate change are not a concern!

The North European Plain is bordered by glaciated lowlands and hill lands in Finland and eastern Sweden and by rugged, ice-scoured mountains and spectacular fjords in western Sweden and most of Norway (•**Figure 4.7**). The British Isles also have extensive areas of glacially scoured hill country and low mountains, along with lowlands where glacial deposition occurred.

Wherever they call home, most Europeans have a special relationship with nature, with particularly strong bonds to the seas and the mountains. Europe is relatively small; it has an outstanding transportation infrastructure of roads, rails, and ferries; it has city-dwellers who long for the weekend or seasonal getaway; and it has beautiful beaches and mountains. Even if Europeans don't live by these natural wonders, they can reach them quickly. Europe's natural gems inevitably succumb to their own successes: some of its beach and mountain resorts are notoriously crowded and, many would say, spoiled;

• **Figure 4.6** The maximum extent of glaciation in Europe about 18,000 years ago.
Source: After Hoffman, 1990.

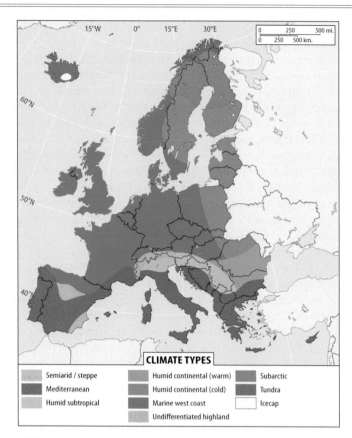

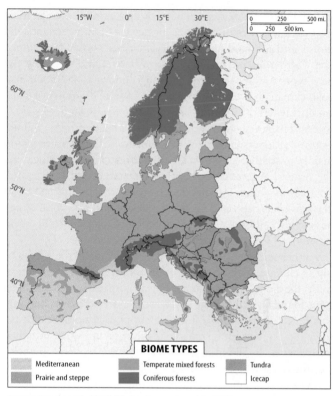

• **Figure 4.8** Climate types (left) and biomes (right) of Europe.
Source: Based on Rand McNally's Classroom Atlas, 2003.

Source: Based on World Wildlife Fund ecoregions data, 1999.

but others have managed to maintain their special charm through careful planning and management. Switzerland's resort of Zermatt is one example (see Figure 4.5).

Diversity of Climate and Vegetation

Despite its relatively small size, Europe also has remarkable climatic and biotic diversity (•**Figure 4.8;** see also Figures 2.4 and 2.5). A marine west coast climate extends from the coast of Norway to northern Spain and inland to west central Europe. Comparable to the climate found in the U.S. Pacific Northwest, the main characteristics are mild winters, cool summers, and ample rainfall, with many drizzly, cloudy, and foggy days. Throughout the year, changes of weather follow each other in rapid succession as different air masses temporarily dominate or collide with each other along weather fronts. In lowlands, winter snowfall is light, and the ground is seldom covered for more than a few days at a time (in the first decade of this century, however, there were several freakish, massive snow events that blanketed much of the British Isles and the European continent). Summer days are longer, brighter, and more pleasant than the short, cloudy days of winter, but even in summer, there are many chilly and overcast days. The frost-free season of 175 to 250 days is long enough to support typical middle-latitude crops, but most areas have summers that are too cool for heat-loving crops like corn (maize). Agricultural and other land uses are depicted in •**Figure 4.9**.

Inland from the coast, in western and central Europe, the marine climate gradually changes. Winters become colder

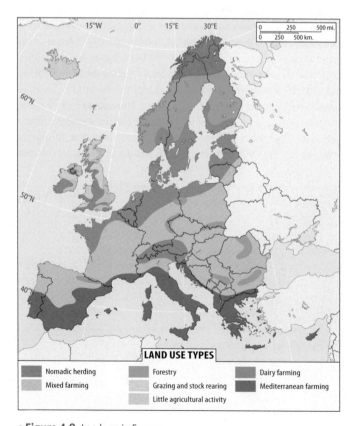

• **Figure 4.9** Land use in Europe.
Source: The Oxford School Atlas edited by Patrick Wiegand (OUP, 1997), copyright © Oxford University Press. Used by permission of Oxford University Press.

and summers are hotter; cloudiness and annual precipitation decrease. Influences of maritime air masses from the Atlantic diminish and are modified by continental air masses from inner Asia. Farther inland, conditions are different, and two new climate types are apparent: the humid continental short-summer (cold) climate in the north (principally in southern parts of Sweden and Finland, Estonia, Latvia, Lithuania, Poland, Slovakia, eastern Hungary, and northern Romania) and the humid continental long-summer (warm) climate in the warmer south (mainly in southern Romania, Serbia, and northern Bulgaria). The natural vegetation is mostly temperate mixed forest, and soils vary in quality. Among the best soils are those formed from alluvium (soils deposited by rivers and streams) and loess along the roughly 1,800-mile (3,000 km) valley of the Danube River in central and Eastern Europe (see Figure 4.1 and page 77).

Southernmost Europe has a distinctive climate: the Mediterranean climate, with a pattern of dry summers and wet winters (this is the archetype for Mediterranean climates worldwide, like those found in California; see Figure 2.4 for their distribution). This pattern results from seasonal shifts in atmospheric belts. In winter, the belt of westerly winds shifts southward, bringing precipitation. In summer, a belt of high atmospheric pressure over the Sahara shifts northward, bringing extremely dry and warm or hot conditions. Winters are mild, and frosts are rare.

Drought-resistant trees originally covered Mediterranean lands, but little of that forest remains, having been depleted by centuries of human use—especially the clearing of land for farming, and the pressure of goats and other grazing livestock. George Perkins Marsh, one of the early great American geographers, wrote about these landscape changes in the Mediterranean in his classic nineteenth-century book *Man and Nature*. Forests have been replaced by the wild scrub that the French call *maquis* (called *chaparral* in the United States and Spain) or by cultivated fields, orchards, or vineyards (•Figure 4.10). Much of the land consists of rugged, rocky, and eroded slopes where thousands of years of deforestation, overgrazing, and excessive cultivation have taken their toll. The region's subtropical temperatures can support a great variety of crops, but irrigation is a must in the dry summers.

Anastasios71/Shutterstock.com

• **Figure 4.10** This landscape on the Greek island of Corfu is covered with typical Mediterranean vegetation, and shows off the physical beauty so characteristic of much of the Mediterranean region.

This climate is also a draw for summer travelers, and tourism has immense economic importance in Mediterranean Europe.

Some northerly sections of Europe experience the harsh conditions associated with subarctic and tundra climates. The subarctic climate, characterized by long, severe winters and short, cool summers, covers most of Finland, much of Sweden, and parts of Norway. A short frost-free season, along with thin, leached, and acidic soils, makes agriculture difficult. Human settlement is sparse, and forest of needleleaf conifers (the taiga described on page 25), such as spruce and fir, covers most of the land. In the tundra climate of northernmost Norway and much of Iceland, cold winters combine with brief, cool summers and strong winds to create conditions hostile to tree growth. An open, windswept landscape results, covered with lichens, mosses, grass, low bushes, dwarf trees, and wildflowers (a typical Norwegian tundra landscape may be seen in Figure 2.6b on page 25). Human inhabitants are few, and agriculture usually impossible, on the tundra.

The higher mountains of Europe, like high mountains in other parts of the world, have an undifferentiated highland climate, varying with elevation and exposure to sun, wind, and precipitation. The variety can be startling. The Italian slope of the Alps, for example, ascends from subtropical conditions at the base of the mountains to tundra and ice cap climates at the highest elevations. The ice cap climate experiences temperatures that historically have averaged below freezing every month of the year. At the higher latitudes, these extreme conditions are found even at sea level. Europe's glaciers are following the almost worldwide trend of retreat in the past half century due mainly, most climatologists believe, to increasing levels of greenhouse gases in the Earth's atmosphere and the accompanying global warming. Probably because of their interior continental location, the Alpine glaciers are melting much faster than the world average. 250, 358, 428

Rivers and Waterways

Europe has many rivers evoking deep cultural and historical meanings; think of the Blue Danube, the left and right banks of the Seine in Paris, and the castles on the Rhine, for example. These waterways are important for transport, water supply, electricity generation, and recreation. Rivers were a critical part of Europe's transport system at least as far back as Roman times, and they still make it possible to move cargo at low cost. The most important rivers for transportation are those of the highly industrialized areas in the core area of west north central Europe. An extensive system of canals connects and supplements the rivers. Some are quite old; the Romans built canals throughout Northern Europe and Britain, mainly for military transport.

The Dutch developed the pound (pond) lock for canals in the late fourteenth century, leading to the expansion of regional connections of waterways through canals and to the use of canals to link farmlands with the coasts. Starting in the late 1700s, the British built canals to move raw materials toward factory locations. By the mid-nineteenth-century, the expanding influence of railroads and the steam engine competed with canals. Although canals continue to play a significant transport role, since the 1950s railroads and highways have been the main means to transport goods throughout Europe.

Important seaports developed along the lower courses of many rivers, and some became major cities. London on the Thames, Antwerp on the Scheldt, Hamburg on the Elbe, and Rotterdam in the **delta** (the usually triangular-shaped alluvial area at the mouth of a river) of the Rhine are outstanding examples.

The Rhine and the Danube are particularly important European rivers, touching or crossing the territory of many countries (see Figure 4.1). The Rhine countries include Switzerland, Liechtenstein, Austria, France, Germany, and the Netherlands. Highly scenic for much of its course, the Rhine is Europe's most important inland waterway. Along and near it, an axis of intense urban industrial development and high population density developed. At its North Sea end, the Rhine connects to world commerce at one of the world's most active seaports, Rotterdam, in the Netherlands. The Danube, on its journey from the Black Forest of southwestern Germany to the Romanian shore of the Black Sea, touches or crosses more countries than any other river in the world. The river is an important artery for the flow of goods and is also held in deep esteem by the peoples that share its waters. The Danube delta region of eastern Romania is one of the world's most important wetlands for waterfowl, and it draws growing numbers of ecotourists.

4.3 Cultural and Historical Geographies

The apparent crowding of so many countries into the relatively small land area of Europe is an indication of the region's extraordinary cultural diversity. Travelers in Europe experience this richness. A brief train ride, for example, takes the visitor from French-speaking western Switzerland eastward into the country's mainly Germanic region and southward through mountain tunnels into the Italian world. Today, almost all of Europe's peoples coexist harmoniously, but this harmony was a long time in coming.

Linguistic and Ethnic Groups of Europe

This section will help you appreciate Europe's diverse cultures. Europe emerged from prehistory as the homeland of many different peoples. In ancient and medieval times, some of them expanded with vigor, diffusing their languages and cultures widely. The first millennium B.C.E. witnessed a great expansion of the *Greek* and *Celtic* (pronounced *kel'*-tik) peoples. In peninsulas and islands bordering the Aegean and Ionian Seas, the early Greeks developed a civilization that reached previously unsurpassed heights of philosophical inquiry and literary and artistic expression. Greek adventurers, traders, and colonists used the Mediterranean Sea as their highway to spread classical Greek civilization and its language along much of the Mediterranean shoreline. Evidence of the geographic range and influence of the **Greek** language and culture is apparent in the many Greek elements in modern European languages. But over time, the use of Greek in most areas disappeared as new peoples and languages were introduced and expanded. Several language families are represented in Europe today, and as in all the world regions described in this book, language is one of the most important components of this region's ethnicities and cultures (see •Figure 4.11).

Europe's **Celtic languages** expanded at roughly the same time as Greek and, like Greek, are represented today only by remnants. Preliterate Celtic-speaking tribes radiated from a culture hearth in what is now southern Germany and Austria, eventually occupying much of continental Europe and the British Isles. Conquest and cultural influence by later arrivals, mainly Romans and Germans, eventually eliminated the Celtic languages except for a few traces that survive today as Gaelic (or Goeidelic) and Brythonic tongues in isolated pockets in the British Isles (notably in Wales, Scotland, and Ireland) and in French Brittany.

In present-day Europe, the overwhelming majority of the people speak Romance, Germanic, and Slavic languages. Like Greek and Celtic, these are **Indo-European languages.** The **Romance languages** evolved from *Latin,* originally the language of ancient Rome and a small district around it. In the few centuries before and just after the beginning of the Common Era, the Romans subdued territories extending through Western Europe as far west as Great Britain, and the use of Latin spread to this large empire. It is often said that the Mediterranean Sea became the "Roman Lake." Latin had the greatest impact in the less developed and less populous western parts of the empire. Over a long period, well beyond the collapse of this part of the empire in the fifth century, regional dialects of Latin survived and evolved into *Italian, French, Spanish, Catalan* (still spoken in the northeastern corner of Spain), *Portuguese, Romanian,* and other Romance languages of today.

It is important to point out that language frontiers do not always coincide with political frontiers. For example, the French language extends to western Switzerland and also to southern Belgium, where it is known as *Walloon.* French was Europe's dominant language as recently as the eighteenth century, when France was the continent's intellectual capital. French is still widely spoken outside the handful of countries in which it is the leading language and is second only to English as a leading tongue in European affairs.

In the middle centuries of the first millennium C.E., the power of Rome declined, and a prolonged expansion by Germanic and Slavic peoples began. Germanic (Teutonic) peoples first appeared in history as groups inhabiting the coasts of Germany and much of Scandinavia. They later expanded southward into Celtic lands east of the Rhine. They repelled Roman attempts at conquest, and the Latin language had little impact in Germany. In the fifth and sixth centuries, Germanic incursions overran the western Roman Empire, but in many areas, the conquerors were eventually absorbed into the culture and language of their Latinized subjects.

However, the *German* language expanded into, and remains, the language of present-day Germany, Austria, Luxembourg, Liechtenstein, the greater part of Switzerland, the previously Latinized part of Germany west of the Rhine, and parts of easternmost France (Alsace and part of Lorraine). The **Germanic languages** of Europe include many tongues other than German itself. In the Netherlands, *Dutch* developed as a language closely related to dialects of northern Germany, and *Flemish*—virtually identical to Dutch—became the language of northern Belgium. *Danish, Norwegian, Swedish,* and *Icelandic* descended from the same ancient Germanic tongue.

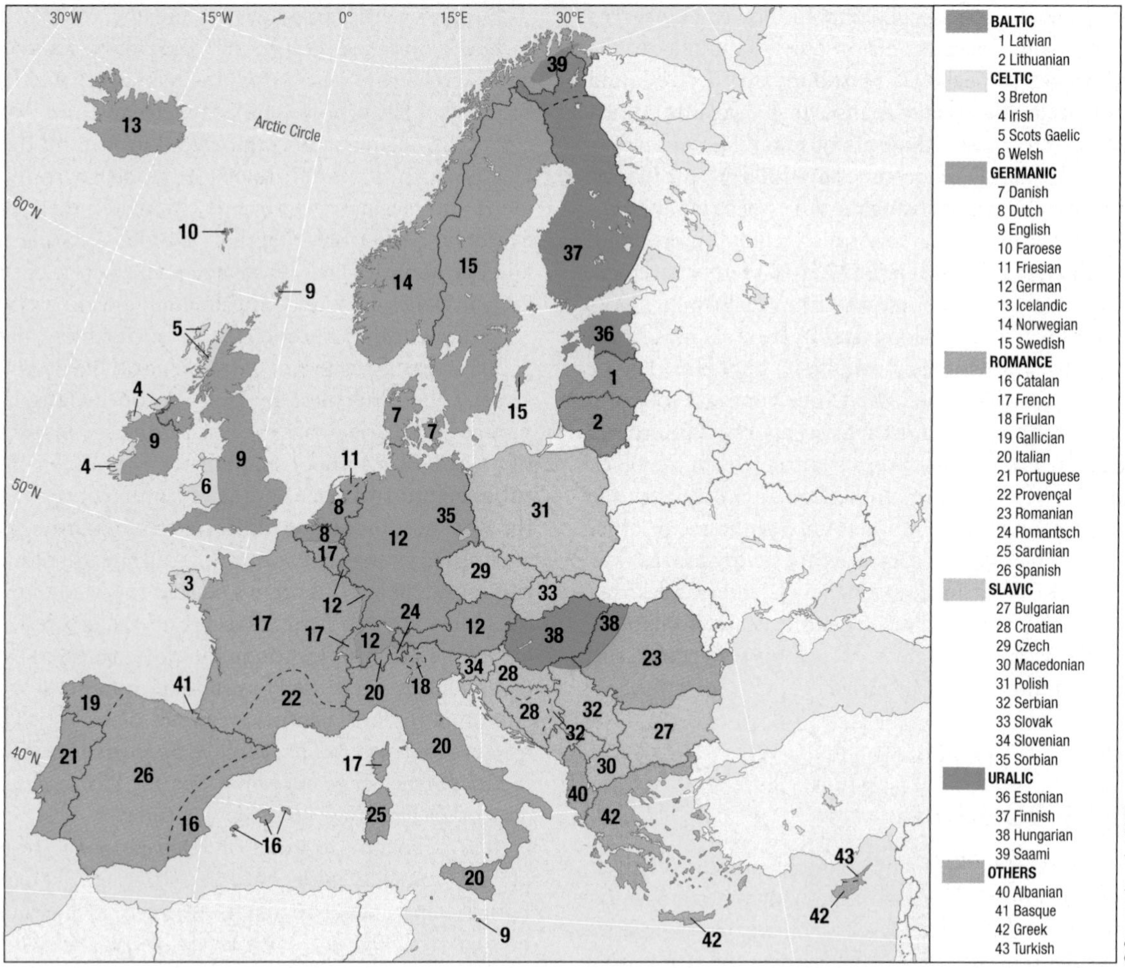

• **Figure 4.11** The languages of Europe.

English is a Germanic language, but it has many words and expressions derived from French, Latin, Greek, and other tongues. Originally, English was the language of the Germanic tribes known as Angles and Saxons who invaded England in the fifth and sixth centuries C.E. The Norman Conquest of England in the eleventh century established French for a time as the language of the English court and the upper classes. Modern English retains the Anglo-Saxon grammatical structure but borrows much vocabulary from French and other languages. English is now the principal language in most parts of the British Isles, having been imposed by conquest or spread by cultural diffusion to areas outside England.

Although the European Union has 23 official languages, English is becoming the de facto *lingua franca* of this 27-nation organization, with about two-thirds of all EU documents drafted in English. By EU law, and at great expense, all those documents must be translated into all those other languages, which most recently have come to include Ireland's Gaelic. Although some of the key institutions of the European Union are in French-speaking cities, French usage across the bloc is declining. One reason is the recent accession of EU member countries in Eastern Europe, where about 60 percent of the population speaks English as a second language, compared with only 20 percent speaking French as a second language. English is by far the most widely taught second language in non-English-speaking European countries, where it is studied by over 90 percent of secondary school students (compared with 33 percent studying French and 13 percent German).

Slavic languages—also in the Indo-European language group—are dominant in most of Eastern Europe. These include *Russian* and *Ukrainian*, the primary languages of Europe's adjoining region of Russia and other former republics of the Soviet Union along Europe's edge; *Polish, Czech*, and *Slovak*, which are spoken in central Europe; and *Serbian, Croatian*, and *Bulgarian*, which prevail in the Balkan Peninsula. They apparently originated in Eastern Europe and Russia, spreading and differentiating from each other during the Middle Ages as people traded and migrated. Closely related to the Slavic languages are the **Baltic languages** of *Latvian* and *Lithuanian*.

A few languages in present-day Europe are not related to any of the groups just discussed. Some are ancient languages that have persisted from prehistoric times in isolated, usually mountainous, locales. Two outstanding examples are *Albanian* (of the Indo-European language family) in the Balkan Peninsula and *Basque* (an indigenous European language without Indo-European roots), spoken in and near the western Pyrenees

Mountains of Spain and France. Some languages unrelated to others in Europe have relatives in Russia and reached their present locales through migrations of peoples westward. The prime examples are *Finnish, Saami, Estonian,* and *Hungarian* (also called Magyar). These are all **Uralic languages,** thought to have originated in the Ural Mountain region of Russia and Kazakhstan. *Turkish,* an **Altaic language** that originated in central Asia, is spoken by the ethnic Turkish inhabitants of northern Cyprus. The **Roma (Gypsies)** speak *Romany,* in the **Indic** branch of the Indo-European languages.

108,
225

What patterns can be seen in the rich and complex linguistic mosaic of Europe? Keeping in mind that language is a key component of ethnic identity, that question may be best answered by appreciating the seven ethnic patterns of Europe, as identified by David Levinson:[5]

1. Several nations are populated overwhelmingly by a single ethnic group—for example, the Germans of Germany, the Poles of Poland, and the Dutch of the Netherlands.

2. There are also substantial populations of ethnic groups that originated in other nations living in "foreign" countries (for example, the Poles of Lithuania), where they are known as **national minorities.**

3. There are groups distinguished more by culture than by ethnicity; they share a common culture but have many different ethnic origins (for example, the French and the Italians).

4. There are a few ethnolinguistic minorities, like the Saami (Lapps) of Scandinavia and the Ladin of Italy, who have resided in an area for a very long time but have retained cultural distinctiveness.

5. There are also regional ethnic minorities that have a major presence in a particular nation, where they are culturally very different from the country's other ethnic groups; examples are the Basques and Catalans in Spain and the Bretons in France. Such groups often call for the devolution of power from the host country to their ethnic subregion (see Insights, page 80).

6. There are recent immigrants from outside Europe, including guest workers mainly from former European colonies in Asia, Africa, and the Americas; examples are the Pakistanis and Indians of Britain and the North Africans of France.

7. Finally, there are ethnic groups with members all across Europe but without a homeland there; the Jews, who originated in modern Israel and adjacent areas (see Chapter 6, page 170) and the Roma (Gypsies), who probably originated in India, are the most prominent of these. The characteristics, contributions, and conflicts of Europe's ethnic groups are discussed in more detail later in the chapter.

Europeans' Religious Roots

Religion is one of the most important attributes of culture. On first glance, Europe's religious geography seems relatively straightforward: one of this region's principal culture traits is the dominance of **Christianity** (see the map of Europe's faiths in •Figure 4.12; the discussion of Christianity in Chapter 6,

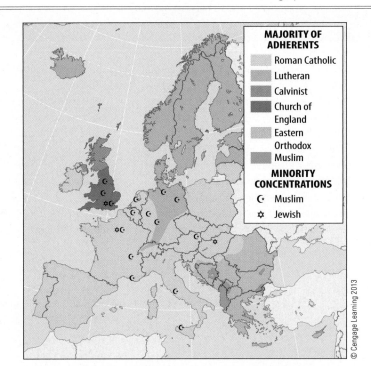

• **Figure 4.12** The religions of Europe. Sects of Christianity dominate, but Islam has a strong presence in the Balkans and, along with Judaism, in cities throughout Europe.

pages 172–173; and a note on the importance of religion in culture on page 4). But Christianity has a broad spectrum in Europe, where Christian encounters with other believers became part of the continent's troubled history. Initially outlawed by the Romans, Christianity was embraced as the faith of the realm by the Emperor Constantine in the fourth century, and it became a dominant institution in the late stages of the empire. It survived the empire's fall and continued to spread, first within Europe and then to other parts of the world. In total number of adherents, the **Roman Catholic Church,** headed by the pope in Vatican City in Rome, is Europe's largest religious group (with 280 million followers, or 45 percent of Europe's Christians), as it has been since the Christian church was first established (•Figure 4.13). Today, it remains the principal faith of a

• **Figure 4.13** The Vatican in the "Eternal City" of Rome is the seat of the Roman Catholic Church.

A notable political process in modern Europe, often related to issues of linguistic, religious, and ethnic identity, is **devolution**, the dispersal of political powers to ethnic minorities and other subnational groups. Like that of many other world regions, Europe's historical geography is rife with dominant groups occupying the traditional lands of smaller minority groups. Minorities are sometimes absorbed into the larger culture and political system, but more often than not, they retain some degree of distinctiveness and a desire to control their own affairs.

This book relates numerous examples from around the world of strife and war taking place because the dominant power refuses to yield any measure of authority to minority ethnolinguistic groups. But in Europe, the process of devolution is occurring peacefully for the most part and more energetically than ever. This may be because the European countries, after centuries of strife, are finally self-assured of their sovereignty and well-being and are so focused on international cooperation that demands for autonomy among their minorities are no longer perceived as threats. Devolution is an important issue in Wales, Scotland, Northern Ireland, Spain, and elsewhere; •Figure 4.B is a map of the major locales where devolution has occurred or is being appealed for in modern Europe.

• **Figure 4.B** Devolutionary areas of Europe. Power is being devolved from central governments to regional authorities in many parts of Europe. This map depicts places like Scotland and Wales, where the transfer has taken place, and others like Corsica, where devolution is anticipated, declared, or sought.

highly secularized Europe. The main areas that are predominantly Roman Catholic are Italy, Spain, Portugal, France, Belgium, the Republic of Ireland, large parts of the Netherlands, Germany, Switzerland, Austria, Poland, Hungary, the Czech Republic, Slovakia, Croatia, and Slovenia.

During the Middle Ages, a center of Christianity evolved in Constantinople (now Istanbul, Turkey) as a rival to Rome. From that seat of power, the **Eastern Orthodox Church** spread to Greece, Bulgaria, Romania, Serbia, Montenegro, and Macedonia, as well as to Ukraine and Russia in the east.

In the sixteenth century, the **Protestant Reformation** took root in various parts of Europe, and a subsequent series of religious wars, persecutions, and counterpersecutions left **Protestantism** dominant in Great Britain, northern Germany, the Netherlands, Denmark, Norway, Sweden, Iceland, and Finland. Except for the Netherlands, where a higher Catholic birth rate has since reversed the balance, these areas are still mainly Protestant, with the *Church of England* and *Calvinism* the strongest sects in Britain and *Lutheran Protestantism* dominant in Germany and Scandinavia. Among the three Baltic countries, Lutheran Protestantism is the dominant faith in Estonia and Latvia, while most Lithuanians are Catholic.

The fact that Christians form a vast majority is a somewhat misleading characterization of religion in Europe. Mainstream Christianity is in decline as Europeans have become increasingly secularized in recent decades. Historic legacy plays a strong role in the trend: over a period of centuries, Europeans paid a costly toll in lives by fighting for causes associated with religion, and they increasingly associate peace with secularism. Recent polls indicate that almost 75 percent of Europeans believe in God, but in most countries, even larger majorities shun traditions like attending church. In predominantly Catholic France, about one person in 20 attends a religious service every week (compared with one in three in the United States). In Poland, 95 percent of the people identify themselves as Catholic, but only about 40 percent attend church on Sunday. Various polls suggest that somewhere between 15 and 33 percent of Italians attend church in mainly Catholic Italy. Polls indicate that overall, about 20 percent of all Europeans regard religion as "very important" to them (compared to about 60 percent of Americans). The Christian churches that are growing—even surging—in attendance tend to be independent *Pentecostal churches* attended by immigrants from Nigeria, Sierra Leone, and other African countries.

The faith of **Islam** (practiced by **Muslims**) has the fastest-growing number of adherents of any world religion and is the fastest-growing religion in Europe (for a description of Islam, see Chapter 6, pages 173–175). Muslims are a majority in just three European countries: Albania, Kosovo, and Bosnia and Herzegovina on the Balkan Peninsula. Islam was once widespread in the Balkans (as that region is often called), where the Ottoman Turks established it during their period of rule from the fourteenth to nineteenth centuries. It was also the religion of the **Moors,** who ruled the Iberian Peninsula from the eighth century until the end of the fifteenth, when Spain's Catholic monarchy ousted them. Islam is growing throughout the continent, particularly among immigrant communities in Western Europe. Estimates vary, but France has as many as 4.7 million Muslims, followed by Germany with about 4.1 million, Britain with 2.8 million, and the Netherlands with almost 1 million.

As related in the earlier discussion of immigration, Muslims in Europe are often received with fear, suspicion, or hostility by "native" Europeans. But nothing can compare with the experience of the Jews as a minority in Europe. One of humankind's greatest tragedies is the story of their plight on the continent (see also Chapter 6, page 172). The Romans forcibly dispersed these followers of **Judaism** to Europe from Palestine in the first century C.E., and over time, they became more numerous in Europe than anywhere else. In 1880, before many emigrated to the United States, 90 percent of the world's Jews lived in Europe. In 1939, as World War II erupted, there were 9.7 million Jews, 60 percent of the world total, in Europe. By the end of the war, systematic murder by the Nazis in the **Holocaust** reduced their numbers to 3.9 million. About half of those subsequently emigrated to Israel and elsewhere, and there are about 1 million Jews in Europe today.

European Colonialism and Its Consequences

Far more than any other region, Europe has shaped the human geography of the modern world. Before the late fifteenth century, Europe played only a minor role in world trade patterns: goods moved from southern France and northern Italy northwest to Britain, for example, and more active trade connected Italy with the margins of the Mediterranean Sea and beyond to southwestern Asia. But the most important global trade routes were much farther east and south. The longest link was the 5,000-mile (8,000 km) **Silk Road** that moved goods overland (and on the Mediterranean Sea) from Chang-an (Xi'an) in China to Venice.

The balance of world affairs started to shift to Europe with the beginning of the Age of Discovery in the fifteenth century. European sailors, missionaries, traders, soldiers, and colonists burst onto the world scene. Although these agents had different motives, they shared a conviction that European ways were superior to all others, and by various means they sought to bring as much of the world as possible under their sway. Economic benefits for the European homelands were among their main motivations. By the end of the nineteenth century, Europeans had created a world in which they were economically dominant and exercised great influence on indigenous cultures.

The process of exploration and discovery by which Europeans filled in the world map began with Portuguese expeditions down the west coast of Africa. In 1488, a Portuguese expedition headed by Bartholomeu Dias rounded the Cape of Good Hope at the southern tip of Africa and opened the way for European voyages eastward into the Indian Ocean. In 1492, North America was brought into contact with Europe when a Spanish expedition commanded by a Genoese (Italian) man named Christopher Columbus (Cristóbal Colón) crossed the Atlantic to the Caribbean Sea. Less than half a century later, these early feats of exploration culminated in the first circumnavigation of the globe by a Spanish expedition commanded initially by a Portuguese man, Ferdinand Magellan. Worldwide exploration continued apace for centuries, with the Dutch, French, and English eventually wresting leadership from the Spanish and Portuguese.

The explorers were the vanguards of a global European invasion that would in turn bring the missionaries, soldiers, traders, settlers, and administrators. They gradually built the frameworks of European colonization of the Americas, Africa, Asia, and the Pacific. The colonial patterns are described on pages 44–45 in Chapter 3 and are not repeated in detail here. Briefly, their main manifestations were the transfer of wealth from the colonies to Europe, the movement of great numbers of indigenous laborers to serve European interests in other

178, 230

41

Try It   Where Did That Come From?

Take a moment to reflect on the momentous impacts of the Columbian exchange.

Imagine the United States and the rest of the Western Hemisphere without coffee; there was no "morning Joe" for anyone in the Americas before 1492. Imagine Europe, Asia, and Africa without cigarettes; just about 500 years ago, there was no tobacco except in the Americas.

The other thing to ponder is something we seldom give attention to: every single one of the plants and animals we depend on for our sustenance (and indulgence) had roots in geographic space; its ancestors thrived in a particular environment in ancient times and then was domesticated and diffused through human activities.

Think about one or more of your favorite foods (plant or animal), pets, or domestic animals. Where did it originally come from? If it is already in Table 4.3, dig deeper ("Old World" and "New World" cover a lot of territory). If you have a sweet tooth, what about chocolate or sugar? Where did the crops that yield them originally come from? Answering these questions will take you into the geographic realms of biogeography, diffusion, natural and cultural landscapes, and historical geography. Try it.

enterprise in the region as people took greater risks to gain wealth. Energetic entrepreneurs mobilized capital into large companies and used (or exploited) both European and overseas labor.

Europeans reached a level of technology generally superior to that of non-Europeans with whom they came in contact during the early Age of Discovery. Achievements in shipbuilding, navigation, and the manufacture and handling of weapons gave them decided advantages. Nations such as China, once technologically superior, fell quickly behind in terms of the power and geographical reach they projected (as we will see throughout the book, China has certainly changed that balance in recent decades!). Europe gained almost free rein in control and expansion of technological innovation for several centuries. The foundations of modern science were also constructed almost entirely in Europe, and the great names in science between about 1500 and 1900 were overwhelmingly European.

These material, cultural, and intellectual forces combined to launch the Industrial Revolution in Europe, which was the first world region to evolve from an agricultural into an industrial society. In the first half of the eighteenth century, primarily in Britain, industrial innovations made water power and then coal-fueled steam power increasingly available to turn machines and drive gears in new factories. James Watt's invention of the steam engine in 1769 made coal a major resource and greatly increased the amount of power available.

lands, the settlement of Europeans in agriculturally productive colonies (where their descendants are great majorities or significant minorities today), and the attempted subjugation of indigenous cultures to European ones. These are among the most prominent issues in world geography; every chapter of this book deals with the impacts of European colonization on cultures and landscapes beyond Europe.

Of great importance in reshaping the world's **biogeography** (the distribution of living things on the Earth) was the transfer of plants and animals from one place to another, following Europe's conquest of the Americas, the so-called **Columbian Exchange.** Major examples include the introduction of hogs and cattle and the reintroduction of horses to the Americas from Europe; of tobacco and corn (maize) from the Americas to Europe and other parts of the world; of rubber from the Americas to Asia; of the potato from South America to Europe, where it became a major food; and of coffee from Africa to Latin America, where it became the main cash crop in a number of countries. More examples are listed in •Table 4.3, which shows some of the most significant foodstuffs diffused from the New World to the Old World after 1492, and the major Old World foods introduced to the New World after that time (see the Try It feature on this page).

4.4 Economic Geography

Awash in Cash and Talent

Europe had significant material and cultural riches, and the colonial system built on that endowment to make Europe the world's wealthiest region for several centuries. The European capitalist system, and the huge strides in science and technology that accompanied it, helped launch the region into economic supremacy. Parts of Europe had already developed capitalist institutions by the end of the Middle Ages, and the colonial era saw further rapid development of economic

Table 4.3	Diffusion of Agricultural Products between the New World and the Old: The Columbian Exchange
Diffused from Old World to New World	**Diffused from New World to Old World**
Wheat	Maize
Grapes	Potatoes
Olives	Cassava (manioc, yuca)
Onions	Tomatoes
Melons	Beans (lima, string, navy, kidney, etc.)
Lettuce	Tobacco
Rice	Squash, including pumpkins
Soybeans	Sweet potatoes
Coffee	Peanuts
Bananas	Cacao
Yams	Chili pepper

New processes and equipment, including the development of **coke** (coal with most of its volatile constituents burned off), made iron smelting cheaper and more abundant. The invention of new industrial machinery made of iron and driven by steam engines multiplied the output of manufactured products, initially textiles. Then, in the nineteenth century, British inventors developed processes that allowed steel, a metal superior to iron in strength and versatility, to be made inexpensively and on a large scale for the first time. These developments soon diffused to new hearths of massive, mechanized industrial production and brought unprecedented changes to landscapes and ecosystems around the world (for more on the revolutionary nature of the Industrial Revolution, see Chapter 3, page 41). Among the most immediate and most significant processes accompanying industrialization were rapid population growth and a dramatic shift of people from rural areas into cities.

By 1900, the people and machines of Western European cities created about 90 percent of the world's manufacturing output. Europe led the world in the production of textiles, steel, ships, chemicals, and other industries. But in the twentieth century (especially after 1960), Europe's preeminence in world trade and industry diminished to about 25 percent of the world's manufacturing output, for several reasons.

Europe Displaced

One was warfare (•Figure 4.14). Europe suffered enormous casualties and damage in World Wars I and II, which were initiated and fought mainly on that continent. Recovery was eventually achieved, in large part with assistance from the United States, but the region did not regain its former supremacy.

During the twentieth century, rising **nationalism**—the quest by colonies and ethnic groups to possess their own homelands—brought the European colonial empires to an end. Taking advantage of a weakened Europe and mounting disapproval of colonialism in the world at large, one European colony after another gained independence quickly in the decades following the end of World War II. Perhaps ironically, opposition to continued European control was often spearheaded by colonial leaders who had been educated in Europe and had acquired nationalistic ideas there. Later, we will meet some of these leaders, including Vladimir Lenin of Russia and Ho Chi Minh of Vietnam. 128, 258

Europe's predominance was also seriously eroded by the rising economic and political stature of the United States and the Soviet Union. Each far larger than any European nation, these countries outpaced Europe in military power, economic resources, and world influence, particularly in the postwar years 1950–1970.

A major shift took place in global manufacturing patterns as well. Europe once enjoyed a near monopoly in exports of manufactured goods, but manufacturing accelerated in many countries outside Europe in the twentieth century. The United States reached industrial maturity by 1950 and became a vigorous competitor of Europe. More recently, the Asian countries of Japan, South Korea, China, and Taiwan have become industrial forces capable of competing with Europe in markets all over the world.

• **Figure 4.14** In the twentieth century, two devastating world wars originated in and centered on Europe, causing enormous suffering and eclipsing the region's global economic dominance. These are the ruins of the Reichstag in Berlin in 1945, photographed by the author's grandfather. Employed by a firm in Pittsburgh, Edmund Rhodes visited Germany a chemical engineer to assess the reconstruction of a Germany in ruins.

Finally, Europe's leadership was weakened by a new dependence on outside sources of energy. The region's traditional reliance on coal was superseded by technological developments and environmental concerns, and the energy balance shifted to petroleum and natural gas. Despite the development of North Sea oil and gas resources (which mainly belonged to the United Kingdom and Norway), Europe became, and remains, highly dependent on imported Middle Eastern oil.

⁹⁷

An Imbalance of Wealth

A striking pattern in Europe's economic geography is that Western Europe is wealthier than Eastern Europe. This trend dates to at least the 1870s, when per capita incomes in the west were twice those in the east. The gap grew even wider throughout the twentieth century. After World War II, Eastern European countries were in effect colonized by the Soviet Union, serving as vassal states that gave up human and material resources to service the motherland. Only recently, with the dissolution of the Soviet Union in 1991 and the admission of Eastern European countries to the European Union, have the eastern countries been able to hope that they may eventually catch up with their western counterparts. Membership in the European Union is structured to help redistribute wealth from west to east, through mechanisms such as agricultural subsidies that could help to breathe new life into the faltering farms and factories of the east. As we will see however, the newer member states have not been automatically granted all the privileges enjoyed by the older, wealthier members.

Living off the Land and Sea

Despite the dominance of a postindustrial, urban economy, some European countries continue to rely significantly on the productivity of land and sea. Agriculture was the original foundation of Europe's economy, and it is still important. Food provided by the region's agriculture allowed Europe to become a populous area at an early time. After about 1500, a period of steady agricultural improvement began. Introduction of important new crops such as the potato played a part, but so did such practical improvements as new systems of crop rotation and scientific advances that produced better knowledge of the chemistry of fertilizers. The continuing expansion of industrial cities provided growing markets for European farmers, who received protection through **tariffs** (tax penalties imposed on imports) or direct **subsidies** (payments made to domestic producers) to encourage production and support rural incomes (these are also discussed on page 420 in Chapter 11, in the context of similar protective measures by the United States). As a result of European Union policies, surpluses of many products are common, including grains, butter, cheese, olive oil, and table wines. But Europe is not agriculturally self-sufficient and depends on trade for much of its foodstuffs.

• **Figure 4.15** Cod fishing in the North Atlantic Ocean.

© Jeffrey L. Rotman/CORBIS

Europe, like the United States, is finding itself increasingly at odds with the rest of the world as global free-trade agreements insist that its doors be open to more (and generally cheaper) food imports. Europe wants to protect its farmers, particularly those who raise what it calls the "sensitive" products (beef, chicken, and sugar), with high tariffs on imported foods and with subsidies that allow farmers to "dump" their surplus production onto the world market at very low prices. The European Union says it is doing enough to open its doors to imports because it imposes no tariffs on imports from the world's 50 poorest countries, but free-trade advocates insist that more trade barriers must fall.

Throughout history, fishing has been an important part of the European food economy, and the coasts still have many busy fishing ports (•**Figure 4.15**). At times, control of fishing grounds has been a major commercial and political objective of nations, even resulting in warfare. Fisheries are particularly important in shallow seas that are rich in plankton, the principal food of herring, cod, and other fish of commercial value. The Dogger Bank (to the east of Great Britain) in the North Sea and the waters off Norway and Iceland are major historical fishing areas, and Norway, Iceland, and Denmark are Europe's leading nations in total catch. Fishing has long been especially crucial to countries such as Norway and Iceland, which have little agriculturally productive land. However, overfishing of cod, the world's most popular food fish, sometimes known as the "beef of the sea," has grown so severe, particularly in the North Sea, that since 2006 the European Union and Norway have legislated cuts of cod fishing by as much as 25 percent. Environmentalists have cried foul; organizations, including the World Wildlife Fund and Greenpeace, have insisted on a complete ban on cod fishing. Already, plummeting fish catches have caused economic hardship in many traditional fishing communities on North Sea shores. ²⁹

Postindustrialization

Europe's economy today, like its population trend, is best characterized as **postindustrial.** In the European context, this means that since about 1960, the region has experienced **deindustrialization,** shifting by choice and by force away from energy-hungry, labor-costly, and polluting industries toward an economy based on production of high-tech goods and on services. The high-technology products include electronic devices, data processing hardware, **robotics,** and telecommunications equipment. Other industries like drug and pesticide manufacturing make use of these technological products to create their own, so there is an important interconnectedness centered on high technology. Europe is also renowned for the production and export of high-quality, expensive luxury brands, including Ferrari automobiles and Rolex watches. In the service sector, growing proportions of Europeans are employed in transportation, energy production, health care, banking, retailing, wholesaling, advertising, legal services, consulting, information processing, research and development, education, and tourism.

Deindustrialization has had important spatial consequences for Europe. As we have seen, cities and industries grew up where important natural resources like coal and iron ore were located. Now most of these geographical correlations and barriers have been lifted. With the capabilities of the Internet and the advantages of fast and convenient transport of goods, it is possible for entrepreneurs to begin new enterprises almost anywhere. European cities are generally cleaner than they once were, and as the book describes in North America, neglected and decaying cores of formerly industrial cities are being rehabilitated and gentrified. Many European cities are exceptionally attractive places to live and visit, providing amenities that encourage socializing, good health, and enjoyment of the museums and historical legacies that endow each city and give it its unique character. The pedestrian and bike-only street in Amsterdam seen in •**Figure 4.16** exemplifies the people-friendly character of many European cities.

Neither high-tech nor service industries employ as many people as the old manufacturing sector did, and like the United States, the European countries are grappling with problems resulting from industrial unemployment. There are concerns over where new jobs will come from and how to make the transition from manufacturing to service roles without too much social dislocation. Many European nations for a long time fit the model of the **welfare state,** using the resources collected through high taxation rates to provide generous social services to citizens. That "social model" of the European state is, however, under tremendous pressure from the forces of globalization, the costs of absorbing immigrants, the effects of the global financial crisis of 2008–2009, and the debt crisis that erupted in Europe soon after that.

The European Union, Built on a Market

The European Union, which maintains its headquarters in the Belgian capital of Brussels, is Europe's most important **supranational organizations** (an organization in which member countries are united beyond the authority of any single national government and are planned and controlled by a group of nations). The countries of the European Union are depicted in •**Figure 4.17**.

The European Union began in 1957, when the Treaty of Rome created the **European Economic Community (EEC),** also known as the **Common Market,** made up of France, West Germany, Italy, Belgium, Luxembourg, and the Netherlands. By 1996, three years after it acquired the name European Union, the organization had an additional nine members: the United Kingdom, Denmark, Ireland, Greece, Portugal, Spain, Austria, Finland, and Sweden.

Much of the European Union's reason for being is economic. The EEC was designed, as the European Union is today, to secure the benefits of large-scale production by pooling the resources—natural, human, and financial—and markets of its members. Tariffs (taxes) were eliminated on goods moving from one member state to another, and restrictions on the movement of

• **Figure 4.16** to come.

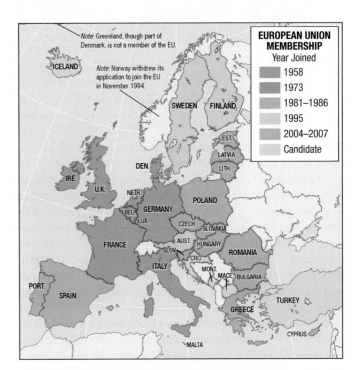

• **Figure 4.17** Members of the European Union.

Source: Adapted from "European Union Survey," The Economist, May 31, 1997, p. 5.
© The Economist Newspaper Limited, London 1997. All rights reserved. Reprinted with permission. Further reproduction prohibited.

INSIGHTS | Genetically Modified Foods and "Food Fights"

The United States is Europe's leading trading partner, but the exchange is not always open or friendly, especially when it comes to food. Countries often pay lip service to the concept of free trade while actually limiting trade in some products to protect their own industries and interests. Nations on both sides of the Atlantic use these trade barriers, sometimes precipitating **"food fights"** and other kinds of **trade wars.** In the late 1990s, for example, the United States wanted to sell bananas that U.S. corporations grew in Latin America to the countries of the European Union. The Europeans, who wanted to protect their investments in banana farming in former European colonies in Africa, the Caribbean, and the Pacific, refused to buy, and the so-called **"banana war"** ensued.

An ongoing trade dispute involves **genetically modified (GM) foods,** also called **genetically modified organism (GMO) foods,** produced in the United States. These are products such as rice, soybeans, corn, and tomatoes that have been genetically manipulated through biotechnology to be more productive and resilient (•**Figure 4.C**). European farmers and consumers, with their long-established food cultures, generally disdain GM foods, citing uncertain long-term health and environmental risks (for example, they fear that a GM plant cross-pollinating neighboring plants might have unpredictable ecological

consequences). For a long time, the European Union practiced a "zero tolerance" policy toward GM foods.

The United States argued that the EU ban on GM foods violated the free-trade principles of the World Trade Organization (WTO). Under pressure in international trade talks, the European Union gradually made concessions, allowing some GM foods and livestock feeds to be imported from the United States and elsewhere. But the European Union insisted that GM foods had to be labeled. (In the United States, they are usually not). And European consumers, who derisively refer to GM food as **"Frankenfoods"** (as in Frankenstein), have not shown much appetite for these imports. Some member countries, notably Greece, have defied EU guidelines and imposed outright bans on GM foods and crops.

Globally, the Europeans will find themselves increasingly isolated on the issue of GM foods and less able to compete in the international agricultural marketplace, especially as China and other large food producers and consumers embrace GM products. Some other countries do share the European perspective, however; for example, some African nations have actually refused to accept international food aid in the form of genetically modified corn.

Joe Hobbs

• **Figure 4.C** Biotechnological processes can manipulate the genetic makeup of mung beans (seen here in Vietnam with University of Missouri plant scientists Jerry Nelson, left, and Henry Nguyen) and other food crops to make them more productive, increasing yields—and also, say many Europeans, endangering local plant varieties and evolutionary processes.

---

labor and capital between member states were eased (in theory, a citizen of any EU country may work in any other EU country, but the process is not simple; a work permit must be obtained from the host country). Monopolies that restricted competition were discouraged. A common set of external tariffs was established to regulate imports from other countries, and a common system of price supports for agriculture replaced the individual systems of member states (see Insights on this page for European Union's policy on importing genetically modified foods). The founders expected that free trade within such a populous and highly developed bloc of countries would stimulate investment in mass-production enterprises, which could sell freely to all member countries. This trade would encourage productive geographic specialization, with each part of the union expanding lines of production for which it was best suited. Each country might thereby achieve greater production, larger exports, lower costs to consumers, higher wages, and a higher standard of living than it could achieve on its own.

Bring on the Euro

The European Union, under provisions of the *Maastricht Treaty of European Union* that went into force in 1993, began pursuing even greater new steps toward unification. A single EU currency, the **euro,** was launched in 1999 as the centerpiece of the *European Economic and Monetary Union (EMU).* This common currency was based on economic theories suggesting that sharing a currency across borders had many advantages, such as lower transaction costs, more certainty for investors, enhanced competition, and more consistent pricing. A common currency could theoretically restrain public spending, reduce debt, and tame inflation. In practical terms, consumers are able to spend money in euro-using countries just as Americans do at home; they can compare the price of a car, for example, in Germany and Italy, and decide to buy the cheaper car, just as an American might buy a car in Utah instead of Colorado. Besides sharing a currency, all

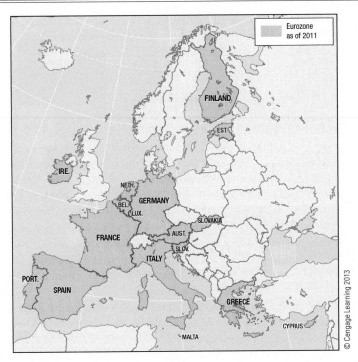

© Cengage Learning 2013

Eurozone
as of 2011

• **Figure 4.18** The "Eurozone," countries that use the euro as a common currency.

countries adopting the euro have agreed to allow the strong European Central Bank to make critical decisions on issues such as interest rates, which would be identical for all euro-using nations.

A certain amount of euphoria accompanied the creation of the euro: this would be yet another symbol of a strong and united Europe. But from the beginning the common currency had detractors. The Swedes, Danes, and British rejected the currency. Sweden feared a loss of financial and monetary independence. Danes worried that the euro would threaten their generous welfare benefits or even their national unity. Britain's economy has recently been stronger than the economies of Germany and France, suggesting to the British that they might as well shun the euro. The new Eastern European members of the European Union decided to adopt the euro, but had to make difficult economic reforms to meet the requirements of adopting this common currency. The number of euro-using nations grew to 17; these countries comprise the so-called **Eurozone** (•Figure 4.18). As we will see, a major crisis developed in the Eurozone in 2010 in part because of the economic burdens imposed by the expansion of the common currency and the expansion of the European Union itself.

Europe's "Big Bang"

Embracing the Less Wealthy

Ten Eastern European nations joined the European Union all at once in what was known as the **"big bang"** of 2004. The newcomers were Poland, the Czech Republic, Hungary, Slovakia, Estonia, Latvia, Lithuania, Slovenia (the first former Yugoslav state to join the European Union), Malta, and officially all of Cyprus, although de facto EU membership applies only to the Greek portion of the divided island nation. Turkey's application to join was rejected. The big bang created a mega-Europe of 450 million people, stretching from the Atlantic to Russia, with an economy valued at almost $10 trillion, nearly as strong as the U.S. economy. The European Union grew enormously, with the addition of 75 million people (about 20 percent, almost half of them in Poland) and a geographic expansion by almost one-third of its prior size. A more modest expansion took place in 2007 when Romania and Bulgaria joined as the European Union's poorest member nations. This growth is accompanied by some clear advantages for the organization and its new members, but also by some problems. Some thorny issues had to be resolved before the dozen new countries could join, and even thornier problems lie ahead.

Consistent with Europe's historical patterns of west and east, and core versus periphery, the most outstanding difference between the old and new members is in their economies. The old EU member nations have some of the highest per capita gross domestic product purchasing power parity (GNI PPP) levels in the world—including the very highest, Luxembourg, with $59,970. The new states, by contrast, are much less affluent, and some are classified statistically as LDCs with a GNI PPP of less than $15,000 (see Figure 3.5 and Table 4.1). Bulgaria has a per capita GNI PPP of just $13,260. The old EU countries have 95 percent of the continent's wealth, and the new ones only 5 percent. When the big bang countries joined in 2004, the European Union's average wealth per person fell by 13 percent.

Naturally, questions have arisen about the European Union's ability to afford its new members. There is particular concern with the costs of the European Union's agricultural policies. The European Union provides generous agriculture subsidies to its member states. Prior to the big bang, $50 billion, fully half of the European Union's budget, was spent every year on these subsidies. For example, the European Union pays an Austrian farmer $28,000 per year in subsidies for 30 hectares' (c. 75 acres') production of seed potatoes and milk. That farmer, like many in the European Union, would probably not be able to farm profitably without EU subsidies. The new EU members are upset that they began membership receiving only one-fourth of the agricultural subsidies provided to the older members (with an annual 5 percent increase in subsequent years). They fear that their farmers will go out of business, unable to compete with the lower prices from the subsidized, heavily mechanized farms to the west.

Feeding the PIGS

For its part, the older member states of the European Union argue that the new countries do not have the high production costs and quality controls of the old countries. Mainly, though, the older European Union members fear that providing full subsidies to the new countries would bankrupt the organization. Germany is especially fearful. It is the wealthiest member of the European Union and carries the largest cost

burden. But Germany had its own economic pains from absorbing the costs of reunion and integration with the poor
former East Germany and is concerned about a far more expensive integration on a much grander scale. Despite hopes
among many in the newly joining states that membership will
bring them prosperity, there will also be at least short-term
pain. Food prices have risen sharply, while salaries have increased slowly. Many western Europeans fear that the lower
salaries and taxes of the new eastern member countries will
lure industries and jobs away from the west, contributing to
the decline of their home economies.

A similar set of problems revealing the strains within a united
Europe has plagued the Eurozone. In this case, the European
Union's strongest economies, those of Germany (which is seen
as the "economic locomotive" of the continent) and France,
have come under pressure to help out the weaker countries—
especially Portugal, Ireland, Greece, and Spain, known collectively by the unfortunate acronym **PIGS**. After they joined
the common currency, these countries embarked on spending
and borrowing sprees, building up huge deficits and debts
(and apparently hoping that in case of any troubles the larger
economies of the Eurozone would bail them out). In the debt
crisis that followed, beginning in 2010, a number of economic
bubbles in these countries burst—most viciously in Spain's real
estate. Some countries had to introduce severe economic austerity programs, including salary cuts, which led to growing unemployment, poverty, and social unrest.

The euphoric vision of the euro has faded away, and new
words have crept into common usage: Euromess, Europhobia,
Eurocrisis, and especially the **European debt crisis** or "sovereign debt crisis" among them. The European debt crisis began
early in 2010 when it appeared that Greece might default on its
loans. There was no way to address that problem quickly (in
fact, a tentative bailout for Greece's debt was not put together
until mid-2011), because all 17 of the Eurozone's members had
to be consulted for input. The European Central Bank does not
have much authority to make binding decisions for all its members. Individual countries in the Eurozone have different goals
and expectations. The wealthier ones, like Germany, do not
want their taxes used to bail out countries like Greece that are
heavy borrowers. The poorer indebted countries like Greece do
not want wealthier outsiders like the Germans to dictate how
their government taxes their citizens and spends their money.

The Eurozone's finance ministers began studying the way
the United States runs its financial structures and started considering a more centralized approach like the United States
has. The Europeans may finally end up financially as a kind
of "United States of Europe," with a central finance body,
like the U.S. Treasury Department, with power over national
budgets and tax regimes. The organization would enable the
Eurozone countries to make common fiscal policies for all
17 members. The notion of a fiscal union is extremely unpopular among the public in many of these countries, but economic crisis may push them inevitably toward greater unity.
The Eurozone countries will have to draft and ratify a new
treaty, a process that will take years, but the alternative may

be the fragmentation of the union. In the midst of the crisis in
2011, the prevailing motto among European finance ministers
was "centralize or die."[6]

4.5 Geopolitical Issues

In the past 100 years, Europe's geopolitical situation changed
more profoundly and more violently than that of any other
world region. Europe experienced two world wars that
wrought unprecedented devastation. World War I (1914–1918),
known until 1939 as the Great War, saw Europe fracture on
fault lines of military alliances. Germany and the Ottoman
Empire (based in what is now Turkey) were solidly defeated
by an alliance led by Britain, France, Russia (early on), and
the United States (later). The war accomplished little and destroyed much, and the peace treaty placed a costly burden of
reparations on the defeated Germany.

The economic and social humiliation of Germany helped
plant the seeds of the rise of Adolf Hitler's fascist Nazi (National Socialist) Party and the outbreak of World War II (1939–
1945). Hitler's Germany surged in strength and allied with
Italy, the Soviet Union (early), and Japan (later) in a military
quest to establish European and Far Eastern empires. Early in
the war, Germany advanced swiftly on the European mainland
and Scandinavia and set its sights on an occupation of Britain. When Japan bombed a U.S. naval base at Pearl Harbor,
in the Hawaiian Islands, in 1941, the United States was drawn
into the war firmly on Britain's side. Soon joined by the Soviet
Union, these forces, assisted by many smaller ones, began to
turn the tide against Germany. In 1945, with German cities
in ruin and two Japanese cities nearly obliterated by the first
atomic bombs, the war ended. In the European theater, about
50 million people had died. "What is Europe now?" the British
prime minister, Winston Churchill, asked. "It is a rubble heap,
a charnel house, a breeding ground of pestilence and hate."[7]

Postwar Europe

The wounds of that gigantic conflagration have healed. For-
ward-looking Europe is intent on establishing, in the form of
the European Union, that "virtual" United States of Europe
with a federation of nations similar to the states of the United
States. Remarkably, George Washington said "Some day, following the example of the United States of America, there will
be a United States of Europe". The European Union is the latest and largest of the various postwar European supranational
organizations. There are many reasons, largely the economic
ones discussed above, for this push toward greater unity. One
of the most important but understated reasons is simply that
Europeans want their continent to be warproof, with its member countries so intertwined in economics, foreign policy, and
other ways that war among them would be all but impossible.

Europe is still recovering and reorganizing from another
momentous geopolitical shift: the end of the **Cold War**, which
came with the collapse of the Soviet Union in 1991. When

World War II ended, Soviet forces and political institutions moved in on most of the countries of Eastern Europe, including part of defeated Germany. As Churchill memorably described it, an **"Iron Curtain"** had descended from Poland southward to the Balkan Peninsula. West of it, the capitalist democracies led by Britain, France, West Germany, Italy, and the United States prevailed; to the east, across a 4,000-mile (6,400 km) line of concertina wire, walls, minefields, and guard towers, the Soviet Union controlled authoritarian, state-run regimes in the countries that came to be known as its European "satellites," including East Germany, Poland, Czechoslovakia, Hungary, Romania, and Bulgaria. The United States nurtured Western Europe's interests and from 1947–1952 pumped billions of dollars into the reconstruction of the devastated European infrastructure in a program known as the **Marshall Plan.**

The United States was also the linchpin of the military alliance known as the **North Atlantic Treaty Organization (NATO),** founded in 1949 between the United States, Canada, most of the European countries west of the Iron Curtain, and Turkey. NATO faced off against the **Warsaw Pact,** an alliance of the Soviet Union and its Eastern European satellites. On more than one occasion, notably in the Cuban Missile Crisis of 1962 and in the Arab-Israeli War of 1973, these military alliances came all too close to apocalyptic nuclear war.

178, 373

In 1989 the Berlin Wall, which separated the German city that epitomized the Cold War, was dismantled in a popular and peaceful uprising. The Soviet grip on East Germany and its other satellites had been slipping for years, and now went limp. The Soviet Union itself disbanded in 1991 (see Chapter 5, page 131). An exhilarated public throughout Eastern Europe embraced the prospects of democracy, capitalism, and material improvement. The Warsaw Pact was dissolved. The nuclear arsenals of the respective alliances were reduced substantially. Plans were made to turn the path of the Iron Curtain into the **European Greenbelt,** a mosaic of national parks and other protected areas stretching across 18 countries, from the Barents Sea to the Black Sea. There is an analogous place in the world where a protected natural area may emerge from a former war front: the demilitarized zone (DMZ), separating North and South Korea.

274

Since the end of the Cold War, NATO has continued to serve an important peacemaking role. In the Balkans during the 1990s, NATO forces reversed the advance of the Serbs against their neighbors in the former Yugoslavia (see page 107). More recently, NATO members deployed their military assets against the regime of Muammar Gaddafi in Libya. In that conflict, the United States preferred that NATO take a lead role, especially to prevent the United States from having only a militaristic reputation in the Middle East. NATO's membership has grown to 28 with the addition of nine former Soviet allies (Croatia, Estonia, Latvia, Lithuania, Slovakia, Slovenia, Romania, and Bulgaria; see •Figure 4.19). Russia resented this Western expansion into its former sphere of influence, but its opposition was tempered by its own growing alignment with Western defense and security concerns.

194

• **Figure 4.19** European members of the North Atlantic Treaty Organization (NATO).

Sources: EU map based on "European Union Survey," The Economist, May 31, 1997, p. 5. Data from EUROSTAT (Statistical Office of the European Communities); World Factbook, CIA.

Welcome to Schengenland?

In theory, the European Union would like to move toward a situation in which there were no passport, visa, or other control issues at any internal land, sea, and airport frontiers of its member countries. This would be an extraordinary geopolitical bloc, considering Europe's fractious history. A framework that began outside the European Union and now operates within it has attempted this kind of integration. Known as the **Schengen Agreement,** it allows free circulation of people between the nations that signed the agreement; these are the EU countries (except for Britain, Ireland, Romania, and Bulgaria), plus non-EU Iceland, Norway, and Switzerland. To maintain internal security, within this greater **Schengenland,** the member states are supposed to exercise common visa, asylum, and other policies at their external borders. Some countries, notably the United Kingdom, are skeptical of the effectiveness of control of external borders and have stayed out of Schengen.

Truly open borders are probably still far in the future. Anti-immigrant fears that rose with the exodus of refugees from North Africa during 2011's Arab Spring have caused some countries, notably France and Denmark, to backpedal on their commitment to the principle of open borders. Even two countries that recently joined the European Union—Romania and Bulgaria—have been rejected for membership in Schengenland. Western European fears of a tidal wave of cheap eastern labor, terrorists, or human traffickers have produced special rules for these eastern countries, long regarded as abodes for these shady activities. Like the Euro, Schengenland is one of the big European goals that is difficult to proclaim as a success.

72, 192

Differences between Europeans and Americans

Many of us instinctively equate the United States with Europe when we speak about global patterns; after all, we share many common cultural roots, share a great deal of the world's wealth, and have many of the same measures of quality of life. But there are differences, and these seem to have grown in recent years. "Americans are from Mars and Europeans are from Venus," wrote policy analyst Robert Kagan.[8]

Here are a few differences—and please keep in mind that these are not hard facts, but generalizations that reflect policies and public opinions. The concept of social justice, in which steps are taken to reduce unemployment and poverty among the least privileged, is a key value in Europe, whereas it lags in the United States. European governments and people have historically had a stronger sense of the social compact between the state and its citizens. Governments have been expected to provide "free" public services like education and health care, while the people have understood they must foot the bill for these "free" services by paying very high taxes. (Incidentally, undergraduate public education in most European nations is free. However, per capita spending on public education is higher in the United States, and an American graduate degree is widely regarded as superior to that obtainable in Europe.) Europeans accept high taxes on gasoline as an incentive to drive small cars and conserve energy, whereas for an American politician, advocating higher fuel taxes is political suicide; as a result, gas prices in Europe are roughly double those in the United States. Europeans complain that U.S. cultural industries such as Hollywood films humiliate other cultures and marginalize languages other than English. For awhile, there was strong resistance, especially in France, to the proliferation of McDonalds and other American fast-food restaurants in Europe; but that has largely faded, and McDonalds has "Europeanized" its restaurants in both menus and architecture. Europeans are less inclined than Americans to allow questions of spirituality into political debate; in fact, they are repelled by personal expressions of religiosity in public, which they consider indiscreet. Europeans abhor the death penalty, which is outlawed in all EU countries.

Europeans tend to apply the **precautionary principle**—the notion that it is appropriate to attempt to reduce or eliminate any risky practice—to things they think may be harmful. Thus there may be risks from GM foods, so many people would like to ban them; climate change is probably occurring, so most people want to take action to reduce greenhouse gases. The United States has adopted more of a "show me" attitude—for example, with respect to global warming; compared with Europeans, a large percentage of Americans think there is not enough scientific proof, so it is better not to impose reductions in carbon dioxide emissions. There are differences on the geopolitical front too. For Americans, September 11, 2001, was the watershed time when it reevaluated its place in the world and began to feel less secure. For Europeans, the fall of the Berlin Wall in 1989, leading to the collapse of the Soviet bloc, was the watershed that made them start to feel more secure.

4.6 Regional Issues and Landscapes

In this section, we take a closer look at some problems and issues of interest in the European subregions described on page 65. We begin by looking at some of the elements that made the European Core the continent's strongest subregion, and then examine characteristic or exceptional qualities of Northern, Southern, and Eastern Europe.

The European Core

Properties of the European Core

In Chapter 1, we saw that core and peripheral locations are among the simplest and most common properties of the Earth's countries and regions (see page 10). Europe has a recognizable core, a subregion that has long played a dominant role in the continent's political, economic, and cultural development. Today, the European Core is recognizable as a prominent subregion of Europe with the following traits:

- Densest, most urbanized population
- Most prosperous economy
- Lowest unemployment
- Most productive agriculture
- Most conservative politics
- Greatest concentration of highways and railroads
- Highest levels of crowding, congestion, and pollution[9]

Defined in this way, the core consists mainly of three of Europe's major countries and of smaller nations in the British Isles and the west central portions of the European mainland (•Figure 4.20). The United Kingdom (U.K.) and the Republic

• **Figure 4.20** The political geography of Europe's core.

of Ireland share the British Isles. The countries of west central Europe are Germany and France and their smaller neighbors, the Netherlands, Belgium, Luxembourg, Switzerland, and Austria. Also sharing these traits of the core are roughly the northern half of Italy, Denmark, southern Sweden, and parts of Poland, but other characteristics of these countries place them more appropriately in the European Periphery, discussed later. Because they are nestled among the core countries, the microstates of Liechtenstein, Monaco, and Andorra are included here.

The European Core includes some of the world's most geopolitically and economically significant countries (see Table 4.1, pages 67–68). As we saw in Chapter 1, geography is very concerned with appreciating how people have reshaped the face of the Earth. This European subregion became prosperous in part through the exploitation of its own natural resources. Today, it has one of the world's most transformed landscapes, shaped by thousands of years of productive agriculture and hundreds of years of industrial development and affluence (•Figure 4.21). European countries do not have the natural advantage enjoyed by one of their affluent peers, the United States: vast size. As we will see when we get to North America in Chapter 11, there is actually more forest in the United States now than there was when European settlement began (see page 398). European countries have not enjoyed the luxury of allowing vast parts of their hinterlands to reclaim something similar to their "presettlement" environments.

The European Core is one of just four regions in the world classified as a **major cluster of continuous settlement**, meaning that all habitations lie no more than 3 miles (5 km) from other habitations in at least six different directions. In addition,

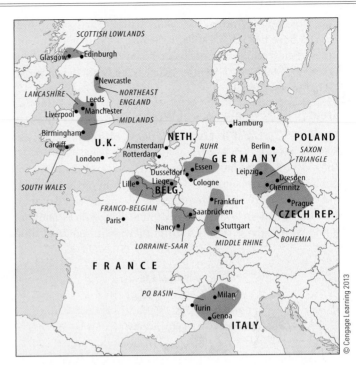

• **Figure 4.22** Historical industrial concentrations, cities, and seaports of the European core. Older industries such as coal mining, heavy metallurgy, heavy chemicals, and textiles clustered in these congested districts. Local coal deposits provided fuel for the Industrial Revolution in most of these areas, which have shifted increasingly to newer forms of industry as coal has lost its economic significance and older industries have declined.

roads or railroads lie no farther than 10–20 miles (16–32 km) away in at least three directions.[10]

Europe's cultural landscape is unusually intricate. Different layers of history appear at every turn, and there are sharp differences in human geographies and national perspectives of its peoples. Environments and resources also vary widely. Local resources (especially **coal** and iron ore) and transportation opportunities were the basis for industrial and urban development in past centuries—that correlation is seen in •Figure 4.22—but the modern European Core has a pronounced postindustrial geography. There are distinct areas of declining and growing industrial activity. Resources and industries that put many major cities on the map are played out or are in serious decline.

This process of deindustrialization hit Britain especially hard, but other regions in the European Core, notably the Ruhr in Germany, also suffered. High unemployment is a major consequence of this trend, but a positive result is that the cities are cleaner and more livable. Many cities are replacing old industrial areas with greenbelts and attractive architecture. Improved environments should lure in high-tech and other clean industries, services, and tourists that would drive a new economy (I emphasize *should* and *would* because these trends were underway until the gigantic forces of economic recession stopped them in their tracks, at least temporarily, after the economic crisis erupted in 2007). These greener industries tend to occupy smaller areas than the older, more polluting industries. In a geographical play on

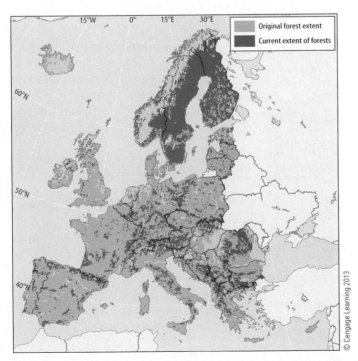

• **Figure 4.21** Human activities over long periods of time have transformed the natural landscapes of Europe, especially in the core region. This map depicts areas that once sustained temperate mixed forests and coniferous forests but have lost that tree cover.

words evoking California's Silicon Valley, parts of the British Isles have been dubbed "Silicon Bog," "Silicon Glen" and Silicon Fen," while the continent boasts Germany's "Silicon Saxony," France's "Scientific City," and northern Italy's "Techno City."[11]

Great Britain: Aging Seat of World Power

The islands of Great Britain and Ireland, off the northwestern coast of Europe, are known as the British Isles (•**Figure 4.23**). They are home to the Republic of Ireland, with its capital at Dublin, and the much larger United Kingdom of Great Britain and Northern Ireland (often referred to simply as "Britain"), with its capital at London. The United Kingdom is made up of the entire island of Great Britain, with its political units

of England, Scotland, and Wales; the northeastern corner of Ireland, known as Northern Ireland; and most of the smaller outlying islands.

The six counties that make up Northern Ireland separated from Ireland in 1922 and maintained their link with the United Kingdom, while the remaining counties achieved full independence as the Republic of Ireland. Altogether, the United Kingdom encompasses about four-fifths of the area and over 90 percent of the population of the British Isles.

Lying only 21 miles (34 km) from France at the closest point, England is the largest of the United Kingdom's four main subdivisions. England was originally an independent political unit and twice conquered mainly Catholic Ireland, including Northern Ireland.

• **Figure 4.23** Principal features of the British Isles. The "Tees-Exe" line connecting the mouths of rivers having those names (not shown) is the general dividing line between Highland Britain to the northwest and Lowland Britain to the southeast.

Scotland was first joined to England when a Scottish king inherited the English throne in 1603, and the two became one kingdom under the Act of Union in 1707. Nearly 300 years later, in 1997, the U.K. government yielded to increasing Scottish demands and allowed for the devolution (dispersal of political power and autonomy—see page 80) of many administrative powers (including agriculture, education, health, housing, taxation, and lawmaking, but excluding defense and social security) to a new Scottish parliament. In a groundbreaking 2007 election, Scottish nationalists won political dominance in the parliament and promised a referendum on whether Scotland should secede from the United Kingdom and declare independence. Scots are extremely proud of their distinctive culture, history, and identity, and bristle when they are identified as "English," as happens all too often.

England conquered Wales in the Middle Ages, but like the Scots, the Welsh managed to preserve their cultural distinctiveness and hold it high—especially in staying true to their Welsh language (a member of the Celtic subfamily). Any thought that the Welsh are English withers quickly when one considers this Welsh place name (known as the longest place name in the world): Llanfairpwllgwyngyllgogerychwyrndrobwllllantysiliogogogoch, meaning "The Church of Saint Mary in the hollow of white hazel trees near the rapid whirlpool by Saint Tysilio's of the red cave." A 1997 referendum voted into existence a Welsh assembly dedicated to promoting Welsh interests. Although this body has some powers independent of the United Kingdom, growing numbers of Welsh are calling for a stronger devolution of power that would transfer more authority from the national government to a full-fledged parliament, as Scotland has now.

Between the defeat of Napoleonic France in 1815 and the outbreak of World War I in 1914, the United Kingdom was the world's strongest country. Its overseas empire covered a quarter of the Earth at its maximum extent; "the sun never sets on the British Empire" was a proud boast. The influence of English law, education, and culture spread still farther. The British take great pride in much of that legacy; the Magna Carta, an English charter written in 1215, granted certain human rights that centuries later would be enshrined in the constitutions of other democracies, including that of the United States. However, Britain was one of the main forces behind the cruel industry of the enslavement of Africans. Until the late nineteenth century, the United Kingdom was the world's leading manufacturing and trading nation. Britain's Royal Navy dominated the seas, and its merchant marine moved half or more of the world's ocean trade. London became the center of a free-trade and financial system that invested its profits from industry and commerce around the world.

London continues to be one of the world's greatest financial centers. Unlike many of the United Kingdom's other large cities, London did not owe its existence to the proximity of nearby coal or other natural mineral resources. The Thames River has provided its port with access to worldwide ocean trade, and London has been a magnet for entrepreneurs, artists, tourists, and vagabonds. It is by all measures a **"world city,"** commonly defined as a city that is a center of economic and political power in the global economy.

The United Kingdom's economic and political decline began late in the nineteenth century. World War I damaged its free-trade and financial system, and the downward trend accelerated after World War II. All the large colonies of the former British Empire gained independence by the 1960s, and Britain's remaining overseas possessions are scattered and small.

The United Kingdom is, however, still a very important country on the world stage, in large part because of its lasting impacts from colonial times. It plays a major role in the European Union and is associated with many of its former colonies in the worldwide **Commonwealth of Nations.** The Commonwealth is a voluntary association of 54 countries that nominally recognizes the British monarch as its head. Its strong allegiance to the United States, even in times of great adversity such as during the wars in Iraq and Afghanistan, is another of Britain's characteristic traits in the international arena. Britain's involvement in those wars has, however, been costly in lives and in money spent, and has deepened the impacts brought on the country by the global economic financial crisis. There is a growing sense among the British that their country is experiencing declinism, with Britain taking a more passive place in world affairs.[12]

Much world culture has British roots; English is by far the leading language of the World Wide Web and serves as a *lingua franca* in many countries with diverse cultures (including the European Union; see page 78). In the 1960s, the Beatles, four lads from Liverpool (a city recently crowned a "European Capital of Culture"), revolutionized music in their home country, moved on to conquer a generation in the United States, and left an indelible mark on music everywhere. British pop culture and more traditional fare—Scottish reels, fish and chips, Buckingham Palace Beefeaters—draw fans and visitors from around the world. For many foreigners, the images and artifacts of Britain may become distilled into something homogeneous, but a Briton of Anglo origin is quick to speak about his or her unique piece of the country and what makes it distinct. And then there is multicultural Britain, another legacy of its colonial past, with a bit of Jamaica, India, Kenya, and scores of other lands infused in every corner of the country.

Ireland: Troubles and Resilience on the Emerald Isle

Ireland (known as *Eire* in the indigenous Celtic language) is a land of hills and lakes, marshes and peat bogs, cool dampness, and the verdant grassland that earned its epithet the "Emerald Isle" (•Figure 4.24). The island consists of a central plain surrounded on the north, south, and west by hills and low rounded mountains. The Republic of Ireland, which occupies a little over four-fifths of the island, is far into its transition from an agricultural to an industrial economy.

Ireland ranked low among Western European countries for many decades, but recent economic growth changed this pattern. High-tech industries such as electronic products and software boosted Ireland's economy. The size of the economy doubled in the 1990s, earning Ireland the nickname **Celtic Tiger** and placing it on par with the United Kingdom in per capita GNI PPP (see Table 4.1, page 67). Unemployment dropped

Joe Hobbs

• **Figure 4.24** A typical Irish landscape, seen from the Blarney Castle. There is a tradition that kissing the so-called Blarney Stone in one of the castle battlements will make one eloquent in speech.

sharply. This development surge was accomplished by a government program luring foreign-owned industries with inexpensive labor, tax concessions, and help in financing plant construction. Hundreds of industrial factories, owned mainly by U.S., British, and German companies, began to produce computers, electronics, foods, textiles, office machinery, organic chemicals, and clothing, and Ireland's economy became more than 50 percent reliant on exports of these products to overseas markets. Most of these plants sprang up around the two main cities, Dublin and Cork, where they make up an economic region dubbed "Silicon Bog," and near Shannon International Airport in western Ireland.

Unfortunately, Ireland is the "I" in the PIGS economies described on page 87. Much of Ireland's strong economic growth—with an annual rate of about 7 percent between 1999 and 2007—was the result of a borrowing and spending spree prompted by very low interest rates. Housing and banking bubbles developed, and inevitably—as elsewhere in the Eurozone, but with greater consequences—they popped. In 2008, the country adopted economic austerity measures including increased taxes and cuts in salaries and budgets. Ireland fell into a deep recession as industrial and service companies

shuttered completely or cut way back on staff. As happened so many times in Ireland's history, the Irish began fleeing their home island in search of better lives abroad. But this time, the Irish emigrants took skills in IT, law, and financial services with them—they were among Ireland's best and brightest, not menial laborers looking for construction jobs abroad (this is one manifestation of the "brain drain" problem described in Chapter 3; see page 56). Ironically, the deep sacrifices made by the Irish may help the country come out of recession more quickly, as low wages and prices should attract more foreign investment.

Many Troubles

In the past, Ireland had an agrarian economy that made the country ripe for suffering. England conquered Ireland in the seventeenth century and made the island a formal part of the United Kingdom. Ireland was effectively an English colony. Most of the land was expropriated and divided among English landlords into large estates, where Irish peasants worked as tenants. The income from this enterprise generally found its way to Britain, and Ireland became a land of poverty, sullen hostility, and periodic violence.

A crop disease known as blight caused the notorious **potato famine** of 1845–1851. Nearly 10 percent of Ireland's pre-famine population of 8 million died of starvation or disease during these years. An even larger number emigrated to England, Wales, Scotland, Australia, and North America. The large Irish American community in the United States traces much of its ancestry to this migration. Today, Americans of Irish descent are just a few of the eager consumers abroad of Irish culture, especially dance and music.

Northern Ireland, now a majority Protestant part of the United Kingdom, has had a particularly difficult struggle that has colonial roots. One outcome of the second English conquest of Ireland, in the seventeenth century, was the settlement of English and Scottish Presbyterians and other Protestants in the north, where they became numerically dominant. These newcomers were supposed to form a loyal population in an otherwise hostile, conquered, largely Roman Catholic country. In 1921, when the Irish Free State (later to become the Republic of Ireland) was established in the southern, mainly Catholic part of the island, for economic and religious reasons the predominantly Protestant north chose to remain with the United Kingdom. Northern Ireland (also known as Ulster) was given much autonomy, including its own parliament in Belfast. Northern Ireland's Catholic minority felt increasingly marginalized by the pro-British Protestant majority and in the late 1960s began agitating—often violently—for change. The British responded by imposing **direct rule** (in which the United Kingdom's central government in London makes all major policy decisions for Northern Ireland), and Catholic discontent soon focused on the perceived British occupation. The **Irish Republican Army (IRA)** began a campaign of bombings, shootings, and arson, sometimes into the heart of England, ostensibly designed to drive the British army from Northern Ireland. **The Troubles,** as the locals refer to this period of strife, involved an often violent struggle between **Catholic Republicans** and **Protestant Unionists** in Northern Ireland.

In 1998, by which time more than 3,400 Irish on both sides had died in the conflict, the U.S.-brokered **Good Friday Agreement** was signed. Approved by the IRA (and its political counterpart, **Sinn Fein,** pronounced "shin fayn") and the Unionists, the agreement called for the devolution of British power. Direct rule by Britain was to be replaced by the *Northern Ireland Assembly,* in which Protestants and Catholics would share power. Violent challenges to the peace agreement persisted from both sides, however, leading the British to halt devolution and reimpose direct rule in 2002. In 2005, the IRA announced that it would permanently cease military operations. Despite the refusal of three small splinter groups to lay down arms, Unionists agreed to share power with the IRA, and unprecedented negotiations between the two sides continued. If it lasts, peace could help bring prosperity to Northern Ireland, which has had periods of strong economic growth based on shipbuilding and linen textile industries, concentrated around the main industrial and residential city of Belfast.

Paris as a Primate City

A **primate city** is a city that, according to a convenient definition, is larger than the second and third largest cities in that country combined. Paris, for example, has a metropolitan area population of 10.4 million, far exceeding the combined populations of France's next largest cities, Lille (with 1 million people) and Lyon (with 1.5 million people). In practical terms, a primate city dwarfs all others in its demographic, economic, political, and cultural importance. The primate city produces and consumes a disproportionately high share of the country's goods and services, and many national resources—such as bureaucratic offices and educational opportunities—are available only in the primate city. The primate city therefore acts as a great magnet for rural-to-urban migration and tends to absorb more investment than other communities.

Urban primacy is often bad for a country's development because funds that could be spent to improve services and quality of life in smaller cities and rural areas are often diverted instead to the primate city. With greater opportunities available in the primate city than in the smaller cities of the hinterland, the most educated and talented people abandon the hinterland for the primate city. This internal brain drain contributes further to underdevelopment outside the primate city. Primate cities are rare in the more developed countries but are quite typical of the LDCs, where regional development is a critical issue. Cairo, Karachi, and Mexico City are exemplary primate cities of Africa, Asia, and the Americas.

Paris is the greatest urban and industrial center of France, a primate city completely overshadowing all other cities in both population and economic activity. It is by far the largest city on the mainland of Europe. Paris is located at a strategic point on the Seine River relative to natural lines of transportation; therefore its geographic *situation,* as described earlier in the chapter, was important in the city's development. In France, it would be appropriate to say "all roads lead to Paris" (see

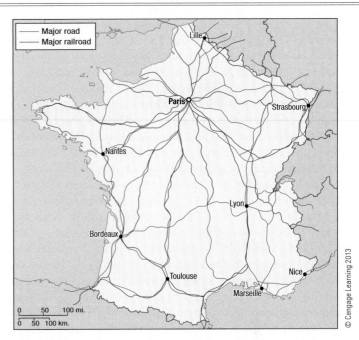

• **Figure 4.25** Paris is the primate city at the hub of France's transportation network.

•Figure 4.25). Paris' growth was also the product of the growth and centralization of the French government and the transportation system it created. Like London, it has no major natural resources for the industry in its immediate vicinity, yet it is the greatest industrial center of the country. Despite its economic clout, Paris maintains its international reputation as the romantic and monument-studded **"City of Light"** (many of these attractions are on the map in •**Figures 4.26**). It is the world's leading urban tourist destination, and France attracts more tourists than any other country in the world.

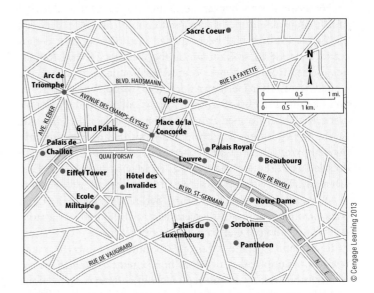

• **Figure 4.26** Map of downtown Paris and some of its world-famous attractions.

In the Middle Ages, Paris became the capital of a succession of kings who gradually extended their effective control over all of France. As the French monarchy became more absolute and centralized, its seat of power grew in size and came to dominate the cultural and political life of France. As national road and rail systems were built, their trunk lines were laid out to connect Paris with outlying regions. The result was a radial pattern with Paris at the hub, and this is essentially the pattern that exists today. As the city grew in population and wealth, it became an increasingly large and rich market for goods.

The local market, plus transportation advantages and proximity to government offices, provided the foundations for a huge industrial complex. This development has involved two major classes of industries. Paris is the main producer of the high-quality luxury items (fashions, perfumes, cosmetics, and jewelry, for example) for which that city and France have long been famous. Paris is also the country's leading center of engineering industries, secondary metal manufacturing, and diversified light industries, concentrated in a ring of industrial suburbs that sprang up in the nineteenth and twentieth centuries. Paris inverts the usual pattern of the industrial city by maintaining a core of low-profile historic monuments, while the skyscrapers of the modern economy tower away from the city center. Parisian urban planners have carefully cultivated and protected the "crown jewels" of this beautiful city's architecture.

Divided and Reunified Germany

One of the most important geopolitical events of the late twentieth century—and one that has lingering effects in the new millennium—was the reunification of the two Germanys. If you go on to study Korea in this book, you will see that one of the big questions for this century is whether or not the divided Koreas might follow the German example. In both cases, economic issues play a prominent role; in each case, there is a stronger, wealthier country that would have to assume much of the cost of merging with a weaker and poorer counterpart.

The Federal Republic of Germany (•Figure 4.27) reappeared on the map of Europe as a unified country in 1990. Between 1949 and 1990, it had been divided into two nations: democratic West Germany (known as the German Federal Republic), formed from the zones occupied by Britain, France, and the United States after Germany's defeat in World War II in 1945, and Communist East Germany (the so-called German Democratic Republic), formed from territory occupied after the war by the Soviet Union. The former capital, Berlin, was similarly divided. West Berlin was a part of West Germany for those decades, even though the city was entirely surrounded by East Germany. The withdrawal of Soviet support for East Germany in 1989 and the destruction of the Berlin Wall led to the rapid collapse of its Communist government in 1990 and to the jubilant reunification of Germany's 16 states (*Länder*).

Reunified Germany is Europe's dominant country in many respects. Its population of 82 million is much greater than that of any other nation in the region. Politically it is seen, along with France, as the cornerstone of the European Union. Economic considerations are even more important; Germany has

• **Figure 4.27** Principal features of Germany. Note which parts of the country belong to what formerly were East and West Germany.

the world's fourth largest economy, after the United States, China, and Japan. Germany is among the top three countries, along with China and the United States, in exports of goods. Germany must also bear a substantial part of the economic burden of problems in the Eurozone and in the European Union.

The former West Germany was Europe's leading industrial and trading country during the latter half of the twentieth century. The former East Germany was an advanced country by Soviet bloc standards but was an economic disaster by West German standards. Many billions of dollars and much time have been required to rehabilitate Germany's east, and the task is still incomplete.

Germany's urban geography is unusual. Industrial and urban centers are widely scattered. Many cities are large, but none is as nationally dominant as London or Paris. This dispersed pattern came about because Germany was divided for a long time into petty states, many of which eventually became internal states within German federal structures. These often had their own capitals, which would in time become Germany's large cities.

In World War II the western Allies, led by the United States, heavily bombed many of these industrial cities, which also had large civilian populations. Perhaps the most tragic bombing was of Dresden in Germany's Saxony region, in 1945. The firestorm in this city, known as the "Florence on the Elbe" because of its cultural treasures, was immortalized in Kurt Vonnegut's novel *Slaughterhouse Five.*

Allied bombing did not eliminate Germany's industrial capacity. West Germany was reconstructed rapidly in the postwar years, thanks to several factors. The massive U.S.-funded aid package known as the Marshall Plan (see page 89) supplied billions of dollars in capital goods between 1948 and 1952. This aid was strategically aimed at rebuilding the country to stave off the Communist threat from the east. Skilled workers were eager to work, even for low wages, and markets were hungry for the industrial products that Germans could make. Germany's economic growth continued through the 1960s. Labor vacancies created by upwardly mobile Germany workers were filled by foreign workers, especially Turks and Kurds from Turkey.

East Germany, Still German

East Germany was Communist ruled and Soviet dominated, but it was still German. Even Soviet exploitation and bureaucratic inefficiency were unable to destroy the traditional German ethics of hard work, efficiency, attention to detail, and high standards. East Germany was the most productive of the Soviet satellite countries that lay between Russia and Western Europe, but its economy lagged far behind West Germany's. When East Germany collapsed in 1989, a major migration stream began flowing from East to West Germany. After 1989, more than 4 million people fled westward from what Germans call the "new states" of the former East Germany. The emigrating easterners left behind what one German professor described as "a series of towns and enclaves for senior citizens."[13] The urban landscapes of the east changed to accommodate an aging population. Hundreds of the old Soviet-style apartment blocks were demolished and new parks and other green spaces developed in their wake. The ongoing urban architectural experiment in eastern Germany, known as *Ruckbau,* is a cross between demolition and reconstruction.

The pool of young and educated people that remained in the east began to attract employers looking for lower-wage workers, and the region's economy may benefit from their presence. More than 3 million people have moved from the west to the east to take advantage of these opportunities. There is a growing semiconductor industry in what local promoters like to call "Silicon Saxony" (one of those plays on words modeled on Silicon Valley, as described on page 92), especially in Dresden and Leipzig. The automakers Volkswagen and BMW have opened plants in Dresden and Leipzig, respectively. Taking advantage of the skilled and inexpensive labor, scores of U.S. chip-making companies have set up operations in the region.

Billions of dollars of investment are helping to raise economic standards in the east and to clean up the severe environmental damage inflicted during Communist rule there. To complete that transition, the former West Germans are bearing much of the tax and social burden of bringing the former East Germans up to the standards long enjoyed in the west. More than 20 years after reunification, western Germans (known as

"Wessies" by the east Germans) continue to resent paying to support the eastern economy and believe the Easterners (whom they call "Ossies") are ungrateful for that help.[14]

Economically, Germany was one of the characteristic welfare states that subsidized its citizens generously. The costs for that support contributed to Germany's economic downturn early in the 2000s, when economic growth was the slowest of all EU countries and unemployment surged to 10 percent. In the area of former East Germany, conditions were worse, with the rate of unemployment more than twice that in the west (that ratio continued even after conditions became better). Germany continued to work on its painful transition from a welfare state to one in which citizens increasingly must pay for health care and other vital services. Other difficult measures, such as cutting jobs, extending work hours without extra pay, and outsourcing work to other countries, helped Germany's economy rebound after 2005 and grow even more sharply after 2009 in the wake of the global recession. Germany's climb out of its economic hole was remarkable, and it raised hopes for an economic rebound throughout the EU community.

But is Germany really committed to the European community and to its partnership in the Atlantic alliances, including NATO? These became persistent questions after 2010. Germany is still considered the economic locomotive of Europe, and other EU member countries look to Germany for leadership and aid (many of the poorer EU countries came to believe that Germany was their "paymaster" in times of need). Germany is, however, increasingly cultivating economic partnerships with countries to the east. Fewer German manufactured goods like precision machinery, cars, chemicals, and electronics are going to France, whereas more are being exported to China, India, and Russia. France continues to be China's leading trade partner but is nervous that Germany will "decouple," going its own way and serving its own interests rather than those of the European Union. Germany also broke ranks politically with its NATO allies, including France, the United Kingdom, and the United States, in refusing to support military action against Muammar Gaddafi's Libya in 2011.

Europe Paves the Way on Alternative Energy

While we are on the subject of French and German differences, let us consider this: France is the world champion of nuclear energy, getting about 80 percent of its electricity from nuclear power. Germany, on the other hand, has vowed to phase out its nuclear power program (which supplies a quarter of its electricity) completely by 2022. What explains the difference? The devastating tsunami in Fukushima, Japan, and the series of nuclear power plant accidents in its wake. Awed by what happened in Japan, Germany's leaders and citizens decided that nuclear power was simply too risky to maintain. France, on the other hand, is too deeply committed to nuclear power to consider alternatives.

Germany's decision to abandon nuclear power puts it in a difficult position to attain the European Union's collective goal of reducing greenhouse gases to 20 percent *below* 1990 levels by 2020. One of nuclear power's advantages is that it produces no

(a)

(b)

(c)

(d)

•**Figure 4.28a–d** Europe has a wide variety of energy sources, with Poland almost completely reliant on coal to produce electricity and France nearly so on nuclear power. Traditional and modern sources of energy in Europe include **(a)** peat from a Scottish bog, **(b)** coal in Poland, and windmills, both **(c)** old (in the Netherlands) and **(d)** new (in Denmark).

greenhouse gas emissions, so maintaining the nuclear program would be helpful to Germany. Some EU leaders think that goal is not ambitious enough and are pushing to cut carbon dioxide emissions to 30 percent *below* 1990 levels by 2020. Britain vows to derive 20 percent of its total energy needs from renewable sources by 2020 and to cut carbon dioxide emissions to 60 percent *below* their 1990 levels by 2050. I emphasize *below* 1990 levels to draw attention to these extraordinary ambitions. Imagine the serious and fundamental changes needed to turn the clock back on the relentless forward movement of Europe's fossil-fuel driven economies and lifestyles. And compare these ambitions with those of the United States. Under Democratic and Republican administrations alike, the United States has consistently backed away from making serious commitments to reduce greenhouse gas emissions.

Although the precise goals of EU countries differ, the underlying assumption is rock steady: the European Union has taken an official stance that global climate change is a serious problem and must be confronted through dramatic changes in the ways energy is produced and consumed (•Figure 4.28). Officially, the European Union's stated target is to get 20 percent of its energy from renewable sources by the year 2020. To achieve this goal, EU members will promote fuel efficiency in automobiles,

encourage the use of public transportation, expand the cap-and-trade mechanisms of carbon emissions trading, and bring a vast array of alternative energies—hydroelectric power, solar power, tidal power, wave power, geothermal power, and even trash power—online. France and some other counties will also add nuclear power into this mix, but nuclear is not considered a renewable alternative energy source.

One of the most popular renewable alternative energies in Europe is wind power. Denmark uses wind power to meet more than 20 percent of its electricity needs—the highest rate in the world from this source. Germany has established itself as the world's leading manufacturer of wind turbines. More than 12,000 of these mills dot Germany itself, providing about 7 percent of the country's electricity needs. That share is projected to rise to 20 percent by 2020, after 40 offshore wind farms as far as 70 miles from the coast come online. Britain is building more than 1,000 wind turbines off England's coast, mainly in the Thames estuary, the east coast area, and along the northwest coast.

Wind energy produces no greenhouse gas emissions, and you might think Greens would embrace it. But environmentalist Britons, recalling William Blake's image of the "dark satanic mills" of early English industry, lament that the "silver satanic blades" will sully the storied Lake District of the northwest.[15] The **NIMBY** ("not in my back yard") attitude often wins the day. This viewpoint, combined with the fact that Europeans are very short on land—compared with Americans especially—and the characteristic geography of Western Europe's numerous islands, peninsulas and seas, and enormous amount of coastline, makes a perfect recipe for offshore wind power.

Some of the technical challenges of providing wind power are impressive. Have a look at the North Sea in Figure 4.1 on page 66. Its waters can be turbulent, with winds often howling at 60 miles per hour and waves cresting higher than 30 feet (9 m). As many as 6,000 gigantic wind turbines will be installed in these treacherous North Sea and other British waters by 2020. Fortunately, British engineers have had the know-how of installing oil rigs in the North Sea since the 1970s. The plan is to put that fossil fuel experience to work on alternative energies.

The European Periphery

Properties of the European Periphery

Just as there is a recognizable European Core, where the region's demographic, economic, and political weight is centered, there is also a **European Periphery.** This "rimland" is made up of countries whose interests are tied closely to those of the core and are strongly influenced by the core countries. But they have less clout in the political and economic systems than the core countries do and tend to be dependent on the core. As we have seen in the context of the European Union, for example, many of the countries of the periphery hope that the core countries Germany and France will bail them out of their economic troubles.

Here we use the tool of geographic regions to subdivide these countries into three categories: those of Northern, Southern, and Eastern Europe (•Figure 4.29). Within these subgroups, the

countries have distinct national identities, histories, and experiences. In this section, we explore some problems and issues of these countries outside the European center. The major traits of each nation are listed in Table 4.1 on pages 67–68.

Northern Europe

Save the Whales . . . for Dinner   One reason that both Norway and Iceland (see •Figure 4.30) have refused to join the European Union is that both have economically vital fishing industries, and they fear that common EU policies on fishing might diminish their fishing profits. The two countries, along with Japan, also are distinguished by their unique perspective on another marine resource: the world's whales. People in all three countries maintain that populations of some whale species, especially the minke (pronounced "ming'-key"), have rebounded to levels that should allow regular, limited harvesting for human consumption. One of their arguments is that growing populations of these whales would feed on huge amounts of commercially important fish stocks. Whale meat has also long been a prized food for people in these countries. "Save the whales for

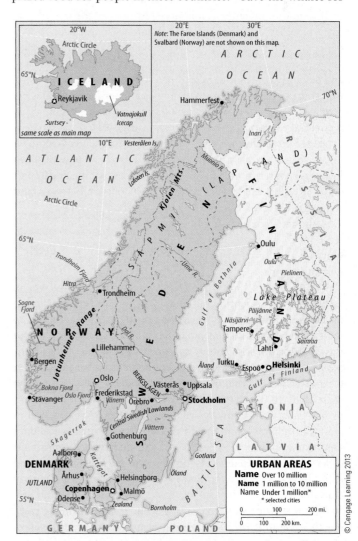

• **Figure 4.30** Principal features of Northern Europe. Note Iceland in the inset. See the map of Figure 4.1 on page 66 to appreciate the isolated location of Iceland in the North Atlantic. Iceland and Norway are two of the world's three countries still practicing whaling; the third is Japan.

• **Figure 4.29** The subregions and countries of the European periphery.

• **Figure 4.31** Norwegian whalers haul in a minke whale.

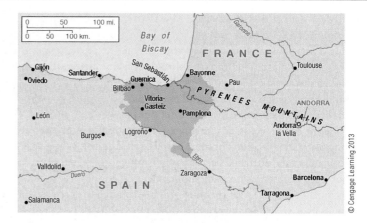

• **Figure 4.32** The Basque country of Spain and France.

the dinner table" is the ironic Norwegian response to the popular Western bumper sticker "Save the whales" (•**Figure 4.31**).

In a landmark agreement honored by most of the world's marine countries, the International Whaling Commission (IWC) banned commercial whaling worldwide in 1986. The Norwegians never accepted the ban, calling it "cultural imperialism," and defiantly continue to take about 500–600 minkes each year, well short of their self-imposed quota of about 1,300 whales per year. Norwegians eat whale meat without remorse, arguing that the rest of the world is foolish in its insistence that the whale is a "sacred cow." Even Norway's famously staunch environmentalists see no problem with the whale harvest. But with video footage of water stained red by the blood of writhing and apparently suffering whales, environmentalists outside Norway have stepped up pressure on Norway to stop whale hunting.

Although Norway simply defies the IWC whaling ban and since 1993 has hunted whales commercially for consumption in Norway and for export abroad (especially to Japan), the other whaling countries at least pretend to be considerate of the ban. The 1986 ban forbids commercial whaling but allows whales to be killed for scientific purposes and for subsistence (the latter clause allows Native Americans in Alaska, for example, to hunt whales). Iceland and Japan have exploited the loophole of the scientific purpose and in the process put whale meat on the table. They argue that there is a legitimate need for the science: Iceland estimates that its territorial waters are inhabited by more than 40,000 whales that eat perhaps 1 to 2 million tons of fish each year. Only by studying the whales' stomach contents can reliable estimates of the whales' impact on Iceland's vital fishing industry be evaluated. The studies accomplish two other goals: They reduce the pressure on the fish resource, and they allow whale meat to find its way to markets for ordinary consumers.

Killing whales does carry some costs. Iceland resumed commercial whaling in 2006, establishing quotas of 30 minkes and 9 fin whales each year. Most Icelanders support the hunt, but others argue that whale watching—which previously had drawn 90,000 tourists yearly—was the more economically

sensible way to manage the resource. As the European economy hit hardest by the global economic downturn, Iceland can ill afford to lose tourists. And with "Save the whales" an international mantra, Iceland, Norway, and Japan are subjected to intense criticism from around the world.

Southern Europe

The Basque Country The 2.3 million **Basques** of Spain and 300,000 Basques of France (see Figure 4.B and •**Figure 4.32**) have a unique ethnicity and culture completely unrelated to those of their host country majorities. Like other minorities, they have often been the targets of discrimination and violence. This is especially true in Spain, where during the Spanish civil war in 1937 a German bombing (called for by Spanish fascists) led to tragic loss of life among Basque civilians in the city of Guernica. Artist Pablo Picasso painted a mural of the suffering of Guernica's civilians that brought international attention to the Spanish civil war and ever since has served as a symbol of the tragedy of war (•**Figure 4.33**).

By the 1960s, Basque desire for independence led to the emergence of the militant group **ETA** (an acronym for **Basque Homeland and Liberty** in Euskara, the Basque language). Spanish authorities gradually allowed considerable autonomy for the Basques, and they now control their own affairs in taxation, education, and policing. For more than four decades, these concessions were not enough for ETA, which carried out attacks on Spanish interests and triggered a vigorous crackdown by Spanish security forces. But in 2011, ETA vowed to cease violence with its declaration of a "permanent, general and verifiable" ceasefire. The European Union and the United States regard ETA as a terrorist organization and are wary of this ceasefire because of ETA's history of breaking earlier ceasefires. It is not clear whether in the long term ETA will settle for what they have now, a devolutionary status comparable to Scotland's autonomy within the United Kingdom (see page 93), or will push for outright independence as Kosovo did (described later in this chapter).

Independence for Basques in Spain or France seems very unlikely, and most Basques are not pushing for it or expecting it. The Basques both in Spain and France have worked hard at retaining their unique customs. In Spain, there are Basque

• **Figure 4.33** During the Spanish Civil War, Spanish Republicans commissioned Pablo Picasso to create an image of the suffering inflicted by Spanish Nationalists on Spanish civilians. Picasso's *Guernica* went on display at the 1937 World's Fair in Paris.

• **Figure 4.34** North and South in Italy, as perceived by Italians. This is a vernacular map, so there are no "true" boundaries separating north and south.

language television broadcasts. French Basques would like to have the kind of autonomy their counterparts in Spain have achieved.

North versus South in Italy All Italians are united by their pride in the heritage bestowed by civilization from the Romans to the Renaissance and beyond, and Italians live on some of the most picturesque and storied landscapes on Earth. But within Italy, there is a longstanding vernacular distinction between the north and the south Northerners tend to see themselves as more sophisticated and cosmopolitan than the more earthy and provincial southerners (see •**Figure 4.34**). Southern Italians, whose region is known as the **Mezzogiorno** (a fine example of a vernacular or perceptual region, as described on page 6 in Chapter 1), often acknowledge their agrarian roots as the source of their superior kinship values and enjoyment of life. Statistics underscore the economic discrepancies between the two Italys—for example, northern Italy has labor shortages while the south has unemployment; northern labor and industries (such as luxury automobile, jewelry, textile, leather, ceramic, and furniture industries) are more productive, and income levels in the north are much higher.

The feeling among many northerners that they would be better off if they did not have to subsidize the south has even led to a secessionist movement, championed by an organization called the **Northern League**, that calls for the creation of an autonomous or independent state of **Padania** in the north (see Figure 4.B). The Northern League wants to maintain the isolated affluence of northern Italy in part by keeping out immigrants, particularly Muslims and Africans. [71] Another organization, **Liga Veneta**, wants to create an autonomous region of Veneto, centered on the affluent city of Venice (•**Figure 4.35**).

In the toe of Italy's boot many illegal immigrants, mainly black Africans, live in squalid conditions. They began coming to Italy to pick oranges and tangerines when farming was still profitable because of EU agricultural subsidies. But by 2010, when EU subsidies for farming were curbed, there was no work for them in a weakening agricultural sector. The xenophobic Northern League denounced the "excessive tolerance" of illegal immigration in Italy. In Italy's toe, some white Italians

© Jonathan Blair/CORBIS

• **Figure 4.35** Venice from the air. The amphibious and vulnerable setting of the city is most apparent from this perspective.

engaged in the "hunting of blacks." Both whites and blacks ignited ugly race riots that took many lives and destroyed property. The climate of economic hardship provided a perfect breeding ground for this kind of racism and violence.

North versus South in Cyprus Keep your eyes on Cyprus. It is another one of those situations where there are divisions between north and south, and questions about reunification or continued divisiveness. In the case of Cyprus, there is much at stake beyond the confines of this Mediterranean island. What happens there will play a big role in the question of whether Turkey will join the European Union, changing the character and complexity of the European Union much as the "big bang" did.

The large Mediterranean island of Cyprus (see •Figure 4.36 and Figure 4.1 for its location relative to the continents), located near southeastern Turkey, came under British control in 1878 after centuries of Ottoman Turkish rule. In 1960, it gained

© Cengage Learning 2013

• **Figure 4.36** Cyprus.

independence as the Republic of Cyprus. The critical problem on the island is the division between the Greek Cypriots, who are Greek Orthodox Christians and make up about three-fourths of the estimated population of 1 million, and the Turkish Cypriots, who are Muslims and make up about one-fourth of the population. Agitation by the Greek majority for union (**enosis**) with Greece led in the 1950s to widespread terrorism and guerrilla warfare by Greek Cypriots against the occupying British. Violence also erupted between Greek advocates of *enosis* and the Turkish Cypriots, who feared a transfer from British to Greek sovereignty.

A major national crisis erupted in 1974 when a short-lived coup by Greek Cypriots, led mainly by Greek military officers, temporarily overthrew Cyprus's President Makarios, who had embraced a conciliatory policy toward the Turkish minority. Turkey then launched a military invasion that overran the northern part of the island. Cyprus was soon partitioned between the Turkish north and the Greek south. A buffer zone (the **Attila Line** or **Green Line**) sealed off the two sectors from each other, and even the main city of Nicosia was divided. A separate government was established in the north, and in 1983, the Turkish Republic of Northern Cyprus was proclaimed. Only Turkey recognized this state. Meanwhile, the internationally recognized Republic of Cyprus functioned in the Greek Cypriot sector, which makes up about three-fifths of the island. Both republics have their capitals in Nicosia.

The north had dominated the economy prior to the partitioning of Cyprus, but since then, the north has had economic problems, while Greek Cyprus has prospered. Most nations have refused to trade directly with Turkish Cyprus since the invasion, and economically weak Turkey has not been able to boost the north's prospects. The depressed north remains tied to Turkey, while the Greek sector makes effective use of economic aid from Greece, Britain, the United States, and the United Nations. Tourism has flourished, and new businesses have thrived.

Events leading to the European Union's expansion in the Big Bang of 2004 provided a strong incentive for the two sides to resolve their differences and join the union. The United Nations devised a plan for the two halves to vote separately in a referendum on reunification. Had both sides voted in favor of reunification, a united Cyprus would have taken up membership in the European Union. However, the United Nations and European Union agreed that should either the north or the south, or both, reject the referendum, only the Greek south would actually join the European Union. In the vote, the Turkish north voted overwhelmingly in favor of reunification, while the Greek south overwhelmingly rejected it. The somewhat perplexing result is that in EU terms, "Cyprus" nominally refers to both parts of the island, but only the Greek south is a de facto member of the European Union. This allows the United Nations, the European Union, and other international bodies to continue their policy of not officially recognizing the north as a legitimate, separate political entity. Turkey, long seen as the obstacle to reunification of Cyprus, won rare acclaim in

Europe because of the Turkish Cypriots' strong vote in favor of reunification. Turkey's prospect of joining the European Union seemed to strengthen, but in 2006, it refused to open its ports to Greek Cypriot vessels. The European Union halted member-ship discussions with Turkey over this issue.

Relations between Turkey and the European Union hit an-other rough patch in 2011, when Turkey threatened to send warships against the Republic of Cyprus. The Republic had come to an agreement with Israel over seabed resources in the eastern Mediterranean and began exploratory drilling for oil and gas. Turkey and the Turkish Cypriots argued that any maritime agreements or drilling should take place only after reunification of the island.

Eastern Europe

Wrenching Reforms in the Shatter Belt of Eastern Europe   The Eastern European countries (•Figures 4.37 and 4.38) are in the process of reinventing themselves after more than four de-cades of direct or indirect control by the Soviet Union. Prior to German reunification in 1990, the former East Germany was also part of Eastern Europe, and it is brought into the present discussion when appropriate. Three of the countries—Estonia, Latvia, and Lithuania—were incorporated into the Soviet Union during World War II, while many others became **Soviet satellites,** with local Communist governments effec-tively controlled from Moscow. Exceptions were Albania and the former Yugoslavia (consisting of Slovenia, Croatia, Bosnia and Herzegovina, Serbia, Montenegro, and Macedonia), where national Communist resistance forces took power on their own as German and Italian power collapsed.

• **Figure 4.38** The northern portion of Eastern Europe.

• **Figure 4.37** The southern portion of Eastern Europe.

Majority Slavic ethnicity, former Communist status, and subjugation to Soviet interests were among the few unifying themes of the region prior to the end of the Cold War. Now the true complexity of the region is more apparent.

The extension of Soviet power into Eastern Europe after World War II was a replay of history. In the Middle Ages, several peoples in the region—the Poles, the Czechs, Magyars, Bulgarians, and Serbs—enjoyed political independence for long periods and at times controlled extensive territories outside their homelands. Their situation deteriorated as stronger powers—Germans, Austrians, Ottoman Turks, and Russians—pushed into east central Europe and carved out empires. These empires frequently collided, and the local peoples were caught in wars that devastated great areas, often resulted in a change of authority, and sometimes brought about large transfers of populations from one area to another. In geopolitical terms, Eastern Europe is a classic **shatter belt**—a large, strategically located region composed of conflicting states caught between the conflicting interests of great powers (see the evolution of this region in •Figure 4.39).

Peoples of this region have been on the move in large numbers since the beginning of World War II. Jews, Germans, Poles, Hungarians, Italians, and others were uprooted by Nazi and Soviet authorities, often without notice, losing all their possessions, and were dumped as refugees in so-called "homelands" that many had never seen. During the war, Nazi Germany systematically killed approximately 6 million Jews and perhaps as many as 1.5 million Gypsies (Roma) and other "undesirables" in the Holocaust. The prewar populations of Germans in Poland and Czechoslovakia were expelled at the end of World War II and forced into East and West Germany. Many died in this process of forced migration overseen by the Soviet Union. Ethnic minorities now make up only 2 percent of the population in Poland, which transferred most of its German population to Germany and whose territories containing Lithuanians, Russians, and Ukrainians were absorbed by neighboring countries. A number of countries still have large minorities, including the Roma or Gypsies (see page 108). These population transfers helped create an ethnic map of Eastern Europe that was overwhelmingly and purposefully Slavic.

Communism in Ideals and Practice   In becoming satellites of the Soviet Union, the nations of Eastern Europe (and East Germany) were reformed by **communism.** You should consider the principal traits of communism in order to understand their legacy in today's geography of Eastern Europe and of the former Soviet Union, which is the subject of the next chapter. Communism had these principal traits:

- One-party dictatorial governments
- National economies planned and directed by organs of the state
- Abolition of private ownership (with some exceptions) in the fields of manufacturing, mining, transportation, commerce, and services
- Abolition of independent trade unions
- Varying degrees of **socialization** (state ownership) of agriculture

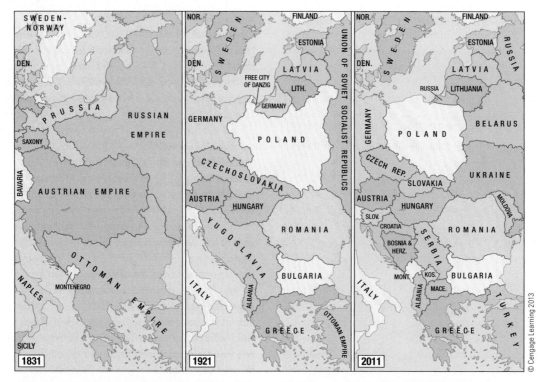

• **Figure 4.39** Positioned between stronger powers to the east and west, Eastern Europe is a classic shatter belt with a tumultuous past, reflected in its shifting borders.

Soviet military force crushed attempts by East Germany in 1953, Hungary in 1956, and Czechoslovakia in 1968 to break away from Soviet control. In agriculture, the new Communist governments liquidated the remaining large private holdings and in their place introduced programs of collectivized agriculture on the Soviet model. Some farmland was placed in large state-owned farms on which the workers were paid wages, but most was organized into collective farms owned and worked jointly by peasant families who shared the proceeds after operating expenses of the collective had been met. This **collectivization**—the bringing together of individual landholdings into a government-organized and government-controlled agricultural unit—met with strong resistance. Conversion back to private farming after four decades of collectivization presented painful obstacles, and progress has been slow. This process is still ongoing, which is why it is important that you understand the legacies of communism and socialism. Despite recent attempts to diversify and intensify agriculture, in Eastern Europe agriculture is focused on a narrow range of products, especially corn, wheat, pigs, and sheep.

130, 146

As it did in agriculture, communism in Eastern Europe inaugurated a new era in industry based on a Soviet model. Most industries were taken out of private hands, and national economic plans were developed. Communist planners did not aim at a balanced development of all types of industry. Instead they emphasized those industries seen as essential to the development of a greater Soviet empire: mining, iron and steel, machinery, chemicals, construction materials, and electric power, for example. These were favored while consumer-type industries were neglected.

The central planning agencies of the Communist governments maintained rigid control over individual industries. Plant managers were order to produce certain goods in quantities determined by state governmental planners. The plant's success was judged by its ability to meet production targets rather than its ability to sell its products competitively and at a profit. Plant managers were expected to conform to the central plan and not ask questions. As you can imagine, this was not an economic environment that fostered creativity and entrepreneurship. There were no visionaries like Apple's Steve Jobs—or if there were, they would have been branded as nonconformist and been alienated.

This system resulted in shoddy goods that were in short supply or in mountainous oversupply because production was not driven by consumer demand and satisfaction. The planned expansion of mining and industry in Eastern Europe increased the output of minerals, manufactured goods, and energy. But it also produced inefficient, overstaffed industries that were ill-prepared for competition in the world markets of the post-Communist period. This was a problem shared by all the economies of all the successor states of the Soviet Union, as we will also see in the next chapter.

Under communism, the economies of Eastern Europe were closely tied to that of the Soviet Union. During the 1970s and 1980s, however, the region's trade relationship with the Soviet Union weakened. Trade and financial relations with Western countries, companies, and banks expanded rapidly. With relatively little to export, the Eastern European countries borrowed massively from Western governments and banks to pay for imports from the West. Poland and Romania pursued this course especially strongly. New industrial plants and equipment acquired in this way were supposed to be paid for by goods that would be produced for export to the West, by industries that had learned efficient production and marketing techniques from the West.

But then the Western economies began to slump, lowering the demand for imports. This led to a situation by the 1980s in which large debts to Western governments and banks needed to be repaid if countries were to maintain any credit at all. However, with the help of mismanagement by Communist bureaucrats, exports with which to pay were not being produced or, if produced, could not be sold. The result was a severe impediment to economic expansion and falling standards of living as the governments squeezed out the needed money from their people. In Poland, social action through the formation of an independent trade union called **Solidarity** paralyzed the country and threatened to upset Communist dominance. These conditions, along with the Soviet Union's own growing economic and political distress, set the stage for the withdrawal of Soviet control and decommunization in 1989 and 1990.

As late as the early summer of 1989, East Germany and the Eastern European countries seemed to be firmly in the control of totalitarian Communist governments. Then, suddenly, the Communist order began to crumble. Public demands for freedom, democracy, and a better life gathered momentum in one country after another. Communist dictators who had ruled for many years were forced out, and reformist governments took charge. By mid-1991, democratic multiparty elections had been held in all countries. This liberalizing process continued throughout the 1990s and into the 2000s.

In 1989 and 1990, the Soviet Union withdrew its backing for the region's Communist order. That support was too costly for the Soviet Union, whose leaders were forced to recognize the inevitable victory of "people power" in Eastern Europe. Several Communist regimes (including East Germany's) quickly collapsed and were replaced by democratically elected non-Communist governments.

You many not have witnessed these momentous events that occurred in Europe and the Soviet Union in 1989. But you have seen the extraordinary upheaval in the Middle East and North Africa in the Arab Spring of 2011. These jaw-dropping events 192 give you a good idea of what 1989 was like. And what happened in Eastern Europe may provide clues to how long and how difficult the restructuring of the Arab World will be.

Governments in Eastern Europe have been struggling ever since 1989 to build democracies with capitalist economies out of the economic wreckage of unproductive, unprofitable, uncompetitive, state-owned enterprises. Many countries pursued reforms aggressively to meet qualifications for joining the European Union. Economic restructuring has required many difficult steps, including the **privatization** (shift to nongovernmental ownership) of state-owned enterprises, an increase in the efficiency of state-run enterprises by allowing noncompetitive enterprises to fail and removing support of poor performers, and encouragement of the development of new private enterprises, often with foreign capital and management playing a role. In addition, economic restructuring has seen the termination of price controls so that prices reflect competition in the market; the development of new institutions required by a market-oriented economy (such as banks, insurance companies, stock exchanges, and accounting firms); the fostering of joint enterprises between state-owned firms and foreign firms; and the elimination of bureaucratic restrictions on the private sector. Internationally convertible currencies have also been created to increase trade and enhance competition, and access to international communications media of all sorts has been expanded to help local entrepreneurs learn from foreign examples.

235 One of the most remarkable economic trends has been Western European and Chinese **outsourcing** of investments that take advantage of relatively cheap labor in Eastern Europe. In the late 1990s, Austrian and German electronics firms established large factories in east central Europe, especially in Hungary. Even China opened television and other electronics factories in Eastern Europe, because China's labor costs were rising.

The Eastern European countries enjoyed strong economic growth for about a decade (1997–2007). But when the global economic crisis hit Europe in 2007, it hit the former communist countries hardest.

Why They Call It "Balkanization" You may have heard someone use the term *balkanize*. This word comes from the real-life geopolitical events in a part of southeastern Europe known as the Balkans or the Balkan Peninsula. Here we will see what these events were and what their impact has been on the modern geography of the Balkan states, which includes some of the world's youngest countries.

Although economic progress characterized most of Eastern Europe in the 1990s, Yugoslavia (seen on the map of the Balkan Peninsula in •Figure 4.40) was plagued by ethnic warfare following the dissolution of the Yugoslav federal state in 1991. Previously, under the Communist regime instituted by Marshal Josip Broz Tito during World War II, Yugoslavia ("Land of the South Slavs") had been organized into six "socialist people's republics" (Serbia, Croatia, Slovenia, Bosnia and Herzegovina, Montenegro, and Macedonia) and two "autonomous provinces" within Serbia (Kosovo and Vojvodina).

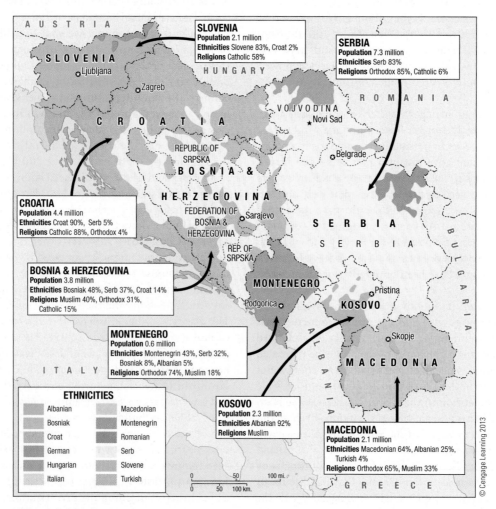

• **Figure 4.40** Ethnic composition of the Yugoslav successor states.

As long as Tito and his successors could retain a firm grip, these groups could coexist within the artificial boundaries of a single nation. However, this apparent unity belied many underlying tensions and conflicts based on the linguistic, ethnic, and religious distinctions you read about earlier in the chapter: Orthodox Serb versus Muslim Kosovar, Catholic Croat versus Orthodox Serb, and so on.

As the Iron Curtain dissolved across Europe, Yugoslavia began to fracture along its ancient ethnic fault lines (**Figure 4.39**). In 1988, Serbia took direct control of Kosovo and Vojvodina as part of its push (under an elected Communist president, Slobodan Milosevic) for greater influence in federal Yugoslavia. The quasi-independent states within Yugoslavia initially demanded a looser federation with more autonomy for each republic. But in 1991 and 1992, Croatia, Slovenia, Macedonia, and Bosnia insisted on independence. When Croatia declared its independence, fighting broke out between ethnic Croats and Serbs in the new country. Croatian Serbs took over one-third of Croatia and embarked on a policy of **ethnic cleansing,** the forced emigration or murder of one ethnic group by another within a certain territory. Although Bosnia's Muslims (known as *Bosniaks*) and Croats voted for independence from Yugoslavia, ethnic Serbs in Bosnia and the Yugoslav government of Slobodan Milosevic opposed this.

War erupted in 1992 between the Bosnian government and local Serbs. Supported by Serbia and Montenegro, the Bosnian Serbs fought to partition Bosnia along ethnic lines and join Serb-held areas to a "greater Serbia." Bosnian Serbs laid siege to the mainly Bosniak city of Sarajevo. The United Nations withdrew recognition of Yugoslavia because of the government's failure to halt Serbian atrocities against non-Serbs in Croatia and Bosnia. In 1994, the Bosniaks and Croats agreed to create the joint federation of Bosnia and Herzegovina.

In 1995, the presidents of Bosnia and Herzegovina, Croatia, and Serbia signed the **Dayton Accord** (it was brokered by U.S. diplomats at an airbase near Dayton, Ohio). This peace agreement retained Bosnia and Herzegovina's international boundaries and created a joint multiethnic, democratic government charged with conducting foreign and economic policies. The agreement also recognized a second tier of government composed of two similarly sized entities charged with overseeing internal functions: the Bosniak-Croat Federation of Bosnia and Herzegovina and the Bosnian Serb–led Serb Republic (Republika Srpska). To help implement and monitor the agreement, NATO fielded a 60,000-strong implementation force known as IFOR. It was later replaced by a smaller NATO force, in turn replaced by EU peacekeepers numbering just 7,000.

While the Bosnian situation quieted, in 1998 Serbia turned its sights on Kosovo, a majority ethnic Albanian and Muslim region with a minority of ethnic Serbs in the north. Here again, Serbs aspired to ethnically cleanse Kosovo of non-Serbs. Serbs asserted that Kosovo was the cradle of their culture and the site of the decisive battle of Kosovo Polje in 1389, which Serbs lost to the Ottoman Turks. NATO responded to the Serb offensive with 11 weeks of bombing. When the bombing stopped, the United Nations took control of Kosovo and oversaw a tense

• **Figure 4.41** On February 17, 2008, the Albanian majority of Kosovo celebrated their self-proclaimed newborn independence in the city of Pristina. Kosovo has yet to gain recognition as independent by UN security members Russia and China.

truce between its Serb and Muslim inhabitants. The fate of more than 200,000 ethnic Serbs who fled during the conflict from Kosovo to Serbia and Montenegro has not been resolved. Nominally still a part of Serbia, but with a population of only 5 percent ethnic Serb and 95 percent ethnic Albanian, Kosovo came under United Nations jurisdiction.

Kosovo declared its independence in 2008, naming itself the Republic of Kosovo. The ethnic Albanian majority poured out into the streets of the capital Pristina to joyously proclaim their country as "newborn" (•**Figure 4.41**). It has been recognized as independent by more than 80 countries, including the United States, most EU member countries (22 out of 27), and other nations, and some countries within the critical United Nations Security Council, but not by other countries, most notably the Security Council members Russia and China. United Nations membership is the ultimate stamp of a country's sovereignty, but Russia and China (which have veto power in the UN) uphold Serbia's claim that Kosovo is a Serbian province, so Kosovo may never receive that recognition. In this book, we recognize Kosovo as an independent nation.

The ouster of Serbian leader Milosevic in 2000 and his capture and subsequent transfer to The Hague to stand trial for war crimes (he died before the trial was completed), opened the way for a period of stable and peaceful borders in the 2000s. The international community insisted on no further redrawing of Balkan boundaries. The region had undergone a profound and violent process of **balkanization,** which is political-geographic shorthand for fragmentation into ethnically based, contentious units that took its name from the characteristic disharmony of this region.

The Balkan conflicts unfolded on television and other media through the 1990s. They reminded Europeans and Americans of the earlier world wars on the continent and seemed a horrible anomaly in a world region accustomed to peace and prosperity. Western military and political institutions eventually seemed to extinguish the fires of conflict, and the Balkan countries were set on the back burners of Europe.

Joe Hobbs

• **Figure 4.42** This Roma woman and her children pass their days begging on a street corner in Szczecin, Poland.

By all measures of quality of life and economic prosperity, the Balkans are the poor stepchildren of the continent, and most cannot expect to join the European Union for many years.

The Roma One of Europe's largest ethnic minorities is the Gypsies—properly known as the Roma—who number about 12 million (•Figure 4.42). Romania has the highest number—as many as 2.5 million—but estimates of the Roma population vary widely because the Roma live on the road and at the margins of society.

The Roma began their great odyssey in what is now India, and their Romany language is very close to languages still spoken on the Indian subcontinent. Having in early times traveled thousands of miles from their homeland, and still often moving in caravans, they have come to be the archetypal people on the move (in informal use, we often refer to inveterate travelers as "gypsies.") Throughout the Roma realm, host governments and majority populations have for centuries regarded the Gypsies with disdain. They are typically depicted as a rootless, lawless, and violent people. They are "the other," apart, not deserving of the educational and economic opportunities offered to majority populations. Wherever they are, they have more children than the majority populations. Sterilizations of Roma women without their consent have been reported in the Czech Republic, Slovakia, Hungary, Romania, and Bulgaria. The Roma are poorer than the majority populations; in Hungary, for example, the Roma unemployment rate is five times that of non-Roma, and in some Roma communities the unemployment rate is as high as 90 percent. They were the first to lose their jobs when the Eastern European countries gained economic freedom from the Communists.

The Roma typically live in shantytowns without water, sewage, or other services and make their living on scant child benefit payments, minor mechanical work, begging, foraging for food, and petty crime. In 1999, one Czech city built a wall between Gypsy and Czech neighborhoods, insisting it was needed to protect other townspeople from Gypsy criminals, noise, and visual squalor. Most Roma children do not attend schools because their parents do not believe that education will improve their job prospects, given the discrimination against them. Many that want to attend school are funneled into institutions for the mentally disabled. There is also violence; Gypsies are often the targets of attacks by skinheads. These assaults, especially in the Czech Republic and Slovakia, have resulted in recent waves of Roma emigration to Canada, France, Britain, and Finland.

Romany activists have responded to discrimination by demanding more rights to welfare, pensions, and other benefits offered to regular citizens. Already feeling overwhelmed by the economic burden of their welfare services and by the migrants described on pages 71 and 72, the wealthier countries of the European Core are worried that the new EU membership of Eastern European countries will unleash a tide of Roma immigration. The latest 14 EU members have almost 5 million Roma. Some Roma have taken advantage of the free passage allowed within Schengenland to travel to France in search of work. In 2010, France began deporting many Roma back to their home counties, especially Romania. France said, as we heard earlier in this chapter (page 89), that its policy about immigrants was that they could settle in France only if they could support themselves. After three months in France, they had to prove they were working or studying, and had sufficient funds and health care. If they could not provide proof—and many Roma could not—France would deport them, and it did. The deported Roma found themselves back in Romania with scant hope of work there—which is what drove them to France in the first place. Economic recession closed the mines and other enterprises where the Roma used to find work in Romania.

When Romania and other Eastern European countries joined the European Union, the European Union earmarked funds to be spent in those countries on projects that would help Roma communities. These funds got caught up in bureaucratic red tape, and the projects never got off the ground.

The stigma associated with being Roma has made successful Roma reluctant to reveal their ethnic identity for fear of losing their jobs. When someone takes a public stance in favor of the Roma they are often silenced or dismissed. For example, at a concert in Bucharest, Romania, the pop singer Madonna urged greater toleration of the Roma. She was booed.

This concludes the survey of Europe. We did not cover every issue in each of Europe's diverse countries, but you should have a good foundation for appreciating land and life in Europe. You should be in a good position to read news coming out of Europe and be able to connect that to your foundation in this chapter. Let us try to achieve those goals as we move eastward to the world's largest country, Russia, and many of the countries within its realm of great influence.

summary

- Europe is physically part of the great continent of Eurasia but is generally labeled a continent and is treated as a separate region. Europe's population is about twice that of the United States, and Europe is more densely settled. Europe is demographically postindustrial, with a slowly declining and aging population.

- Among Europe's most distinctive physical geographic traits are its northerly location, temperate climate, and varied topography.

- The North European Plain is a major belt of settlement and agricultural productivity. The marine west coast climate, continental climates, and Mediterranean climate are characteristic. The Rhine and the Danube are the two most important rivers.

- European languages derive primarily from Indo-European roots and include Romance, Germanic, and Slavic languages. English is a Germanic language, and like most European languages, it is enriched by many other tongues. The dominant religion in Europe is Christianity, with major followings of Protestant, Roman Catholic, and Eastern Orthodox churches. There are significant populations of Muslims (most of them recent immigrants) and of Jews (whose population is a small remnant of the pre-Holocaust community).

- Immigration is enriching Europe's ethnic mosaic but also presenting economic and security dilemmas for the European countries.

- From the beginning of the sixteenth century until late in the nineteenth century, Europe was at the center of global patterns of colonization and foreign settlement, long-distance trade, and agricultural and industrial innovation. During this time, Europeans diffused crops and animals between the Old and New Worlds.

- The Industrial Revolution originated in Europe, with energy derived from coal and factory technology focused on textiles and iron. Industrialization and colonization launched Europe to global economic and political supremacy.

- Recent decades have seen a global shift in power away from Europe. War dislocation, rising nationalism, the ascendancy of the United States, a shift in world manufacturing patterns, and new energy sources have combined to diminish Europe's global centrality. Europe is nevertheless a very strong force in world economic, political, and social affairs, and its peoples are among the most prosperous in the world.

- Europe's economy is postindustrial, making the transition from energy-hungry, labor-costly, and polluting industries to leaner high-tech industries and services. This shift has caused unemployment.

- European nations often try to protect their domestic industries and have been involved in trade wars with the United States, which likewise wants to protect its industries. Europe is under pressure from international trade agreements to reduce its trade barriers and subsidies that protect its domestic agriculture and other industries.

- Eastern Europe has long been much poorer than Western Europe.

- In recent decades, Europe has been reorganizing itself to ensure that nothing like the two world wars will happen again and to strengthen its economies. Its principal military alliance, NATO, is growing in membership and redefining its focus toward peacekeeping.

- The most important development is the growth of the 27-member European Union (EU), a supranational organization pooling the economic and human resources of its member countries. There have been obstacles to its achievement of common EU policies in money matters, defense, and foreign affairs.

- The Euro, the common currency used by many members of the European Union, was endangered by overspending in the "PIGS" countries. Germany and France had to come to the rescue of Europe in its debt crisis.

- The U.S. war in Iraq highlighted major differences among EU members, and overall there are several marked differences in the ways Europeans and Americans view the world.

- The nations that make up the European Core are the United Kingdom, Ireland, France, Germany, the Low Countries (the Netherlands, Belgium, and Luxembourg), Austria, and Switzerland. The giant nations in this region are Germany, the United Kingdom, and France. They are the primary engines for the economic momentum in this part of the world. Three microstates—Andorra, Monaco, and Liechtenstein—are also included here.

- The United Kingdom is made up of England, Scotland, Wales, and Northern Ireland. It forms the greater part of two major islands: Great Britain and Ireland, which are called the British Isles. The island of Great Britain has distinctive highlands and lowlands regions.

- Most of the eighteenth- and nineteenth-century industrial and urban development of Britain related to the location of coal and iron ore, the two leading resources of the Industrial Revolution. London arose on the Thames estuary with neither coal nor iron as a local resource and is today a major global financial center.

- Although the Republic of Ireland has recently experienced strong economic growth based on high-tech industries, Northern Ireland's growth has been restrained by a long-standing but perhaps nearly resolved conflict between historically indigenous Roman Catholics and more recent residents of Protestant, English background.

- The EU countries have established an ambitious schedule to replace fossil fuel energy with wind and other renewable energy alternatives. Britain seeks leadership in alternative energy technology and cuts in greenhouse gas emissions.

- Germany, restored when West Germany reunified with East Germany in 1990, has the largest economy and population of all Europe. German urban centers have a much broader distribution than the industrial cities of France and even Britain. As in Britain, these cities were generally located by the availability of coal and other resources and of rivers for transport, and as in Britain, they have been deindustrializing in recent decades. Germany's economy has had to bear the costs of reunification and the burdens of being a welfare state. The former East Germany remains poorer than the West, but new industrial investment is drawn there by low labor costs.

- The core region of Europe is ringed by a periphery of three subregions: Northern, Eastern, and Southern Europe. These have historically been less integrated into the Core, and the eastern and southern countries have been less prosperous.

- There is an imbalance between northern and southern Italy, with the north far more prosperous. Some northerners want to develop an autonomous or independent region of Padania.

- As the result of a 2004 referendum in which both sides of divided Cyprus voted on whether they should be reunited, the Greek south (which rejected reunification) was allowed to join the European Union, whereas the Turkish north (which favored reunification) was not allowed to join.

- Eastern Europe is made up of Estonia, Latvia, Lithuania, Poland, the Czech Republic, Slovakia, Hungary, Romania, Bulgaria, Slovenia, Croatia, Bosnia and Herzegovina, Serbia, Montenegro, Albania, and Macedonia. These countries have been shaped powerfully by struggles between stronger countries and make up a geopolitical shatter belt. Most critical was their role as satellite states of the Communist Soviet Union between World War II and 1990.

- Collectivization under Communist rule during the Soviet era from the late 1940s until the 1980s was the agricultural model in Eastern Europe. Privatization of farmland has taken place since the fall of communism.

- Mining, iron and steel, machinery production, construction materials, and electrical power were the highlights of the industrial effort during the years of Soviet domination, with the majority of raw materials coming from Russia. The shift to a market economy left many firms uncompetitive, and many were abandoned. Both Western European countries and China have taken advantage of inexpensive labor in Eastern Europe to outsource some production in the region, and many economies have rebounded.

- Politically, the disintegration of the former Yugoslavia was the most disabling phenomenon of the post-Soviet era in Eastern Europe. Yugoslavia was dismantled and replaced with six countries, some with precarious rivalries among ethnic groups. NATO and the UN have used various military and diplomatic means to prevent further balkanization of the countries. Kosovo declared its independence in 2008, but it is not recognized by Russia and some other countries.

Key Terms + Concepts

Altaic languages (p. 79)
Attila Line (p. 102)
"baby bounty" (p. 70)
balkanization (p. 107)
"banana war" (p. 86)
Basques (p. 100)
Benelux (p. 65)
"big bang" (p. 87)
biogeography (p. 82)
"birth dearth" (p. 69)
Catholic Republicans (p. 95)
Celtic Tiger (p. 93)
Christianity (p. 79)
"City of Light" (p. 95)
coal (p. 91)
coke (p. 83)
Cold War (p. 88)
collectivization (p. 105)

Columbian Exchange (p. 82)
Common Market (p. 85)
Commonwealth of Nations (p. 93)
communism (p. 104)
Dayton Accord (p. 107)
deindustrialization (p. 85)
delta (p. 77)
devolution (p. 80)
direct rule (p. 94)
enosis (p. 102)
ETA (Basque Homeland and Liberty) (p. 100)
ethnic cleansing (p. 107)
Eurabia (p. 71)
Euro (p. 86)
European Core (p. 65)
European debt crisis (p. 88)

European Economic Community (EEC) (p. 85)
European Greenbelt (p. 89)
European Periphery (p. 99)
European Union (EU) (p. 70)
Eurozone (p. 87)
"food fights" (p. 86)
"Frankenfoods" (p. 86)
genetically modified (GM) foods (p. 86)
genetically modified organism (GMO) foods (p. 86)
glacial deposition (p. 74)
glacial scouring (p. 74)
glaciation (p. 73)
Good Friday Agreement (p. 95)
Gulf Stream (p. 73)
Gypsies (p. 79)

Holocaust (p. 81)
Ice Age (p. 74)
Indo-European languages (p. 77)
  Baltic languages (p. 78)
  Celtic (p. 77)
  Germanic languages (p. 77)
  Greek (p. 77)
  Indic languages (p. 79)
  Romance languages (p. 77)
  Slavic languages (p. 78)
Irish Republican Army (IRA) (p. 94)
Iron Curtain (p. 89)
Islam (p. 81)
Islamophobia (p. 71)
isostatic rebound (p. 74)
Judaism (p. 81)
Liga Veneta (p. 101)

loess (p. 73)

major cluster of continuous settlement (p. 91)

Marshall Plan (p. 89)

Mezzogiorno (p. 101)

microstates (p. 65)

Moors (p. 81)

Muslims (p. 81)

nationalism (p. 83)

national minorities (p. 79)

NIMBY (p. 99)

North Atlantic Drift (p. 73)

North Atlantic Treaty Organization (NATO) (p. 89)

Northern League (p. 101)

North European Plain (p. 73)

outsourcing (p. 106)

Padania (p. 101)

PIGS (p. 88)

postindustrial (p. 84)

potato famine (p. 94)

precautionary principle (p. 90)

primate city (p. 95)

privatization (p. 105)

Protestant Reformation (p. 80)

Protestant Unionists (p. 95)

Protestantism (p. 80)

Roma (Gypsies) (p. 79)

Roman Catholic Church (p. 79)

Schengen Agreement (p. 89)

Schengenland (p. 89)

shatter belt (p. 104)

Silk Road (p. 81)

Sinn Fein (p. 95)

site (p. 70)

situation (p. 70)

socialization (p. 104)

Solidarity (p. 105)

Soviet satellites (p. 103)

subsidies (p. 84)

supranational organization (p. 85)

tariffs (p. 84)

the Troubles (p. 94)

trade wars (p. 86)

undocumented workers (p. 72)

Uralic languages (p. 79)

Warsaw Pact (p. 89)

welfare state (p. 85)

westerly winds (p. 73)

"world city" (p. 93)

Review Questions

1. Why is Europe usually treated as a separate region?

2. What terms and trends best describe Europe's population today?

3. Why is immigration so critical to Europe's demographic future, and what problems does immigration pose?

4. What is unusual about Europe's coastline?

5. What are some of Europe's other major physical and environmental characteristics? Why are they consistently described as diverse?

6. What roles have rivers played in Europe's development, and which rivers are the most important ones? Where is industrial and other economic activity concentrated in Europe and why?

7. What are the dominant languages, religions, and other ethnic traits of Europe?

8. What factors led to Europe's global dominance in economic and political affairs? What impacts did that dominance have on other peoples and environments?

9. Explain the origins and significance of the Industrial Revolution, including the role of resources in that revolution. How does Europe's present urban pattern reflect its industrial past?

10. What are Europe's main economic traits? How have these changed in recent decades?

11. What happened to Europe in the twentieth century? How are Europe's political and economic institutions of today trying to create a different Europe?

12. What are the main goals and principles of the European Union? What successes and difficulties has the organization and the Eurozone had?

13. What are some of the differences between the Europeans and Americans?

14. What are the countries of the European Core, and on what basis have they been designated as belonging to the core?

15. What was the spatial relationship between coalfields and cities in industrial Britain and mainland Europe?

16. What are the political affiliations of Ireland, Northern Ireland, England, Scotland, and Wales?

17. What are "the Troubles" of Northern Ireland?

18. What is a primate city? What examples exist in Europe and elsewhere?

19. What was the impact of German reunification on the country's economy?

20. What threatens Italy's economic future?

21. What are the major countries and ethnolinguistic groups of Eastern Europe?

22. Why is Eastern Europe considered a shatter belt in geopolitical terms?

23. How did communism shape agriculture and industry in Eastern Europe? What economic processes have occurred there since 1990?

24. What are the main forces behind the breakup of Yugoslavia and the current borders and ethnic components of its successor countries?

25. Who are the Roma and what are their unique attributes? Why and how have other ethnic groups in Europe discriminated against them?

Notes

1. Elisabeth Rosenthal, "European Union's Plunging Birthrates Spread Eastward." *New York Times,* September 4, 2006, p. A3.

2. Quoted in Judy Dempsey, "New Interior Minister Revives a Debate: Can Muslims be True Germans?" *New York Times*, March 7, 2011, p. A5.

3. Quoted in Dwight Garner, "A Turning Tide in Europe as Islam Gains Ground." *New York Times*, July 30, 2009, pp. C-1, 4.

4. Simon Kuper, "The Myth of Eurabia." *Financial Times Life and Arts*, October 4, 2009, p. 2.

5. David Levinson, *Ethnic Groups Worldwide* (Westport,Conn.: Oryx Press, 1998), p. 1.

6. Quoted in Steven Erlanger, "Europe Steers into a Zone of Uncertainty." *New York Times*, September 8, 2011, pp. 1, 8.

7. Quoted in A. A. Byatt, "What Is a European?" *New York Times Magazine*, October 13, 2002, p. 58.

8. Robert Kagan, *Of Paradise and Power: America and Europe in the New World Order* (New York: Knopf, 2003), p. 1.

9. Terry G. Jordan and Bella Bychova Jordan, *The European Culture Area: A Systematic Geography* (Lanham, Md.: Rowman and Littlefield, 2002), p. 402.

10. Ibid., p. 163

11. Ibid., p. 148

12. Philip Stephens, "Shrunken Ambitions." *Financial Times*, April 28, 2010, p. 9.

13. Quoted in Kevin J. O'Brien, "Last Out, Please Turn Off the Lights: Poor Economy Is Driving East Germans from Home." *New York Times*, May 28, 2004, pp. W1, W7.

14. These perceptions of Germans are discussed in Quentin Peel, "An Unequal Union." *Financial Times*, October 1, 2010, p. 11.

15. Ian Cowell, "Menacing the Land, but Promising to Rescue the Earth." *New York Times*, July 4, 2005, p. A4.

Online Resources

 CourseMate: Make the most of your study time by accessing everything you need to succeed in one place. Read your textbook, take notes, review flashcards, watch videos, complete activities, take practice quizzes, and more—online with CourseMate. Log in at **www.cengagebrain.com**.

"The Russian drama began at the end of 1991, when the Soviet Union mercifully ended. Russia and 14 other new countries emerged from the ruins of the Soviet Union. Every one of those 15 new states faced a profound historical, economic, financial, social and political challenge."

—Jeffrey Sachs, Director, The Earth Institute

C. Mayhew & R. Simmon (NASA/GSFC), NOAA/NGDC, DMSP Digital Archive

Joe Hobbs

Above, the great landmass of Russia and neighboring lands at night. Note the string of lights along the Trans-Siberian Railway. Left: The beautiful onion domes of Russian Orthodox churches are among the most characteristic features of Russian architecture.

Russia and the Near Abroad

5

Take a moment to appreciate the size of the region you are about to study. Russia and the countries along its perimeter make up an area of breathtaking scale. To get started, use the maps within this book's front and back covers, Google Earth, a globe, or the chapter-opener image of this region. You are about to explore not only a huge region, but also one of enormous consequence in world affairs.

This vast geographic realm is made up of Russia; the kindred Slavic countries of Ukraine, Belarus, and Moldova; the decidedly non-Russian but heavily Russian-influenced "stans" of central Asia—Kazakhstan, Uzbekistan, Turkmenistan, Kyrgyzstan, and Tajikistan; and the countries of Georgia, Armenia, and Azerbaijan in the fractious Caucasus. These 12 countries, known here as "Russia and the Near Abroad," are also the member states of the **Commonwealth of Independent States (CIS)**, an economic and political association anchored by Russia.

chapter objectives

This chapter should enable you to

- Appreciate the environmental obstacles to development in vast areas of the world's largest country and nearby nations.

- Become familiar with the ethnic complexity of a huge region until recently held together as a single country.

- Learn the significant milestones in the historical and geographic development of Russia and the Soviet Union, some of them accompanied by unimaginable loss of life.

- Recognize the differences between command and free-market economies and the post-Soviet difficulties in shifting from one to the other.

- Understand the reasons for the reversal of Russia's progress through the demographic transition.

- Come to know the geopolitical and ethnic forces threatening the unity of Russia and pitting various groups and countries within and outside the region against one another.

- Recognize core and peripheral subregions of Russia and the Near Abroad.

- Identify geographic obstacles (for example, vast distances) and opportunities (for example, transpolar routes) for Russia in east–west trade.

- Appreciate the importance of fossil fuels in Russia's economy, and how overdependence on natural resources poses risks to the country's development.

- Examine the geopolitical considerations that go into planning for oil and natural gas pipelines in this region.

- See how free enterprise in former Communist countries has introduced a new set of environmental problems.

- View the transformation of the Aral Sea as a legacy of Soviet colonialism in Central Asia.

Geographers do not have an easy task in defining and naming regions that have emerged from the breakup of the Soviet Union. As discussed in Chapter 1, regions are organizing tools, not facts on the ground. There is no acknowledged best way to classify this region. Some geographers have chosen to cleave the central Asian "stans" from Russia, establishing them as an entirely separate region, or to include them as part of a greater Middle Eastern region. Some still consider the Baltic countries (Estonia, Latvia, and Lithuania) as part of the greater Russian realm. The variety of geographers' names for the region suggests how unsettled the terminology is: "Post-Soviet Region," "Russia and Its Neighbors," "Russia and the Newly Independent States," "Russia and Neighboring Countries," "Former Soviet Union" (with the acronym FSU), and this book's "Russia and the Near Abroad" are among them (see Insights on page 115). The must-read book on the geography of this region is Mikhail Blinnikov's *A Geography of Russia and Its Neighbors* (Guilford, 2011).

From 1917 to 1991, the huge region known in this text as Russia and the Near Abroad, plus the three Baltic countries, made up a single Communist-controlled country known as the Union of Soviet Socialist Republics (USSR) or Soviet Union (•Figure 5.1). The Russian-dominated government in Moscow controlled the affairs of many non-Russian peoples in the country, and after World War II also effectively controlled its Communist satellite countriesin Eastern Europe (those satellite countries were discussed in Chapter 4). For four decades, the Cold War between the Soviet bloc of nations and the U.S.-led Western bloc dominated world politics. The Union of Soviet Socialist Republics and the United States vied in a geopolitical struggle for the allegiance of newly independent countries and other nations around the world following World War II.

The stakes in this contest were high, and at times the chilly relations of the Cold War threatened even to bring on a nuclear conflict with unimaginable consequences. Each side had massive nuclear stockpiles ready to be unleashed at a moment's notice. The 11th hour of such a nuclear showdown was reached in the Cuban Missile Crisis of 1962, when Russians and Americans prepared for the end (see page 373). In recently released audio tapes, American First Lady Jackie Kennedy can be heard pleading with her husband, U.S. President John Kennedy, to be together with their children at the White House when the Soviet nuclear warheads detonated there. Each side regularly denounced the other with hostile rhetoric. American President Ronald Reagan attached moral significance to the struggle, dubbing the Soviet Union the **"Evil Empire"** in 1983.

Remarkably, less than a decade after President Reagan used that term, he and a Soviet leader named Mikhael Gorbachev stood together shaking hands and smiling in an agreement to reduce their nuclear stockpiles. And almost unbelievably, in 1991, while Gorbachev was still the Soviet leader, the giant country he ruled split into 15 independent countries (see •Table 5.1). Russia—now officially known as the Russian Federation—remained by far the largest of these in area, population, and political and economic influence (see Insights, page 115).

Russia emerged from the fragmentation of the Union of Soviet Socialist Republics as a great power, but it was no longer a superpower. Russia suddenly found itself struggling to maintain influence in the other countries that emerged from the ashes of the Soviet Union. From the Russian perspective, the other 14 countries constituted the **Near Abroad**, a zone in which Russia had to preserve and exert its special interests and influence.

The choice to use "Russia and the Near Abroad" in this book was made not to impose a Russian-centered view of the region, as the name might imply, but to reflect the enormous power that Russia wields (or would like to wield) there. The

| **Regional Names of Russia and the Near Abroad**

Here are some useful geographic terms to know when studying this region. The region of Russia and the Near Abroad consists of what used to be the Union of Soviet Socialist Republics, also known as the Soviet Union or USSR, minus the Baltic countries of Estonia, Latvia, and Lithuania. The Union of Soviet Socialist Republics came into existence in 1922, following the overthrow of the last Romanov tsar in the Russian Revolution of 1917 and the subsequent civil war. Prerevolutionary Russia is known as Old Russia, tsarist Russia, Imperial Russia, or the Russian Empire.

The name *Russia* now refers to the independent country of Russia. It is known politically as the Russian Federation, which was the largest of the 15 Soviet Socialist Republics (Union Republics, or SSRs) that made up the Soviet Union. The full name of the Russian Federation during the Communist period was the Russian Soviet Federated Socialist Republic, or RSFSR; the name appears on many older maps.

The loosely aligned Commonwealth of Independent States (CIS) was formed by 12 of the 15 former Union Republics late in 1991. Estonia, Latvia, and Lithuania are not members of this organization. They distanced themselves strongly from Russia, and in 2004 joined the European Union. They are characterized as European in Chapter 4.

The area west of the Ural Mountains and north of the Caucasus Mountains has been known historically as European Russia. The Caucasus and the area east of the Urals have been called Asiatic Russia, Soviet Asia, or the eastern regions. Transcaucasia is the region of the mountains plus the area south of the Caucasus Mountains. Siberia is the general name for the huge area between the Urals and the Pacific; and Central Asia is the arid area occupied by the five countries with large Muslim populations immediately east and north of the Caspian Sea. They were known collectively in pre-Soviet times as "Turkestan" and are known informally today as "the stans" because all the country names end in -*stan*, Persian for "land of."

KEY TO NUMBERED COUNTRIES
1 Moldova
2 Georgia
3 Armenia
4 Azerbaijan
5 Kyrgyzstan
6 Tajikistan

• **Figure 5.1** Russia and the Near Abroad.

Table 5.1   Russia and the Near Abroad: Basic Data

Political Unit	Area (thousands; sq mi)	Area (thousands; sq km)	Estimated Population (millions)	Estimated Population Density (sq mi)	Estimated Population Density (sq km)	Annual Rate of Natural Increase (%)	Human Development Index	Urban Population (%)	Per Capita GNI PPP ($US)
Slavic States and Moldova	**6,919.1**	**17,920.4**	**202.1**	**29**	**11**	**−0.3**	**0.747**	**72**	**15,000**
Belarus	80.2	207.7	9.5	118	46	−0.3	0.756	75	12,740
Moldova	13.0	33.6	4.1	315	122	−0.1	0.649	42	3,010
Russia	6,592.8	17,075.3	142.8	22	8	−0.2	0.755	74	18,330
Ukraine	233.1	603.7	45.7	196	76	−0.4	0.729	69	6,180
Caucasus Region	**71.8**	**185.9**	**16.6**	**224**	**86**	**0.8**	**0.711**	**55**	**7,230**
Armenia	11.5	29.8	3.1	270	104	0.4	0.716	64	5,410
Azerbaijan	33.4	86.5	9.2	275	106	1.1	0.700	54	9,020
Georgia	26.9	69.6	4.3	160	62	0.4	0.733	53	4,700
Central Asia	**1,542.3**	**3,994.5**	**63.3**	**38**	**15**	**1.8**	**0.665**	**40**	**5,000**
Kazakhstan	1,049.2	2,717.4	16.6	16	6	1.4	0.745	54	10,320
Kyrgyzstan	76.6	198.4	5.6	73	28	1.9	0.615	35	2,200
Tajikistan	55.3	143.2	7.5	136	52	2.4	0.607	26	1,950
Turkmenistan	188.5	488.2	5.1	27	10	1.4	0.686	47	6,980
Uzbekistan	172.7	447.3	28.5	165	64	1.9	0.641	36	2,910
Summary Total	**8,533.2**	**22,100.8**	**282.0**	**33**	**13**	**0.2**	**0.726**	**64**	**12,300**

Sources: World Population Data Sheet, Population Reference Bureau, 2011; Human Development Report, United Nations, 2011; World Factbook, CIA, 2011.

Table 5.2   Russia and the Near Abroad: Metropolitan Populations, in Millions

1	Moscow, Russia	14.9
2	St. Petersburg, Russia	4.8
3	Tashkent, Ukraine	3.2
4	Kiev, Ukraine	3
5	Baku, Azerbaijan	2.3
6	Minsk, Belarus	1.8
7	Nizhny Novgorod, Russia	1.8
8	Donetsk, Ukraine	1.6
9	Kharkiv, Ukraine	1.6
10	Volgograd, Russia	1.5

Population in millions.

© Cengage Learning 2013

former Soviet countries have enduring and often uncomfortable strategic and economic associations with Russia. For example, the Soviet-era oil refinery and pipeline system still links Russia with many countries of the Near Abroad. As we will see, Russia has periodically shut off the pipelines supplying oil and natural gas, or has increased the prices of these commodities,

to obtain political concessions from some of these nations. The needs of many of the Near Abroad countries to buy Russian energy or export their own energy resources across Russian territory give the countries an incentive to maintain peaceful political relations with Russia. The countries of the Near Abroad generally do not want to trigger Russian military intervention, as this would potentially open the door to Russia re-exerting its control. In this chapter, you will get acquainted with many of these issues in their geographical context, which is simply enormous.

5.1   Area and Population

With an area of 8.5 million square miles (22.1 million sq km), the region of Russia and the Near Abroad is the largest in this book. Russia is about twice as big as the United States, including Alaska. You can get a sense of this comparison by looking at Figure 5.3 on page 120. A good indication of its staggering size is the fact that this region (and even Russia by itself) spans 11 time zones; in comparison, the United States, from Maine to Hawaii, spans 7. For most Russians, having a land mass that sprawls unbroken across 11 time zones has been a matter of national pride, rather like the way Texans or Alaskans feel about the size of their states. Many Russians were dismayed when, in

2009, their president proposed (in the interest of the country's economy) that Russia should cut back to just four time zones. This would make it easier for businesses in Moscow, for example, to communicate with offices and customers in the Russian Far East on the same business day.

The eventful geopolitical history of Russia and the Near Abroad has given this region land frontiers with 15 countries in Eurasia (Figure 5.1). Between the Black Sea and the Pacific, the region borders Turkey, Iran, Afghanistan, China, Mongolia, and North Korea. Pakistan and India also lie close by. In the Pacific Ocean, narrow water passages separate the Russian-held islands of Sakhalin and the Kurils from Japan. In the west, the region has frontiers with Romania, Hungary, Slovakia, Poland, Lithuania, Latvia, Estonia, Finland, and Norway.

Although the region of Russia and the Near Abroad is huge, it does not have a lot of people. Much of this vast region is sparsely populated (•Figure 5.2). Great stretches of economically unproductive terrain separate populated areas from one another. The mountainous frontiers in Asia have few inhabitants. As we will see, shrinking populations pose many challenges for Russia and, to a lesser extent, some of its neighbors. Mindful of the maxim that "possession is nine-tenths of the law," Russia worries that there are not enough Russians in its far eastern region along the frontier with overcrowded China. You can see the population distributions clearly in Figure 5.2. In contrast with the lightly populated east, you can see that to the west, on the frontier between the Black and Baltic Seas, international boundaries pass through populous lowlands. These have long been disputed territory between Russia and other countries.

Russia has the region's largest population, with about 143 million people. The next largest country is Ukraine, with about 46 million. Uzbekistan is the most populous central Asian nation, with about 28 million people. The regions' other countries trail far behind, with populations generally under 15 million apiece. Population growth rates are highest, around 1.8 percent

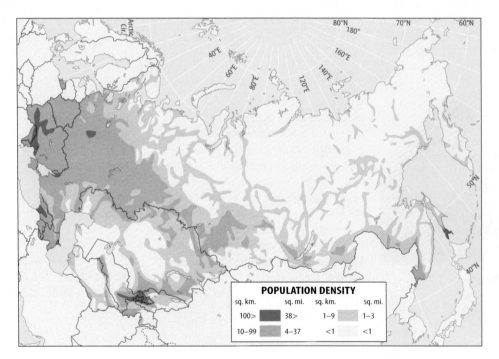

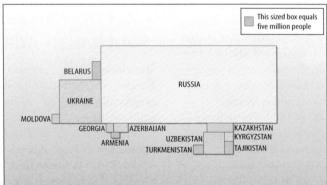

• **Figure 5.2** Population distribution (above) and population cartogram (below) of Russia and the Near Abroad.

Source: The Oxford School Atlas edited by Patrick Wiegand (OUP, 1997), copyright © Oxford University Press. Used by permission of Oxford University Press.

Demographically, Russia ought to be a postindustrial country like the countries of Western Europe or like the United States. It should have passed into the fourth state of the demographic transition, where both birth rates and death rates have fallen parallel to one another (like the United States in Figure 5.B). But it didn't happen that way. Look at the trend lines for Russia in •Figure 5.B, where a startling intersection and trend have taken place.

Russia's economic health plummeted after the breakup of the Soviet Union, and so did human health. Along with the economic downturn came more unemployment, a decline in health and other services, increasing crime rates, and a growing sense of despair at the individual level. These gave rise to some very unhealthful habits among Russians, which have been stubbornly persistent for more than two decades.

• **Figure 5.A** Alcoholism is a major killer in Russia.

One of the biggest problems is an increase in alcoholism. Russians do not imbibe much in beer and wine, the customary adult beverages of Western Europe. Instead they go for the harder stuff, especially vodka (Figure 5.A). I worked as a tour leader on Russian ships every summer in the 1990s and found my Russian companions to be invariably warm and hospitable. Socializing includes good-spirited ritual drinking of vodka. The alcohol is usually served ice-cold in small shot glasses, in so many rounds that the unaccustomed drinker may easily end up under the table. One toast is followed by another, and then another—and not always just in the late afternoon or evening. Vodka at lunch and sometimes even at breakfast and in between. . . .

The figure for annual consumption of hard liquor is an eye-blearing 5 gallons per person (that statistic would include every man, women, and child in the country). Researchers working in Siberian cities report that alcohol is the cause of more than half of all deaths of people ages 15–54, from accidents, violence, and alcohol poisoning. Alcohol poisoning kills an estimated 40,000 Russians (mainly men) each year, compared with about 300 in the United States. One statistic after another points to alcohol-related problems. Perhaps the worst one is the Russian government's own estimate that 500,000 Russians die every year from causes directly related to or aggravated by alcohol. That figure represents more than half of all premature deaths in Russia. Russia's government has repeatedly tried, and failed,

to crack down on alcohol abuse through public awareness campaigns, higher taxes, and restricted sales. There is always a black market producing cheap vodka.

There are other problems too. Smoking has long been a national pastime; an estimated 60 percent of Russian men smoke. Organized and petty crime have resulted in an alarmingly high incidence of physical violence. Infections of HIV/AIDS, often accompanied by tuberculosis, have surged through the population of intravenous drug users and their sexual partners. Russia's geography is part of the problem: it is uncomfortably close to the world's largest producer of heroin, Afghanistan. Highly potent and inexpensive heroin makes its way swiftly out of Afghanistan and through the central Asian countries into Russia.

Heroin addicts are left to their own devices to wean themselves off the drug through the rather primitive "cold turkey" method. This is not very effective; the relapse rate is high. The West's common treatment of addiction with methadone is not practiced in Russia, where the health care system suffers from a vast array of problems. The universal system of health care under the Soviets was not efficient, but a health care system close to collapse replaced it, and hospitals are ill-equipped to help stem a rising tide of deaths from a variety of causes.

After 1990, Russia's death rate rose about 25 percent, and it has persisted as one of the world's highest levels (14 per 1,000 in 2011), surpassed only by many countries in Africa,

annually, among the predominantly Muslim populations of the central Asian countries. At the other end of the spectrum, Russia, Ukraine, and Belarus are losing population at a rate of up to 0.4 percent per year. Russia's population is declining at an astonishing 500,000 people each year. Russia's annual population loss is the highest in the world and has seldom been seen on Earth except in times of warfare, famine, and epidemic.

But there is no war and no shortage of food. The depopulation is instead due to an imbalance between fertility and mortality, which has its roots in the declining quality of life that occurred after the breakup of the Soviet Union (see Insights, this page). In an atmosphere of economic and political uncertainty, couples have been choosing to have fewer children, and there have been more broken marriages. More disturbingly, death rates have been soaring.

5.2 Physical Geography and Human Adaptations

Stretching nearly halfway around the globe in northern Eurasia, the region of Russia and the Near Abroad is geographically gifted by its sheer size. But it is also geographically challenged. Most of the region has some combination of cold temperatures, infertile soils, marshy terrain, aridity, and ruggedness. These natural conditions are more comparable to those of Canada than to those of the United States. Four-fifths of the total area is farther north than any point in the conterminous United States (•Figure 5.3). Interaction with this demanding environment was a major theme in Russian and Soviet expansion and development.

and, outside Africa, only by Afghanistan and by other countries of the former Soviet Union and Eastern Europe where people have similar unhealthful habits. This clearly is not the benign demographic transition, in which a country's population declines because of falling birth rates that reflect growing affluence (see the graph and discussion on pages 52–53). Instead, Russia is doing something completely unprecedented in the world's demographic experience: it is moving *backward* through the transition. Not only have death rates soared but, as quality of life has deteriorated, birth rates have also plummeted. The conditions that would promote higher fertility are in short supply. Fully 80 percent of Russian marriages end in divorce. A physician cited "male/female estrangement and a loss of family cohesion" among the reasons for

Russia's plummeting birth rate.[1] When women do become pregnant, many choose abortion as a common method of birth control.

The disturbing trend lines of increasing death rates and falling birth rates in Russia intersected in the mid-1990s, forming the grim graphic that demographers dubbed the **"Russian cross"** (•Figure 5.B). It is an ominous intersection. The odd shape of Russia's age structure diagram in •Figure 5.C also depicts the declining population. Many projections put Russia's population at close to 100 million in 2050, down nearly 30 percent from 2011.

Alarmed, the government in Moscow has been trying to incentivize higher fertility through cash payouts to mothers, extended maternity leaves, and child care benefits, similar to the incentives offered in Italy and Japan. In 2007, Russia's President introduced Family Contact Day, a new national holiday on which Russians were encouraged to stay home and make babies. These efforts to grow Russia's fertility have shown few signs of success. In order to grow the population, Moscow is also trying to lure Russian expatriates home.

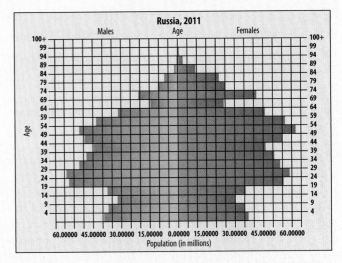

• **Figure 5.C** Russia's age structure diagram. Russia's population topped out at 149 million when the Soviet Union broke up in 1991 and has been falling ever since. Note the relatively low numbers of the country's young people. World War II's impact can be clearly seen in the imbalance between men and women at the top of the diagram; the men were killed.

Source: Based on U. S. Census Bureau, International Database.

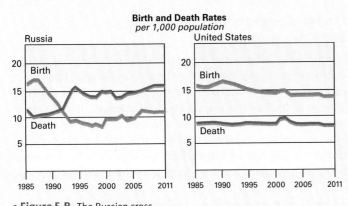

• **Figure 5.B**   The Russian cross.

Source: Data from World Factbook, CIA, 2011.

The Roles of the Climates and Vegetation

The region of Russia and the Near Abroad has a harsh climatic setting. Severe winter cold, short growing seasons, drought, and hot, crop-shriveling winds are major disadvantages. But there are also advantages in the form of good soils, the world's largest forests, natural pastures for livestock, diverse wild fauna, and great reserves of fossil fuels. Many of these resources are hard to reach because of the logistical difficulties posed by climatic extremes and vast distances.

Most parts of the region have continental climatic influences, with long, cold winters, short, warm summers, and low to moderate precipitation. Severe winters, for which Siberia—roughly the eastern two-thirds of Russia—is particularly infamous, prevail because the region is generally at a high latitude

and has few of the moderating influences of oceans. As we have seen in Chapter 4, Europe benefits from those moderating influences. Westerly winds from the Atlantic reach only the western portion of Russia and the Near Abroad, and their moderating effects become weaker toward the east, where it is much colder.

The most extreme continental climate of the world is here. Russia cannot boast the lowest temperatures on Earth; those are claimed by places in the high mountains of Antarctica. But Russia can claim the lowest temperature ever recorded in an inhabited region. The lowest temperature recorded in the Northern Hemisphere reached *minus 90°F* (minus 68°C) in the Siberian settlement of Verkhoyansk (located at 67°N, 135°E). That location is well inland from the shores of northern Siberia, so even in this context you can appreciate the moderating

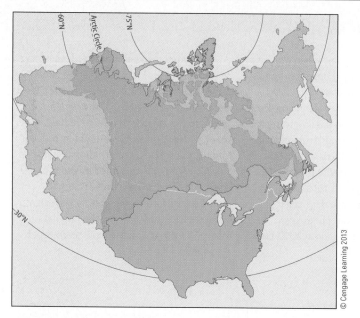

• **Figure 5.3** Russia and the Near Abroad compared in latitude and area with the continental United States and Canada.

influence of ocean waters. Even the Arctic Ocean is relatively warm, so Siberia's northern coast is not as cold as the region's interior, further to the south.

Russia and the Near Abroad has a large amount of arable land; Russia has the third largest amount of the world's farmable land, behind only the United States and India. But several factors make agriculture difficult in large parts of this region. The average frost-free season of 150 days or less in most areas (except the extreme south and west) is too short for many crops to mature. Aridity and drought make problems too. The annual average precipitation is less than 20 inches (50 cm) nearly everywhere except in the extreme west, along

the eastern coast of the Black Sea and the Pacific coast north of Vladivostok, and in some of the higher mountains. Most places are relatively warm during the brief summer, and the southern steppes and deserts are hot. Thankfully, while the summer is short, the high latitudes bring longer days in the summer, and these encourage plant growth. Summer is also a time for more outdoor activities. Russians are passionate campers and sun worshipers, and they take every opportunity to "get away from it all" to enjoy the fleeting pleasures of summer. Urban residents never lose the traditional Russian passion for nature, and many retreat as much as possible to their rural cabins called *dachas*. These have an important place in Russia's agriculture; see page 146.

There are five main climatic belts in the region of Russia and the Near Abroad: tundra, subarctic, humid continental, steppe, and desert. In •**Figure 5.4**, you can see how each of these belts with its associated vegetation and soils succeeds the other from north to south. There are also smaller scattered areas of subtropical, Mediterranean, and undifferentiated highland climates. Most human activities take place in the subarctic, humid continental, and steppe climatic zones. •**Figure 5.5** depicts land uses all across the vast region of Russia and the Near Abroad.

Unlikely Sources of Global Warming in Russia's Wilderness

Much of the tundra and subarctic climatic zones are permanently frozen to an average depth of three meters. The frozen ground, called **permafrost**, makes construction difficult. Heat generated by buildings melts the upper layers of permafrost and causes foundations and walls to sink and tilt. Most buildings are elevated on pilings to reduce this risk, like the building in the photograph of •**Figure 5.6**. Pipelines carrying crude oil (which is hot when it comes from the ground) would likewise

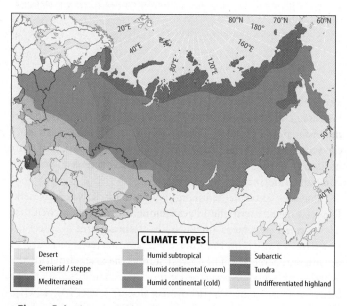

CLIMATE TYPES

Desert	Humid subtropical	Subarctic
Semiarid / steppe	Humid continental (warm)	Tundra
Mediterranean	Humid continental (cold)	Undifferentiated highland

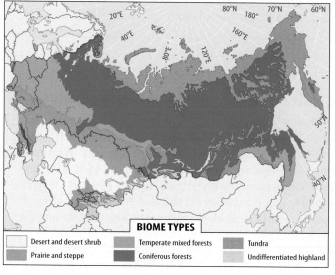

BIOME TYPES

Desert and desert shrub	Temperate mixed forests	Tundra
Prairie and steppe	Coniferous forests	Undifferentiated highland

• **Figure 5.4** Climates (left) and biomes (right) of Russia and the Near Abroad.

Sources: (left) Based on Rand McNally's Classroom Atlas, 2003. (right) Based on World Wildlife Fund ecoregions data, 1999.

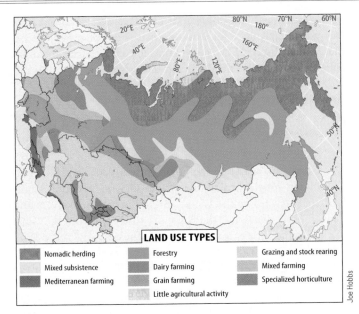

LAND USE TYPES

- Nomadic herding
- Mixed subsistence
- Mediterranean farming
- Forestry
- Dairy farming
- Grain farming
- Little agricultural activity
- Grazing and stock rearing
- Mixed farming
- Specialized horticulture

• **Figure 5.5** Land use in Russia and the Near Abroad.

Source: The Oxford School Atlas edited by Patrick Wiegand (OUP, 1997), copyright © Oxford University Press. Used by permission of Oxford University Press.

melt the permafrost and sink, and so they are heavily insulated or built on elevated supports.

The subarctic climate zone corresponds with the Russian taiga, or northern coniferous forest (also known as boreal forest). This is not an agriculturally productive zone; the soils are mainly acidic, infertile **spodosols** (from the Greek word for "wood ash" and known in Russian as *podzols*). But conditions are ideal for coniferous tree growth. Representing 20 percent of the world's forested land, this is the largest forest on Earth. Half of the world's evergreen trees are here, with enormous stretches of pine, spruce, and fir. Economically, this is a region of major timber reserves. Russia exports raw logs to markets in Sweden, Finland, Japan, South Korea, and increasingly, China. Many observers fear that Russia, in its

• **Figure 5.6** In the High Arctic, buildings must be erected on pilings so that they do not melt the permafrost below. This is the Russian coal-mining settlement of Barentsburg in Norway's Svalbard (Spitsbergen) Archipelago.

• **Figure 5.7** The taiga in Russia is the Earth's largest continuous forest biome, and Russia's economic growth is based in part on its exploitation.

drive to advance the economy, will deplete the taiga as an easy cash crop (•Figure 5.7).

There would be global consequences for selling this commodity. Russia's blanket of coniferous forest acts as a major carbon dioxide "sink" that helps absorb human-made greenhouse gases, so its removal would contribute to global warming.

As discussed in Chapter 2 (page 32), global warming is likely to be most intense in the high latitudes and polar regions. We looked at a "tipping point" scenario in which warming temperatures trigger other mechanisms that will warm the atmosphere irreversibly. We saw how warm temperatures melt arctic sea ice, leading to more warming. Another positive feedback loop of this kind involves the permafrost underlying the subarctic and tundra zones, most extensively in Russia but also in Northern Europe, northern North America, and Greenland. The carbon locked up in this organic matter of permafrost is more than double the carbon already in Earth's atmosphere. If warmer temperatures thaw this frozen ground, bacteria and fungi will begin to break down the carbon contained in the organic matter and release it into the atmosphere as carbon dioxide and methane, both of which are greenhouse gases. This would lead to even more atmospheric warming.

Steppes and Good Soils

A humid continental climate occupies a triangular area south of the subarctic climate, narrowing eastward from the region's western border to the vicinity of Novosibirsk (see Figure 5.4). This area has the cold short-summer subtype of humid continental climate, comparable to that of the Great Lakes region and the northern Great Plains of the United States and adjacent parts of Canada. Both the climate and the soils are more favorable for agriculture in the humid continental climate than in the subarctic.

South of Russia's humid continental climate zone, and in most of Ukraine, Moldova, and Kazakhstan, is the drier climatic zone of the steppe. The average annual precipitation is 10 to 20 inches (c. 25 to 50 cm). The steppe's characteristic natural vegetation is short grass with deep roots. Pastoralists, including the Scythians who in the fourth century B.C.E. produced astonishing artwork

• **Figure 5.8** A grazier with his cattle on the steppe near the Don River in southern Russia.

of gold in the area that is now Ukraine, grazed their herds from an early time on the treeless steppe grasslands.

The steppe corresponds with the **black-earth belt**, which today is the most important area of crop and livestock production in the region of Russia and the Near Abroad. Wheat is the main crop, supplemented by sugar beets and sunflowers. Cattle are the principal livestock (•**Figure 5.8**). The main soils of this belt are known as *chernozem*, which is Russian for "black earth." Among the best soils to be found anywhere in the world, these are the thick and productive soil type known as **mollisols**. A similar belt of mollisols occupies the eastern portion of the Great Plains of North America. Their great fertility is due to an abundance of **humus** (decomposing plant and animal matter) in the topsoil.

Growing crops is difficult in the desert climate areas east and just north of the Caspian Sea in the central Asian countries. Cotton is the most important crop where irrigation water is available from the Syr Darya, Amu Darya, and other rivers flowing from

high mountains to the south and east. The Syr Darya and Amu Darya flow into the Aral Sea—or what is left of it; on page 151, there is discussion of Soviet era management of agriculture in this area and its legacy in the disrupted ecosystems of Central Asia.

The Role of Rivers

In the early history of Russia, rivers formed natural passageways for trade, conquest, and colonization. They were especially crucial in settling Siberia, which is drained by some of the greatest rivers on Earth: the Ob, Yenisey, Lena, and Kolyma, all flowing northward to the Arctic Ocean, and the Amur, flowing eastward to the Pacific. By following these rivers and their lateral tributaries, the Russians advanced from the Urals to the Pacific in less than a century. (Incidentally, there is no substance to the widespread belief that rivers *should* run south. There are quite a number of northward-flowing rivers in the world, and Russia's stand out.)

The Moscow region lies on a low upland from which a number of large rivers radiate like the spokes of a wheel (see Figure 5.1). The longest ones lead southward: the Volga to the landlocked Caspian Sea, the Dnieper to the Black Sea, and the Don to the Sea of Azov, which connects with the Black Sea through the strategic Kerch strait (see The Geography of World's Great Rivers, page 123). Shorter 116 rivers lead north and northwest to the Arctic Ocean and Baltic Sea. These river systems are accessible to each other by portages.

A major link in the inland waterway system is the Volga-Don Canal, tying the two rivers together where they approach each other near the city of Volgograd (see the canal in •**Figure 5.9** and its location in Figure 5.23). The canal allowed the White Sea and Baltic Sea in the north to be linked to the Black Sea and Caspian Sea in the south in a single water transport system.

The Role of Topography

Most of the big rivers in Russia and the Near Abroad wind slowly for hundreds or thousands of miles across large plains.

• **Figure 5.9** A lock in the Volga-Don Canal, a vital link for trade between the Volga watershed and the Black Sea.

GEOGRAPHY OF THE WORLD'S GREAT RIVERS | The Volga

Russians think of the Volga as their most important river. They call it *Matushka*—"Mother." It rises about 200 miles (320 km) southwest of Saint Petersburg and flows 2,300 miles (3,680 km) to the Caspian Sea. Before railroads supplemented river traffic, the Volga was Russia's most important commercial artery. Wheat, coal, and pig iron from Ukraine; fish from the Caspian Sea; salt from the lower Volga; and oil from Baku, on the western shore of the Caspian, traveled upriver toward Moscow and the Urals. Timber and finished products moved downriver to the lower Volga and Ukraine (•Figure 5.D). Boatmen towed barges upstream on a 70-day journey from Astrakhan, near the Volga mouth, to Kazan, on the middle Volga. The steamboat arrived in the late 1800s, bringing an end to the way of life recalled in the famous Russian song "The Volga Boatmen."

The Volga was difficult to navigate until the latter half of the twentieth century, when the **Great Volga Scheme** transformed the river. The goal was to control the flow of the river completely with a stairway of huge reservoirs, each of which reaches upstream to the dam forming the next reservoir. These bodies of

• **Figure 5.D** Cargo moving along the Volga, Russia's "Mother" river. Russian literary depictions of life along the river are sometimes reminiscent of Mark Twain's Mississippi. Down along the Volga wharves at Kazan, Maxim Gorki found "a whirling world where men's instincts were coarse and their greed was naked and unashamed."

water allow shipping during the 6 months when the river is ice-free, and they supply hydroelectric power and water for irrigation. These reservoirs are so vast that they are often labeled as "seas" on maps.

The "taming" of rivers for power, navigation, and irrigation was an important component of Soviet economic development. Large dams were raised, and reservoirs formed behind them, on main courses and major tributaries of the Dnieper, Don, Kama, Irtysh, Ob, Yenisey, Angara, and other rivers.

Such plains and low hills compose nearly all the terrain from the Yenisey River to the western border of the region (see Figure 5.1). The only mountains in this lowland are the Urals, a low and narrow range (with an average elevation less than 2,000 ft [600 m]) separating Europe from Asia and European Russia from Siberia. The Urals trend almost due north–south. A wide lowland gap between the southern end of the Urals and the Caspian Sea allows uninterrupted east–west movement by land. Cut by river valleys offering easy passageways, the Urals are not a serious barrier to transportation.

Between the Urals and the Yenisey River, the West Siberian Plain is one of the flattest areas on Earth. Immense wetlands, through which the Ob River and its tributaries slowly wind their way, cover much of this vast flatland. This waterlogged country, underlain by permafrost that blocks downward seepage of water, is a major barrier to land transport and discourages settlement. Tremendous floods occur in the spring when the breakup of ice in the upper basin of the Ob releases great quantities of water while the river channels farther north are still frozen and act as natural dams. Russian aircraft sometimes bomb ice jams on the rivers to prevent worse flooding.

The hilly Central Siberian Uplands, rising 1,000 to 1,500 feet (c. 300 to 450 m) above sea level, sprawl across the area between the Yenisey and Lena Rivers. Mountains dominate the landscape east of the Lena River and Lake Baikal, a natural gem described on page 149. Extreme northeastern Siberia is an especially bleak and difficult country for people to live in.

Low mountains rim the region from Lake Baikal to the Pacific, and high mountains flank the southern frontiers of Russia and the Near Abroad from the Black Sea to Lake Baikal (see Figures 5.1 and 5.10). In the Caucasus Mountains

• **Figure 5.10** The rugged Tien Shan Mountains.

of extreme southern Russia, Mt. Elbrus rises to 18,510 feet (5642 m). Great mountains rise east of the Caspian Sea. Their local names suggest their majesty: the Pamirs are the "Feet of the Sun;" the Tien Shan are the "Celestial Mountains" or the "Mountains of the Spirits." The Soviet Union named the highest mountain in its borders "Mt. Communism, "at 24,590 feet (7495 m). When it gained independence in 1991, Tajikistan renamed this peak in the Pamir Mountains "Is-mail Somoni," after a tenth-century leader of the Samanid Dynasty. Outside of the settled southwestern core land written about in pages 143–146, "rugged," "wild," and "vast" describe much of the physical geography of Russia and the Near Abroad.

5.3 Cultural and Historical Geographies

A Babel of Languages

Russia and the Near Abroad have a complex cultural and linguistic mosaic. The region's peoples belong to about 30 major ethnic groups and speak more than 100 languages (see •Figure 5.11). In Russia, Belarus, and Ukraine, the majority are Slavs who originated in east central Europe as speakers of an ancestral Slavic language (a member of the Indo-European language family) and who now speak *Russian, Belarusian,* and *Ukrainian,* respectively. Along the upper Volga River, in the

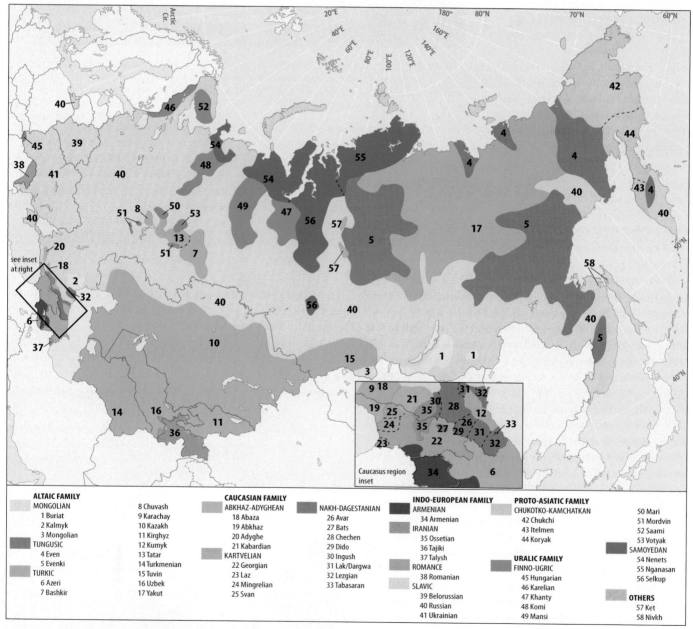

• **Figure 5.11** Ethnolinguistic distributions in Russia and the Near Abroad, where there is a close correlation between languages and the ethnic groups using them. The Soviet Union had difficulties trying to hold together such a vast collection of culture groups. The Russian Federation faces many of the same challenges.

Source: Adapted with permission from Bernard Comrie, et al., The Atlas of Languages, Revised Edition (New York: Facts On File, 2003). Reprinted with permission of the Publisher.

small regions of Tatarstan, Chuvashia, and Mari El, there are majority speakers of **Finno-Ugric** (a subfamily of the Uralic languages) and **Turkic** (in the **Altaic** language family). Another outpost of Turkic speakers is Yakutia (the Sakha Republic) in northeastern Russia. In Moldova, the majority ethnic Moldovans speak *Moldovan*, which is the same language as Romanian, a Romance (Indo-European) language.

The Caucasus is particularly diverse in its ethnicity and languages; that is why it needs an inset in the language map of Figure 5.11. This area has long been a magnet for linguistic researchers, and for good reason: about 40 indigenous languages are spoken here, more than any part of the world except for Papua New Guinea. Many researchers argue that the rugged terrain of the Caucasus serves a natural laboratory for the evolution and isolation of distinct languages. Throughout history, the mountains have kept invaders and their languages at bay.

The *Armenians* have a unique language within the Indo-European group. Their neighbors in the Caucasus—the Georgians and Azerbaijanis—speak, respectively, a **Kartvelian** (**South Caucasian**) and a Turkic language in the Altaic family. Most smaller ethnic groups in the Caucasus speak languages unrelated to Georgian, belonging to the **Abkhaz-Adyghean** and **Nakh-Dagestanian** families.

The "stan" countries are less complex linguistically. Turkic languages are spoken by most of the ethnic groups in the central Asian countries of Turkmenistan, Kazakhstan, Uzbekistan, and Kyrgyzstan. In Tajikistan, however, the dominant group of *Tajiks* speak an **Iranian** language in the Indo-European family. In far northeastern Russia, there are speakers of the **Chukotko-Kamchatkan** languages, including *Chukchi* and *Koryak*, belonging to the **Proto-Asiatic** language family and related to some of the languages that ancestors of Native Americans carried eastward into the Americas. It is worth remembering that North and South Americas were populated by Asian cultures that presumably crossed over the Bering land bridge (now the Bering Strait) from Russia to Alaska.

This is just a brief overview of the region's tongues; there are many languages spoken by smaller populations. This rich multi-ethnicity is part of the great cultural wealth of Russia and the Near Abroad. This diversity has too often been a source of discord between peoples.

The cultural histories of the minority ethnic peoples have long been influenced and confronted by the Russians, whose origins reach more than a thousand years into the past. As we will see, in Soviet times Russians tried to assert their power by having speakers of non-Russian languages give up their tongues in favor of the Russian language. Language is a powerful political and cultural tool. The story of Russia's political and cultural dominance began with Slavic peoples who colonized Russia from the west, interacted with many other peoples, stood off or outlasted invaders, and acquired the giant territory that would become Russia.

Vikings, Byzantines, and Tatars

Slavic peoples have inhabited European Russia since the early centuries of the Christian era. During the Middle Ages, Slavic tribes living in the forested regions of western Russia came under the influence of Viking adventurers from Scandinavia known as **Rus** or **Varangians**. The newcomers carved out trade routes, planted settlements, and organized principalities along rivers and portages connecting the Baltic and Black Seas. In the ninth century, the principality of Kiev (now in Ukraine), ruled by a mixed Scandinavian and Slavic nobility, achieved mastery over the others and became a powerful state. The culture it developed was the foundation on which the Russian, Ukrainian, and Belarusian cultures later arose.

Contacts with Constantinople (modern Istanbul, Turkey) greatly affected Kievan Russia. Located on the straits connecting the Black Sea with the Mediterranean, Constantinople was the capital of the Eastern Roman or Byzantine Empire, which endured for nearly 1,000 years after the collapse of the Western Roman Empire in the fifth century C.E. Constantinople became an important magnet for Russian trade, and the Russians borrowed heavily from its culture. In 988, the ruler of Kiev, Grand Duke Vladimir I, formally accepted the Christian faith from the Byzantines and had his subjects baptized. Following its cleavage from the Roman Catholic Church in 1054, Orthodox Christianity became a permanent feature of Russian life and culture (•Figure 5.12 is the map of the region's religions). Moscow eventually came to be known as the **"Third Rome"** (after Rome itself and Constantinople) for its importance in Christian affairs.

The Bolshevik Revolution in 1917 began a 75-year period of official repression and neglect of the Orthodox church and other religions. Since the breakup of the Soviet Union, however, there has been a renaissance in religious observance—not just among Christians but also among Muslims, Buddhists, and others—across the vast region of Russia and the Near

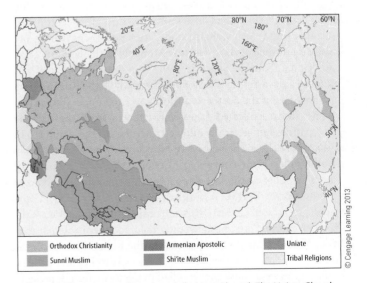

Orthodox Christianity	Armenian Apostolic	Uniate
Sunni Muslim	Shi'ite Muslim	Tribal Religions

© Cengage Learning 2013

• **Figure 5.12** Religions of Russia and the Near Abroad. The Uniate Church of northwestern Ukraine is an Eastern Catholic Church that accepts the Catholic dogma and the primacy of the pope in Rome but is not subject directly to the pope's control. Sunni and Shiite Islam are described on page 174 in Chapter 6, and the Armenian Church is discussed on page 149.

• **Figure 5.13** Some of the faithful in a Russian Orthodox church in Vyborg, Russia. Since the breakup of the USSR, there has been a resurgence of organized worship throughout the region.

Abroad (•**Figure 5.13**). There has even been a small reversal in the emigration of Jews to Israel and the West. Ten million Jews lived in the region before the Nazi Holocaust killed 3 million of them, and perhaps another million assimilated to escape official Soviet anti-Semitism. In the decade following the breakup of the Union of Soviet Socialist Republics, half of Russia's Jews fled the country. Some are coming back to today's more tolerant Russia, which now recognizes four official religions: Orthodox Christianity, Islam, Buddhism, and Judaism.

Yet another cultural influence reached the Russians from the heart of Asia. The steppe grasslands of southern Russia had long been the habitat of nomadic horsemen of Asian origin. During the later days of the Roman Empire and in the Middle Ages, these grassy plains, stretching far into Asia, provided the Huns, Bulgars, and other nomads a passageway into Europe. In the thirteenth century, the **Tatars** (also known as Tartars) of central Asia took this route. Many steppe and desert peoples of Turkic origins were in their ranks, with the **Mongols** in the lead. In 1237, Batu Khan ("Batu the Splendid"), the grandson of Genghis Khan, launched a devastating invasion that brought all the Russian principalities except the northern one of Novgorod under Tatar rule.

The Tatars collected taxes and tributes from the Russian principalities but generally allowed the rulers of these units to be autonomous. Even the princes of Novgorod, who were not technically under Tatar control, paid tribute to avoid trouble. The Tatars established the khanates (governmental units) of Kazan (on the upper Volga), Astrakhan (on the lower Volga), and Crimea (on the Black Sea). Russians called the Kazan Tatars the **"Golden Horde,"** after the brightly colored tents in which they lived. When Tatar power declined in the fifteenth century, the rulers of the Moscow principality were able to begin a process of territorial expansion that resulted in the formation of present-day Russia. Oil-rich Tatarstan, with its capital at Kazan on the upper Volga River, became one of Russia's most important "autonomous" political units. Its distinct ethnic Tatar population

has proved remarkably resilient to centuries of Russian supremacy. We will have a closer look at the Tatars and their culture when we look at Russia's struggle to maintain control over diverse peoples within Russia's borders, on page 134.

The Empire of the Russians

The Russian monarchy reached outward from Muscovy, its original domain in the Moscow region. From the fifteenth century until the twentieth, the tsars created an immense Russian empire by building onto this core. This imperialism by land, in the era when the maritime powers of Western Europe were expanding by sea, brought a host of alien peoples under tsarist control (see Insights on page 127, to understand how Russia built a "land empire"). Russian motivations for expansion were diverse. Quelling raids by troublesome neighbors, particularly nomadic Muslim peoples in the southern steppes and deserts, was one objective. Many Russian cities, such as Volgograd (originally named Tsaritsyn and then Stalingrad) on the Volga River, were founded as fortified outposts on the steppe frontier. In the wilderness of Siberia, the search for valuable furs and minerals, especially gold, stimulated early expansion, and the missionary impulse of Orthodox priests also played a role. Land hunger and a desire to escape serfdom and taxation led to the flight of many peasants into the fertile black-earth belt of southwestern Siberia, and many landlords also moved there with their serfs.

The initial outward thrust from Muscovy under Ivan the Great (reigned 1462–1505) was mainly northward. Ivan the Great annexed the rival principality of Novgorod, near present-day Saint Petersburg. This secured a Russian domain that extended northward to the Arctic Ocean and eastward to the Ural Mountains. Later tsars pushed the frontiers of Russia westward toward Poland and southward toward the Black Sea. Peter the Great (reigned 1682–1725) defeated the Swedes under Charles XII, to gain a foothold on the Baltic Sea, where he established Saint Petersburg as Russia's capital and its **"Window on the West."** At the expense of Turkey, Catherine the Great (reigned 1762–1796) secured a frontage on the Black Sea. It has been a foreign policy principle of Russia (and during its existence, the Soviet Union) to warn other powers against threatening its "warm water ports" on the Black Sea and its access through the Turkish straits to the Mediterranean Sea and Atlantic Ocean. It is ironic that despite its great coastlines, 138, 139 Russia faces geographic obstacles to seaborne trade and power.

Russia's eastward conquest was very impressive. Ivan the Terrible (reigned 1533–1584) added large new territories in conquering the Tatar khanates of Kazan and Astrakhan, thus giving Russia control over the entire Volga River. At the end of Ivan the Terrible's reign, traders and pioneers were already penetrating wild and lonely Siberia, where there were only scattered indigenous inhabitants. Among the new settlers were the **Cossacks**, peasant-soldiers of the steppes who originally were runaway serfs and others fleeing from the rule of the tsars. They eventually gained special privileges as military communities serving the tsars.

A Cossack expedition reached the Pacific in 1639, but Russian expansion toward the east did not stop at the Bering

Russia, and later the Soviet Union, developed as a **land empire**. Rather than establishing its colonies overseas, as imperial powers such as Spain and Britain did, Russia established colonies in its own vast continental hinterland 44 (•**Figure 5.E**). Many of the colonized peoples had little in common with the ethnic Russians ruling from faraway Moscow. Like former colonies of overseas empires such as Britain, non-Russian regions of the periphery of Russia and later the Soviet Union were drawn into a relationship of economic dependence on the imperial Russian core. The central Asian republics of the Soviet Union, for example, followed Moscow's demands to grow cotton, which was shipped to Moscow to be manufactured into value-added clothing that was subsequently 151 sold throughout the empire. This pattern contributed to the growth of the Union of Soviet Socialist Republics' economy but often inhibited local development. Other entities characterized historically as land empires include the 262, 407 United States, China, and Brazil, as well as the Ottoman, Mogul, Aztec, and Inca empires.

• **Figure 5.E** The development of Russia's land empire.

Source: Based on R. R. Milner-Gulland, Cultural Atlas of Russia and the Former Soviet Union (New York: Checkmark Books, 1998).

409 Strait. It continued down the west coast of North America as far as northern California, where the Russian trading post of Fort Ross was active between 1812 and 1841. In 1867, however, Russia sold Alaska to the United States (for 2 cents per acre) and withdrew from North America. The relentless quest of Russian fur traders for sable, sea otter, and other valuable pelts brought great cultural changes to aboriginal peoples in Siberia and Pacific North America. These effects became increasingly pronounced, especially in the last decades of the Russian Empire, when millions of Russians moved into Siberia. Russia's expansionism bore similarities to the American sense of fulfilling a "manifest destiny" in acquiring territory, and to Chinese efforts (ongoing now) to settle ethnic Chinese in the non-Chinese 262, 408 areas along China's western and southern borders.

There was one stumbling block in Russia's eastern expansion. In the Amur River region near the Pacific, the Russians were unable to consolidate their hold for nearly two centuries because of opposition by strong Manchu emperors who claimed this territory for China. Russians were thus barred from the Siberian area best suited to growing food for the fur-trading enterprise, and best endowed with good harbors for maritime expansion. This situation changed in 1858–1860 following the victory of European sea powers over China in the Opium Wars. As a result of two treaties, Russia was able to add the Amur region to its earlier gains in the Ob, Yenisey, and Lena basins. Finally, in a series of military actions during the nineteenth and early twentieth centuries, Russian tsars annexed most of the Caucasus region and Turkestan, the central Asian region east of the Caspian Sea.

The Soviet state that succeeded tsarist Russia had to grapple with difficult questions of how to govern and interact with a mosaic of scores of languages and cultures. Soviet authorities permitted the different ethnolinguistic groups to retain their own languages and other elements of their traditional cultures, and even created alphabets for those that previously had no written language. Politically, the Soviet Union officially recognized many of the groups as non-Russian nationalities. The Union of Soviet Socialist Republics established 16 autonomous Soviet socialist republics (ASSRs) as homelands for large ethnic minorities, in theory endowing them with limited autonomous (self-governing) powers. Smaller nationalities were organized into politically subordinate autonomous units, including five autonomous regions (*oblasts*) and 10 autonomous areas (*okrugs*). The Soviets' labeling of these units as "autonomous" came back to haunt the Russian nation in the 1990s. 131, 134

In spite of such apparent concessions to non-Russians, the Russian-dominated regime implemented a deliberate policy of **Russification**, an effort to implant Russian culture and language in non-Russian regions and to make non-Russians more like Russians. Local peoples could retain their own languages as 137 long as they were at the same time educated in Russian language. They were compelled to use Russian in public life, media, and publications. Russian was to be the "second mother language" of non-Russian ethnic peoples. Ethnic Russians migrated to work in the factories and state farms in non-Russian republics. Russians were prominent in positions of responsibility even within the non-Russian republics. Throughout the Union of Soviet Socialist Republics, Russians held the majority of top posts in the Communist Party, the government, and the military. But most of these ethnic groups that the Russians tried to Russify were very resilient and carried on their own distinct cultural practices. As we will see, Russian ways had little impact on non-Slavic cultures such as the Georgians and Armenians. Even Slavic kinspeople of the Russians, such as the Ukrainians, insisted on retaining their cultural distinctiveness and their original mother tongues. In general, the policy of Russification failed because it underestimated the strength of ethnicity and culture.

Russia and the Soviet Union: Tempered by Revolution and War

Russia and the Soviet Union repeatedly triumphed over powerful invaders, notably the Swedish forces led by King Charles XII in 1709, the French and their allied European forces under Napoleon I in 1812, and the German and other European forces sent by Hitler into the Soviet Union in World War II. In each case, the Russians lost early battles and much territory but eventually inflicted a crushing and decisive defeat on the invaders. The success of Old Russia and the Soviet Union in withstanding invasions by such formidable armies was due in part to the environmental rigors (particularly, the brutal Russian winter) that invaders faced, the overwhelming distances of a huge country with poor roads, and the defenders' love of their homeland. It was also due to talented Russian military leadership and to the willingness of good and poor generals alike to lose great numbers of soldiers in combat. Finally, the successful defense used **scorched earth** tactics to protect the motherland; rather than leave Russian railways, crops, and other resources to fall into the invaders' hands, the defenders destroyed them.

The **Russian Revolution** of 1917, which set the stage for the formation of the Soviet Union, was really two revolutions that occurred against the backdrop of World War I. In 1914, when a Serb in Sarajevo (in modern Bosnia and Herzegovina) assassinated the heir to the throne of Austria-Hungary, a complicated series of alliances required Tsar Nicholas II to commit Russian troops to fight with Serbia, France, and Britain against Austria-Hungary and Germany in World War I. The first revolution in Russia began early in 1917 as a general protest against the terrible sacrifices of Russian forces on the Eastern Front during this war. That revolt overthrew Nicholas II, the last of the Romanov tsars.

The second was the **Bolshevik Revolution** that came later that year. Led by Vladimir Ilyich Lenin (1870–1924), the Bolshevik faction of the Communist Party seized control of the government. The new regime made a separate peace with Germany and its allies and survived a difficult period of civil war and foreign intervention between 1917 and 1921. Lenin presided over the establishment of Russia's successor state, the Soviet Union, in 1922. To say that Lenin was the "George Washington of the Union of Soviet Socialist Republics" would only begin his biography. He was the head of the pantheon of

• **Figure 5.14** "Under the Banner of Lenin We Will March to Complete Victory!" This 1944 propaganda poster was designed to inspire the final push by the Red Army into the heart of the Nazi homeland.
Source: Aurora Art Publishers. The Great Patriotic War, 1941–1945: Posters. Leningrad, USSR, 1985.

Soviet leaders. Lenin's ideology and name were evoked at every opportunity to inspire the Soviet peoples (•**Figure 5.14**). His portraits and statues were omnipresent throughout the Union of Soviet Socialist Republics. Some survive, but when the Soviet Union fell, so did many of his likenesses—often quite savagely, with hammers—in quarters where he was resented.

World War II found the Russians again allied with France and Britain in a far more ferocious war against Germany. The Soviet Union's success in withstanding the German onslaught that began as **Operation Barbarossa** in June 1941 surprised many outside observers, who had predicted that the Soviets would prove too weak and disunited to resist for more than a few weeks or months. The crucially important cities of Leningrad (now Saint Petersburg) and Moscow held out through brutal sieges. Late in 1942, Soviet forces halted the German push eastward at the Volga River in the huge Battle of Stalingrad (see Geography of Sacred Space, page 129). This engagement was the turning point in the war, when Soviet and Allied forces began to reverse Nazi advances.

The failure of powerful Germany to conquer the Soviet Union was a clear indication that the strength of the Soviet Union had been underrated and that the country's power would henceforth

GEOGRAPHY OF SACRED SPACE | Stalingrad

A **sacred space** or **sacred place** is any location that people hold in reverence. Sacred places in the realms of ordinary experience include places of worship such as synagogues, churches, and mosques. Cemeteries too are perceived and managed as sacred sites. Places where people have lost their lives defending their homeland, or where they died at the hands of an enemy, are among the most revered and enduring sacred spaces—the Gettysburg Battlefield and Manhattan's Ground Zero are examples in the United States. Russia has the world's most extensive network of sacred places associated with the loss of life in defense of a country.

Perhaps in no other country is the memory of a war etched so vividly in the national consciousness as it is in Russia, which lost the greatest number of people, perhaps 23 million (including civilians) in World War II. Even today, pilgrims visit a large number of war monuments and cemeteries in an effort to heal deep emotional wounds and to keep alive the memory of Russia's costly wartime resistance. These rites of visitation and commemoration pass to each new generation. Immediately after the wedding ceremony, for example, it is common for a newlywed couple to place flowers at the local tomb of the unknown soldier. Contemporary Russian pride and nationalism have strong roots in wartime sacrifice.

During the Second World War, superiors urged Russian soldiers to fight to the death for the motherland: Russia was sacred ground to be defended at any cost. Today, the battlegrounds where those soldiers fell are sacred places. They are kept hallowed by the continuous ritual visitation of veterans, war widows, and three generations of descendants of war survivors.

Stalingrad (now named Volgograd), site of the most ferocious battle on World War II's eastern front, is the greatest of all these sites. Germany's Nazi leader Adolf Hitler had a geographic rationale for sending his forces against the city in August 1942. Located on the border of the steppe and semidesert where the Volga and Don Rivers are closest together, it was strategically situated. Take a close look at the geographical *situation* of Volgograd/Stalingrad in Figure 5.23 on page 144. It was a grain and livestock center with railway and water connections (through the Volga-Don canal described on page 122 and in the photo of Figure 5.9) to the Don Valley and the Caucasus region. Hitler wanted the vital oilfields of the Caucasus—the same ones that are still prized in the global oil trade and that are described on page 130. This vast energy supply would fuel the German war effort. But first, Stalingrad had to be wrested from Soviet control. Although vital strategically, Stalingrad was to Hitler as much a symbolic as a military prize: Because it bore the name of the Soviet leader, its fall would be of great propaganda value to the German war effort. Equally, its salvation from the invader was a goal of nearly religious significance for the Stalin regime.

"Not one step backward," Stalin ordered his troops. Soviet resistance at Stalingrad is an extraordinary chapter in the history of warfare. The invaders were unprepared for the resolve of the Red Army and for the ferocity of the Russian winter. The 5-month battle ended in a devastating defeat for the Germans: Two armies consisting of 24 generals, 2,000 officers, and 90,000 soldiers were taken prisoner. One-quarter of the German army's war material was lost. Two years earlier, the Germans could not have imagined such a defeat.

But victory for the Soviet Union exacted an unimaginable cost. Stalingrad lay in ruins; Soviet authorities dubbed it a "city without an address" and, in honor of its defenders, a "hero town." Fifty years after the battle, Russian military authorities finally released figures on the number of dead. In this sacred ground lay the bodies of 3.5 *million* soldiers and civilians, of whom 2.7 million were Soviet citizens, mainly Russians.

The visitor to Volgograd cannot help but feel the pain of war that has lingered through the decades. All over the city are monuments to remind the living of the dead. The devastated shell of a mill stands as the only physical artifact of the past, but the monuments built after the war rekindle the emotional losses most strongly. Volgograd's central memorial is the complex on Mamayev Hill, a mecca for Russians. Old soldiers, still wearing their medals, look war-weary even now as they shuffle through. Grandmothers lead small children to place flowers at the feet of statues. Over the mass grave of an estimated 300,000 Soviet and German soldiers, a huge hand raises an eternal torch. The honor guard changes in goose-step once each hour. This sacred place is dominated by the world's largest statue, a female sword-wielding figure called *The Russian Motherland* (•**Figure 5.F**).

Joe Hobbs

• **Figure 5.F** The Mamayev Hill monument to the memory of the defenders of Stalingrad. Within the hill lie the bodies of hundreds of thousands of the city's defenders and attackers. This sword-wielding figure, known as *The Russian Motherland*, is the world's largest statue, standing 279 feet (85 m) high and weighing 8,000 tons.

be a major feature of world affairs. But the war's impact on the Union of Soviet Socialist Republics would linger. The German invasion took an estimated 20 million or more Soviet lives and caused the relocation of millions of people. It did enormous damage to settlements, factories, and livestock. As a strategic precaution during the war, Lenin's successor, Josef Stalin, directed major Soviet industries to relocate eastward away from the front, a move that has had a lasting imprint on the region's economic geography.

5.4 Economic Geography

The Soviet Union's collapse dramatically changed the global political landscape and had far-reaching economic consequences. It was in fact the failing economic system that brought the giant country down. This section discusses the economic events leading up to and in the wake of the Soviet Union's demise and then focuses on the modern economy of the largest successor state, Russia.

The Communist Economic System

The Soviet Union's Communist economic system was an attempt to put into practice the economic and social ideas of the nineteenth-century German philosopher Karl Marx. According to Marx, the central theme of modern history is a struggle between the capitalist class ("the **bourgeoisie**") and the industrial working class ("the **proletariat**"). He forecast that exploitation of workers by greedy capitalists would lead the workers to revolt, overthrow the capitalists, and turn over ownership and management of the means of production to new workers' states. In the classless societies of these states, there would be social harmony and justice, with little need for formal government.

104

Marx's utopian vision did not materialize anywhere, but it did provide guidelines for anti-government revolts and Communist political systems in many countries. The ideas of Marx and of Lenin (who died in 1924), as Lenin's successor Josef Stalin interpreted and implemented them, provided the philosophical basis for the Soviet Union's centrally planned **command economy**. Beginning in 1928, a series of 5-year economic plans demanded the fulfillment of quotas for the nation: types and quantities of minerals, manufactured goods, and agricultural commodities to be produced; factories, transportation links, and dams to be constructed or improved; and residential areas to be built for industrial workers. The goals were to abolish the old aristocratic and capitalist institutions of tsarist Russia and to develop a strong socialist state equal in stature to the major industrial nations of the West.

In the command economy, an agency in Moscow called **Gosplan (Committee for State Planning)** formulated the national plans, which were then transmitted downward through the bureaucracy until they reached individual factories, farms, and other enterprises. This was an unwieldy and inefficient process in several ways. First, the planners in Moscow were essentially required to act as CEOs of a giant corporation, effectively "USSR, Inc.," that would manage the economy of an area larger than North America—a gargantuan, impossible task. Further, the Soviet planning bureaucracy had no free

market to guide it, so the system produced goods that people would not buy or failed to produce goods that people would have liked to buy. And Gosplan stated production targets in quantitative rather than qualitative terms, churning out abundant but substandard products. There was often an obsession with fulfilling huge quotas or implementing grandiose schemes, a Soviet preoccupation sometimes known as **gigantomania**. Finally, fearing reprisals from people higher up the ladder, no one wanted to suggest ways to increase quality and efficiency.

Although cumbersome, this system succeeded in propelling the Soviet Union to superpower status, improved the overall standard of living, prompted rapid urbanization and industrialization, and altered the landscape profoundly. The principal goal of Soviet national planning after 1928 was a large increase in industrial output, with emphasis on heavy machinery and other capital goods, minerals, electric power, better transportation, and military hardware. Masses of peasants were converted into factory workers. New industrial centers were founded, and old ones were enlarged. There were huge investments in defense, and the country grew strong enough to survive Germany's onslaught in World War II. After the war, the Soviets maintained large armed forces and accumulated a massive arsenal of conventional and nuclear weapons in the **arms race** against the world's only other superpower, the United States. By the height of the Cold War in the early 1980s, 15 to 20 percent of the country's GDP was dedicated to the military (in contrast to less than 10 percent in the United States), representing an enormous diversion of investment away from the country's overall economic development.

Farmers and farming had troubled careers under the Soviet system of **collectivized agriculture**. Between 1929 and 1933, 105 about two-thirds of all peasant households in the Soviet Union were collectivized. In this process, their landholdings were confiscated and reorganized into two types of large farm units: the **collective farm (kolkhoz)** and the factory-type **state farm (sovkhoz)**. The consolidation of individual farmsteads and villages into fewer but larger communities on the collective farms was supposed to permit the government to administer, monitor, and indoctrinate the rural population and provide services, including education, health care, and electricity, more cheaply and efficiently.

But the rural people fiercely resisted collectivization. In their own version of scorched earth, peasants and nomads slaughtered millions of farm animals and burned crops to avoid turning them over to the "socialized sector." Government reprisals followed, including wholesale imprisonments and executions, together with confiscation of food at gunpoint (often including the peasants' own food reserves and seed). The more prosperous private farmers, known as *kulaks*, were killed, exiled, sent to labor camps, or left to starve. Famine took untold millions of lives—perhaps 10 or 12 million. Soviet leaders—most notoriously, Josef Stalin—disregarded these costs, and by 1940, virtually all of the Soviet Union's farmed land was collectivized.

The drive to increase the national supply of farm products demanded an enlargement of cultivated area. In the 1950s, the Soviet Union began a program to increase the amount of grain (mainly spring wheat and spring barley) produced by bringing

• **Figure 5.15** Soviet agricultural expansion into the so-called virgin and idle lands, 1954–1957.

tens of millions of acres of what were called **virgin and idle lands** or **new lands** into production in the fertile black earth steppes of northern Kazakhstan and adjoining sections of western Siberia and the Volga region (•**Figure 5.15**).

Soviet enterprises in both agriculture and industry harnessed the energies and resources of the entire country to achieve specific objectives. The government called on the people to sacrifice and to make the country strong, especially by working on so-called **"hero projects"** such as the construction of tractor plants, dams, railways, and land reclamation. The government in turn operated as a collectivized welfare state, providing guaranteed employment, low-cost housing, free education and medical care, and old-age pensions. As we will soon see, when the Soviet Union fell apart, so did these benefits that people counted on for sustenance. Social services were often minimal but in some sectors were quite successful; the literacy rate, for example, rose from 40 percent in 1926 to 99 percent in 1959.

Economic Roots of the Second Russian Revolution

The Soviet Union's ambition to achieve and maintain military-industrial superpower status was generally achieved at the expense of the country's consumers. Many of the needs and wants of Soviet citizens were overlooked in favor of heavy metallurgy and the manufacture of machinery, power-generating and transportation equipment, and industrial chemicals. Consumers lined up in stores to purchase scarce items of clothing and everyday conveniences. The exasperation of shoppers confronted by long lines and empty shelves was an important factor generating dissatisfaction with the economic system and demands that it be reformed.

The flow of goods and services ebbed to a trickle in the 1980s. People began taking to the streets in protest. In large demonstrations and strikes, the public expressed their anger at a political and economic system that was sliding rapidly downhill. Their outpouring of dissent was unprecedented in Soviet

history. They openly challenged the Communist system on the grounds that it failed to provide a good living for most people, stifled democracy, and blocked the ambitions of the country's many ethnic groups for a greater voice in running their own affairs. Meanwhile, the economy turned from bad to worse.

The revamping of the economic system became an urgent priority during the regime of Mikhail Gorbachev, which began in 1985. Gorbachev had a leadership style that was also unprecedented in Soviet history: he was a reformer. He introduced new policies of **glasnost** ("openness") and **perestroika** ("restructuring") to allow a more democratic political system, more freedom of expression, and a more productive economy with a market orientation. Little did he know that opening the door slightly to allow free market, capitalist-style enterprise would lead quickly to a flood of public demands for more reforms and set the Union of Soviet Socialist Republics on a course of rapid unraveling. During Gorbachev's rule, the various republics and ethnic groups organized by the Soviet system into nominally "autonomous" units took advantage of new freedoms to resurrect old quarrels and demand greater autonomy. Fighting among ethnic groups erupted in several republics. In all of the republics, declarations of sovereignty and in some cases outright independence challenged the authority of the central government. In 1991, the three Baltic republics of Estonia, Latvia, and Lithuania unilaterally declared their independence. The empire's disintegration had begun.

At the center of the crumbling empire, having failed to reverse the downward slide of the economy, Gorbachev faced growing public demands to scrap the command economy and move as rapidly as possible to an entirely market-oriented economy. Gorbachev's go-slow approach to the economy and his insistence that the Soviet Union should survive as a political entity, even after the Baltic republics fell away, was increasingly unpopular.

Matters came to a head with an attempted (but failed) coup against the Soviet leaders. Gorbachev resigned his position as head of the Communist Party and began to work with Boris Yeltsin, a reform-minded political leader, to reconstruct the political and economic order. But his efforts to preserve the Soviet Union and a modified form of the Communist economic system could not stand against the growing tide of change. During the autumn of 1991, the Communist Party was disbanded, and on December 25, 1991, Gorbachev resigned the presidency. The national parliament voted the Soviet Union out of existence the following day. A powerful empire had quickly and quietly faded away, to be replaced by 15 independent countries, in what was dubbed the **Second Russian Revolution**.

Russia's Road to Misdevelopment

The unprecedented experiment to transform a colossal Communist state into a bastion of free-market democracy produced strange and often unfortunate results in the largest successor state of the Union of Soviet Socialist Republics: the vast Russian Federation, from here on referred to simply as "Russia."

Boris Yeltsin emerged as the first leader of Russia in the post-Soviet era. Yeltsin introduced a program of rapid economic reform, known as **economic shock therapy**, designed to replace the Communist system with a free-market economy. It threw out all tenets of the command economy, removing all price controls and encouraging the privatization of businesses (up to this point, the Soviet government controlled all sectors of the economy).

Shock therapy proved to be too much too fast for the Russian people. Former employees of the Communist Party resented the loss of their jobs and privileges. Poor people living on fixed incomes struggled to survive high prices for basic necessities. A new consumer-oriented society developed along class lines. Unemployment and homelessness increased, and the gap between rich and poor widened (•Figure 5.16a, b). There was a bewildering explosion of new goods and services that most people simply could not afford. But some people could, particularly the well-heeled "oligarchs" who moved in to buy up formerly state-controlled sectors of the economy and profit handsomely from them.

One of the major components that emerged in Russia's new economic geography was the **underground economy**, also known as the *countereconomy* or *second economy* (or economy *na levo*, meaning "on the left"). Rampant black marketeering developed. Although much of this exchange was illegal, there was a general tendency to overlook such transactions because they were essential to the economy. Widespread **barter**—the exchange of goods and services instead of cash—resulted from the declining value of Russia's currency, the ruble. Many people resorted to selling personal possessions to buy high-priced food and other necessities. Most Russians grew economically worse off. Many privately owned industries were too unproductive to pay their employees, or else they paid workers "under the table" to avoid taxation. Unpaid

workers could not pay taxes to the government, of course, and workers paid under the table had no incentive to pay taxes. Russia's gross domestic product plummeted, shrinking by almost half in the 1990s. This was the largest fall in production that any industrialized country had ever experienced in peacetime.

Russia's new private entrepreneurs came to include a large criminal "mafia" that preyed on government, business, and individuals. Organized crime and corruption became pervasive and made "free enterprise" far from free. A company wanting to build a factory, for example, would have to pay off officials at every stage of construction; without the payouts, organized crime would terminate the project. Russia became a **kleptocracy**, an economic and political system based on crime. Corruption was rampant. And more than 20 years after the collapse of the Soviet Union, it still is. Russians accept corruption as part of everyday life. Here is an example: Rather than endure hours to pay or contest a traffic violation, most Russians will discreetly hand the equivalent of $15 over to the traffic policeman, and the matter is resolved. Like any other commodity, bribes are subject to inflation: in 2009, the cost of an average business-related bribe tripled to $1,000 from the previous year's cost. On the corruption index maintained by Transparency International, only 24 of the world's 178 countries were listed as more corrupt than Russia in 2011.

Tracking figures like these, some observers classified post-Soviet Russia not as a less developed or more developed country, but as a **"misdeveloped country."** The disturbing demographic trend of a plummeting population due to rising death rates—the Russian cross described earlier—is perhaps the strongest indicator of Russia's backward momentum in the years following independence. From the technical standpoint of GNI PPP used to define more and less developed countries, Russia is an MDC—but on the lower end, with a value of $18,330, not far above the $15,000 cutoff. [43]

As the largest percentage of Russia's public slid backward, there were those who prospered. The biggest beneficiaries of

• **Figure 5.16a** Having fallen through the cracks in Russia's transition from a command economy to free enterprise, these women are begging for help on a Saint Petersburg sidewalk. Their placards say, in essence, "Help us dear brothers and sisters. Please give us money for food, for God's sake. We are invalids with cancer. We have given of ourselves but are now forgotten by the state. We are poor and hungry, with no protection, just left to rot."

Joe Hobbs

OLIVER LANG/AFP/Getty Images

• **Figure 5.16b** While some beg on Russia's streets, others indulge in Western-style overconsumption. This Maserati sports car, out of reach for all but Russia's super-rich, was on display in Moscow at the 2009 "Millionaire Fair."

the new economic system were the so-called **oligarchs**, Russia's leading businesspeople (*oligarchy* is literally the "rule of the few"). In the 1990s, a controversial privatization program transferred an immense amount of wealth to these tycoons; by some estimates, they acquired control over about 70 percent of Russia's economy.

The oligarchs wielded enormous political power and helped usher Vladimir Putin, a former KGB (state security) officer of the Soviet Union, into Russia's highest office, the presidency, in 2000. Putin's authoritarian style, reminiscent of former Soviet ways, proved popular with voters, who awarded him a succession of terms as president and prime minister that may continue to stretch all the way to 2024. He enjoyed broad popular support among Russians, sometimes with an approval rating of over 80 percent. Even with the secondary office title of prime minister, he was the "power behind the throne" of President Medvedev. For many years, his consolidation and retention of power, with little public opposition, seemed to reaffirm what that most powerful Soviet leader Josef Stalin often said: "The Russians need a tsar." Not only did Putin remind many people of the relative stability and security of the Soviet state, but much more important, he was credited with pulling Russia out of the economic black hole it fell into when the Union of Soviet Socialist Republics broke apart. But he may have taken his power and popularity too much for granted. In 2011, when he and President Medvedev announced Putin's plan to return to the presidency, many Russians took to the streets in protest. They wanted real rivals to Putin, and they wanted a verifiably fair election. They did not get either, and Putin won.

Putinomics

Vladimir Putin had a simple plan for Russia. He would export the country's natural resources at full tilt to flood the country with wealth, starting to rebuild its economy and reinstilling a sense of national pride in Russians.

It worked. The steady rise in global energy demand and prices after 1999 (with a setback in 2008–2009) was a big stimulus to the Russian economy. Russia experienced a new period of self-assurance and economic renewal. Gross domestic product and wages more than doubled in the decade of the 2000s. Russia came to be classified as a member of an elite group: the so-called **"BRIC economies,"** with the acronym standing for Brazil, Russia, India, and China. According to the Goldman Sachs outlook, these four countries (which together have a quarter of the world's land mass and two-fifths of the world's population) would have the world's largest economies by 2050. Russia and Brazil would establish dominance by exporting raw materials, while India and China would do so through manufacturing and services. Putin's plan was for Russia to be world's fifth largest economy by 2020.

Russia has enormous energy reserves; its proven oil reserves represent 6 percent of the world total, and its natural gas, 27 percent of proven global reserves, the largest in any single country. Russia also has the world's second largest coal reserves, after the United States. Russia frequently vies with Saudi Arabia as the world's largest oil producer and exporter. Its reserves, however, are just a fraction of those of Saudi Arabia, and such high levels of production can only be maintained until about 2020. Oil production will then start trailing off, and Russia will have to rely on a more diversified economy. Producing raw materials is no substitute for the economic diversification that can protect a country's economy in a variety of economic conditions.

The biggest danger to Russia's economy now lies in excessive dependence on natural resources including fossil fuels, metals, and timber. Energy represents about two-thirds of the value of Russia's exports, and oil sales fund half of Russia's federal budget. Should prices for these commodities slump—as the price of oil did in 2008—revenues would also plummet and drag the entire economy into a recession. If, on the other hand, revenues from oil and other natural resource exports continue to be high, the profits will be rolled into manufacturing and high-tech industries, and the country will enjoy a more stable, diversified economy. This is the scenario that Vladimir Putin is counting on, so much so that this economic rationale has come to be known as "Putinomics." Disturbed by the scenario of Putin leading Russia for another dozen years, detractors argued that his model was bad for Russia's development. One of them, Laza Kekic, wrote that Russia was "on the way to becoming a third-world petrokleptocracy."[2]

Russia will have an uphill climb in preparing for a future that is not driven by natural resources exports. Russia has not promoted the "knowledge economy" as post-industrial European and North American countries have been doing to replace heavy industry in major urban areas. Heavy industries, which had been so vital during the Soviet era, went idle when the recent oil boom began. The country's industries and industrial resources are concentrated in four areas: the regions of Moscow and Saint Petersburg, the Urals, the Volga, and the Kuznetsk region of southwestern Siberia; see Figure 5.22. Part of the problem is that much of the knowledge fled Russia after the Soviet Union collapsed, in one of the most unfortunate brain drains ever.

Why did the best minds start looking for the exits? Russians are justifiably proud of their past achievements in literature, art, music, and a variety of fields in the humanities, and in science and technology. The Union of Soviet Socialist Republics had a healthy lead against the United States at the start of the "space race" of the 1960s, for example. Enormous resources were deployed to advance research and development. There were research parks such as "Academy Town" (Akademgorodok) where the Union of Soviet Socialist Republics' best scientists came to work. Soviet scientists and their research and publication were esteemed, and many developing countries sent their best students to earn degrees in Soviet universities. But in the economic downdrafts that followed the breakup of the Soviet Union, knowledge was devalued. In Russia, the salary of a senior researcher with a Ph.D. decreased by a factor of 10 between 1989 and 1999.[3] By 1999, the salary of a bus driver was five to seven times more than that of a professor. Not surprisingly, the "best and the brightest" fled the country or went into fields other than science and academia.

5.5 Geopolitical Issues

Among those Russians longing for the "good old days" of the Soviet Union is none other than Vladimir Putin. He describes the collapse of the Soviet Union as "the greatest geopolitical catastrophe of the century."[4] Today, there are three concentric spheres of major geopolitical concern in the former Soviet realm: the unity of Russia itself, Russia's relationships with its Near Abroad (the other countries that were once part of the Soviet Union), and the relations between this region and the rest of the world. There are many critical and compelling geopolitical issues to consider, and getting to know them is worth the effort. "I cannot forecast to you the action of Russia," Britain's Prime Minister Winston Churchill said in a radio address to his nation in 1939. "It is a riddle, wrapped in a mystery, inside an enigma; but perhaps there is a key. That key is Russian national interest."[5] Here we will see whether, early in the twenty-first century, that key is to be found in Russia's geopolitical contexts.

Geopolitics within Russia

Internally, Russia is struggling to maintain a cohesive whole that is fashioned from diverse and sometimes volatile ethnic units. As mentioned earlier, the Soviet Union included a number of "autonomous" units based on ethnicity. When Russia emerged from the Soviet Union, it kept those designations within its borders. The Russian Federation ended up with 83 "subjects" divided into six categories: 48 *oblasts* (regions), 7 *krais* (territories), 21 autonomous republics, 4 autonomous *okrugs* (ethnic subdivisions of oblasts or krais), 2 federal cities, and 1 autonomous oblast. You can see these political units in •**Figure 5.17** (many of them were reorganized in 2000, when the Putin government created a system of seven federal districts into which all the units were placed). The 26 **autonomies** (nationality-based republics and lesser units: the 21 republics, plus the 4 *orkugs* and the single Jewish *oblast*) in total occupy more than 40 percent of Russia's total area and are home to about 20 percent of the country's population.

Nearly half of the people in Russia's 26 autonomies are ethnic Russians. The other peoples are ethnically diverse. Some, such as the Karelians, Mordvins, and Komi, are Uralic (also known as Finnic, related to the Finns and Magyars); others are Turkic (for example, Tatars, Bashkirs, and Yakuts), Mongolian (for example, Buriats near Lake Baikal), and members of many other ethnic groups. Their largest autonomous units form a nearly solid band stretching across northern Russia from Karelia, bordering Finland, to the Bering Strait. The Karelian, Komi, and, the largest of all, Sakha (Yakut) republics are in this group. This band also includes several other large but thinly inhabited units of aboriginal peoples. It is useful for you to look at the correlation of some of the autonomous regions in **Figure 5.17** with the ethnolinguistic regions in

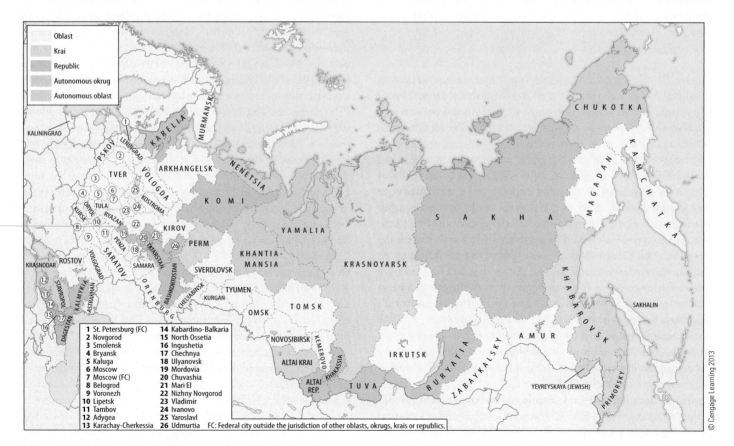

• **Figure 5.17** Subject units in the Russian Federation. Russia fears that independence for any of the republics, particularly Chechnya, might fracture the country.

Figure 5.11. Look, for example, at the large Sakha Republic in northeastern Russia in Figure 5.17, and then at the Yakut ethnolinguistic region (#17) in Figure 5.11. Likewise, look at the Republic of Tatarstan in Figure 5.17 and the Tatar ethnolinguistic region (#13) in Figure 5.11; and the Republic of Chechnya (#17) in Figure 5.17 and the Chechen ethnolinguistic region (#28) in Figure 5.11.

What gives some of these political and ethnolinguistic regions their exceptional geopolitical significance is the presence of major mineral resources within their boundaries: oil and gas in the Volga-Urals fields in Tatarstan and Bashkhortostan, high-grade coal deposits at Vorkuta in the Komi Republic, and a quarter of all the world's diamonds in Sakha. After the breakup of the Union of Soviet Socialist Republics, many of the former Soviet ASSRs (see page 127), recalling Moscow's long-standing promises of self-rule for them, issued declarations of sovereignty, asserting their right to greater self-direction of their internal affairs and greater control over their own resources. Eighteen of Russia's 21 republics signed the 1992 Federation Treaty, which granted them considerable autonomy. The treaty called for the devolution of power centralized in Moscow and for more cooperation between regional and federal governments. Each republic of the Russian Federation was legally entitled to have its own official language, constitution, president, budget, tax laws and other legislation, and foreign and domestic economic partnerships. But many soon complained that Moscow did not honor the treaty, and they began looking for regional solutions to their economic and other problems.

Moscow gave in to some of these demands. The Sakha (formerly Yakut) Republic, for example, won the right to keep 45 percent of hard-currency earnings from foreign sales of Sakha diamonds, compared with only a small fraction during the Soviet era. In turn, Sakha would pay for the government subsidies that Moscow previously gave to local industries. Without credits from Moscow, Sakha no longer paid taxes to Moscow. Salaries and other indexes of living standards in the Sakha Republic rose after this agreement was implemented. The majority-Muslim, oil-rich Tatarstan, which did not sign the Federation Treaty in 1992, earned "limited sovereignty" with its own constitution, parliament, flag, and official language, and it has since signed the treaty. The autonomous republic of Chechnya—of particular geopolitical importance because of its crossroads location for oil pipelines and its links with militant Islamist movements in the region and abroad—would not sign the treaty. Chechnya's future, along with that of neighboring Dagestan, became critical to the integrity of the Russian Federation (see Problem Landscape, page 136).

As Russia sees it, these regional demands for greater self-rule threatened the unity of the Russian Federation and threatened to deprive Russia of valuable resources. Independence for either Tatarstan or Chechnya could inspire the oil-rich Turkic-speaking Bashkirs and the Chuvash to push for their own independence. This could begin a process that would virtually cut Russia in half. Although some analysts argue that this process of devolution in Russia may actually bring more stability and

prosperity to the country, the government in Moscow worries that what happened to the Union of Soviet Socialist Republics (its breakup) might happen to Russia itself.

To keep the country's vast periphery tied to its core, Moscow created a new layer of bureaucracy with seven administrative regions (Northwestern, Central, Northern Caucasus, Volga, Urals, Siberian, and Far Eastern) appointing a presidential envoy (known as a governor general) to manage national defense, security, and justice matters in each. In addition, the presidents of Russia's republics, formerly elected, are now appointed by Moscow. Under Putin, the central government rolled back the powers of autonomy and now tightly controls regional affairs. Moscow drafts laws and keeps a lid on dissent. Putin established a new "social contract" with the regions: he will ensure that standards of living will rise, as long as people pay taxes and do not demand greater freedoms.[6] The message is clear: although there are nominally autonomous units within Russia, Moscow in fact rules them all.

Geopolitics in the Near Abroad

And then there are geopolitical intrigues in the Near Abroad, which President Medvedev called Russia's "zone of privileged interests." With the collapse of the Soviet Union, the Russian Federation took over the property of the Union of Soviet Socialist Republics within the federation's borders. This included Moscow's Kremlin, Russia's historical seat of power, a compound with a stunning array of powerful architectural buildings that backs up to Red Square and the mausoleum of Vladimir Lenin (•Figure 5.18).

From the Kremlin, Russia sought to be the anchor of all the newly independent successor states of the Soviet Union. In 1991, the new countries of Russia, Ukraine, and Belarus (formerly Byelorussia) formed a loose political and economic organization, the Commonwealth of Independent States (CIS). Except for Estonia, Latvia, and Lithuania, the other former Soviet republics eventually joined the CIS. Russia

• **Figure 5.18** The Kremlin, the seat of political power of vast Russia, and formerly the Soviet Union, occupies roughly the left half of the photo. Beneath the spire just left of center is the Lenin Mausoleum. A long line of people stretching across Red Square awaits entry to view the Soviet leader's body.

Nominally part of the Russian Federation, the Caucasus autonomy of Chechnya (population 1 million; for its location, see Figure 5.20) insisted on independence after the Union of Soviet Socialist Republics broke up. The Chechens, who are Sunni Muslims, have long resisted Russian rule, and Moscow has fought back. Russian troops attempted but failed to seize control over Chechnya soon after its 1991 declaration of independence. Beginning a second offensive in 1994, and at the cost of an estimated 30,000 to 80,000 lives on both sides, Russian troops succeeded in exerting physical control over most of Chechnya. But in 1996, Chechen forces recaptured the capital city of Groznyy. At that time, the Chechen war was deeply unpopular among Russians, and the prospect of further Russian losses led to a peace agreement with the Chechens.

The cessation of hostilities was short-lived. Chechen rebels kidnapped and tortured hundreds of Russian citizens in southern Russia and in 1999 carried their campaign to the heart of Russia. They detonated bombs in Moscow and other civilian centers, killing more than 300 Russians. The Chechen resistance by this time had become internationalized, with Islamic militant groups from abroad (including from Osama bin Laden's al-Qa'ida organization based in Afghanistan) supplying men and war materials to their Islamic guerrilla brethren in Chechnya. No longer controlled by Chechnya's president, in August 1999, these Chechen Islamists invaded the neighboring province of Dagestan with the hope of creating an independent Islamic state there.

The 1999 Dagestan invasion and the terrorist bombings in Moscow led to a surge in anti-Chechen sentiment among Russians and to calls for resolute action against Chechnya. Russia's newly appointed prime minister, Vladimir Putin, decided to pursue an all-out second Chechen war. In September 1999, Russian forces launched a furious attack that this time put them firmly in control of most of Chechnya.

In a Russian-sponsored referendum in 2003, Chechens voted that Chechnya should remain part of Russia, and in a subsequent election chose a new president, Akhmad Kadyrov. Behind the scenes, Russia ran a rigged election to ensure the victory of their handpicked candidate, who most Chechens revile as a traitor (in the first Chechen war, he fought the Russians, but he switched sides during the second war). He was assassinated the following year. His son Ramzan Kadyrov became Chechnya's president in 2007 with the blessings of Vladimir Putin. With Kadyrov, Putin implemented his policy of "Chechnization," using a handpicked and highly paid local ruler to carry out Moscow's will in Chechnya, and especially to quell the insurgency against Russia.

This Russian government poured money into Kadyrov's Chechnya, investing in the reconstruction of roads and the urban infrastructure of shattered Groznyy. Chechnya's capital now boasts skyscrapers and glitzy shopping malls (•Figure 5.G). Underneath the glamour and the veneer of peace, Chechen resentment against Russia remains. Chechen guerrilla resistance has continued with periodic terrorist attacks against Russian civilians. Many of these are suicide bombings carried out by Chechen or Dagestani women whose fathers, sons, or brothers were killed by Russians. They are the so-called "Black Widows" who are very effective at concealing explosive devices in their conservative Muslim clothing.

Islamist groups are still active in Dagestan, a volatile area with 37 distinct and fractious ethnic groups described by one observer in these terms: "a seemingly intractable knot of clan and business rivalries, religion, separatism, crime and extortion, feuds and vendettas whose roots runs deep into its region's

history."[7] How dangerous is Dagestan? It is so dangerous that in 2011 a Russian oligarch bought an Italian soccer player to play on a Dagestani team. In order to play, he flew in and out from his new home in Moscow on match days, never daring to spend a night in Dagestan. As in Chechnya, but with less success, Moscow pours subsidies into Dagestan to try to keep the area loyal to Russia. Dagestan ought to be prosperous, with its Caspian Sea beaches attracting tourists, its oil and gas reserves flowing, and its fisheries and vineyards yielding harvests. But violence or the potential for violence keeps investors and tourists away and keeps resources from being utilized.

Many Russians harbor strong prejudices against ethnic minorities and immigrants, particularly those from the Caucasus. Just as there is Islamophobia in Europe and in America, so there is in Russia. About 2.5 million Muslims live in Moscow, where they frequently encounter discrimination. About 20 million Russians—or 15 percent of the country's population—are Muslim.

• **Figure 5.G** Fireworks celebrating Chechen leader Ramzan Kadyrov in Groznyy, 2011.

took a strong leadership role in the CIS, although it was officially headquartered in Minsk, Belarus. To date, the CIS has been concerned mainly with establishing common policies on economic concerns. Russia has proposed that its members unite on other issues, particularly security, with the formation of an antiterrorism alliance. Russia is especially keen to prevent any CIS countries from becoming full members in the western military alliance of NATO. Russia has made it clear that if Ukraine or Georgia joined NATO, they would pay a heavy price.

Energy as Russia's Carrot and Stick

As noted earlier, Russia is an energy powerhouse. Russia is using fossil fuels to achieve maximum political clout in the Near Abroad. A major economic link is the vital flow of Russian oil and gas to other states, notably Ukraine (which is dependent on Russia for 80 percent of its energy), Belarus, the Baltic countries, and Kyrgyzstan.

Under Vladimir Putin, the Russian government began in 2004 to increase its control of Russia's energy sector, a process of **renationalization** that reversed the trend toward privatization that had been under way. The world's largest natural gas producer is Russia's largest company, its state-owned **Gazprom**. With its state powers in the critical energy sector, Russia has repeatedly wielded fossil fuel as a political weapon. Loyal former Soviet countries get the "carrot" of large subsidies, whereas disloyal ones get the "stick."

A good example of the carrot is Russia at one time charging Belarus just one-third of what it charged Latvia and Estonia for natural gas. This was in the 1990s, at a particularly good time in relations between the two countries, when they were actively pursuing a **union state** in which their economic systems, political systems, and defense networks would be integrated.

It rankles Russia when Western clout grows in the Near Abroad, particularly in critical nations like Ukraine. With independence from the Union of Soviet Socialist Republics, Ukraine gave up its nuclear weapons, and the United States rewarded that move with an increased foreign-aid package. Ukraine's orientation toward the West grew with its membership in NATO's "Partnership for Peace" program, and with its application to become a member of the European Union. The plot thickened during Ukraine's 2005 presidential election. Russia's favored candidate lost, following an eventful campaign in which his party poisoned but failed to kill the eventual winner, pro-Western Viktor Yushchenko.

Russia answered with the energy stick. In what was widely seen as retaliation for the so-called **Orange Revolution** that brought Yushchenko to power, Gazprom briefly cut natural gas supplies through a pipeline to Ukraine. Russia later cut oil supplies to Belarus when its authoritarian president, Alexander Lukashenko (known as "Europe's last dictator"), angered and embarrassed Vladimir Putin.

Ethnicity

Ethnic issues are also important in this geopolitical realm. About 25 million ethnic Russians lived in the 14 smaller Soviet states at the time of independence. Although some have migrated to Russia since 1991, most stayed, and many have had difficulties finding housing and employment. When Russians in the 14 non-Russian nations have complained that host governments and peoples discriminate against them, the Russian government has said that it has a right and a duty to protect these Russian minorities.

Russia encourages these Russian minorities to maintain and strengthen their ethnic identities, creating prospects for future political secession from the host countries. In certain areas, notably eastern Ukraine and Crimea, many ethnic Russians are causing political instability because they want to secede and form their own state or join Russia. (see •**Figure 5.19**). In political geography terms, such a movement by an ethnic group in one country to revive or reinforce kindred ethnicity in another country—often in an effort to promote secession there—is known as **irredentism**.

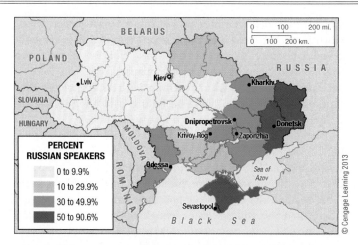

• **Figure 5.19** An important feature in Ukraine's domestic and foreign interests is the distribution of ethnic groups. Ethnic Russians dominate eastern Ukraine and the Crimean Peninsula.

Since independence in 1991, many of the governments and ethnic majorities of the post-Soviet countries have asserted their non-Russianness and tried to sweep away the remains of Moscow's Russification program. These sentiments are especially strong in the Baltic countries of Estonia, Latvia, and Lithuania, which were in effect colonized by Russia and where several generations of colonizing Russians have lived. Russian language is banned from government offices in Lithuania. In Estonia and Latvia, people must pass language tests in Estonian and Latvian to obtain citizenship; this compels ethnic Russians in these countries to learn the local languages or suffer the consequences of not knowing them.

There are similar sentiments elsewhere. In Georgia, Russian language is banned from most television and radio programs. Russian language schools and newspapers have been shut down in Turkmenistan. In Ukraine, the Ukrainian language must be taught in schools—even in the country's eastern region, where Russian has long been the first language for most people (see Figure 5.19). In the western part of the country, and among leaders elected from that region, the sentiment is to advocate Ukrainian language and culture while cultivating closer ties with Western Europe and the United States rather than with Russia. The Ukraine's pro-Western President Yushchenko promoted a deeper understanding of Ukrainian language as a means of distancing the country from Russia; he said, "With our native language, we preserve our culture. That greatly contributes to preserving our independence. If a nation loses its language, it loses its memory, its history, and its identity."[8] Much to the delight of Moscow, a pro-Russian president from eastern Ukraine (Viktor Yanukovych) succeeded Yushchenko.

Territorial Issues

Ukraine's post-independence struggles with Russia also involve control of strategic territory. An early issue of contention was the Crimean Peninsula, a picturesque and verdant resort region on the Black Sea (see Figure 5.24). Russia's Catherine the

Great annexed the Crimea in 1783, but Soviet leader Nikita Khrushchev returned it to Ukraine in 1954 (mainly a symbolic gesture, because Ukraine at the time belonged to the Soviet Union). After the Soviet Union collapsed, Russian irredentists agitated to get back the Crimea with its 70 percent Russian ethnic population. Ukraine yielded a bit by granting the Crimea special status as an autonomous republic.

For now, Russia recognizes Crimea as belonging to Ukraine but continues to be at odds with Ukraine over a strategic strait separating the Crimea from Russia's Taman Peninsula to the east. This Kerch Strait is the vital sea connection between the Azov Sea and the Black Sea—and therefore between southern Russia and the wider world through the warm water access it covets (see page 126 and Figure 5.24 on page 146). Russia wants to share sovereignty of the strait with Ukraine, but Ukraine claims most of the waters as its own and charges Russia several million dollars per year in transit fees.

Russia has a strategic interest in the Crimean port of Sevastopol, where it bases its Black Sea naval fleet. After Ukraine gained independence in 1991, it charged Russia a hefty fee to use the port. Recently the two countries struck a deal by which Ukraine allows Russia to use the naval facility at Sevastopol, and in exchange Russia gives Ukraine a 30 percent discount for buying Russian natural gas.

Near Abroad in the Caucasus

Russia has a number of geopolitical concerns with countries and ethnic groups of the Caucasus south of Chechnya (•Figure 5.20; see also pages 149–150). The Armenians embrace Russia as a strategic ally against their historic enemy, the Turks. The Georgian government is strongly anti-Russian,

but several nominally Georgian provinces have populations that are strongly pro-Russian. Russia is keen to nurture these pro-Russian elements and to keep the national governments friendly to Russia.

For its part, Georgia wants to loosen ties with Russia in favor of better relations with the West and its allies (for example, it wants to join the European Union and the North Atlantic Treaty Organization). Together with the Ukraine, Azerbaijan, and Moldova, Georgia has formed the regional grouping known as **GUAM** (an acronym for its four members) to promote economic integration, democratic reform, and increasing orientation toward Europe—and away from Russia. For years, as a member state, Georgia was able to block Russian accession to the World Trade Organization (WTO), the organization that sets the ground rules for globalization.

Russia has retaliated. In 2001 and again in 2006, when Georgia insisted that it was time for Russia to scale back its military presence in the region, Russia responded by cutting off supplies of natural gas to Georgia. Russia imposed punitive tariffs against imports of Georgian wine and other products, and threatened to send home a million Georgian guest workers whose remittances are vital to the Georgian economy. Most road and rail links between Russia and Georgia were severed, and Turkey replaced Russia as Georgia's main trading partner. Georgia also wants to reestablish its historically important position in overland trade with the central Asian countries, particularly by serving as an outlet for the export of oil from the Caspian Basin.

For many years, a critical issue between Georgia and Russia was the fate of Batumi, the Black Sea oil-shipping port (see Figure 5.20). It lies in the rebellious province of Ajaria (Adzharia), home to another Russian military base and to a population

140

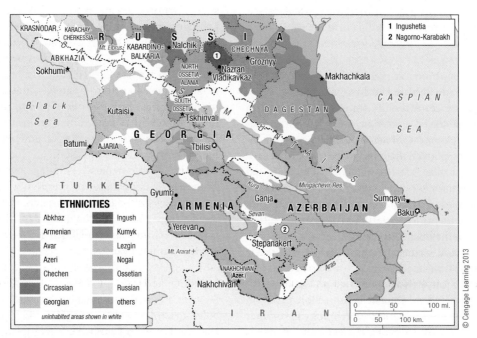

• **Figure 5.20** The Caucasus.

bitterly opposed to the rule of Georgian President Mikhail Saakashvili, who ousted former President Eduard Shevardnadze in the peaceful **Rose Revolution** of 2003. Saakashvili vowed to reassert firm Georgian control over all of the separatist regions within Georgia, thereby risking a direct military confrontation with Russia. His forces secured Batumi and the rest of Ajaria in 2004, and his attention turned to the even more dangerous South Ossetia and Abkhazia (see Figure 5.20).

Russia supports the aspirations of the primarily Muslim South Ossetians, who would like to free themselves from Georgian control and establish an Ossetian nation (almost 100 percent of the population voted to do so in a 2006 referendum). The mostly Christian North Ossetians, whom they wish to join, live adjacent to them in Russia in a region the North Ossetians call Alania. (North Ossetia was the setting of the horrific civilian deaths at Beslan in 2004 after Islamists from nearby Ingushetia seized hostages at a school.) Not far from Ossetia is the southern Russian city of Sochi, the venue for the 2014 Winter Olympics that observers fear would be a magnet for terrorism.

Also seeking independence from Georgia, with Russian help, is another Muslim people, the Abkhazians, who make up about 2 percent of Georgia's population and are concentrated in the province known as Abkhazia (see Figure 5.20). In 1993, Abkhazian separatists captured the Georgian Black Sea port of Sokhumi in Abkhazia. Russian forces initially aided them, both to regain access to Black Sea resorts and to take revenge on Georgia's President Shevardnadze, the former Soviet foreign minister whom many Russians held partly responsible for the breakup of the Union of Soviet Socialist Republics. Russia was also putting pressure on Georgia to rejoin the CIS. Shevardnadze responded by having Georgia rejoin the CIS and agreeing to allow four Russian military bases to remain for a while on Georgian soil and Russian troops to be stationed on Georgia's border with Turkey. Russian forces then put a stop to the Abkhazian offensive but did not drive the rebels from the territory they had captured. Russia began granting Russian citizenship to the majority of Abkhazians—a step that Shevardnadze's successor Saakashvili regarded as Russian annexation of Georgian territory.

For Russia, a friendly Abkhazia is a geopolitical prize, as it affords Russia even more warm waterfront of the Black Sea, from where Russia can project military and commercial power. A friendly Abkhazia would help Russia to safeguard the vulnerable gas and oil pipelines crossing the area and would secure a wider buffer around the Winter Olympics venue of Sochi. In August 2008, Russia saw its chance to jump on an opportunity.

Determined to assert control over South Ossetia, in August 2008, Georgian president Saakashvili sent Georgian troops against South Ossetian rebels. Directed by Russian leaders Putin and Medvedev, Russian forces invaded and quickly secured control over both South Ossetia and Abkhazia. While Georgia scrambled to secure its fractured borders, Russia recognized both regions as newly independent countries (only a handful of other countries in the world recognized them as independent). This conflict was very much an old-style Cold War engagement of proxy forces: The United States and the European Union backed the Georgia side, whereas Russia propped up South Ossetia and Abkhazia.

What Are Russia's Objectives?

Some observers contend that Russia's diplomatic, economic, and military involvement in the countries of the Near Abroad is an effort to establish a buffer zone between Russia and the "Far Abroad." Russia has a military presence in more than half the former Soviet countries. Some contend that above all Russia wants access to important resources and economic assets in its former republics, such as uranium in Tajikistan, aviation plants and the vital BTC oil pipeline (see Geography of Energy, page 140) in Georgia, military plants in Moldova, and the Black Sea coast and naval fleet in Ukraine's Crimea. Within Russia, there is a body of public opinion favoring reassertion of Russian control over its former empire, but the strength of this feeling is unknown. Vladimir Putin spoke of Russia's urgent priority to draw the countries that were republics of the former Soviet Union closer to Russia.

Such reintegration could, of course, take place through peaceful diplomatic and economic means. Within days of announcing his intention to run for president again in 2012, Putin unveiled plans for building a "Eurasian Union" of former Soviet states: Russia, Belarus, and Kazakhstan would be joined by Kyrgyzstan and Tajikistan, and then others. The Eurasian Union, Putin said, would become "one of the poles of the modern world," rivaling the United States and the European Union.

The Far Abroad

Finally, there are outstanding geopolitical issues in relations between this region and the "Far Abroad," the rest of the world. In the years immediately following the breakup of the Union of Soviet Socialist Republics, the most crucial issues involved the development of a new model of relations between East and West and a peaceful succession to the Cold War. After 9/11, these issues were complemented by others related to the "war on terrorism," as perceived by both the United States and Russia. Another phase began after 2004, with a popular and confident Vladimir Putin, buoyed especially by Russia's soaring energy revenues, putting more distance between Russia and the West and constantly reminding Russians of their greatness.

It is important to keep in mind—and Russians especially want the world to know—that Russia is one of the world's great powers. Russia has, since 1997, been a member of what is known as the **Group of Eight (G-8)**, the world's eight most economically powerful and politically influential countries (along with Canada, France, Germany, Italy, Japan, the United Kingdom, and the United States). Russia is a member of the UN Security Council and along with its other permanent members (China, France, the United Kingdom, and the United States) has the enormous veto power that position affords. Russia can therefore have a large impact on world affairs.

To follow this discussion, use the map in •Figure 5.H. The Caspian Sea and the land around it, making up the Caspian Basin, has some of the world's largest reserves of oil and natural gas. Look at the number of countries that border the Caspian Sea. As you might imagine, each one of these five wants to maximize its share of the fossil fuels under the seabed.

The Caspian is called a sea, but looking at it, you can see clearly that it is a large lake. Nevertheless, the countries of the Caspian Basin have squabbled over the geographic naming of the Caspian, with some arguing that it is a lake, and the others, that it is a sea. The reason is that the resolution of this question determines what share of the fossil fuels each country has the right to exploit.

There is not much oil or gas just offshore of Iran's and Turkmenistan's Caspian waterfronts. Iran and Turkmenistan therefore argue that the Caspian should be regarded as a lake. This designation would mean that lakebed resources are common property, with shares divided equally among the five surrounding countries. Russia, Kazakhstan, and Azerbaijan have signed treaties recognizing the Caspian not as a lake, but as a sea. This designation establishes separate sovereign territorial waters, with each country having exclusive access to the reserves in its domain. In this arrangement, Azerbaijan has the largest share. These three countries are actively exploiting their fossil fuel reserves within their designated portions of the seabed.

Have a look at the geopolitics involved in getting the biggest share of Caspian oil to market: Azerbaijan's. Azerbaijan is landlocked, and its oil must cross the territories of other nations by pipeline to reach the Mediterranean Sea. On the map, that looks easy enough, but any export route crosses territories that are embroiled in the political unrest left in the wake of the collapse of the Union of Soviet Socialist Republics.

Russia's position is clear: Azerbaijan's oil should be pumped by pipeline through Russian territory to Russia's Black Sea port of Novorossiysk. In keeping with Moscow's principle that the Near Abroad is Russia's "zone of privileged interest," this route (which can be seen in Figure 5.H) would maximize Russia's influence over Azerbaijan's economic affairs. Russia would also earn revenue from transit fees it would charge to allow oil to pass through Russian territory.

Azerbaijan's leaders, however, want to build political and economic ties with the West (Europe and the United States). The Western nations want to be able to obtain Caspian Sea oil without relying on the goodwill of Russia. Azerbaijan therefore decided to export its oil through a route that does not transit Russia at all. This is the BTC (Baku-Tbilisi-Ceyhan) pipeline, which is strategically routed to block either Russian or Iranian assertion of control over the oil resource. Russia and Iran are of course displeased with this route. The BTC runs across Azerbaijan from Baku through Georgia and into Turkey. Note how it skirts Armenia, which is technically still at war with Azerbaijan over Nagorno Karabakh (see page 149). It continues across Turkey to the Mediterranean port of Dörtyol, near Ceyhan. Why does the BTC take a sudden 90° turn in central Turkey? As you will see in Chapter 6 (on page 202), Turkey has a domestic security problem with its large Kurdish population, located mainly in eastern Turkey. The BTC route detours around the volatile Kurdish region.

Azerbaijan's oil exports through these pipelines have brought considerable new wealth to the country. Its capital of Baku on Azerbaijan's Caspian Sea coast was a windblown, bleak place with blocks of Soviet-era apartments. Now it is beginning to glisten, Dubai-style, with ostentatious urban wealth.

• **Figure 5.H** Numerous physical and political obstacles stand in the way of oil exports from the Caspian region.

168

Energy Issues

Countries of the Far Abroad are asking serious questions about how reliable a business partner Russia is, and how far it might go in flexing its muscle in the countries of the former Soviet Union. In ramping up oil production, Russia is trying to sell itself to the United States and other major industrial powers as a more stable source of energy than the volatile Persian Gulf countries. But that claim rings hollow when Russia manipulates fuel supplies and prices in its Near Abroad and has a downstream effect on supplies to Western Europe.

Western Europe frets because of its growing dependence on gas and oil from Russia through a cat's cradle of politically vulnerable pipelines. Remember how Russia cut off its oil to Belarus and its natural gas to Ukraine? The pipeline that carries Russian oil to Belarus continues on to Germany, so Western Europe's oil supplies were temporarily disrupted •Figure 5.21. The pipeline that carries Russian natural gas to Ukraine likewise continues on to Germany, so the suspension of gas delivery to Ukraine also meant that Western European markets were not getting their supply of natural gas. Russia supplies about one-fourth of Europe's natural gas, and 80 percent of the Russian natural gas delivered to the European Union passes by pipeline through Ukraine.

Western European countries naturally began to regard Russia as an unreliable energy supplier. These countries sought

137

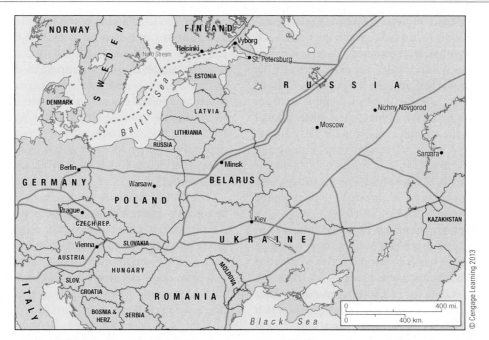

© Cengage Learning 2013

• **Figure 5.21** Is Russia a reliable provider of fossil fuels to Europe, or is it too ready to turn off the tap for political reasons?

ways to reduce their reliance on Russian gas and oil and, in some cases, to find new pipeline routes for Russian oil and gas that would bypass countries with which Russia had disagreements.

The Nord (North) Stream pipeline was built with enormous expense and technical challenges to carry Russian natural gas across the Baltic seafloor to Germany, the largest foreign consumer of Russian gas. Why not build a cheaper, easier pipeline above ground? It would have to cross the territories of Estonia, Latvia, and Lithuania, which are unfriendly to Russia. Its prospective southern counterpart, either a South Stream pipeline or the Nabucco pipeline, likewise would be strategically routed: the South Stream, for Russian gas to bypass Ukraine, which is often at odds with Russia; and the Nabucco, for Azerbaijan's gas to bypass Russia (just as Azerbaijan's oil does in the BTC pipeline described on page 140) and decrease Europe's dependence on Russian gas from 70 to 35 percent.

These projects ultimately do nothing to reduce Western Europe's reliance on Russian and other foreign sources of energy. The European countries are now looking under their own feet to see whether they can find energy independence. One possibility getting much attention these days is the production of shale gas from Europe's old coal fields—the same coal fields that fuelled the Industrial Revolution but then appeared to be played out and were forgotten in Europe's development.

Weapons Proliferation Issues

In addition to energy concerns, the West frets about other problems related to a stronger Russia, such as Russia's assistance to the nuclear and would-be nuclear weapons powers North Korea and Iran; its sales of military aircraft to Sudan, Myanmar, and Venezuela; and its support of the Islamist Hamas organization in the Palestinian Territories.

As important as these issues are, they pale in comparison to one that has thankfully been put to rest: the specter of a nuclear war of annihilation between East and West. The Warsaw Pact, a military alliance formed around the Soviet Union to counter the West's North Atlantic Treaty Organization (NATO) has dissolved. NATO's mission is no longer centered on a confrontation with the East. Both sides have reduced their nuclear arsenals, with Russia taking the lead in calling for even deeper cuts. When President Obama took office, he and his Russian counterpart agreed to "reset" U.S.-Russian relations from a friendly new starting point.

Although no longer preoccupied by the prospect of a conventional nuclear war, the United States and its European allies are very fearful that low budgets and lax security at nuclear facilities throughout the region could help divert nuclear weapons or nuclear fuel to terrorists. There have already been hundreds of thefts of radioactive substances at nuclear and industrial institutions in the former Union of Soviet Socialist Republics. The destination points of these materials, known as **"loose nukes,"** remain largely unknown. The West also fears that underpaid nuclear scientists will sell their know-how to governments or organizations such as al-Qa'ida with nuclear strike intentions. Former Soviet nuclear experts are known to have worked in Iran, Iraq, Algeria, India, Libya, and Brazil, and it is possible that some have cooperated with stateless organizations like al-Qa'ida. The late Osama bin Laden repeatedly stated his intention to acquire and use nuclear weapons. Most analysts think that al-Qa'ida would try first to acquire loose nukes in the region of Russia and the Near Abroad.

Central Asia: Geopolitics Astride the Silk Road

Since 9/11, and particularly since the U.S. invasion of Iraq in 2003, the geopolitical significance of the central Asian "stan" countries has increased enormously. This region, also now empty of nuclear weapons, borders unsettled Afghanistan, the potential flashpoint of Iran, and the great power of China. Central Asia is itself rich in mineral resources. Diverse kinship, spiritual, economic, and political relations exist between the central Asian Muslims and their neighbors on the other side of the international frontiers. Along with internal ethnic conflicts and dissatisfaction over living conditions, these international affiliations may make the region a major political problem area in the future. The popular uprisings of the Arab Spring have not yet spread into non-Arab Central Asia, but some of the same ingredients for the Arab Spring—autocratic regimes controlling a large population of disaffected youth foremost among them—are in Central Asia.

Both Iran and Turkey are vying for increased influence in central Asia. Except in Tajikistan, support for Iran is largely lacking, in part because the majority of central Asians practice Sunni rather than Shi'ite Islam (these branches of Islam are described on page 174). Ethnicity is also important; the four Turkic states are inclined to orient with Turkey, with which they have already established cultural ties such as educational exchanges and shared media. Turkey also has extended economic assistance and established small business ventures and large construction contracts in central Asia. Turkmenistan has good relations with both Turkey and Iran, and has built a rail link with Iran. Turkmenistan also tries to maintain good ties with Russia.

The region has become a focus for rivalry between the historical enemies Turkey and Russia. Russians are concerned that Turkey is winning too much influence in the young Muslim countries of Central Asia. For their part, some Turks believe Russia will try to take over the Transcaucasus region again and then turn to Central Asia. Some Turks uphold a dream of **pan-Turkism**, uniting all the Turkic peoples of Asia from Istanbul to the Sakha Republic in Russia's Siberia region—a prospect that Moscow of course does not like.

The twin scourges of terrorism and narcotics—often closely linked—have put central Asia on the map of geopolitical hot spots. Uzbekistan has erected concrete and barbed-wire barriers and sown land mines on its borders with Tajikistan and Kyrgyzstan. The Uzbek government is trying to prevent the incursion of what it sees as two huge threats: an epidemic in heroin use and a militant Islamic insurgency. Both issues have preoccupied the region's governments. The major routes for exports of heroin from Pakistan and Afghanistan are north through Central Asia and then into Russia and Western Europe. Along this pathway, large numbers of central Asians have fallen prey to inexpensive, pure heroin. The associated health and social problems, including HIV infection, hepatitis, and prostitution, have also taken root. This extremely valuable commodity corrupts people at every level, all the way up to governments. The United Nations estimates that about 40 percent of Tajikistan's economy is drug-related. As we will see later (page 242), drug sales fund the war effort of the Taliban against the United States and its NATO allies.

Several powers are alarmed by what they perceive as an Islamist threat to Central Asia. Russia has expressed great concern about an Islamist wave spreading from Afghanistan and Pakistan into Central Asia and Russia itself. Tajikistan and Kyrgyzstan have allowed Russia to deploy thousands of Russian troops in their countries, in part to stave off such a threat. Some observers believe Russia is using the Islamist threat as a pretense to maintain influence in the region, or even to counter the growing U.S. military presence there since 9/11. And human rights groups in the West argue that authoritarian, oppressive regimes in Central Asia exaggerate the threats from Islamist militants to justify their harsh treatments of political opponents.

The United States is genuinely worried about the Islamist danger. Its greatest fears are focused on Tajikistan, Central Asia's poorest country. Islamist rebels with links to both al-Qa'ida and Afghanistan's Taliban are intent on disrupting U.S. military supply lines from Uzbekistan into Afghanistan and have also been attacking Tajikistan's soldiers. Those confrontations could lead to the reemergence of civil war, which plagued Tajikistan from 1992 to 1997.

Civil war also threatens Kyrgyzstan, where a minority population of ethnic Uzbeks (who make up 15 percent of the country's population) are seeking greater autonomy. They frequently engage in bloody clashes with ethnic Kyrgyz. Uzbekistan is fearful that the dominant Kyrgyz people will ethnically cleanse the country's Uzbeks. Ethnic Kyrgyz for their part fear that Uzbekistan may take advantage of instability to occupy land in Kyrgyzstan. Russian president Medvedev expressed his concern that Kyrgyzstan would soon become "a second Afghanistan."

Tajikistan was home to a group called the **Islamic Movement of Uzbekistan (IMU)**, which has recently been pressured southward into the Afghanistan-Pakistan border region. For many years, its principal aims were to overthrow the Uzbek government of President Islam Karimov and establish an Islamic state in the Fergana Valley, which runs through Uzbekistan, Tajikistan, and southern Kyrgyzstan (see Figure 5.31). Following a 1999 attempted assassination blamed on the rebels, Karimov cracked down hard on suspected Islamists and even began to restrict personal symbols of being Muslim, such as beards on men and headscarves on women. Tajikistan's government is also employing these repressive tactics, which many observers believe will only help grow Islamist anti-government sentiments.

Anxious to repress any external support for Islamist insurgency within its borders and to curry political and economic favor with Washington, Uzbekistan initially welcomed the U.S. military to use its land and air space in attacks against al-Qa'ida and the Taliban in 2001 and 2002. Uzbekistan shut down this access in 2005, after the United States and its European partners in NATO, citing human rights abuses, slashed economic assistance to the country. Soon, however, the United States' escalating engagements in Afghanistan led Washington to overlook poor human rights records in both Uzbekistan and Kyrgyzstan. American and NATO military supplies began pouring by air and rail into Afghanistan from both countries, and Uzbekistan built a 45-mile-long rail line into the Afghan

243 city of Mazar-e-Sharif, mainly for the transport of these military goods. The U.S. airbase at Manas in Kyrgyzstan has served as a critical airbase in the West's military campaign in Afghanistan. The Russian military also operates bases in both Kyrgyzstan and Tajikistan. In 2009, Kyrgyz president Bakiyev double-crossed Russia, promising to shut down U.S. access to the Manas airbase in return for a large economic aid package from Russia. Once Bakiyev secured the Russian aid, he collected another large sum from the United States to renew its lease of Manas. Bakiyev was ousted in a coup the following year, and Kyrgyzstan headed down the road toward a "failed state" status.

264 Last but certainly not least as we consider the Far Abroad is that country whose influence is on the rise in every region covered in this book: China. China has sought to build security relationships with the Central Asian countries, especially to blunt the aspirations of Muslim insurgents in China's far western region. In a proclaimed effort to confront terrorism, separatism, and extremism (and perhaps to counterbalance NATO), China and Russia have recently formed a confederacy with all of the central Asian countries except Turkmenistan, known as the **Shanghai Cooperation Organization (SCO)**.

Not surprisingly, given its geographic proximity to the region, China has growing economic interests in the Central Asia. Chinese exports to Central Asia are surging. China is upgrading its roads leading into Kyrgyzstan and cooperating with Kazakhstan in building a railway across Kazakhstan that will ultimately link producers and markets in China and Europe. China has made multibillion dollar loans to Turkmenistan and Kazakhstan in exchange for guaranteed future access to oil and gas supplies in those countries. Natural gas flows from Turkmenistan to western China through a 1,140-mile-long pipeline crossing three countries. Notably, Russia is not one of these: when the pipeline began operation in 2009, it was the first major export route for the region's natural gas that did not pass through Russia. This pipeline is also notable as the first to run east out of the region. If all roads once led to Rome, they now lead to China.

Following are more insights into the land and life in the subregions and countries that make up Russia and the Near Abroad.

5.6 Regional Issues and Landscapes
Peoples and Resources of the Coreland

In this book's first chapter, you were introduced to the geographic concepts of core and periphery. If you read the previous chapter on Europe, you know that there are distinctive core and periphery elements of that region. This is also true of Russia and the Near Abroad.

An outstanding feature in the geography of this huge region is that resources are distributed unevenly, favoring development in certain subregions and countries. Agricultural and industrial resources are in fact clustered in a core region encompassing five countries partially or entirely: western Russia,

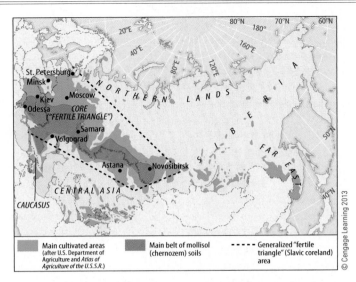

• **Figure 5.22** Major regional divisions of Russia and the Near Abroad, as discussed in the text.

northern Kazakhstan, Ukraine, Belarus, and Moldova. Nearly three-fourths of the region's people—the great majority of them Slavic—and an even larger share of the cities, industries, and cultivated land of this immense region are packed into this roughly triangular **Fertile Triangle** comprising about one-fifth of the region's total area. Also known as the **Agricultural Triangle** and the **Slavic Core**, this is the distinctive functional hub of the region (•Figures 5.22 and 5.23).

Slavic peoples are the dominant ethnic groups in most of the region of Russia and the Near Abroad in both numbers and political and economic power. The major groups are the Russians (discussed extensively earlier in the chapter; the Slavs and other ethnolinguistic groups are introduced on page 124), Ukrainians, and Belarusians (Byelorussians). Most of these Slavs live in the core region extending from the Black and Baltic Seas to the neighborhood of Novosibirsk in Siberia. Moscow, with a population of about 15 million, and Saint Petersburg (formerly Leningrad), with almost 5 million people, are the largest cities, followed by 3 million people living in the Ukrainian capital, Kiev.

The Ukraine

Ukrainians, the second largest ethnic group in the Slavic Core, are closely related to the Russians in language and culture; "Russia writ small" is how one author described Ukraine.[9] The name Ukraine translates as "at the border" or "borderland," and this region has long served as a buffer between Russia and neighboring lands (•Figure 5.24). Here armies of the Russian tsars fought for centuries against nomadic steppe peoples, Poles, Lithuanians, and Turks, until the eighteenth century, when the Russian Empire finally absorbed Ukraine. For three centuries, Ukrainians were subordinate to Moscow, first as part of the Russian Empire and then as a Soviet republic. Ukraine's industrial and agricultural assets were always vital to the Soviet Union; the Bolshevik leader, Vladimir Lenin, once declared, "If we lose the Ukraine, we lose our head!"

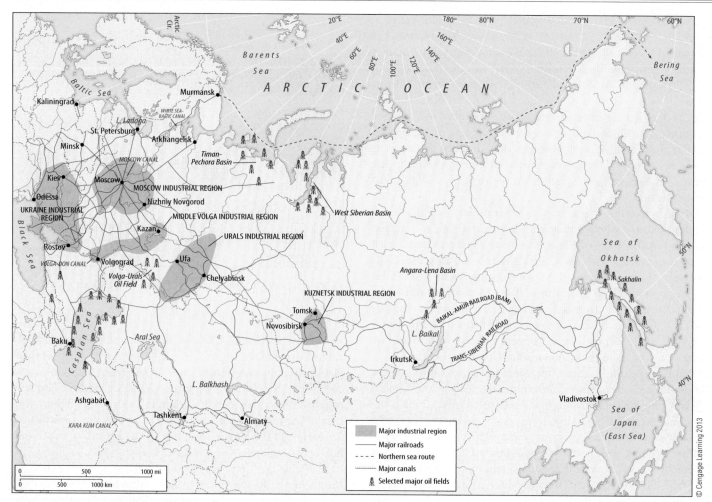

• **Figure 5.23** Major economic zones and transportation routes of Russia and the Near Abroad. The map reveals that the Fertile Triangle is also the industrial core of the region. Moscow's function as a transportation hub is evident. It is similar to the depiction of Paris as a hub for France, in **Figure 4.25**; however, Russia does not have a primate city as France does.

Slightly smaller than Texas, Ukraine lies partly in the forest zone and partly in the steppe. On the border between these biomes, and on the banks of the Dnieper River, is the historic city of Kiev, Ukraine's political capital and a major industrial and transportation center. Ukraine is blessed with an abundance of fertile black earth soils, especially in the steppe region south of Kiev. Historically, Ukraine has been a great "breadbasket" of wheat, barley, livestock, sunflower oil, beet sugar, and many other products. Ukraine also has a generous endowment of industrial raw materials, especially coal and iron ore. Agricultural and industrial outputs plummeted after Ukraine gained independence from the Soviet Union in 1991, but are beginning to pick up again. Many of the social problems that plagued Russia after the breakup of the Union of Soviet Socialist Republics also troubled Ukraine: increasing mortality, falling birth rates, and broken families are among them. In Ukraine, Moldova, Belarus, and Russia, increasing numbers of women fell victim to sex trafficking as they struggled to improve their lives (**see A Closer Look on page 145**).

Chernobyl

North of Kiev is **Chernobyl**, where an explosion at a nuclear power station in 1986 rendered parts of northern Ukraine and adjoining Belarus and parts of Russia incapable of safe agricultural production for years thereafter. Encircling the station is still an 18-mile (29 km) "exclusion zone" where farming is prohibited. Recent studies suggest that 100,000 to 200,000 people are still severely affected by the Chernobyl disaster and that 4,000 deaths will ultimately be attributed to it. However, the toll on human health and life is well short of what had originally been predicted. The psychological trauma to the region's inhabitants has been profound, and the three affected countries have spent large sums to provide benefits to Chernobyl's victims. With financial assistance from Russia, the European Union, and the United States, Ukraine is decommissioning all of its Chernobyl-type nuclear plants. Until recently, the Chernobyl disaster was a singular event. Now it is joined by Japan's Fukushima catastrophe in the world's imagination of nuclear power gone wrong.

A **closer** LOOK

Human Trafficking and the Sex Trade

One of the outcomes of the collapse of communism and the dissolution of the Soviet Union has been the unleashing of long-restrained expressions of sexuality. Entrepreneurs in Russia and other former Soviet countries quickly adopted an axiom of Western advertising: sex sells. Russian television has surpassed even the spiciest fare on standard American television.

Many sexual awakenings in the region are applauded or accepted as good, clean fun, but others have a decidedly dark side, and some send ripples far abroad. Far too many women of the former Soviet Union have become commodities (•**Figure 5.1**). Russian, Ukrainian, and Kazakh partners or brides can in effect be purchased by anyone in the world with access to the Internet. Dot-com agencies act as brokers for these New Age "mail-order brides." Some of the subsequent unions end happily, of course, but others collapse with physical abuse and heartache. Frequently, the woman ends up abandoned and destitute in a strange land.

The sex trade takes the greatest toll. Primarily in this industry, women become victims of what is known as **human trafficking**, often unwittingly and almost always out of poverty. Following is the classic pattern, played out over and over, not only in this region but also around the world. Criminal gangs use false promises of conventional employment to lure young women into the trade. The typical victim is an 18- to 30-year-old Russian, Ukrainian, Belarusian, or Moldovan woman—often with children to support—who is approached by a trafficker promising her a job as a waitress, barmaid, babysitter, or maid in Eastern or Western Europe. She pays a fee for travel and visa costs. Arriving in Bosnia and Herzegovina, the Czech Republic, or Germany, she is met by another trafficker who confiscates her passport and identity papers and then sells her for several thousand dollars to a brothel owner. She is escorted to her guarded apartment. There she is compelled to reside with several other women who have taken a similar journey, and she is informed of her real job: prostitute.

Berlin papers advertise the merchandise: "The Best from Moscow!" "New! Ukrainian Pearls!" The working women meet their clients in clubs or brothels or are driven to their homes. Of the rates charged for their services, typically less than 10 percent goes to the women. Their pimps and other gang members take the lion's share, and the women use their meager earnings to buy food and pay rent. Most end up in debt. Unable to pay off their debts, lacking identification papers, fearful of going to the police, and terrified the gang might harm family members back home if they try to escape, the women are trapped. Reports of depression, isolation, sexually transmitted disease, beatings, and rape abound.

This is a big business with striking demographics. The sex industry in Europe generates revenues of several billion dollars per year. Ukrainian sources estimate that since the country's independence in 1991, more than half a million Ukrainian emigrant women have ended up in the European sex trade. Figures vary widely, but tens and perhaps even hundreds of thousands of women of the former Soviet Union immigrate to Europe each year, the great majority of them becoming prostitutes.

There is an overflow into other regions. These so-called "Natashas" work in Turkey, Israel, the United Arab Emirates, and other countries in the Middle East in similar deplorable circumstances. Many of the world's countries, including the United States, have human trafficking problems (see page 394).

89,
158,
325

Joe Hobbs

• **Figure 5.1** Window shopping in Amsterdam's Red Light District. An estimated 85 percent of the working women here are trafficked, mainly from the former Soviet countries.

• **Figure 5.24** Belarus, Ukraine, and Moldova were among the most productive republics of the Soviet Union and are now struggling to develop their economies independently or in association with Russia.

• **Figure 5.25** Fluorescent lighting and artificial climate controls allow year-round harvests in Russian greenhouses.

Ukraine paid a high economic cost for independence. Previously, the Soviet economy subsidized the provision of Russian oil, gasoline, natural gas, and uranium to Ukraine. As discussed earlier (page 137), Ukraine is now subject to the whims of Russia where oil prices and supplies are concerned. Ukraine's chemical plans and steel mills are natural gas guzzlers, and Ukraine cannot afford to run out. Like the Western European countries, Ukraine is looking to extract natural gas in its own territory through fracking of its shale deposits. Ukraine is well endowed with other mineral resources, including large deposits of coal, iron, manganese, and salt. Since the breakup of the Union of Soviet Socialist Republics, Ukraine has remained far more reliant on heavy industrial production than Russia.

Farming the Fertile Triangle

Most of the Fertile Triangle is within Russia, which still faces huge problems in transforming state-run into free-market farming. Throughout the second half of the twentieth century, the Fertile Triangle was a global-scale producer of farm commodities such as wheat, barley, oats, rye, potatoes (Russia's leading staple food), sugar beets, flax, sunflower seeds, cotton, milk, butter, and mutton. The overall high output, however, did not reflect high agricultural productivity per unit of land and labor. There were many inefficiencies in the state-operated and collective farms that dominated the agricultural sector. On the state-operated farms, workers received cash wages in the same manner as industrial workers. Bonuses for extra performance were paid. Workers on collective farms received shares of the income after obligations of the collective had been met. As on the state farms, there were bonuses for superior output.

These methods, however, failed to provide enough incentives for highly productive agriculture. The patterns of agricultural failure parallel those in the industrial sector in the post-Soviet era. Farm machinery stood idle because of improper maintenance; since no individual owned the machine, the incentive to repair it was diminished. There were also shortages of spare parts. Poor storage, transportation, and distribution facilities, plus wholesale pilfering, caused alarming losses after the harvest. Young people deserted the farms for urban work, often living with the family in the countryside but commuting to their city jobs. With the shortage of younger male workers, the elderly, women, and children did an increasingly large share of the farmwork.

In an effort to reverse declines in the agricultural sector, all of the countries of the Fertile Triangle are promoting land reform by privatizing collective and state farms and developing more independent farming. Russia has been slow to privatize farming. Private farms produce only about 10 percent of Russia's agricultural output. The former state farms control about 75 percent of Russia's cultivated land and produce more than half of the country's agricultural output. Russia has made revival of the agricultural sector one of its "national projects," but like almost all the countries in this region, Russia remains a net food importer.

Surprisingly, even in the dead of Russia's winter, fresh vegetables find their way from greenhouses onto supermarket shelves (•Figure 5.25). And everywhere Russians have an opportunity to grow crops on the land they privately own, they make the best of it. Dachas, the usually modest second homes that many Russians own in the countryside (see page 120), make a large contribution to the country's overall agricultural output.

Russia's Far East and Northern Lands

The Far East

The Far East is Russia's mountainous Pacific edge (see Figure 5.1). Most of it is a thinly populated wilderness in which the only settlements are fishing ports, lumber and mining camps, and the villages and camps of aboriginal peoples. Port functions, fisheries, and forest industries provide the main support for most Far Eastern communities, and until recently the output of coal, oil, and a few other minerals was small. Most of the Russians and Ukrainians who make up the majority of its people live in cities along two transportation arteries: the Trans-Siberian Railroad and the lower Amur River. At the south, on the Sea of Japan, is the port of Vladivostok, which is kept open throughout the winter by icebreakers. About 50 miles (80 km) east of the city, the main commercial seaport area of the Far East has developed at Nakhodka and nearby Vostochnyy (East Port). Both ports are nearly ice-free and well positioned for trade in goods manufactured in Korea, China, and Japan. Seaborne shipments of cargo containers from northeast Asia to Scandinavia require 40 days, but the cargo can make the 6,000-mile (9,600 km) journey from Vostochnyy to Finland by rail in only 12 days. Business in Vostochnyy is picking up. A new oil terminal is being built at nearby Krylova Cape to support exports of oil, to be brought in initially on the Trans-Siberian Railroad, to Japan and elsewhere.

The diversified industrial and transportation center of Khabarovsk is located at the confluence of the Amur and Ussuri Rivers, where the main line of the Trans-Siberian Railroad turns south to Vladivostok and Nakhodka and the Amur River turns north toward the Sea of Okhotsk (see Figure 5.23). Before World War II, the Soviet Union and Japan held the northern and southern halves, respectively, of the large island of Sakhalin. The island has much pristine wilderness, with some forest, fishing, and coal mining, but is particularly important for its offshore petroleum and natural gas. Russian energy officials are describing it as a **"Second Kuwait,"** with about 1 percent of global oil reserves and enough natural gas to supply the U.S. import market for about 25 years. Some of the liquefied natural gas will make its way to the United States from a terminal in northern Mexico, but most of the gas will be exported to China, Japan, and South Korea.

As World War II came to a close, the Union of Soviet Socialist Republics annexed southern Sakhalin and the Kuril Islands and sent their Japanese residents back to Japan. Russian control of these former Japanese territories continues to be a problem in relations between the two countries and has kept them from reaching a formal post–World War II treaty; technically, Russia and Japan are still at war! The countries have, however, been trying to break the impasse, negotiating about how to divide the islands the Russians call the southern Kurils and the Japanese know as the Northern Territories.

Both countries are trying to put these differences aside in favor of economic cooperation, particularly in the energy realm. Japan negotiated contracts to buy natural gas from Sakhalin and to help build an oil pipeline from the eastern Siberian fields near Irkutsk, to the port of Nakhodka (•Figure 5.26). Russia is happy to have the foreign investment, and Japan, which has almost no oil, is happy to have the energy from a source outside the volatile Middle East. With its booming economy, China is desperate for more energy, especially

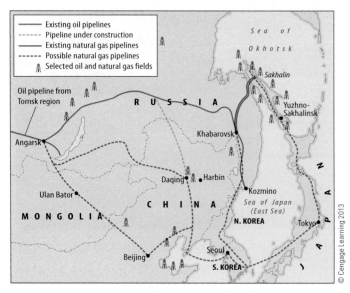

• **Figure 5.26** Existing and proposed pipelines for eastern Siberian and Far Eastern oil and gas exports.

from closer and therefore cheaper sources, and is also negotiating with Russia to obtain eastern Siberian oil through a new pipeline. Russia is diversifying the geography of its oil exports. Until recently, all of Russia's energy exports were geared toward western markets. Now Russia is opening export routes in the east. If one or the other is threatened for any reason, there will be a backup.

At the north, the Kurils Islands approach the mountainous peninsula of Kamchatka (total population 300,000, with about 60 percent living in Petropavlovsk). Like the Kurils, Kamchatka is located on the Pacific Ring of Fire—the geologically active perimeter of the Pacific Ocean—and has 23 active volcanoes (•Figure 5.27). Soviet authorities protected Kamchatka for decades as a military area, and no significant development activities occurred there. A California-size land of rushing rivers, extensive coniferous forests, and gurgling hot springs, Kamchatka is now one of the last great wilderness areas on Earth, resembling the U.S. Pacific Northwest landscape of a century ago. A struggle has been underway between those who want to develop Kamchatka's gold and oil resources and those who wish to set the land aside in national parks and reserves where ecotourism would be the only significant source of revenue.

Much to the surprise of environmentalists abroad, Russia's conservationists seem to have prevailed. Fossil fuel development will take place, but so will an ambitious effort to set aside seven wilderness tracts totaling more than 9,400 square miles (c. 24,400 sq km), or about three times the size of Yellowstone National Park in the United States. The main goal will be to have multiple uses from Kamchatka's rich salmon stocks—for food, sport, study, and profit—and also protect this resource from the kind of overfishing that has laid waste to the cod fisheries of the Atlantic and North Seas and to Russia's sector of the nearby Bering Sea.

Image courtesy of the Image Science & Analysis Laboratory, NASA Johnson Space Center

• **Figure 5.27** Kamchatka's spectacular volcano-studded landscape, as seen from space.

• **Figure 5.28** The Russian nuclear-powered icebreaker *Yamal* at the North Pole. The former Soviet icebreaker fleet still escorts commercial ships on the Northern Sea Route but during the summer carries Western tourists such as these to the High Arctic. To be honest, we were 12 miles from the Pole (south of the Pole, of course) when I took this photograph. The North Pole was ice-free in August 2006 (see page 32). The ship's Captain brought the gigantic vessel to 90° north latitude, and we all celebrated being there. But we could not touch the North Pole. Twelve miles from the Pole, the captain "beached" the *Yamal*, and all the passengers came down the gangway to spend some time on the ice.

Joe Hobbs

The Wild North

North and east of the Fertile Triangle and west of (and partially including) the Pacific coast lie enormous stretches of coniferous forest (taiga) and tundra extending from the Finnish and Norwegian borders to the Pacific (see Figure 5.4). These outlying wilderness areas in Russia may, for convenience, be called "The Northern Lands," although parts of the Siberian taiga extend to Russia's southern border.

The Arctic port of Murmansk (population 300,000, the world's largest city within the Arctic Circle) is located on a fjord along the north shore of the Kola Peninsula west of Arkhangelsk. It is the headquarters for important fishing trawler fleets that operate in the Barents Sea and North Atlantic. Murmansk also has a major naval base and cargo port and is homeport to the icebreakers that escort cargo vessels and carry Western tourists to the North Pole and other high Arctic destinations (•Figure 5.28). It is connected to the Russian Core by rail. The harbor is open to shipping all year thanks to the warming influence of the North Atlantic Drift, an extension of the Gulf Stream. Murmansk and Arkhangelsk played a vital role in World War II, when they continued to receive supplies by sea from the Soviet Union's Western allies after Nazi forces captured and closed the other ports of the western Soviet Union.

Murmansk and Arkhangelsk are western terminals of the **Northern Sea Route**, a waterway the Soviets developed to provide a connection with the Pacific via the Arctic Ocean (see Figure 5.23). Until recently, navigation along the whole length of the route was possible for only up to 4 months per year, despite the use of powerful icebreakers (including several nuclear-powered vessels) to lead convoys of ships. Areas along the route provide cargoes such as the timber of Igarka and the metals and ores of Norilsk. Some supplies are shipped north on Siberia's rivers from cities along the Trans-Siberian Railroad and then loaded onto ships plying the Northern Sea Route for delivery to settlements along the Arctic coast.

The Northern Sea Route is now open to international transit shipping, and if not for the capricious sea ice, it would be an attractive route. In recent winters, typical sea ice coverage in the Arctic Ocean has retreated substantially—one of the most accelerated trends in overall global warming, as described on page 32—and the Russian shipping industry is hoping this trend will continue. The distance between Hong Kong and all European ports north of London is shorter via the Northern Sea Route than through the Suez Canal, and Russia could earn valuable transit fees from this route. Russia is also anxious to develop the Northern Sea Route for its own exports because Russian exports by land to Western Europe must now pass over the territories of Ukraine, Belarus, and the Baltic countries, which collect their transit fees from Russia.

Countries that have territories in the Arctic are making plans to exploit resources made more accessible by the forces of climate change. The race for Arctic riches is described on pages 426–427. Here is a preview of that discussion: in 2007, Russians working in a mini-submarine planted the Russian flag on the seabed at the North Pole. This was a not-so-subtle message that Russia was laying claim to the oil and gas wealth that may lie here.

There is growing support in Russia, the United States, and Canada to construct a rail link between Siberia and North America. This would require laying another 2,000 miles (c. 3,200 km) of track from the Baikal-Amur Mainline (BAM) to the Bering Strait (see the rail lines in Figure 5.23). An underwater tunnel would carry the track across the Bering Strait, where another 750 miles (c. 1,200 km) of track would be laid to Fairbanks, Alaska. New lines would connect Alaska with western Canada and the lower 48 United States. Proponents of this route say that a Russia–North America railway link would reduce trans-Pacific shipping times by weeks and that it could carry 30 billion tons of cargo each year.

Baikal

The BAM passes through Bratsk, site of a huge dam and hydroelectric power station on the Angara River. Lake Baikal, which the Angara drains, lies in a mountain-rimmed rift valley. It is the deepest body of freshwater in the world, more than 1 mile (1.6 km) deep in places, and may contain as much as one-fifth of the world's unfrozen freshwater (•**Figure 5.29**). Said to be the oldest lake in the world at 30 million years of age, the lake contains an estimated 1,800 endemic plant and animal species. "Nature as the sole creator of everything still has its favorites, those for which it expends special effort in construction, to which it adds finishing touches with a special zeal, and which it endows with special power," wrote Valentin Rasputin, a native of Siberia. "Baikal, without a doubt, is one of these. Not for nothing is it called the pearl of Siberia."[10]

There is an environmental controversy over wastewater containing dioxin and sulphuric compounds that is discharged into Lake Baikal from the Baikalsk Pulp and Paper Mill. The mill had been shuttered after environmentalists won a case arguing that the waste jeopardizes Baikal's unique natural history. In 2010, Prime Minister Putin signed a decree allowing the resumption of the discharge. Russian environmentalists are upset with this outcome.

But Russia's government did take other steps to protect Lake Baikal. Proposed routes for exporting oil eastward from the Angarsk region (see Figure 5.26) also threatened this natural wonder. Each route would have crossed a number of rivers flowing into Lake Baikal. Responding to fears that an earthquake in this seismically active region could result in calamity, then-president Putin in 2006 decreed that the proposed northern route should be moved still farther north, away from Baikal's watershed. As major oil production comes online in the eastern Siberian oilfields around Irkutsk, there will be new environmental concerns.

• **Figure 5.29** Remote sensing image of Lake Baikal, Earth's deepest lake, in winter.

MODIS/NASA

The Caucasus

A Cauldron of Inter-Ethnic Conflict

The Caucasian isthmus has been an important north–south passageway for thousands of years, and the population includes dozens of ethnic groups that have migrated into this region (see Figure 5.11). Most of these are small ethnic populations confined to mountain areas that became their refuges in past times. The tiny republic of Dagestan, within Russia on its southeastern border with Georgia and Azerbaijan, is by itself home to 34 different ethnic groups; as we saw earlier, they do not always live comfortably together. Some mountain villages, particularly in southeastern Azerbaijan, have achieved international fame as the abodes of the world's longest-lived people—up to the challenged claim of 168 years!

In addition to Russians and Ukrainians, most of whom live north of the Greater Caucasus Range, the most important nationalities are the Georgians, Azerbaijanis, and Armenians, each represented by an independent country. Throughout history, all have defiantly maintained their cultures in the face of pressure by stronger intruders. These nationalities have ethnic characteristics and cultural traditions that are primarily Asian and Mediterranean in origin. Their religions differ. The Azerbaijanis (also known as Azeri Turks) are Shiite Muslims. Most Georgians belong to one of the Eastern Orthodox churches. The Armenian Apostolic Church is a very ancient, independent Christian body that is an offshoot of Eastern Orthodoxy; in 301 C.E., Armenia became the world's first Christian country when its government pronounced the young religion as its national faith (•**Figure 5.30**).

There has been a history of animosity between the Armenians and Azeri Turks (including a war between them in 1905), growing out of the Turks' persecution of Armenians in the Ottoman Empire prior to and during World War I. Turkey and Azerbaijan still have not accepted responsibility for their roles in the **Armenian genocide**, in which an estimated 1.5 million Armenians died between 1915 and 1918. The descendants of Armenians who fled this holocaust have established themselves in many places around the world (for example, there is a large population in Hollywood's film industry) and today outnumber Armenians living within the country by almost two to one. One indicator of Armenia's declining economic fortunes following independence in 1991 is that an estimated 40 percent of the country's population has moved abroad since then.

Turkey and Armenia finally established diplomatic relations in 2009. If Turkey were to tolerate a dialog on the Armenian genocide, it would help Turkey's application to join the European Union. At least that was the reasoning before the European Union became so troubled that Turkey began to lose interest in it.

Nagorno-Karabkh

There are some outstanding problems among the Caucasus countries, in addition to the international geopolitical issues discussed earlier (pages 138–139). Early in the 1990s, historical animosity flared into massive violence between Armenians and Azeris over the question of Nagorno-Karabakh, a predominantly Armenian

• Figure 5.30 The Armenian Church of the Holy Cross on the island of Akhtamar, now in eastern Turkey. Armenia was the world's first Christian country, and its ancient borders included Mount Ararat and other land later lost to Turkey. In 2010, in a symbolic act of growing religious tolerance, Turkish authorities allowed Armenians to pray inside the Church of the Holy Cross for the first time since the collapse of the Ottoman Empire in 1915.

enclave of 1,700 square miles (4,420 sq km) within Azerbaijan, governed at the time by Azerbaijan but claimed by Armenia (see Figure 5.20). From 1992 until 1994, when Armenian forces finally secured the region (which is still technically part of Azerbaijan), fighting over Nagorno-Karabakh took 30,000 lives and created nearly a million refugees. Large numbers of Armenians fled from Azerbaijan to Armenia as refugees. There was also a flight of Azeris from Armenia to Azerbaijan, where hundreds of thousands continue to live as refugees.

Armenia and Azerbaijan have not yet made peace. It is possible that they will go to war again over Nagorno-Karabakh. But diplomats from France, Russia, and the United States have been trying to develop a framework for a peace agreement. In this proposal, the Armenians would return to Azerbaijan six of the seven regions they captured. Nagorno-Karabakh and the adjacent Lachin region would be granted self-governing status. Azerbaijan's compensation would be the construction of an internationally protected road linking Azerbaijan with its enclave of Nakhichevan, currently cut off from Azerbaijan by Armenian territory.

Central Asia

The Shrinking Aral Sea

The five central Asian "stans" are mainly flat, with plains and low uplands, except for Tajikistan and Kyrgyzstan, which are spectacularly mountainous and contain the highest summits in the region of Russia and the Near Abroad in the Pamir and Tien Shan Ranges (•Figure 5.31; see also Figure 5.10). Central Asia

• Figure 5.31 Principal features of Central Asia.

is almost entirely a region of interior drainage. Only the waters of the Irtysh, a tributary of the Ob, reach the ocean; all the other streams either drain into enclosed lakes and seas or gradually lose water and disappear in the central Asian deserts.

Historically, many of this region's people were pastoral nomads who grazed their herds on the natural forage of the steppes and in mountain pastures in the Altai and Tien Shan Ranges. Over the centuries, they slowly drifted away from nomadism. The Soviet government accelerated this process by forcibly collectivizing the remaining nomads and settling them in permanent villages. During the twentieth century, Russians and Ukrainians poured into central Asia, mainly into the cities. Meant to Russify this non-Russian area, they included political dissidents banished by the Communists, administrative and managerial personnel, engineers, technicians, factory workers, and in the north of Kazakhstan, farmers in the "new lands" wheat region. Most of the Slavic newcomers lived apart from the local Muslims, as most of those who remain today still do.

Most people today live in irrigated valleys at the base of the great mountains in the region's south. There, the wind-blown loess and other soils are fertile, the growing season is long, and rivers flowing from mountains provide irrigation water. The principal rivers in the heart of the region are the Amu Darya (Oxus) and Syr Darya, both of which empty into the enclosed Aral Sea.

Only northern Kazakhstan, which extends into the black-earth belt, receives enough moisture for dry farming (unirrigated agriculture). Irrigation is needed everywhere else, providing water to mulberry trees (grown to feed silkworms), rice, sugar beets, vegetables, vineyards, and fruit orchards.

Cotton began to be cultivated here during the American Civil War, when U.S. exports abroad virtually ceased. The Soviets later fostered a "plantation economy" of their own in Central Asia, exploiting the region's potential to produce large quantities of this valuable crop (known for a time as **"white gold"**) for the central Russian textile mills. Cotton remains the region's major crop. Production flows to textile mills in Tashkent, the Moscow region (at levels far below those of Soviet times), and locations outside the region.

There was a significant environmental price to pay for the development of central Asian agriculture. Driven by the need for foreign exchange from export sales, Soviet planners reshaped the basins of the Syr Darya and Amu Darya with 20,000 miles (c. 32,000 km) of canals, 45 dams, and more than 80 reservoirs. As the area under irrigated cotton grew, the rivers and the Aral Sea became highly polluted with agricultural chemicals draining from the fields. Diversion of water from the rivers caused the Aral Sea to shrink rapidly in volume and area, virtually destroying a vast natural ecosystem and an important regional fishery. Since the 1950s, the Aral Sea has lost 60 percent of its surface area and 80 percent of its volume and has divided into three small lakes (see the former and present shorelines in Figure 5.31 and remote sensing images of the lake's transformation in •Figure 5.32). With assistance from the World Bank, Kazakhstan recently built a dam to contain the smaller lake, known as the North Aral Sea, completely within its borders. It is engineered to reduce salinity in the lake and divert any excess water to the larger, parched Aral Sea. The North Aral Sea's waters have risen faster than anticipated, and there is much optimism about a revitalization of its fishing industry.

Both Kazakhstan and Uzbekistan use tremendous quantities of water to flood paddies to produce rice—an extraordinary land use decision in such a thirsty region. Rice cultivation is one reason Uzbekistan is desperate for water and accuses Kyrgyzstan of hoarding it. Kyrgyzstan insists it needs to store water for hydroelectric power because the downstream countries will not provide it with the fossil fuels it needs. Meanwhile, Turkmenistan has declared that it will harness Amu Darya waters to create a new "Lake of the Golden Turkmen" to irrigate an additional 1,875 square miles (4,860 sq km) of land. Where all the water will come from is not clear, especially since the five "stans" signed an agreement in the 1990s, promising to freeze in place their Soviet-era water distribution allotments.

(a)

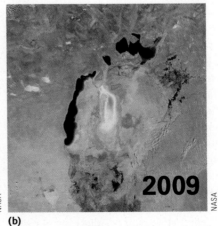
(b)

• **Figure 5.32** The Aral Sea in decline. In this remotely-sensed image sequence, **(a)** was taken in 2000, and **(b)** in 2009.

Another environmental problem—this one with international security dimensions—existed on the former island of Vozrozhdeniye, on Uzbekistan's side of the Aral Sea. In 1988, as the Mikhail Gorbachev's Soviet Union hastened to build new bridges with the West, Soviet biological warfare specialists buried hundreds of tons of living anthrax bacteria, encased in steel canisters, on the island. As the Aral Sea continued to shrink, the island became a peninsula, raising fears that these more accessible biohazards could be exposed and blown into populated areas or be carried off by would-be terrorists. Following a number of incidents of anthrax-tainted letters reaching U.S. Senate offices and other facilities in the wake of the 9/11 attacks, the U.S. government provided the funds needed to clean up "Anthrax Island"; all anthrax burial sites were decontaminated in 2002.

The central Asian countries are like those of the Middle East in many respects, particularly in their overwhelmingly Muslim populations, their traditions of pastoral nomadism and irrigated agriculture, and their petroleum endowments. The text now turns to that critical world region. The discussion of Islam in Chapter 6 provides especially useful background for further appreciation of central Asia and the Caucasus.

summary

- From the Russian perspective, the other 14 countries of the former Soviet Union make up the region of the "Near Abroad," a special realm of policy and interaction. The three Baltic countries have struggled to divorce themselves from this orbit. The remaining countries, along with Russia, are members or associate members of the Commonwealth of Independent States (CIS). All have significant ties with Russia, and Russia aspires to increase its influence on them—sometimes by force, as seen in its 2008 invasion by Georgia.

- The area west of the Ural Mountains and north of the Caucasus Mountains is known as European Russia; the Caucasus and the area east of the Urals is Asiatic Russia. Siberia is the area between the Urals and the Pacific. Central Asia is the name for the arid area occupied by the five countries immediately east and north of the Caspian Sea.

- Russia and the Near Abroad span 11 time zones. Stretching nearly halfway around the globe, the region has formidable problems associated with climate, terrain, and distance.

- In the years of economic uncertainty after 1991, birth rates fell and death rates rose dramatically in Russia, leading to steep declines in population. Growing revenues from oil have begun to improve standards of living, however.

- Vast stretches of the region are at very high latitudes, making agriculture difficult. Much of the tundra and subarctic climate zones are underlain by permafrost. If this "permanently" frozen grown should thaw, large amounts of greenhouse gases would be released into the atmosphere, accelerating a trend of global warming. The subarctic taiga or coniferous forest acts as a carbon "sink" and helps offset carbon dioxide emissions, so there are fears about deforestation in Siberia. The most productive agricultural area is the steppe region south of the forest in Russia and in Ukraine, Moldova, and Kazakhstan. In early Russia, rivers formed natural passageways for trade, conquest, and colonization. In Siberia, the Russians followed tributaries to advance from the Urals to the Pacific. More recently, alteration of rivers for power, navigation, and irrigation became an important aspect of economic development under the Soviet regime.

- Large plains dominate the terrain from the Yenisey River to the western border of the country. The area between the Yenisey and Lena Rivers is occupied by the hilly Central Siberian Uplands. In the southern and eastern parts of the region, mountains, including the Caucasus, Pamir, Tien Shan, and Altai ranges, dominate the landscape.

- Slavic and Scandinavian peoples were prominent in the early cultural development that was the foundation for the Russian, Ukrainian, and Belarusian cultures. Although Slavs came to dominate economic and political life, the region has large and diverse populations of non-Slavic peoples.

- Early contacts with the Eastern Roman or Byzantine Empire led to the establishment of Orthodox Christianity in Russian life and culture. Although it was repressed after the Bolshevik Revolution, this and the region's other religions have been experiencing a rebirth since the breakup of the Soviet Union.

- From the fifteenth century until the twentieth, the tsars built an immense Russian empire around the small nuclear core of Muscovy (Moscow). This imperialism brought huge areas under tsarist control. Russian expansion extended to the Pacific, across the Bering Strait, and down the west coast of North America. Russia has often triumphed over powerful foreign invaders. These achievements have been due in part to the environmental obstacles the invaders faced, the overwhelming distances involved, and the defenders' willingness to accept large numbers of casualties and implement a scorched-earth strategy to protect the motherland.

- Policies of the Soviet Union permitted most ethnic groups to retain their own languages and other elements of traditional cultures. However, the Soviet regime implemented a deliberate policy of Russification in an effort to implant Russian culture in non-Russian regions. Millions of Russians settled in non-Russian areas, but local cultures were not converted to Russian ways.

- To meet the needs of its people, the Union of Soviet Socialist Republics became a collective welfare state. However, military and industrial superpower status was generally achieved at the expense of ordinary consumers.

- The sudden transition from a command economy to capitalism in Russia and other countries in the region led to a widening gap between the rich and the poor, with growing ranks of poor people. Organized crime and an underground economy emerged. Agricultural and industrial production fell dramatically. However, high oil prices in the early 2000s improved Russia's economic outlook. This overdependence on a single commodity could prove risky for Russia. However, Vladimir Putin's plan is to propel Russia to being the world's fifth largest economy by 2020.

- There are three important realms of geopolitical concern in the region. Within Russia, the aspirations of non-Russian people like the Chechens and Tatars appear to pose a danger to the unity of the Russian Federation. Between Russia and the Near Abroad countries, there are challenging issues including energy shortages and supplies, desires of Russians outside Russia to achieve their own rights and territories, stationing of Russian troops, and Islamist terrorism. Between this region and the rest of the world—the "Far Abroad"—there are concerns about the fate of Soviet-era nuclear materials, whether the oil-rich central Asian countries are more sympathetic to Russia or Turkey, and what should be done to stem the tide of narcotics and terrorism.

- The Russian government has said that it has the right and duty to protect Russian minorities in the other countries. Some observers contend that Russia is attempting to reestablish the Near Abroad as a buffer between Russia and the Far Abroad.

- The Caspian Sea and adjacent areas make up one of the world's greatest petroleum and natural gas regions. Russia, Iran, and the West, particularly the United States, are vying for influence in this increasingly important area. Of particular concern to the rivals is how the energy supplies should be routed via pipeline to reach ocean terminals. There are similar questions about how Russian natural gas pipelines should be routed to reach Western European markets.

- Because of nuclear weapons and other critical issues, relations between Western countries and the countries of the former Soviet Union, particularly Russia, are vital to global security.

- Nearly 75 percent of the people, and an even larger share of the cities, industries, and cultivated land of the region of Russia and the Near Abroad, are packed into the Fertile Triangle (Agricultural Triangle or Slavic Core), comprising one-fifth of the total area. The rest (mostly in Asia) consists of land where environmental obstacles limit settlement.

- Ukraine is one of the most densely populated and most productive areas of the region. Ukraine's industrial and agricultural assets were vital to the Soviet Union. Post-Soviet Ukraine relies more on heavy industry than most of the other successor countries.

- Under Soviet rule, Belarus became an industrial power. Belarus's trade is now almost exclusively with Russia, much of it in the form of barter. Belarus and Russia technically formed a union state to more closely link the two predominantly Slavic countries, but political problems have prevented its implementation.

- Young women from Moldova, Ukraine, Russia, and other successor states of the Soviet Union are among the main participants in and victims of sex trafficking to Europe.

- In an effort to reverse the inefficiencies left by the state-run system, all of the countries are promoting land reform by privatizing collective and state farms and developing more independent or peasant farming. After a period of decline following independence in 1991, Russian agricultural production has improved.

- Russian industrial production plummeted after independence. The industrialized area around Moscow has a central location within the populous western plains and is functionally Russia's most important industrial region.

- Most of the Russian Far East is a thinly populated wilderness with few settlements. Several small to medium-sized cities form a north–south line along the Trans-Siberian Railroad and the lower Amur River, the two main transportation arteries. The coastal ports of Vladivostok, Nakhodka, and Vostochnyy are coming to life as shipments of oil destined for Japan and other foreign markets have picked up. Oil from the eastern Siberian fields will flow through new pipelines to China and Vostochnyy. The island of Sakhalin is becoming an important oil and gas producer for the Japanese market.

- The Northern Lands comprise one of the world's most sparsely populated areas, with coniferous forest (taiga) and tundra extending from the Finnish and Norwegian borders to the Pacific. Murmansk and Arkhangelsk are ports on the Northern Sea Route, a transpolar route for ocean shipping that can be kept open in winter with icebreakers.

- The world's deepest lake, Lake Baikal, is a unique natural ecosystem apparently threatened by effluent from paper and pulp mills. However, oil pipelines have been rerouted to reduce threats to the lake.

- Siberia's Northern Lands have rich mineral resources that began to be exploited in Soviet times, often with forced labor. Now it is expensive for Russia to maintain populations in some of the northern settlements, and the government is trying to attract people to more southerly latitudes.

- The far southern Caucasus region borders Russia between the Black and Caspian Seas. It includes the Caucasus Mountains, a fringe of foothills and level steppes to the north, and the area south known as Transcaucasia. Russians and Ukrainians are the majority north of the Greater Caucasus Range. To the south, the important nationalities are the Georgians, Armenians, and Azerbaijanis, each represented by an independent country and maintaining their unique cultures. Numerous smaller ethnic groups also inhabit the region, which has been a cauldron of conflict since independence from the Soviet Union. One of the outstanding internal problems is the fate of Nagorno-Karabakh, a region claimed by Armenia but controlled by Azerbaijan.

- Soviet-era colonization of central Asia for a cotton plantation economy had pronounced and lingering effects on the environment. Most notable was the diversion of water for irrigated agriculture and the subsequent shrinkage of the Aral Sea. The post-Soviet countries of the region are locked in a critical competition over the region's water supplies from the Amu Darya and Syr Darya Rivers.

Key Terms + Concepts

Abkhaz-Adyghean language family (p. 125)
Agricultural Triangle (p. 143)
Altaic language family (p. 125)
    Turkic subfamily (p. 125)
Armenian genocide (p. 149)
arms race (p. 130)
autonomies (p. 134)
barter (p. 132)
black-earth belt (p. 122)
Bolshevik Revolution (p. 128)
bourgeoisie (p. 130)
BRIC economy (p. 133)
Chernobyl (p. 144)
chernozem (p. 122)
collective farm (*kolkhoz*) (p. 130)
collectivized agriculture (p. 130)
command economy (p. 130)
Commonwealth of Independent States (CIS) (p. 113)
Cossacks (p. 126)
economic shock therapy (p. 132)

"Evil Empire" (p. 114)
Fertile Triangle (p. 143)
Gazprom (p. 137)
gigantomania (p. 130)
glasnost (p. 131)
"Golden Horde" (p. 126)
Gosplan (Committee for State Planning) (p. 130)
Great Volga Scheme (p. 123)
Group of Eight (G-8) (p. 139)
GUAM (p. 138)
hero projects (p. 131)
human trafficking (p. 145)
humus (p. 122)
Iranian language subfamily (p. 125)
irredentism (p. 137)
Islamic Movement of Uzbekistan (IMU) (p. 142)
Kartvelian (South Caucasian) language (p. 125)
kleptocracy (p. 132)
land empire (p. 127)

"loose nukes" (p. 141)
"misdeveloped country" (p. 132)
mollisol soils (p. 122)
Mongols (p. 126)
Nakh-Dagestanian language family (p. 125)
Near Abroad (p. 114)
new lands (p. 131)
Northern Sea Route (p. 148)
oligarchs (p. 133)
Operation Barbarossa (p. 128)
Orange Revolution (p. 137)
pan-Turkism (p. 142)
perestroika (p. 131)
permafrost (p. 120)
podzols (p. 121)
proletariat (p. 130)
Proto-Asiatic language family (p. 125)
renationalization (p. 137)
Rose Revolution (p. 139)
Rus (p. 125)
"Russian cross" (p. 119)

Russian Revolution (p. 128)
Russification (p. 127)
sacred space (sacred place) (p. 129)
scorched earth (p. 128)
"Second Kuwait" (p. 147)
Second Russian Revolution (p. 131)
Shanghai Cooperation Organization (SCO) (p. 143)
Slavic Core (p. 143)
spodosols (p. 121)
state farm (*sovkhoz*) (p. 130)
Tatars (p. 126)
"Third Rome" (p. 125)
underground economy (p. 132)
union state (p. 137)
    Finno-Ugric subfamily (p. 125)
Varangians (p. 125)
virgin and idle lands (p. 131)
"white gold" (p. 151)
"Window on the West" (p. 126)

Review Questions

1. What are the main climatic belts and corresponding vegetation types of Russia and the Near Abroad? Which are most and least productive for agriculture?

2. What are the region's major river systems?

3. Why has Russia's population been declining?

4. Why is Russia known as a land empire? How did it acquire its empire?

5. What significant conflicts have occurred in this region? Why has Russia so often won them? What was the geographic significance of Stalingrad in World War II?

6. What were the perceived advantages of collectivized agriculture?

7. What major Soviet Communist projects changed the landscape of this region?

8. What were some of the successes and failures of efforts to industrialize the Soviet Union?

9. What were some of the results of Russia's "shock therapy" economic reforms and the subsequent conditions of the country and its people? What accounts for the recent turnaround in Russia's economy?

10. Which of the non-Russian ethnic groups in Russia are particular security concerns to Russia today?

11. How does Caspian Sea oil reach markets abroad? What factors go into decision making about those exports?

12. Where, outside Russia, are Russian military troops stationed? What are the reasons for their presence? What events in 2008 stirred new concerns about Russia's intentions?

13. What resources and boundaries are associated with the Fertile Triangle?

14. To what extent are Ukraine and Belarus dependent on Russia or interested in developing further ties with Russia? Which are most and least like Russia? What problems in relations have occurred between Russia and each of these counties?

15. What are the resources and economic development prospects of Sakhalin and Kamchatka? Why are the Kurils important in Russian-Japanese relations?

16. What major dispute between two groups is focused on Nagorno-Karabakh, and what is the source of that dispute?

17. What environmental problems do the Central Asian countries face, especially with water resources?

Notes

1. Murray Feshbach, quoted in C. J. Chivers, "Putin Urges Plan to Reverse Slide in the Birth Rate." *New York Times,* May 11, 2006, pp. A1, A6.

2. Quoted in Andrew K. Kramer. September 28, 2011. "For Investors, Russia's Putin is Good For Business." *New York Times,* p. B4.

3. From Mikhail Blinnikov, *A Geography of Russia and Its Neighbors.* (New York: Guilford, 2011).

4. Quoted in C. J. Chivers, "For Putin and the Kremlin, a Not So Happy New Year." *New York Times,* January 3, 2006, p. A8.

5. The Phrase Finder, www.phrases.org.uk/meanings/31000.html.

6. Charles Clover, "Strife on the Edge." *Financial Times,* March 25, 2010, p. 8.

7. Seth Mydans, "Lured by Oligarch's Cash, Soccer Star Enlivens Dagestan." *New York Times,* September 12, 2011, p. A10.

8. Quoted in Clifford J. Levy, "Retreat of the Tongue of the Tsars." *New York Times,* September 13, 2009, p. wk3.

9. 'The Ukrainian Question." *Economist,* November 29, 1999, p. 20.

10. Valentin Rasputin, "Baikal." In *Siberia on Fire: Stories and Essays,* G. Mikkelson, M. Winchell, & V. Rasputin (Eds.). (De Kalb: Northern Illinois University Press, 1989), pp. 188–189.

Online Resources

 CourseMate: Make the most of your study time by accessing everything you need to succeed in one place. Read your textbook, take notes, review flashcards, watch videos, complete activities, take practice quizzes, and more—online with CourseMate. Log in at **www.cengagebrain.com**.

"Knowledge is light, and all who contribute bring light to our world."

—BEDOUIN PROVERB

C. Mayhew & R. Simmon (NASA/GSFC), NOAA/ NGDC, DMSP Digital Archive

Joe Hobbs

Above: The Middle East and North Africa at night. Note especially the population concentrations along the great river valleys of the Nile, Tigris and Euphrates. Left: The Monastery of St. Katherine at the foot of Jebel Musa in Egypt's Sinai Peninsula. This is widely believed to be Mount Sinai, where God gave Moses the Ten Commandments.

The Middle East and North Africa

6

Welcome to the Middle East. What other part of the world outside your own is in the news every single day? What part of the world seems so threatening to your own, even if you feel dependent upon it just to drive across town? What part of the world seems to have so many problems for which no solutions seem to be in sight?

The good news is that there *is* good news coming out of the Middle East, and there always has been. This is the fountainhead of many faiths, civilizations, and sciences. Also good is that there *are* solutions to the region's seemingly intractable problems (and in fact, peace and stability in the Middle East will make your life better). This chapter will equip you with enough knowledge that you should feel capable of attempting to solve the Palestinian-Israeli conflict in this chapter's "Try It" feature.

chapter objectives

This chapter should enable you to

- Understand and explain the mostly beneficial relationships between villagers, pastoral nomads, and city dwellers in an environmentally challenging region.

- Know the basic beliefs and sacred places of Jews, Christians, and Muslims.

- Recognize the importance of petroleum to this region and the world economy.

- Identify the geographic chokepoints and oil pipelines that are among the world's most strategically important places and routes.

- Appreciate the problems of control over freshwater in this arid region.

- Know what al-Qa'ida and other Islamist terrorist groups are and what they want.

- Understand the major issues of the Arab-Israeli conflict and the obstacles to their resolution.

- Learn about the Arab Spring, how it evolved, and what can be expected in its wake.

- Balance the pros and cons of Egypt's Aswan High Dam and other large water engineering projects.

- Recognize how Islamist terrorism transformed American geopolitical perspectives and priorities.

- Trace the trajectory of Persian Gulf (Arabian Gulf) countries from underpopulated, wealthy welfare states to populous, vulnerable states overly dependent on a single commodity.

- See the U.S. wars on Iraq as the result of tragic miscalculations by Saddam Hussein, combined with strategic American interests in the region and, many analysts say, by American miscalculations.

- Recognize Turkey as an "in-between" country, a less developed Asian nation aspiring to be a European power.

The *Middle East* is a fitting designation for the places and the cultures of this vital world region: they really are in the middle. This is a physical crossroads, where the continents of Africa, Asia, and Europe meet, and the waters of the Mediterranean Sea and the Indian Ocean mingle. Its peoples—Arab, Jew, Persian, Turk, Kurd, Berber, and others—express in their cultures the coming together of diverse influences. They have bestowed on humankind a rich legacy that includes the ancient civilizations of Egypt and Mesopotamia and the world's three great monotheistic faiths of Judaism, Christianity, and Islam.

People tend to overlook contributions like these and instead associate the region with war and terrorism. These negative connotations are often accompanied by muddled understanding.

This chapter makes the complex and misunderstood Middle East and North Africa more intelligible for you by illuminating their geographic context.

6.1 Area and Population

What is the Middle East, and where is it? The term itself applies to a Eurocentric vernacular region invented by the British, who placed themselves in the figurative center of the world. They began to use the term before the outbreak of World War I, when the *Near East* referred to the territories of the Ottoman Empire in the eastern Mediterranean region, the *East* to India, and the *Far East* to China, Japan, and the western Pacific Rim. With *Middle East,* they designated as a separate region the countries around the Persian Gulf (known to Arabs as the Arabian Gulf and in this text often simply the Gulf). Gradually, the perceived boundaries of the region grew.

Sources today vary widely in their interpretation of which countries are in the Middle East. For some, the Middle East includes only the countries clustered around the Arabian Peninsula. For others, the Middle East spans a vast 6,000 miles (9,700 km) west to east, from Morocco in northwest Africa to Iran in central Asia, and a north–south distance of about 3,000 miles (4,800 km) from Turkey, on Europe's southeastern corner, to Sudan, which adjoins East Africa (•Figure 6.1). This larger area is the region covered in this chapter, where it is referred to as the Middle East *and* North Africa, and often by the acronym MENA, which is popular in many regional studies. North Africa is added to the Middle East because the North African peoples of Morocco, Algeria, and Tunisia generally do not consider themselves Middle Easterners; they are, rather, from what they call the *Maghreb,* meaning "western land."

Defined with these borders, the Middle East and North Africa include 21 countries, the Palestinian territories of the West Bank and Gaza Strip, and the disputed Western Sahara (•Table 6.1), occupying 5.6 million square miles (14.7 million sq km) and inhabited by about 503 million people in 2011. This area is about 1.8 times the size of the lower 48 United States and is generally situated at latitudes equivalent to those between Boston, Massachusetts, and San Jose, Costa Rica (see •Figure 6.2).

These half-billion people are not distributed evenly across the region but are concentrated in major clusters (•Figure 6.3). Three countries contain the lion's share of the region's population: Turkey, Iran, and Egypt, each with more than 70 million people. There is a very strong correlation between the distribution of population and the distribution of water resources. Where water is abundant in this generally arid region, so are people. Egypt has the Nile River, and parts of Iran and Turkey have bountiful rain and snow. Conversely, where rain seldom falls, as in the Sahara of North Africa and in the Arabian Peninsula, people are few.

There is a considerable range in the region's rates of population growth. Bahrain, the United Arab Emirates, Turkey, and Tunisia had the lowest rate of population growth, 1.2 percent in 2011. The highest is 3.1 percent in Yemen, followed by

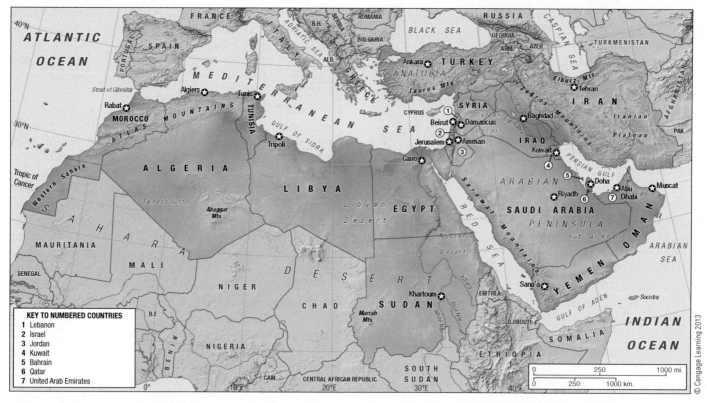

• **Figure 6.1** The Middle East and North Africa.

3.0 percent in Iraq and 2.8 percent in the Palestinian territories (consisting of the West Bank and Gaza Strip). These are some of the highest population growth rates in the world, on par with many countries in Africa South of the Sahara. Yemen's high population growth is related to its great poverty. In the Palestinian territories, rapid growth may be attributed in part to the Palestinians' poverty and probably also to the wishes of many Palestinians to have more children to counterbalance the demographic weight of their perceived Israeli foe. As we will see later in the chapter (on pages 187–189), there is an important demographic component to the Palestinian-Israeli conflict.

Between these extremes are countries with modest rates of population growth. Their governments have regarded most of these countries as too populous for their resources and economic bases, and have encouraged family planning. They have been successful in lowering birth rates, which bodes well for their development.

At the other end of the spectrum, some oil-rich but population-poor countries of the Gulf nations have been encouraging citizens to give birth to more citizens. If there were more resident nationals, there would be less need to import foreign laborers and technicians, and these countries would be more self-sufficient in their development. Some oil-rich countries of the Gulf region have far more foreigners than citizens living in them; about 80 percent of the working-age population of the United Arab Emirates, for example, is nonnative, mainly from South Asian countries. Many of these workers complain of dreams broken by poor living conditions and low wages,

and in the worst cases, of physical and emotional abuse shared by victims of human trafficking.

The Middle East and North Africa as a whole have a medium rate of population growth, suggesting that the region is moving out of the second and into the third phase of the demographic transition. Death rates have been on the decline for decades, and recently birth rates have begun falling. The result is a declining rate of population growth. The average annual rate of population change for the 21 countries, the Palestinian territories, and Western Sahara was 1.8 percent in 2011. This is higher than the world average of 1.2 percent but is a significant decline from the high of 3.0 in 1980.

These numbers seem promising for progress: as we saw in Chapter 3, a lowering of population in the LDCs is a fundamental requirement for their sustainable development. On closer examination, however, we can see a problem. The number of people in the lower age groups is so high that even with the declining rate of growth, very large numbers of people are young. This is the **youth bulge** present in all but a handful of the Middle East and North African countries. Sixty percent of MENA's people are less than 25 years old. The number of youth (people ages 15–24) in the region is estimated to grow until peaking at 100 million in 2035.

If there were jobs for such large numbers of people, there would be little reason for concern. Unfortunately, this region has the highest unemployment rate of all the world's regions (officially, 10% overall, with unemployment among young people over 25%, but the actual numbers may be considerably

Table 6.1 Middle East and North Africa: Basic Data

Political Unit	Area (thousands; sq mi)	Area (thousands; sq km)	Estimated Population (millions)	Estimated Population Density (sq mi)	Estimated Population Density (sq km)	Annual Rate of Natural Increase (%)	Human Development Index	Urban Population (%)	Per Capita GNI PPP ($US)
Middle East	**2,378.9**	**6,161.4**	**298.5**	**125**	**48**	**1.8**	**0.680**	**70**	**12,060**
Bahrain	0.3	0.8	1.3	4,333	1,673	1.3	0.806	100	33,690
Iran	630.6	1,633.3	77.9	124	48	1.3	0.707	70	11,470
Iraq	169.2	438.2	32.7	193	75	3.0	0.573	67	3,330
Israel	8.1	21.0	7.9	975	377	1.7	0.888	92	27,010
Jordan	34.4	89.1	6.6	192	74	2.6	0.698	83	5,730
Kuwait	6.9	17.9	2.8	406	157	1.6	0.760	98	29,200
Lebanon	4.0	10.4	4.3	1,075	415	1.5	0.739	87	13,400
Oman	82.0	212.4	3.0	37	14	2.6	0.705	73	24,530
Palestinian Territories	2.4	6.2	4.2	1,750	676	2.8	0.641	83	N/A
Qatar	4.2	10.9	1.7	405	156	1.0	0.831	100	N/A
Saudi Arabia	830.0	2,149.7	27.9	34	13	1.8	0.770	81	24,020
Syria	71.5	185.2	22.5	315	122	2.2	0.632	54	4,620
Turkey	299.2	774.9	74.0	247	95	1.2	0.699	76	13,500
United Arab Emirates	32.3	83.7	7.9	245	94	1.2	0.846	83	23,990
Yemen	203.8	527.8	23.8	117	45	3.1	0.462	29	2,330
North Africa	**3,037.2**	**7,866.4**	**204.9**	**67**	**26**	**1.8**	**0.606**	**51**	**5,680**
Algeria	919.6	2,381.8	36.0	39	15	1.5	0.698	67	8,110
Egypt	386.7	1,001.6	82.6	214	82	2.0	0.644	43	5,680
Libya	679.4	1,759.6	6.4	9	4	1.8	0.760	78	16,400
Morocco	172.4	446.5	32.3	187	72	1.3	0.582	56	4,400
Sudan	718.7	1,861.5	36.4	51	20	2.4	0.408	41	1,990
Tunisia	63.2	163.7	10.7	169	65	1.2	0.698	68	7,810
Western Sahara	97.2	251.7	0.5	5	2	2.3	N/A	82	N/A
Summary Total	**5,416.1**	**14,027.8**	**503.4**	**93**	**36**	**1.8**	**0.650**	**62**	**9,460**

Sources: World Population Data Sheet, Population Reference Bureau, 2011; Human Development Report, United Nations, 2011; World Factbook, CIA, 2011.

higher). The conditions of high population growth and lack of economic opportunities have long made this region ripe for unrest. As we will see, in the spring of 2011 the Middle East and North Africa exploded in a wave of popular revolutions. This was the "Arab Spring" that is presented on pages 192–197 of this chapter.

Many developing countries have economies largely dependent on subsistence agriculture, with low percentages of urban inhabitants. Perhaps surprisingly, however, the Middle East and North Africa have more urbanites than country folk. The average urban population among the 23 countries and territories is 62 percent. The most prosperous countries are also the most urban. Essentially city-states, Bahrain, Qatar, and Kuwait are 100 percent urban. The other oil-wealthy Gulf countries have urban populations over 70 percent. Consistent with its profile as a Western-style industrialized country without oil resources, Israel is 92 percent urban. At the other end of the spectrum, desperately poor Yemen has an urban population of 29 percent.

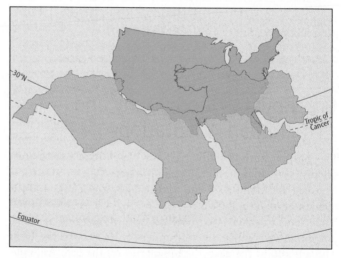

• **Figure 6.2** The Middle East and North Africa compared in latitude and area with the conterminous United States.

Source: Based on John Haywood, ed., Atlas of World History (Barnes and Noble Books, 1997).

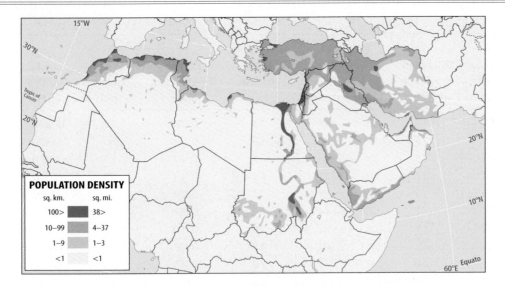

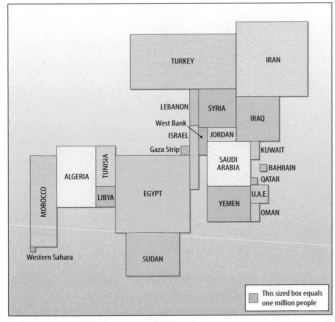

• **Figure 6.3** Population distribution (top) and population cartogram (bottom) of the Middle East and North Africa.

Source: The Oxford School Atlas edited by Patrick Wiegand (OUP, 1997), copyright © Oxford University Press. Used by permission of Oxford University Press.

6.2 Physical Geography and Human Adaptations

The margins of the Middle East and North Africa are mainly oceans, seas, high mountains, and deserts (Figure 6.1). To the west lies the Atlantic Ocean; to the south, the Sahara and the highlands of East Africa; to the north, the Mediterranean, Black, and Caspian Seas, together with mountains and deserts lining the southern land frontiers of Russia and the Near Abroad; and to the east, the mountains on the Iran-Afghanistan frontier and the Baluchistan Desert straddling Iran and Pakistan. The land is composed mainly of arid plains and plateaus, together with large areas of rugged mountains and

isolated "seas" of sand and "islands" or ribbons of fertility known as oases. Despite the environmental challenges, this region has given rise to some of the world's oldest and most influential ways of living.

A Region of Stark Geographic Contrasts

Aridity dominates the Middle East and North Africa (•Figure 6.4). At least three-fourths of the region has average yearly precipitation of less than 10 inches (25 cm), an amount too small for most types of dry farming (unirrigated agriculture). Sometimes, however, localized cloudbursts release moisture that allows plants, animals, and small populations of people—the Bedouin, Tuareg, and other pastoral nomads—to live in the desert. Even the vast Sahara, the world's largest desert, supports a surprising diversity and abundance of life. Plants, animals, and even people have developed strategies of **drought avoidance** and **drought endurance**. Migrating to avoid drought is a coping strategy that pastoral nomads and many animals use. Other organisms, such as trees, must endure drought by such adaptations as having very deep roots and small leaves. Populations of people, plants, and animals are all but nonexistent in the region's vast **sand seas**, including the Great Sand Sea of western Egypt and the Empty Quarter of the Arabian Peninsula (•Figure 6.5).

The region's climates have the comparatively large daily and seasonal ranges of temperature characteristic of dry lands. Desert nights can be surprisingly cool. Most days and nights are cloudless, so the heat absorbed on the desert surface during the day is lost by radiational cooling to the heights of the atmosphere at night. Summers in the lowlands are very hot almost everywhere. The hottest shade temperature ever recorded on earth, 136°F (58°C), occurred in Libya in September 1922. Many places regularly experience daily maximum temperatures over 100°F (38°C) for weeks at a time.

Human settlements located near the sand seas often experience the unpleasant combination of high temperatures and

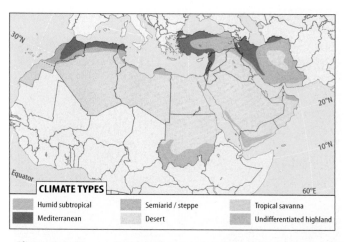

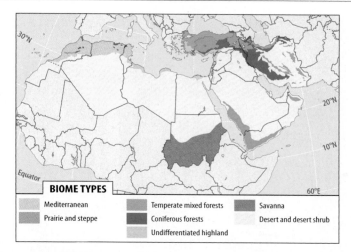

• **Figure 6.4** Climates (left) and biomes (right) of the Middle East and North Africa.

Sources: (left) Based on Rand McNally's Classroom Atlas, 2003. (right) Based on World Wildlife Fund ecoregions data, 1999.

hot, sand-laden winds, creating the sandstorms known locally by such names as *simuum* ("poison") *sirocco*, and *haboob* (the latter, ushered in by an ominous, towering wall of sand, is also familiar to the people of southern Arizona). Only in mountainous sections and in some places near the sea do higher elevations or sea breezes temper the intense midsummer heat. The population of Alexandria, on the Mediterranean Sea coast, explodes in summer as Egyptians flee from Cairo and other hot inland locations. In Saudi Arabia, the government relocates from Riyadh to the highland summer capital of Taif to escape the lowland furnace.

Lower winter temperatures bring relief from the summer heat, and the more favored places receive enough precipitation for dry farming of winter wheat, barley, and other cool season crops. Land uses across the region are depicted in •Figure 6.6. In general, winters are cool to mild. But very cold winters and snowfalls occur in the high interior basins and plateaus of Iran, Afghanistan, and Turkey (there are even glaciers in Iran and Turkey!). These locales generally have

a steppe climate. Only in the southernmost reaches of the region, notably in southern Sudan, do temperatures remain consistently high throughout the year. A savanna climate and biome prevail there.

Most areas bordering the Mediterranean Sea have 15 to 40 inches (38 to 102 cm) of precipitation a year, falling almost exclusively in winter, whereas the summer is dry and warm—a typical Mediterranean climate pattern. Throughout history, people without access to perennial streams have stored this moisture to make it available later for growing crops that require the higher temperatures of the summer months. The Nabateans, for example, who were contemporaries of the Romans in what is now Jordan, had a sophisticated network of limestone cisterns and irrigation channels (•Figure 6.7).

Along the southern and northern margins of the region, there is enough rainfall to support dry (nonirrigated) farming during the summer. There are some places that defy the usual perception of the region as bone-dry. The Black Sea side of Turkey's Pontic (Kuzey) Mountains, for example, is lush and moist

• **Figure 6.5** Seas of sand cover large areas in Saudi Arabia, and parts of the Sahara. This is the edge of the Empty Quarter (al-Rub' al-Khali) in the United Arab Emirates.

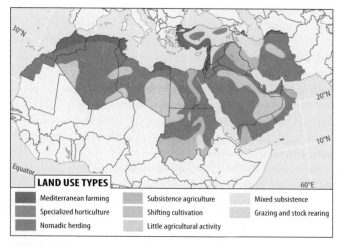

• **Figure 6.6** Land use in the Middle East and North Africa.

Source: The Oxford School Atlas edited by Patrick Wiegand (OUP, 1997), copyright © Oxford University Press. Used by permission of Oxford University Press.

• **Figure 6.7** The Treasury, a temple carved in red sandstone, probably in the first century c.e., by the Nabateans at their capital of Petra in southern Jordan. The English poet Dean Burgen called Petra "a rose red city, half as old as time." The Nabateans built sophisticated networks for water storage and distribution. For scale, note the man at the lower right.

• **Figure 6.8** Rainfall is heavy on the Black Sea side of Turkey's Pontic (also known as Pontus or Kuzey) Mountains. Note that roofs are pitched to shed precipitation; in the region's drier areas, most roofs are flat. The crop growing here is corn.

in the summer, and tea grows well there (•**Figure 6.8**). In the southwestern Arabian Peninsula, a monsoonal climate brings summer rainfall and autumn harvests to Yemen and Oman, probably accounting for the Roman name for the area: *Arabia Felix* ("Happy Arabia"). Mountainous areas in the region, like the river valleys and the margins of the Mediterranean, play a vital role in supporting human populations and national economies. Due to orographic or elevation-induced precipitation, the mountains receive much more rainfall than surrounding lowland areas.

There are three principal mountainous regions of the Middle East and North Africa (see Figure 6.1). In northwestern Africa between the Mediterranean Sea and the Sahara, the Atlas Mountains of Morocco, Algeria, and Tunisia reach over 13,000 feet (3,965 m) in elevation. Mountains also rise on both sides of the Red Sea, with peaks up to 12,336 feet (3,760 m) in Yemen. These mountains rose with the tectonic processes that are pulling the African and Arabian plates

apart at the northern end of the Great Rift Valley. The hinge of this crustal movement is the Bekaa Valley of Lebanon, where the widening fault line follows the Jordan River Valley southward to the Dead Sea (•**Figure 6.9**). This valley is the deepest depression on the earth's land surface, lying about 600 feet (183 m) below sea level at Lake Kinneret (also known as the Sea of Galilee or Lake Tiberius) and nearly 1,300 feet (400 m) below sea level at the shore of the Dead Sea, the lowest point of land on the planet. The rift then continues southward to the very deep Gulf of Aqaba and Red Sea, before turning inland into Africa at Djibouti and Ethiopia, and continuing all the way down the Great Rift Valley of Africa (see page 317). By the way, should you visit the Dead Sea one day, just to try to sink. It's impossible—as this lake has no outlet, the waters are so salty at 32 percent that you will bob like a cork.

In tectonic processes, one consequence of rifting in one place is the consequent collision of the earth's crustal plates elsewhere. There are several such collision zones in the Middle East and North Africa, particularly in Turkey and Iran, where mountain

Image Science and Analysis Laboratory, NASA-JSC.

• **Figure 6.9** It looks peaceful from space, but the land around the Jordan Rift Valley is one of the most contested areas on Earth. The Jordan River runs into and out of Lake Kinneret (the Sea of Galilee) near the top of the image, emptying into the Dead Sea near the bottom of the image. If political boundaries were drawn on the image, they would include parts of Israel, Lebanon, Syria and Jordan, and all of the West Bank, Gaza Strip, and Golan Heights.

building has resulted. These are seismically active zones—meaning earthquakes occur there—and rarely does a year go by without a devastating quake rocking Turkey or Iran.

A large area of mountains, including the region's highest peaks, stretches across Turkey and Iran. The loftiest mountain ranges in Turkey are the Pontic, Taurus (•**Figure 6.10**), and Anti-Taurus, and in Iran, the Elburz and Zagros Mountains. These chains radiate outward from the rugged Armenian Knot in the tangled border country where Turkey, Iran, and the countries of the Caucasus meet. In Turkey, the glacier-fringed summit of the extinct volcano Mt. Ararat soars to 16,804 feet (5122 m)

over the frontier between Turkey and Armenia. Armenians seethe in anger and resentment that this mountain, so sacred to them, fell into Turkish hands. Some people believe that Mt. Ararat is where Noah's Ark lies (and even go in search of it), for in Genesis it is written that the vessel came to rest "in the mountains of Ararat." (Incidentally, like so many things, the accounts of Noah are shared by all three Abrahamic traditions of Judaism, Christianity, and Islam.)

Extensive forests existed in early historical times in the Middle East and North Africa, particularly in these mountainous areas, but overcutting and overgrazing have almost eliminated them. Since the dawn of civilization in this area, around 3000 B.C.E., people have cut timber for construction and fuel faster than nature could replace it. Lebanon, described in ancient times as "an oasis of green with running creeks" and "a vast forest whose branches hide the sky," illustrates the result. If not for human impacts, forests would cover large parts of Lebanon today. On many mountainsides, the dominant tree would be the cedar, the tree that is the centerpiece of Lebanon's flag. This hardwood was highly valued throughout the Mediterranean Basin. The Phoenician civilization, with its base in Lebanon, grew wealthy by trading cedarwood. Ancient Egyptians were among those who prized the cedar. The solar boat of King Cheops, dating to the third millennium BCE, would carry this pharaoh's soul in the afterlife, and it had to be built of the best wood: cedar of Lebanon. King Tutankhamen's funerary shrines and Solomon's Temple in Jerusalem were built of cedar of Lebanon (•**Figure 6.11**). Only a few isolated groves of cedar remain in Lebanon. Some of these are on the grounds of monasteries and survive because they grow in these sacred places. One grove is called "The Cedars of God."

The "primordial" or potential vegetation of Morocco's Atlas Mountains can be easily reconstructed by looking at the vegetation in the untouched sacred space surrounding saints' tombs.[1] This is a fine example of the importance of

Joe Hobbs

• **Figure 6.10** The Taurus Mountains of southeastern Turkey.

Joe Hobbs

• **Figure 6.11** Ancient Egyptians believed that the solar boat of King Cheops of Egypt (c. 2500 B.C.E.), builder of the Great Pyramid, would carry the pharaoh's spirit through the firmament. It had to be made of the best wood—cedar from Lebanon. The solar boat was interred next to the pyramid and now stands in a specially constructed museum.

environmental perception in geography. The principle of environmental perception is that people, societies, and cultures perceive and use the environment in different ways, depending on their age, gender, socioeconomic standing, language, religion, nationality, and other variables. Often these perceptions have an important role in shaping the cultural landscape. In this case, Islamic beliefs about the sacred space surrounding saints' tombs have protected the plants and animals living there, offering us a window into the environmental history of the Atlas Mountains.

Villager, Pastoral Nomad, Urbanite: The Ecological Trilogy

In the 1960s, the American geographer Paul English developed a useful model for understanding relationships between the three ancient ways of life that, although highly modified, still characterizes the Middle East and North Africa today: villager, pastoral nomad, and urbanite. Each of these modes of living is rooted in a particular physical environment. Villagers are the subsistence farmers of rural areas where dry farming or irrigation is possible; pastoral nomads are the desert peoples who migrate through arid lands with their livestock, following patterns of rainfall and vegetation; and urbanites are the inhabitants of the large towns and cities, generally located near bountiful water sources but sometimes placed for particular trade, religious, or other reasons. Describing these ways of life as components of the **Middle Eastern ecological trilogy**, English explained how each of them has a characteristic, usually mutually beneficial, pattern of interaction with the other two (•**Figure 6.12**).

The peasant farmers of Middle Eastern and North African villages (the **villagers**) represent the cornerstone of the trilogy. They grow the staple food crops such as wheat and barley that feed both the city dweller and the pastoral nomad of the desert. Neither urbanite nor nomad could live without them. The village also provides the city, often unwillingly, with tax revenue, soldiers, and workers. And before the mid-twentieth century,

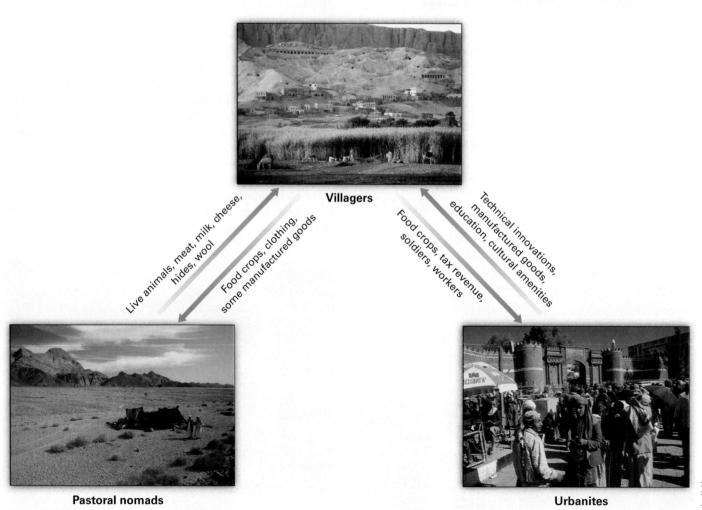

Villagers

Live animals, meat, milk, cheese, hides, wool

Food crops, clothing, some manufactured goods

Food crops, tax revenue, soldiers, workers

Technical innovations, manufactured goods, education, cultural amenities

Pastoral nomads

Urbanites

Joe Hobbs

• **Figure 6.12** People, environments, and interactions of the ecological trilogy (here, villagers harvesting sugarcane in upper Egypt, Bedouin at camp in Egypt's Eastern Desert, and shoppers at the main gate to the historic city of San'a, Yemen). This relationship is generally symbiotic, although historically both urbanites and pastoral nomads preyed on the villagers, who are the trilogy's cornerstone.

villages provided **pastoral nomads** with plunder as the desert dwellers raided their settlements and caravan supply lines. Generally, however, the exchange is mutually beneficial.

The nomads provide villagers with livestock products, including live animals, meat, milk, cheese, hides, and wool, and with desert herbs and medicines. Educated and progressive **urbanites** provide technological innovations, manufactured goods, religious and secular education and training, and cultural amenities (today, including films and music).

There is little direct interaction between urbanites and pastoral nomads, although some manufactured goods such as clothing travel from city to desert, and some desert folk medicines pass from desert to city. Historically, the exchange was sometimes violent, as urban-based governments have sought to control the movements and military capabilities of the elusive and sometimes hostile nomads. Pastoral nomads once plundered rich caravans plying the major overland trade routes of the Middle East and North Africa. Governments did not tolerate such activities and often cracked down hard on the nomads they were able to catch.

Later, in the 1970s, English wrote an article marking the "passing of the ecological trilogy." He noted that cities were encroaching on villages, villagers were migrating into cities and giving some neighborhoods a rural aspect, and pastoral nomads were settling down—thus the trilogy no longer existed. In reality, although the makeup and interactions of its parts have changed, the trilogy model is still valid and useful as an introduction to the major ways of life in the Middle East and North Africa. It is especially significant that a given man or woman in the region strongly identifies as a villager, a pastoral nomad, or an urbanite. This perception of self has an important bearing on how these people of very different backgrounds interact, even when they live in close proximity. Urban officials may work in rural village areas, but they remain at heart and in their perspectives city people and usually live apart from farmers. Extended families of pastoral nomads may settle down and become farmers, but they continue to identify themselves by affiliation with the nomadic tribe. Many continue to harvest desert resources on a seasonal basis and retain marriage and other ties with desert-dwelling relatives.

Let's take a closer look at each of the components of the ecological trilogy.

The Village Way of Life

Agricultural villagers historically represented by far the majority populations in the Middle East and North Africa; only in recent decades have urbanites begun to outnumber them. In the region's typically dry environments, the villages are located near reliable water sources with cultivable lands nearby. They are usually made up of closely related family groups, with many fields owned by an absentee landlord. Villagers typically live in closely spaced homes made of mud brick or concrete blocks. Production and consumption focus on a staple grain such as wheat, barley, or rice. As land for growing fodder is often in short supply, villagers keep only a small number of sheep and goats and rely in part on nomads for pastoral produce. Residents of a given village usually share common ties of kinship, religion, ritual, and custom, and the changing demands of agricultural seasons regulate their patterns of activity.

These patterns of village life have been increasingly exposed to outside influences since the mid-eighteenth century. European colonialism brought significant economic changes, including the introduction of cash crops and modern facilities to ship them. Improved and expanded irrigation, financed initially with capital from the West, brought more land under cultivation. Recent agents of change have been the countries' own government-supported doctors, teachers, and land reform officers. Modern technologies such as motor vehicles, gasoline-powered water pumps, radio, television, and—most recently—the Internet and cell phones have modified old patterns of living. The educated and ambitious, as well as the unskilled and desperately poor—motivated, respectively, by pull and push factors (see Chapter 3)—have been drawn to urban areas. Improved roads and communications have in turn carried urban influences to villages, prompting villagers to become more integrated into the national society.

The Pastoral Nomadic Way of Life

Pastoral nomadism emerged as an offshoot of the village agricultural way of life not long after plants and animals were first domesticated in the Middle East (about 7000 B.C.E.). Rainfall and the wild fodder it brings forth, although scattered, are sufficient resources to support small groups of people who migrate with their sheep, goats, and camels (and in some locales, cattle) to take advantage of this changing resource base (see Perspectives from the Field, next page). In addition to selling or trading livestock to obtain food, tea, sugar, clothing, and other essentials from settled communities, pastoral nomads hunt; gather; work for wages; and where possible, grow crops. Many now work in the tourist industry, as Westerners hungry for insight into traditional cultures seek them out. Their multifaceted livelihood has been described as a strategy of **risk minimization** based on the exploitation of multiple resources so that some will support them if others fail.

Although renowned in Middle Eastern legends, pastoral nomads have been described as "more glamorous than numerous." It is impossible to obtain adequate census figures on the number living in the deserts of the Middle East and North Africa, but they are very small relative to the numbers of villagers and urbanites. Recent decades have witnessed the rapid and progressive settling down, or **sedentarization**, of the nomads—a process attributed to a variety of causes. In some cases, prolonged drought virtually eliminated the resource base on which the nomads depended. Traditionally, they were able to migrate far enough to find new pastures, but modern national boundaries now inhibit such movements. Some have returned with the rains to their desert homelands, but others have chosen to remain as farmers or wage laborers in villages and towns. In the Arabian Peninsula, the prosperity

PERSPECTIVES FROM THE FIELD   **Way-Finding in the Desert**

The research for my doctoral dissertation in geography at the University of Texas–Austin was an 18-month journey with the Khushmaan Ma'aza, a clan of Bedouin nomads living in the northern part of Egypt's Eastern Desert. I studied their perceptions, knowledge, and uses of natural resources, and tried to understand how their worldviews and kinship patterns apparently helped them devise ways of protecting these scarce resources. I have worked with these people from the early 1980s to the present; I am now doing a multiyear study of how they carefully protect acacia trees, which offer them many resources when drought prevents the growth of annual plants. I have always been astonished by how different they are from me in perceiving the world around us, and how much better they are in their ability to "read" the ground and recall landscape details. In the following passage from my book *Bedouin Life in the Egyptian Wilderness*, I marvel at their way-finding abilities.

> The process by which the Khushmaan nomads have developed roots in their landscape, fashioning subjective "place" from anonymous "space," and the means by which they orient themselves and use places on a daily basis deserve special attention: these are essential parts of the Bedouins' identity and profoundly influence how they use resources and affect the desert ecosystem.
>
> The Bedouin have little need or knowledge of maps. On the other hand, when shown a map or aerial image of countryside they know, the nomads accurately orient and interpret it, naming the mountains and drainages depicted. Saalih Ali (•**Figure 6.A**) especially enjoyed my star chart. He would tell me what time the

Pleiades would rise, for example, and ask me what time the chart indicated. The two were almost always in agreement.

Many desert travelers have been astonished by the nomads' navigational and tracking ability, calling it a "sixth sense." The Khushmaan are exceptional way-finders and topographical interpreters able, for instance, to tell from tracks whether a camel was carrying baggage or a man; whether gazelle tracks were made by a male or female; which way a car was traveling and what make it was; which man left a set of footprints, even if he wore sandals; and how old the tracks are. Bedouins are proud of their geographical skills, which, they believe, distinguish them from settled people. Saalih told me, "Your people don't need to know the country but we do, to know exactly where things are in order to live." Pointing to his head, he said, "My map is here."

The difference in the way-finding abilities of nomads and settled persons may be due to the greater survival value of these skills for nomads and to the more complex meanings they attribute to locations. Bedouin places are rich in ideological and practical significance. The nomads interpret and interact with these meanings on a regular basis and create new places in their

• **Figure 6.A**   Saalih Ali, a Bedouin of the Khushmaan Ma'aza, with the author in 2010, 30 years after they began fieldwork together.

lifetimes. Theirs is an experience of belonging and becoming with the landscape, whereas settled people are more likely to inherit and accommodate themselves to a given set of places.

The nomads' homeland is so vast, and the margin of survivability in it so narrow, that topographic knowledge must be encyclopedic. Places either are the resources that allow human life in the desert or are the signposts that lead to resources. A Khushmaan man pinpointed the role of places in desert survival: "Places have names so that people do not get lost. They can learn where water and other things are by using place names." Tragedy may result if the nomad has insufficient or incorrect information about places: many deaths by thirst are attributed to faulty directions for finding water.*

Source: From BEDOUIN LIFE IN THE EGYPTIAN WILDER-NESS by Joseph J. Hobbs, Copyright © 1989. Courtesy of the University of Texas Press.

---

and technological changes prompted by oil revenues made rapid inroads into the material culture—and then the livelihood preferences—of the desert people; many preferred the comforts of settled life. Some governments, notably those of Israel and prerevolutionary Iran, were frustrated by their inability to count, tax, conscript, and control a sizable migrant population and therefore forced the nomads to settle.

Pastoral nomads of the Middle East and North Africa identify themselves primarily by their tribe, not by their nationality. The major ethnic groups from which these tribes draw

are the Arabic-speaking Bedouin of the Arabian Peninsula and adjacent lands, the Berber and Tuareg of North Africa, the Kababish and Bisharin of Sudan, the Yörük and Kurds of Turkey, and the Qashai and Bakhtiari of Iran. Members of a tribe claim common descent from a single male ancestor who lived countless generations ago; their kinship organization is thus a patrilineal descent system. It is also a segmentary kinship system, so called because there are smaller subsections of the tribe, known as clans and lineages, that are functionally important in daily life. Members of the most closely related families comprising the lineage, for example, share livestock, wells, trees, and other resources. Both the larger clans, made up of numerous lineages, and the tribes possess territories. Members of a clan or tribe typically allow members of another clan or tribe to use the resources within its territory on the basis of **usufruct**, or nondestructive mutual use. That arrangement has often proved to be lifesaving in an environment where water often is in dangerously low supply.

Although some detractors have depicted pastoral nomads as the "fathers" rather than "sons" of the desert, blaming them for wanton destruction of game animals and vegetation, there are numerous examples of pastoral nomadic groups that have developed indigenous and very effective systems of resource conservation. Most of these practices depend on the kinship groups of family, lineage, clan, and tribe to assume responsibility for protecting plants and animals.

The Urban Way of Life

The city was the final component to emerge in the ecological trilogy, beginning in about 4000 B.C.E. in Mesopotamia (modern Iraq) and 3000 B.C.E. in Egypt. Though they resembled villages in many ways, the early cities were distinguished by their larger populations (more than 5,000 people), the use of written languages, and the presence of monumental temples and other ceremonial centers. The early Mesopotamian city and, after the seventh century C.E., the classic Islamic city, called the **medina**, had several structural elements in common (•Figure 6.13). The medina had a high surrounding wall built for defensive purposes. The congregational mosque and often an attached administrative and educational complex dominated the city center. Although Islam is often characterized as a faith of the desert, religious life has always been focused in, and diffused from, the cities. The importance of the city's congregational mosque in religious and everyday life is often emphasized by its large size and outstanding artistic execution.

A large commercial zone, known as a *bazaar* in Persian and a *suq* in Arabic and recognizable as the ancestor of the modern shopping mall, typically adjoined the ceremonial and administrative heart of the city (•Figure 6.14). Merchants and craftspeople selling various commodities occupied separate spaces within this complex, and visitors to an old medina today can still find sections devoted exclusively to the sale of spices, carpets, gold, silver, traditional medicines, or other particular goods. Smaller clusters of shops and workshops were located at the city gates.

Residential areas were differentiated as quarters, not by income group but by ethnicity; the medina of Jerusalem, for example, still has distinct Jewish, Arab, Armenian, and non-Armenian Christian quarters (see Figure 6.B). Homes tended to face inward toward a quiet central courtyard, buffering the occupants from the noise and bustle of the street. The narrow, winding streets of the medina were intended for foot traffic and small animal-drawn carts, not for large motor vehicles, a fact that accounts for the traffic jams in some old Middle Eastern and North African cities today and for the wholesale destruction of the medina in others.

• **Figure 6.14** A typical Middle Eastern *suq or bazaar* (market). This one is in Cairo, Egypt.

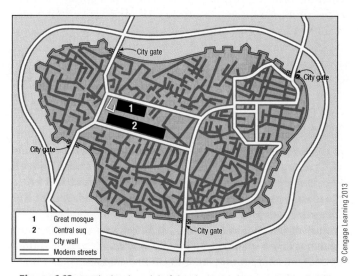

• **Figure 6.13** An idealized model of the classic medina, or Muslim Middle Eastern city.

1	Great mosque
2	Central suq
	City wall
	Modern streets

Table 6.2	Middle East and North Africa: Metropolitan Populations Data	
1	Cairo, Egypt	16.4
2	Istanbul, Turkey	14.3
3	Tehran, Iran	13.2
4	Baghdad, Iraq	11.8
5	Khartoum, Sudan	10.1
6	Algiers, Algeria	6.7
7	Riyadh, Saudi Arabia	4.9
8	Alexandria, Egypt	4.6
9	Basra, Iraq	4.4
10	Ankara, Turkey	3.8
11	Aleppo, Syria	3.5
12	Tel Aviv, Israel	3.4
13	Casablanca, Morocco	3.3
14	Jiddah, Saudi Arabia	3.2
15	Isfahan, Iran	3.1
16	Izmir, Turkey	2.9
17	Mosul, Iraq	2.9
18	Damascus, Syria	2.8
19	Kuwait City, Kuwait	2.6
20	Mashhad, Iran	2.5

© Cengage Learning 2013

Population in millions.

The medinas that survive today are gently decaying vestiges of a forgotten urban pattern. Periods of European colonialism and subsequent nationalism changed the face and orientation of the city. During the colonial age, resident Europeans preferred to live in more spacious settings at the outer edges of the city, and later, the national elite followed this pattern. In recent times, independent governments have adopted Western building styles, with broad traffic arteries cutting through the old quarters, and large central squares near government buildings. This opening up of the cityscape has spread commercial activity along the wide avenues, diluting the prime importance of the central bazaar as the focus of trade.

Rural-to-urban migration and the city's own internal growth contribute to a rapid rate of urbanization that puts enormous pressure on services in the region's poorer countries. Governments often build high-rise public housing to accommodate the growing population, contributing to a cycle in which the urban poor move into the new dwellings, only to leave their old quarters as a vacuum to draw in still more rural migrants. In Cairo, millions of former villagers now live in the "City of the Dead," an extraordinary urban landscape composed of multistory dwellings erected above graves—a last resort for the poor who have no other place to go. The overwhelmingly largest city, or primate city, so characteristic of Middle Eastern and North African capitals,

• **Figure 6.15** In Dubai, one of the city states of the United Arab Emirates, wealth created initially on oil revenues has made it possible for some to build almost anything, including islands in whimsical shapes. The islands, which are available for sale, have begun to erode and are sinking.

NASA/GSFC/MITI/ERSDAC/JAROS/ASTER

thus grows at the expense of the smaller original city. Two of the primate cities of MENA, Cairo and Istanbul, are also **megacities**, defined as having more than 10 million inhabitants. The region's major metropolitan areas and their populations are listed in •**Table 6.2**.

Much of the rural-to-urban migration and subsequent urban gridlock and squalor could probably be avoided if governments invested more in the development of villages and smaller cities. The oil-rich countries with relatively small populations generally enjoy an urban standard of living equaling that of affluent Western countries. Modern industrial cities such as Saudi Arabia's Jubail and others founded on oil wealth were built virtually overnight, providing fascinating contrast to the region's colorful, complex ancient cities. With the recent resurgence of oil revenue into Saudi Arabia and the other Gulf countries, some of the cities boast the most elegant, sophisticated, and expensive amenities and architecture to be found anywhere in the world. There are no constraints in the over-the-top Persian Gulf emirate of Dubai. There is an indoor ski "resort" in a mall, and developers have even created islands in the shapes of the world's continents and of palm trees (•**Figure 6.15**).

6.3 Cultural and Historical Geographies

Peoples of the Middle East and North Africa have made many fundamental contributions to humanity. During the Agricultural Revolution, many of the plants and animals on which the world's agriculture is based today were first domesticated in the Middle East between 5,000 and 10,000 years ago. The list includes wheat, barley, sheep, goats, cattle, and pigs, whose wild ancestors were processed, manipulated, and bred until their physical makeup and behavior changed to suit human needs. The interaction between people and the wild plants and animals they eventually domesticated took place mainly in the well-watered **Fertile Crescent**, the arc of land stretching from Israel to western Iran.

By about 6,000 years ago, people sought higher yields by irrigating crops in the rich soils of the Tigris, Euphrates, and Nile River Valleys. Their efforts produced the enormous crop surpluses that allowed civilization—a cultural complex based on an urban way of life—to emerge in Mesopotamia (literally, "the land between the rivers," the Tigris and Euphrates) and in Egypt. Accomplishments in science, technology, art, architecture, language, mathematics, and other areas diffused outward from these centers of civilization. Egypt and Mesopotamia are thus among the world's great culture hearths.

Ethnicities and Languages

The Middle East and North Africa are sometimes mistakenly referred to as the "Arab World"; in fact, the region has huge populations of non-Arabs. It is true that a majority of the region's inhabitants are Arabs. An **Arab** is best defined as a person of Semitic Arab ethnicity whose ancestral language is *Arabic*, a **Semitic language** spoken by about 250 million, or 49.7 percent, of the region's people. Originally, the Arabs were inhabitants of the Arabian Peninsula, but conquests after their majority conversion to Islam took them, their language, and their Islamic culture as far west as Morocco and Spain.

The region is also the homeland of the **Jews**, whose definition today is complex. Originally, Jews were both a distinct ethnic and linguistic group of the Middle East who practiced the religion of Judaism. It is possible (although difficult) for non-Jews to convert to Judaism (the convert is known as a proselyte, or "immigrant"), so a strictly ethnic definition does not apply today. Many Jews do not practice their religion but still consider themselves ethnically or culturally Jewish. Whatever debate exists about what defines a Jew, Jewish identity is strong and resilient.

It is important to recognize that although political circumstances made them enemies in the twentieth century, Jews and Arabs lived in peace for centuries, and they share many cultural traits. Both recognize Abraham as their patriarch. Arabic is a Semitic language in the same **Afro-Asiatic language family** as *Hebrew*, which is spoken by most of the 5.7 million Jewish inhabitants of Israel (see the language map, •Figure 6.16).

There are other very large populations of non-Semitic ethnic groups and languages in the region. The greatest are the 60 million **Turks** of Turkey, who speak *Turkish,* a member of the Altaic

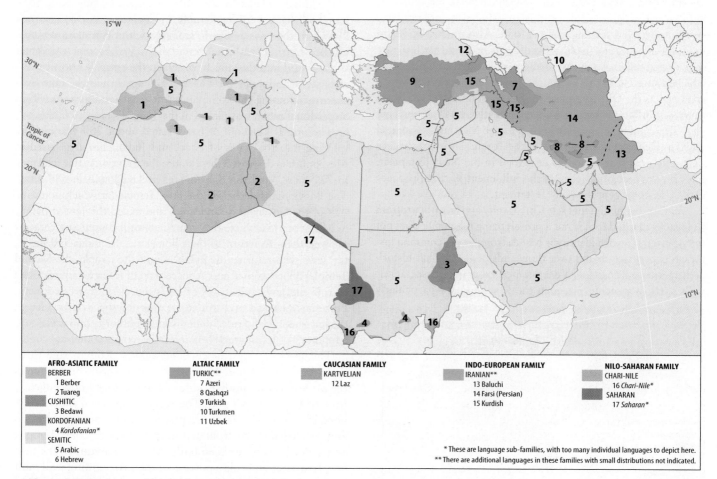

AFRO-ASIATIC FAMILY
BERBER
  1 Berber
  2 Tuareg
CUSHITIC
  3 Bedawi
KORDOFANIAN
  4 *Kordofanian**
SEMITIC
  5 Arabic
  6 Hebrew

ALTAIC FAMILY
TURKIC**
  7 Azeri
  8 Qashqzi
  9 Turkish
  10 Turkmen
  11 Uzbek

CAUCASIAN FAMILY
KARTVELIAN
  12 Laz

INDO-EUROPEAN FAMILY
IRANIAN**
  13 Baluchi
  14 Farsi (Persian)
  15 Kurdish

NILO-SAHARAN FAMILY
CHARI-NILE
  16 *Chari-Nile**
SAHARAN
  17 *Saharan**

* These are language sub-families, with too many individual languages to depict here.
** There are additional languages in these families with small distributions not indicated.

• **Figure 6.16** Languages of the Middle East and North Africa.
Source: Based on Bernard Comrie, et al., The Atlas of Languages, Revised Edition (New York: Facts on File, 2003).

language family, and the 41 million Persians of Iran, most of who speak **Persian** (*Farsi*) in the Indo-European language family. Both Persian and Arabic are written in Arabic script and so appear related, but they are not. Turkish was also written in Arabic script until early in the twentieth century, but since then it has been written in a Latin script. About 30 million **Kurds**—a people living in Turkey, Iraq, Iran, and Syria—speak *Kurdish*, which is also an Indo-European language. Many people in North Africa speak **Berber** and **Tuareg** (in the Afro-Asiatic language family).

The Abrahamic Faiths

Out of the Middle East came the closely related monotheistic faiths of Judaism, Christianity, and Islam. The region's human and political geographies are closely tied to these religions and their holy places (•Figure 6.17). All three are Abrahamic traditions; they regard Abraham as their patriarch, and they share many fundamental beliefs. The first to emerge was the faith of Judaism.

The Promised Land of the Jews

Judaism, the first significant monotheistic faith as far as we know, is today practiced by about 14 million people worldwide, mostly in Israel, Europe, and North America. September 2011 marked the start of the year 5772 in the Jewish calendar, but unlike its kindred faiths—Christianity and Islam—Judaism does not have an acknowledged starting point in time. Also unlike Christianity, Judaism does not have a fixed creed or doctrine. Jews are encouraged to behave in this life in compliance with God's laws, which, according to the Torah (the Jewish holy scripture, which is known as the Old Testament in the Christian Bible) God gave to Moses on Mount Sinai as a covenant with God's "chosen people." The coming of a savior known as the Messiah ("Anointed One" in Hebrew) is prophesied in the Torah; Christians believe that Jesus was that savior, as described in the New Testament; Jews do not recognize Jesus as the fulfillment of that prophecy and so do not accept the New Testament.

156

Another distinction that sets Christianity and Islam apart from Judaism is that Judaism is not a proselytizing religion; it does not seek converts. Jewish identity is based strongly on a common historical experience shared over thousands of years. That historical experience has included deep-seated geographic associations with particular sacred places in the Middle East—particularly with places in Jerusalem, capital of ancient Judah (Judea), the province from which Jews take their name. Tragically, the Jewish history also has included unparalleled persecution.

Depending on one's perspective, the Jewish connection with the geographic region known as Palestine, the area now composed of Israel, the West Bank, and the Gaza Strip (see Figure 6.33), is most significant on a time scale of either about 4,000 years or about 100 years. According to the Bible, around 2000 B.C.E., God commanded Abraham and his kinspeople, known as **Hebrews** (later as Jews), to leave their home in what is now southern Iraq and settle in Canaan. God told Abraham that this land of Canaan—geographic Palestine—would belong to the Hebrews after a long period of persecution. The Bible says that the Hebrews did settle in Canaan until famine struck that land. At the command of Abraham's grandson Jacob, the Hebrews— known then as **Israelites**—relocated to Egypt, where grain was plentiful. That began the long sojourn of the Israelites in Egypt, which, according to the Bible, ended in about 1200 B.C.E. when Moses led them out in the journey known as the **Exodus**.

According to Jewish history, the prophecy of Abraham was first fulfilled when the Israelites settled once again in Canaan, their **Promised Land**. The Jewish King Saul unified the 12 tribes that descended from Jacob into the first united Kingdom of Israel in about 1020 B.C.E. In about 950 B.C.E. in Jerusalem—the capital of the kingdom enlarged by Saul's successor, David—King Solomon built Judaism's **First Temple**. He located it atop a great rock known to the Jews as *Even HaShetiyah*, the **Foundation Stone**, plucked from beneath the throne of God to become the center of the world and the core from which the entire world was created (see Geography of Sacred Space, next page). The Ark of the Covenant, containing the commandments that God gave to Moses atop Mount Sinai, was placed in the Temple's Holy of Holies.

The united kingdom of Israel lasted about 200 years before splitting into the states of Israel and Judah. Empires based in Mesopotamia destroyed these states: the Assyrians attacked Israel in 721 B.C.E., and the Babylonians sacked Judah in 586 B.C.E. The Babylonians destroyed the First Temple (at which point the Ark of the Covenant disappeared) and exiled the Jewish people to Mesopotamia, where they remained until conquering Persians allowed them to return to their homeland. In about 520 B.C.E., the Jews who returned to Judah rebuilt the temple (the **Second Temple**) on its original site. A succession of foreign empires came to rule the Jews and Arabs of Palestine: Persian, Macedonian, Ptolemaic, Seleucid, and around the time of Jesus, Roman. Herod, the Jewish king who ruled under Roman authority and was a contemporary of Jesus, greatly enlarged the temple complex.

The Jews of Palestine revolted against Roman rule three times between 64 and 135 C.E. The first revolt broke out as the profoundly monotheistic Jews refused to acknowledge the Roman emperor as a god. The Romans quashed these rebellions in a series of famous sieges, including those of Masada and Jerusalem. The Romans destroyed the Second Temple, and a third has never been built. All that remains of the Second Temple complex is a portion of the surrounding wall built by Herod. Today, this **Western Wall**, known to non-Jews as the Wailing Wall, is the most sacred site in the world accessible to Jews (Figures 6.B and •6.18). Some religious traditions

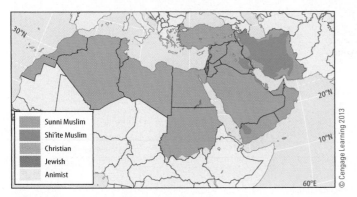

Sunni Muslim
Shi'ite Muslim
Christian
Jewish
Animist

© Cengage Learning 2013

• **Figure 6.17** Religions of the Middle East and North Africa.

GEOGRAPHY OF SACRED SPACE | **Jerusalem**

The old city of Jerusalem is filled with sacred places and is one of the world's premier pilgrimage destinations. Over thousands of years, Jerusalem has been coveted and conquered by people of many different cultures and faiths. In the process, a place held sacred by one group has often come to be held sacred by a second and even a third.

The most important thing to notice on this map is that the location of the First and Second Jewish Temples is identical to that of the Muslims' Dome of the Rock. Judaism, Islam, and Christianity are very close in their origins and many of their basic precepts, but their faiths are not conjoined or syncretized in practice, as, for example, the Maya and Catholic faiths have been in Mesoamerica (see page 365), or Buddhism and other traditions have been in East Asia (see page 228). In each faith, there has been a long tradition of quarreling over ownership of sacred space. Nowhere in the world does contested sacred space have more potential to ignite violence and even war than in Jerusalem (which means, in Hebrew, "City of Peace"). The later section on Regional Issues and Landscapes offers more insight into the role of Jerusalem's holy places in the region's turbulent history (see page 187). •Figure 6.B depicts the major places sacred to Jews, Christians, and Muslims, and also the city's ethnic quarters; the walled old city is a classic medina (see page 167).

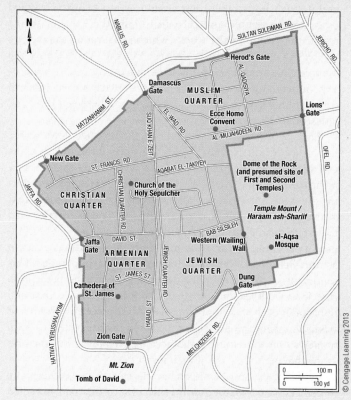

• **Figure 6.B** Sacred sites and ethnic quarters of the old city of Jerusalem.

• **Figure 6.18** Jerusalem's Western Wall, Temple Mount, and Noble Sanctuary. In this view are some of Judaism's and Islam's holiest places. At left, below the golden dome, is the Western Wall, which is all that remains of the structure that surrounded the Jews' Second Temple. The dome is the Muslims' Dome of the Rock. At far right, with the black dome, is the al-Aqsa Mosque, another very holy place in Islam.

prohibit Jews from ascending the **Temple Mount** above, the area where the temple actually stood, because it is too sacred. After the temple's destruction, that site was occupied by a Roman temple and then in 691 by the Muslim shrine called the **Dome of the Rock**, which still stands today (also in Figures 6.B and 6.18). The mostly Muslim Arabs know the Temple Mount as *al-Haraam ash-Shariif*, the **Noble Sanctuary**. Supercharged with meaning, this place has in modern times often been the spark of conflagration between Jews and Palestinian Arabs.

The victorious Romans scattered the defeated Jews to the far corners of the Roman world. Thus began the Jewish exile, or **Diaspora**. In their exile, the Jews never forgot their attachment to the Promised Land. One Jewish prayer, recited during the weeklong Passover holiday, ends with the words "Next year in Jerusalem!" In Europe and Russia, Jews were subjected to systematic discrimination and persecution and were forbidden to own land or engage in a number of professions. Known as **anti-Semitism**, hatred of Jews developed deep roots. This sentiment in part grew out of the long-simmering perception that Jews were responsible for the murder of Jesus Christ and in part out of the fact that Christians were generally prohibited from practicing moneylending. Jews were permitted to loan money to the Christians, who accepted the loans but then resented having to pay the Jews interest.

In the 1930s, anti-Semitism became state policy in Germany under the Nazis, led by Adolf Hitler. Many German Jews fled to the United States, and others emigrated to Palestine in support of the **Zionist movement**, which, since the late 1890s, had aimed at establishing a Jewish homeland in Palestine, with **Zion** (Jerusalem) as its capital. Most Jews were not as fortunate as the Zionist emigrants. Within the boundaries of the Nazi empire that conquered most of continental Europe during World War II, Hitler's regime implemented its "Final Solution" to the "Jewish problem" by shipping to prison camps and eventually murdering all the Jews they could round up. The Nazi Germans and their allies killed an estimated 6 million Jews, along with millions of other minorities they deemed "inferior," including Roma (Gypsies) and homosexuals. It was this Holocaust that prompted the victorious allies of World War II, from their powerful position in the newly formed United Nations, to create a permanent homeland for the Jewish people in Palestine. A seemingly endless cycle of violence, discussed beginning on page 185, ensued.

Christianity: Death and Resurrection in Jerusalem

Nearly a thousand years after Solomon established the Jews' First Temple, a new but closely related monotheistic faith emerged in Palestine. This was Christianity, named for Jesus Christ (*Christ* is Greek for "Anointed One," the equivalent of the Hebrew word for Messiah). Jesus, a Jew, was born near Jerusalem in Bethlehem, probably around 4 B.C.E. Tradition relates that when he was about 30 years old, Jesus began spreading the word that he was the Messiah, the deliverer of humankind long prophesied in Jewish doctrine. A small group of disciples accepted that he was the Messiah, the Son of God and a living manifestation of God Himself. They followed him for several years as he preached his message. He taught that the only path to God was through Him, and that faith in Christ as God's Son and as the redeemer of humanity's sins was the key to salvation.

Christianity is built upon a Jewish foundation: the common scriptures of Judaism and Christianity are known as the Hebrew Bible. Theologically, Christians would argue that Jews have the correct foundation but did not accept God's complete message, which continues beyond the Hebrew Bible to the new covenant, or the New Testament.

Jesus Christ's own teachings denied the validity of some Jewish doctrines, and around 29 C.E., a growing chorus of Jewish protesters called for his death. Palestine was then under Roman rule, and Roman administrators in Jerusalem placated the mob by ordering that he be put on trial. He was found guilty of being a claimant to Jewish kingship, and Roman soldiers put him to death by the particularly degrading and painful method of crucifixion. The cornerstone of Christian faith is that Jesus Christ was resurrected from the dead on the third day after his crucifixion and later ascended into heaven. For Christians, this was fulfillment of the scriptures: Jesus had come to redeem humanity's sins through His own death on the cross and through His resurrection. Christians believe that He continues to intercede with his Father on their behalf and that He will come again on Judgment Day, at the end of time.

After a period of relative tolerance, the Romans began actively persecuting Christians. Nevertheless, Christian ranks and influence grew. The turning point in Christianity's career came after 324 C.E., when the Roman Emperor Constantine embraced Christianity and established it as the official religion of the empire. The Christian Byzantine civilization that developed in the **"New Rome"** that Constantine established—Constantinople, now Istanbul, Turkey—created fine monuments at places associated with the life and death of Jesus Christ. These include the place long acknowledged as the center of the Christian world, Jerusalem's **Church of the Holy Sepulcher** (•Figures 6.19 and 6.B). This extraordinary, sprawling building, now administered by numerous separate Christian sects, contains the locations where tradition says Jesus Christ was crucified and buried. Incidentally, several of these Christian denominations have a long history of squabbling over turf within the church, and the key to this place is in the hands of a trusted, neutral party: a Muslim family.

Christianity has seldom been the majority religion in the land where it was born; only until Islam arrived in Palestine in 638 C.E. was the region primarily Christian. From then until the twentieth century, most of Palestine's inhabitants were Muslim, and since the mid-twentieth century, Muslims and Jews have been the major groups. Between the eleventh and fourteenth centuries, European Christians dispatched military expeditions to recapture Jerusalem and the rest of the Holy Land from the Muslims. These bloody campaigns, known as the **Crusades**, resulted in a series of short-lived Christian administrations in the region.

Joe Hobbs

• **Figure 6.19** Jerusalem's Church of the Holy Sepulcher, containing the locations where many Christians believe Jesus Christ was crucified and buried.

There are significant minority populations of Christians throughout the Middle East, including members of distinct sects such as the **Copts** of Egypt and the **Maronites** of Lebanon. Their share of the population has generally been declining, both because of emigration and lower birth rates than those of the majority Muslims. Christians have been persecuted in some countries, which is one of the push factors in their emigration. Nevertheless, the Middle East remains the cradle of their faith for Christians the world over, and Jerusalem and nearby Bethlehem are the world's most important Christian pilgrimage sites. Whatever your faith, you would surely find Jerusalem to be one of the most compelling places on Earth. I hope you will have an opportunity to travel there.

The Message of Islam

Islam is by far the dominant religion in the Middle East and North Africa; only Israel. within its pre-1967 borders, has a non-Muslim majority. Because of Islam's powerful influence not merely as a set of religious practices but as a way of life, an understanding of the religious tenets, culture, and diffusion of Islam is vital for appreciating the region's cultural geography.

Islam is a monotheistic faith built on the foundations of the region's earliest monotheistic faith, Judaism, and its offspring, Christianity. Indeed, Muslims (people who practice Islam) call Jews and Christians **People of the Book**, and their faith obliges them to be tolerant of these special peoples. Muslims believe that their prophet, Muhammad, was the very last in a series of prophets who brought the Word of God to humankind. Thus they perceive the Bible as incomplete but not entirely wrong— Jews and Christians merely missed receiving the entire message (just as Christians would insist that Jews missed the entire message). Muslims do not accept the Christian concept of the divine Trinity (God manifested in the form of the Father, his son Jesus, and the Holy Spirit) and regard Jesus as a prophet rather than as God.

Muhammad was born in 570 C.E. to a poor family in the western Arabian (now Saudi Arabian) city of Mecca. Located on an important north–south caravan route linking the frankincense-producing area of southern Arabia (now Yemen and Oman) with markets in Palestine (now Israel) and Syria, Mecca was a prosperous city at the time. It was also a pilgrimage destination because more than 300 deities were venerated in a shrine there called the **Ka'aba** (the Cube; •Figure 6.20). Muhammad married into a wealthy family and worked in the caravan trade. Muslim tradition holds that when he was about 40 years old, Muhammad was meditating in a cave outside Mecca when the Angel Gabriel appeared to him and ordered him to repeat the words of God that the angel would recite to him. Over the next 22 years, the prophet

Nabeel Turner/Getty Images

• **Figure 6.20** The black-shrouded cubical shrine known as the Ka'aba (just right of center) in Mecca's Great Mosque is the object toward which all Muslims face when they pray and is the centerpiece of the pilgrimage to Mecca required of all able Muslims. Note the carpet of humanity spread across this image: this is the height of the *hajj* (pilgrimage), when millions of the devout converge on the holy places of Mecca and Medina.

A schism occurred very early in the development of Islam, and it persists today. The split developed because the Prophet Muhammad had named no successor to take his place as the leader (caliph) of all Muslims. Some of his followers argued that the person with the strongest leadership skills and greatest piety was best qualified to assume this role. These followers became known as **Sunni**, or orthodox, Muslims. Others argued that only direct descendants of Muhammad, specifically through descent from his cousin and son-in-law Ali, could qualify as successors. They became known as **Shia**, or **Shiite**, Muslims.

The military forces of the two camps engaged in battle south of Baghdad at Karbala in 680 C.E., and in the encounter Sunni troops caught and brutally murdered Hussein, a son of Ali (Karbala is thus sacred to Shiites, as is nearby al-Najaf, where Ali is buried). The rift thereafter was deep and permanent. The martyrdom of Hussein became an important symbol for Shiites, many of whom still today regard themselves as oppressed peoples struggling against cruel tyrants, including some Sunni Muslims.

Today, only three of the region's countries, Iran, Iraq, and Bahrain, have Shiite majority populations. Significant minority populations of Shiites are in Syria, Lebanon, Yemen, and the Arab states of the Persian (Arabian) Gulf (see Figure 6.17).

related these words of God (whom Muslims call **Allah**) to scribes who wrote them down as the Qur'an (or Koran), the holy book of Islam.

During this time, Muhammad began preaching the new message, "There is no god but God (Allah)," which the polytheistic people of Mecca viewed as heresy. As much of their income depended on pilgrimage traffic to the Ka'aba, they also viewed Muhammad and his small band of followers as an economic threat. They forced the Muslims to flee from Mecca and take refuge in Yathrib (modern Medina), where a largely Jewish population had invited them to settle. There were subsequent skirmishes between the Meccans and Muslims, but in 630, the Muslims prevailed and peacefully occupied Mecca. The Muslims destroyed the idols enshrined in the Ka'aba, which became a pilgrimage center for their one God.

The Ka'aba is Islam's holiest place, and Mecca and Medina are its holiest cities. Jerusalem is also sacred to Muslims. Muslim tradition relates that on his **Night Journey**, the Prophet Muhammad ascended briefly into heaven from the great rock now beneath the Dome of the Rock (the same rock Jews regard as the Foundation Stone). Nearby on the Temple Mount (al-Haraam ash-Shariif) is **al-Aqsa Mosque**, a sacred congregational site. The proximity, even the duplication, of holy places between Islam and Judaism came to be the most difficult issue in peace negotiations between Palestinians and Israelis, as explained on page 190.

After Muhammad's death in 632, Arabian armies carried the new faith far and quickly. The two decaying empires that then prevailed in the Middle East and North Africa—the Byzantine or Eastern Roman Empire, based in Constantinople (Istanbul), and the Sassanian Empire, based in Persia (Iran) and adjacent Mesopotamia (Iraq)—put up only limited military resistance to the Muslim armies before capitulating. Local inhabitants generally welcomed the new faith, in part because administrators of the previous empires had not treated them well, whereas the Muslims promised tolerance. Soon the Syrian city of Damascus became the center of a Muslim empire. Baghdad assumed this role in 750 C.E.

Arab science and civilization flourished in the Baghdad immortalized in the legends of *The Thousand and One Nights*. Important accomplishments and discoveries were made in mathematics, astronomy, and geography. Scholars translated the Greek and Roman classics, and if not for their efforts, many of these works would never have survived to become part of the modern European legacy. It was an age of exploration, when Arab merchants and voyagers visited China and the remote lands of southern Africa. Many important discoveries by the Arab geographers were recorded in Arabic, a language unfamiliar to contemporary Europeans, and had to be rediscovered centuries later by the Portuguese and Spaniards. Arab merchants carried their faith on the spice routes to the East. One result, surprising to many today, is that the world's most populous Muslim country is not in the Middle East and North Africa, and its people are not Arabs; it is Indonesia, 5,000 thousand miles (8,000 km) east of Arabia.

Whether in Arabia or Indonesia, whether they be Sunni Muslims or Shiite Muslims (see Insights, above), all believers are united in support of the five fundamental precepts, or **Pillars of Islam**. The first of these is the **profession of faith**: "There is no god but God, and Muhammad is His Messenger." This expression is often on the lips of the devout Muslim, both in prayer and as a prelude to everyday activities.

The second pillar is **prayer**, required five times daily at prescribed intervals. Two of these prayers mark dawn and sunset. Business comes to a halt as the faithful prostrate themselves before God. Muslims may pray anywhere, but wherever they are, they must turn toward Mecca. There also is a congregational prayer at noon on Friday, the Muslim Sabbath.

The third pillar is **almsgiving**. In earlier times, Muslims were required to give a fixed proportion of their income as charity, similar to the concept of the tithe in the Christian church. Today, the donations are voluntary. Even Muslims of very modest means give what they can to those in need.

The fourth pillar is **fasting** during Ramadan, the ninth month of the Muslim lunar calendar. Muslims are required to abstain from food, liquids, smoking, and sexual activity from dawn to sunset throughout Ramadan. The lunar month of Ramadan falls earlier each year in the solar calendar and thus periodically occurs in summer. In the torrid Middle East and North Africa, that timing imposes special hardships on the faithful, who, even if they are performing manual labor, must resist the urge to drink water during the long, hot days.

The final pillar is the *hajj*, or **pilgrimage to Mecca**, Islam's holiest city. Every Muslim who is physically and financially able is required to make the journey once in his or her lifetime. A lesser pilgrimage may be performed at any time, but the prescribed season is in the 12th month of the Muslim calendar. Those days witness one of earth's greatest annual human migrations as about 3 million Muslims from all over the world converge on Mecca. Hosting these throngs is an obligation the government of Saudi Arabia fulfills proudly and at considerable expense. However, there has been some trepidation in recent years because of the security threat foreign visitors may pose to the host country and because accidents such as stampedes and tent city fires have cost numerous lives. Many pilgrims also visit the nearby city of Medina, where Muhammad is buried. Most Muslims regard the *hajj* as one of the most significant events of their lifetimes. All are required to wear simple seamless garments, and for a few days, the barriers separating groups by income, ethnicity, and nationality are broken. Pilgrims return home with the new stature and title of *hajji* but also with humility and renewed devotion.

All Muslims share the Five Pillars and other tenets, but they vary widely in other cultural practices related to their faith, depending on the country they live in; whether they are from the desert, village, or city; and their education and income. The governments and associated clerical authorities in Saudi Arabia and Iran insist on strict application of **Islamic law** (known as *sharia*) to civil life; in effect, there is no separation between church and state. The Qur'an does not state that women are required to wear veils, but it does urge them to be modest, and it portrays their roles as different from those of men. Clerics in Saudi Arabia insist that women wear floor-length, long-sleeved black robes and black veils in public, that they travel accompanied by a male member of their families, and that they not drive cars. In Egypt, by contrast, Muslim women are free to appear in public unveiled if they choose. However, in most Muslim countries, conservative ideas about the role of women are still very strong: they should be modest, retiring, good mothers, and keepers of the home, although they can work. The Qur'an portrays women as equal to men in the sight of God, and in principle, Islamic teachings guarantee the right of women to hold and inherit property.

Most Muslim women argue that what others often see as "backward" cultural practices are in fact progressive. For example, they say, their modest dress compels men to evaluate them on the basis of their character and performance, not their attractiveness. They argue that segregation of the sexes in the classroom makes it easier for both women and men to develop their confidence and skills (•**Figure 6.21**). A married woman

Joe Hobbs

• **Figure 6.21** Education of women in the Muslim countries has increased dramatically in recent decades, but strong traditions keep most women out of the workforce. This is a geographic information systems (GIS) class in the Department of Geography, United Arab Emirates University.

retains her maiden name. These apparent advantages can be weighed against the drawbacks that make women generally subordinate to men in public affairs.

6.4 Economic Geography

Overall, the Middle East and North Africa is a poor region; per capita GNI PPP for the 23 countries and territories averages only $9,460 (see Table 6.1). This may seem surprising in view of the "rich Arab" stereotype. Only the oil-endowed states around the Persian Gulf (Arabian Gulf) deserve the reputation for wealth, and by most measures other than per capita wealth, only non-Arab Israel is truly a more developed country (MDC). Israel's prosperity comes from its innovation in computer and other high-technology industries, the processing and sale of diamonds, large amounts of foreign (mostly U.S.) aid, and investment and assistance by Jews and Jewish organizations around the world.

Here we will examine the role of oil in the region's economies, and the lagging economic indicators and human conditions that should have served as the writing on the wall for the "Arab Spring" (see page 192).

"The Prize"

Vital to the industrialized countries as a source of fuels, lubricants, and chemical raw materials, petroleum is one of the world's most important natural resources. Daniel Yergin, recognized as the world's preeminent authority on oil, wrote the definitive book on this essential commodity. His title sums up what oil has come to represent in economic and geopolitical terms: *The Prize*.

• **Figure 6.22** Petroleum facilities in the interior wilderness of Arabia's Empty Quarter (al-Rub' al-Khali). Most of Saudi Arabia's production is from vast, shallow reservoirs of oil that are much closer to marine terminals on the Persian Gulf.

A crucial feature of world geography is the concentration of about two-thirds of the world's proven conventional petroleum reserves in a few countries that ring the Gulf. In the second half of the twentieth century, this geographical endowment transformed a handful of inconsequential countries and shaikhdoms into some of the most critical locations in the global economic system (•**Figure 6.22**).

By coincidence, the countries rich in oil tend to have relatively small populations, whereas the most populous nations have few oil reserves; Iran is an exception. All but about 1 percent of the Gulf oil region's proven reserves of crude oil are located in Saudi Arabia, Iraq, Kuwait, Iran, and the United Arab Emirates (UAE), with smaller reserves in Oman, Bahrain, and Qatar. Saudi Arabia, by far the world leader in reserves, has about 18 percent of the proven crude oil reserves on the globe. Venezuela is in second place, with about 14 percent of the world's proven oil; Canada is third, with 12 percent (not in the form of easily extractable crude oil, but as tar sands; see page 417); the next four are all along the Gulf: Iran (9%), Iraq (8%), Kuwait (7%), and the United Arab Emirates (6%). Libya also has significant oil reserves. The world's largest oil consumer, the United States, has 2 percent. China, the second largest consumer by a narrow margin, has 2 percent.

In the Gulf area, the great thickness of the region's oil-bearing strata and high reservoir pressures have made it possible to secure an immense amount of oil from a small number of wells. The productivity of each well makes each barrel inexpensive to extract, and it is simple to increase or reduce production quickly in response to world market conditions. Gulf oil is thus cheap to produce unless expenditures to maintain huge military forces and infrastructure to defend it are factored in (in economic terms, unless the "external costs" of oil are internalized). Some analysts—including a former U.S. Navy secretary—prefer to calculate the "real price" of oil with military expenditures calculated. If one

accepts the premise that maintaining a costly military presence in the region is part of the United States' long-term strategy to maintain the flow of Gulf oil, that oil is no longer cheap by any measure.

Production, export, and profits of Middle Eastern and North African oil were once firmly in the hands of foreign companies. That situation changed after 1960, when most of the Gulf countries and other exporting nations formed the **Organization of Petroleum Exporting Countries (OPEC)**, with the aim of taking joint action to demand higher profits from oil. It changed again after 1972, when the oil-producing countries began to nationalize the foreign oil companies. OPEC was relatively obscure until the Arab-Israeli war of 1973, after which the organization began a series of dramatic price increases. In 1980, the organization's price reached $37 (U.S.) per barrel, compared with $2 a barrel in early 1973. Adjusted for inflation, that $37 would be the equivalent of about $100 per barrel in 2011, so the price hikes were truly titanic.

These events had enormous repercussions for the world economy. Immense wealth was transferred from the more developed countries to the OPEC countries to pay for indispensable oil supplies. The skyrocketing cost of gasoline and other oil products helped cause serious inflation in the United States and many other countries and contributed to the 1973 **energy crisis** in the United States. Desperately poor, less developed countries (LDCs) found that high oil prices not only hindered the development of their industries and transportation but also reduced food production because of high prices for fertilizer made from oil and natural gas. In the Gulf countries, the oil bonanza produced a wave of spending on military hardware, showy buildings, luxuries for the elite, and ambitious development projects of many kinds. Per capita benefits to citizens were greatest in the Arabian Peninsula, where small populations and immense inflows of oil money made possible the abolition of taxes, the establishment of comprehensive social programs, and heavily subsidized amenities such as low-cost housing and utilities, including freshwater distilled from the salty Gulf in desalination plants.

Then, as the 1980s began, the era of continually expanding OPEC oil production, sales, and profits seemed to come to an end. After 1973, the high price of oil stimulated oil development in countries outside OPEC. Oil conservation measures such as a shift to more fuel-efficient vehicles and factories were introduced. Substitution of cheaper fuels for oil increased. Coal replaced oil in many electricity-generating stations. Meanwhile, the world entered a period of economic recession due in part to high oil prices. Decreased business activity reduced the demand for oil. Profits of the world oil industry (and taxes paid to governments) were severely cut, large numbers of refineries had to close, and much of the world tanker fleet was idled.

Oil prices have risen and fallen, with predictable consequences for producers and consumers. But the immense oil and gas reserves still in the ground guarantee that the Gulf region will continue to have major long-term influence in world affairs and will remain prosperous as long as these finite resources are in demand—especially in the established industrialized countries and in the booming economies of China and India.

There are, of course, other resources and industries in the region's economic geography—**remittances** (earned income sent home by guest workers) in the oil-rich countries; revenues from ship traffic through the Suez Canal; and exports of cotton, rice, and other commercial crops, for example—but oil dominates the region's economy and is central to the global economy.

This depiction of MENA's oil wealth, which is concentrated in a few countries, does not consider the economic poverty so characteristic of the region's other countries and the great majority of its peoples. The worsening conditions in many of the Arab countries—the poverty, crowding, unemployment, and the growing gap between the great mass of the poor and the increasingly wealthy rich—were the economic ingredients of the Arab Spring described later.

6.5 Geopolitical Issues

The Middle East and North Africa have long been vital in world affairs, with their resources and geographical assets such as the Suez Canal coveted by outside interests. From very early times, overland caravan routes, including the great Silk Road, crossed the Middle East and North Africa with highly prized commodities traded between Europe and Asia. The security of these routes was vital. Geopolitical concerns are now focused on narrow waterways, access to oil, access to freshwater, and terrorism.

Chokepoints

One of the striking characteristics of the geography of the Middle East and North Africa is how many seas border and penetrate the region. In many cases, these bodies of water are connected to one another though narrow straits and other

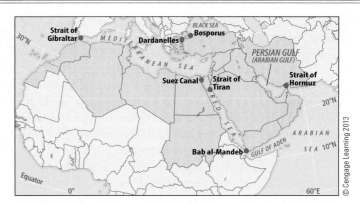

• **Figure 6.23** Chokepoints in the Middle East and North Africa.

passageways. In geopolitical terms, such constrictions are known as **chokepoints**—strategic narrow passageways on land or sea that may be easily closed off by force or even the threat of force (•**Figure 6.23**). Chokepoints must be unimpeded if world commerce is to carry on normally. Keeping them open is therefore usually one of the top priorities of regional and external governments. Likewise, closing them is a priority to a combatant nation or a terrorist entity seeking to gain a strategic advantage. Many notable events in military history and the formation of foreign policy in the Middle East focus on these strategic places.

One of the world's most important chokepoints is the Suez Canal, which opened in 1869. Slicing 107 miles (172 km) through the narrow Isthmus of Suez, the British- and French-owned Suez Canal linked the Mediterranean Sea with the Indian Ocean, saving cargo, military, and passenger ships a journey of many thousands of miles around the southern tip of Africa (•**Figure 6.24**). Keeping the Suez Canal in friendly hands was one of Britain's major military concerns during World War II. That objective led to a hard-fought and eventually successful British and Allied campaign against Nazi Germany in North Africa. Egyptian President Gamal Abdel Nasser's nationalization of the canal in 1956 led immediately to a British, French, and Israeli invasion of Egypt and a conflict known as the **Suez Crisis** and the **Arab-Israeli War of 1956**. Egypt effectively won

• **Figure 6.24** The strategically vital Suez Canal zone saw bitter fighting in the Middle East wars of 1956, 1967, and 1973.

that war when international pressure caused the invading forces to withdraw, leaving the canal in Egypt's hands.

Nearby, another chokepoint played a critical role in the most important Arab-Israeli war, the Six-Day War of 1967. One of the events that precipitated the war was President Nasser's closure of the Strait of Tiran (at the southern end of the Gulf of Aqaba) to Israeli shipping. Israel had won the right of navigation through the strait after its war of independence and would not accept its closure; therefore, it attacked Egypt.

On either side of the Arabian Peninsula are two more critical chokepoints. One is exceptionally vital to the world's economy: the Strait of Hormuz, connecting the Persian-Arabian Gulf with the Gulf of Oman and the Arabian Sea. Much of the world's oil supply passes through here in the holds of giant supertanker ships. Closure of the Strait of Hormuz would have devastating effects on the world's industrial and financial systems. Iran's plans to station Chinese-made missiles on the Strait of Hormuz and thus threaten international oil shipments led to a new level of U.S. involvement late in the Iran-Iraq War of 1980–1988. That war also prompted a flurry of new oil pipeline construction designed to bypass the Strait of Hormuz chokepoint (see Geography of Energy, next page). One of these pipelines, the Petroline route running east–west across the Arabian Peninsula, was built to bypass another important chokepoint, the Bab el-Mandeb, which connects the Red Sea with the Gulf of Aden and the Indian Ocean. Iran threatened to close the Strait of Hormuz when Western powers imposed sanctions against Iran for ramping up its nuclear program in 2012.

Turkey controls two more chokepoints that together are known as the Turkish Straits. The northernmost strait is the Bosporus, which cleaves the city of Istanbul into western (European) and eastern (Asian) sides, and the southern strait is called the Dardanelles. Their security has long been critical to the successful passage of goods between Europe and Asia and even more so to the successful passage of vessels between Russia (and the Soviet Union) and the rest of the world. As discussed on page 126, throughout the twentieth century one of the Soviet Union's constant strategic priorities was the right of navigation through the Turkish Straits.

Finally, the Strait of Gibraltar, connecting the Mediterranean Sea with the Atlantic Ocean, is also a chokepoint. Have a look at the map of Figure 6.23, and consider how geographic good luck gave the lands of the Mediterranean Sea access to the rest of the world through Gibraltar. Britain's insistence on maintaining control over its enclave of Gibraltar, decades after the decolonization of most of the world, is an excellent indicator of how critical this chokepoint is.

Access to Oil

The Suez Canal and other chokepoints, the cotton of Egypt's Nile Delta, and the strategic location of the region were important during colonial times and have remained so ever since. But oil has been and will remain (as long as fossil fuels drive the world's economies) what keeps the rest of the world interested economically in the Middle East and North Africa. The region's oil flows to many countries, but most of it is marketed in Western Europe, China, and Japan. The United States also imports large amounts of Gulf oil but has a smaller relative dependence on this source than Japan and Europe do; about 18 percent of its imported oil came from the Gulf in 2011. However, the Gulf region is very important to the United States because of the great dependence of close American allies on Gulf oil and because of the importance of the oil as a future reserve. American companies are also heavily involved in oil operations and oil-financed development in the Gulf countries. Gulf "petrodollars" are spent, banked, and invested in the United States, contributing significantly to the U.S. economy. Maintaining a secure supply of Gulf oil has therefore been one of the long-standing pillars of U.S. policy in the Middle East.

The United States has long had a precarious relationship with the key players in the Middle Eastern arena. The U.S. has pledged unwavering support for Israel, while at the same time courting Israel's oil-rich Arab foes, notably Saudi Arabia. That Arab kingdom and its neighbors around the Persian/Arabian Gulf possess more than 60 percent of the world's proven oil reserves. Thus they are vital to the long-term economic security of the Western industrial powers, China, India, and Japan. The United States and its Western allies made it clear they would not tolerate any disruption of access to this supply when Iraqi troops, under orders from Iraqi President Saddam Hussein, occupied Kuwait on August 2, 1990. U.S. President George H.W. Bush drew a "line in the sand," proclaiming, "We cannot permit a resource so vital to be dominated by one so ruthless—and we won't." In what came to be known as the **Gulf War**, the United States and a coalition of Western and Arab allies mounted a massive array of military might that ousted the Iraqi invaders within months and secured the vital oil supplies for Western markets. When the United States invaded and occupied Iraq in 2003, ostensibly to extinguish Iraq's ability to develop and deploy **weapons of mass destruction (WMD)**, including chemical, biological, and nuclear weapons, many critics insisted that this was just another example of America's determination to control Middle Eastern oil.

The Gulf War was not the first time the United States expressed its willingness to use force to maintain access to Middle Eastern oil. In 1979, in the wake of the Islamic revolution in Iran, the Soviet Union invaded neighboring Afghanistan. U.S. military analysts feared that the Soviets might use Afghanistan as a launching pad to invade oil-rich Iran. The United States deemed this prospect unacceptable, and President Jimmy Carter issued the foreign policy statement that came to be known as the **Carter Doctrine**: "An attempt by any outside force to gain control of the Persian Gulf region will be regarded as an assault on the vital interests of the United States of America, and such an assault will be repelled by any means necessary, including military force." "Vital interests" meant oil, and defending them with military force meant that the United States was willing to go to war with the Soviet Union, presumably nuclear war.

Middle Eastern wars had already become proxy wars for the superpowers, with oil always looming as the prize. In the 1967 and 1973 Arab-Israeli wars, for example, Soviet-backed Syrian forces fought U.S.-backed Israeli troops. American support of Israel in this war prompted Arab members of OPEC to impose an **oil embargo**, refusing to sell their oil to the United States, thus precipitating the nation's first energy crisis. During

242

415, 440

GEOGRAPHY OF ENERGY | **Pipelines and Chokepoints in the Middle East**

The oil-exporting countries of the Middle East, often with the financial support of the United States and other leading oil-consuming countries, have invested enormous resources to ensure the safe passage of oil to world markets. The Middle East's oil region is not landlocked like the oil-rich "stans" of central Asia, but some of the same kinds of geopolitical issues are involved in oil export planning. These concerns often point toward pipelines as the best means of routing oil shipments (•Figure 6.C). Pipelines are attractive because they shorten the time and expense involved in seaborne transport, and they bypass chokepoints. However, they are vulnerable to disruption. Weapons as small as grenades can disrupt supplies. A major challenge is how to route a pipeline so that it will not cross through a potential enemy's territory, and another is how to maintain friendships with the countries the oil crosses.

The following paragraphs will not be interesting at all if you don't use the map in Figure 6.C. Please use that map, and you will get a feel for the compelling intersection of economics, military thinking, and the consequences of war in pipeline geography.

One of the first major pipelines in the region was the 1,100-mile (1,760-km) Trans-Arabian Pipeline (Tapline), leading from the oil-producing eastern province of Saudi Arabia to a terminal on the Mediterranean coast of Lebanon. It has the advantage of bypassing three chokepoints (Strait of Hormuz, Bab el-Mandeb, and Suez Canal). However, at the time it was completed (1950), it could not have been anticipated that it would cross two of the worst impending conflict zones in the Middle East: the Golan Heights of Syria (which fell to Israel in the 1967 war) and southern Lebanon (a major theater of the Lebanese civil war of the 1970s and the subsequent Israeli–Lebanese Shiite struggles). Tapline was knocked out early by its unfortunate geography.

Another Middle East conflict, the 1967 Six-Day War between Israel and its neighbors, led to the closure of the Suez Canal (for about a decade) and thus to the construction of two new pipelines to bypass Suez: one across southern Israel from the Gulf of Aqaba to the Mediterranean Sea and another across Egypt

from the Gulf of Suez to the Mediterranean (thus its acronym, SUMED).

No Middle Eastern country has been more dependent on pipelines than Iraq, and none has so systematically experienced the liabilities of pipelines. The Iran-Iraq war of 1980–1988 had major impacts on pipeline geography. Saudi Arabia supported Iraq in the war and so tried to ensure it could get oil to market without being threatened by Iran; this meant bypassing the Strait of Hormuz. That led to the construction of Petroline (opened in 1981), the east–west pipeline across the Arabian Peninsula. Iraq built a pipeline to ship some of its oil south to Petroline, thereby bypassing the Strait of Hormuz as well. Beginning in 1961, Iraq was also able to ship oil through pipelines almost due west though Syria to the

Mediterranean Sea. But when it attacked Iran in 1980, Iraq lost that route because Syria was Iran's ally.

Iraq quickly responded to that setback by building a new pipeline leading almost due north to bypass Syrian territory and then making a sharp 90-degree turn westward through southern Turkey to the Mediterranean Sea. But in 1990, Iraq attacked Kuwait, threatened Saudi Arabia, and prompted a U.S.-led counterattack. Saudi Arabia responded by closing the southern link with Petroline. Turkey, a NATO ally of the United States, reacted by closing the northern link. The United Nations responded by issuing strict controls on Iraqi oil exports. In sum, by his military adventurism, Saddam Hussein did almost everything imaginable to deprive his country of oil exports.

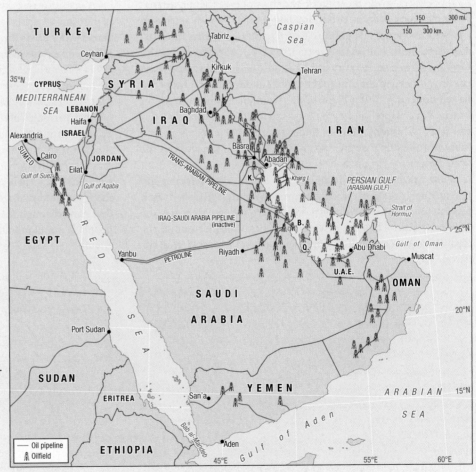

• **Figure 6.C** Principal oilfields and pipelines in the heart of the Middle East. Vulnerable chokepoints and volatile political relations have led to the construction and often indirect routing of many pipelines.
Source: Based on National Geographic's Atlas of the Middle East (National Geographic, 2003).

the 1973 war, the United States put its forces on an advanced state of readiness to take on the Soviets in a nuclear exchange if necessary. All of these events illustrate that the Middle East and North Africa form, as Eastern Europe did, a shatter belt— a large, strategically located region composed of conflicting states caught between the conflicting interests of great powers.

Access to Freshwater

"The next war in the Middle East will be over water."

—Jordan's late King Hussein

Some of the most serious geopolitical issues in the Middle East and North Africa relate to **hydropolitics**, or political leverage and control over water. In this arid region, where most water is available either from rivers or from underground aquifers that cross national boundaries, control over water is an especially difficult and potentially explosive issue. An estimated 90 percent of the usable freshwater in the Middle East crosses one or more international borders.

Water is one of the most problematic issues in the Palestinian-Israeli conflict. Freshwater aquifers underneath the West Bank supply about 40 percent of Israel's water. Palestinians point to Israel's control over West Bank water as one of the most troubling elements of its occupation. The average Jewish settler on the West Bank uses 74 gallons (278 l) per day, whereas the average West Bank Palestinian uses 19 (72 l). (The World Health Organization calculates that 13 gallons, or 50 liters per person per day, is needed for minimal health and sanitation standards.) Israeli policies prohibit Palestinians from increasing their water usage. Many Israeli policymakers insist that water resources in the West Bank must remain under strict Israeli control and on these grounds oppose the creation of a Palestinian state in the West Bank.

More promisingly, Jordan and Israel are working on agreements to share waters from the Jordan River (which forms a portion of their common border) and its tributary, the Yarmuk River. They are discussing a joint venture to build the **Red–Dead Peace Conduit**, which would send seawater from the Gulf of Aqaba to the Dead Sea via a network of canals and pipelines (see Figure 6.33). This would replenish the Dead Sea, which has retreated by about 3 feet (c. 1 m) per year for the last 25 years. The gravity flow of seawater to the Dead Sea would run generators to produce electricity, some of which would be used to desalinate the water. The two nations, and potentially the Palestinian Authority, could share this water and power.

Water is a critical issue blocking a peace treaty between Israel and Syria. If Syria were to recover all of the Golan Heights (which it lost to Israel in the 1967 war), it would have shorefront on Lake Kinneret and therefore, presumably, rights to use its water. That prospect is unacceptable to Israel. This body of water, also known as Lake Tiberias and the Sea of Galilee, is Israel's principal supply of freshwater, feeding the National Water Carrier system that transports water south to the Negev Desert (see Figure 6.33). In occupying the Golan Heights, Israel also controls some of the northern bank of the Yarmuk River on the border with Jordan. For many years, Israel has stated that it would never allow Syria and Jordan to construct their proposed Unity Dam, intended to store waters of the Yarmuk River to be shared between those countries. Israel thus implied it would bomb the dam rather than allow it to deprive Israel of water from the Jordan River.

A useful way to think about the geography of hydropolitics is in terms of **upstream** and **downstream countries**. Simply because water flows downhill, an upstream country is usually able to maximize its water use at the expense of a downstream country (a situation described as a **zero-sum game**, where any gain by one party represents an equivalent loss to the other). However, the situation between Israel and the countries upstream on the Yarmuk shows that this is not always true. Although downstream, Israel is far more powerful militarily and can use the threat of force to wrest more water out of the system.

Historically, Egypt has also defied the norm of upstream countries exercising their power over downstream countries. Egypt is the ultimate downstream country, at the mouth of a great river that runs through five countries and sustains about 170 million people (a population that is expected to double in about 20 years; •Figure 6.25). However, it has long been the strongest country in

• **Figure 6.25** Recent and proposed water developments in the Nile Basin. Note that this map is oriented such that north is at the left rather than at the top.

REGIONAL PERSPECTIVE | The Aswan High Dam

Egypt's enormous Aswan High Dam (•**Figure 6.25**; and see location in Figure 6.25) is located on a stretch of the Nile known as the First Cataract (the Nile's many **cataracts** are areas where the valley narrows and rapids form in the river). Its reservoir, Lake Nasser, stretches for more than 300 miles (480 km) and reaches into northern Sudan. Like all giant hydrological projects, the Aswan High Dam has generated benefits and liabilities, and its construction was controversial. On the plus side:

- By storing water in years of abundant rain for use in years of low rainfall, the High Dam provides a constant water supply to the Egyptian Nile Valley's 11,000 square miles (28,500 sq km) of irrigated fields, thus extending the cultivation period.

- Year-round irrigation has significantly boosted crop production in Egypt.

- The dam made possible the "reclamation" of previously uncultivated land on the desert fringe adjoining the western Nile Delta; there is enough surplus water to irrigate lands beyond the river's flood plain.

- Because water levels can be adjusted, there are fewer threats to ships navigating the Nile.

- The dam provides hydroelectricity to Egypt's industries and citizens.

- Excess waters stored in Lake Nasser help Egypt to avoid the damaging impacts of drought.

- The dam can hold back overabundant water, helping Egypt to avoid the damaging impacts of flooding.

The Aswan High Dam has also had drawbacks:

- The dam has caused the water table to rise, making it harder for irrigated soil to drain properly. When farmers use too much water, standing water evaporates and leaves a deposit of mineral salts—a problem known as **salinization**—that causes soil to lose its productivity.

- Canal waters, which are now high throughout the year, are ideal breeding grounds for the snail that hosts the parasite that causes schistosomiasis (bilharzia), a debilitating disease affecting a large proportion of Egypt's rural population.

- Mediterranean sardine populations, now deprived of the rich silt that nurtured their food supply of phytoplankton, have plummeted off the Nile Delta. The lucrative sardine fishing industry has declined by more than 90 percent.

- The dam prevents rich volcanic silt washed down the Nile from Ethiopia from reaching the floodplain below the dam. Without this "free fertilizer," Egyptian farmers must buy expensive artificial fertilizers.

- Because silt is no longer replenishing the soils of the Nile Delta, there is coastal erosion along the Mediterranean seafront. This is allowing seawater to destroy productive farmland.

- The Nubian people who lived in the area now inundated by the reservoir behind the dam had to be relocated from their homeland to an alien environment.

- Many of the archaeological treasures of ancient Egypt were drowned by the reservoir (but others were relocated).

- If the dam burst because of an earthquake or a bombing, the lives of nearly 70 million people downstream would be at risk. One analyst wrote:

If a bomb hit the top of the Dam, the water pouring out uncontrolled would make the Dam burst in its complete breadth within a few hours. The masses of water sweeping across the whole of Egypt (they would be at least the discharge of two years within hours or days) would cause utter devastation from Aswan to Damietta, destroy all cultivated land, and most probably about 99 per cent of the total population would lose their lives.[2]

Whenever a dam is built, there are questions about whether its advantages outweigh its drawbacks. When great dams are built in the LDCs—like the Aswan High Dam in Egypt or the Three Gorges Dam in China—such questions tend to be secondary to the symbolic importance of the dam. These massive engineering projects announce that a developing country is developed enough to master its environment and marshal the extraordinary engineering and other resources needed to build the dam. The dam announces that a country has "come of age."

the Nile Basin and has threatened to use its greater force if it does not get the water it wants. In 1926, when the British ruled Egypt and many other colonies in Africa, 10 countries located on the Nile and its tributaries upstream of Egypt were compelled to sign the **Nile Water Agreement**. This guaranteed Egyptian access to 75 percent of the river's flow, even though barely a drop of the Nile's waters actually originates in Egypt. Sudan was given access to 25 percent of the water, and Ethiopia was given none. The treaty forbids any projects that might threaten the volume of water reaching Egypt, prohibits use of Lake Victoria's water without Egypt's permission, and gives Egypt the right to inspect the entire length of the Nile to ensure compliance.

In recent years, however, one country after another has defied the treaty, calling it an outmoded legacy of colonialism. Kenya and Tanzania have plans to build pipelines to carry Lake Victoria waters to thirsty towns and villages inland. Uganda is building its controversial Bujagali Dam on the Nile, mainly for hydroelectricity production. With Chinese assistance, Ethiopia is building the huge Tekaze Dam, for hydropower and irrigation, on a tributary of the Blue Nile. Sudan is building the Merowe (Hamdab) and Kajbar Dams on its northern stretch of the Nile. South Korea, China, and India have purchased huge swaths of land in Sudan and Ethiopia to grow wheat and rice through irrigation. This will result in a massive new drawdown of Nile waters. Predictably, Egypt has had a bellicose response to these developments. Egypt has even called Kenya's stated intention to withdraw from the Nile Water Agreement an "act of war."

Meanwhile, Egypt's demands on Nile waters are increasing. Following on the ambitious Aswan High Dam (see Regional Perspective), Egypt recently built the multibillion-dollar

Egypt's Aswan High Dam image appears here, with caption below.

• **Figure 6.26** Egypt's Aswan High Dam, on the First Cataract of the Nile River.

Toshka Canal, which transports water from Lake Nasser over a distance of 100 miles (160 km) to the Kharga Oasis of the Western Desert (see Figure 6.25). Proponents of the canal insist that it will result in the cultivation of nearly 2,350 square miles (c. 6,100 sq km) of "new" land and provide a living for hundreds of thousands of people. Critics argue that it is a waste of money and that salinization and evaporation will take a huge toll on the cultivated land and the country's water supply.

The familiar specter of population growth will play a big role in the hydropolitics of the Nile. The combined populations of Egypt, Sudan, and Ethiopia, at about 210 million now, are predicted to grow to 270 million by 2025 and 360 million by 2050. Every person's direct needs for water plus production of food for him or her means growing demands for Nile waters and the potential for conflict. The solution? Lester Brown of Worldwatch proposes three steps: Reduce population growth, grow crops that are less demanding of water, and go back to the negotiating table for a new Nile Water Agreement.[3]

Turkey—the source of four-fifths of Syria's water and two-thirds of Iraq's—exercises its upstream advantage on the Tigris and Euphrates rivers (•Figure 6.27). Turkey's position has long been that water in Turkey belongs to Turkey, just as oil in Saudi Arabia belongs to Saudi Arabia. Not surprisingly, downstream Syria and Iraq reject this position and are distraught by the diminished flow and quality of water resulting from Turkey's comprehensive **Southeast Anatolia Project**. When completed, the project is expected to reduce Syria's share of the Euphrates waters by 40 percent and Iraq's by 60 percent. Also increasing the likelihood of serious future tension is a history of strained relations among Turkey, Syria, and Iraq, accompanied by the fact that no commonly accepted body-of-water law governs the allocation of water in such international situations.

Turkish leaders have said they will never use water as a political weapon, but Turkey has already wielded water to its advantage. In 1987, for example, Turkey increased the Euphrates flow into Syria in exchange for a Syrian pledge to stop support of Kurdish rebels inside Turkey. Turkey now says it wants to use water to promote peace in the Middle East by shipping it in converted supertankers for sale to such thirsty (and therefore potentially combative) countries as Israel, Jordan, Saudi Arabia, Libya, and Algeria.

Terrorism

During the two terms of U.S. President George W. Bush, almost all of the geopolitical issues related to this region—oil, economic

• **Figure 6.27** The Tigris and Euphrates Rivers rise in Turkey, giving this non-Arab country control over a resources vital to the lives of millions of Arabs in downstream Syria and Iraq. This waterfall is on a tributary of the Tigris in far eastern Turkey.

development, trade, aid, the Arab-Israeli conflict, and more—were subsumed beneath the broader rubric of the president's declared **War on Terror**. Even the 2003 invasion of Iraq was justified as a response to that war. The open-ended, unconventional War on Terror would destroy the enemies who attacked the United States on 9/11 and use preemptive strikes if needed to protect the U.S. homeland against future attacks. President Obama much more frequently employed preemptive attacks on suspected terrorist targets, while avoiding the term *War on Terror*. One of his favorite tactics was the use of unmanned "predator drones" that fly quietly overhead and unleash deadly force on ground targets. Their use is controversial because they often result in civilian casualties.

197, 238

Since 9/11, Americans have become accustomed to heightened security measures to provide protection against **terrorism**, which by the U.S. State Department definition is "premeditated, politically motivated violence perpetrated against noncombatant targets by subnational groups or clandestine agents, usually intended to influence an audience."[4] The terrorists pursued by the United States are almost without exception Islamist militants—best known as **Islamists**—and so it is useful to understand the nature of Islamic "fundamentalism" and radicalism. Not all Islamists are militant or terrorist, but they all reject what they view as the materialism and moral corruption of Western countries and the political and military support these countries lend to Israel. Both Sunni and Shiite Muslims have advanced a wide range of Islamist movements, notably in Iran, Lebanon, Egypt, Sudan, Algeria and the Palestinian Territories.

Although nominally religious, the more radical of these movements have political and cultural aims, particularly the destabilization or removal of U.S. and Israeli interests in the region and abroad. In 1993, followers of the radical Egyptian cleric Sheikh Umar Abdel-Rahman bombed New York City's World Trade Center as a protest against American support of Israel and Egypt's pro-Western government. In an attempt to destabilize and replace Egypt's government, which it viewed as an illegitimate regime too supportive of the United States, another Egyptian Islamist group attacked and killed foreign tourists in Egypt in the 1990s. In the 1980s, members of the pro-Iranian **Hizbullah**, or Party of God, in Lebanon kidnapped foreign civilians to use as bargaining chips for the release of comrades jailed in other Middle Eastern countries.

Within Israel and the autonomous Palestinian territories of the West Bank and Gaza Strip, Palestinian members of **Hamas** (an Arabic acronym for the Islamic Resistance Movement) and another organization called the al-Aqsa Martyrs Brigade carried out terrorist attacks on Israeli civilians and soldiers in a largely successful effort to derail implementation of the peace agreements reached in the 1990s between the Israeli government and the Palestine Liberation Organization (PLO).

189

The Iranian Threat

In Western capitals, concern about "state-sponsored terrorism" has long focused on Iran. Iran has extended both open and clandestine assistance to a variety of Islamist terrorist groups, including Hizbullah and Hamas. There is great concern about Iran's nuclear weapons potential because such weapons might find their way to terrorist groups or be delivered by Iran itself on its own missiles against Israel or another target. Iran insists that its nuclear program is aimed only at electricity generation and medical research, and denies that it is developing weapons. American and Israeli intelligence agencies, however, believe Iran is on course to create nuclear weapons some time before 2016. The United States, Israel, and the United Nations' **International Atomic Energy Agency (IAEA)** are particularly watchful of Iran's acquisition of centrifuges, which are used to enrich uranium. The enrichment process is necessary for the production of fuel for civilian use in nuclear power production, which Iran is permitted to do, but can also be expanded for the production of nuclear weapons.

141

The United States was weighing three difficult options in response to Iran's nuclear weapons threat. As U.S. Defense Secretary Robert Gates noted, "There are no good options on Iran."[5] First, it could engage in diplomatic dialogue with Iran, as it had done with North Korea, offering a package of incentives and security guarantees in exchange for Iran's pledge to halt weapons development. The United States and Iran do share some common interests, notably stability in Iraq; Iran does not want to deal with strong new Sunni or Kurdish states there, or with the tide of refugees that more Iraqi conflict would send its way. Second, the United States could impose even harsher sanctions than it has already against Iran, especially to cut off investment and finance (the United Nations, European Union, and several countries also have levied sanctions against Iran).

Finally, with or without coordination with Israel, United States could launch military strikes on Iran's suspected nuclear weapons development sites, beginning at facilities northeast of the Shiite holy city of Qum, at nearby Natanz and Arak, and near the Gulf city of Bushehr (•Figure 6.28). Most analysts

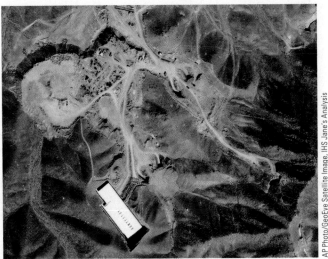

• **Figure 6.28** Iran's suspected gas centrifuge uranium enrichment facility northeast of the holy city of Qum, in the rugged Zagros Mountains of northwest Iran. Just above and just below the center of the image you can see tracks leading into tunnel entrances. The critical equipment is presumably accessed from these tunnels, and may be so deep that "bunker buster" bombs cannot damage it.

AP Photo/GeoEye Satellite Image, IHS Jane's Analysis

believed that the military option would have far-ranging and dangerous consequences. Iran's nuclear program would be set back a few years, but the country would portray itself as a victim of U.S. aggression, galvanizing support across the Muslim world (even among Iran's traditional enemies, the Sunni Arabs), perhaps inciting its Hizbullah allies in Lebanon and Hamas allies in the Palestinian territories to attack Israel, encouraging al-Qa'ida–style terrorism against the West, and discouraging Iranians (particularly the younger ones) who favor peaceful relations with the United States. A military strike would probably also lead to a sharp reduction in oil exports from the Gulf, perhaps triggering a global economic crisis. Russia and China would meanwhile have an opportunity to enhance their geopolitical interests by offering diplomatic support and perhaps more to Iran throughout such a crisis. Both countries have considerable leverage in Iran.

Israeli officials have said publicly that Israel (which has its own nuclear weapons, a fact it never acknowledges officially) regards the Iranian bomb as an "existential threat" and will prevent its development, presumably by a single surprise air strike like the one Israel carried out against Iraq's nuclear facility in 1981. It is more likely that Israel rather than the United States would strike Iran. Again, Israel would delay the development of an Iranian bomb, but the regional and global consequences would be enormous. The United States has repeatedly warned Israel against attacking Iran because of the probable consequences discussed above.

The United States and Israel may have found a way to circumvent both diplomatic and military means to blunt Iran's nuclear weapons program: sabotage. In 2010, a computer worm (probably planted by Israeli and U.S. agents) called Stuxnet apparently did significant damage to centrifuges at Iran's nuclear facilities. The most sophisticated cyberweapon ever developed, the Stuxnet malware took over the control systems at Natanz and Bushehr, sending the centrifuges spinning wildly out of control while also issuing a sound that made control room staff think that systems were operating normally. By some intelligence estimates, the Stuxnet attacks may have set Iran's nuclear weapons program back by two years.

Al-Qa'ida

In 1998, the world began to hear about Osama bin Laden, a former Saudi businessman living in exile in Afghanistan, whose **al-Qa'ida** organization bombed U.S. embassies in Kenya and Tanzania as part of a struggle, or *jihad*, against American imperialism and immorality. (In Islam, though, *jihad* has a very wide range of interpretations, from expressing one's commitment to a personal conviction, to an all-out war against nonbelievers.) Bin Laden's organization was also responsible for the 2000 bombing of the American naval destroyer *U.S.S. Cole* in Yemen's harbor of Aden. However shocking those assaults were, they pale in comparison to al-Qa'ida's attacks against targets in the United States on September 11, 2001. In the most ferocious terrorist actions ever undertaken to that date, members of al-Qa'ida "cells" (small groups of agents) in the United States hijacked four civilian jetliners and succeeded in piloting

two of them into the twin towers of New York City's World Trade Center and one into the Pentagon, home of the U.S. Department of Defense, outside Washington, D.C. Passengers on a fourth plane succeeded in aborting the terrorists' likely mission of destroying the White House or damaging the U.S. Capitol, forcing the plane to crash in rural Pennsylvania. More than 3,000 people, mainly civilians, perished. Bin Laden and his followers cheered the carnage as justifiable combat against an infidel nation whose military troops occupied the holy land of Arabia, where Mecca and Medina are located.

As the United States counterattacked, ousting al-Qa'ida from its bases in Afghanistan and cracking down hard on its leadership around the world, al-Qa'ida evolved into a far more geographically diffuse organization. It carried out or supported attacks in Indonesia, Morocco, Algeria, Tunisia, Turkey, Spain, Uzbekistan, Egypt, Saudi Arabia, Jordan, the United Kingdom, Yemen, and Pakistan. A succession of al-Qa'ida videotapes and audiotapes promised an unrelenting and costly continuation of *jihad* against the United States and its allies. Even with Bin Laden dead, it will be many years before Americans and Europeans might be able to emerge from the shadow of this threat (•Figure 6.29). Al-Qa'ida clearly will not hesitate to use the most devastating weapons, even against large numbers of civilians; this deadly goal is a tactical priority. There is also growing consensus that al-Qa'ida is no longer a solitary severe threat to U.S. interests; instead, the *jihadist* cause has spread to smaller groups and to "lone wolves" that are also keen to take on the West. There is little evidence that al-Qa'ida's message has any broad appeal, and it had virtually no role in the tumultuous Arab Spring (see page 196). But the danger remains.

Using GIS and a variety of other geographic tools, many geographers—along with U.S. intelligence agencies, of course—are at work interpreting the geographic dimensions of terrorism. One of the most critical questions for geographers is where future al-Qa'ida attacks are likely to take place. Rohan Gunaratna, probably the greatest authority on al-Qa'ida, estimates that al-Qa'ida maintains a reserve of at least 100 targets worldwide.[6] Where are most of these? Suleiman Abu Ghaith, one of bin Laden's top aides, offered a clue, along with a geographic answer to the question of what al-Qa'ida wants: "Let the United States know that with God's permission, the battle will continue to be waged on its territory until it leaves our lands."[7]

In 9/11 and other atrocities, a tiny minority of Muslims carried out terrorist actions that the vast majority of Muslims condemned. Islamic scholars and religious authorities point out that murder of civilians is prohibited in Islamic law and that the attacks have no legitimate religious grounds. Mainstream Islamist movements are not military or terrorist organizations and have distinguished themselves through public service to the needy.

The following section offers more insight into the peoples and nations of this vital region. It begins with a continuation of the geopolitical theme, examining one of the world's most problematic, persistent, and influential conflicts, the one between Israel and its Arab neighbors.

• **Figure 6.29** Bookends to Osama bin Laden's engagement with the United States. In the 9/11/2001 attack on the World Trade Center **(a)**, note United Flight 175 about to strike the South Tower. Belgian newspapers trumpet bin Laden's death on May 3, 2011 **(b)**.

6.6 Regional Issues and Landscapes

Israel and Palestine

The Arab-Israeli Conflict

The Arab-Israeli conflict persists as one of the world's most intractable disputes. It has not been resolved, in part because the central issues are closely tied to such life-giving resources as land and water, and to deeply held and unyielding religious beliefs. When you understand these issues, you should be proud and prepared. Among other things, you will have a strong foundation from which to follow the news about this conflict for years to come.

The Arab-Israeli conflict is above all a conflict over who owns the land—sometimes very small pieces of land—and is therefore of extreme interest in the study of geography. It is also a conflict that has repercussions far beyond the boundaries of the small countries and territories involved. As long as it simmers or boils, there are other countries and entities that will use the unresolved Arab-Israeli conflict to advance their interests at the expense of others; both al-Qa'ida and Iran, for example, derive much beneficial propaganda value from it. A United Nations–sponsored group called the Alliance of Civilizations argues that the Palestinian-Israeli conflict is *the largest force behind global tensions*. Let's look at this in different terms: when it is resolved, the world will be a much more safer place.

A geographic understanding of this conflict requires familiarity with the events leading up to the creation of the state of Israel and with the wars that have followed, particularly as they have rearranged the boundaries of nations and territories.

The modern state of Israel was carved from lands whose fate had been undetermined since the end of World War I. The Ottoman Empire, based in what is now Turkey, had ruled Palestine (roughly the area now made up of Israel and the Palestinian territories) and surrounding lands in the eastern Mediterranean since the sixteenth century. After the British and French defeated the Ottoman Turks in World War I and destroyed their empire, they divided the region between themselves (•**Figure 6.30**). The British received the "mandate" (authority to establish a government) for Palestine, Transjordan (modern Jordan), and Mesopotamia (Iraq), and the French received the mandate for Syria (now Syria and Lebanon).

During World War I, British administrators of Palestine had made conflicting promises to Jews and Arabs. They implied that they would create an independent Arab state in Palestine and yet at the same time vowed to promote Jewish immigration to

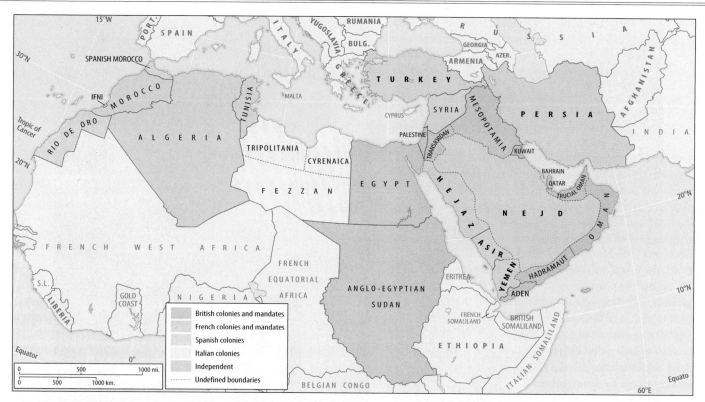

•**Figure 6.30** The Middle East and North Africa in 1920. The victorious allies of World War I carved up the Middle East among themselves. Growing difficulties of administration would drive them from the region within a few decades.
Source: Based on John Haywood, ed., *Atlas of World History* (Barnes and Noble Books, 1997).

Palestine with an eye to the eventual establishment of a Jewish state there. The **Palestinians**—Arabs who historically formed the largest majority of the region's inhabitants—did not welcome the ensuing Jewish immigration and rioted against both the migrants and the British administration. Militant Jews attacked British interests in Palestine, hoping to precipitate a British withdrawal.

Placing themselves in a no-win position with these conflicting promises and under increasing pressure from both Jews and Arabs, in 1947 the British decided to withdraw from Palestine and leave the young United Nations with the task of determining the region's future. The United Nations responded in 1947 with the **two-state solution** to the problem of Palestine. It established an Arab state (which would have been called Palestine) and a Jewish state (Israel).

The two-state plan was deeply flawed. The states' territories were long, narrow, and fragmented, giving each side a sense of vulnerability and insecurity (**Figure 6.29**). When Israel declared itself into existence in May 1948, the armies of the neighboring Arab countries of Transjordan, Egypt, Iraq, Syria, and Lebanon mobilized. In what Israelis call "the War of Independence" and Palestinians call "the Catastrophe" (*al-Nakba*), the smaller but better-organized and more highly motivated Israeli army defeated the Arab armies, and Israel acquired what have come to be known as its **pre-1967 borders** (the 1949 Armistice Agreement borders shown in •**Figure 6.31**). The boundary separating Israel from the West Bank would later come to be known as the "Green Line" (seen in **Figure 6.36**).

Prior to and during the fighting of this **1948–1949 Arab-Israeli war**, approximately 800,000 Palestinian Arabs chose or were forced to flee from the new officially Jewish state of Israel to neighboring Arab countries (•**Figure 6.32**). The United Nations established refugee camps for these displaced peoples in Jordan, Egypt, Lebanon, and Syria. Little was done to resettle them in permanent homes, and both the local Arab governments and the refugees themselves continued to insist on the right of the refugees to return to Israel and the restoration of their property there. This Palestinian **right of return** is one of the central problems of the modern peace process.

Other countries assumed control of those parts of the proposed Palestine that Israel did not absorb. Egypt occupied the Gaza Strip, a piece of land on the Mediterranean shore adjacent to Egypt's Sinai Peninsula that was inhabited mostly by Palestinian Arabs. Transjordan, renaming itself simply Jordan, occupied the predominantly Arab hilly region of central Palestine on the west side of the Jordan River, known as the West Bank, Fig. 6.31 and the entire old city of Jerusalem, including the Western Wall and Temple Mount, Judaism's holiest sites. The Palestinian Arab state envisioned in the UN partition plan was thus stillborn.

The **Six-Day War** of 1967 fundamentally rearranged the region's political landscape in Israel's favor, setting the stage for subsequent struggles and the peace process. This conflict was precipitated in part when Egypt, by positioning arms at the Strait of Tiran chokepoint, closed the Gulf of Aqaba to Israeli shipping. Egypt's President Nasser and his Arab allies took several other belligerent but nonviolent steps toward a war they

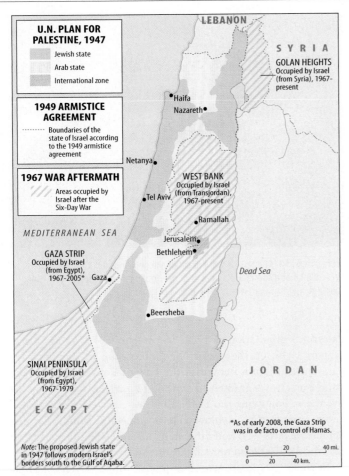

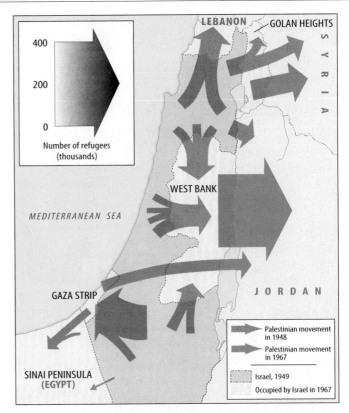

• **Figure 6.32** Palestinian refugee movements in 1948 and 1967. Many who fled in the first conflict relocated in the second.

Source: The Middle East and North Africa: A Political Geography, by A. Drysdale and G. Blake. 1985 Oxford University Press. Based on The Wandering Arabs, Geographical Magazine XLV (3) 1972 page 180.

• **Figure 6.31** The 1947 UN partition plan for Palestine and Israel's original (pre-1967) borders. The 1948–1949 war, which began as soon as Britain withdrew from Palestine and Israel proclaimed its existence, aborted the UN plan and created a tense new political dynamic in the region.

Source: The Middle East and North Africa: A Political Geography by Drysdale & Blake (1985) Fig.9.7 The 1947 UN Partition Plan for Palestine and Israel's Original (Pre-1967) Borders p.280. By permission of Oxford University Press, Inc.

were ill prepared to fight. Israel elected to make a preemptive strike on its Arab neighbors, virtually destroying the Egyptian and Syrian air forces on the ground. Israel gave Jordan's King Hussein an opportunity to stay out of the conflict. However, Jordan went to war and quickly lost the entire West Bank and the historic and sacred Old City of Jerusalem. The entire nation of Israel was transfixed by the news that Jewish soldiers were praying at the Western Wall. The Israeli army (Israeli Defense Forces, or IDF) also seized the Gaza Strip, Egypt's Sinai Peninsula (shutting down the Suez Canal), and the strategic Golan Heights section of Syria overlooking Israel's Galilee region. These three pieces of land would henceforth be known to the world as the **Occupied Territories**. Israel had tripled its territory in six days of fighting (•**Figure 6.34**).

The Palestinian refugee situation became more complex as a result of the war. Israel took over the areas where most of the camps were located. Many people in the camps again took flight, and Palestinians fleeing villages and towns in the newly Occupied Territories joined them as refugees (see Figure 6.32). Some later returned to their homes, but an estimated 116,000

either did not attempt to return or were denied permission by Israel authorities to do so.

The **1973 Arab-Israeli War** was a multifront Arab attempt to reverse the humiliating losses of 1967. On October 6, 1973, the Jewish holy day of Yom Kippur, Egypt and Syria launched a surprise attack on Israel, with the hope of rearranging the stalemated political map of the Middle East. Egypt's army initially showed surprising strength, penetrating deep into the Sinai Peninsula and overturning Israel's image of invulnerability. Israeli troops soon reversed the tide and surrounded Egypt's army, but in the ensuing disengagement talks Egypt won back the eastern side of the Suez Canal and by 1975 was able to reopen it to commercial traffic. In 1979, Israel agreed to return Sinai to Egypt under the U.S.-sponsored **Camp David Accords**, and Egypt recognized Israel's rights as a sovereign state. The Gaza Strip, West Bank, and Golan Heights continued to be held by Israel. Different Israeli administrations regarded them as lands that rightfully belong to Jews or as cards to be played in the regional game of peacemaking.

Arabs and Jews: The Demographic Dimension

In addition to issues of land, water, politics, and ideology, the Palestinian-Israeli conflict is about sheer numbers of people. Each side has wanted to maximize its numbers to the disadvantage of the other. To realize the Zionist dream of establishing a Jewish state, Jews began immigrating to Palestine around the start of the twentieth century. Jews made up 11 percent of

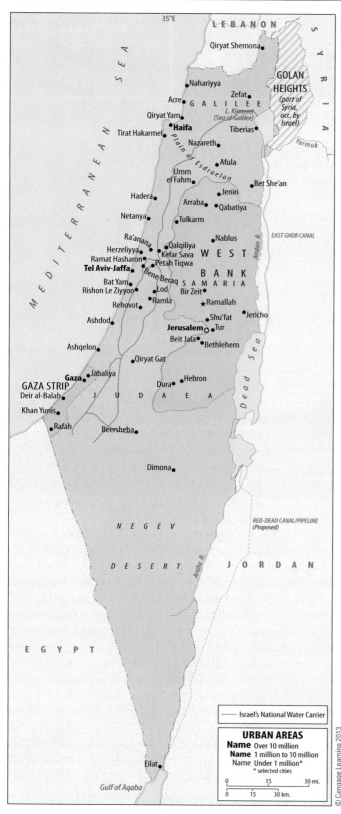

• **Figure 6.33** Israel and Arab territories as of 2012. Israel occupied the Sinai Peninsula in the 1967 war but returned it to Egypt following the Camp David Accords of 1979. Israel withdrew settlers and soldiers from the Gaza Strip in 2005 but kept them in the West Bank and the Golan Heights.

• **Figure 6.34** The Jewish settlement of Maaleh Adumim in the occupied West Bank has a commanding position overlooking the strategic West Bank road linking Jerusalem with the Dead Sea rift valley.

Palestine's population in 1922, 16 percent in 1931, and 31 percent in 1946, on the eve of Israel's creation. In keeping with national legislation known as the **Law of Return**, the state of Israel has always granted citizenship to any Jew who wishes to live there. Following the 1948–1949 war, waves of new immigrants from Europe (**Ashkenazi Jews**) joined the Middle Eastern **Sephardic Jews** who had inhabited Palestine and other parts of the Middle East (and until 1492, Spain) since early times.

In the 1990s, more waves of Jewish immigration followed the dissolution of the Soviet Union and a change of government in Ethiopia, where an ancient Jewish group called **Falashas** had lived in isolation for thousands of years. In 2011, Jews made up 76 percent of the population of 8 million people living within Israel's pre-1967 borders. About 1.5 million Palestinian and non-Palestinian Arabs, or 19 percent of the country's population, are Israeli citizens and residents; they are known as "Israeli Arabs." Palestinian Arabs without Israeli citizenship, along with a variety of other ethnic groups, comprise the balance of the population within Israel's pre-1967 borders.

Another 4 million people live in the West Bank, Gaza Strip, and Golan Heights—areas occupied by Israel in the wars of 1967 and 1973. About 300,000 of those are Jewish settlers, and most of the rest are Palestinian Arabs (mainly in the West Bank and Gaza Strip), **Druze** (members of a small offshoot of Shiite Islam, living mainly in the Golan Heights), and other Arabs. Another 200,000 Jewish settlers live in East Jerusalem, which is also part of the territories Israel conquered in 1967. There are approximately 6 million more Palestinians living outside Israel and the Occupied Territories. Jordan has the largest number (3 million), followed by Syria and Lebanon (with about 500,000 each). The remainder are in other Arab countries and in numerous nations around the world, comprising a sizable diaspora of their own that Palestinians call the *ghurba*. In this respect, the Palestinians are in much the same situation as the Jews prior to Israel's creation.

After 1967, and especially after the election of its conservative **Likud Party** to power in 1977, Israel moved to strengthen its grip on the Occupied Territories through security measures and the government-sponsored establishment of new **Jewish settlements**. Most of these settlements were, and still are, spread through the West Bank (see Figure 6.36). Much smaller numbers of Jews settled in the troubled Golan Heights, where about 19,000 remained in 2012. These are generally not frontier farming settlements but communities inhabited by relatively prosperous middle-class Jews who commute to jobs in Israel. They were not built as temporary encampments but as permanent fixtures meant to create what have been described as **"facts on the ground"**—an Israeli presence so entrenched that its withdrawal would be almost inconceivable (Figure 6.34).

Some of the motivation for the settlements has been ideological, as the West Bank is composed of the biblical lands of **Judaea and Samaria**, which many devout Jews regard as part of Israel's historical homeland. The Jews believe strongly that God Himself declared that his "chosen people" should inhabit these lands forever. Put yourself in their shoes and imagine being told, as they were, that your occupation of this land was illegal, or even that you must move away as part of a peace deal.

The proliferation of Jewish settlements worsened relations between Israel and its Arab neighbors and even drove a wedge into the Jewish population of Israel itself. Many Israelis strongly oppose further settlement in the Occupied Territories and annexation of these territories to the Jewish state because if the territories were annexed, Israel would become a country whose population was only about 60 percent Jewish. With an Arab minority that has a far higher birth rate than that of the Jews, Jews could eventually become the minority population in their own country. The Israeli advocacy group Peace Now leaked official Israeli government documents showing that about 40 percent of the land on which Jewish settlements in the West Bank were built is privately owned by Palestinians. There is also international pressure against the settlements because they violate Geneva Conventions on activities allowed in territories captured in war. The International Court of Justice in the Hague has declared them illegal. **United Nations Resolutions 242 and 338** (passed after the wars of 1967 and 1973, respectively) called on Israel to withdraw completely from the Occupied Territories. The United States routinely calls the Jewish settlements "an obstacle to peace" but is loath to push Israel hard about them.

The fate of the Jewish settlements in the Occupied Territories is one of the key issues in the resolution of the Palestinian-Israeli conflict. It is a difficult problem to solve, but it is not insolvable.

The settlements and other aspects of Israeli occupation have fostered widespread and long-lasting Palestinian resistance against Israel. In 1987, Palestinians instigated a popular uprising in the Occupied Territories known as the *Intifada* (Arabic for "shaking"). Initially, they relied on rocks and bottles to engage Israeli troops, who answered with rubber and metal bullets. International media coverage of a Palestinian "David" fighting an Israeli "Goliath" did much to damage Israel's image abroad, and many Israelis came to question the justice and importance to security of Israel's continued occupation. Popular opinion and, by 1992, a new Israeli government led by the liberal **Labor Party** began to consider the previously unthinkable: giving the Palestinians at least some control over land and internal affairs in the Occupied Territories.

Land for Peace

They loom large in the headlines and in world affairs, but a closer look reveals that Israel and the Occupied Territories are almost unfathomably small. Within its legally recognized, pre-1967 borders, Israel's area is only 8,019 square miles (20,770 sq km), about the size of New Jersey or Slovenia (•Figure 6.35). Israelis often cite the small size of their country, nestled within the vastness of the Arab Middle East and North Africa, to highlight their vulnerability on the world stage. When and if Palestinians acquire their country, Palestine will be even smaller. The question "Whose lands are these?" has always been at the heart of the conflict between Israel and its neighbors.

President Anwar Sadat of Egypt and Prime Minister Menachem Begin of Israel set the precedent in 1979: Arabs could make peace with Israel on the formula of **"land for peace,"** with Israel swapping Arab lands it occupied in 1967 in exchange for peace with its neighbors. The United States, Russia, and Norway initiated negotiations in the 1990s that made such a prospect appear possible. The process began in earnest in 1993, when Israel recognized the legitimacy of and began to negotiate with the **Palestine Liberation Organization (PLO)**, long recognized by the Palestinians as their legitimate governing body. In return, the PLO, under the leadership of Yasser Arafat, recognized Israel's right to exist and renounced its long-standing use of military and terrorist force against Israel.

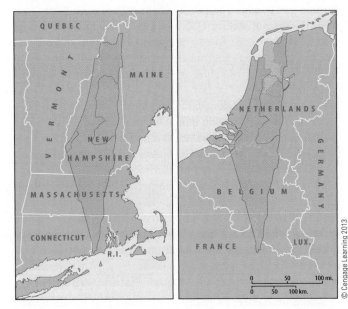

• **Figure 6.35** Israel and the Occupied Territories compared in size with New England and the Benelux countries.

U.S. President Bill Clinton orchestrated a historic handshake between Arafat and the Israeli prime minister, Yitzhak Rabin, and the leaders signed an agreement known as the **Gaza-Jericho Accord** (in its implementation, it came to be known as the **Oslo I Accord**, after the Norwegian capital where it was negotiated). It was designed to pave a pathway to peace by Israel's granting of limited autonomy, or self-rule, in the Gaza Strip and in the West Bank town of Jericho. Autonomy meant that the Palestinians were responsible for their own affairs in matters of education, culture, health, taxation, and tourism. Israel also pulled its troops out of these areas. Palestinians were allowed to form their own government, known as the **Palestinian Authority (PA)**, and they elected Yasser Arafat as its first president.

The accord established a five-year timetable for the resolution of much more difficult matters, the so-called **final status issues.** These included the political status of Jewish and Muslim holy places in Jerusalem and of the city itself, the possible return of Palestinian refugees (the "right of return"), the future of Jewish settlements in the Occupied Territories, and Palestinian statehood (independence). To this day, the final status issues are the ones that prevent the Palestinian-Israeli conflict from being settled. You will be asked to grapple with them as if you were a peacemaker, so please be acquainted with them.

In the wake of a deal known as the Wye Agreement, or **Oslo II Accord**, three successive Israeli prime ministers (Yitzhak Rabin, Benjamin Netanyahu, and Ehud Barak) promised to transfer more West Bank lands from Israeli to Palestinian control, and Arafat promised increased Palestinian efforts to crack down on Palestinian terrorists and so guarantee the security of Israelis in the Palestinian territories and in Israel. If finally implemented (it had not been as of 2012; see •**Figure 6.33**), this arrangement would have brought the total of West Bank lands under complete Palestinian control to 17 percent, leaving 57 percent completely in Israeli hands and 26 percent under joint control (•**Figure 6.36**).

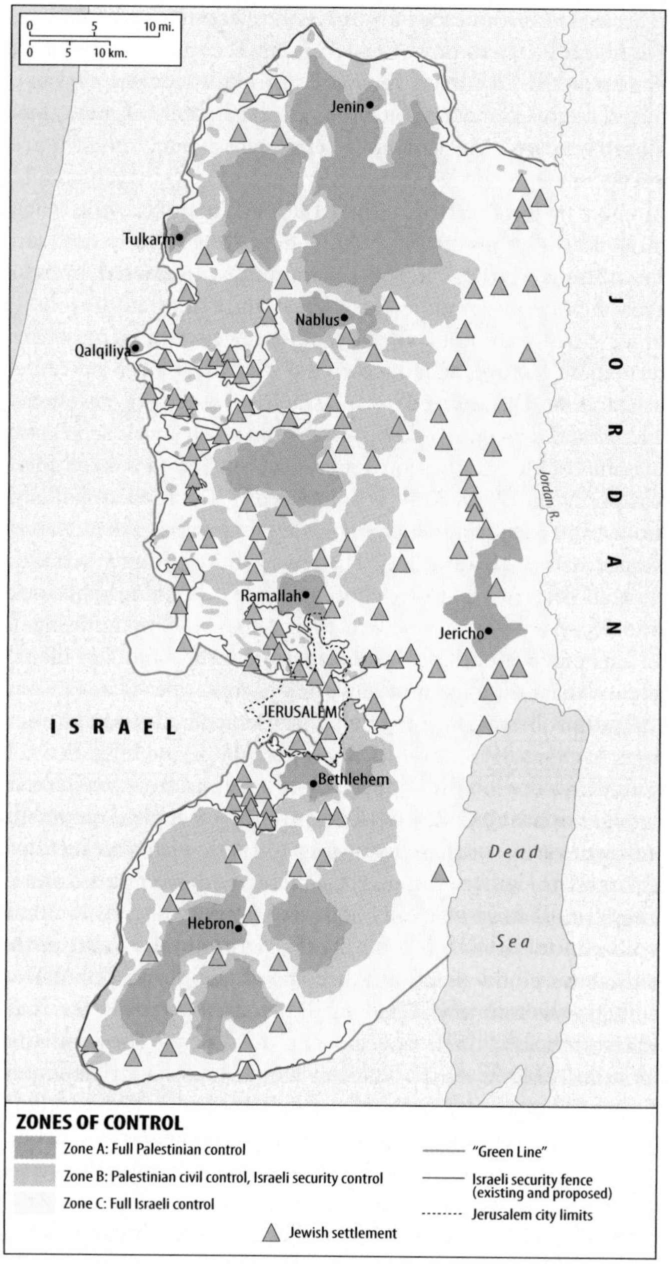
0 5 10 mi.
0 5 10 km.
Jenin
Tulkarm
Nablus
Qalqiliya
Ramallah
Jericho
JERUSALEM
Bethlehem
Hebron
ISRAEL
JORDAN
Jordan R.
Dead Sea
ZONES OF CONTROL
Zone A: Full Palestinian control
Zone B: Palestinian civil control, Israeli security control
Zone C: Full Israeli control
Jewish settlement
"Green Line"
Israeli security fence (existing and proposed)
Jerusalem city limits

• **Figure 6.36** The West Bank. Israeli and Palestinian areas of control were delimited in the peace process of the 1990s, but because of recurrent violence, almost all areas are effectively controlled by Israel. Also depicted are Jewish settlements, the completed and planned portions of the Israeli-built security fence, and the Green Line delimiting the internationally recognized border between sovereign Israel and the occupied West Bank.

Sources: Zone control data from Foundation for Middle East Peace, 2002; security wall route from IDF maps.

On the Brink of Peace During his final year in office in 2000, U.S. President Clinton sought to solidify his legacy as peacemaker by brokering a historic final settlement between Israelis and Palestinians. PLO Chairman Arafat and Israeli Prime Minister Barak huddled with Clinton and his advisers in the presidential retreat at Camp David, a site chosen because of its historical significance in Middle East peacemaking. Over weeks of tough negotiations, mostly over the "final status" issues, the two sides came close to a deal. It would have included the following:

- The creation of an independent Palestinian country in the Gaza Strip and the West Bank.
- A "land swap:" the Palestinian state would include 95 percent of the West Bank. The other 5 percent of the West Bank would be clusters of Jewish settlements that would remain under Israeli control. Palestine would receive 5 percent of Israel's land in exchange.

However, the talks broke down over two thorny issues:

- The status of Palestinian refugees abroad; Israel was unwilling to permit the "right of return" of the large numbers demanded by the Palestinians.
- The deal breaker, Jerusalem, which both Israelis and Palestinians proclaim as their capital. Israel agreed to have the Old City of Jerusalem divided, with Israel assuming sovereignty over the Jewish and Armenian quarters and Palestinians assuming sovereignty over the Muslim and Christian

quarters (see Figure 6.B). These issues were not as problematic as those involving the places held holy by each side. President Clinton put an interesting compromise on the table: neither side would have sovereignty over the Jewish Temple Mount/Muslim Noble Sanctuary (*al-Haraam ash-Shariif*). Instead, these would belong to God. This proposal, which came to be known as Divine Sovereignty, was rejected by both sides. Israel agreed to allow the Palestinians "soft sovereignty" over the Temple Mount/Noble Sanctuary, meaning that Palestinians could administer the area and fly a flag over it, but Israel would technically "own" it. Israel saw this point as essential because the First and Second Temples had stood on this spot. The Palestinians, however, insisted on full sovereignty over the Temple Mount and walked out on the negotiations.

Losing Ground Again   Tragically, within weeks, this historic opportunity for peace evaporated and was replaced by a state of war. On September 28, 2000, the right-wing Israeli opposition leader (and subsequently prime minister) Ariel Sharon walked up to the Temple Mount under heavily armed escort. He meant to make the point that this sacred precinct belonged to Israel. Many observers claim he also meant to disrupt the peace process. Whether or not it was his intention, a disruption in the peace process resulted. Beginning that day and for the next few years, Palestinians took to the streets in the **al-Aqsa *Intifada*** to challenge Israeli forces with rocks, bottles, and guns. Hundreds of Palestinians and Israelis died.

Following the collapse of the Camp David talks and an escalation of suicide bombings of Israeli civilians, Israel chose to effectively suspend the peace process. Prime Minister Sharon branded Arafat an untrustworthy partner for peace negotiations, isolating him physically in the West Bank town of Ramallah and systematically dismantling the Palestinian Authority he headed. Israel also persuaded the United States to stop negotiating with Arafat, who died a forlorn figure in 2004. Israel and the United States insisted on the creation of the new post of Palestinian prime minister, hoping it would be filled by someone they could negotiate with. The United States led an effort joined by the other members of the so-called Quartet—the European Union, the United Nations, and Russia—to establish what they called a **"road map for peace"** between the Israelis and the Palestinian Authority. It was based largely on the 2000 Camp David formula and, if successful, would presumably have led to an independent Palestinian state consisting of the West Bank and the Gaza Strip.

Prime Minister Sharon, while resuming dialogue with the Palestinian Authority, decided unilaterally to withdraw Israeli settlers and troops from the Gaza Strip in August 2005 and to build an immense **security fence** separating the West Bank from Israel (the fence is depicted in Figure 6.36). Israel's stated goal in constructing the new barrier was to prevent Palestinian suicide bombers from reaching Israel. Noting that the security fence did not follow the internationally recognized **Green Line** of Israel's pre-1967 border but instead penetrated significant portions of the West Bank to pick up Jewish settlements there,

Palestinians condemned it as a "land grab." The International Court of Justice concluded that the wall violated customary international law and several conventions on human rights to which Israel is a signatory.

Work on the barrier continued after Sharon lapsed into a coma and was succeeded as prime minister by a Likud Party colleague, Ehud Olmert, in 2006. Olmert vowed to continue Sharon's consolidation of Israeli territory, holding on to three major blocs of Jewish settlements in the West Bank and to "united Jerusalem." On the Palestinian side, a political earthquake occurred that year with the surprising majority vote for parliamentary candidates from the Islamist group Hamas, giving Hamas control of the Palestinian Authority. Palestinians had grown disillusioned with the showy wealth of many elected Palestinian Authority officials and by the PA's failure to deliver an independent country or any other tangible benefits. Hamas, meanwhile, won hearts and minds through its network of social services, often providing what the Palestinian Authority failed to, including education, medicine, and food rations.

Hamas's stunning victory in the parliamentary elections added another layer of uncertainty and discord to the region's political landscape. In the subsequent power-sharing agreement, Palestinian president Mahmoud Abbas of the PLO retained his post, but a Hamas leader, Ismail Haniyeh, took office as the Palestinian prime minister presiding over the Hamas-dominated parliament. Because the Hamas charter does not recognize Israel and in fact calls for its destruction, Israel and the Quartet suspended financial aid and discontinued dialogue with the Palestinian Authority. These sanctions would be lifted on three conditions: the Hamas-led PA must recognize Israel's right to exist, renounce violence, and abide by previous agreements reached between Israel and the Palestinians.

As the Palestinian economy deteriorated even further, clashes erupted between Hamas and PLO factions led by Abbas's Fatah Party. Hamas emerged victorious in fighting to control the Gaza Strip in 2007. Expelled from the territory, Abbas withdrew his recognition of Haniyeh as prime minister, and the Palestinian Authority government was dissolved. Abbas's Fatah Party consolidated its hold in the West Bank. In the winter of 2008–09, there was yet another conflict, the **Gaza War**, during which Israel sought to destroy as much of the weaponry and manpower of Hamas as possible.

In the wake of that conflict, U.S. President Barack Obama vowed to put the weight of his office behind the resolution of the Palestinian-Israeli conflict. The American leader put unprecedented political pressure on Israel, then led by conservative Likud Prime Minister Benjamin Netanyahu, to stop the construction of Jewish settlements in East Jerusalem and the West Bank. Netanyahu refused, and peace talks failed to get off the ground. Obama put the conflict on the back burner for nearly two years, then called for talks to restart—again insisting that Israel halt the settlements. Netanyahu again refused. With the U.S. political campaign for the 2012 presidential election moving into full swing, Obama again moved away from peacemaking in the volatile Middle East.

Try It Solve the Palestinian Arab Conflict

You know the problems now. Are you ready to propose a solution to the Palestinian-Israeli conflict? Try It.

Because many of these issues are sensitive, you may wish to remain anonymous if you are sharing your solution with anyone but yourself.

Many but not all of the issues you may wish to consider are the "final status" issues described earlier. You can start with those.

Palestinian statehood. Should the Palestinians have independence? Where and under what conditions? What concessions might be made in creating the country's boundaries? Hamas controls the Gaza Strip and calls for the destruction of Israel. Should Hamas participate

in the creation of this state, and if so, under what conditions?

Palestinian refugees. Many Palestinians took flight in the 1948 and 1967 wars. Should they and their descendants have a "right of return" to an independent Palestine?

Jewish settlements. Many Jewish settlements have been built within the Green Line or pre-1967 borders of the West Bank. What should be done with them?

Jerusalem. Should the historic Old City of Jerusalem be divided between Israelis and Palestinians? How? Don't forget to use **Figure 6.B**. Who should control the sacred sites of the Western Wall and the Temple Mount/Noble Sanctuary (*al-Haraam al-Sharif*)?

A frustrated Mahmoud Abbas then took a risky but calculated political step. Putting aside the Americans, the Quartet, and all the conventions used since the peace process began, he went to the UN General Assembly in September 2011, appealing for the recognition of an independent Palestinian country inside the "pre-1967" (the Green Line) borders of East Jerusalem, the West Bank, and the Gaza Strip. The UN General Assembly is the same political body that, in November 1947, fulfilled the Zionist dream by portioning Palestine into two countries, one Jewish and one Arab.

You have been very patient in sinking your teeth into the complexities of the Arab-Israeli conflict. I hope you will share my view that there are no shortcuts when it comes to understanding this critical problem. I hope you will read or view the news about this part of the world with much appreciation and interest. It is my hope that you will come to your own conclusions about how to solve the conflict. Are you ready to try? **See the Try It feature on this page.**

The Arab Spring

On December 17, 2010, in the Tunisian city of Sidi Bouzid, a 26-year-old vegetable vendor named Muhammad Bouazizi was shaken down for a bribe by a city inspector. Muhammad had paid the $7 to the inspectors many times before, but this time he protested. The inspector and her companions took apples from his rolling cart and confiscated his scale. The young man went to the local governor's office to lodge a complaint but was denied entry. Later in the day, he returned to the governor's office, shouting from the street, "How do you expect me to earn a living?" He poured paint thinner over his body and lit a match. Flames consumed his body.

This was the beginning of the Arab Spring.

Stirring from Stagnation

News of Bouazizi's self-immolation spread quickly throughout Tunisia and the Arab World by way of YouTube, Facebook, Twitter, and al-Jazeera television. This was an unfamiliar event, quite unlike suicide bombings or anything known even in the worst of times.

The young man's death struck a nerve among much of Tunisia's largely young and disaffected population. Tunisian resentment against the rule of the autocratic President Zine al Abidine Ben Ali had been growing throughout his more than two decades in power. Ben Ali's family was estimated to have directly or indirectly controlled half of the country's economy.

Protestors poured into the streets of Tunisian cities, bearing signs reading "We are all Muhammad Bouazizi!" and calling for Ben Ali's resignation. Demonstrators soon numbered in the hundreds of thousands, and Ben Ali began to make political concessions. It was too late. Within a month of the start of what became known as the **Jasmine Revolution**, Ben Ali's own generals turned against him. Ben Ali and his family fled to Saudi Arabia. The first domino had fallen.

Protests soon erupted in a dozen other countries across the Middle East and North Africa. The experiences in each of these countries were diverse, and it would take an entire book to describe them! Instead, I will discuss briefly some of the unique issues in the countries affected, focus mainly on the most important country affected—Egypt—and identify some patterns underlying the dissent.

The Pharaoh Falls

The next domino was the Middle East's demographic and political heavyweight, Egypt. There are several identifiable factors leading to this most important revolution in the region in half a century.

The first are two problems you know about: the general trend of "people overpopulation," and MENA's own youth bulge that is one of its consequences. Over the past decade, the number of people of working age in Egypt, Jordan, Lebanon, Morocco, Syria, and Tunisia have grown by 2.7 percent, faster than any other world region except sub-Saharan Africa.

Egypt has MENA's biggest youth bulge: 60 percent of Egypt's 83 million people are under 25. In Arabic the word *shabaab* means "young people" and represents that youth bulge. It sometimes has the connotation of "bad boys." Bad or not, these boys have palpable frustrations. Without treading too much into psychosocial analysis, it is clear that the most basic desires and needs of young men that include courtship and marriage are extremely difficult to fulfill. They are expected to have steady work and to provide a home and other amenities to their brides, and this is hard given the economic challenges they face.

Unemployment and underemployment were the second set of ingredients in Egypt's Arab Spring. Egypt's annual rate of population growth is 2 percent. At this rate, Egypt's population will double in 35 years. Will Egypt double the number of jobs for its people in 35 years? That prospect is remote. In recent decades, South Korea, China, and other successful developing countries nurtured the growth of private sector employment, while Egypt and other MENA countries generally stayed with a Soviet style model of state-controlled enterprises that was unresponsive to people's needs and wants, and failed to grow productive jobs. Outside the agricultural sector, 70 percent of working Egyptians are employed by the government. Private sector jobs are scarce.

Throughout MENA, the number of college graduates far exceeds positions to hire them. Egypt cranks out huge numbers of college graduates, but the quality of their education is questionable. All Egyptian college graduates are guaranteed jobs in the public sector. They are stuffed into already bloated bureaucracies where 10 people do the work that could be done by 1. Egypt and many other MENA countries suffer in this way from **underemployment** (the underutilization of labor). Their wages cannot possibly sustain a family, and most people "moonlight" with second or even third jobs. Under such circumstances, the visionary Steve Jobses of MENA are unlikely to ever be heard of, unless they flee with their brains to the West or the East.

Third, Egyptians suffered from government repression and a lack of individual freedoms. I heard the same complaints in Egypt in 1977 and on the eve of the Arab Spring. Egyptians had no control over their lives. They could not express dissatisfaction with government policies they were unhappy with, which were many. There was no real democratic political system. Every few years, Egyptians endured the sham of elections, in which an amazing 99 percent of Egyptians always chose President Hosni Mubarak's party to lead them. They had no freedom of the press. They lived under emergency law, which allowed authorities to arrest people without charges, detain prisoners indefinitely, and limit freedoms of expression and assembly. Their ranks were penetrated by the countless plainclothes secret police, the dreaded *mukhabarat* whose duty was to look for any sign of sedition. Egyptian security forces had torture down to a fine art (one of the reasons why the United States sent al-Qa'ida suspects to Egypt under the Bush and Obama programs of "extraordinary rendition").

The fourth factor in Egypt's uprising was the yawning gap between the haves and the have-nots, another problem of LDCs described in the book's introduction (see page 47). There was a veneer of prosperity. Egypt had a gleaming culture of conspicuous consumption for the relative few who could afford to consume. The gap between rich and poor grew widely in Mubarak's Egypt, and 20 percent of Egyptians were classified as poor at the time of the Arab Spring. For the wealthy and the well-connected, getting things done was easy: you just had to pay bribes and use your favors. To know someone with influence always helped. Corruption, known as *wasta*, was just the way it was. And if you didn't know anyone or could not afford to pay bribes, you could feel just like Muhammad Bouazizi did.

The results of unleashed *shaabab* power can be dramatic, as I have seen. In January 1977, Egypt's government under the soon-to-be-assassinated president Anwar Sadat removed many of the subsidies that made it possible for Egypt's poor to buy bread, cooking oil, and many other basic needs. Egyptians were furious, and the *shabaab* took to the streets in Egypt's infamous "food riots." In the country's main cities, young men broke businesses' windows, overturned and burned vehicles, and attacked government buildings. A large, loud mob assembled in Cairo's Tahrir Square and marched on a nearby train station, determined to raze it. A curious college student, I was on their periphery. Riot police with batons and shields advanced on us and began firing tear gas. Many of the demonstrators picked up the fuming canisters and threw them back at the riot police, who did not have gas masks and were themselves in as much discomfort as we. I picked up a canister that had burned out, and it read "Made in USA. For outdoor use only."

The police began firing rubber bullets, which are painful and can be fatal. The crowd continued to advance, pushing back the police. The police finally escalated to live ammunition, and when demonstrators began falling dead and wounded, the crowd evaporated. That was the end of it. Authorities imposed a two-day curfew prohibiting mass assemblies—defined as more than six persons—and answered the angry public's demands by restoring most of the subsidies. The food riots were over, but real hatred against Egyptian authorities continued to seethe, especially among the country's huge population of poor people. That hatred persisted for decades, as did Egypt's miserable conditions. Conditions just before the spring of 2011 didn't seem like anything special to most observers.

I go into such detail to underscore why I, as well as so many Middle East analysts, was taken off guard by Egypt's revolution in the Arab Spring. I expected only the *shabaab* to take to the streets in Egypt—and then more to let off steam rather than articulate meaningful dissent. I expected them to melt away when the riot police that you saw on television began using living ammunition. I expected the Mubarak regime to hold on and for Egypt's disaffected people to soldier on. I was wrong on all counts.

One of the opposition parties trying to organize in the last months of Mubarak's rule called itself Kifaya, meaning "Enough." Enough of all of this. The people of Egypt—not just the *shabaab*, but the older generations (including the *shabaab* I had seen rioting in 1977); girls and women; Muslims and Christians; poor, middle class, and even the well-to-do who wanted democracy and freedom of expression—were inspired by Tunisia's Jasmine Revolution to challenge the government—a government led by an 80-year-old man who died his hair jet black and who was grooming his son to succeed him.

Egypt's revolution played out all over the country, but its epicenter was in Cairo's Tahrir ("Liberation") Square

• **Figure 6.37** Egyptian masses protesting against the government of Hosni Mubarak in Cairo's Tahrir Square, February 2011.

(•**Figure 6.37**). Egypt's despised secret police used deadly force even when there was no need. Do you remember the scenes of armed men mounted on camels and horses charging against the demonstrators in Tahrir Square (which became known as the "Battle of the Camels")? Rather then disperse, over a period of 18 days the protestors held their ground in Tahrir and Egypt's other public squares and outside government buildings. Without arms, the demonstrators took casualties of dead and wounded. In the end, their people power prevailed. The army, an institution that Egyptians long revered, took up their cause. Mubarak offered political concessions to Egyptians, but it was too late. As the army assumed power, he fled from Cairo, and was soon imprisoned and put on trial.

The Libyan Domino

A worse fate would befall Libya's dictator, Muammar Qaddafi. Libya was not like Egypt: with a population of just 6.4 million, Libya did not have an unmanageable youth bulge, nor was it propped up by American aid. With its considerable oil reserves, more than 3 percent of the world total, Libya should have been a prosperous country. For more than 40 years, it was instead the toy of Muammar Qaddafi, an eccentric man (what other world leader would call for Switzerland to be abolished?). Colonel Qaddafi ran the country as if it were his fiefdom, favoring and distributing money to those who were loyal to him. Tribal and ethnic politics were decisive. Qaddafi loyalists drew mainly from his own tribe on the central coast and from other tribes he favored in the center and west. King Idris I, whom Qaddafi overthrew in 1969, had favored tribes in the eastern region known as Cyrenaica, especially around the city of Benghazi. Those tribespeople never forgave Qaddafi for ousting their leader, and Qaddafi made sure they never earned any graces from him. Qaddafi also treated the ethnic Berber tribes of Libya's western mountains (whose livelihood but not ethnicity is much like that of the Bedouin) as second-class citizens.

The people of Cyrenaica along with the Berbers of the west were inspired by the tide of the Arab Spring to rise up against Qaddafi, and a brief civil war emerged between them and

Qaddafi loyalists. The opposition to Qaddafi was advantaged by the commitment of NATO air power, weapons, and intelligence, and it was only a matter of time until Qaddafi and his power base would be eliminated. Qaddafi was captured and killed execution-style in October 2011, eight months after the uprising began. [89]

Syria's Minority Dynasty versus the People of Syria

As in Bahrain and Libya, Syria's leaders opted for violent repression against demonstrators inspired by the Arab Spring; at least 9,000 and probably many more died. The Syrian uprising began in January 2011 after another self-immolation.

In Syria, a minority Shiite group (making up 7 percent of population) called the **Alawites** and led by president Bashir al-Assad, son of the late president Hafez al-Assad, ruled the majority (74 percent) Sunni. Resentment of Alawite rule was nothing new in Syria. Four decades of the repressive al-Assad regime rankled the Sunni majority, the Christian minority, and others. In 1982, a brutal army assault against a Sunni Muslim Brotherhood uprising in the city of Hama resulted in tens of thousands of deaths. The 2011/12 revolt involved much broader participation in more cities (•**Figure 6.38**). Arab Spring demonstrations in most of the Arab countries were loudest in the capital cities, but Syria was an exception. The fighting came to Damascus rather late.

The challenge of taking down the Alawite regime in Syria was great. The Alawites controlled the key unit of the Syrian military, so it would be difficult to secure a broad military uprising against the regime. NATO and the United States would be far less inclined to become involved than they did in Libya; the military challenge would be much greater. Western countries stepped up their sanctions against Syria. The United Nations tried to pass resolutions against Syria, but Russia and China vetoed these. The Arab League moved to oust Syria from the organization and to impose its own sanctions. Turkey was more involved, providing shelter to forces of the Free Syrian Army, made up mainly of Sunni defectors

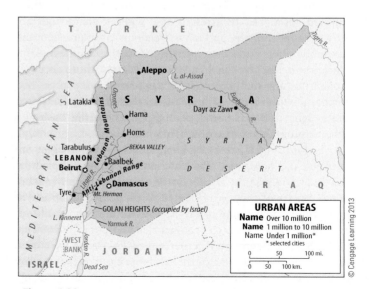

• **Figure 6.38** Principal features of Syria and Lebanon.

from the Syrian army. Israel did not become involved. Iran had much at stake: it would lose its only Arab ally should the Syrian regime fall.

Syria tried to keep foreign reporters out of the country, but a few sneaked in and were able to verify the authenticity of the video footage posted daily on YouTube. The savagery with which the Syrian army took on armed and unarmed civilians was extraordinary. Soldiers seemed to have no compassion even for children and women. It was unrestrained carnage, and there appeared to be no way to bring it to a stop.

Bahrain: A Pearl Is Crushed

In Bahrain, a small but oil-rich Gulf island shaikhdom (Bedouin monarchy) linked to eastern Saudi Arabia by a causeway, there is a Shiite majority of 70 percent that has long been ruled 174 by a Sunni monarchy (•Figure 6.39). Like Egyptians, Bahrain's Shiites were inspired by the Jasmine Revolution to challenge King Hamad bin Isa al-Khalifa.

Bahrain was the Arab Spring's best example of how repression can backfire by fueling resentment. The mainly Shiite protestors began by simply expressing wishes for democracy, public participation, and justice. They asked that King Khalifa take his place in a new constitutional monarchy. The monarch could have responded as a benevolent despot, yielding some ground to his subjects.

Instead, he ordered government forces to crush the rebellion with deadly force. As the casualties mounted, demonstrators' cries changed from reasonable petitions to "Death for Khalifa!" Worried that Bahrain's Shiites might overthrow the monarchy and put Bahrain into the embrace of his adversary Shiite Iran, the king called on Saudi Arabian forces to drive across the causeway to Bahrain and helped put down the uprising. 203 Government forces destroyed the demonstrations' epicenter, Pearl Square, with the country's iconic pearl sculpture in the capital city of Manama, and turned it into a traffic intersection to ensure that crowds would not gather again. King Khalifa's opponents retreated, their hopes unrequited. Below we will consider Bahrain's future, especially in its geopolitical context.

Revolt in Yemen's Mountainous Redoubt

Although located on the Arabian Peninsula (see Figure 6.39), Yemen has only a small bounty of oil reserves. A stunningly beautiful country graced by mountains and deserts, and a distinctive urban architecture looking like gingerbread houses,

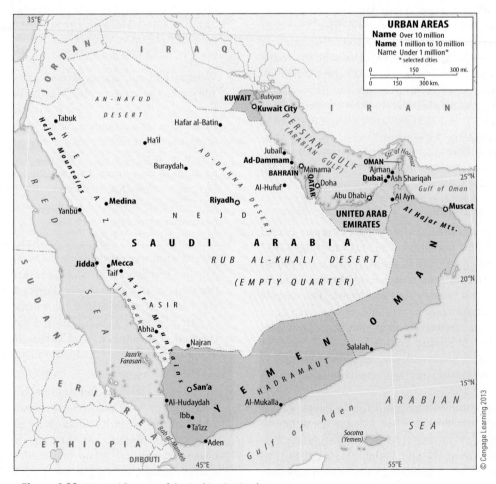

• **Figure 6.39** Principal features of the Arabian Peninsula.

• **Figure 6.40** The palace of the Imam in Wadi Dhahr, Yemen.

Yemen is cursed by poverty (•**Figure 6.40**). Loyalties are to clan and tribe, and Yemen has no natural sense of cohesion as a nation state. The country was previously divided into North and South Yemen. Al-Qa'ida has a strong foothold here; Yemen is in fact the ancestral homeland of Osama bin Laden's clan.

With the Arab Spring, a widespread revolt against the long-time ruler Ali Abdullah Saleh erupted. After surviving an assassination attempt, and just before resigning office, President Saleh offered concessions to Yemeni protestors. The country's established opposition political parties pleaded with the young people to stop their protests. Yemen's *shabaab* ignored that advice. They are a force to be reckoned with; the demographic profile of Yemen's *shabaab* is very large. With a population of 23.8 million, Yemen has one of the world's highest population growth rates, a staggering 3.1 percent (with a doubling time of 22 years). Poverty, unemployment, and repression of dissent made Yemen ripe for revolt. Its geopolitical context is also discussed here.

Hallmarks of the Revolution

In the Arab Spring's aftermath, Middle East observer Fareed Zakaria recommended a book on his Sunday morning "GPS" show on CNN. It was the United Nations' *Arab Development Report 2002*. Why recommend a decade-old book? To make the point that little had changed in those years. The chapter titles that so well summarized the region's conditions in 2002 did the same a decade later: "Development not engendered is endangered ... Bridled minds, shackled potential ... As development management of stumbles, economies falter ... The curse of poverty: denying choices and opportunities, degrading lives."[8]

The Arab Spring rumbled through Tunisia, Egypt, Libya, Bahrain, Syria, and Yemen—described in some detail here—and in Sudan, Algeria, Mauritania, Morocco, Jordan, the Palestinian Territories, Lebanon, and Oman. Many of Egypt's problems that we have considered—population growth,

unemployment and underemployment, lack of political representation, and oppression by authorities—were shared by these countries. Here are some of the other notable characteristics of the uprising and its aftermath in the countries of the Arab Spring:

- The revolutions were much facilitated by Facebook, Twitter, 24/7 satellite television news (notably from Qatar's *al-Jazeera* network), and other social media. Governments tried to crack down, bringing down the Internet and suspending cell phone service, but to no avail. By the time the authorities woke up, detailed instructions on how, where, and when to gather and protest in public spaces had been widely posted and tweeted.

- Although new media made their debut as tools of revolution in the Arab Spring, the more traditional geography of revolt in public spaces played a prominent role. In the traditional Middle Eastern *medina*, with the exception of the mosque there was no large public space in which large numbers of people could gather. However, spacious traffic hubs that doubled as public squares were established in the post-colonial modern Middle Eastern cities. In the Arab Spring, Cairo's Tahrir Square, Tripoli's Green Square in Libya (renamed Martyrs' Square in recognition of the lives lost there in protest), and Bahrain's Pearl Square, served as central places where social media called upon demonstrators to gather. Many of the key events leading to success (and in Bahrain's case, the failure) of the revolutions took place in these locations.

- Women had an unprecedented strong role in the Arab Spring. In the conservative Muslim societies of the Middle East and North Africa, there is strict segregation of men and women (in some countries more so than in others) in public space. But in Tahrir Square and elsewhere, that convention was thrown aside as women and men stood shoulder to shoulder in protest. In a second round of revolt against Egyptian rule—this one against the Egyptian army late in 2011—the army struck back against women protestors with particular savagery.

- The traditional Islamic classification of sacred time played an important role in the Arab Spring. As often in Muslim history, Friday noon prayers and sermons incorporated politics with prayer. Prayer time gave way to protest time, often on such a huge scale that the authorities dreaded Fridays. This traditional day of prayer was often tweeted, Facebooked, and otherwise announced as the "Day of Rage" for which demonstrators should prepare themselves.

- Remarkably, religion and militant Islamism did not otherwise feature prominently in the Arab Spring. Mainstream and radical Islamists alike were conspicuous by their absence. In Egypt, the Muslim Brotherhood, a well-organized political group that could have promoted itself and its agenda, kept a low profile. Throughout the region, no demonstrators evoked Osama bin Laden or al-Qa'ida. In fact, the Arab Spring succeeded nonviolently in doing what al-Qa'ida aimed to accomplish through terrorism: to overthrow the autocratic, Western-oriented regimes of the Middle East and North Africa. Bin Laden must have been dismayed by what he saw in the last months of his life.

What Now? The Arab Fall

The Arab Spring transitioned into an autumn, a period of danger and uncertainty that is likely to persist for years. What will take the place of the relative stability that the region's repressive, autocratic regimes guaranteed? What will happen to the dominoes that have yet to fall?

Amidst the hope and such remarkable scenes as Tunisians turning out in October 2011 for their first-ever democratic elections, there were worrisome trends and incidents. Divisiveness and polarization based on major faith, minority sect, and tribal affiliation were of special concern.

Euphoria gave way to hard realities in Egypt. Instability reigned while the army held power in anticipation of democratic elections. Violent sectarian clashes broke out between Muslims and Coptic Christians, leaving scores of people dead. The economy tanked. At least 10 percent of Egypt's economy relies on tourism, and for awhile almost no one wanted to visit Egypt. Foreign investment dried up. The terrible economic conditions that brought people out on the street to begin with only became worse. People began talking about going back to the streets for another revolution, and by late November they were again in Tahrir Square in great numbers, this time challenging the army. Egyptians began voting in truly democratic parliamentary elections that brought members of the Muslim Brotherhood and other "modern Islamists" to power. These peaceful, democratic elections were a first for Egypt and might be an inspiration for many in the Arab World. Egypt's critical peace treaty with Israel was tested on several fronts: Egyptian rioters sacked the Israeli embassy in Cairo, skirmishes broke out between Egyptian and Israeli forces on the Sinai frontier, and Libyan weapons began making their way overland through Egypt into the Palestinian Gaza Strip.

What about Uncle Sam?

The problems between Egypt and Israel raise an intriguing geopolitical question about the Arab Spring's aftermath: How much instability in this troubled, oil-rich region will the United States tolerate without intervention? With economic aid, the United States had effectively "bought" the peace between Egypt and Israel (not counting Iraq for military expenditures, these two countries consistently rank as the top two countries in the value of American foreign aid). Throughout the region, the United States made "deals with the devil," helping to prop up authoritarian regimes like Egypt's that would maintain strategic and political interests in Middle Eastern oil, the region's chokepoints, and Israel's security, to name a few.

Another interest was to prevent the ascendance of militant Islamists. Many of these one-party states appealed to the United States for aid, saying that the alternative to their rule was that al-Qa'ida or the like would find a footing. That was an effective ploy in Yemen, Egypt, and elsewhere. Certainly these are valid concerns:

- The well-organized Muslim Brotherhood will grow much stronger in post-Mubarak Egypt, but the Brotherhood does not advocate violence.

- The United States would be loath to "lose" Bahrain to the rule of the majority Shiites, who could help Iran gain traction just miles from Saudi Arabia. Bahrain is also home to the U.S. Fifth Fleet, whose ships patrol the Gulf to ensure safe passage of ships carrying oil through the Strait of Hormuz and to confront Iran if necessary.

- An anti-American Islamist regime, or Islamist insurgents, could find home in post-Qaddafi Libya, or more likely and ominously in Yemen, on Saudi Arabia's doorstep. It is difficult to envision anything other than at least an ongoing low-level war waged by the United States against al-Qa'ida affiliates in Yemen. Those predator drones will be busy.

- With its oil wealth, Saudi Arabia represents the ultimate red line in the sand. Would the United States "allow" the Saudi Arabian monarchy to fall in a popular uprising? That scenario of revolution in Saudi Arabia, perhaps beginning with an uprising by the Shiites in the Eastern Province (•Figure 6.41) is the ultimate nightmare for U.S. foreign policy decision makers.

You have witnessed the Arab Spring, one of the region's most momentous events in centuries. These issues will be prominent in world affairs for a long time to come.

• **Figure 6.41** Saudi Arabia's strict interpretation of Sunni Islam (Wahhabism) casts the Shiite minority of the country's Eastern Province as heretics. These Shiites have long been the country's poorest and most marginalized group. Already perceived as potentially destabilizing sympathizers with Iran, these Shiites could in fact be among the first to challenge the monarchy. These Shiites are at a market in the Eastern Province town of Qatif.

The Gulf Oil Region

GIS Helps Turn an Arabian Mirage into Reality

A futuristic city is taking shape on the desert sands of Abu Dhabi, one of the United Arab Emirates (location in **Figure 6.39**). Called **Masdar**, meaning "the Source" in Arabic, this city has the extraordinary goal of being "carbon-neutral." That means that unlike any other settlement in the world, Masdar will minimize the carbon dioxide it produces and neutralize or offset the small emissions it does make (**•Figure 6.42**). Automobiles will be banned, and the city's 45,000–50,000 residents will be required to use low-emissions public transport. Residents will work in nearby Abu Dhabi city or in the clean technology companies based in Masdar itself. Abu Dhabi's government is bearing the $20 billion price tag to build the city, which is supposed to become a magnet for domestic and foreign investment in high-tech research and development firms. Masdar's first residents are students and faculty of the Masdar Institute, an educational center staffed by professors from the Massachusetts Institute of Technology (MIT). Its graduates are meant to help meet the world's growing demand for scientists and visionaries with knowledge of green technology, green cities, "smart" cities, and renewable energy.

Energy for Masdar City is to be provided by renewable resources, especially photovoltaic energy, a solar tower, and other techniques of solar power that, as you might imagine,

• **Figure 6.42** Artist's rendition of Masdar, the carbon-neutral city under construction in Abu Dhabi, UAE.

has abundant potential there in Arabia. Wind farms, geothermal energy, and the world's largest hydrogen power plant will contribute energy as well. Masdar will be zero-waste, with its waste products reused, recycled, or turned into yet more energy. Water will be provided by desalinization of the very salty Gulf waters, an energy-intensive process that will also use solar power. Eighty percent of the water will be recycled, and greywater—water that has been used but does not contain human waste—will be used for irrigation.

In 2007, I was employed by a foundation in the UAE to come up with the first draft of an environmental master plan for the country. I visited each of the seven emirates and spoke with government officials, environmental nongovernment organizations, CEOs, foreign workers, citizens, and others to identify the country's environmental challenges and then proposed specific methods for dealing with them. One of most memorable visits was with Masdar's Chief Executive Officer, Dr. Sultan al-Jaber. While we sipped tea, Dr. al-Jaber spoke at length about his vision of the futuristic city. As I prepared to leave, he said, "Come with me—I have something to show you." He led me to a conference room and opened the door. Inside, seated around a long table, were about 20 persons: engineers, urban planners, GIS technicians, and professors from a variety of fields at the Massachusetts Institute of Technology (MIT). They were on a site visit to see whether they would commit to a five-year leave from MIT, relocating to Abu Dhabi with their families to help build Masdar. You know the expression "the smartest guy in the room." This was a room full of the smartest guys in the room, and Dr. al-Jaber needed them to turn his zero-carbon vision into reality.

As I learned, GIS has an important role in Masdar's urban planning and management. Using ESRI's ArcGIS software (see page 15), Masdar's GIS crew was able to carefully consider the city site's geography of sun angles, wind patterns, street widths, building density, and building heights. "Building a city like this has never been done before," GIS consultant Shannon McElvaney said, "and GIS is proving to be an absolutely critical tool."[9] ESRI's Derek Gliddon wrote:

> Data layers contained in the geodatabase include information such as transportation, vegetation, drainage, structures, boundaries, elevation, biodiversity, buildings, and utilities, as well as terrain elevation, bathymetric data, and remotely sensed imagery. A sophisticated Web browser–based virtual city visualization and navigation tool is used to visualize the construction of the city over time. Construction managers can navigate anywhere in the city; "play" the project timeline; and identify spatiotemporal clashes, accessibility problems, and other logistical issues. On a fast-paced, high-density development, these issues are very important. ArcGIS introduced the spatial analysis and modeling necessary for the most efficient placement of facilities at the city. Water and sewage treatment plants, recycling centers, a solar farm, geothermal wells, and plantations of various tree species were placed using traditional planning principles modeled with ArcGIS. Questions—Is there enough physical space available? How much are the buildings shading each other? How much space is needed between a facility and the residents?—are modeled, and the best answer chosen through GIS.[10]

Masdar City: Abaca/Newscom

The global financial crisis has derailed the city's ambitious construction schedule. When Phase I is completed—tentatively in 2015—Masdar will open for settlement. It won't be cheap to live there. Some critics have decried Masdar as a luxury development and the ultimate gated community. Whatever you call it, there is nothing quite like Masdar.

Iraq and the United States: A Deadly Dance

Iraq (population 32.7 million) occupies a broad, irrigated plain drained by the Tigris and Euphrates rivers, together with fringing highlands in the north and deserts in the west (•Figure 6.43). Known since ancient times as Mesopotamia, meaning "the land between the rivers," this landscape of today's Iraq has a complex geopolitical history. It has been a cradle of civilization, a seat of empires, a target of conquerors, and in the twentieth and early twenty-first centuries, a focus of oil development and political and military contention.

A critical feature of modern Iraq's geography is its ethnic and religious composition, with three major groups present: the Shiite Arabs, mainly in the south, who make up about 60 percent of the country's population; the Sunni Arabs, about 35 percent of the total and living mainly in the center (especially in the so-called Sunni Triangle); and the Kurds, mainly in the north, most of whom are Sunni Muslims and who represent 15 to 20 percent of the total (•Figure 6.44; see also the discussion of the Kurds on page 203). There are numerous smaller minorities, including Yazidi and Sunni Turkoman.

On the map of Iraq in Figure 6.43, you can see a fundamental geographic disadvantage of this country. It has a tiny, 35-mile (55 km) window on the Gulf. Exports of Iraqi oil by oil tanker are hampered by poor port facilities; this narrow access to the sea does not include natural deepwater ports that would facilitate exports. One of the main reasons Iraq attached Kuwait in 1990 and Iran in 1980 was to widen Iraq's access to the Gulf.

Gulf War I Iraq and the United States became entangled in a deadly dance in 1990, when Iraqi forces invaded Kuwait—a New Jersey-sized country that has the world's sixth largest oil reserves. Claiming that it was historically part of Iraq and calling it Iraq's "19th province," Saddam Hussein's regime attempted to annex Kuwait. Saudi Arabia and other Gulf oil states seemed in danger of invasion. Worldwide opposition to Iraq's

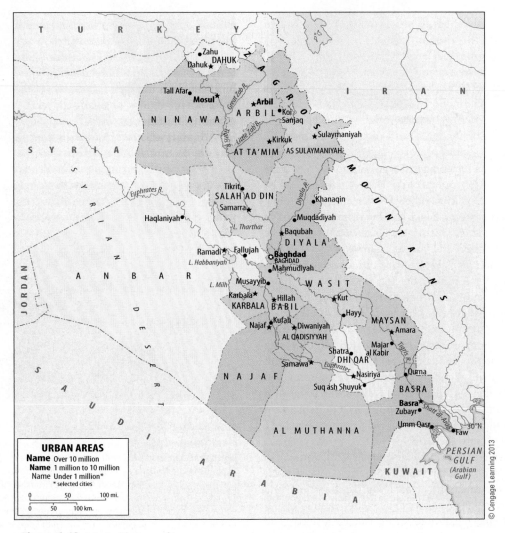

• **Figure 6.43** Principal features of Iraq.

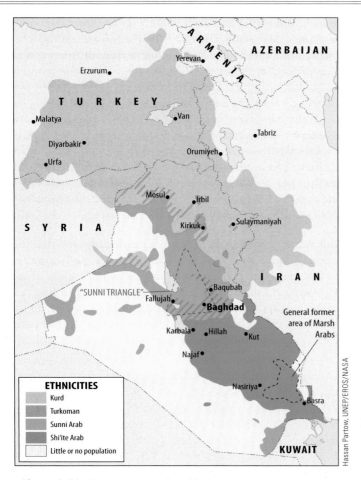

• **Figure 6.44** Ethnicity in Iraq and neighboring regions.

surviving forces then mercilessly put down rebellions by Shiite Arabs in southern Iraq and by Kurds in the north. To deter Iraq's military and provide a safe haven for the Kurds in the north and the Shiites in the south, the United Nations established "no-fly zones," where Iraqi aircraft were prohibited from operating. But Iraqi forces had no difficulties in mounting ground campaigns against the Shiites and Kurds. Coalition forces did not intervene on behalf of the Shiites and Kurds, who felt betrayed: the coalition had encouraged them to rise up against Saddam Hussein but then abandoned them and left them at the mercy of Saddam's vengeful forces. Iraq's Shiites, both in their urban strongholds, including Basra, and in the countryside, fared especially badly.

One of history's great environmental and cultural tragedies was Iraq's systematic destruction after 1991 of the marshes adjoining the southern floodplains of the Tigris and Euphrates rivers (•**Figure 6.45**; see also the ethnicity map in Figure 6.44). This

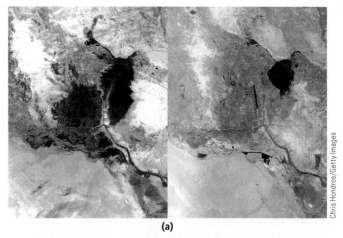

(a)

(b)

• **Figure 6.45** **(a)** Remote sensing of Iraq's southern marshes in 1971 and 2000. This is a "false color" image; the red is the real-life green color of marsh vegetation. Iraqi engineers built embankments to cut off the flow of Tigris and Euphrates waters into these wetlands, laying waste to a unique ecosystem and the way of life of the "Marsh Arabs." **(b)** Marsh Arab life in the marshes in 1974, before their destruction. Note the unique architecture of dwellings and community buildings, all made from marsh reeds. Some Biblical scholars have proclaimed that this marsh was the Garden of Eden, where Adam and Eve begat humankind.

aggression resulted in a sweeping embargo on Iraqi foreign trade imposed by the Security Council of the United Nations. Led by the United States, a coalition of 28 UN members (including several Arab states) staged a rapid buildup of armed forces in the Gulf region. The stage was set for what would come to be known as "the Gulf War," and later as "Gulf War I."

Beginning in January 1991, the coalition's crushing air war drove the Iraqi air force from the skies, severely damaged Iraq's infrastructure, and hammered the Iraqi forces in Kuwait and adjacent southern Iraq with intensive bombardment. In the ensuing ground war, the Iraqis were driven from Kuwait within 100 hours. Overall, an estimated 110,000 Iraqi soldiers and tens of thousands of Iraqi civilians were killed, while the technologically superior coalition forces suffered 340 combat deaths. As they withdrew from Kuwait, the Iraqis also made war on the environment, setting fire to hundreds of oil wells in Kuwait and allowing damaged wells to discharge huge amounts of crude oil into the Gulf.

A formal cease-fire to the Gulf War came in 1991, with harsh conditions imposed on Iraq by the UN Security Council. UN-sponsored economic sanctions created shortages of food and medical supplies, bringing great hardship to the country's population. Iraq agreed to pay reparations for the damage its forces had caused, and it renounced its claims to Kuwait. Saddam Hussein's

area is the homeland of the ancient culture of the **Ma'adan**, or **Marsh Arabs**. These people established a way of life seen nowhere else in the Middle East, using the resources of the marshland to sustain almost all their needs. To quell any opposition these Shiite Muslims might pose, the Iraqi regime destroyed their marshy habitat and forced some 300,000 Ma'adan to flee. Engineers built massive embankments and canals to drain the marshes, and later set fire to the dry vegetation and to Ma'adan settlements.

After the Iraqi regime was taken down in 2003, the embankments were demolished, and restoration of the marshlands began. More than 50 percent of the marshes have come back to life, albeit with less biodiversity than before. The fires changed the soil chemistry so that many of the soils are now claylike and unsuited for the vegetation that grew there previously. Less than one-third of Ma'adan who fled have returned to restore their traditional way of life. As extraordinary as their adaptation to this environment was, their life in the marshes was one of poverty. The task of rebuilding communities in the marshes is an ongoing social challenge. Much damage was done to a culture whose social fabric and economy was woven into the reeds and waters of the marsh.

Throughout the 1990s, the regime of Saddam Hussein played a cat-and-mouse game with UN inspectors whose duty was to eliminate Iraq's potential to produce weapons of mass destruction. Iraq refused to allow full access by the UN inspectors. United States and allied forces responded with air strikes and more stringent economic sanctions. International support for the sanctions against Iraq weakened greatly, especially in the Arab World. President Saddam Hussein endured, much to the chagrin of his foes, and might have been able to hang on to power indefinitely if not for 9/11.

Gulf War II Immediately after the Gulf War and during the Clinton administration of 1993–2001, the reasoning in Washington was that a unified Iraq under Saddam Hussein was better than the alternative: its possible fragmentation into the three entities of a Shiite south, a Sunni center, and a Kurdish north, some or all of which might be hostile to U.S. interests in the region. Although unsavory, Saddam Hussein could be contained. Under the new Bush administration, and especially after 9/11, the downfall of Saddam Hussein became an urgent priority, ranking alongside the War on Terrorism. The Bush administration asserted that Saddam Hussein had linkages with al-Qa'ida. There was no basis for this association, however, as al-Qa'ida had an unwavering contempt for Saddam.

Publicly, Washington built the case that Iraq was harboring and continuing to develop weapons of mass destruction that might be used against the United States and its allies. The United States was able to prompt further United Nations pressure on Iraq, which finally allowed UN weapons inspectors to resume their work. When these inspections failed to turn up weapons, the United States insisted that Iraq was hiding them and that only the use of force could remove this threat. It was necessary to effect "regime change" in Baghdad. To take on Saddam Hussein, President Bush assembled what he called a "coalition of the willing," consisting mainly of Britain, Australia, Denmark, and Poland. It fell far short of the coalition assembled during the Gulf War. France, Germany, and other traditional allies of the United States opposed the impending war against Iraq, and even within the United States there was considerable dissent against what was perceived as a war of choice rather than one of necessity.

The U.S.-led ground and air assault on Iraq (dubbed **Operation Iraqi Freedom**) that began in March 2003 led quickly and with relatively few casualties to the downfall of the Iraqi regime. With about 130,000 troops, concentrated mainly in the Sunni-dominated center of the country around Baghdad, the United States settled in for a long and troubled occupation of the country. The United States handpicked what it called the Governing Council of Iraqis, broadly representative of the country's ethnic composition, and ran many of Iraq's security and other affairs with its own Coalition Provisional Authority (CPA) until June 2004, when it turned over sovereignty to an interim Iraqi government.

The United Nations, with strong U.S. support, organized a January 2005 election for a 275-member transitional national assembly. In this, Iraq's first-ever free election, Iraq's largest ethnic population unsurprisingly won the lion's share of the seats: the Shiites, who turned out in huge numbers to vote, won 48 percent for their main political party. Anticipating the Shiite success and condemning the election as an illegitimate U.S. action, most Sunnis boycotted the vote. That made it possible for the main political party of the Kurds, a smaller ethnic group, to capture another large bloc of the assembly votes—26 percent. Iraq's Sunnis ended up with just 2 percent of the assembly seats. Having been the dominant force in Iraq for decades under Saddam Hussein (who was convicted of war crimes by an Iraqi court and hanged in 2006), the Sunnis now found themselves deprived of power and increasingly angry about the American occupation. The pro-U.S. government meanwhile faced a perplexing array of challenges in cementing a constitution for the fragile country: how much autonomy the Kurds and Shiites might receive within the country, how oil revenues would be split among the geographic and ethnic regions, and what the role of Islam should be in national affairs, for example.

Sunni resistance against the United States grew. Using roadside bombs, car bombs, small arms, and rocket-propelled grenades, Sunni militants were able to inflict American casualties on an almost daily basis. Predictably, al-Qa'ida used the U.S. occupation of Iraq as a rallying cry for intensified resistance against the United States. A well-organized insurgency officially linked with al-Qa'ida started up, under the leadership of Jordanian-born Abu Mus'ab al-Zarqawi. His al-Qa'ida in Iraq organization kept up a steady stream of bloody attacks on U.S. targets even after al-Zarqawi was killed in 2006. Also in the crosshairs were the Iraqis perceived to be collaborating with the Americans, particularly those who joined the new Iraqi army. Sunni insurgents relentlessly targeted Shiites in

(Left): Chris Hondros/Getty Images
(Right): AP Photo/Hameed Rasheed

• **Figure 6.46** In February and June of 2006, Sunni militants—possibly belonging to al-Qa'ida in Iraq—bombed the holy site of the Imam al-Askari shrine in Samarra, north of Baghdad. Shiite Islam's holiest sites outside Saudi Arabia are in Iraq. Sunni insurgents used the tactic of bombing these sites to sow sectarian strife in Iraq.

particular, even twice bombing one of their holiest sites, the golden-domed Askariya Mosque in Samarra (•**Figure 6.46**).

These Sunni assaults on Shiite sacred space helped push Iraq into civil war. Increasingly, this internecine violence took the form of ethnic cleansing, in which one group systematically murdered members of the other. The more militant anti-Sunni and anti-U.S. Shiite factions, rallying chiefly behind a cleric named Moqtada al-Sadr and his Mahdi Army (Mahdi Militia), ascended. Moderate Shiite leaders such as Ayatollah Ali al-Sistani were no longer able to restrain a Shiite backlash. As Shiites retaliated against Sunnis, a ceaseless cycle of revenge killings ensued.

At home, the Bush administration found itself in an increasingly difficult position. Costly in lives and dollars—with thousands of soldiers and supporting civilians killed and a price tag of over $100 billion per year—the war grew more unpopular among Americans, and the president's approval rating slid. Many opposition politicians, academics, and others spoke increasingly of Iraq as "Vietnam," referring to a previous war that proved untenable and unwinnable for the United States. The Bush administration faced a dilemma: "stay the course," as President Bush insisted, with about 140,000 U.S. soldiers still on the ground in Iraq, or pull out (as the United States did in Vietnam) and risk Iraq's plunge into further chaos. Fearing anarchy, the Bush administration in 2007 opted for a "surge" in U.S. troops to maintain order, even as its main ally, Britain, drew down its military presence (primarily in Shiite southern Iraq). The surge, which appears to have been an effective tactic in assisting Iraqi troops in securing the volatile Anbar province, took the number of U.S. troops in Iraq up to nearly 170,000. By this point in the war, al-Qa'ida in Iraq had become deeply unpopular among Iraqis, including the Sunnis who were intended to become al-Qa'ida's main supporters. "Moderate" Sunnis and U.S. forces joined up to push the militants out of Anbar Province.

As a candidate for the presidential office, Bush's successor Barack Obama promised a speedy withdrawal of U.S. troops from Iraq, effectively handing the country's problems back to the Iraqis. His Iraq policy helped Obama clinch the election in 2008. The troop drawdown began in 2009. In August 2010,

President Obama announced that U.S. combat operations in Iraq had ceased. He declared that Operation Iraqi Freedom was over, and that Operation New Dawn was beginning. All but some American military "advisors" were withdrawn from Iraq in 2011. To keep a foothold in the unstable Gulf region, the U.S. military beefed up its forces in Kuwait. It would be especially vigilant toward Iran, which would surely seek to fill the vacuum left with the withdrawal of U.S. forces.

The Iraq War is now being viewed through historical lenses. The conflict was seen in a new light after the controversial release in 2010 of 400,000 Wikileaks documents about the war. These documents reinforced and fleshed out the major findings of the official U.S. government investigation into the 9/11 attacks, establishing the following narrative. The Iraq war was a miscalculation by the United States. The claim that there were weapons of mass destruction in Saddam Hussein's Iraq was a costly intelligence failure. The allegation of Saddam Hussein's connection with al-Qa'ida was false. In the wake of 9/11, the Gulf War squandered American lives and military resources in Iraq, while the United States lost its focus on Osama bin Laden and al-Qa'ida in Afghanistan and neighboring Pakistan. In invading Iraq, the United States lost most of the international sympathy and goodwill that grew out of 9/11. The United States inadvertently fulfilled many of Iran's ambitions in Iraq. And finally, the United States created enemies in Islamic countries who perceived that the United States was at war with Muslims. The United States inadvertently played into the hands of the conspiracies theories created by Osama bin Laden and other terrorists.

About 4,400 U.S. soldiers and perhaps 150,000 Iraqi soldiers died during the war. About 5 million Iraqis became refugees, with about half fleeing abroad (mainly to countries in the Arab World) and half displaced internally.

The Kurds

A mostly Sunni Muslim people of Indo-European origin, the Kurds are the largest of several non-Arab minorities in Iraq and the largest minority group in Turkey. They have long revolted against Iraqi and Turkish authorities. Kurds today remain the world's largest ethnic group without a country. European powers promised an independent state of Kurdistan at the end of World War I, but the governments concerned never took steps to create it. The Kurdish Question—How are the political wishes of these people to be accommodated?—has international implications because the area occupied by an estimated 30 million Kurds extends from Turkey and Iraq into Iran, Syria, and the southern Caucasus region (see Figure 6.44); an estimated 2 million more Kurds live outside the Middle East.

A book about the Kurds is appropriately titled *No Friends but the Mountains*.[11] Turkish officials have long downplayed the identity of Kurds (who make up about 20 percent of Turkey's population), often referring to them as "Mountain Turks," and even banned or restricted Kurdish-language media in Turkey until 2002. A Kurdish rebellion against Turkish rule, rising in part from the great poverty of the Kurds relative to the rest of Turkey's population, has smoldered and sometimes flamed since 1984. The main Kurdish resistance to Turkish rule comes from the **Kurdistan Workers Party**, known by its Kurdish acronym **PKK**.

Under Saddam Hussein, Iraq tried to "Arabize" its Kurds by forcing them to renounce their ethnicity and sign forms saying they were Arabs. However, after a no-fly zone and "safe haven" established after Gulf War I put them beyond Saddam Hussein's reach, Iraq's Kurds came to enjoy relative autonomy and prosperity. Even in the tumultuous years after the U.S.-led invasion in 2003, the Kurdish region of the north escaped most of the bloodshed and economic chaos characteristic of the south.

Hotspot Iran

Iran, the easternmost country of the Middle East, was known formerly as Persia (•Figure 6.47). From the city of Persepolis in southern Persia, Persian kings ruled a great empire in the fifth century B.C.E. Alexander the Great destroyed Persepolis, but Persia's grandeur would return. Some of the world's great architectural wonders were commissioned by Shah Abbas the Great in sixteenth-century Esfahan. Persia became a backwater in subsequent centuries and was a poor and undeveloped country a century ago. In the 1920s, a military officer of peasant origin, Reza Khan, seized control of the government and began a program to modernize Iran and free it from foreign domination. He had himself crowned as Reza Shah Pahlavi, the founder of the new Pahlavi dynasty. Influenced by the modernizing efforts of Mustafa Kemal Atatürk in neighboring Turkey, the new shah (king) introduced social and economic reforms. In 1941, his young son took the throne as Mohammed Reza Shah Pahlavi. A decade later, intervention by the U.S. Central Intelligence Agency (CIA) helped to solidify the shah's absolute rule of Iran. The shah then set out to expand the program of modernization and Westernization begun by his father. In the 1970s, Iran played a leading role in raising world oil prices. Mounting oil revenues underwrote explosive industrial, urban, and social development.

In the countryside, the shah's government tried to upgrade agricultural productivity and rural standards of living through land reform and better technologies. Despite the fact that Iran was a cradle of plant and animal domestication, dry and rugged habitats make agriculture difficult over large parts of the country. Agricultural reforms generally failed, and poverty-stricken families from the countryside poured into Tehran and other cities. From this devoutly Muslim group of new urbanites came much of the support for the revolution that ousted the shah in 1978–1979. Dissent also grew because the regime spent vast sums on its military, mainly for U.S.-made weapons, thus undercutting benefits to the people.

The revolution that overthrew the shah took Iran abruptly out of the American orbit. The central figure among the revolutionaries was an **ayatollah** ("sign of God") named Ruhollah Khomeini, an elderly critic of the regime whom the shah had forced into exile in 1964. From Paris, Khomeini sent repeated messages to Iran's Shiites that helped spark the revolution. A furious tide of Shiite religious sentiment rose against corruption, police heavy-handedness, modernization, Westernization, Western imperialism, the exploitation of underprivileged people, and the monarchy, which Shiite beliefs regard as illegitimate. Growing numbers of people staged strikes and demonstrations in 1978, forcing the shah to flee the country and abdicate his throne. After hospitalization in the United States and life in exile in Panama and Egypt, the shah died in Cairo in 1981 and is buried there. The Ayatollah Khomeini returned in triumph to Iran in 1979. His followers seized and held 52 U.S. diplomatic personnel as hostages in Tehran for 444 days. U.S. President Jimmy Carter proved unable to resolve this crisis, which contributed to his defeat in his 1980 bid for reelection.

Under Khomeini (who died in 1989), the country became an Islamic republic governed by Shiite clerics, who included a handful of revered and powerful ayatollahs at the head of the religious establishment, and an estimated 180,000 priests

© Cengage Learning 2013

• **Figure 6.47** Principal features of Iran.

called **mullahs**. Khomeini served as the Guardian Theologian, an infallible supreme leader enshrined by the principle of **divine rule by clerics** (*velayet-e-faqih*). These religious leaders continue to supervise all aspects of Iranian life and perform many functions allotted to civil servants in most countries. They base their authority on Shiite interpretations of Islam (about 90 percent of Iran's 78 million people are Shiites).

In contrast to their status during the shah's reign, when they were encouraged to take on larger public roles, women in postrevolutionary Iran have lost many freedoms. Under the new regime, "polluting" influences of Western thought and media were purged. However, recent years have seen a general softening of restrictions on personal freedom of expression, and the Internet and other media have introduced broader worldviews to Iran's youth, a very important cohort, as about 60 percent of the population is under the age of 30. Young people led an unprecedented series of prodemocracy protests in 2002. Five years later, Iranians of different generations took to the streets to protest government hikes on gasoline prices; ironically, this oil-rich country has a low oil-refining capacity and needs to import one-third of its gasoline. Iran imports more gasoline than any country in the world except for the United States. The country is fortunate to have the world's fourth largest reserves of oil, and natural gas reserves that are second only to those of Russia.

Before 2005, there was tug-of-war between the moderate and reform-minded president Muhammad Khatami and the less tolerant supreme religious leader, Ayatollah Ali Khamanei (Khomeini's successor), who has lifetime control over the military and the judicial branch of government. There was speculation that the country's younger and more Western-oriented sectors might win the day and that relations with the wider world—the United States in particular—would improve. But the winner of Iran's 2005 presidential election was Mahmoud Ahmadinejad. This hardliner outspokenly opposed American policies in the region. He goaded Israel by refusing to recognize its existence, insisting that the Holocaust never happened and calling for the country's destruction. He accelerated Iran's 183 nuclear power program. The level of rhetoric and tension between Iran and the United States escalated.

Here are a few things to keep in mind about Iran in its regional context. First, remember that Iranians (Persians) are 170 not Arabs; their ethnolinguistic roots are Indo-European. Iran (and previously Persia) is a historic enemy of most Arabs. Its big Arab neighbors Iraq and Saudi Arabia despise Iran, and vice-versa. Even the Shiite Muslims in Iraq do not naturally ally themselves with Iran; their ethnicity (Arab) seems to outweigh their faith where relations with Iran are concerned. In other contexts, however, Iran's identity as a Shiite state has important effects on its relations with Arabs. Iran claims sovereignty over the Gulf Arab country of Bahrain, where 195 70 percent of the population is Shiite. During the Arab Spring of 2011, while Saudi Arabia came to the aid of the minority Sunni monarchy, Iran appealed to the Shiites to overthrow the regime.

The Sunni/Shiite equation is reversed in Syria, with Sunnis the majority and Shiites the minority. When (Arab) Iraq attacked (Persian) Iran in 1980, the Arab nation of Syria allied itself with Iran. Why? In part because, as you have seen, Syria's Sunni Muslim majority is ruled by a family practicing an offshoot of the Shiite faith known as Alawites. Iran is strongly allied with the militant Shiite Arabs of Hizbullah, in Lebanon, and is able to support Hizbullah logistically because of strong ties with Lebanon's neighbor Syria. 183 Iran also supports Hamas in the West Bank, not because its members are Shiites (they are Palestinian Sunni Muslims), but because Hamas is set on Israel's destruction, which Iran would like to see. 183

Turkey: Where East Meets West

Like Iran, Turkey is a demographic heavyweight (population 74 million people) and has long had an enormously important presence in the region (•Figure 6.48). The Turks who organized the Ottoman Empire beginning in the fourteenth century originated as pastoral nomads from central Asia, where Turkic cultures still prevail. From the sixteenth 142 century to the nineteenth, their empire, based in what is now Turkey, was an important power. Modern Turkey was created from the wreckage of the old empire after World War I. Its founder, Mustafa Kemal Atatürk, was determined to Westernize the country, raise its standard of living, and make it a strong and respected national state. He inaugurated social and political reforms designed to break the hold of traditional Islam and open the way for modernization and Turkish nationalism.

Islam had been the state religion under the Ottoman Empire, but Atatürk divorced church from state, and to this day Turkey is the only Muslim Middle Eastern country to officially separate them. The wearing of the red cap called the *fez*, an important symbol of piety under the Ottoman caliphs, was prohibited, and state-supported secular schools replaced the all-pervasive religious schools. To facilitate public education and remove further traces of Muslim culture, Latin characters

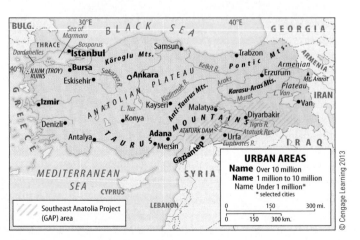

• **Figure 6.48** Principal features of Turkey.

replaced the Arabic script. Slavery and polygamy were outlawed, and women were given full citizenship. Legal codes based on those of Western nations replaced Islamic law, and forms of democratic representative government were instituted, although Atatürk ruled in a dictatorial fashion.

Turkey's economic standing is near the top of the LDCs, close to MDC status. Not long ago, poverty and prominently rural characteristics marked Turkey. But the country embarked on a course of change that is modernizing agriculture, expanding industry, and raising standards of living. Its economic growth has been impressive. The value of output from manufacturing (mainly textiles, agricultural processing, cement manufacturing, metal industries, and automobiles) is greater than that from agriculture, which is considerable in this fertile, well-watered country. Some of Europe's most recognizable brand names, like Bosch, Renault, and Fiat, have factories in Turkey. Turkey does produce some oil.

Early in the 1990s, Turkey began an impressive $32 billion agricultural effort called the **Southeast Anatolia Project** (known by its Turkish acronym **GAP**). Its aim is to double the country's irrigable farmland, converting Turkey's semiarid southeastern into the "breadbasket of the Middle East," especially with surplus production of cotton, wheat, and vegetables for export. The centerpiece of this massive irrigation project is the Atatürk Dam on the Euphrates River. About 20 other dams on the Turkish Euphrates and Tigris are also part of the ongoing project to provide hydroelectricity as well as water (•**Figures 6.48** and **6.49**).

Like most great dam projects, including the Aswan Dam described earlier, The GAP is controversial. It is inundating historic settlements and priceless archaeological treasures and is compelling people to move. It is being developed in the part of the country where a poor and restive Kurdish population is seeking development, human rights, and autonomy. Many Kurds do not have confidence in government promises that the project will boost their prospects. Tens of thousands of Kurds

202

are being relocated from areas inundated by reservoirs, and there are fears that local people will succumb to waterborne diseases. In addition, Turkey's Tigris and Euphrates waters flow downstream into neighboring states, raising serious questions about downstream water allocation and quality.

182

Is Turkey European?

Turkey is an "in-between" country. Economically, it is near the line between MDCs and LDCs. Culturally, it is between traditional Islamic and secular European ways of living. Part of Turkey—the section called Thrace—is actually in Europe, and Turkey as a whole aspires to become more European. It has been a member of the North Atlantic Treaty Organization (NATO) since 1952. It is an associate member of the European Union and is a candidate for full membership. The European Union, however, insists that it would first like to see Turkey become more European politically, for example, by diminishing the army's influence in government and by improving its human rights record. Turkey insists it has made huge strides

87

toward satisfying the minimal democratic and human rights norms for EU member countries. It has abolished the death penalty and eased restrictions it imposed on the cultural expressions of its minority Kurdish population.

At expansion talks, EU leaders nevertheless have repeatedly rejected Turkey's bid and would not promise a date for membership negotiations. They argued that Turkey has still not gone far enough with democratic and human rights reforms. The Europeans probably fear the costs of absorbing Turkey's mainly low-income population into the union—a fear that is justified, considering the EU's struggles with its own lower income countries like Greece, which have threatened the very existence of the EU.

88

Many Turks feel that their membership in the European Union was denied because they are Muslims. It is indeed possible that behind the scenes, EU decision makers rejected EU membership for Turkey because they feared that Islamist terrorists would easily make their way from an EU Turkey into the heart of continental Europe. The European Union's rejection of Turkey contributes to the sentiment in the Islamic world that the West is waging a systematic campaign against Muslims. Turkish public opinion has become decidedly more anti-European, even as the prospect of joining the troubled European Union, with its sovereign debt crises and other problems, looks increasingly unattractive. A Turkey shunned by Europe may seek alliances eastward in Russia, India, and China.

Or Is Turkey a Great Middle Eastern Power?

Turkey's standing in Middle Eastern affairs has changed dramatically in recent years. Not long ago, diplomatically and otherwise, the Arabs reviled Turkey, just as they had for centuries (remember that Turks are not Arabs and that The Ottoman Empire that subjugated much of the Middle East from the sixteenth century until World War I was Turkish). Turkey's closest ally in the Middle East was Israel, which is also reviled in much of the Arab World. That association made Turkey deeply unpopular among Arabs.

• **Figure 6.49** Hasankeyf, on the Tigris River. Important archaeological sites are to be inundated by a reservoir that will form behind a dam that is part of the Southeast Anatolia Project.

Joe Hobbs

A single incident in 2010 dramatically rearranged Turkey's regional priorities and its standing in the Arab Middle East. Israel has a blockade that prohibits any unauthorized access to the Palestinian-controlled Gaza Strip. On May 10, 2010, a so-called "Gaza Freedom Flotilla" of ships carrying humanitarian aid and construction materials to the Gaza Strip was confronted at sea by Israeli commandoes. The Israeli forces boarded several of these ships belonging to the Free Gaza Movement and the Turkish Foundation for Human Rights and Freedoms and Humanitarian Relief. An altercation ensued, and a number of Turkish civilians as well as Israeli commandoes were killed. Turkey condemned Israel and demanded that Israeli apologize, pay compensation, and prosecute those responsible for the Turkish fatalities. Israel refused.

The incident accelerated an ongoing reshuffling of the Middle East's political landscape. Turkey's Prime Minister, Recep Tayyip Erdoğan, denounced the raid as Israeli "state terrorism." Turkey recalled its ambassador to Israel, suspended military trade with Israel, and otherwise indicated that Israel was no longer an ally. Turkey had long promoted a national policy of "zero problems with the neighbors," but Prime Minister Erdoğan now portrayed Israel as a regional problem and offered himself and Turkey as champions of Arab causes. Amidst the tumult of the Arab Spring in 2011, Turkey appealed to the United States to recognize it, rather than Israel, as the best ally of the United States in the Middle East. Prime Minister Erdoğan visited the countries most transformed by the Arab Spring—Egypt, Tunisia, and Libya—and vowed to assist in their reformations. He spoke of a new axis of power, based on Turkey's alliance with Egypt, the Arab World's strongest country. He indicated that Turkey would use military force if needed to advance the interests of its Arab allies and of itself—for example, in preventing the Republic of Cyprus from working with Israel to tap oil and gas reserves in the Mediterranean Sea. 102

Domestically as well, Prime Minister Erdoğan shook up Turkey. After coming to power in 2003, he began restraining the powers of the military. Turkey's constitution invests the military with the responsibility of protecting the country's secular government, and the army has used its powers to overthrow four elected governments since 1960. Erdoğan's growing power and popularity blunted any prospect that he would be ousted by the military. Turkey's economy improved markedly under his rule—most notably, it became a powerhouse of automobile manufacturing—and trade, transport, and communication with the Arab countries accelerated. As the leader of secular Turkey, Prime Minister Erdoğan has had to restrain his personal convictions. Prior to his election, he had described himself as a political Islamist and an advocate of Islamic law or *shari'a*. Some analysts worry that he will violate Turkey's rigid separation of church and state, and advocate Islamist domestic and foreign policies.

This concludes our journey through the Middle East and North Africa. We now turn eastward into the realm of South and East Asia.

summary

- Misleading stereotypes about the environment and people of the Middle East and North Africa are common, as people outside the region often associate the region solely with military conflict and terrorism.

- The region has bestowed on humanity a rich legacy of ancient civilizations and the three great monotheistic faiths of Judaism, Christianity, and Islam.

- Middle Easterners include Jews, Arabs, Turks, Persians, Berbers, people of sub-Saharan African origin, and other ethnic groups who practice a wide variety of ancient and modern livelihoods.

- Arabs are the largest ethnic group in the Middle East and North Africa, and there are also large populations of ethnic Turks, Persians (Iranians), and Kurds. Islam is by far the largest religion. Jews live almost exclusively in Israel, and there are minority Christian populations in several countries.

- Population growth rates in the region are moderate to high. Oil wealth is concentrated in a handful of countries, and as a whole, this is a developing region.

- The Middle East has served as a pivotal global crossroads, linking Asia, Europe, Africa, and the Mediterranean Sea with the Indian Ocean. These countries have historically been unwilling hosts to occupiers and empires originating far beyond their borders.

- The margins of this region are occupied by oceans, high mountains, and deserts. The land is composed mainly of arid and semiarid plains and plateaus, together with considerable areas of rugged mountains and isolated "seas" of sand.

- Aridity dominates the environment, with at least three-fourths of the region receiving less than 10 inches (25 cm) of yearly precipitation. Great river systems and freshwater aquifers have sustained large human populations.

- Many of the plants and animals on which the world's agriculture depends were first domesticated in the Middle East.

- The Middle Eastern "ecological trilogy" consists of peasant villagers, pastoral nomads, and city dwellers. The relationships among them have been mainly symbiotic and peaceful, but city dwellers have often dominated the relationship, and both pastoral nomads and urbanites have sometimes preyed on the villagers, who are the trilogy's cornerstone.

- About two-thirds of the world's oil is here, making this one of the world's most vital economic and strategic regions.

- Since World War II, several international crises and wars have been precipitated by events in the Middle East. Strong outside powers depend heavily on this region for their current and future industrial needs. Unimpeded access to Persian-Arabian Gulf oil is one of the pillars of U.S. foreign policy.

- The region of the Middle East and North Africa is characterized by a high number of chokepoints, strategic waterways that may be shut off by force, triggering conflict and economic disruption.

- Oil pipelines in the Middle East are routed both to shorten sea tanker voyages and to reduce the threat to sea tanker traffic through chokepoints, but are themselves vulnerable to disruption.

- Access to freshwater is a major problem in relations between Turkey and its downstream neighbors, Egypt and its upstream neighbors, and Israel and its Palestinian, Jordanian, and Syrian neighbors.

- Al-Qa'ida and affiliated Islamist terrorist groups aim to drive the United States and its allied governments from the region and to replace them with an Islamic empire. Al-Qa'ida is an apocalyptic group that seeks to inflict mass casualties on its enemies, particularly on Americans in their home country.

- The United Nations Partition Plan of 1947 attempted a two-state solution to the dilemma Britain had created by promising land to both Arabs and Jews in Palestine. It envisioned geographically fragmented states, making each side feel vulnerable to the other. War prevented the plan's implementation. The Arab-Israeli conflict has continued from that time, with Israel gaining more territory and the indigenous Palestinians failing to acquire a country of their own. Principal obstacles to peace between Israelis and Palestinians are the status of Jerusalem, the potential return of Palestinian refugees, the future of Jewish settlements in the Occupied Territories, and Israeli construction of a security barrier that penetrates into portions of the Palestinian West Bank.

- Israel's eastern neighbor, Jordan, has a majority Palestinian population.

- The Arab Spring of 2011 was a wave of revolutions against autocratic regimes across the region, notably in Tunisia, Egypt, Libya, Syria, Bahrain, and Yemen.

- Masdar is a "carbon-neutral" city under construction in Abu Dhabi, one of the United Arab Emirates.

- Two-thirds of the world's proven petroleum reserves are concentrated in a few countries that ring the Persian-Arabian Gulf. Saudi Arabia controls more than one-fifth of the world's oil.

- Under Saddam Hussein, Iraq squandered its oil wealth on military misadventures, including invasions of Iran and Kuwait. The Kuwait invasion led to an enormous U.S.-led counterattack that devastated the country's infrastructure and subjected it to years of economic sanctions. The U.S.-led invasion of Iraq in 2003 resulted in Saddam Hussein's downfall and death, and an uncertain future for this ethnically diverse country (with large Shiite Arab, Sunni Arab, and Kurdish populations) engaged in civil war.

- Iran's oil revenues were used to modernize and Westernize the nation during the Pahlavi dynasty. Rapid social change and uneven economic benefits precipitated a revolutionary Islamic movement, which forced the shah to abdicate and flee in 1979. This theocratic state has had difficult relations with the United States, most recently over its alleged nuclear weapons program. Iran's young population is agitating for change, but the ruling clerics are resisting.

- Turkey was formerly the seat of power for the Islamic Ottoman Empire. It is a secular nation with membership in NATO and possible future membership in the European Union. It is blessed with freshwater resources that it has mustered with the hope of creating the "breadbasket of the Middle East."

Key Terms + Concepts

Review Questions

1. What countries constitute the Middle East and North Africa? Which are the three most populous? Which encourage and discourage population growth?

2. What are the major climatic patterns of the Middle East? Where are the principal mountains, deserts, rivers, and areas of high rainfall?

3. Why can this region be described as a culture hearth? What major ideas, commodities, and cultures originated there?

4. What are the major ethnic groups, and in which countries are they found? What is an Arab? A Jew? A Turk? A Kurd? A Persian? A Muslim?

5. What are the principal beliefs and historical geographic milestones of Jews, Christians, and Muslims?

6. What is the difference between Shiite and Sunni Islam?

7. Where is oil concentrated in this region?

8. What is the Carter Doctrine?

9. What options does the United States weigh in dealing with Iran's suspected nuclear weapons development program?

10. What are upstream and downstream countries, and which are usually the more powerful? What are the exceptions to this rule?

11. What are Hizbullah, Hamas, and al-Qa'ida?

12. What precipitated the various conflicts between Israel and its Arab neighbors? How did each of these conflicts rearrange the political map?

13. What lands did the peace process of the Oslo Accords yield to Palestinian control? What is the current situation in those areas? Who controls the Gaza Strip and the West Bank?

14. According to Lester Brown, what three steps can be taken to avert conflict over Nile waters?

15. What are the pros and cons of Egypt's Aswan High Dam?

16. What were essential elements of the Arab Spring, especially in Egypt, and why is the United States wary about the "Arab Fall?"

17. What are the objectives of the MASDAR project?

18. What happened when Saddam Hussein's troops invaded Kuwait? What happened when they invaded Iran? What critical decisions must the United States make in Iraq?

19. What is the ethnic composition of Iraq? How did that consideration cause the United States to avoid invading the country in 1991?

20. Who are the Kurds, and where do they live?

21. What were some of the grievances against the Pahlavi dynasty that led to revolution in Iran? What changes has the Islamic regime brought to the country?

22. What three options has the United States weighed in response to Iran's nuclear weapons program?

23. In what ways is Turkey an "in-between" country, and how is Turkey's role in the MENA changing?

Notes

1. Ulrich Deil, Heike Culmsee and Mohamed Berriane. 1980. "Sacred Groves in Morocco: A Society's Conservation of Nature for Spiritual Reasons." *Silva Carelica 49*, 185–201.

2. Lester Brown, "When the Nile Runs Dry." *New York Times*, June 2, 2011, p. A23.

3. Fouad Ibrahim and Barbara Ibrahim, *Egypt: An Economic Geography.* (London: I.B. Tauris, 1983).

4. From the U.S. State Department website at http://www.state.gov/s/ct/rls/crt/2000/2419.htm.

5. Quoted in Daniel Dombey, James Blitz and Najmeh Bozorgmehr, "Atomic Agitation." *Financial Times*, January 8 2010, p. 7.

6. Rohan Gunaratna, *Inside al Qaeda: Global Network of Terror* (New York: Cambridge University Press, 2002), p. 188.

7. Suleiman Abu Ghaith, "Al-Qa'ida Statement (October 10, 2001)." In *Anti-American Terrorism and the Middle East,* Barry Rubin and Judith Colp Rubin, eds. (Oxford University Press, 2002), p. 251.

8. United Nations Development Programme Report 2002. http://www.arab-hdr.org/publications/other/ahdr/ahdr2002e.pdf

9. Derek Gliddon, "Building an Oasis in the Desert." *Arc News, 31*(3) Fall 2009, 26–27.

10. Ibid.

11. John Bullock, *No Friends but the Mountains* (London: Oxford University Press, 1993),

Online Resources

 CourseMate: Make the most of your study time by accessing everything you need to succeed in one place. Read your textbook, take notes, review flashcards, watch videos, complete activities, take practice quizzes, and more—online with CourseMate. Log in at **www.cengagebrain.com.**

"A book holds a house of gold."

—CHINESE PROVERB

Southern Asia at night. Note especially how the Himalaya separate the densely-populated Indian subcontinent from the Tibetan wilderness; the belts of urban-industrial activity in southeast China and in Japan; and the contrast between the two Koreas. Left: boats on the New Year pilgrimage to the Perfume Pagoda, Vietnam.

C. Mayhew & R. Simmon (NASA/GSFC), NOAA/ NGDC, DMSP Digital Archive

Joe Hobbs

South and East Asia

7

South and East Asia make up a great triangle that runs from Afghanistan in the west to Japan in the northeast and the island of New Guinea in the southeast (•**Figure 7.1**). In this land area of about 8 million sq mi (21 million sq km—a little more than twice the size of the United States [•**Figure 7.2**]), or less than one-quarter of the earth's landmass—lives more than half of the world's population (•**Figure 7.3**). The cultural makeup of these 3.8 billion people is as diverse and expressive as the physical environments they inhabit. The world's highest mountain peaks, some of the longest rivers, and the earth's most highly transformed and densely settled river plains are the settings for some of the oldest civilizations and the most modern economies. Just as this region has played a major role in human development over thousands of years, it is a major force in world affairs in the twenty-first century. Some experts say this will be known as the **Asian Century** for the ways in which the region will reshape the globe.

chapter objectives

This chapter should enable you to

- Recognize the role of seasonal monsoon winds and rains in people's livelihoods and perceptions.
- Appreciate China and India as the demographic giants and surging economies of early-twenty-first-century Asia.
- Learn how unique spatial and spiritual considerations have influenced the traditional layout of Asian settlements.
- Know the basic beliefs of Hinduism, Buddhism, Confucianism, and Daoism.
- Become familiar with the pros and cons of the Green Revolution.
- Understand the geopolitical dimensions of the tensions between India and Pakistan, North Korea and the West, and Islamists and governments in Pakistan and Indonesia.
- Consider how Afghanistan has become known as the "Graveyard of Empires."
- Recognize the tensions between Muslims and Hindus in South Asia that often give rise to violent conflict.
- Appreciate how the partition of India created lasting problems in political relations, resource use and allocation, and industrial development.
- See that food production has kept pace with enormous population growth in India.
- Recognize the serious consequences of political insurgencies in Kashmir, India, Pakistan, and Sri Lanka.
- Appreciate the catastrophic reach of tsunamis and some of the region's other natural hazards.
- Understand some of the major political obstacles and environmental consequences associated with large dams and other water management projects.
- Recognize China as a land empire in which a single ethnic group added vast peripheral areas to its eastern core.
- Appreciate the constraints imposed by China's Communist system on personal, political, and economic freedoms, and the recent substantial easing of those restrictions.
- Balance the pros and cons of China's hugely ambitious plans to transform its natural environment through dams, canals, forest restoration, and other projects.
- Consider the economic discrepancies between China's rural and urban populations and the forces behind large-scale rural-to-urban migration.
- Understand the regional and geopolitical risks of Taiwan's aspirations for, and identity independent of, China.
- Appreciate how countries burdened by a lack of natural resources can become prosperous by marshaling their human resources.
- Understand the unique challenges associated with a postindustrial society that has a prosperous and aging population.
- See how a country's geographic location can make it a target of division and conquest.
- Appreciate how different political and economic systems can produce dramatically different results for almost identical peoples.

The countries of eastern Eurasia were at one time referred to as the Orient, from a Latin word meaning "facing the east" or "facing the sunrise." (The West was then known as the Occident, derived from a word meaning "facing the sunset.") But this terminology reflected the view from Europe and became associated with Western stereotypes and misconceptions about Asia. These terms have fallen out of favor, but in older literature you are sure to read about the Orient.

7.1 Area and Population

South and East Asia encompass the following subregions, which in some geography books are classified as separate regions:

- East Asia, which includes Japan, North and South Korea, China, Mongolia, Taiwan, and many near-shore islands
- South Asia, which includes Afghanistan, Pakistan, India, Sri Lanka, Bangladesh, the mountain nations of Bhutan and Nepal, and the island country of the Maldives
- Southeast Asia, which includes both the peninsula jutting out from the southeast corner of the Asian continent, on which are located Myanmar (Burma), Thailand, Laos, Cambodia, Vietnam, Malaysia, and Singapore, and the island world that rings this peninsula, which includes the countries of Indonesia, the Philippines, Brunei, and Timor-Leste (•Table 7.1; see also Figure 7.1)

Demographic Heavyweights of South and East Asia

Perhaps the most critical feature in the geography of South and East Asia is that this region is home to about 55 percent of the world's population (Figure 7.3; see also Figures 3.13 and 3.14). Two countries alone, India and China, together have 2.6 billion people, or 37 percent of the world total. Their extraordinary demographic weight reflects thousands of years of human occupation of productive agricultural landscapes, combined with twentieth-century advances in health technology that reduced death rates. The highest population densities are found in the areas of most abundant precipitation or surface water and with the most productive soils; rivers and coastal plains are especially densely settled. Rugged mountains; almost waterless deserts; and until recently, tropical rain forest vegetation have limited populations elsewhere.

Some of the world's highest urban densities are found in the cities of Hong Kong, Macao, and Singapore, but in Mongolia and Nepal population densities are extremely low. Singapore is one of the few nations to claim a 100 percent urban

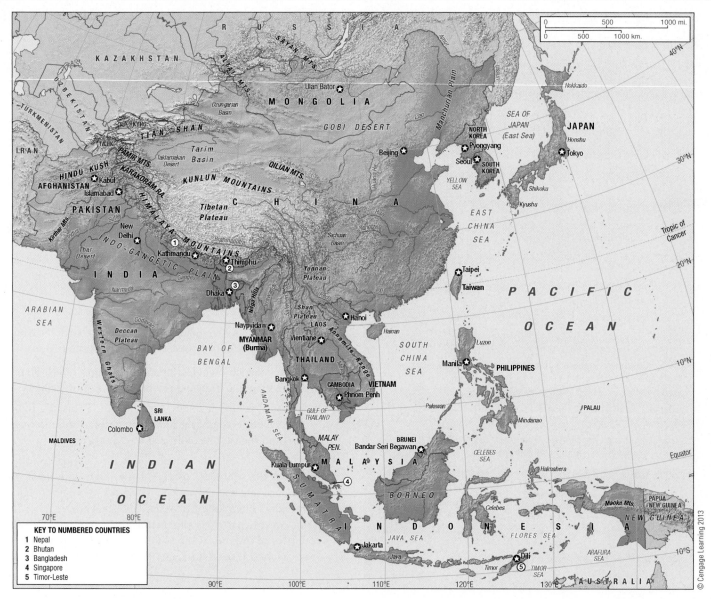

• **Figure 7.1** South and East Asia.

population; in contrast, only 17 percent of Nepal's and 15 percent of Sri Lanka's population are urban. A mere 1 percent of Mongolia is arable, whereas farmers in India and Bangladesh cultivate more than 50 percent of their lands.

Population Growth Patterns

It is not possible to generalize about population growth in South and East Asia. Growth rates range from about zero percent growth (ZPG) in Japan to a very high 3.1 percent per year in Timor-Leste. The region reflects the general worldwide trends of poorer countries having higher growth rates and wealthier countries having lower growth rates. The region is demographically and otherwise recognizable as one of mainly less developed countries (LDCs), with some outstanding exceptions. Japan is a classic postindustrial country and, like Italy, must worry about a declining, aging population in which

young people will have to shoulder an increasing economic burden to support their elders. As discussed later in this chapter, there are other strong economies in the region.

Migration is discussed in many parts of this chapter, but a brief note on the Philippines has a place here. The country is unique in the world for its strong tradition of sending family members to work abroad. One quarter of the country's workforce works abroad. There is an acronym for them: **OFWs**, for **overseas foreign workers**. They send as much money home as possible, so much that these remittances account for more than 10 percent of the Philippines' gross domestic product. There is a price to pay for the many years that a Filipino mother or father spends working abroad: loneliness; isolation; the pain of being far from one's children; broken families; and ultimately, divorce are common.

One of the big anomalies in Asia's population equation is China. As a developing country with a per capita GNI PPP of just $6,890, China might be expected to have a rather high

Table 7.1   East and South Asia: Basic Data

Political Unit	Area (thousands; sq mi)	Area (thousands; sq km)	Estimated Population (millions)	Estimated Population Density (sq mi)	Estimated Population Density (sq km)	Annual Rate of Natural Increase (%)	Human Development Index	Urban Population (%)	Per Capita GNI PPP ($US)
South Asia	**1,984.4**	**5,139.6**	**1,653.7**	**833**	**322**	**1.6**	**0.535**	**29**	**2,990**
Afghanistan	251.8	652.2	32.4	129	50	2.8	0.398	22	860
Bangladesh	55.6	144.0	150.7	2,710	1,046	1.5	0.500	25	1,550
Bhutan	18.1	46.9	0.7	39	15	1.4	0.522	33	5,290
India	1,269.3	3,287.5	1,241.3	978	378	1.5	0.547	29	3,280
Maldives	0.1	0.3	0.3	3,000	1,158	2.0	0.661	35	5,250
Nepal	56.8	147.1	30.5	537	207	1.9	0.458	17	1,180
Pakistan	307.4	796.2	176.9	575	222	2.1	0.504	35	2,680
Sri Lanka	25.3	65.5	20.9	826	319	1.3	0.691	15	4,720
Southeast Asia	**1,735.3**	**4,494.4**	**602.0**	**347**	**134**	**1.3**	**0.619**	**42**	**4,930**
Brunei	2.2	5.7	0.4	182	70	1.3	0.838	72	N/A
Cambodia	69.9	181.0	14.7	210	81	1.8	0.523	20	1,820
Indonesia	735.4	1,904.7	238.2	324	125	1.3	0.617	43	3,720
Laos	91.4	236.7	6.3	69	27	2.2	0.524	27	2,200
Malaysia	127.3	329.7	28.9	227	88	1.6	0.761	64	13,710
Myanmar	261.2	676.5	54.0	207	80	1.2	0.483	31	N/A
Philippines	115.8	299.9	95.7	826	319	1.9	0.644	63	3,540
Singapore	0.2	0.5	5.2	26,000	10,039	0.5	0.866	100	49,780
Thailand	198.1	513.1	69.5	351	135	0.5	0.682	31	7,640
Timor-Leste	5.7	14.8	1.2	211	81	3.1	0.495	22	4,730
Vietnam	128.1	331.8	87.9	686	265	1.0	0.593	30	2,790
East Asia	**4,545.6**	**11,773.1**	**1,573.5**	**346**	**134**	**0.4**	**0.711**	**55**	**9,760**
China	3,696.1	9,572.9	1,345.9	364	141	0.5	0.687	50	6,890
Japan	145.9	377.9	128.1	878	339	−0.1	0.901	86	33,400
Korea, North	46.5	120.4	24.5	527	203	0.6	N/A	60	N/A
Korea, South	38.3	99.2	49.0	1,279	494	0.4	0.897	82	27,240
Mongolia	604.8	1,566.4	2.8	5	2	1.8	0.653	61	3,330
Taiwan	14.0	36.3	23.2	1,657	640	0.1	N/A	78	N/A
Summary Total	**8,265.3**	**21,407.1**	**3,829.2**	**463**	**179**	**1.0**	**0.621**	**42**	**6,080**

Sources: World Population Data Sheet, Population Reference Bureau, 2011; Human Development Report, United Nations, 2011; World Factbook, CIA, 2011.

population growth rate like India, which has a similar economic ranking. But in the twentieth century, China's population was so alarmingly large and rapidly growing that its Communist leaders decided to restrain its growth drastically. China adopted a coercive "one-child policy" that has brought its growth rate to just 0.5 percent per year, representing a doubling time of 140 years (see Geographic Spotlight, page 216). That is a remarkable turnabout from the situation in 1964, when China had a growth rate of 3.2 percent and a doubling time of 22 years.

To reduce its poverty, India has made great strides in reducing its population growth, but not in China's coercive fashion. India is projected to overtake China as the world's most populous country in 2040, when it will have an estimated 1.5 billion people. A wild card in India's population deck is HIV/AIDS. The number of Indians infected with HIV was estimated at 2.4 million in 2011. The potential for a **pandemic** in the world's second largest country is uncertain, and much will depend on how India's government chooses to fight the disease. About 80 percent of India's HIV infections are spread by heterosexual contact, and public health authorities have so far been shy to mount a strong campaign about sexual ethics and practices. In China, the number of people infected was estimated at 740,000 in 2011.

Some Chinese parents want to produce a son if their first child was a girl, and many girls are given to orphanages (China is the world's leading source country of adopted children, about 90 percent of them girls). Female feticide is increasingly associated

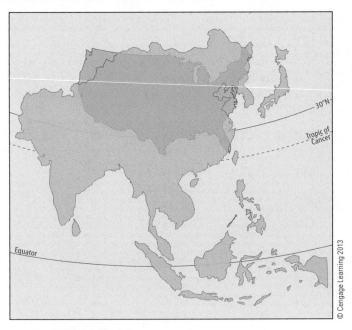

© Cengage Learning 2013

• **Figure 7.2** South and East Asia compared in latitude and area with the conterminous United States.

with the desire to have a son. Couples often insist on an ultrasound test to determine the baby's gender, and about 90 percent of girls detected in this way are aborted. In some provinces where the practice is widespread, three boys are born for every two girls. Countrywide, there are 118 male newborns for every 100 females. As in India, this practice is beginning to have serious repercussions, reflected in a shortage of eligible brides. Rural women have been abducted and trafficked as virtual slaves to other parts of the country, where they become brides. On the positive side, the relative shortage of women in China has helped enhance their status, particularly in urban areas, where they are also becoming more influential in the workplace.

250

145

7.2 Physical Geography and Human Adaptations

The physical geography of South and East Asia may be described broadly as consisting of three concentric arcs, or crescents, of land: an inner western arc of high mountains, plateaus, and basins; a middle arc of lower mountains, hill lands, river plains, and basins; and an outer eastern arc of islands and seas (see Figure 7.1).

The inner arc includes the world's highest mountain ranges, interspersed with plateaus and basins. In the south, the great wall of the Himalaya, Karakoram, and Hindu Kush Mountains overlooks the north of the Indian subcontinent. In the north, the Altai, Tien Shan, Pamir, and other towering ranges separate this region of Asia from the countries of central Asia. Between these mountain walls lie the sparsely inhabited Tibetan Plateau, at over 15,000 feet (c. 4,500 m) in average elevation, and the dry, thinly populated basins and plateaus of Xinjiang (Sinkiang) and Mongolia.

The middle arc, between the western inner highland and the sea, is occupied by river floodplains and deltas bordered and separated by hills and low mountains. The major features of this area are the immense alluvial plain of northern India, built up through ages of meandering and deposition by the Indus, Ganges, and Brahmaputra Rivers; the hilly uplands of peninsular India, geologically an ancient plateau; the plains of the Irrawaddy, Chao Praya (Menam), Mekong, and Red Rivers in the Indochinese Peninsula of Southeast Asia, together with bordering hills and mountains; the uplands and densely settled small alluvial plains of southern China; the broad alluvial plains along the middle and lower Chang Jiang (Yangtze River) in central China and the mountain-girded Red Basin on the upper Chang Jiang; the large delta plain of the Huang He (Yellow River) and its tributaries in northern China, backed by loess-covered hilly uplands; and the broad central plain of northeastern China (Manchuria), almost enclosed by mountains.

The outer arc is an offshore fringe of thousands of islands, mostly grouped in great **archipelagoes** (clusters of islands) bordering the mainland. On these islands, high interior mountains (including volcanoes) are flanked by coastal plains, where most of the people live. Most of the islands are in three major archipelagoes: the East Indies, the Philippines, and Japan. Sri Lanka, Taiwan, and Hainan are large, densely populated islands outside these archipelagoes. Between the archipelagoes and the mainland lie the East and South China seas and, to the north, the Sea of Japan (known in Korea as the East Sea). At the southwest, the Indian peninsula projects southward between two immense arms of the Indian Ocean: the Bay of Bengal and the Arabian Sea.

Climate and Vegetation

South and East Asia are characterized generally by a warm, well-watered climate, but there are nine distinct types of climate in the region: tropical rain forest, tropical savanna, humid subtropical, warm humid continental, cold humid continental, desert, steppe, subarctic, and undifferentiated highland (•Figure 7.4, left; see also Figure 2.5). These are generally associated with predictable patterns of vegetation, which have been heavily modified by millennia of human use (Figure 7.4, right; see also Figure 3.4). Remember that discussion of biomes and vegetation considers what ought to be growing in the absence of people. From this baseline, we will look at what people have done to the living landscape (see Perspectives from the Field, page 218).

The tropical rain forest climate zone and vegetation types are typically found within 5° to 10° of the equator and are characteristic of most parts of the East Indies, the Philippines, and the Malay Peninsula. Precipitation is spread throughout the year so that each month has considerable rain. The amount of precipitation is at least 30 to 40 inches (c. 75 to 100 cm) and often 100 inches (c. 250 cm). Average temperatures vary only slightly from month to month; Singapore, for example, has a difference of only 3°F between the warmest and coolest months. High temperatures prevail year-round, although some relief is afforded by a drop of 10°F to 25°F (6°C to 14°C) at

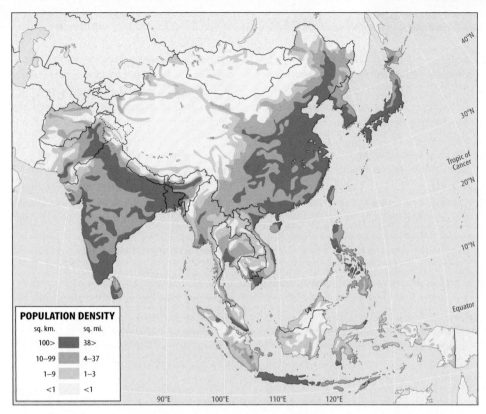

• **Figure 7.3** Population distribution (top) and population cartogram (bottom) of South and East Asia.

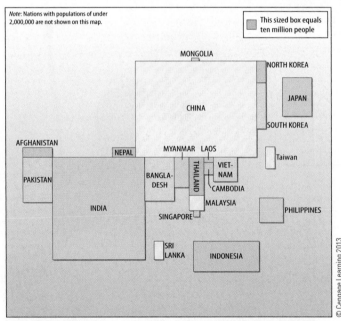

Note: Nations with populations of under 2,000,000 are not shown on this map.

This sized box equals ten million people

© Cengage Learning 2013

night, and sea breezes refresh coastal areas. Winter in climates like this, it can be said, "comes only at night."

The tropical savanna climate, like the tropical rain forest climate, is characterized by high temperatures year-round but is generally found in areas farther from the equator, and the average temperatures vary more from month to month. The savanna has a better-defined dry season, lasting in some areas as long as six or eight months each year, creating problems for agriculture. Tropical savanna climates occur in southern and

central India, in most of the Indochinese Peninsula, and on eastern Java and the smaller islands to the east.

In Asia, the characteristic natural vegetation associated with this climate is a deciduous forest of trees smaller than those of the tropical rain forest. Tall, coarse grasses such as those found on African and Latin American savannas grow only in limited areas.

The humid subtropical climate zone prevails in southern China, the southern half of Japan, much of northern India, and a number of other countries. It is characterized by warm to hot summers, mild or cool winters with some frost, and a frost-free season lasting 200 days or longer. The annual precipitation of 30 to 50 inches (c. 75 to 125 cm) or more is well distributed throughout the year, although monsoonal trends produce a dry season in some areas. The natural vegetation is a mixture of evergreen hardwoods, deciduous hardwoods, and conifers.

The humid continental climates characterize the northern part of eastern China, most of Korea, and northern Japan. They are marked by warm to hot summers, cold winters with snow, a frost-free period of 100 to 200 days, and less precipitation than the humid subtropical climate. Most areas experience a dry season in winter. The main natural vegetation is a mixture of broadleaf deciduous trees and conifers, although prairie grasses probably formed the original cover in parts of northern China. Extreme northeastern China and northern Mongolia have a subarctic climate associated with coniferous trees.

Steppe and desert climates are found in Xinjiang and Mongolia and in parts of western India, Pakistan, and Afghanistan. Due to limitations on agriculture, along with very high mountains, they are the region's least inhabited climate zones. The undifferentiated

GEOGRAPHIC SPOTLIGHT | China's One-Child Policy

China's population of 1.3 billion represents nearly one-fifth of the world's people and is increasing by 6.7 million each year. Population is a very serious matter for a less developed country whose area of about 3.7 million square miles (9.4 million sq km) is only slightly smaller than the United States but whose inhabitants outnumber the U.S. population by about five to one. But a glimpse at China's age structure diagram, with its wide middle ages but tapering youth, reveals that China is one of the world's great success stories in terms of population management (•**Figure 7.A**).

The drive for smaller families was motivated by the fact that China's population was surging while per capita food output was increasing at only a very modest rate. In 1976, China's Communist regime instituted one of the most stringent and most controversial programs of birth control ever attempted. It culminated in the **one-child campaign,** which aims to limit the number of children per married couple to one (there are some exceptions, for example, for ethnic minorities) (•**Figure 7.B**). The government tracks individual family birthing patterns—especially in the cities—and maintains surveillance through local authorities. It dispenses free birth control devices, free sterilization operations, free hospital care in delivery, free medical care for the child, free education for the child, an extra month's salary each year for parents who comply, and other favors and preferences to induce compliance with the one-child-per-family norm.

Those who violate the norm are subjected to constant social and political pressure, denial of the privileges accorded one-child families, pay cuts, and fines. A woman who becomes pregnant after having had one child is pressured to have a free state-supplied abortion, with a paid vacation provided. China's government hopes to continue to decrease childbearing so that population will stabilize before reaching the projection of 1.5 billion by 2025. The number may be higher because of widespread disregard of the "one couple, one child" policy. Reasons for noncompliance vary. In some cases, parents feel wealthy enough to pay the fines of about $6,200 or more per extra child. Many rural dwellers think their extra children will escape detection by authorities, who are most effective in the cities. Anticipating their years as elders, some couples want the "safety net" of additional children—especially boys. If a young urban couple has just one child and it is a girl, it is likely that she will marry and become, by Chinese tradition, such a part of the husband's family that she may not be available to care for her parents as they get older. China has traditionally used the son's family as the "old folks' home." On the other hand, rising property prices in cities may affect this preference for boys: parents have to provide a married son with an apartment, and that is becoming prohibitively expensive in cities.

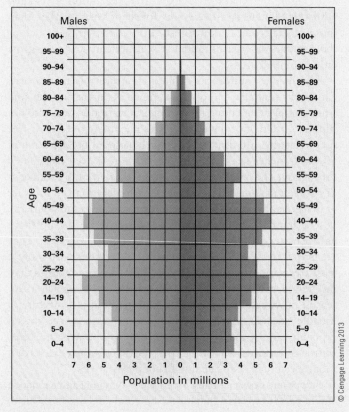

Males | Females

Age

Population in millions

© Cengage Learning 2013

• **Figure 7.A** The age structure diagram of China (2011) reveals the country's success in slowing population growth.

Joe Hobbs

• **Figure 7.B** Strong disincentives mean one child is the norm for Chinese couples.

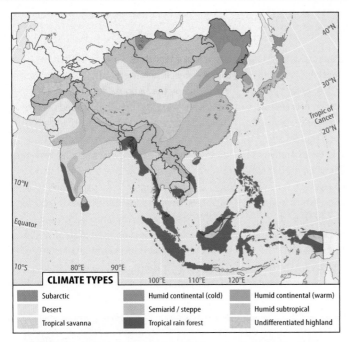 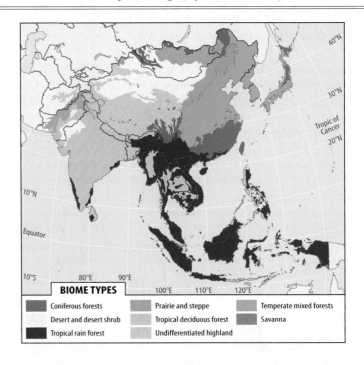

• **Figure 7.4** Climates (left) and biomes (right) of South and East Asia.

Sources: (left) Based on Rand McNally's Classroom Atlas, 2003. (right) Based on World Wildlife Fund ecoregions data, 1999.

highland climate is most extensive in the Tibetan Highlands and adjoining mountain areas. It is made up of a broad range of montane microclimates that limit plant growth and vary from place to place according to elevation, solar exposure, and latitude.

The Monsoons

The **monsoons** are the prevailing sea-to-land and land-to-sea winds that are the dominant climatic concern for people living in this world region (•**Figure 7.5**). They play a significant role in both wet and dry environments and are especially influential in the coastal plains and lowlands of South Asia, the peninsula and islands of Southeast Asia, and the eastern third of China. The ways in which the sea and land absorb **insolation** (heat from the sun) are different, causing the instability in air masses that creates monsoonal winds. If coastal waters and the adjacent coastline receive about equal amounts of warmth from the sun, the land becomes warm or hot much more quickly than the seawater. As a result, air over the land begins to become unstable and to rise. This ascending air creates a low-pressure attraction for the air masses over the water, and marine winds begin to blow toward the land. In Asia, this is the **summer monsoon**, characterized by high humidity, moist air, and generally predictable rains. Because of the moisture carried by such air masses, these wind shifts are the sources of major rainfall in the late spring, summer, and early fall seasons. Agriculture and patterns of human activity are keyed to these incoming rains.

If there is even a little bit of elevation on the landscape—hills or mountain flanks—the moist air is driven higher, where it cools and releases even more rain in the pattern known as *orographic* precipitation. If there is a lot of elevation—as there is with the

giant mountain walls of the Himalaya and Karakoram—the orographic effect of wringing out the moisture-laden atmosphere is impressive, with tremendous rainfall and snowfall events.

In the **winter monsoon**, the land loses its relative warmth while the sea and coastal waters stay warm longer. As a result, the wind shifts and air masses begin to flow from the inland areas of Asia toward the sea. However, because there is hardly any moisture in the source areas over land for the winter

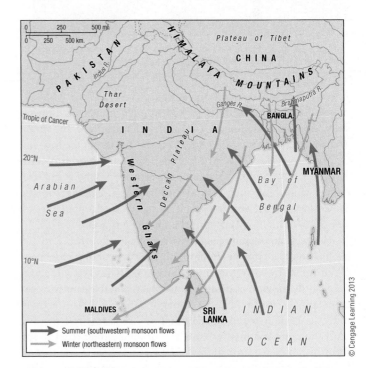

• **Figure 7.5** This map illustrates how the monsoons work, with prevailing winds blowing from the sea (wet summer monsoons) and toward the sea (dry winter monsoons).

The pace of landscape change in this region in recent decades has been truly breathtaking. We could perhaps say the same is true in all world regions except Antarctica, but the change is particularly striking in Asia because of the deforestation described on page 251.

Much of the geographer's focus is on transformation of natural or rural environments, but we are also interested in urban environmental change. When we look at Asian cities, we often see both nature and culture transformed. Nowhere else in the world is nature so prominent in urban life. Next time you watch a television reporter in Karachi or New Delhi, listen for the hooded crows.

As a child, I lived in the desert of Saudi Arabia. In 1968, my parents took my brother, sister, and me across Asia to Bangkok. In this verdant paradise of a city, it was love at first sight for this animal-loving youngster. Behind our hotel sprawled a vast lawn literally seething with reptiles and amphibians (many flew home to Saudi Arabia with me; others did not make it when my dad saw the sides of my carry-on bag pulsating). This was a green, wet city: canals ran every which way, and rain forest and fruit trees grew wherever they had an opportunity. It seemed like a forest with an enchanting city tucked into it (•**Figure 7.C**).

In 2006, I returned to Bangkok for the first time. How was it possible this was the same city? I looked for trees—any trees. And the canals—any canals. There were some remaining, including the famous "floating market," but most were gone, cut down and paved over in Thailand's rapid modernization, urbanization, and economic growth. Bangkok is often gridlocked, with some of the worse traffic anywhere on Earth (•**Figure 7.D**). One morning I had a 9:00 A.M. appointment across town and took the locals' advice to leave the hotel at 6:00 for the 30-minute drive. Any later and it would have been a two- or even three-hour drive. In some of those long commutes, I had plenty of time to reflect: The verdant paradise of Bangkok was just a child's memory.

• **Figure 7.C** Bangkok, 1968

• **Figure 7.D** Bangkok, today

monsoon, little moisture is picked up, and much less rain results. A monsoon climate, overall, is characterized by spring and summer precipitation and a long dry season in the low-sun (winter) cycle. Japan, however, experiences an unusual pattern of winter precipitation related to the monsoon wind shifts. The winter winds blow out of China and East Asia, sweep across the Sea of Japan (East Sea), and pick up moisture, dropping heavy, wet snow on the west coast of the Japanese islands.

The wet summer monsoon is of critical importance to the livelihoods of a large portion of humanity, particularly the people of India. By early May in India, the land has typically been parched by months of drought, and the rising temperatures become almost insufferable. No thought can be given to farming until the rains come. Anticipation and discomfort rise steadily until the monsoon finally "breaks," bringing great relief in the form of much needed rain and lower temperatures (**Figure 7.6**). Sometimes the onset of the rain comes with great ferocity. India and Bangladesh have experienced particularly heavy early monsoonal rains for the past three decades, with significant loss of

• **Figure 7.6** Torrential monsoon rains are regular and usually welcome features of land and life in much of the region. This is a flooded road in south central Sri Lanka.

Many Asian farmers, particularly in tropical rain forest regions, practice **shifting cultivation** (also known as **swidden cultivation** or **slash-and-burn cultivation**), or land rotation between crop and fallow years (●**Figure 7.E**). The procedure begins with people clearing forest cover and then burning the dried vegetation during the dry season. Ash from the burn provides short-lived fertility to the soil in which farmers plant their subsistence crops. Most tropical soils lose their natural fertility quickly when they are cropped, and after two or three harvests must be rested for several years (often 10 to 15 years) before they will again produce a crop. During this **fallow** period, the land reverts to wild vegetation.

The system of shifting agriculture is widespread in the tropics. Although it is not very productive, it does provide a bare living for people too poor to afford fertilizer or farm machinery. Traditionally, this system has minimized soil erosion because most of the land is not in cultivation at any given time. Unfortunately, there has been a recent widespread trend for farmers to reduce the length of the fallow period in the cycle of shifting cultivation or to abandon it altogether. This is a direct result of a surging growth in population: with more mouths to feed, in many places there is simply not enough land to provide the "luxury" of letting it lie fallow.

The tragic consequence is that short-term overuse of the land is eroding the soil. On the **lateritic soils** characteristic of many tropical regions, excessive exposure of the land to the sun's ultraviolet radiation and to repeated wetting and drying turns the ground to a bricklike, iron-rich substance called **laterite** or **plinthite**, which is essentially impossible to cultivate. In fact, this transformed soil is so tough that people used it in the construction of some of Asia's greatest architecture, including the temples of Angkor Wat in Cambodia (●**Figure 7.F**). When the best available lands are exhausted, farmers are often marginalized to places unsuitable for farming, such as semiarid lands or slopes that are too steep to till without causing disastrous erosion. Such dilemmas of land use are typical of the world's tropics from Indonesia to West Africa and Amazonia.

47

345,
384

Joe Hobbs

● **Figure 7.E** Shifting cultivation (swidden or slash-and-burn cultivation) is widespread in the world's tropical regions and can be sustained as long as the land is given sufficient time to recover. Part of the cycle is harvesting crop residue (here, cornstalks) to use as fuel or fodder. This Hmong woman's backbreaking work is part of the greater fuelwood crisis scenario described on page 48.

Joe Hobbs

● **Figure 7.F** Laterite or plinthite, created by the transformation of tropical soils, is hard enough to provide foundations for buildings. This is one of the outstanding examples: the twelfth-century temple complex in the Cambodia site of Angkor Wat.

life. Pakistan has had fewer events of this kind, but unusually heavy, sustained monsoon rains in 2010 caused catastrophic flooding all along the Indus floodplain. Inevitably, some scientific fingers point to global climate change as the culprit.

Agricultural Adaptations

The wet and warm climates associated with the monsoons might be expected to offer excellent opportunities for agriculture. However, the high temperatures and heavy rains promote rapid leaching of mineral nutrients and decomposition of organic matter, so many soils are infertile. The lush vegetation cover in these low-latitude areas is deceiving because most of the system's nutrient matter is in the vegetation itself rather than in the underlying soil. When Asian farmers clear the tree cover, they often find that the local soils will not support more than one or two poor harvests (see Insights, this page). Agricultural and other land uses in South and East Asia are depicted in ●**Figure 7.7**. 384

The Importance of Rice

Agriculture in South and East Asia—with the principal exception of Japan—is characterized by the steady input of arduous manual labor. Known as **intensive subsistence agriculture**, it is built around the growing of cereals, especially rice, which is the premier staple of South and East Asia and the crop of choice in areas with adequate rainfall or where irrigation waters are available (see the areas of rice subsistence in Figure 7.7).

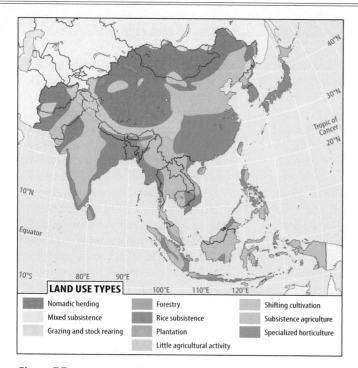

• **Figure 7.7** Land use in Asia.

Source: The Oxford School Atlas edited by Patrick Wiegand (OUP, 1997), copyright © Oxford University Press. Used by permission of Oxford University Press.

• **Figure 7.8** Harvesting wet rice from an irrigated terrace in Ha Giang, Vietnam

Farmers use organic fertilizers from both their animals and their own latrines and irrigate to the extent that they can or need to. Where natural conditions are not suitable for irrigated rice, grains such as wheat, barley, soybeans, millet, sorghum, or corn (maize) are raised. Where possible, farmers grow cash crops for local markets. Large-scale, commercial plantation agriculture is common in Southeast Asia, the subregion of South and East Asia in which cash cropping has had the strongest role.

Shifting cultivation represents a precarious adaptation to local soil and climate conditions, capable of sustaining only small populations for brief periods in particular locales. At the other end of the spectrum of human adaptation to Asian lands is **wet rice cultivation**, capable of producing two or three crops per year and sustaining large populations over long periods of time. Wet rice agriculture takes place both in lowland floodplains and on carefully constructed upland **terraces** requiring enormous inputs of human labor (the term **paddy**, also spelled **padi**, may refer to the rice itself or the field in which it is grown). Terraces have been built almost everywhere in Asia where there is a combination of sloping land, available irrigation water, and productive soil (•**Figure 7.8**). The dramatic stair-step terraces of the island of Luzon in the Philippines are the most famous, but dramatic cultural landscapes of this type can be found all across the region.

Terraces have the important function of preventing or dramatically slowing soil erosion, as the terrace walls allow heavy rainfall to run off without carrying topsoil away. Doing what the Chinese call "teaching water," people use canals and pipes to direct irrigation water from a source to the terrace, which acts like a kind of dam to hold the water in place. Wet rice is actually immersed in standing water for extended periods, creating an environment unfavorable to many crop pests. The productivity of wet rice agriculture is sometimes exceptionally high, supporting as many as 500 people per square mile (200/sq km) on the volcanic soils of Java and Bali in Indonesia. This extraordinary output is a contributing factor to the high densities and overall populations of some of the Asian countries and regions, including Java. One anthropologist has observed that most of Asia's rice is eaten within walking distance of where it is grown.[1]

Are There Correlations between Agriculture and Culture?

Anthropologists have observed some unique cultural adaptations to opportunities for terraced wet rice. Group welfare takes precedence over individual interests, for example. Individual farmers cannot cultivate at will but must work with the community to ensure that crops are staggered over time. This allows water to be shared fairly and efficiently, and gives some fields the chance to dry out while others are irrigated. This agricultural system also prevents the use of tractors and other mechanical technology. Over much of Asia today, just as in ancient times, wet rice cultivation is the product of intensive work by many human hands, mostly women's. They spend much of their lives in very difficult labors, bent over to plant and harvest in challenging heat and humidity.

Geographers and other social scientists have also observed a remarkable set of cultural adaptations to harsh environments on high elevation steep slopes in the region, especially in Southeast Asia; see the Regional Landscapes feature "Zomia."

REGIONAL LANDSCAPES | Zomia

Stretching from Vietnam in the east to India in the west, at elevations over 1,000 feet (300 m) is a region known as **Zomia** (•**Figure 7.G**). Evocative of imaginary Asian realms like Shangri-La, the name *Zomia* comes from a word meaning "highlander" in Burma, Bangladesh, and India. Although first used in the scientific literature by a Dutch scholar, Zomia was popularized by Yale University professor James C. Scott in a groundbreaking book entitled *The Art of Not Being Governed*. Scott's thesis is that in China, Vietnam, Cambodia, Laos, Thailand, and Burma, upland peoples including the Hmong, Akha, Karen, Lahu, Mien, and Wa have found something perhaps all of us yearn for: an escape from tax collection and other unpleasantries of government rule. They have done so by choosing to live in the highest, most difficult, and least appealing lands of Southeast Asia—cold, steep, infertile places where no one else wants to live. They wish to be left alone, ungoverned, and undisturbed.

Scott's thesis is controversial, but after reading his book it is impossible to visit these places, as I did on a journey along the China-Vietnam frontier, without thinking, "I am in Zomia!" In the foothills below Zomia, the ethnic Tay people enjoy a prosperous livelihood growing wet rice (they are depicted in Figure 7.8). Higher in the mountains, ethnic Yao and Nung are clearly poorer than the Tay, growing dry rice and maize (corn) on slopes that they must maintain carefully with terraces to ward off erosion. Finally, on the steepest and most forbidding slopes near the mountaintops are the ethnic Hmong. The Hmong are the poorest of these ethnic groups, and it is easy to see why: it is all but impossible to productively farm these lands. The Hmong must resort to slash-and-burn farming, cutting the forest and planting maize without the benefit of terracing, which is impossible because the slopes are too steep. Without terraces, soils wash away and leave a desert-like environment in their wake. The Hmong move on and clear more forest. The result is breathtaking: an impoverished, lunar landscape where these people scour the hillsides for the last patches of forest to clear and plant (•**Figure 7.H**).

At first glance, it appears that the Hmong are the ultimate illustration of marginalization, where more powerful groups push weaker ones to the least hospitable lands. If Scott is correct, however, the Hmong are marginalized by choice: they prefer to live in Zomia because no one else—including governments and other powerful entities—would want to live in and control such daunting places.

Many Hmong have resettled from Southeast Asia to the United States, especially in rural California and urban Minnesota and Wisconsin. Here too the Hmong have largely succeeded in being left alone, at least culturally. Their American neighbors are often baffled why Hmong have not assimilated as other immigrants have. There is an exceptional account of the culture clash between the traditional Hmong and modern America—especially in the realm of medicine. (The book's focus is on a Hmong girl afflicted with epilepsy, which the Hmong attribute to the work of evil spirits. They struggle to keep her out of the care of doctors with their powerful anti-seizure medications.) This is Anne Fadiman's *The Spirit Catches You and You Fall Down*.

47

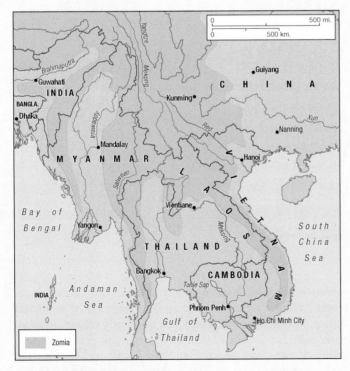

• **Figure 7.G** Map of Zomia.
Source: Based on James C. Scott, 2009, The Art of Not Being Governed: An Anarchist History of Upland Southeast Asia. New Haven: Yale University Press, p. 17.

• **Figure 7.H** Landscapes of the Hmong up to an elevation of about 6600 feet (2000 m) in the Ha Giang province, northern Vietnam near the Chinese border.

In some cases, population growth has outpaced the ability of even the most productive wet rice farming to feed the people, creating the classic scenario of "people overpopulation." In Nepal, for example, high rates of population growth on limited areas of moderately sloped, terraceable land have been a push factor behind migration. After the 1950s, when malaria began to be eradicated in the lowland Terai region, millions of Nepalis migrated there, cleared the forests, and planted crops. For decades, the Terai served as a virtual "safety valve" for overpopulation in Nepal. Now that the Terai itself is mostly settled, there is nowhere for landless Nepalis to go but up. They clear and cultivate increasingly steep lands on which it is impossible to build terraces. Without terraces in place to hold the topsoil, heavy monsoon rains cause relentless erosion of their farmlands. Locally, the consequences are that the eroded plots cannot be cultivated again, and landslides occur downslope, often causing loss of life.

Much farther away, in the lowlands of India and Bangladesh, the increased sediment load contributed by erosion in Nepal causes rivers to swell out of their banks and bring extensive flooding—again, with losses of crops and lives. This sequence of events, beginning with overpopulation in Nepal and ending with flooding in Bangladesh, has been described as the **theory of Himalayan environmental degradation**.[2]

Where Asians Live

Many of the world's largest cities are in South and East Asia, including the greatest of all, Tokyo, with a metropolitan population of 38 million. About 1.6 billion people, or 42 percent of the region's total, live in urban settlements (•**Table 7.2**).

In Villages

The main unit of Asian settlement is the village. Most of Asia's roughly 2 million villages are essentially groups of farmers' homes, although some villages are home to other occupational groups, such as miners or fisherfolk. Clusters of houses bunched tightly together are typical, and inexpensive and simple structures—often made of local clays and other building materials—are characteristic. Indoor plumbing continues to be the exception in village homes, but electricity lines have been extending steadily throughout Asia. So have cell phones and satellite dishes; it is not unusual today to see what appear to be very "traditional" and even quite poor Asians enjoying some of the latest innovations in global communications (•**Figure 7.9**). In South and East Asia, as elsewhere in the developing world, many people are making the jump directly from nineteenth- to twenty-first-century technology. Wireless technology is especially appropriate where landlines and other infrastructure are difficult and expensive to install and where bureaucracies often slow down applications for access to utilities.

The original site selection for Asian villages was usually closely related to natural conditions and to perceived spiritual circumstances as well (see Geography of Sacred Space, page 224.). With consideration always given to the possibility of flooding

Table 7.2	South and East Asia: Metropolitan Populations Data	
1	Tokyo, Japan	37.7
2	Seoul, South Korea	22.7
3	Mumbai, India	21.9
4	Manila, Philippines	20.6
5	Shanghai, China	19.9
6	Jakarta, Indonesia	19.2
7	Delhi, India	18.9
8	Osaka, Japan	17.4
9	Kolkata, India	15.6
10	Dhaka, Bangladesh	14.3
11	Karachi, Pakistan	13.2
12	Beijing, China	13.2
13	Bangkok, Thailand	10.1
14	Hong Kong, China	9.5
15	Nagoya, Japan	8.9
16	Taipei, Taiwan	8.5
17	Kuala Lumpur, Malaysia	8.1
18	Chongqing, China	7.8
19	Chennai, India	7.4
20	Lahore, Pakistan	7.1
21	Bandung, Indonesia	6.7
22	Shenyang, China	6.6
23	Bangalore, India	6.5
24	Tianjin, China	6.4
25	Hyderabad, India	6.4
26	Singapore	6
27	Gunagzhou, China	5.7
28	Pune, India	5.5
29	Ahmedabad, India	5.4
30	Xi'an, China	5.3
31	Ho Chi Minh City, Vietnam	5.3
32	Hangzhou, China	5.1
33	Yangon, Myanmar	5

Population in millions.

© Cengage Learning 2013

monsoon rains, lowland villages tend to be situated on natural **levees** (raised riverbanks built up by deposition of sediments during floods), dikes, or raised mounds. Early villages in Indonesia were often built in defensible mountain sites. European colonists sometimes rejected traditional settlement patterns to serve their own interests. Dutch colonial administrators in Indonesia, for example, required that villages be built along main roads and trails in the lowlands, making it easier to exercise

• **Figure 7.9** Across South and East Asia, satellite dishes sprout incongruously from traditional dwellings.

Joe Hobbs

control, collect taxes, and draft soldiers or laborers for road-
45 work and other projects.

The most infamous example of foreign intervention in traditional rural settlements came during the Vietnam War in the
258 1960s. The American military wanted to make it more difficult
for insurgent Viet Cong forces to infiltrate South Vietnamese
villages, where they might indoctrinate and blend in with civilians in their fight against the Americans. The military therefore
forced millions of farming families to relocate into fortified communities known as **strategic hamlets**. The project was an utter
failure. The insurgents managed to infiltrate these settlements.
More important, the Americans violated one of the strongest
precepts of Vietnamese culture: to live close to the graves of
228 their ancestors. The U.S. military lost the "hearts and minds"
of Vietnamese civilians in the strategic hamlets program. Today,
U.S. intelligence agencies are still trying to learn from the lessons
of Vietnam by trying to understand the culture of the enemy
(for example, the Pashtun of Afghanistan's Taliban) (see A Closer
Look, page 244.). In a controversial program known previously as
"human terrain" and now remarkably as "human geography",
social scientists with expertise in local cultures are imbedded
with troops. Early on, some of these scholars met gruesome ends
when the enemy captured them and discovered their roles. But
now the program is more effective and less risky.

In Cities

The majority of people in South and East Asia are rural, but
the region's future, in terms of demographic weight, economic
power, and cultural change, is focused on the cities (see the
feature on megacities on page 226, and Table 7.2). Some of the countries are already highly urbanized, notably Japan (86%), South
Korea (82%), and Taiwan (78%) and entities like Singapore
and Brunei that are effectively city-states. The layouts and
architectures of the cities span a continuum from the ancient,
walled fortress city to the modern Western-style metropolis, and many, like Beijing in China and Lahore in Pakistan,
accommodate both. Unfortunately, in the push for urban

modernization, many traditional urban structures and communities uniquely adapted to local environments have been
sacrificed. Urban conservations are struggling to protect some
of the remaining old buildings including those of Dhaka in
Bangladesh and of Shanghai in China.

A striking pattern across the region is the steady stream of
rural-to-urban migration, with people motivated by both push
and pull factors to leave the relative poverty of the countryside in
search of employment or even prosperity in the city. Many Asian 56
cities are hard-pressed to provide accommodation and services to
their swelling populations, and in authoritarian countries (notably China), restrictions on internal migrations are imposed.

7.3 Cultural and Historical Geographies

South and East Asia have been home to some of the most important cultural developments of humankind in landscape
transformation, settlement patterns, religion, art, and political systems. From Korea (not Gutenberg's Germany, as is
widely thought) came the first movable printing type. From
China came gunpowder, paper, silk, and china (porcelain).
From India came the great faiths of Hinduism and Buddhism.
Handmade textiles, artware, bone and leather and paper products, and precious metals and gems have flowed from Asia into
global trade for millennia. Spices, foodstuffs, and exotic plants
and products from the East Indies have been major commodities in Asian-European trade and commerce for more than five
centuries. Some of the world's most important domesticated
plants and animals originated in this region, including rice,
cabbage, chickens, water buffalo, zebu cattle, and pigs.

These countries, and their cultures, histories and geographies, are a source of endless fascination. It is for example difficult to many people, the present author included, to dispute
Ramachandra Guha's claim that "India is the most interesting
country in the world."[3]

Ethnic and Linguistic Patterns

As would be expected of such a vast region, in many places fragmented by islands and mountain barriers, the ethnic and linguistic composition is rich and complex (•Figure 7.10). Central and
South Asia gave birth to the Indo-European language family,
which contains tongues as diverse as English and Hindi. *Hindi*
and *Urdu*, which are major languages in India and Pakistan, respectively, are in the **Indo-Iranian subfamily** that includes such
other major South Asia languages as *Pashto* (closely related to Persian, and spoken in western Pakistan and Afghanistan; Afghanistan's official language is *Dari*, or Afghan Persian). Indo-Iranian
tongues also include *Sanskrit* (which is extinct as a vernacular
language but is still used in Hindu ritual), *Sinhalese* (the majority language of Sri Lanka), *Punjabi* (spoken on both sides of the
northern frontier between Pakistan and India), Nepal's national
language of *Nepali, Bengali* (the language of Bangladesh), and
Romany (the tongue of the Roma, or Gypsies, of Europe).

GEOGRAPHY OF SACRED SPACE | **The Korean Village**

In North America, people are accustomed to communities laid out on a perfect grid of streets intersecting at right angles, with their homes and businesses symmetrically fronting those streets. But in large parts of Asia, settlement and residence considerations are far more varied, especially because they are tied to spirituality, aesthetics, and topography. Villages, and the homes in particular, tend to be arranged for maximum harmony with the natural and spiritual environments—a practice known as **geomancy**—and constitute a special realm that may be considered sacred space. The traditional Korean village illustrates these unique Asian qualities.[4]

From ancient times, Koreans developed ways of determining the most auspicious village sites. The principles of *feng shui* and of Confucianism played strong roles in how they adapted settlements and homes to geographic conditions. **Feng shui** (pronounced "fung-*shway*") is a traditional theory for selecting favorable sites for buildings, homes, and cities and for guiding many other urban and architectural planning decisions.[†] It is essentially an ordering device for environmental planning and for daily living. *Feng shui* (the words mean "wind" and "water") attempts to balance **yin** and **yang** to achieve harmony. *Yin* (conceived of as negative, dark, and feminine) and *yang* (representing the positive, bright, and masculine) stand for all the dualistic, opposing forces that may be reconciled harmoniously, including female and male, night and day, moon and sun, cold and hot, and right and left. With cultural diffusion and growing interest in Asian lifestyles, *feng shui* principles are spreading in American suburban home architecture, as in houses built to conceal overhead beams and sharp protrusions (which are disharmonious) and provided with windows opening to the east (a harmonious situation).

The theory of *feng shui* and the Confucian view of nature have some common characteristics that affect site, settlement, and residence designs. Confucianism reflects the sociopolitical philosophy of the Chinese philosopher known in the West as Confucius (see page 231). One of the main Confucian principles is that architecture and the lived environment should be in harmony with nature. This can be seen in the Confucian institutes for higher learning, called *seowon,* which in Korea peaked in the Joseon Dynasty, around 1400 c.e. Each *seowon* had a lecture hall where Confucian scholars gathered to learn philosophy. The buildings were laid out

in such a way that from the inside of the complex, the scholars could see distant landscapes. The scholars' minds and spirits were enhanced as they gazed at the mountains and contemplated nature's grandeur. They believed that the earth's power was an invisible force that would help them become men of dignity and eminence.

The principles of *feng shui* and Confucianism are also apparent in the design of Korean villages. Each village is laid out to achieve harmony with the mountains—an important consideration in a land that is 70 percent mountainous. Rising behind many villages is a mountain, called *jinsan,* which is a spiritual focus for the villagers. *Jinsan* serves as the village symbol and guardian. Typically, a small stream flows in front of the village, and beyond it, there is a flat expanse of farmland. Beyond those fields is another mountain, called *ansan,* facing the village (●Figure 7.1).

These geographic conditions are economically advantageous to the community, providing it with productive fields and a good supply of water. But just as important, this situation is psychologically, spiritually, and symbolically beneficial to the villagers. The natural features are meant to embrace the village, as Koreans say a mother would hold a child. Korean historical records describe the traditional village as a place of comfort and serenity that safeguards against negative natural forces.

Symbolic guardians are associated with the topography. Ideally, there is the Green Dragon to the east, the White Tiger to the west, the Black Tortoise to the south, and the Red Bird to the north. These guardians, named for figures in Chinese astronomy, could have varied manifestations. The Green Dragon, for example, could be

a hill or a river to the east of the village, while the White Tiger could be a mountain or a road on the west side. One enters the village through a symbolic portal. For many villages, a pair of "spirit posts" stands at the foot of a nearby mountain. The posts mark the preliminary boundary to the outside, and they ward off evil spirits.

Not only is the village site selection process different from that in the West, but the settlement pattern is uniquely Asian too. Unlike in Western settlements, no more than two houses should be built side by side, they should not stand parallel to each other, and their gates should never face each other. The side street should curve naturally between the home lots. Overall, then, Koreans choose to emphasize the surrounding topography rather than geometric order in the construction of their communities and homes.

The traditional Korean home is adapted to the natural environment, embracing its surroundings while offering protection, comfort, and order. As in the *seowon,* the layout brings outdoor scenery into the house itself. Homes are also adapted to the environment in their construction materials. A thatched roof and earthen walls effectively block heat in the summer and insulate in winter, a perfect combination for Korea's sweltering summers and raw winters. The mulberry paper covering lattice windows diffuses sunlight, providing a pleasant natural light while absorbing sound and allowing some ventilation. The home's eaves are built so that sunlight penetrates deep into the house during the cold weather but never during summer—a rather recent innovation in the West, where it is known as "passive solar" design.

● **Figure 7.1** The Korean village of Yangdong is nestled between mountains that are separated by a river and lush rice fields.

Joe Hobbs

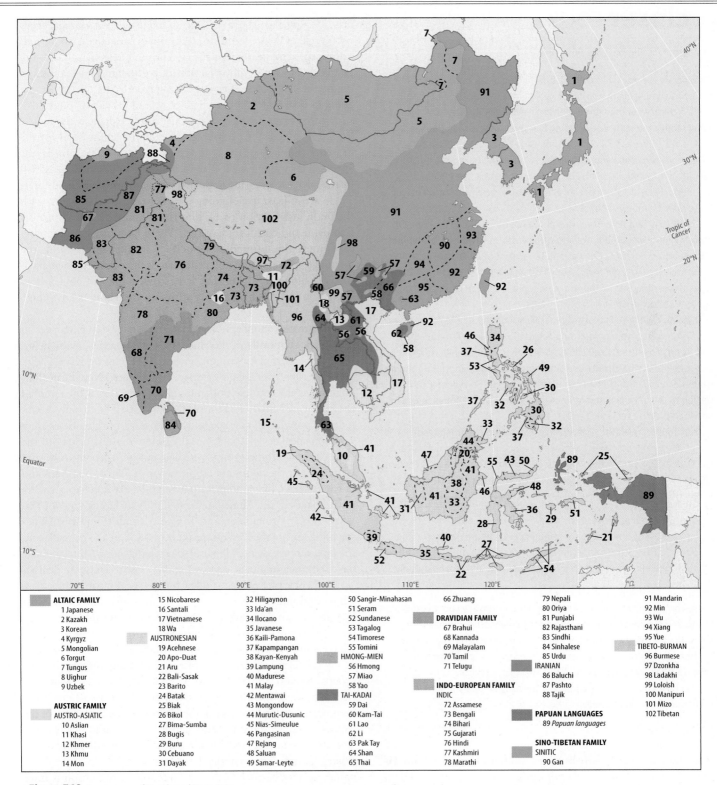

• **Figure 7.10** Languages of South and East Asia.

Source: Based on Bernard Comrie, et al., *The Atlas of Languages, Revised Edition* (New York: Facts on File, 2003).

ALTAIC FAMILY
1 Japanese
2 Kazakh
3 Korean
4 Kyrgyz
5 Mongolian
6 Torgut
7 Tungus
8 Uighur
9 Uzbek

AUSTRIC FAMILY
AUSTRO-ASIATIC
10 Aslian
11 Khasi
12 Khmer
13 Khmu
14 Mon

15 Nicobarese
16 Santali
17 Vietnamese
18 Wa
AUSTRONESIAN
19 Acehnese
20 Apo-Duat
21 Aru
22 Bali-Sasak
23 Barito
24 Batak
25 Biak
26 Bikol
27 Bima-Sumba
28 Bugis
29 Buru
30 Cebuano
31 Dayak

32 Hiligaynon
33 Ida'an
34 Ilocano
35 Javanese
36 Kaili-Pamona
37 Kapampangan
38 Kayan-Kenyah
39 Lampung
40 Madurese
41 Malay
42 Mentawai
43 Mongondow
44 Murutic-Dusunic
45 Nias-Simeulue
46 Pangasinan
47 Rejang
48 Saluan
49 Samar-Leyte

50 Sangir-Minahasan
51 Seram
52 Sundanese
53 Tomini
54 Timorese
55 Tomini
HMONG-MIEN
56 Hmong
57 Miao
58 Yao
TAI-KADAI
59 Dai
60 Kam-Tai
61 Lao
62 Li
63 Pak Tay
64 Shan
65 Thai
66 Zhuang

DRAVIDIAN FAMILY
67 Brahui
68 Kannada
69 Malayalam
70 Tamil
71 Telugu

INDO-EUROPEAN FAMILY
INDIC
72 Assamese
73 Bengali
74 Bihari
75 Gujarati
76 Hindi
77 Kashmiri
78 Marathi

79 Nepali
80 Oriya
81 Punjabi
82 Rajasthani
83 Sindhi
84 Sinhalese
85 Urdu
IRANIAN
86 Baluchi
87 Pashto
88 Tajik

PAPUAN LANGUAGES
89 *Papuan languages*

SINO-TIBETAN FAMILY
SINITIC
90 Gan

91 Mandarin
92 Min
93 Wu
94 Xiang
95 Yue
TIBETO-BURMAN
96 Burmese
97 Dzonkha
98 Ladakhi
99 Loloish
100 Manipuri
101 Mizo
102 Tibetan

When the first of the Indo-European languages was carried from central Asia into India—apparently by the ethnic group known as the **Aryans**, between 2000 and 1000 B.C.E., languages in the ancient and indigenous **Dravidian language family** were being spoken there by peoples who had much darker skin than the newcomers. The Dravidian languages, today found almost exclusively in southern India and northern Sri Lanka (with the exception of a small pocket in Pakistan), include *Tamil, Malayalam,* and *Telegu.* Along the eastern side of the Himalayan mountain wall separating South Asia from China, people speak

A **closer** LOOK

Asia, Motherland of Megacities

A **megacity** is a city with more than 10 million inhabitants. We are accustomed to living in a world of such gigantic cities. Yet as recently as 1975, there were only three of them: New York City, Mexico City, and Tokyo. Today there are 23, including 6 in the Americas, 3 in the Middle East, 2 in Europe, and 1 in Africa. The rest are in South and East Asia. India has 3, and China and Japan have 2, including the world's greatest city, Tokyo, with its 38 million souls. South Korea, the Philippines, Indonesia, Pakistan, Bangladesh, and Thailand each have 1, and each of these is that country's primate city.

South and East Asia have a lot of urban growth ahead of them. On a worldwide scale, urban dwellers surpassed rural dwellers for the first time in history in 2008. Only about 40 percent of Asians are urbanites, so just to catch up with the global trend Asians will be moving into cities. Three-quarters of the world's people will live in cities by 2050—again, with the greatest numbers in Asia.

It is impossible to make any generalizations about "the Asian megacity." They range from prosperous Tokyo to Bangladesh's poor Dhaka, and from planned, vertical, and tidy Shanghai to unplanned, sprawling, and squalid Mumbai. In site and situation too, they span the continuum from the auspicious, chosen layout of the original Forbidden City in Beijing to the unfortunate but necessary placement of Tokyo atop a network of active tectonic faults.

Having described the Asian village as a sacred nurturing cradle, it is tempting to contrast it with secular fleshpots of the modern Asian city. But such a depiction of the Asian megacity would be wrong. As Edward Glaeser writes of the Asian city in his book *Triumph of the City,*

> Cities are at the apex of human endeavour. High-density cities are creative, thrilling and less environmentally destructive than sprawling car-based suburbs typical of America. Cities are passports from poverty. They attract poor people, rather than creating them. They are where humans are at their most artistically and technologically creative.[5]

As for the sacred, one need not wander far from any spot in the Asian city to find people praying, whether in the neighborhood mosque or below the towering *goporums* of a South Indian Hindu temple complex. At the larger pilgrimage sites, many of the devotees are villagers who have travelled far to petition for better harvests or harmonious marriages of daughters and sons. Many of those daughters and sons will eventually move into the city.

languages in the **Tibeto-Burman subfamily** of the **Sino-Tibetan language family**. *Tibetan,* spoken farther north in Tibet, and *Burmese,* spoken to the east in Myanmar (Burma), are also in the Tibeto-Burman subfamily. So is *Koro,* a language spoken by only 800 people in India's northeasternmost state. Koro was unknown to the outside world until 2008, when it was discovered by a team of researchers working for National Geographic's Enduring Voices Project.

With so few speakers, Koro is an endangered language, and like so many other languages, seems doomed for what linguists call **language death**, which occurs when a tongue's last speaker

dies. The last fluent speaker of a language dies, on average, every two weeks. Of the world's 6,909 known languages, about half are expected to die in this century. The main reason is that young people are drawn by globalization and other forces to learn more dominant languages. Koro-speaking youth, for example, are going to boarding schools where they study Hindi or English.

In other cases, there are stringent efforts to protect languages. Indian languages, for example, have served as the basis for the drawing of many of the country's state boundaries. India has 14 official languages, and the country's political map is still in the process of being redrawn along ethno-linguistic lines. There are at least five ongoing efforts to create new states, with the main justification being that these boundaries will protect linguistic and cultural identities. These languages are in many cases mutually unintelligible; neither speaker understands the language spoken in an adjacent state. One economic benefit of British colonialism was that it bestowed English as a *lingua franca* on India. This has given India a tremendous advantage in the marketplace of outsourced call centers.

The greatest numbers of people in China speak languages in the **Sinitic subfamily** of the Sino-Tibetan languages, overwhelmingly *Chinese,* and its variants, including *Mandarin, Xiang, Min,* and *Yue.* The Chinese government has adopted Mandarin (Putonghua) as the country's official dialect—making it the national tongue of one-fifth of the world's population—but only about half the population can speak it. Almost all of these more than 1 billion speakers of about 1,500 Chinese dialects belong to the **Han Chinese** ethnic group, which is culturally, politically, and economically the dominant ethnic group of China. Northern and northwestern China, along with Mongolia, are home to speakers of languages in the Altaic language family including *Uighur, Kazakh, Turkic, Mongolian,* and *Tungus.* Cut off from their Altaic-speaking kin are the *Korean* speakers of the Korean Peninsula and the *Japanese* speakers of Japan. Japan's indigenous **Ainu** language has no known linguistic relatives and is close to language death.

The people of Taiwan include Chinese-speaking migrants from China and indigenous speakers of an unrelated language. Most linguists assign this tongue to the **Austronesian subfamily** of the more widespread **Austric language family**. Extreme southern China and adjacent portions of Vietnam and Laos are linguistically complex, reflecting the presence of many minority ethnic groups. Several are kin to the peoples of peninsular Southeast Asia, including the *Hmong* of southern China, Laos, and Vietnam, who speak a language in the **Hmong-Mien subfamily** of the Austric language family. Most people in Laos and neighboring Thailand speak *Lao* and *Thai* in the **Tai-Kadai subfamily** of the Austric family. The majority languages of Cambodia and Vietnam, respectively, are *Khmer* and *Vietnamese,* in the **Mon-Khmer** branch of the **Austro-Asiatic subfamily** of the Austric languages.

Southward and eastward into peninsular Malaysia, Indonesia, and the Philippines, the linguistic map is exceptionally rich and fragmented. The eight languages of the Philippines belong to the Austronesian group. Also in the Austronesian branch is a multitude of languages spoken in Indonesia, including *Malay, Dayak, Kayan-Kenyah, Barito, Javanese, Bugis, Tomini,* and *Timorese.* In many cases, a single language is spoken almost exclusively by an ethnic group inhabiting a single island. In other cases, as will

Joe Hobbs

• **Figure 7.11** Tamil youth near Ipoh in West Malaysia. Making up about 10 percent of Malaysia's population, the Tamils are descendants of laborers brought from India to (what was then called) Malaya by the colonial British.

be seen later, different ethnolinguistic groups sometimes live in conflicted proximity to one another on a single island. The large island of New Guinea and some of its smaller island neighbors are home to the 750 different **Papuan languages**, which are not 291 related to Austronesian tongues.

Colonialism and regional economic enterprises led to the infusion of distant tongues into many parts of East and South Asia. As well as bringing English to India, British colonists introduced Tamil-speaking Dravidians from south India into what is now Malaysia, where they worked the rubber plantations and tin mines and where their descendants remain as an important mi-295 nority today (•**Figure 7.11**). Another small but economically prosperous minority in several countries (and a majority in Singapore) is ethnic Han Chinese. Some people in Indonesia speak Dutch, reflecting that country's former colonial status, and the Philippines' former status as a U.S. commonwealth is reflected in the fact that English remains one of that country's two official languages.

Religions and Belief Systems

South and East Asia are, along with the Middle East, one of the world's two great hearths of religion. Hinduism, Buddhism, and related traditions of Confucianism and Daoism—collectively practiced or observed by an enormous 2 billion people (or more than a quarter of the world's population) originated in this region. The basics of these belief systems are introduced here. The precepts of Islam and Christianity, also practiced in the region, are described in Chapter 6. The faiths with small numbers of adherents, like Zoroastrianism and Sikhism, are described later in the contexts of particular countries. Religion is sometimes,

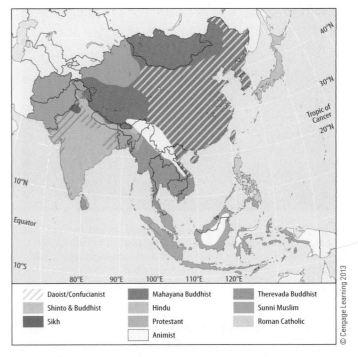

© Cengage Learning 2013

Daoist/Confucianist	Mahayana Buddhist	Therevada Buddhist
Shinto & Buddhist	Hindu	Sunni Muslim
Sikh	Protestant	Roman Catholic
	Animist	

• **Figure 7.12** Religions of Asia.

unfortunately, at the root of domestic or international conflict, and these cases are also discussed in the appropriate places.

The geographic distribution of the region's faiths is depicted in •**Figure 7.12**. As this map reveals, Hinduism is dominant in India, although many other religions are practiced there as well. Islam prevails in Afghanistan and Pakistan, Bangladesh, most parts of the East Indies, parts of the southern Philippines, parts of western China, and among the Malays of western Malaysia. Indonesia, in the East Indies, has the world's largest number of Muslims; 90 percent of the 238 million Indonesian people are Muslim. However, Hinduism and Buddhism thrived in Indonesia until Islam arrived in the fifteenth century, and many aspects of these religions permeated Indonesia's unique Islamic culture. Various forms of Buddhism are dominant in Myanmar (Burma), Thailand, Cambodia, Laos, Tibet, and Mongolia. Most people in Sri Lanka and Nepal are Buddhists or Hindus. The religious patterns of the Chinese, Vietnamese, and Koreans are more difficult to describe. Among them, Buddhism, Confucianism, and Daoism have been influential, often in the same household.

Growing prosperity among the Chinese has been accompanied by a renewed interest in spiritual matters. China's government officially recognizes and even encourages five faiths (Buddhism, Islam, Daoism, Catholicism, and Protestantism) but discourages or forbids some sects and belief systems, including Tibetan Buddhism; the quasi-Buddhist movement known as **Falun Gong** (or Falun Dafa, a practice of calisthenics and meditation, that the government calls an "evil cult") and "house church" Protestantism (•**Figure 7.13**). In Japan, religious affiliations often overlap, with an estimated 84 percent of the Japanese population sharing Buddhism with the strongly nationalistic religion of **Shintoism**, itself a blend of nature reverence, Japanese folklore, and Chinese ritual. The Philippines and Timor-Leste, with large Roman Catholic majorities, are

• **Figure 7.13** Falun Dafa (Falun Gong) is outlawed in China. In many countries you will see followers sharing information about their movement with passersby. This posting is near the Sydney Opera House. After I took this photo, the Falun Dafa followers eagerly greeted Chinese tourists stepping off a tour, bus, telling them about the human rights abuses against their compatriots back home in China.

the only predominantly Christian nations in Asia, although Christian groups are found in all other countries in this region. About one-third of South Koreans are Christian, and Korean missionaries work in many countries of the world.

Other customs include **ancestor veneration**, which is especially prominent among Chinese, Japanese, Vietnamese, and Koreans. Most homes in these countries contain small shrines with photographs and other memorabilia of deceased ancestors at which family members place fresh fruits and other offerings on a daily basis. It is crucial in these cultures that the ancestors never be forgotten (•**Figure 7.14**). Many members of hill tribes in Southeast Asia believe that natural processes and objects possess souls, a doctrine known as **animism**. Such indigenous animistic beliefs have been incorporated into a number of mainstream religions, creating vibrant and highly varied belief systems by Western standards. This

• **Figure 7.14** Ancestor veneration is not a religion per se but an important spiritual tradition that requires the living to remember and pay tribute to their forebears. This is a shrine to the deceased parents of an elderly woman in a southern Thai village.

• **Figure 7.15** Worship at the Main Temple of the Holy See of the Cao Dai religion in Tay Ninh, southern Vietnam. Note the "All-Seeing Left Eye of God" at the upper right. Tay Ninh is like the Vatican City of the Cao Dai faith.

blending, known as **syncretism**, has parallels elsewhere in the world, for example, among the Maya of Central America.

The ultimate syncretic faith is the **Cao Dai** religion, which originated in Vietnam during the French colonial era and persists today despite constraints imposed by the communist Vietnamese government (•**Figure 7.15**). It recognizes and blends the precepts and iconography of Buddhism, Daoism, Confucianism, Judaism, Christianity, and Islam. Prophets and saints of the Cao Dai faith include Moses, Louis Pasteur, Victor Hugo, and Albert Einstein.

This kind of layering and mixing in the spiritual realm is typical of most belief systems in South and East Asia. Most of the faithful put stock in what many Westerners might dismiss as "superstitions." Numerology, shamanism, amulets, horoscopes, and fortune telling have an important place in peoples' lives. Marriage, travel, site selection for homes and businesses, and countless other aspects of life are frequently influenced by auspicious and inauspicious omens. In Thailand, Laos, Cambodia, and Myanmar, the most visible sign of beliefs in a multifaceted spiritual realm is the "spirit house" (•**Figure 7.16**). The spirit house is placed in an auspicious spot of a home of business, where it serves as an abode for spirits that people petition for special favors or appease to avoid bad fortune.

Hinduism is a regionally varied faith and a belief system so complex that it is extremely difficult to summarize. However, it

• **Figure 7.16** Spirit House in Laos.

has the following characteristic elements. It lacks a definite creed or theology. It is very absorptive, encompassing an unlimited pantheon of deities (all of which, most Hindus say, are simply different aspects of the one God, Brahman) and an infinite range of types of permissible worship (•**Figure 7.17**). Most Hindus recognize the social hierarchy of the caste system, deferring authority to the highest (Brahmin) caste. They practice rituals to honor one or another of the principal deities (Brahman the creator, Vishnu the preserver, and Shiva the destroyer) and thousands of their manifestations (*avatars*) and lesser deities, attending temples to worship them in the form of sanctified icons in which the gods' presence resides. They believe in reincarnation and the transmigration of souls and revere many living things (**see A Closer Look**).

• **Figure 7.17** Hinduism has a complex and very colorful pantheon. This is Vishnu, preserver god, and his consort, Lakshmi, goddess of wealth and prosperity.

A **closer** LOOK

The Sacred Cow

Many attitudes, beliefs, and practices associated with cattle make up the world-famous but often poorly understood "sacred cow" concept of India. There are nearly 200 million cattle in India, representing about 15 percent of the world total and the largest concentration of domesticated animals anywhere on earth. India's dominant religion of Hinduism forbids the slaughter of cows but allows male cattle and both male and female water buffalo to be killed. Reverence for the cow is well founded. Indians favor cow's milk, ghee (clarified butter), and yogurt over dairy products from water buffalo. They value cows as producers of male offspring, which serve as India's principal draft animals. Both cows and bullocks provide dung, an almost universal fuel and fertilizer in rural India.

Reverence for the cow in particular and cattle in general pervades Hindu religion and mythology. The bull Nandi is associated with the Hindu god Shiva, the cow is linked to the goddess Lakshmi, and Krishna (an incarnation of the god Vishnu) was a cowherd. People allow cattle to roam the streets of Indian cities, and simply accommodate them in vehicle and pedestrian traffic (•**Figure 7.J**). As a symbol of fertility, cows are associated with several deities. The mother of all cows, Surabhi, was one of the treasures churned from the cosmic ocean. Hindus honor cows at several special festivals. They use cow's milk in temple rituals. They believe that the "five products of the cow"—milk, curds, ghee, urine, and dung—have unique magical and medicinal properties, particularly when combined. In India, there are about 4,000 *gaushalas*, or "old folks' homes," for aged and infirm cattle.

Remarkably, however, the subcontinent countries of Pakistan, Bangladesh, and India are major exporters of leather and leather goods. In India, the leather industry is mainly in the hands of Muslims, who do not share Hindus' restrictions on killing or eating cows. Muslims are forbidden to eat pork; therefore, pigs, a major food resource in many developing countries, are of little importance here. They are eaten mainly by Christians, very low-caste Hindus (some of whom are pig breeders), and tribal peoples.

• **Figure 7.J** People and cows share street space throughout India.

They are supposed to be tolerant of other religions and ideas. They participate in folk festivals to commemorate legendary heroes and gods. To earn religious merit and to struggle toward liberation from the bondage of repeated death and rebirth, they make pilgrimages to sacred mountains and rivers.

The Ganges is a particularly sacred river to Hindus, who believe it springs from the matted hair of the god Shiva, and who call it "Mother" and "Goddess." The pilgrimage to celebrate the festival of Ardh Kumbh Mela at Allahabad, at the confluence of the Ganges and Yamuna rivers, drew 75 million pilgrims in January 2007. It was the largest pilgrimage, and probably the largest gathering of any kind, in human history. The most enduring Hindu pilgrimage destination is the Ganges city of Varanasi (Benares) in the state of Uttar Pradesh. Many elderly people go to die in this city and be cremated where their ashes may be strewn in the holy waters. Indian authorities have passed laws, so far ineffective, to clean up the Ganges, polluted in part by incompletely cremated corpses; many of the faithful poor cannot afford to buy the fuel needed for thorough immolation. Scavenging water turtles released into the river at Varanasi help dispose of the cadavers.

Buddhism is based on the life and teachings of Siddhartha Gautama, known as the Buddha or the Enlightened One (•Figure 7.18). The Buddha was born a prince in about 563 B.C.E. in northern India.[6] Although presumably born a Hindu, he came to reject most of the major precepts of Hinduism, including caste restrictions. When he was about 29 years old, he renounced his earthly possessions and became an ascetic in search of peace and enlightenment. The Buddha eventually settled for a "middle path" between self-denial and indulgence, and he meditated through a series of higher states of consciousness until he attained enlightenment (*bodhi*). From that point, he traveled, preached, and organized his disciples into monastic communities. In his sermons, he described the "Four Noble Truths" revealed in his enlightenment: life is suffering; all suffering is caused by ignorance of the nature of reality; suffering can be ended by overcoming ignorance and attachment; and the path to the suppression of suffering is the Noble Eightfold Path, made up of right views, right intention, right speech, right action, right livelihood, right effort, right mindedness, and right contemplation.

An important Buddhist concept is the doctrine of **karma**, a person's acts and their consequences. A person's actions lead to rebirth, in which good deeds of the previous life are rewarded (for example, by which one could be reborn a human) and bad deeds punished (for example, by being reborn as a resident of hell). The goal of the practicing Buddhist is to attain **nirvana**, a transcendent state in which one is able to escape the cycle of birth and rebirth and all the suffering it brings.

Buddhism evolved into two separate branches: **Theravada** ("Way of the Elders") and **Mahayana** ("Great Vehicle"). (Mahayanists sometimes apply the term *Hinayana*, "Lesser Vehicle," in a derogatory fashion to the Theravada.) Theravada Buddhism is strongest in Sri Lanka, Myanmar, Laos, Cambodia, and Thailand. Theravada Buddhists claim to follow the true teachings and practices of the Buddha. Mahayana Buddhism originated in India and then diffused along the Silk Road to

•**Figure 7.18** The Buddha has many manifestations. This temple of the Theravada faithful is in Chiang Mai, Thailand.

central Asia, Tibet, and China and eventually into Vietnam, Korea, Japan, and Taiwan. Mahayana Buddhists accept a wider variety of practices than those espoused by the Theravada, have a more mythological view of the Buddha, and are interested in broader philosophical issues.

Confucianism is not as much a belief system as it is a sociopolitical philosophy that serves on its own or is blended with other religions, particularly among ethnic Han Chinese. It is based on the writings of Confucius (Kung Fu-tzu, or "Master Kung," who lived from 551 to 479 B.C.E.), collected primarily in his *Analects*. Mencius (c. 371–288 B.C.E.) later became a major force in the widespread diffusion of this belief system throughout China. Following a period in which Communist authorities disavowed them, Confucian ideals are making a major comeback in China today.

The major tenets of Confucianism are embodied in a system of ethical precepts for the proper management of society, emphasizing honor of elders and other authorities, hierarchy, and education. Confucius never proclaimed his beliefs to constitute a religion; he was simply attempting to create a social contract between different classes central to Chinese government and society. He viewed his philosophy as secular, and its diffusion to Korea, Japan, and Vietnam has been part of the cultural baggage taken abroad by a steady stream of Chinese emigrants to those places.

Another important Chinese school of thought, second only to Confucianism and also widely blended with Buddhism among ethnic Han Chinese, is **Daoism** (sometimes spelled **Taoism**). Its philosophy comes from a body of work known as the *Daodejing (Tao-te Ching),* or in English, *Classic of the Way and Its Power,* ascribed to Chinese Lao-tzu (Laozi), who lived in the sixth century B.C.E. Daoism encourages the individual to reject Confucian-style social conformity and seek to conform only to the underlying pattern of the universe, the "Way" (Dao, or Tao). To follow that way, the individual should "do nothing," meaning nothing unnatural or artificial. One can become rid of all doctrines and knowledge to achieve unity with the Dao and thereby gain a mystical power. With that power, the individual can transcend everything ordinary, even life and death. Daoists hold the simple earthly life of the farmer in high esteem.

There are about 25 million Christians in India, most of whom live in the south of the peninsula; Buddhists, Jains, Parsis, and members of a variety of tribal religions make up the numerous smaller remaining religious minorities. Most of the estimated 7 million Buddhists are recent converts from among India's lowest castes. Numbering perhaps 75,000 and concentrated mainly in Mumbai, the famously entrepreneurial **Parsis** have attained wealth and economic power far out of proportion to their number. Their religion is the pre-Islamic Persian faith of **Zoroastrianism**, a monotheistic faith known mainly for its reverence of fire. Zoroastrians follow the teachings of the prophet Zarathustra, who was born around 600 B.C.E. in Persia (modern Iran). He preached that people should have good thoughts, speak good words, do good deeds, and believe in a single god, Ahura Mazda ("Wise Lord"). Persia's Zarathustras, as the Zoroastrians are also known, fled Persia in two waves, first during the conquest by Alexander the Great around 300 B.C.E. and then after Islam swept into Persia in

the seventh century. They settled in China, Russia, and Europe, where they became assimilated with other cultures. In India, they maintained a separate existence until quite recently; now about one in three takes a spouse of another religion and abandons Zoroastrianism. Many scholars consider theirs an endangered religion, with only 124,000 to 190,000 followers worldwide.

India's 6 million **Jains** also have influence beyond what their numbers suggest, as they control a significant share of India's business. They live mainly in the western state of Gujarat. **Jainism**, founded on the teachings of Vardhamana Mahavira (a contemporary of the Buddha, c. 540–468 B.C.E.), is renowned for its respect for geographic features and animal life. Jains believe that souls are in people, plants, animals, and nonliving natural entities such as rocks and rivers. Jainism has taken the principle of nonviolence (*ahimsa,* also present in Hinduism) to mean they should not even harm microbes. Therefore, Jain worshipers often wear masks to prevent inhalation of microscopic organisms, and Jains cannot be farmers because they would have to destroy plant life and living organisms in the soil.

Effects of European Colonization

As in much of the developing world, Western colonialism reshaped many traditional geographic patterns in South and East Asia (•Figure 7.19). By the end of the fifteenth century, Portugal and Spain had begun to extend economic and political control over some islands and mainland coastal areas of South and Southeast Asia. In the eighteenth and nineteenth centuries, the pace of colonization and economic control quickened, and large areas came under European domination. By the end of the nineteenth century, Great Britain was supreme in India, Burma (modern

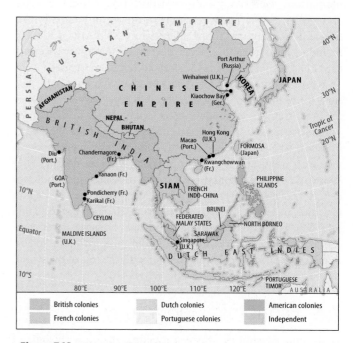

• **Figure 7.19** Colonial realms and independent countries in South and East Asia at the beginning of the twentieth century.

Source: Based on 'The World in 1900,' in R. R. Palmer, ed., Atlas of World History (Chicago: Rand McNally, 1957), pp. 169–171.

Myanmar), Ceylon (today's Sri Lanka), and Malaya and northern Borneo (now parts of Malaysia); the Netherlands possessed most of the East Indies (Indonesia); France had acquired Indochina (the region of modern Vietnam, Laos, and Cambodia) and small holdings on the coasts of India; and Portugal had Goa and Diu in India, Macau in China, and Timor (today's Timor-Leste) in the East Indies. Although retaining a semblance of territorial integrity, China was forced to yield possession of strategic Hong Kong to Britain in the 1842 Treaty of Nanking and to grant special trading concessions and extraterritorial rights to various European nations and the United States through the latter half of the nineteenth century. In 1898, the Philippines, held by Spain since the 1500s, came under the control of the United States.

A few Asian countries escaped domination by the Western powers during the colonial age. Thailand (historically known as Siam) formed a buffer between British and French colonial spheres in peninsular Southeast Asia. Japan withdrew into almost complete seclusion in the mid-seventeenth century but emerged in the late nineteenth century as the first modern, industrialized Asian nation and soon acquired a colonial empire of its own. Korea also followed a policy of isolation from foreign influences until 1876, when Japan began to colonize it. It is noteworthy that Japan, South Korea, and Thailand would emerge as very or relatively prosperous countries, whereas most of the former colonies lagged in economic development. Dependency theorists point to these discrepancies to affirm their argument that colonization hampered development.

Asia was an extraordinarily profitable region for the colonizers. Western nations extracted vast quantities of tropical agricultural commodities and in turn found large markets for their manufactured goods. Westerners also invested heavily in plantations, factories, mines, transportation, communication, and electric power facilities. Some of the region's most important cities, including Shanghai in China, Kolkata (Calcutta) in India, and Singapore, were developed mainly by Western capital as seaports serving Western colonial enterprises.

Western domination of Asia ended in the twentieth century for a variety of reasons. The two world wars weakened the West's ability to conduct its colonial affairs in the region; Japan rose to great power status and challenged the West early in World War II; and effective anti-colonial movements arose in nearly all areas subject to European control. In the decades after World War II, all colonial possessions in Asia gained independence. The last to revert were Hong Kong (returned by Britain to China in 1997) and Macau (returned by Portugal to China in 1999). Until this era of independence, however, twentieth-century Asia was marked by revolution, war, turmoil, and an inability to gain an economic footing with the rest of the world.

7.4 Economic Geography

With some exceptions, this is a dynamic region with some of the world's strongest and fastest-growing economies. Their economic performance is often described as miraculous, given the region's tumultuous history. At the end of World War II,

South and East Asia were poorer than Africa and Latin America. "The depleted states of East Asia, including Korea, Taiwan and Singapore, with few sources of money and practically no existing industry, seemed especially hopeless," writes journalist Michael Schuman in his book *The Miracle*. "Yet it was in these countries, supposedly the bottom of the international economic barrel, that the Miracle was born."[7]

If not literally a product of divine intervention, what explains Asia's extraordinary economic success? Many of the answers—government policies that nurtured the growth of industries, and government investments in industrial growth rather than in public amenities and services, for example—are explained in the following pages, in studies of the individual countries, including Japan and South Korea. It is also useful to look at the level of the individual and family to appreciate Asia's economic geography and economic success. To generalize, Asians are savers (whereas Americans are spenders!). These individual traits arguably played a significant role in the recent "Great Recession:" Americans went deeply into debt to finance consumption, while Asians stepped up their savings and exported more goods to Western consumers.[8] Savings are also important because most of the Asian countries provide few "safety nets" such as retirement pensions and health care. And most important of all, savings help to take care of the family, which is virtually sacred in many Asian cultures. It is always impressive to see how young people working even for low wages manage to save a significant share of those earnings to send home as remittances to Mom and Dad and Grandma. Needless to say, one of the great hopes in the West is that Asians will cut back on savings, loosening purse strings especially to buy Western exports—thus pulling Western economies and (the West argues) the world economic system out of the abyss.

The world's second and third largest economies are in East and South Asia: China and Japan, respectively. The recent global economic downturn slowed them down somewhat, but the economies of China and India have still been surging, with both goods and services inundating the global economic system. Their clout, their accomplishments, and their ambitions are enormous. Some economists and equities marketers have taken to conflating the two countries and calling them "Chindia," a double-barreled engine of economic growth.

It must be noted, however, that this growth is occurring against the backdrop of extraordinary poverty, and many hundreds of millions of people remain desperately poor. Although the region as a whole has taken great strides at reducing poverty rates, the gap between rich and poor is growing in many countries. This is particularly true in India, where the overall economy has grown about 8 percent annually since 2000; yet in that same period, there has been almost no reduction in childhood malnutrition, which afflicts almost half of India's young. Three-quarters of India's people live on less than $2 per day, which is the World Bank's benchmark for poverty. About 70 percent of Indians live in rural areas, but agriculture produces only about 20 percent of India's growth domestic product; no wonder then that rural people are poor, as they are a majority who live off a small portion piece of the economic pie. Poverty

tends to be deepest in the rural areas, fueling the classic push
56 factor of rural-to-urban migration. China overtook Japan to
become the world's second largest economy in 2010 (and is
projected to be the largest by 2020), yet on a per capita basis
the country remains poor, with over a third of the people living
on less than $2 per day. "For the first time, you have this odd
combination," an economist in Hong Kong wrote as China
reached this milestone. "One of the world's largest economies
is also one of the world's poorest economies."[9]

The milestone of China becoming the world's second largest
economy is one of the many that is described as a "China moment." With its galloping growth, China reaches one of these
milestones after another. The world awaits the China moment
of the country becoming the world's largest economy. There
was a China moment recently when the country consumed so
much coal that for the first time it began to import coal. There
was the recent China moment of China becoming the world's
second largest oil importer, after the United States. And so on.

Despite being generally a region of less developed countries,
South and East Asia are home to the strong, industrialized,
export-oriented economies of South Korea, Taiwan, Hong
Kong, and Singapore, which late in the twentieth century
came to be known as the **Asian Tigers** (or "Little Dragons").
Some economists consider the rapidly industrializing Southeast Asian countries of Thailand, Malaysia, Indonesia, the
Philippines, and Vietnam the **New Asian Tigers**, or **Tiger
Cubs**, following the path of rapid economic growth blazed by
their neighbors. Indonesia's economy has been doing so well
that some economists would like to include it as a second "I"
in the exclusive "BRIC" club, which would then be Brazil,
133 Russia, India, Indonesia, and China, or "BRIIC."

Japan was the first Asian country to develop modern cities
and modern types of manufacturing on a large scale. It has
been a major industrial power for more than a century. China
and India are much better supplied with mineral resources
than is Japan. With the world's largest populations, they also
have cheaper labor. But Japan's labor force is, at least for the
time being, generally more skilled, and it leads Asia in high
value-added manufacturing—the process of refining and fabricating more valuable goods from raw or semi-processed materials. Now, however, there are signs that Japan's preeminence
in innovation is waning as China's intellectual resources grow.
The biggest economic story in East and South Asia is the rise
of China. Some highlights of that story are told elsewhere in
Section 7.6.

China's Surging Economy

The statistics portray China as a steamroller. Among the China
moments, China became the world's largest exporter, overtaking Germany in 2010. China overtook the United States as
the world's largest energy consumer in 2010. China overtook
Japan to become the second largest economy that same year.

China's economic growth soared at an average annual rate
of 10 percent between 1990 and 2011. China's exports (along
with imports) increased about 1,000 percent during those

Joe Hobbs

• **Figure 7.20** Mechanized silk production in China. Note the white cocoons
in the vats below the worker's hands; the silk is spun from these.

years, growing an average of 50 percent a year. China's share
of world exports of goods grew 8 percent between 1981 and
2010 (compared with a decline of 3.5 percent in the United
States during the same period).

Just a few years ago, China was notable for making low-
value-added products—cheap toys, for example—that the
richer countries really had no interest in making. There were
shirts and shoes too, most of them manufactured in more than
20,000 factories in the Special Economic Zone (SEZ) of Shenzhen, adjoining Hong Kong. Now China is making at least a 267
little bit of almost everything, becoming the "workshop of the
world" (•**Figure 7.20**). The country is making especially rapid
progress in manufactures and exports of information technology (IT) hardware. Already China is the world's third largest
manufacturer of personal computers, sold under such brand
names as Dell and IBM. The roots of China's boom in IT
may be traced to the dot-com bubble that burst in the United
States in 2000. Many of the talented Chinese who worked in
California's Silicon Valley at that time went back to China,
taking their skills with them in a kind of "reverse brain drain." 56

Advances in China's global status and in government economic policy have fueled the boom. China joined the World Trade Organization in 2001, prompting a surge in foreign investment. China sweetens the deal for investors with incentives like subsidized loans, tax exemptions, and 50 percent discounts on land prices. The addition of Chinese brawn and brains creates a nearly perfect investment climate. Chinese labor is much less expensive than is typical of most more developed countries, and China has a rapidly growing pool of university-trained engineers and other innovators.

China's Economic Impact

China's economic boom is having enormous consequences for other countries, particularly its Asian neighbors. Japanese electronics companies are cutting costs by moving increasing amounts of production to China. Surging investment in China is linked to disinvestment elsewhere, especially in Southeast Asia. That region's Tigers and Tiger Cubs—notably Malaysia, Singapore, and Thailand—dominated the 1990s with manufactures and exports of medium- and high-tech industrial products. With its growing consumer appetite, China was buying such goods from these countries. But recently the manufacturing shifted to China from those countries. South Korea's Samsung made China the main base for production of its flat screens and computers, for example. Japan's Toshiba makes televisions only in China. Dell recently moved some of its computer-making facilities from Malaysia to China.

With China's increasing gravity, Southeast Asian economies may at least temporarily be relegated to supplying low-end products, especially food and raw materials, to China. In turn, they may end up buying cheap Chinese manufactured goods, harming their own fledgling industries. They may try to counter this trend by developing higher-value niche products to meet demands in China. Thailand, for example, wants to boost manufactures of health care products, while Singapore focuses on biomedical products and financial services. Poorer Indonesia will probably be able to count on helping satisfy China's growing appetite for natural gas to fuel its southern industrial cities. Meanwhile, the large economies of Japan, Taiwan, and South Korea have lost ground to China in the market of foreign investment, as investors have been lured away from those countries by China's lower labor costs and its potential 1.3 billion consumers.

Cheap labor is unlikely to make China a permanent magnet for foreign investment, however. As affluence rises there, so will wages, and China's less prosperous neighbors will again attract investment. These figures are telling: in 2011, the average factory worker in Vietnam made $136 per month, in Indonesia $129 per month, and in China $413 per month. China is no longer cheap. China's one-child policy also means it will have fewer industrial workers in the future. Mindful of this demographic problem, farsighted multinational companies are boosting manufacturing investment in India, where the pool of young workers is likely to remain large and growing for decades to come.

In the long-dominant United States, there are rising fears of China's economic ascendance. Many Americans feel their country is becoming subservient to China because Chinese investments are funding U.S. federal deficits and because the United States is losing ground to China's industrial might. China's rise has been steadily chipping away at U.S. economic dominance in the region. In one country after another (including the most critical, Japan and South Korea), China has eclipsed the United States as Asia's most essential trading partner. The U.S.-China relationship is growing in China's favor. Of China's exports, about 18 percent go to the United States (its largest single trading partner), but its imports from the United States lag far behind. This U.S. trade deficit with China threatens to become a sore point in relations between the countries. Textile and other industries in the United States are pressuring the U.S. government to pass punitive tariffs on Chinese imports, a move that could precipitate a costly trade war. Problems with tainted food and toy imports 420 have already led to bans of some Chinese products in the United States.

The United States is also furious at China's reluctance to stop the counterfeiting or "piracy" of American products, particularly computer software and DVDs of Hollywood films. China is the epicenter of prolific Asian manufacture and trade in pirated products. In some countries like Malaysia, authorities try to crack down on sales of such goods on street corners, but the vendors are usually prepared to disappear at a moment's notice with their goods in tow, only to reappear moments later when police leave. In other countries like Vietnam, no one needs to run: there are countless permanent shops selling the latest counterfeited music, movies, watches, handbags, and more (•Figure 7.21).

Another great concern in the United States is how many jobs and profits are lost to outsourcing, especially to India (see Insights, next page).

• **Figure 7.21** Counterfeited goods made in China and sold throughout Asia bear the brand names of U.S. and other manufacturers but are sold to happy consumers at a fraction of the cost of the genuine product. This $4 digital audio player claims to be an Apple iPod, but you can be sure it is not.

Outsourcing, also known as **offshore outsourcing** or **offshoring**, is the flight of technology and other jobs from countries with high manufacturing and service costs to countries with low manufacturing and service costs. India is the world's largest recipient of these jobs, and the United States is the leading outsourcer. India's outsourcing industry was a natural development after 1990, when the country abandoned its Soviet-style, centrally planned economic model of self-sufficiency and opened its doors to world markets and economic liberalization.

In the United States, there is growing worry and even alarm over the trend of outsourcing jobs to India. Many of these jobs are in computer programming and a wide variety of technical support, from computer problem phone calls to X-ray diagnosis. Many are mundane, like bill processing and order taking. Increasingly, however, high-paying white-collar employment is being outsourced to India, in such fields as investment banking, aircraft engineering, and pharmaceuticals research. Top Western corporations like the United States' Cisco Systems, the world's leading maker of communications equipment, are beginning to relocate substantial numbers of their senior executives to India, where much of the firm's work is actually done. The top destination is the south-central city of Bengaluru (Bangalore), which has come to be known as "India's Silicon Valley."

India, which has earned a reputation of being the West's "back office," is an especially attractive venue for information technology outsourcing for several reasons. It has a large population of well-educated people, most of whom speak excellent English. Those who work the telephone call centers typically receive training in vernacular American English and strive to perfect their American accents while shedding their Indian and British pronunciations. The bottom line is another major advantage: Indian skilled labor is cheap. The average computer programmer in the United States earns $75,000 per year, while his or her counterpart in India earns $9,000 per year.

Finally, the Internet and superb telephone communications, ironically combined with India's physical distance from the United States, provide some remarkable opportunities. While people in the United States sleep, the day shift is on in India. An American doctor can X-ray a patient in the afternoon and by early the next morning have the analysis—performed overnight in India. On the downside, when an American consumer places a technical service call for a computer problem, he or she may be speaking with a very tired Indian employee working the graveyard shift. There are many accounts of fatigue and burnout among Indian call center employees.

What about the loss of American jobs due to outsourcing? There is an expression that one's job may be "Bangalored." Some projections are alarming. A former official of the U.S. Federal Reserve described outsourcing as a "Third Industrial Revolution," predicting it would take jobs from as many as 42 million American workers in the coming years.[10] Outsourcing companies generally argue that reducing the costs of their services reduces inflationary costs for their consumers and improves efficiency and productivity. This rationale is, of course, cold comfort to the American worker whose job has moved offshore. Another trend that Americans may fear is that with its well-educated work force (including many graduates of U.S. universities), India is moving further up the value chain. There is new investment in research and development so that the country's information technology workforce will be innovating, and not just producing goods that Western thinkers invented.

There are surprising new twists in the economy of outsourcing. One is "reverse outsourcing," in which Indian and other companies hire American workers because their qualifications cannot be matched in India and other countries. Secondly, the pride in "made in America" products has grown great enough, and the costs of intellectual capital, wages, raw materials, real estate, and energy in India and China have risen high enough, that many American firms have begun to "onshore" work they had previously offshored. This countertrend, known as **insourcing**, onshoring, inshoring, or homeshoring, is especially popular in the United States at a time when jobs are scarce.

The Green Revolution

Most Asian countries do not want to improve their economies only by enhancing their outputs of services and manufactured goods. They also want to boost agricultural self-sufficiency and crop exports. Revolutionary changes in crops are at the heart of efforts to reshape Asia's agricultural economies. One of the most significant agricultural innovations in Asia during the past century is the development of new seed types and innovations in planting, cultivating, harvesting, and marketing of crops, efforts known collectively as the **Green Revolution**. These developments are not exclusive to South and East Asia, but much of the growth and investment have taken place here.

Ever since 1962, when the International Rice Research Institute (IRRI) was founded in the Philippines, there have been efforts to use science to increase food yields (particularly of rice) to stave off hunger and generate export income. Notable success has been achieved in breeding the new high-yield varieties of seed stock, and there has been a large upsurge of production in certain areas where the new strains have been widely introduced. Many of the new varieties are bioengineered or genetically modified (GM) crops. This means that precise manipulation of the genetic components of a crop can not only produce a higher yield but also a crop more resistant to drought, flood, or pests and which generates a higher amount of a desired nutritional component such as vitamin A. China, second only to the United States in the global biotechnology industry, has produced genetically engineered rice, corn, cotton, tobacco, sweet peppers, petunias, poplar trees, and more.

Because the underlying premise of the Green Revolution is economic—more crops will be produced at lower cost for higher revenue—most countries in the region are racing to increase their biotechnological output, especially in an effort to catch up with China and ensure they do not lose positions in the global marketplace. Malaysia hopes to build Biovalley, a biotechnology research park near Kuala Lumpur, and Indonesia is planning a similar Bioisland. One of Biovalley's goals is to create palm oil trees genetically modified to produce the raw material for specialized plastics used in medical devices. South Korea, Japan, and India have large and well-funded biotechnology research programs. Japan has so far resisted production of GM food crops out of fears of possible health hazards, a concern that most Europeans and their governments share.

For Asian farmers to capitalize fully on the Green Revolution, they must overcome many obstacles. Success requires levels of capital that are often beyond the means of peasant farmers, landlords, and governments. Such expenditures are needed for water supply facilities (for example, the tubewells that now number hundreds of thousands in the Indo-Gangetic Plain of the northern Indian subcontinent and that are severely depleting the groundwater), chemical fertilizers, and chemicals to control weeds, pests, and diseases. And as agriculture becomes more mechanized, considerable increases in the costs of machinery, fertilizers, and fuel have to be borne by farmers who have, in many cases, had little experience with the cash economy. Governments must improve transportation so that the large quantities of fertilizer required by the new seed varieties can be delivered in a timely fashion. Not only must there be these associated infrastructural changes to support this "revolution," but also the crop calendar of the farmer becomes much less forgiving because many of the newest seed grains demand more precise water, fertilizer, and cultivation requirements than traditional grains. Grain storage facilities, now subject to plundering by rats, must be improved.

There are other problems associated with the Green Revolution. Economic dislocations result when rice-importing countries become more self-sufficient, causing hardships for rice exporters. Large infusions of the agricultural chemicals associated with high-yield varieties have negative repercussions on natural ecosystems. India's government subsidizes urea fertilizer, and farmers are naturally tempted to overuse this cheap product. Unfortunately, the overuse of urea degrades the soil, and fertilized soil demands much more water. There is also concern that development of a limited number of high-yield crop varieties grown in vast monocultures will dramatically reduce the genetic variability of crops, in essence interfering with nature's ability to adjust to environmental changes.

7.5 Geopolitical Issues

In the Middle East and North Africa, and in Russia and the Near Abroad, principal geopolitical concerns focus on the production and distribution of fossil fuel energy resources. In South and East Asia, by contrast, some of the most serious geopolitical issues are prospects for what may be done with weapons created from a particular energy source: nuclear energy. There are also concerns about seafloor resources (see Geography of Energy, next page), Islamist terrorism, and the security of shipping lanes. Of these, seabed resources may prove to be the most critical. There are several discussions in this chapter of seemingly insignificant small islands that are extremely critical because of the fossil fuels and other resources that may lie in nearby waters.

The earlier discussion of the mushrooming economic clout of China and India raises a big picture geopolitical reality: Asia is emerging as a center of gravity that will seriously challenge the century-long primacy of the United States in world affairs. Japan long stood alone as a pillar of strength in Asia, but with China and even India surging ahead, regional might has been reshuffled. Eyes in the region and around the world will be focused especially on China, which some analysts fear may use its fast-increasing military power to enhance its fast-increasing economic clout.

Nationalism and Nuclear Weapons

As we will see in Section 7.6, after its independence from Britain in 1947, India emerged as an avowedly secular democratic state. But after the 1998 victory of the Hindu nationalist Bharatiya Janata Party (BJP), hopes for new privileges emerged among India's Hindu majority. Among BJP supporters, there was hope for a "Hindu bomb," a counterweight to a long-feared "Islamic bomb" in neighboring Pakistan.

On May 11, 1998, much to the surprise of U.S. and other Western intelligence agencies, India conducted three underground nuclear tests in the Thar Desert. With the blasts, India's government seemed to be staking India's claim as a great world power, exhibit its military muscle to Pakistan and China, and increasing political support for the BJP. Initial reactions among India's vast populace were highly favorable; 91 percent of residents polled within three days of the event supported the tests. Outside India, there was alarm. India had defied an informal worldwide moratorium on nuclear testing that went into effect in 1996, when 149 nations (not including India and Pakistan) signed the **Comprehensive Test Ban Treaty** (also known as the **Nuclear Nonproliferation Treaty**, or **NPT**), which prohibits all nuclear tests.

The world's eyes turned quickly to India's neighbor to the west. Governments pleaded with Pakistan to refrain from answering India with nuclear tests of its own, arguing that Pakistan would have a public relations triumph it if exercised restraint: it would appear to be a mature and responsible power, whereas India would seem to be a dangerous rogue state. But Pakistan's leaders felt obliged by their population's demands for a tit-for-tat response to India's blasts, and soon followed with six nuclear tests. India and Pakistan thus joined just five other nations—the United States, Russia, China, the United Kingdom, and France—in acknowledging that they possess nuclear weapons. Israel neither confirms nor denies that it has nuclear weapons, but is thought to have about 250.

There is disagreement about what this recent regional and global shift in the balance of power means. Some analysts fear an escalating nuclear arms race that, perhaps ignited by a border skirmish in Kashmir or by a terrorist attack like the

GEOGRAPHY OF ENERGY | The Spratly Islands, Geopolitical Hot Spot in the South China Sea

About 60 islands make up the Spratly Island chain, which lies in the South China Sea between Vietnam and the Philippines (for its location, see Figure 7.37). They are an idyllic tourist destination where divers can hire luxury boats to explore the coral reefs and palm-lined beaches of remote atolls. However, the islands are much more significant for their strategic location between the Pacific and Indian Oceans. During World War II, Japan used the islands as a base for attacking the Philippines and Southeast Asia. Still more significant, as much as $1 trillion in oil and gas may lie beneath the seabed around the Spratlys.

Not surprisingly, many nations covet control of the Spratlys. Six countries claim some or all of the islands, and international law is not equipped to sort out these conflicting claims (see the UN Convention on the Law of the Sea, page 272). China, Vietnam, and Taiwan claim sovereignty over all of them, while Malaysia, Brunei, and the Philippines claim some of them. During the Cold War, the competing nations felt it was too hazardous to push their claims on the islands. As the Cold War drew to a close, however, the situation became more volatile. All the contenders except Brunei placed soldiers, airstrips, and ships on the islands. In 1988, the Chinese navy invaded seven of the islands occupied by Vietnam, killing about 70 Vietnamese soldiers. In 1992, China again landed troops in the islands and began exploring for oil in a section of the seabed claimed by Vietnam. In 1995, China moved to expand its territorial designs on islands already claimed by the Philippines and has since built what it calls "shelters," which look like fortifications to the Filipinos. Late in the 1990s, Malaysia occupied two disputed Spratly Islands and began building on one of them. In 2004, Vietnam began renovations of an airport on one of the islands, saying it was

necessary to boost tourism there. China condemned the construction as a violation of its territorial sovereignty, an accusation it repeated as Vietnam negotiated natural gas and oil development with Indian and other foreign companies in the Spratlys. The situation moved to the brink of all-out conflict in 2011, when Chinese ships sabotaged Vietnamese oil exploration vessels operating within what Vietnam claimed was its own economic zone in the South China Sea (which Vietnam calls the East Sea—the same name the Koreans call the Sea of Japan).

As China's need for energy grows, so does its assertiveness over the Spratlys, the nearby Paracel Islands, and the Senkaku Islands, among others (see •Figure 7.K).

"China has indisputable sovereignty over the South China Sea islands and adjacent waters," a Chinese Foreign Ministry spokesman asserted in 2011.[11] Such a bold claim makes smaller powers in the region fear that China could turn small islands into "permanent aircraft carriers" that China could use to dominate them militarily. China even used a tiny submarine to plant a Chinese flag on the bed of the South China Sea. Not surprisingly, the United States took a position in this geopolitical chess match: in 2010, Washington declared that the South China Sea was of strategic importance to the United States. The United States is solidifying its position to offset China: it has an alliance with the Philippines, and has drawn closer militarily to Vietnam. Clearly, an incident in these remote islands could trigger a much wider and more serious conflict in Asia.

248

276

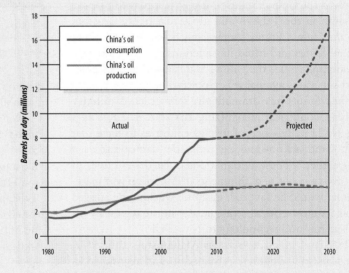

• **Figure 7.K** With China's oil appetite expected to grow enormously, every potential source of oil—like the Spratly Islands'—takes on unprecedented importance.

Source: Based on 'The World in 1900,' in R. R. Palmer, ed., Atlas of World History (Chicago: Rand McNally, 1957), pp. 169–171.

one in Mumbai in December 2008, could lead to the **mutually assured destruction (MAD)** of Pakistan and India. Others, particularly in South Asia, argue that the weapons represent the best deterrent against such a MAD conflict, as they did for decades between the countries of NATO and the Warsaw Pact. However, few people anywhere argue against the view that the nuclear rivalry between India and Pakistan has taken an enormous economic toll in two nations that need to wage war on poverty.

To the United States, both India and Pakistan are **pivotal countries**, defined by several influential historians as countries

whose collapse would cause international migration, war, pollution, disease epidemics, or other international security problems.[12] The disposition of their nuclear arsenals is one of the main reasons for their being considered pivotal. If the government of Pakistan were to fall, Pakistan's nuclear weapons could come into the possession of a rogue element or a new government hostile to the West. One of those influential historians, Francis Fukuyama, declared that Pakistan is "the most dangerous place in the world."[13] India has much to fear from Pakistani scenarios. Either a right-wing or an Islamist government in Pakistan would be far more likely than the current regime to take India on in the disputed Kashmir region. Such a confrontation could set the stage for a nuclear exchange between Pakistan and India, not only decimating those countries but also sending economic shock waves around the world. India's recently surging economy has been a major force in India's recent diplomatic overtures to Pakistan; nothing discourages investment like war or the fear of it.

U.S.-Pakistan Relations since 9/11

Pakistan became even more pivotal with the events of 9/11. Up to that point, the United States had slapped tough economic sanctions against Pakistan (and less severe ones against India) for the nuclear weapons tests. The United States pursued a diplomatic courtship with India, both as a counterweight against China (India's longtime foe in the region) and as a means of expressing displeasure with Pakistan (China's ally) for helping Afghanistan's Taliban harbor Osama bin Laden. But the United States and Pakistan did an abrupt diplomatic about-face and embraced one another after 9/11. Pakistani President Pervez Musharraf instantly dropped support for the Taliban and quietly allowed the United States to use its territory to prepare for the assault on the Taliban and al-Qa'ida in Afghanistan. In return, the United States forgave much of Pakistan's debt to the United States and lifted its sanctions against Pakistan. To avoid isolating India, the United States also lifted its post-nuclear test sanctions against that country.

The United States has dramatically strengthened military ties with both countries, recognizing Pakistan in 2004 as a major non-NATO ally—one of just 14 countries to have that designation—and acknowledging India the same year as a "strategic partner." These designations opened the doors of both countries to major new shipments of American war material. But India would be favored: in 2007, Washington singled out India as its most critical strategic ally in the region by offering nuclear fuel and technology to New Delhi.

Under U.S. and allied attacks in Afghanistan after 9/11, Taliban and al-Qa'ida fighters retreated to and regrouped in western Pakistan, particularly in the semiautonomous federally administered tribal areas (FATA; see Figure 7.29). Here the populace is mainly Pashtun and is sympathetic to the causes of their Taliban ethnic kin and their al-Qa'ida spiritual kin (•Figure 7.22). This is an area where Pakistani government authority has long been kept at bay and where it was considered extremely difficult to insert Pakistani forces on a sustained basis. Bowing to American pressure, Pakistan's government periodically undertakes military operations in the FATA but considers a permanent or ongoing presence there as potentially detrimental to its survival.

The United States is deeply unpopular in Pakistan, so the sustained insertion of U.S. troops in Pakistan (in search of al-Qa'ida, for example) would be very destabilizing, perhaps even result in rebellion or coup against Pakistan's government. This could lead to the rise of an anti-American administration in Pakistan and raise questions about the disposition of Pakistan's nuclear weapons. To avoid that scenario, the U.S.

Joe Hobbs

• **Figure 7.22** The Pashtun of western Pakistan and Afghanistan are a tight-knit ethnic group with a long history of defiance against foreign rule. Most are strongly opposed to American interests in the region.

military has relied heavily on predator drones to stalk and kill al-Qa'ida and Taliban suspects in Pakistan.

Even those unmanned missions are politically risky for the United States, so imagine the risk involved in sending U.S. troops on the mission that killed Osama bin Laden, deep in Pakistani territory. The fact that bin Laden had lived not far from Pakistan's capital, and just around the corner from a Pakistani military academy, raised many questions on the quality and the loyalty of Pakistani intelligence forces. Many analysts and politicians in the United States asserted that elements within Pakistan's intelligence service, the ISI (Inter-Service Intelligence agency), had helped protect bin Laden. There are also suspicions that Pakistani intelligence and armed forces also continue to support the Taliban both in Afghanistan and Pakistan. The American military and intelligence communities often perceive Afghanistan and Pakistan as a single "theater" of operations, and had a neologism for it until the Pakistanis objected: *AfPak*.

What Does North Korea Want?

There are also major concerns about nuclear weapons in Northeast Asia. Ever since suffering Hiroshima and Nagasaki, Japan has had an official policy never to develop or use atomic weapons. But Japan worries about the potential nuclear threat from three adversaries, all of which possess nuclear weapons: Russia, China, and North Korea. The Japanese feel that the West dismissed potential threats from Russia too readily when the Soviet Union dissolved. Japan still has territorial disputes with Russia, particularly involving the four Kuril Islands of Kunashiri, Etorofu, Shikotan, and Habomai (lying just off the northeastern coast of Hokkaido; see Figure 7.56), which the government of Josef Stalin seized at the end of World War II. To show that Russia was intent on maintaining its control, Russian President Medvedev visited one of the islands in 2010 and talked about raising living standards there. The Japanese were furious.

Japan and China also have a territorial dispute concerning the East China Sea, and if, as expected, major oil reserves are discovered there, relations between the two countries could deteriorate. China continued to test nuclear weapons through 1996, adding to tensions in Japan. Finally, Japan fears reunification in Korea, which, as a former Japanese colony from 1910 to 1945, has a historical dislike of Japan. There is speculation that Japan may do the unthinkable and the officially disavowed—develop nuclear weapons—to counter the perceived threat of North Korea's nuclear weapons program.

The world looks nervously at the troubled relations between North and South Korea and between North Korea and the West. A crisis flared in 1994 when North Korea refused to permit full inspection of its nuclear facilities by the International Atomic Energy Agency (IAEA). The country was suspected of separating plutonium that could be used in making nuclear bombs. Over time, the United States, South Korea, and Japan worked out an agreement in which North Korea would agree to freeze development of nuclear weapons in exchange for the others providing fuel oil and assistance in building nuclear power plants. Those nuclear reactors would be of the light water variety, much less likely than North Korea's plutonium-based reactors to be "dual use" for both military and civilian purposes. The Clinton administration trumpeted the agreement as a success, but it had several flaws. One was that North Korea did not

have to dispose of its existing nuclear fuel; it simply had to stow it safely away—meaning that it could quickly be reactivated.

President George W. Bush in 2002 proclaimed that North Korea was one of three countries making up an "axis of evil," along with Iraq and Iran. Late that year, North Korea dropped a virtual bombshell on the United States by admitting—after being confronted with evidence collected by U.S. intelligence agencies—that it did indeed have an active nuclear weapons program (developed, as it turns out, with Pakistani technical assistance). This admission followed earlier strident denials of such a program by North Korean officials.

North Korea's admission could have brought an end to the economic assistance in the form of food and fuel that the United States, South Korea, and Japan had been providing for a decade to the beleaguered country. What, then, was North Korea hoping to accomplish by revealing its nuclear weapons program? There are several possibilities. The least likely is that North Korea was instigating a military confrontation with the United States and its allies South Korea and Japan. In any war scenario, those allies would obliterate North Korea. The United States is obliged to defend South Korea in the event of a war with the North and would effectively wield its huge military advantage (•Figure 7.23). However, that victory would come at an enormous price for South Korea and possibly Japan. North Korea has a standing army of nearly 1 million soldiers, an impressive military arsenal including short- and medium-range missiles, a stockpile of chemical and biological weapons that it might not be shy to use, and presumably at least a few nuclear weapons that it could deploy. The casualties in South Korea would be enormous, whether from a conventional or an unconventional assault from the North.

Joe Hobbs

• **Figure 7.23** A little humor eases the tension for U.S. troops serving along what may be the world's tensest border, the "no-man's-land" separating North and South Korea. Even a minor incident here could touch off a conflagration that might take a huge toll in human lives.

It is much more likely that by disclosing its nuclear weapons program, North Korea sought guarantees that it could avoid war and also gain even more assistance from the West. For years, North Korea had only to mention the prospect of reactivating its nuclear weapons program to get more concessions from the United States and its allies—especially food aid during a succession of droughts and floods in the 1990s. In effect, North Korea was extorting money from the United States and its allies, which have been happy to ante up as a means of containing the rogue nation.

A series of on-again, off-again negotiations known as the Six Party Talks (held between North and South Korea, the United States, China, Japan, and Russia) culminated in 2007 with an agreement: The United States would help unfreeze North Korean funds in a Macau bank, remove North Korea from its list of countries supporting terrorism, lift trade sanctions in place since the Korean War ended in the 1950s, and along with its negotiating partners restart the flow of fuel oil to North Korea; in return, North Korea would close its nuclear plants and allow them to be inspected by international monitors. In the deal, North Korea did not promise to dismantle its nuclear facilities at Yongbyon, but only "abandon" or "disable" them, which it did do by the end of that year. North Korea is not required to destroy any existing stocks of nuclear weapons, and much uncertainty remains. The United States apparently hopes that the involvement of China and the other countries will make North Korea less likely to restart its nuclear program this time. North Korea watchers did not know how the policies of the country's new leader, Kim Jong-Un, might differ from those of his deceased father Kim Jong-Il.

Islands, Sea Lanes, and Islamists

As will be seen in Section 7.6, Indonesia—another pivotal country from the U.S. perspective—is an ethnically complex nation whose integrity has at times been threatened from a variety of secessionist movements. These disturbances range all the way from Aceh (pronounced "a-chay") in the country's far west to Papua (formerly Irian Jaya) in the east (see Figure 7.37). A possible **domino effect** scenario for Indonesia concerns foreign powers: If one province falls or proclaims its independence, many others will, and the largest country in the region will break apart. The United States and other countries fear what would happen to vital international shipping lanes should several new and perhaps militant states emerge from Indonesia's fragmentation. Washington is concerned that many states, some of them anti-Western, could emerge in Indonesia's place and threaten American interests in the country—notably its oil, natural gas, and copper resources and its location astride vital shipping lanes.

There have been scores of pirate attacks on vessels plying the strategic Strait of Malacca (see Figure 7.37), a critical chokepoint through which a quarter of the world's trade passes, including two-thirds of the world's shipment of liquefied natural gas and half of all sea shipments of oil (it ranks second only

to the Strait of Hormuz as an oil shipping lane). Security in the Strait of Malacca is a source of great concern to Japan, which imports 80 percent of its oil on ships that pass through the strait. Aceh's vast natural gas deposits and Papua's copper and gold are also seen as critical in the global economy, so like Indonesia, many world powers are anxious to see that they stay in the "right" hands. The United States gives generous amounts of foreign aid to Indonesia, but some Indonesians resent the United States, perceiving it as anti-Muslim in the wake of 9/11.

U.S. intelligence agencies have kept a watchful eye on Southeast Asia, and Indonesia in particular, fearful that it might emerge as an Islamist terrorist hearth. The country with the largest number of Muslims, Indonesia is a secular rather than a religious state, and a moderate form of Islam prevails in politics and in everyday life. What concerns Western intelligence authorities and Indonesia's moderate leadership are the more militant Islamist organizations that gathered strength after 9/11. One of these is **Laskar Jihad**, created in 2000 to mobilize Indonesian Muslims to fight Christians in Indonesia's Sulawesi and Muluku islands and Americans throughout the country. Another is **Jemaah Islamiah (JI)**, led by Abu Bakar Bashir, whose nominal role was preacher in an Islamic school. Western authorities are much concerned about such religious schools, known as *pesantrens* in Indonesia. Like the *madrasas* of Pakistan and the Arab heartland of the Middle East, these are venues for both religious and secular learning, and some of the schools' curricula are imbued with virulent anti-Western content. Baasyir admitted to having tutored 13 of the men arrested in connection with a plot to blow up Western targets in Singapore.

This concludes an introduction that has set the stage for further exploration of land and life in the subregions and countries of South and East Asia. The journey continues in the region's far west.

7.6 Regional Issues and Landscapes

South Asia

South Asia has an enormous range of ethnic groups, social hierarchies, languages, and religions. This is the most culturally complex area of its size on earth. Its cultures have roots that are deep in antiquity. Its modern engagements with the world are very important in the realms of security, economics, and population, among others.

Afghanistan: Graveyard of Empires

Like Sudan and Turkey, Afghanistan (•Figure 7.24) is a country that is transitional between regions; some geographers place it in Central Asia or the Middle East. It is a land of limited resources, poor internal transportation, and little foreign trade. Afghanistan is one of only five landlocked nations in South and East Asia. As we have seen, most landlocked countries are disadvantaged by their location and are poor. Afghanistan is

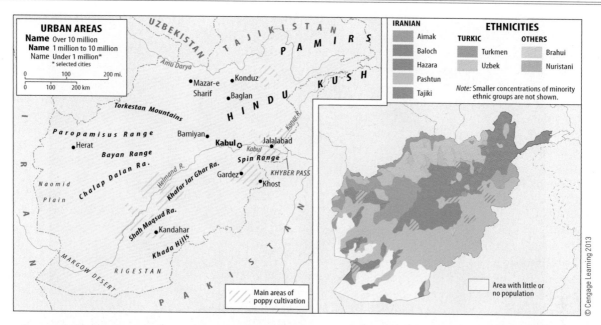

• **Figure 7.24** Principal features and ethnic groups of Afghanistan.

indeed one of the world's poorest countries and also ranks high as a failed state (seventh on that list).

Afghanistan is also a land of much conflict. It has repeatedly drawn the world's most powerful countries into conflict. Why? In large part because of its geographical situation: historically, it has occupied an important strategic location between India and the Middle East, and between India and Central Asia. Major caravan routes crossed Afghanistan, and a string of empire builders sought control of its passes. And although poor in natural resources, Afghanistan is neighbor to oil-rich Iran. This fact was critical to the Carter Doctrine discussed in Chapter 6 and below.

High and rugged mountains dominate Afghanistan. Most of the country's 32 million people live in irrigated valleys around the fringes of the mountains. Northern Afghanistan borders three of the five central Asian countries, and millions of people on the Afghan side are related to peoples of those countries. The most populous area is the northeast, particularly the fertile valley of the Kabul River, where the capital and largest city, Kabul (population 2.3 million; elevation 6,200 ft [1,890 m]), is located. Most of the inhabitants of the southeast are Pashtuns (also known as Pushtuns or Pathans). Their language, Pashto, is related to Persian. The Pashtuns are the largest and most influential of the numerous ethnic groups that make up the Afghan state; the next largest are the Tajik and Hazara, respectively.

The independent-minded tribal Pashtun people have always been loath to recognize the authority of central governments. In 1893, during the days of Britain's Indian Empire, British authorities and the Afghan King drew the "Durand Line" to separate their respective areas of influence. The line sliced right through traditional Pashtun lands, and Pashtun peoples have never accepted it. This border area saw warfare among Pashtun tribes, tribal raids on British-controlled areas, and British punitive expeditions against the tribes. In those days,

the city of Peshawar—on the Indian (now Pakistani) side of the Khyber Pass into Afghanistan—became the hotbed for British and Russian agents playing what came to be known as the **Great Game** of vying (and spying) for strategic influence in Central Asia. The British were especially fearful that the Russians might use Afghanistan as a staging area to invade India, the "jewel in the crown" of Britain's colonies. To this day, Peshawar feels like a wild, lawless frontier city. It plays an important role in regional trade of weapons and drugs, especially heroin. For generations, the people of one of its outlying communities, Daraa, have specialized in replicating whatever foreign weapons a buyer desires, from old British muskets to modern anti-aircraft guns (•**Figure 7.25**).

• **Figure 7.25** Pashtun *mujahadiin* in an Afghan refugee camp outside Peshawar, Pakistan. When I took this photo, these men were taking a break from fighting Soviet troops occupying Afghanistan. They and their sons could well have become Taliban fighting American troops in Afghanistan. Their bought their weapons in Daraa, the nearby factory and market for arms.

THOMAS J. ABERCROMBIE/National Geographic Stock

• **Figure 7.26** The Bamiyan Valley and the Koh I Baba Range of north-central Afghanistan. This photo conveys some of the country's rugged beauty and pastoral character.

Afghanistan is overwhelmingly agricultural and pastoral. Its agriculture relies on traditional methods and produces low yields. The country is so mountainous and arid that only about 12 percent is cultivated (•**Figure 7.26**). The most successful crop has been the opium poppy, which routinely accounts for between one-third and two-thirds of the country's gross domestic product, and more than 90 percent of the world's opium (!). In the context of the global drugs trade, portions of Afghanistan, Pakistan, and Iran have come to be known as the **Golden Crescent**. A succession of rebel armies and then Afghanistan's government under the Taliban used revenues from opium and heroin to obtain their arms. Most of the production takes place in the southern province of Helmand, where the Taliban have retrenched themselves most successfully. Principally to reduce funding to the Taliban, and also to crack down on illicit drugs, Britain and the United States have stepped up efforts to eradicate opium. A policy split has developed: the NATO mission in Afghanistan worries that in the absence of crop alternatives, elimination of the poppy crop could turn destitute farmers into vengeful fighters. One option rarely called for but perhaps most appropriate in Afghanistan's future would be to allow the country to produce opium legally for pharmaceutical needs, as a handful of countries, including Turkey and India, are allowed to do.

Through most of the twentieth century, this highland country was remote from the main currents of world affairs. After the Islamic revolution in Iran in 1979, however, Afghanistan's location next to that oil-rich country made it once again the target of foreign interests. The Soviet military intervention of 1979 and the ensuing devastation catapulted the country into world prominence. The USSR's motives for its invasion of Afghanistan may have included a desire to prevent by force the spread of Iranian-style Islamic fundamentalism into Afghanistan (which is 80 percent Sunni and 19 percent Shiite) or into the central Asian Soviet republics deemed vital to the superpower's security. For its part, with the Carter Doctrine the United States warned the Soviet Union that it would not tolerate further Soviet expansionism.

The Soviet war in Afghanistan brought widespread killing and maiming of civilians, destruction of villages, burning of crops, killing of livestock, sabotage of irrigation systems, and sowing of land mines over vast areas. Soviet ground and air forces caused several million Afghan refugees to flee into neighboring Pakistan and Iran. Arms from various foreign sources filtered into the hands of the *mujahidiin*, the anti-Soviet rebel bands that kept up resistance in the face of heavy odds. The United States was one of the powers supporting the rebels and in that sense waged a proxy war against the Soviet Union in Afghanistan. In its waning years, the USSR recognized that it could not win its "Vietnam War," and its troops withdrew from Afghanistan by 1989. It is estimated that of the 15.5 million people who lived in Afghanistan when it was invaded in 1979, at least 1 million died, 2 million were displaced from their homes to other places within the country, and 6 million fled as refugees into Pakistan and Iran. About two-thirds of these refugees have returned to Afghanistan.

Support that the United States and moderate Arab states such as Egypt and Saudi Arabia lent to the *mujahidiin* and sympathetic Arab fighters soon came back to haunt those countries in a phenomenon that came to be known as **blowback**.

Emboldened by their victory against the Soviets, the most militant Islamists among the fighters turned their attention to the United States and its Middle Eastern allies. Chief among them was Osama bin Laden, who developed "the base" (*al-Qa'ida* in Arabic) of thousands of Afghans, Arabs, and other anti-Soviet war veterans he could call on to wage a wider *jihad*. Bin Laden's organization trained an estimated 10,000 fighters in al-Qa'ida camps in Afghanistan and from this unlikely, remote setting devised spectacular acts of terrorism: an assassination attempt on Egyptian President Hosni Mubarak, the bombing of U.S. military barracks in Saudi Arabia, the bombing of U.S. embassies in Kenya and Tanzania, the attack on the *U.S.S. Cole,* and the 9/11 attacks.

184

The *mujahidiin* succeeded in overthrowing the Communist government of Afghanistan in 1992, but after that, rival factions among the formerly united rebels engaged in civil warfare. During this period, most other countries utterly neglected Afghanistan, inadvertently promoting the growth of militant movements (which tend to thrive in remote, underdeveloped, and ungoverned regions). By 1996, one of the rebel factions, the **Taliban** (backed by Saudi Arabia and Pakistan), gained control of most of the country, including the capital of Kabul. Proclaiming itself the sole legitimate government of Afghanistan, the Taliban imposed a strict code of Islamic law in the regions under its control and gained international notoriety for its austere administration. The Taliban removed almost all women from the country's workforce, forbade public education of girls, and outlawed "un-Islamic" practices such as dancing, flying kites, watching television, keeping birds, and trimming beards.

The Taliban continued to make advances against its opponents inside Afghanistan (particularly the **Northern Alliance**, whose leader, the Ahmed Shah Massoud—to his people a hero known as "the Lion of Panjshir"—was assassinated just days before and in apparent preparation for the 9/11 attacks) and, by 2001, controlled 95 percent of the country's territory. The neighboring central Asian countries, Russia, and even Iran grew increasingly fearful of the spread of the Taliban's extreme interpretation of Islam into their nations. Russia ended up supporting some of the rebels it had fought because they were now fighting the Taliban. Russia's once unlikely ally in supporting those rebels was the United States, which successfully lobbied the United Nations to levy economic sanctions against Afghanistan in 1999. The United States hoped that economic losses would pressure the Taliban into turning over bin Laden for prosecution, and it also announced a $5 million reward for information that would lead to bin Laden's capture or death.

The bin Laden bounty rose into the tens of millions of dollars after September 11, 2001. Named as the mastermind of the attacks against New York and Washington, bin Laden became the "world's most wanted" as U.S. President George W. Bush evoked a Wild West vow to have him captured "dead or alive." In crafting its pronounced war against terrorism (which it promised to carry anywhere in the world necessary), the U.S. administration identified the Taliban as al-Qa'ida's mentor and targeted both organizations for elimination.

184

Within a month of the attacks on the United States, American warplanes and special forces struck in **Operation Enduring Freedom** against Taliban and al-Qa'ida facilities and personnel around Afghanistan. The British military assisted in the air campaign, and most of the ground fighting was carried on by the Northern Alliance, the Taliban's rival and nemesis. The United States successfully persuaded Pakistan to drop its support of the Taliban and join the effort, and much of the military campaign was based on U.S.-Pakistani cooperation. One after another of the Taliban's urban strongholds fell, including Kabul and the Taliban spiritual capital, Kandahar. Withering assaults from U.S. warplanes crushed the tunnels and other hideouts used by al-Qa'ida guerrillas, and many al-Qa'ida prisoners were taken to the U.S. naval base in Guantánamo Bay, Cuba. The elusive bin Laden, however, along with the Taliban leader Mullah Omar, slipped away, presumably seeking safe harbor in Pakistan. 375

The reconstruction of Afghanistan began almost as soon as the Taliban were driven from power. Billions of dollars in aid went into the building of highways and other infrastructure, hospitals, and schools. The numbers of Afghans enrolled in primary schools rose from 25 percent in 1999 to over 90 percent, and girls and women are again being educated. Progress is slow because the country is so devastated, the security situation so tenuous, and the Afghan government so corrupt. Of all those billions of dollars in aid, some unknown amount has been siphoned off to line the pockets of those responsible for distributing it. U.S. Secretary of State Hillary Clinton said it was "heartbreaking" to see that billions of dollars of U.S. aid to the country had been wasted.

American forces and the Afghan government headed by Hamid Karzai effectively took control of only a small area around Kabul, and NATO troops were stationed in the troubled south of the country. Much of the country remains in the hands of warlords whose allegiance is up for grabs. Al-Qa'ida and Taliban fighters operate in the rugged countryside, sometimes carrying fighting into Kabul and other cities.

The mission of this war—the longest in U.S. history—changed substantially over the years. Initially, its purpose was clear: rid Afghanistan of al-Qa'ida and its Taliban supporters. With al-Qa'ida effectively displaced into Pakistan and elsewhere, the focus fell on the Taliban. American and NATO ground troops fought seemingly endless numbers of Taliban fighters, who fought with classic insurgent hit, run, and hide tactics. Western forces became engaged in a conflict that paralleled the long war of Soviets against the *mujahidiin* in Afghanistan. Although the Taliban and other Afghans, including civilians, bore heavy losses, the conflict also cost more than 1,800 American lives and over $400 billion (through 2011). Nearing the end of his first term in office, U.S. President Barack Obama steadily drew down the number of American forces in Afghanistan and announced that in 2014 the mission there would change from one of combat to one of support. U.S. military analysts came to the conclusion that the fight against the Taliban was unwinnable. For the losses of life and retreats suffered in earlier conflicts by the British and Russians

A **closer** LOOK

American Strategic Interests in Your Education

Arguably one of the most important answers to the question *Why has it been difficult for the United States to be involved in Afghanistan?* is something that cannot be seen in an atlas: the fact that the United States has been short on knowledge about the cultures, tribal allegiances, and other social complexities of Afghanistan, and in people fluent in some of the peoples' languages. The U.S. government is playing catch-up, in part by funding undergraduate and graduate-level college students to study the languages and become immersed for long periods in countries deemed critical to national security. See for example the Boren Scholarships, which "provide American undergraduate students with the resources and encouragement they need to acquire skills and experiences in areas of the world critical to the future security of our nation" (http://www.borenawards.org/). Boren Scholarships include language study of four of the five languages of Afghanistan depicted in Figure 7.10: Pashtu, Persian (Dari), Uzbek, and Turkmen. Scholarships like this can open many doors for you. They are also available for graduate study.

In the educational context, national security goes far beyond concerns with military thinking. Consider this definition of national security on the Boren Scholarship program overview: "It draws on a broad definition of national security, recognizing that the scope of national security has expanded to include not only the traditional concerns of protecting and promoting American well-being, but also the challenges of global society, including: sustainable development, environmental degradation, global disease and hunger, population growth and migration, and economic competitiveness." These concerns are virtually a definition of the field of geography. In a sense, your study of geography in this book is contributing to national security.

in Afghanistan, the country had come to be known as the "Graveyard of Empires." Historians began to add America to the list of empires bloodied by Afghanistan.

We can now check our answers to the geographic question that opened the book: *Suggest two or three reasons why it has been difficult for the United States to be involved in Afghanistan.* Even if you knew very little or nothing about Afghanistan, you might have looked at it in this book or in an atlas and proposed that it is mountainous, making supply logistics difficult, and that it is landlocked, also complicating those logistics. Now that you have read this section, you can supply many more details about its geopolitical and other contexts (see A Closer Look, this page).

Faith, Sectarianism, and Strife

The Indian subcontinent is one of the world's culture hearths, rich in a historical legacy and modern diversity of ethnicities, social practices, and faiths. Ironically, this cultural wealth often threatens the well-being and security of the subcontinent's peoples.

There are major social and political divisions within the subcontinent that relate to religion. Religious differences between the two largest religious groups—Hindus (whose religion is described on pages 228–230) and Muslims (whose religion is described in Chapter 6)—are often troublesome. Hinduism was the dominant religion at the time Islam made its appearance in this region. Muslims are significant minorities in the region's non-Muslim countries, including 13 percent of the population of India and 4 percent of Nepal. Muslims make up a seemingly small percentage of India's population, but they number over 160 million people, making India home to the world's third largest Muslim population, behind Indonesia and Pakistan. Muslims are majorities in Pakistan (97 percent) and Bangladesh (90 percent).

Seldom in history have two large groups with such differing beliefs lived in such close association with each other as Hindus and Muslims in the subcontinent. Islam holds to an uncompromising monotheism and prescribes uniformity in religious beliefs and practices. Hinduism is monotheistic for some believers but polytheistic for others, and asserts that a variety of religious observances is consistent with the differing natures and social roles of humans. The exuberant and loud celebrations of the Hindu faith are a striking contrast to the austere ceremonies of Islam. Islam has a mission to convert others to the true religion, whereas most Hindus regard proselytizing as essentially useless and wrong; one is born or reborn as a Hindu, so conversion is not an issue. Islam's concept of the essential equality of all believers is in total contrast to the inequalities of the caste system endorsed in Hinduism (see page 248). Islam's use of the cow for food is an anathema to Hinduism. And then there are religion-based politics, which often turn differences into discord and conflict.

During the colonial era, the colonizing power frequently favored one ethnic or socioeconomic group over another, sowing the seeds for eventual discord. This was the case in British India, where the formerly subordinate Hindus came to dominate the civil service and most businesses. Many Muslims feared the results of being incorporated into a state with a Hindu majority, and their demands for political separation led in 1947 to the creation of two independent countries: the secular but predominantly Hindu nation of India (•Figure 7.27) and the Muslim nation of Pakistan (Figure 7.29). The respective secular and religious identities of both countries remain critical issues today. Pakistan had two parts, West Pakistan and East Pakistan, separated by 1,000 miles (c. 1,600 km) of Indian territory. In 1971, an Indian-supported revolt in East Pakistan led to the birth there of the new independent country of Bangladesh.

Immediately preceding and following partition in 1947, violence broke out between Hindus and Muslims, and hundreds of thousands of lives were lost in wholesale massacres. More than 15 million people migrated between the two countries. Pakistan today has a large population known as **Mohajirs**, or "migrants," the Muslim immigrants and their descendants who poured into the two Pakistans from India when partition occurred. Particularly in Karachi and elsewhere in the southern province of Sind, violent conflict frequently occurs between rival factions among the Mohajirs and between the Mohajirs and other ethnic groups, including Pashtuns (Pathans) and

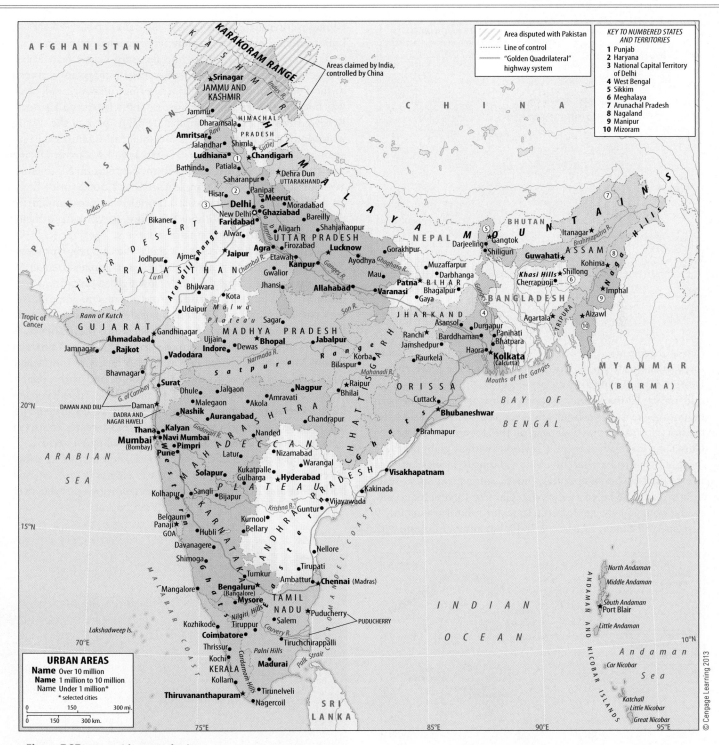

• **Figure 7.27** Principal features of India.

Biharis. Large-scale migrations of ethnic groups like these are part of humanity's troubled experience.

The partition of colonial India precipitated all kinds of problems for the new countries. Some of them are still prominent today. The largest problem is the disputed Kashmir region. Since 1947, India and Pakistan have been in conflict over the status of Kashmir (called Jammu and Kashmir in India), a disputed province straddling their northern border

(•**Figure 7.28**). Before independence, Kashmir was a princely state administered by a Hindu maharaja. About three-fourths of Kashmir's estimated population of 15 million is Muslim, which is the basis of Pakistan's claim to the territory. But under the partition arrangements, the ruler of each princely state was to have the right to join either India or Pakistan, as he preferred. Kashmir's Hindu ruler chose India, which is the legal basis of India's claim.

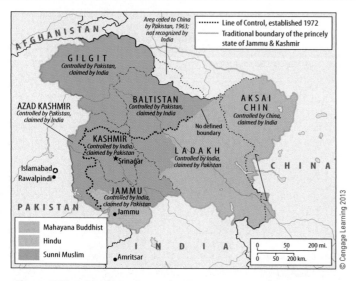

• **Figure 7.28** Political boundaries of Kashmir.

After partition, fighting between India and Pakistan led to the demarcation of a cease-fire line—the "line of control" that still separates the forces—leaving eastern Kashmir, with most of the state's population, in India and the more rugged western Kashmir in Pakistan (see •Figure 7.29). Beginning in the mid-1950s, China pressed its own claims to remote northern mountain areas of Kashmir and occupied some of the border territories. Pakistan ceded part of the occupied territory to China and established friendly relations with its giant northern neighbor, but India rejected China's claims (relations between these countries remain chilly). India's subsequent small-scale military actions failed to dislodge Chinese forces. India now holds about 55 percent of the old state of Kashmir; Pakistan, 30 percent; and China, 15 percent. India has another dispute with

China over China's control of 34,750 square miles (90,000 sq km) in India's northeastern state of Arunachal Pradesh.

Three wars between India and Pakistan have effected little change in Kashmir. In 1965, conflict began in Kashmir, spread to the Punjab, and escalated to a brief but indecisive full-scale war involving tanks, airborne forces, and widespread air raids. Renewed hostilities in Kashmir in 1971, as part of the war in which India supported the revolt of Bangladesh against Pakistan, again did not alter the political landscape of Kashmir. During that conflict, the Siachen Glacier in the Karakoram Mountains, at about 20,000 feet (2,800 m), earned the title of "world's highest battlefield." As recently as 2011, negotiations to demilitarize the glacier have failed. Without this move, it is proving difficult for the two countries to normalize relations.

Kashmir's Muslim majority escalated the campaign for secession from India in 1989, and the strife has continued ever since. India and Pakistan came perilously close to the brink of war over Kashmir in 2002. Late in 2001, armed insurgents attacked India's parliament in New Delhi in a failed attempt to assassinate Indian politicians. India's government identified the assailants as members of two Muslim rebel movements (Lashkar-e-Taiba and Jaish-e-Muhammad) based in Kashmir. A far more spectacular attack—also attributed to these two groups, and with possible links to al-Qa'ida—took place in December 2008. In India's commercial capital of Mumbai, scores of civilians were killed and injured in multiple terrorist strikes against a Jewish center, transport stations, and hotels and restaurants frequented by Westerners.

The upper portion of the Indus River and many of its tributaries are in Kashmir (•Figure 7.30; see location in Figure 7.29), and both India and Pakistan seek as much control as possible over

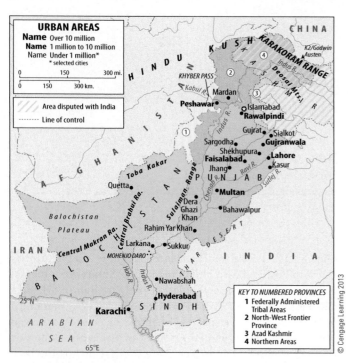

• **Figure 7.29** Principal features of Pakistan.

• **Figure 7.30** Western Kashmir, now in the hands of Pakistan, is a land of massive mountains and grinding glaciers. The world's second highest mountain, K2, is here, as is the "world's highest battlefield" (see page 246). People manage to eke out a livelihood almost all the way to timberline at 13,000 feet. This is the mighty Indus River, not far from K2 and from its source.

these waters. The geopolitical contest for this water threatens to become very dangerous. The militant Lashkar-e-Taiba has threatened to blow up India's dams in its portion of Kashmir. The **Indus Waters Treaty** of 1960 prescribed exactly how much water each side could use (80% to Pakistan, 20% to India), and India's dams have so far not violated that treaty. Pakistan fears, however, that India could use its dams as a weapon to deprive downstream Pakistan of water. Pakistan is sending chilling warnings to India, for example, in a newspaper editorial that read, "Pakistan should convey to India that war is possible on the issue of water and this time war will be a nuclear one."[14]

In the 2002 and 2008 incidents, Indian citizens and many government authorities laid blame on Pakistan's government. Foreign diplomats exerted tremendous efforts to avert war between the South Asian giants. Pakistan banned the militant groups after the 2002 assault (causing many Pakistanis to blame their government for "selling out" to the West following the 9/11 attacks), but the Mumbai attacks revealed how incapable Pakistan might be of preventing dangerous provocations against India. There are also questions about how much control Pakistani leaders have over their own intelligence service, the ISI, which may contain rogue elements intent on continued support of the militants fighting India.

The Kashmir conflict has been very costly to both sides, with more than 65,000 people killed since 1965, about half of those since 1989. India's international reputation suffered because of the country's unyielding position on the region. Pakistan has steadily spent about a third of its budget on defense, mostly focused on a potential engagement with India over Kashmir. There would be economic benefits for both India and Pakistan if lasting peace were achieved. "Kashmir is a Paradise on Earth," wrote Samsar Kour.[15] The stunning snow-crowned peaks and flower-laden valleys of this western Himalayan region (more habitable than the Pakistani-held portion of Kashmir) once lured millions of Indian tourists and ranks of Western trekkers each year, and the tourism development potential is enormous.

Sectarian violence (known in the region as **communal violence**) between Hindus and Muslims is a frequent threat to India's social and political fabric. Contention over places sacred to both faiths sometimes ignites widespread violence. The sixteenth-century Babri Mosque in the city of Ayodhya in Uttar Pradesh (location in **Figure 7.27**) was a place of prayer for Muslims on a spot revered by the Hindus as the birthplace of the god-king Ram, an incarnation of the god Vishnu. Backed by Hindu nationalists in the provincial government, a mob of about 250,000 Hindu fundamentalists demolished the mosque in December 1992 with the intention of building a Hindu temple on the site. The ensuing communal violence between Hindus and Muslims throughout India, even among the cosmopolitan and usually tolerant urbanites of Mumbai, left thousands dead.

Ever since this incident, as a means of garnering votes from Hindu Indians, Hindu politicians have made veiled promises to proceed with construction of the temple. In 1998, the BJP platform included a pledge to build a Hindu temple atop the ruins of the mosque at Ayodhya. His apparent support for the temple helped sweep the BJP leader, Atal Vajpayee, into power as prime

minister that year. A 2010 high court ruling divided the site, with two-thirds going to Hindus and one-third to Muslims. The conflict over this multireligious sacred space in India has parallels with the Temple Mount dispute between Israel and the Palestinians. Politicians in India today are increasingly aware of the threat that Hindu-Muslim discord poses to the country's development.

In India's Punjab, where Sikhs make up 60 percent of the population and Hindus, 36 percent, the 1980s and early 1990s were violent years during which about 20,000 people died in armed clashes. Sikh factions were intent on establishing their own homeland of **Khalistan**, prompted to do so in part by New Delhi's plans to divert water from the Punjab. They challenged Indian authority and turned Amritsar's Golden Temple, the holiest site in the Sikh faith, into their military stronghold. In a controversial move to quell the revolt, Prime Minister Indira Gandhi ordered troops to storm the Golden Temple in June 1984. The resulting deaths and desecration led directly to Gandhi's assassination by her own Sikh bodyguards later that year. Indian authorities accused Pakistan of arming the Sikh militants.

Other trouble spots for India are the northeastern states of Assam, Manipur, Nagaland, and Arunachal Pradesh, an extraordinarily diverse region with more than 200 ethnic groups and followers of many different religions. Several rebel groups have been active in this region for years, agitating for autonomy and independence from India. The government has begun investing more resources in the area, especially infrastructure to boost trade ties with adjacent Myanmar. There are also troubles in half of India's states from Maoist insurgents, known as **Naxalites**, who claim to champion the causes of the rural poor and promise to carry violent rebellion into India's cities and ultimately expand their "compact revolutionary zone" to include the entire country. Some Maoists have sought support among India's roughly 700 tribal groups, whose population is about 85 million.

A conflict over land use has arisen in the eastern state of Orissa, where ancient tribal traditions are in a deadlock with modern industry. Steel companies want to tap the state's deposits of iron ore and coal and develop mills near them. But tribal people living in the forests atop those mineral deposits insist that their identity is tied to the land, forest, and water. If they were displaced by mining, they say, they would lose their cultural and social identities.

Sri Lanka has two major ethnic groups, distinguished from each other by religion and language: the predominantly Buddhist **Sinhalese**, making up about 75 percent of the population, and the Hindu **Tamils**, about 10 percent (**•Figure 7.31**). The light-skinned Sinhalese are an Indo-Aryan people who settled in Sri Lanka about 2,000 years ago. The dark-skinned Tamils, whose main area of settlement is the Jaffna Peninsula and adjoining areas in the north, are descendants of early invaders and more recent imported tea plantation workers from southern India. The migrations of Tamil tea workers from India to Sri Lanka, and their subsequent travails there, are part of the legacy of British colonial rule. Later in this chapter, we will see how the British also relocated Tamils to work the rubber plantations of Malaya. Ethnic discord in modern Malaysia is a legacy of British colonialism.

Sri Lanka's civil conflict reflected antagonisms between ethnic groups and discontent with economic and political conditions,

• **Figure 7.31** Principal features of Sri Lanka.

especially among the minority Tamils. Between 1983 and 2009, more than 70,000 people were killed and many more made homeless as the Tamils of the country's north have fought for autonomy or independence from Sri Lanka's majority Sinhalese government. Tamils complained that the Sinhalese treated them as second-class citizens and deprived them of many basic rights. Tamil fighters called the **Tamil Tigers**, officially the Liberation Tigers of Tamil Eelam (LTTE), wanted to establish their own homeland, Tamil Eelam, along the east, north, and west of the country. The rebels succeeded in holding wide swaths of northern Sri Lanka (see Figure 7.31). Unrest in rebel-held areas made it particularly difficult to rebuild northern sections of the area hit hard by the December 2004 tsunami. Government troops gradually pushed the Tamil forces back, and declared victory over them in 2009 (•Figure 7.32).

254

• **Figure 7.32** The war between Tamils and Sinhalese has ended in Sri Lanka, but Tamil protests against Sinhalese rule continue where they can be practiced. This is a demonstration in Sydney, Australia. The depiction if of the Tamil Tiger founder and leader Velupillai Prabhakaran, who was killed just weeks before this demonstration. The red on the map on the right depicts parts of Sri Lanka held by the Tigers at their maximum advance. The Tamil flag, not fully in view, depicts a tiger jumping through a halo and crossed bayonets.

Arguably, one of the victors in the Sri Lankan civil war was China, which sided with the government and financed the construction of a port in the country's south. Some analysts argue that China will use this port to project its power into the Indian Ocean, where much of the world's oil and other valuable cargo is in transit. They see Sri Lanka as one pearl in China's **"String of Pearls"** strategy to gain critical allies in a line across the region. The largest "pearls" in this part of the string are China's allies Pakistan and Myanmar. India, which wants to see itself projecting power across these sea-lanes, views this as encirclement by China. Keep in mind that China and India are enemies and have been ever since they fought a high-altitude border war in 1962 (which India lost).

Known to early Arab seafarers as Seredib, the "island of Serendipity," Sri Lanka is a breathtakingly beautiful country. The return of peace to Sri Lanka should reopen the doors to international tourism.

The Caste System

One ancient Hindu belief is that every individual is born into a particular **caste**, a social subgroup that determines the individual's rank and role in society. Castes form a hierarchy, with the Brahmin caste at the top (comprising about 5 percent of all Hindus) and three others (Kshatriya, Vaisya, and Sudra) below. Caste membership is inherited and cannot be changed. Particular castes are associated with certain religious obligations, and their members are expected to follow traditional caste occupations. Marriage outside the caste is generally forbidden, and meals are usually taken only with fellow caste members.

At the bottom of the social ladder are the **Dalits**, or *scheduled castes*, once known as **untouchables**, accounting for about 20 percent of all Hindus. They are not part of the caste system but are "outcastes," and according to Hindu belief, they are not reincarnated. The untouchables were so called because they traditionally performed the worst jobs, such as the handling of corpses and garbage; therefore, their touch would defile caste Hindus. I will never forget the horrified look on my taxi driver's face when I shook hands with a man whose job was to cremate bodies on the "ghats" of Mumbai.

This ancient, rigid social system is changing. India's constitution outlawed the caste system in 1950, but a single decree could not undo thousands of years of tradition. Brahmin privileges are now being increasingly challenged and restricted by more laws. Indian law explicitly forbids recognition of Dalits as a separate social group. Confronting the fact that upper castes account for less than one-fifth of India's population but command more than half of the best government jobs, the Indian government has instituted an affirmative action program to allocate more jobs to members of lower castes. Indian leaders of high caste have championed the Dalits' cause for both moral and practical reasons. Politicians use caste divisions to their advantage to bring out voters, promising if elected to act in the interests of their respective large social blocs. In 1997, a Dalit was elected president of India for the first time in the country's history. Since then, Dalits have won more regional and local offices.

The Dalits are voting in greater numbers than ever before, with the conviction they can use India's democratic system to overcome ancient discrimination against them. They are making progress. Disintegration of the caste system has been especially rapid in the cities, where people of different castes must mingle in factories, in public eating places, and on public transportation. In India's vast rural areas, the caste system is more entrenched. There is also a north–south difference: lower caste southern Indians began agitating against the caste system early in the twentieth century and kept up their struggle. In northern India, by contrast, there was less effort to tackle the caste system. Caste is so important in northern politics that it is said that voters don't cast their vote; they vote their caste. Remarkably, partly because of the loosening of caste restraints, economic growth is much stronger in southern than in northern India.

Keeping Malthus at Bay

Poverty and human health are serious problems in South Asia, and the prospect of continued high population growth raises the question of whether growth in food supplies can avert the proverbial Malthusian crisis (see Chapter 3).

India's population has surged since independence, from 352 million in 1947 to 1.2 billion in 2011. In recent decades, the overall trend has been toward lower population growth rates. Fifty years ago, the average Indian woman had six children; now she has three. That rate is still well above the natural replacement rate, however, that would keep the population steady. The population base is already so vast, and the base of its age structure diagram so wide (about 50 percent of the population is younger than 25), that even modest growth will add huge numbers. India is predicted to overtake China as the world's most populous country by 2040. There is uncertainty about how high India's population will grow. Estimates range from 1.5 to 1.9 billion. Most Indian decision makers think that even the low end of that range is too high. Unlike China, India is a democracy and cannot simply declare how many children a couple can have. Indian states rather than the central government establish laws and policies, and there is a wide range of them in the realm of population. Some states offer financial incentives, such as a $100 bonus if couples delay their first child's birth.

For many people, the name *India* invokes an image of grinding poverty, and with good reason: it is estimated that fully 40 percent of the population is "abjectly poor," defined as living on less than $1.25 per day, and almost half of India's children are malnourished. But perhaps in defiance of Malthus, an increased population has so far succeeded in producing more economic resources overall. A few people in India are very wealthy, and there is an emerging middle class of as many as 300 million people—the world's largest middle class (•Figure 7.33). India's economy has grown impressively in recent years. An apparent problem is that the benefits of the growing economy are unevenly distributed. Most of the economic growth benefits those who are already better off, widening the gap between rich and poor.

The issue of population growth is also critical for Pakistan, which has never mounted an effective family planning program, mainly because of opposition from Muslim religious authorities, who regard birth control as an intervention against God's will

• **Figure 7.33** This Indian family visiting the Red Fort in Agra is among the growing ranks of the subcontinent's middle class. Note the Taj Mahal gleaming in the distance.

(Pakistan's birth rate in 2011 was 25 per 1,000, compared with 21 per 1,000 in India). Muslim Bangladesh, along with Nepal and Bhutan, also have high birth rates, but they are much lower in Bangladesh than they once were. In the 1970s, the average woman had seven children, whereas now she has two, and contraception use nationwide has increased from 4 percent then to more than 50 percent now. South Asia's success story in bringing population growth under control is Sri Lanka, with a birth rate of just 17 per 1,000 and an annual population growth of 1.3 percent.

Agricultural output in South Asia has increased since independence. Most notably, despite its huge and growing population, India had achieved self-sufficiency in grain production, with a fourfold increase in output since 1950. Almost half a million "ration shops" sell subsidized food staples to the country's poorest people. The successes of agriculture in the subcontinent have been due mainly to the increased use of artificial fertilizers, the introduction of new high-yield varieties of wheat and rice associated with the Green Revolution, more labor provided by the growing rural population, the spread of education, the development of government extension institutions to aid farmers, and better irrigation. "Mother Nature," however, has the last word. Sixty percent of India's farmland is rain-fed rather than irrigated, and when summer monsoons fail to deliver rainfall, as they did in 2009, the yields of valuable rice, sugarcane, and other crops plummet. This causes immediate difficulties for people (in the form of higher food prices) and the nation (in terms of trade and economic growth).

Many experts in agriculture and economics are calling for a "Second Green Revolution" in India to address agriculture in the context of climate change. The focus would be more nuanced than the Green Revolution of the 1960s. Now there is a need for nationwide cooperation to improve irrigation, capture rainwater and conserve groundwater while boosting crop yields. There is much at stake, not just food security but human well-being and equity. During droughts like the one in 2009, debt-ridden farmers resort to the dire options of selling their wives or daughters (for $83 to $248) or committing suicide.

As the sale of wives and daughters suggests, one of the challenges confronting both rural and urban India is to improve the status of women. Despite a 1961 ban on dowries (money and gifts given by a bride's parents to the groom), the practice is continuing. So is the rate of killing brides who do not provide enough dowry. The burden placed on the bride's family is one reason that many prospective parents choose to abort female fetuses, which can be detected with ultrasound technology. India banned this use of ultrasound in 1996, but the practice continues. Female feticide—particularly prevalent among India's more educated and affluent, who can afford the ultrasound screening—and infanticide have taken tens of millions of young and unborn girls. In some of India's states, especially Haryana, Punjab, Delhi, and Gujarat, one result is a serious shortage of local brides; for every 1,000 men, there are 900 women. The gap is being filled by marriages arranged with women from distant Indian states and Bangladesh.

Low-Lying Bangladesh and Maldives: Canaries in the Climate Change Coal Mine?

Formerly East Pakistan, Bangladesh is about the size of the state of Iowa or the nation of Greece—not an important fact until one considers that it is home to 150 million people (•Figure 7.34). With its limited resources, people-overpopulated Bangladesh is South Asia's third-poorest country on a per capita GNI PPP basis, after mountainous Bhutan and Nepal.

The Bangladesh portion of the Ganges-Brahmaputra Delta ranks, along with the Indonesian island of Java, as one of the two most crowded agricultural areas on earth. This area is subject to catastrophic flooding. Massive tropical cyclones (hurricanes) ravage Bangladesh regularly in September and October. Deforestation upstream on the steep slopes of the Himalayas has increased water runoff and sediment load in the Ganges, also contributing to Bangladesh's flood problems (the region's principal rivers may be seen in Figure 7.34).

Ironically, although there are concerns about climate change–related flooding in Bangladesh, there are even more worries about climate change–related water shortages and droughts. Climate change models point toward accelerating depletion of the

• **Figure 7.35** Melting of glacial ice in the Himalaya and other great mountain ranges of South Asia threatens future water supplies for people downstream. This is Gokyo Glacier, close to Nepal's Mt. Everest.

so-called **"Water Towers of Asia"**—the glaciers and snow cover of the Himalaya, Karakoram, and other great mountain ranges that feed the region's 10 great rivers and scores of smaller ones. The perennial ice mass here is exceeded in volume only by the ice of the Arctic and Antarctic, earning the Tibetan Plateau and environs the name "The Third Pole." The region's roughly 18,000 glaciers, like most others in the world, are melting at a steady, relentless pace (•Figure 7.35). With temperatures on the Tibetan plateau rising much faster than the world average, a respected Chinese glaciologist predicts that two-thirds of them will be gone by 2050. These glaciers supply water to about 40 percent of the world's population! The prospect of their melting away is so alarming that longstanding enemies India and China are putting away their differences and collaborating in scientific research on climate change.

Bangladesh and adjacent portions of India must also keep a wary eye on the potential for the sea level to rise if the world's temperatures increase, as predicted by many climate change models. The country's environment minister has forecast that 20 percent of Bangladesh will be underwater by 2020 if nothing is done to control global warming. The Intergovernmental Panel on Climate Change recognizes this as one the world's most endangered areas, predicting a rise in sea level there of 23 inches (c. 58 cm) by 2100. On the Indian side of the delta, scientists are documenting the shrinkage of the river delta islands known as the Sundarbans. Renowned as habitat for the Bengal tiger, the Sundarbans are succumbing to a combination of sea level rise and deforestation that makes them very prone to flooding and erosion.

The roughly 1,100 islands that make up the Maldives (population 300,000; capital, Male, population 90,000), an independent country since 1965, appear to be the very essence of tropical paradise (•Figure 7.36). So prized are its palm-blessed and coral-fringed beaches that more than half a million tourists, mainly Europeans, visit the country each year. More than 60 percent of the country's foreign-currency earnings come from tourism, with fishing and clothing providing most of the rest. Some sources list the Maldives as South Asia's wealthiest country on a per capita basis. The greatest wealth is in the

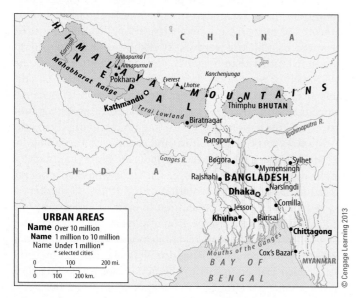

• **Figure 7.34** Principal features of Bangladesh, Nepal, and Bhutan.

tourist sector, where it is concentrated in so few hands that a potentially dangerous rich–poor divide has developed.

There may be other troubles in paradise. More than 80 percent of this country's very limited land area consists of limestone atolls less than 3 feet (90 cm) above the sea. If sea levels were to rise, as predicted by many common climate change scenarios, the entire country could rapidly be submerged. This prospect led to the Maldives' president's famous cry at a climate change conference: "We are an endangered nation!"

Southeast Asia

Deforestation of Southeast Asia

Southeast Asia today is composed politically of 11 countries: Myanmar (formerly Burma), Thailand, Laos, Cambodia, Vietnam, Malaysia, Singapore, Indonesia, Timor-Leste (formerly East Timor), Brunei, and the Philippines (•Figures 7.37 and 7.38). With the exception of Thailand (which was never colonized) and

• **Figure 7.36** This aerial view of Male (pronounced *mah*-lay), the Maldives' capital, reveals how vulnerable the site is to storms and sea level rise. This is the world's most densely populated city.

Peter Essick/Getty Images

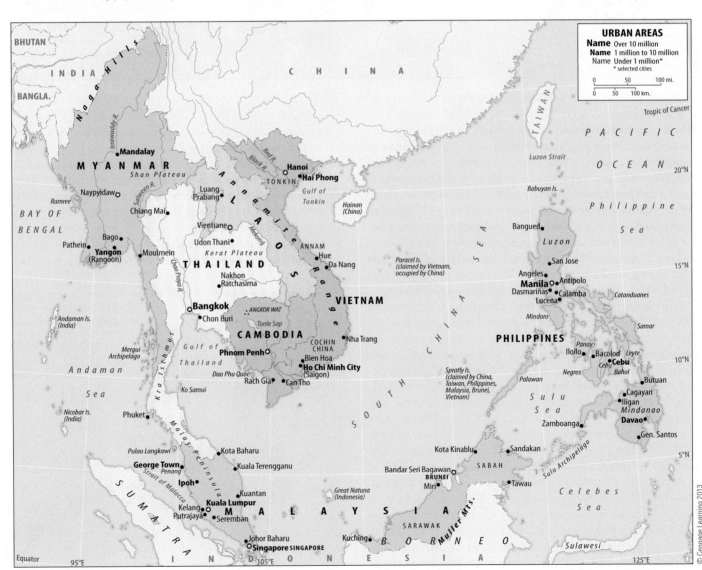

• **Figure 7.37** Principal features of Myanmar, Thailand, Cambodia, Laos, Vietnam, Malaysia, Brunei, Singapore, and the Philippines.

© Cengage Learning 2013

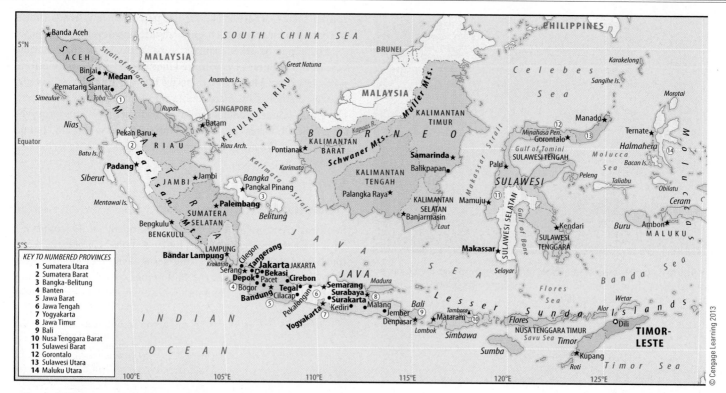

• **Figure 7.38** Principal features of Indonesia and Timor-Leste.

Timor-Leste (which gained independence first from Portugal and then from Indonesia), all of these states became independent from colonial powers after 1946. These 11 nations are fragmented both politically and topographically. Many of the countries are also culturally fragmented and have experienced strife among different ethnic groups. Outside intervention has sometimes complicated and worsened local discord, producing enormous suffering in the region and lasting trauma among many of the foreign soldiers who fought the determined inhabitants of Southeast Asia. The region also must contend with environmental problems related both to its economic progress and its lingering poverty.

Near areas of dense settlement in Southeast Asia, there are still some large areas covered in primary forest, but they are shrinking as population expands and development progresses. Many scientists view the destruction of the region's tropical rain forest as an international environmental problem, especially because of its contributions to climate change (•Figure 7.39). In 2012, only 35 percent of Thailand's original forest cover remained. Poorer neighbor Myanmar had about a 45 percent forest cover. Vietnam lost more than one-third of its forest between 1985 and 2012, by which time only about one-third of the original forest cover remained. Deforestation is advancing most rapidly in Malaysia. Officially, 63 percent of the land area is still forested. However, official statistics include palm oil plantations, so these figures should be evaluated carefully. Only 11 percent of Malaysia's remaining forests are considered "pristine". Deforestation is also progressing quickly in Indonesia, where about half the land is forested. Malaysia is now the world's largest exporter of tropical hardwoods. Virtually all of the primary forests in Peninsular Malaysia have already been cleared, and it is projected

that commercial logging will decimate those remaining outside protected areas in Malaysian Borneo quite soon.

Although Southeast Asia's tropical forests are smaller in total area than those of central Africa and the Amazon Basin, they are being destroyed at a much faster rate. The main forces of deforestation are commercial logging for Japanese and Chinese markets, and clearing of forest to plant oil palm trees, from which palm oil is rendered (•Figure 7.40). The demand for palm oil is soaring for use in biofuels and in household

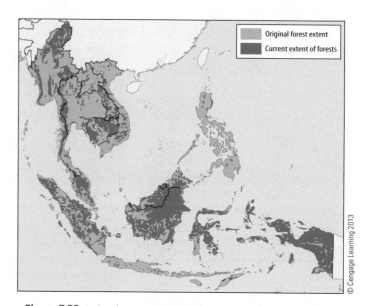

• **Figure 7.39** In Southeast Asia, natural forest cover has been rapidly reduced by many human uses, especially commercial logging and farming.

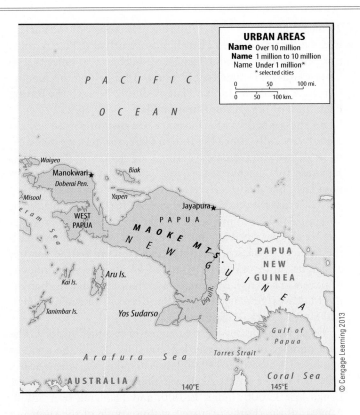

• **Figure 7.40** Over much of the island of Borneo, tropical rain forest has been replaced by palm oil plantations.

products including Vaseline and Dove soap, and food including Kit Kat candy bars. Environmentalists fear the irretrievable loss of plant and animal species and the potential contribution to global warming that deforestation in this region may cause.

Remarkably, the destruction and burning of forests on the islands of Borneo and Sumatra have made Indonesia the world's third largest emitter of carbon dioxide, after China and the United States. There is a new scientific consensus that a fifth of the world's greenhouse gas emissions is due to land use change, with Indonesia and Brazil the most responsible. Indonesia accounts for 40 percent of that; thus the cutting and burning of rain forest mainly to produce palm oil there is responsible for 8 percent of global greenhouse gas emissions. The conversion of peat bogs—some inhabited by Sumatran tigers and bears—to tree plantations for paper and pulp also releases carbon dioxide. As deep as 50 feet, and made up of decomposed trees and plants, the peat stores billions of tons of carbon dioxide.

Due to its large contributions to greenhouse gases through forest conversion, Indonesia is negotiating many projects for the Reducing Emissions from Deforestation and Forest Degradation (REDD) program described in Chapter 2 (see page 35). One major paper and pulp company working in Indonesia is negotiating a REDD deal in which by protecting forests it will receive carbon credits that it can sell on the global carbon exchange for large sums of money. ³⁴

Environmentalists have put pressure on large companies like Cargill, Unilever, and Nestlé to stop buying palm oil produced unsustainably. Nestlé has responded favorably, promising a "zero deforestation" policy. Malaysia and Indonesia have passed laws to slow the destruction, with Indonesia taking the aggressive step of banning the use of fire to clear forests. However, enforcing such legislation has so far proved impossible. Hundreds of Indonesian and Malaysian companies, most of them large agricultural concerns (for example, palm oil plantations and pulp and paper companies) with close ties to the government, continue to use fire as a cheap and illegal method of clearing forests. The ³⁸² World Wildlife Fund estimates that 70 percent of Indonesia's exported trees are illegal, many of them removed by "timber barons" in the country's national parks while bribed park and police officials look the other way. The urgent need for construction materials after the devastating tsunami of 2004 (see p. 254) further increased the pressure on the country's forests.

There is a growing movement in the consuming countries to boycott the tainted timber by selling only certified "good wood" and only palm oil that has been produced from sustainably harvested oil palms. The U.S. company Home Depot, for example, sells only imported woods bearing the Forest Stewardship Council (FSC) logo. Even the source countries are coming on board the certification bandwagon. Malaysia began work on an agreement with the European Union in 2011 to become the first country to sign a voluntary agreement guaranteeing that all timber exports are harvested legally. Malaysia would promise the ³³⁰ European Union that its trees would not be cut from protected forests and that endangered tree species would not be harvested.

Deforestation in Southeast Asia has many actual and potential transboundary consequences. In 1997, a widespread drought attributed to the warming of Pacific Ocean waters by the phenomenon known as El Niño turned the annual July–October burning season into a conflagration of human origin. Deliberately set fires raced out of control over large areas of Sumatra and Borneo, resulting in a choking haze that shut down airports, closed schools, deterred tourists, and caused respiratory distress to millions of people in Indonesia, Malaysia, Singapore, the Philippines, Brunei, and Thailand. Less widespread but still severe smog crises struck Malaysia and Indonesia, particularly the island of Sumatra, in subsequent years, and Indonesian authorities have fought back by revoking the licenses of plantation companies.

Many of the plants and animals in these forests are **endemic species** (found nowhere else on earth). Indonesia, which contains 10 percent of the world's tropical rain forests, is known, after Brazil, as the world's second most important **megadiversity country**, with about 11 percent of all the world's plant species, 12 percent of all mammal species, and 17 percent of all bird species within its borders. The Southeast Asian region is also particularly significant in biogeographic terms because of the so-called **Wallace's Line**—named for its discoverer, the English naturalist Alfred Russel Wallace—that divides it. Nowhere else on earth is there such a striking local change in the composition of plant and animal species in such a small area on either side of this divide. East of the line (separating the Indonesian islands of Bali and Lombok, for example), marsupials are the predominant mammals, and placental mammals prevail west of the line. Similarly, bird populations are remarkably different on either side of the line.

The Great Tsunami of 2004

On December 26, 2004, the most cataclysmic natural event of modern times afflicted a vast portion of the earth. Beneath the Indian Ocean, just off the northwestern coast of the Indonesian island of Sumatra, a gargantuan "megathrust" earthquake measuring magnitude 9.3—with the force of more than 26 megatons of TNT, or 1500 times the power of the nuclear weapon that devastated Hiroshima—struck at 6:58 A.M. local time. In the Java (Sunda) Trench, as much as 750 miles (1,200 km) of fault line slipped 60 feet (over 18 m) along the subduction zone where the Indian Plate is sliding beneath the Burma Plate (part of the greater Eurasian Plate; see Figure 2.1).

As the seabed of the Burma Plate instantly rose several meters vertically above the Indian Plate, the massive displacement of seawater created a series of tsunamis (*tsunami* is Japanese for "harbor wave"; the phenomenon is sometimes popularly but incorrectly referred to as a "tidal wave") that pulsed across the Andaman Sea and the Indian Ocean at up to 500 miles (800 km) per hour. On the open ocean, such waves are barely perceptible, but they break on coastlines with the force of giant storm surges. The tsunami roared ashore, at heights up to 45 feet (14 m), in 14 countries: Indonesia, Malaysia, Thailand, Myanmar, Bangladesh, India, Sri Lanka, the Maldives, Seychelles, Somalia, Kenya, Tanzania, Yemen, and South Africa (•Figure 7.41).

These were the deadliest tsunamis in recorded history. The total number of dead exceeded 200,000. By far the greatest number of deaths (over 130,000) was in Indonesia, where Aceh's principal city of Banda Aceh was obliterated (•Figure 7.42). Sri Lanka lost 35,000; India, 12,000; and Thailand, over 5,000. Almost all the deaths occurred within a mile of the shoreline and were caused by blunt force from debris and by drowning. An estimated one-third of the fatalities were children. There were reports of children running out to investigate fish stranded by retreating seas—a characteristic precursor to a tsunami—only to be struck seconds later by the wall of water.

As many as 2 million people were made homeless by the disaster. An unprecedented international relief effort, designed to feed and house these refugees and to contain the spread of epidemic diseases, began within hours. In many cases, it was clear that recovery efforts were profoundly successful, sometimes in unexpected ways. In India's Tamil Nadu State, for

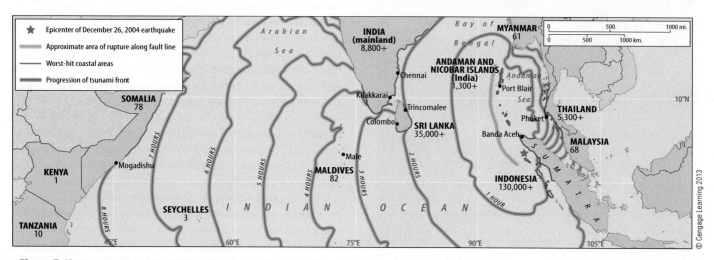

• **Figure 7.41** The epicenter of the 2004 earthquake, which registered magnitude 9.3, was located off the coast of Sumatra, Indonesia. It created a tsunami of tremendous force and scope.

• **Figure 7.42** Banda Aceh before and after the tsunami.

example, there were new roads, housing and utilities, and jobs training. New homes were titled in women's names (because women were deemed more trustworthy than men), giving women an unprecedented rise in status as concerns respect. In the tsunami's wake, the social fabric was transformed.

Tsunami experts concur that countless thousands of lives would have been saved had the affected Indian Ocean region been equipped with the same kind of tsunami early warning system now in place in the Pacific Ocean.[16] That investment had not been made because historically about 90 percent of the world's tsunamis have occurred in the Pacific and because many of the nations afflicted lacked the financial resources for the wave sensors and other infrastructure required by the system. Installing an early warning system in the Indian Ocean basin became an urgent priority, and a system was up and running by 2006. Had this system been operational in 2004, Indonesia would have had just a few minutes' warning, but Thailand would have had two hours, and Somalia, seven hours to prepare for the disaster.

Japan's tragic tsunami in 2011 is discussed on page 16. Its mechanism was similar: an upthrust earthquake along a section of underwater plates. Its magnitude was slightly less (9.0 versus 9.3), and the loss of life in wealthy, tsunami-prepared Japan far less (15,000 versus 200,000). It was such a massive tsunami, however, that Japan's engineering works were simply overwhelmed.

Misrule in Myanmar

Myanmar (population 54 million, Figure 7.37) is centered in the basin of the Irrawaddy River and includes surrounding uplands and mountains.

Britain conquered what was then called Burma in three wars between 1824 and 1885. It became an independent republic outside the British Commonwealth in 1948. Civil war has been almost constant since independence in a country where ethnic minorities make up about 40 percent of the population. Communism, targeted persecution, and ethnic separatism have motivated rebellions by ethnic Karens and Shans, a number of less numerous hill peoples, and Muslims of the Rohingya ethnic group who fear the establishment of a pure

Buddhist state. Beginning around 1999, however, Myanmar's ruling military government reached cease-fire agreements with most of these ethnic groups (except the Karens) and gave these groups what it called **contingent sovereignty**, offering them more civil rights and economic opportunities. The government also encouraged profits made by some of these groups in the drug business to be "invested" in national development.

The economy has been badly damaged by fighting and for years was mismanaged under a form of rigid state control promoted as the **"Burmese Way to Socialism."** Myanmar became the poorest non-Communist country in Southeast Asia, a sad contrast with its reputation in early independence years for having the best health care, highest literacy rate, and most efficient civil service in Southeast Asia. It used to be the world's largest rice exporter; it now exports less than one percent of the world total. Myanmar's government spends comparatively less on health care for its citizens than all but three countries in the world.

In the 1990s, Myanmar began to abandon socialism in favor of a free-market economy and enjoyed considerable economic growth, with China its main trading partner and military ally. Most of the fruits of this economic growth, however, lined the pockets of the ruling military. The **State Law and Order Restoration Council (SLORC)**, the military government that seized power in Burma in 1988 (and changed the country's name to Myanmar), has yielded little to popular pressure for democratization. Myanmar remains one of the world's most repressive places to live. Access to the Internet was prohibited until 1999 and is still strictly regulated. Foreign journalists are banned, and citizens may not allow foreigners into their homes. It is illegal to gather outside in groups of more than five. Giant green billboards across the country read: "Crush all internal and external destructive elements as the common enemy."

In an apparent move to gain strategic depth from its own population and from any outside threats, the government in 2005 established a new capital at Naypyidaw, in the country's remote central region. The move did nothing to quell popular discontent with the regime. Widespread anti-government protests, with broad participation of Buddhist monks, broke out all over the country late in 2007. Authorities responded with deadly force and the insurgency faded—at least temporarily.

In the wake of this political turmoil, Myanmar's isolated leaders turned a natural disaster into a humanitarian catastrophe. On May 2, 2008, a category 4 strength typhoon (hurricane) named Nargis slammed ashore across the Irrawaddy Delta. In this most agriculturally productive and densely populated region of Myanmar, more than 135,000 people lost their lives, and an estimated two and a half million lost their homes and livelihoods. The country's junta largely banned foreign aid agencies from providing medical, food, and other relief supplies. Foreign journalists were also barred entry, so the real impacts of the tragedy may never be known.

The Lady of Burma

A much publicized contest, followed around the world, has been raging between the government and one of its citizens, the Nobel Peace Prize laureate Aung San Suu Kyi, who had been placed under house arrest after leading her antimilitary political party to an overwhelming victory in the 1990 elections (which the military government nullified). In 1995, the government released "The Lady," as Suu Kyi is popularly known. She then called for dialogue with the military junta and appealed for the parliament that was elected in 1990 to be convened. The government responded by again placing her alternatively under house arrest and in prison, triggering strong economic sanctions by the United States.

Aung San Suu Kyi has called for would-be tourists to avoid Myanmar, which has abundant sites of interest but few visitors (•Figure 7.43). She argues that tourism revenue will help strengthen the repressive government. She has also been successful in gaining support abroad for boycotts of U.S. and other Western companies doing business with Myanmar. American corporations including PepsiCo and ConocoPhillips have withdrawn from Myanmar, and the U.S. government has banned

• **Figure 7.43** The huge golden Shwedagon Pagoda is the holiest Buddhist shrine in Yangon, Myanmar, attracting multitudes of the faithful and the few tourists visiting the country. It is said that the relics or bones of three earlier buddhas and the eight hairs of Gautama Buddha are located within this spectacular golden architectural masterpiece.

new investments. The United States regards the country's government as illegitimate and therefore rejects the change of the country's name to Myanmar.

The U.S. government was displeased when the American firm Unocal built a pipeline from the offshore Yadana natural gas field to a port west of Bangkok, Thailand, where the gas is used to generate electricity. Human rights groups complained that slave labor was used to build the pipeline and that profits from gas sales help prop up Myanmar's repressive government. The pipeline has to be well defended in view of the government's many domestic opponents who might try to destroy it as a means of weakening the regime. Other pipelines will serve China's energy appetite. One will carry Myanmar's offshore natural gas across the country to southern China. A parallel pipeline will carry Middle Eastern oil offloaded on the coast of Burma and piped to southern China. This will increase China's energy security, in part by avoiding the congested chokepoint of the Strait of Malacca. India is vying with China for access to Myanmar's natural gas, copper, and other natural resources. 240, 251

The United States has been reevaluating its policies toward Myanmar, especially for an important geographical reason: its geographic *situation* is extremely important, as the country sits right between the two great powers of China and India. By withdrawing from Myanmar and imposing sanctions against it, the United States has been missing out on important commercial and strategic engagements with the country, pushing Myanmar into the arms of its rival China and enemy North Korea, and with sanctions effectively punishing the people rather than the government.

The long isolation of Myanmar may be ending, perhaps because Nargis forced the country to open its doors a little bit to allow food aid. Many changes came in 2011, when the 10 countries of the regional organization known as the **Association of Southeast Asian Nations** (**ASEAN**) agreed to allow its member state Myanmar to chair the bloc in 2014. In 2011, Myanmar made changes to its economy, released hundreds of political prisoners, eased some restrictions on the Internet, media, and political parties, stopped a dam project that was backed by China, and released Aung San Suu Kyi from house arrest again. In 2012, the ruling party allowed Aung San Suu Kyi to run in national elections. She won a seat in Parliament. In the same year the U.S. and Myanmar restored diplomatic relations; they had broken in 1990.

Sex, Drugs, and Health in Southeast Asia

Southeast Asia has long been one of the world's main source areas for opium (sometimes eclipsing Afghanistan) and its highly addictive semisynthetic derivative, heroin. For decades, 242 global concern focused on combating the drug trade originating in this region and on dealing with the problem of heroin addiction in the consuming nations. More recently, owing to the emergence of HIV/AIDS, the Southeast Asian industries of prostitution and tourism have woven drugs into a complex problem of global concern.

The region's drug production is centered in the so-called **Golden Triangle**, a vernacular region comprised of the

Joe Hobbs

• **Figure 7.44** Raw opium oozes from slices in a poppy pod. Heroin, which is derived from opium, has a hearth in Southeast Asia's Golden Triangle region.

borderlands where Laos, Thailand, and Myanmar historically exercised little control over their territories (see Figure 7.37). As in Afghanistan, for a long time the absence of strong government presence together with ideal growing conditions facilitated explosive growth in drug production (•Figure 7.44). But as Thailand and Laos exercised increasing authority in the area, drug production plummeted. Today, Myanmar is the only one of the three countries with a strong opium poppy economy, but even there, production has declined. The main reason is that neighboring China, with an eye to safe, conventional trade with Myanmar, is pressuring authorities there to crack down near their common border. As poppy cultivation slumped in the Golden Triangle, it spiked in the Golden Crescent of Afghanistan. This phenomenon, in which cracking down on a problem in one place causes it to pop up in another, is called the balloon effect and applies to terrorism as well as to drugs.

In Asia, the U.S. Drug Enforcement Administration and other international antidrug authorities focus their concerns on Myanmar, where refining of heroin from opium is done in remote laboratories. In northeastern Myanmar, armed ethnic minority groups have produced both heroin and a potent methamphetamine known as *ya baa*. They smuggle raw opium into both China and India, and amphetamines into Thailand, providing investment capital to contribute to Myanmar's "development." Laundered drug money has paid for the country's recent construction boom, and the government encourages such investment from the drug sector to make up for shortfalls in legitimate export revenues.

Not all of the associated problems are exported. A flood of cheap and potent heroin hit the market in Myanmar after the current regime came to power in 1988. Heroin addiction has soared ever since. Primary delivery is by needles shared in tea stalls. The number of users continues to grow; perhaps 1 percent of the adult

population is addicted. Accompanying the heroin use is an AIDS epidemic, with about 1 percent of the adult population infected. The best available figures indicate that the HIV infection rate among heroin users in Myanmar is around 37 percent.

The epidemic that began among heroin users spread quickly to Myanmar's sex industry. Recent estimates put the HIV infection rate among prostitutes at 26 percent, three times the rate in neighboring Thailand. Many young men with AIDS are shunned by their families and go to monasteries to die. Tragically, their virus has spread among the populations of Buddhist monks (there are now 400,000 in Myanmar), who share razors to shave their heads. AIDS prevention is virtually unheard of. The government had outlawed condoms until 1993. They are now legal, and over 30 million have been distributed by the government since 2001, but this is still short of what is needed. Practically no anti-HIV drugs find their way to Myanmar.

Myanmar's prosperous, cosmopolitan neighbor Thailand also has the scourge, with a 1.3 percent HIV infection rate among adults in 2011. This is a substantial reduction from the peak infection rate the country recorded in 1994. Thailand's liberal social climate has given rise to the unique phenomenon of sex tourism, with package tours catering to an international clientele (•Figure 7.45). Thailand's overall annual income from prostitution is estimated to be up to $4 billion, with prostitutes sending large sums as remittances each year to relatives in the villages from which they migrated. However, the spread of HIV infection among intravenous drug users and among prostitutes in the red-light districts of Bangkok and the popular beach resorts of Pattaya and Phuket threatens both the country's public health and its lucrative tourist trade. The Thai government mounted an all-out war on the drug trade in 2002 that resulted in the deaths of more than 2,000 people, most of them suspected drug dealers.

There are striking differences in how this region's countries are dealing with HIV/AIDS. Thailand, regarded globally as

Taimay Jones

• **Figure 7.45** Bangkok's red-light district was a breeding ground for HIV/AIDS until the government and nongovernmental agencies aggressively promoted a safe-sex campaign.

a leader in efforts to combat the problem, has run an aggressive and increasingly successful anti-AIDS public awareness campaign, resulting in a dramatic decrease in the number of new HIV infections. Thailand distributes antiretroviral drugs to all AIDS patients at extremely low cost. Other Southeast Asian countries lag far behind. The government of the Philippines has been unsuccessful in pursuing an anti-AIDS campaign because of opposition from the country's powerful and anti-contraception Catholic clergy. Church officials have denounced the government's anti-AIDS program as "intrinsically evil" and have set boxes of condoms on fire at anti-government demonstrations. Similarly, in the conservative and largely Islamic nations of Indonesia and Malaysia, Muslim clerics denounce their governments' anti-AIDS campaigns as efforts to encourage promiscuity.

Vietnam Then and Now

The Vietnam War, which is such an important chapter in the American experience, had roots that preceded U.S. interests in the region. This conflict profoundly affected Southeast Asia and has a lingering legacy there today.

France conquered Indochina between 1858 and 1907. The French first extinguished the Vietnamese empire, which covered approximately the territory of today's Vietnam, and defeated the Chinese, whom the Vietnamese called on for help. Later, the French took Laos (in 1893) and western Cambodia (in 1907) from Thailand. France's administration of Indochina, centered in Hanoi and Saigon, left a big cultural imprint. Its major economic success lay in opening the lower Mekong River area to commercial rice production, converting both Vietnam and Cambodia into important exporters of rice.

Japanese forces overran French Indochina in 1941, initiating five decades of warfare in the area. During World War II, a Communist resistance movement led by a Vietnamese nationalist named Ho Chi Minh carried on guerrilla warfare against the Japanese. When French forces attempted to reoccupy the area after the end of war, these forces fought to expel them. In 1954, Ho's forces destroyed an entrenched French garrison at Dien Bien Phu, in the northwestern mountains. With this defeat, France withdrew from Indochina, ending a century of colonial occupation.

Four countries came into existence with France's departure (see Figure 7.37). Laos and Cambodia became independent non-Communist states. North Vietnam, where resistance to France had centered, emerged as an independent Communist state. South Vietnam gained independence as a non-Communist state that received many anti-Communist refugees from the north. Vietnamese Catholics were prominent among those who fled southward.

The partition of Vietnam into two countries was largely the work of U.S. power and diplomacy. South Vietnam was from the outset a client state of the United States, whereas North Vietnam became allied first with Communist China and then with the Soviet Union. Warfare in Vietnam gained momentum. Between 1954 and 1965, the **Viet Cong**, an insurrectionist Communist force supported by North Vietnam, achieved increasing successes in the south in its bid to reunify the country. Increasing intervention by the United States in South Vietnam escalated in

1965 and eventually involved the commitment of half a million U.S. military personnel. Why did the United States get involved on such a massive scale in a small country half a world away? Remember that those were Cold War days, and fighting communism was a foreign policy priority for the United States.

North Vietnam responded to the U.S. military escalation by fielding regular army forces against American and South Vietnamese troops. The United States never invaded North Vietnam, although American air strikes there were devastating, with more ordnance dropped on Vietnam than on all of Europe during all of World War II. American bombs, napalm (an incendiary explosive), and defoliants such as Agent Orange (dioxin) caused enormous damage to the natural and agricultural systems of Vietnam, destroying over 8,400 square miles (nearly 22,000 sq km) of forest and farmland in an effort described in military parlance as **"denying the countryside to the enemy."** Agent Orange also left a legacy of birth deformities among Vietnam's postwar generation.

The United States avoided invading North Vietnam partly because of the risks posed by Chinese and Soviet support of the North. In limiting the theater of ground warfare, however, the United States found itself unable to expel from the South both the Viet Cong and the North Vietnamese army. Those combined forces were determined, skillfully commanded, increasingly better equipped by its allies, accomplished in guerrilla tactics, and willing to bear heavy losses. More than 3 million Vietnamese soldiers and civilians died in the conflict, and about 58,000 U.S. soldiers and support staff perished. In 1973, almost all American forces were withdrawn from the costly war, which was extremely divisive at home as growing ranks of Americans questioned its rationale and protested its conduct. There are still 1,948 Americans listed as missing in action in Indochina.

North Vietnam completed its conquest of South Vietnam in 1975. Saigon was renamed Ho Chi Minh City (although most of its people still call it Saigon), after the country's leader (who died in 1969). Vietnam entered a period of relative internal peace, but there were ongoing conflicts, including Vietnam's 1979 conquest of Cambodia to oust the brutal Khmer Rouge regime, continued warfare against Cambodian and Laotian resistance groups, and a brief border war with China. Within its borders, the Communist government of the reunited Vietnam unleashed a repression against those who had assisted the Americans, including the rebellious hill tribes known collectively as Montagnards and members of the Cao Dai religion. This backlash caused a massive outpouring of refugees, including the "boat people" who risked much to flee. Most of the ethnic Vietnamese living in the United States today are those refugees and their descendants.

Vietnam (population 88 million), with its capital at Hanoi (•Figure 7.46), has restored parts of its war-torn landscape through large-scale reforestation, agricultural reclamation, and nature conservation programs, aided by international organizations such as the World Wildlife Fund. Animal species new to science have been discovered in the rugged Annamite Cordillera (Truong Son Mountains). Such natural treasures are attracting international ecotourists who provide much-needed revenue, and tourism in general is booming.

GEOGRAPHY OF THE WORLD'S GREAT RIVERS | The Mekong

The Mekong River, known as "Water of Stone" in Tibet, "Great Water" in Cambodia, "Nine Dragons" in Vietnam, and "Mother of Waters" (Mae Nam Khong, contracted to Mekong) in Laos and Thailand, is Southeast Asia's great river (its course may be seen in Figure 7.37). From its source on the Tibetan Plateau in the Chinese province of Qinghai, the Mekong snakes along a 2,600-mile (4,200 km) path through China's Sichuan and Yunnan provinces, forms the boundary between Laos and its neighbors Myanmar and Thailand, winds across Cambodia, and fans out across Vietnam's densely populated Mekong Delta region, before empting into the South China Sea.

Unlike the Chao Praya in Thailand, the Red River in Vietnam, and the Irrawaddy in Myanmar, the Mekong is not a great magnet for cities and civilizations. The biggest city on its banks is Cambodia's Phnom Penh, with just over a million people. The main factor limiting more settlement along the Mekong has been the difficulty in navigating the river above Phnom Penh. Historically, very little trade has been carried along much of its length; the oceans and overland routes have provided superior opportunities.

The Mekong is not economically unimportant, however. It is one of the world's most prolific fisheries, yielding about 2 million tons of fish each year, or twice the annual North Sea catch. There are more species of fish in the river (about 1,200) than in any other rivers but the Amazon and the Congo. Unique creatures inhabit its waters, including about 70 Irrawaddy dolphins and the giant catfish that attain 10 feet (c. 3 m) in length and 660 pounds (300 kg) in weight. Overfishing and the construction of dams have all but eliminated this superlative catfish throughout the Mekong Basin.

A big push is being made to capitalize on the Mekong's potential to produce hydroelectric power, with plans to build about 100 dams on the river and its tributaries. The first dam on the river itself (the Man Wan in China) was completed in 1993. China then went to work on five more. Vietnam is building five dams on the Mekong tributary called the Dak Hoyt, which forms part of its border with Cambodia. Thailand has dammed its main tributaries to the Mekong. Laos plans to build the most dams of all—20—and thereby become mainland Southeast Asia's biggest supplier of electricity. Laotian officials boast that their country will become "the battery of Asia." With funding from the World Bank—despite the criticism this organization has received for supporting large dams—Laos built the giant Nam Theun 2 dam. The World Bank insists it went to extraordinary lengths to minimize environmental impacts from this project.

These dams will nevertheless have enormous impacts, some of them unforeseeable.

The fish catch will definitely fall dramatically because the main reason for the Mekong's productive fishery is its regular flooding, which will be dramatically reduced. Fishermen are already complaining of smaller catches since dam building began. There are concerns that dams will ruin the ecosystem of Cambodia's Tonle Sap Lake, which is drained by a river that actually changes direction when downstream Mekong floodwaters inundate its basin. More than a million people make their living by fishing this lake. Inevitably, the downstream countries suffer many ill effects of dams. In this case, for example, Vietnam's Mekong Delta will be deprived of the fertile silt that helps make it an agricultural "breadbasket."

As always with shared river systems, there are serious questions about how the Mekong waters should be divided. Vietnam, Cambodia, Laos, and Thailand have joined to form the **Mekong River Commission**, whose biggest responsibilities are to fix the minimum amount of water each country must discharge downstream on the Mekong and its tributaries, and to agree on rules to ensure water quality. Both quality and quantity of water are threatened by upstream Myanmar and China, which have refused to join the Mekong River Commission. The downstream countries complain that China does not seem to care about what happens to them as a result of China's dam-building and other uses of the Mekong.

181, 265

181, 247, 438

• **Figure 7.46** Typical street scene in the Old Quarter of Hanoi, Vietnam's capital. Urban traffic in Vietnam is a cacophony of motorbikes, bicycles, cars, and pedestrian traffic.

Since its embrace in 1986 of free-market reforms in the policy known as *doi moi*, Vietnam's economy has improved markedly. The poverty rate fell by half in the 1990s, and in recent years Vietnam has had Asia's second-fastest-growing economy, after China's. The Communist government halted collectivized farming in 1988 and turned control of land over to small farmers under 20-year lease agreements. The results have been impressive; Vietnam since 1988 has become a major exporter of rice, second in the world after Thailand. This is a vulnerable commodity, however, especially because production is concentrated along the Mekong River, which is prone to flooding in the May–October rainy season, and to the specter of rising sea levels due to climate change (see following discussions of the World's Great Rivers and GIS). Vietnam is also pinning hopes on exports of its proven reserves of 2.5 billion barrels of oil, especially to energy-hungry China.

Vietnam in 1995 restored full diplomatic relations with the United States. Recently, the United States and Vietnam have been boosting military ties, as they share a joint interest in containing China. U.S. businesses have been scrambling for consumers in this large, promising market of mainly young consumers (about 45% of the Vietnamese are under age 25). After the two countries signed a trade agreement in 2000, Vietnam began exporting shoes, finished clothing, and toys to the United States. Both its imports and its exports boomed when the country joined the World Trade Organization (WTO) in 2007. The dual personalities of North and South are being perpetuated in the new economic climate, with most investment and infrastructural improvement focused in the South. The North is a more austere economic landscape, and northerners dominate the civil service. Along with a growing gap between the rich and the poor, the differences between North and South could slow the country's overall development.

Visualizing Climate Change with GIS

In 2010, I was invited by the government of Vietnam to advise the country's decision-makers on how best to deal with the anticipated impacts of climate change. My focus had to be on adaptation rather than mitigation (see page 34); Vietnam produces only a tiny fraction of the world's greenhouse gases, but has one of the world's longest and most densely populated coastlines. It is commonly ranked as one of the top 10 countries most threatened by climate change, particularly in the forms of rising sea levels and stronger typhoons (hurricanes; •Figure 7.47). How could Vietnam adapt to these threats?

The best way to approach this question is by the basic geographical question: *Where* will the impacts of climate change take place? Using GIS, Vietnam's Ministry of Natural Resources and Environment (MONRE) has developed maps showing exactly where rising sea levels will inundate Ho Chi Minh City and the Mekong Delta (Vietnam's most densely-populated region, and one of the world's "breadbaskets," particularly for rice production). MONRE's Director, Dr. Tran Thuc, told me that IPCC forecasts for sea level rise were far too low, because they did not take into account the melting of the world's great ice sheets in Antarctica and Greenland. Dr. Tran therefore commissioned maps depicting the impacts of 65-, 75-, and 100-centimeter (26-, 30-, and 39-inch) sea level rise by 2100—and told me he "fully expects" Vietnam to experience at least the 100 centimeter (one meter) rise.

MONRE used a GIS method known as "statistical downscaling" to derive large-scale maps of the Mekong Delta and Ho Chi Minh City from small scale global sea level models produced by the IPCC. The maps showing the one meter rise are breathtaking (the Mekong Delta map is shown in •Figure 7.48). Almost 40 percent of the Mekong Delta, and 25 percent of Ho Chi Minh City, would be underwater. About nine million people— 10 percent of Vietnam's population—would have to be relocated. These GIS visualizations are enormously useful, especially in helping Vietnamese leaders appeal for outside foreign aid to help build sea walls, dikes, and other infrastructure to adapt to climate change, and in illustrating the potential impact of climate change on global agriculture and urban infrastructure.

Much more sophisticated GIS work lies ahead. In order for sea walls to be built in just the right places and according to the correct specifications, very detailed large-scale modeling and mapping must be done. Vietnam can look to the Netherlands,

Joe Hobbs

• **Figure 7.47** In recent years, Vietnam has experienced an unprecedented number of "super-typhoons," equating to the category 4 and 5 hurricanes of the western hemisphere. I arrived in the coastal city of Hoi An hours before super-typhoon Xangsane slammed ashore on October 1, 2006, and took this photo 3 days later as the city's residents struggled to recover from extensive flooding and other damage. Many climate change scenarios anticipate more frequent and severe typhoons in the country's future. Just as in Louisiana and in Bangladesh, higher sea levels would magnify the impacts of these storms. Engineering adaptation to climate change in the Mekong Delta must therefore consider sea level rise plus storm surge.

© Cengage Learning 2013

• **Figure 7.48** The impacts of a 1-meter (39-inch) sea level rise on the Mekong Delta, Vietnam. About 40 percent of this extremely productive farmland, one of the world's breadbaskets, would be inundated by the sea by the year 2100.

where Dutch geographers are using a GIS-based process of "spatial planning" to map out exactly where engineering works must be built to stave off the rising waters of the North Sea.

Indonesia: One Country, One People, One Language, 300 Ethnic Groups

The national credo of Indonesia (see map, Figure 7.38), seen on banners and placards throughout the country, is "One country. One people. One language." The government's constitution, in an effort to promote national unity, officially recognizes four faiths: Islam, Christianity, Hinduism, and Buddhism. But the presence of some 300 different ethnic groups, combined with religious frictions, physical fragmentation, and economic problems, has made it difficult to attain peace, order, and unity in Indonesia. There is an official national language (Malay, known locally as Bahasa Indonesian), but more than 200 languages and dialects are in use. The largest ethnic group is the Javanese, who make up 41 percent of Indonesia's population. Eighty-six percent of Indonesians are Muslim, 9 percent are Christian, and 2 percent are Hindu, living mainly in Bali.

Various groups in the outer islands have resented the dominance of the Javanese, who have traditionally asserted their power in colonialist fashion over the larger, more resource-wealthy outer islands. Such animosities have escalated at times to armed insurrections. The Suharto regime (1968–1998) countered these militant expressions with a state ideology called *Pancasila*. Aimed mainly at suppressing militant Muslim aspirations, *Pancasila* was a pan-Indonesian nationalist ideology designed to neutralize all ethnic identities.

For a time, Indonesia's government vowed to break with the repressive ways of Suharto and appeased would-be separatists with promises of **liberal autonomy**, meaning that the provinces would have greater control over local administration and take more profits from the sales of local natural resources. With growing economic problems, pressure from nationalists, and accusations of corruption leveled against Suharto, however, the government shifted to using an iron fist against every Indonesian province that might aspire to follow the path the former enclave of East Timor took.

Now the independent country of Timor-Leste, East Timor was a Portuguese possession that Indonesia occupied after the collapse of the Portuguese colonial empire in the 1970s. The government carried on a long and bitter struggle against an independence movement led by East Timor's Catholics. In 1998, Indonesia and Portugal reached an autonomy agreement that would give the Timorese the right to local self-government.

Indonesia's incoming Wahid administration was bolder, however, and in 1999 allowed the people of East Timor to vote on whether they wanted outright independence. The result was overwhelmingly in favor of independence.

Since its birth, this tiny young country has had many woes: a breakdown of law and order, interethnic violence, and growing unemployment. The economic future of this poor nation could be brightened by resolution of an international legal wrangle over resources. Between Timor-Leste and Australia, in the portion of the Timor Sea known as the Timor Gap, lies a huge field of petroleum and natural gas. Because the distance between Timor-Leste and Australia is less than 400 miles (640 km)—making it impossible to apply the international standard of a 200-mile (320 km) offshore territorial limit—the two countries must come to an agreement on how to divide these undersea riches. Any deal is likely to yield significant wealth to Timor-Leste. 272

From the West ...   Of outstanding concern for Indonesia's unity is the province of Aceh (population 4.5 million), at the northernmost tip of Sumatra. In 1976, the inhabitants of Aceh, who are a predominantly Muslim people of Malayan ethnicity, began seeking independence from Indonesia. The central government had made them a promise of autonomy with Indonesia's 1949 independence from the Netherlands, but failed to keep it. The Acehnese insisted that the Jakarta government was too secular for their Muslim tastes and that too little of the revenue from sales of Aceh's abundant natural gas reserves remained in the province (which also has a wealth of oil, gold, rubber, and timber).

The secessionists' main voice was the **Free Aceh Movement** (known by the acronym **GAM**), which wanted to install a member of its own indigenous royal family as president of a free Aceh. Seeing Aceh as vital to the nation's unity and economic viability and fearful that Aceh might inspire separatism elsewhere in the country, the government refused to discuss independence for the province. Violent clashes persisted, and the death toll grew—to 12,000 between 1976 and 2004. But then an unlikely arbiter appeared. The tsunami of December 2004 concentrated its unspeakable sorrow on Aceh (see Figure 7.42). 255 The international community stepped in with a massive relief effort, and Indonesian authorities listened with unprecedented interest to Acehnese concerns. GAM dropped its demands for independence, and Indonesia agreed to allow GAM to participate in local elections. Nominally secular Indonesia allowed Aceh to become the only one of its 33 provinces to adopt Islamic *sharia* law. Redefining itself as the Aceh Transitional Committee, GAM laid down its weapons, and Indonesia scaled back its military presence in Aceh (•Figure 7.49). The government gave 176

• **Figure 7.49** Following a victory in 2009 local elections, Aceh Party members who are former rebels of the Free Aceh Movement (GAM) hold a press conference in Banda Aceh.

amnesty to the rebels and, critically, pledged that 70 percent of the revenues from Aceh's natural resources would return to the provincial government.

Through a program called Aceh Green, founded by a former rebel, Aceh may emerge as a shining example of sustainable development. In one of Aceh Green's activities, hundreds of former rebels, who know the countryside very well, are being trained by the nongovernmental organization (NGO) Flora and Fauna International as rangers to look out for wildlife poachers and illegal loggers. If the rangers are successful in protecting the Ulu Masen rain forest, Aceh could net $26 million in carbon credits from the REDD funding mechanism.

... to the East   Progress in Aceh raised hopes for a similar future for Papua, at the extreme eastern end of the Indonesian archipelago. Known formerly as Irian Jaya and West Irian, Papua is home to 3 million people of 200 different tribes and speaking 100 different languages. Most are of Melanesian origin, and most are at least nominally Christian. Collectively known as **Papuans**, they have little in common with the Javanese Muslims who control them from 2,500 miles (4,000 km) away in Jakarta. Their homeland on the western half of the island of New Guinea is of great importance to the economies of Indonesia and the United States because it contains the world's largest copper and gold mines. The American company Freeport-McMoRan, which provides more taxes to Indonesia's treasury than any other foreign source, operates these mines. There are also large assets of oil, natural gas, and timber.

Indonesia has never recognized the province's 1961 unilateral declaration of independence. It is little wonder that Jakarta refuses to let its poorest but most resource-rich province go; the Dutch, coveting this wealth, originally refused to relinquish the area when it granted independence to Indonesia. Pressure from the United States led to a UN-sponsored plebiscite on independence in 1969. The vote in favor of the region's becoming part of Indonesia was widely regarded as rigged. Recently, Indonesia has promised some autonomy and a greater share of resource revenues to Papua, but Papuans complain that these promises are unfulfilled.

China

Han Colonization of China's "Wild West"

China's growth as a land empire has involved both subjugation of people who are not ethnic Han and the colonization of those ethnic areas by ethnic Han. There are at least 56 non-Han ethnic groups in China, in total about 100 million people living mainly in the west and southwest of the country. As the Soviets did in the Soviet Union, China's Communist government granted at least token recognition of the distinctive identity and rights of five large minorities by creating **autonomous regions**. These are Guangxi, bordering Vietnam; Nei Mongol (Inner Mongolia) and adjacent, small Ningxia in the north; Xizang (Tibet); and Xinjiang in the far west (see •**Figure 7.50**).

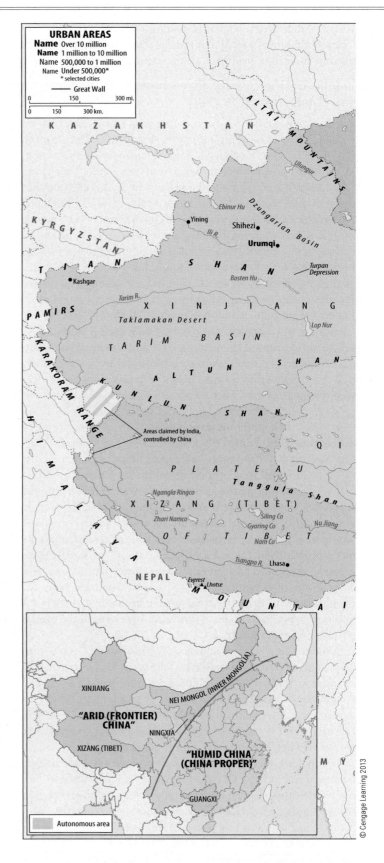

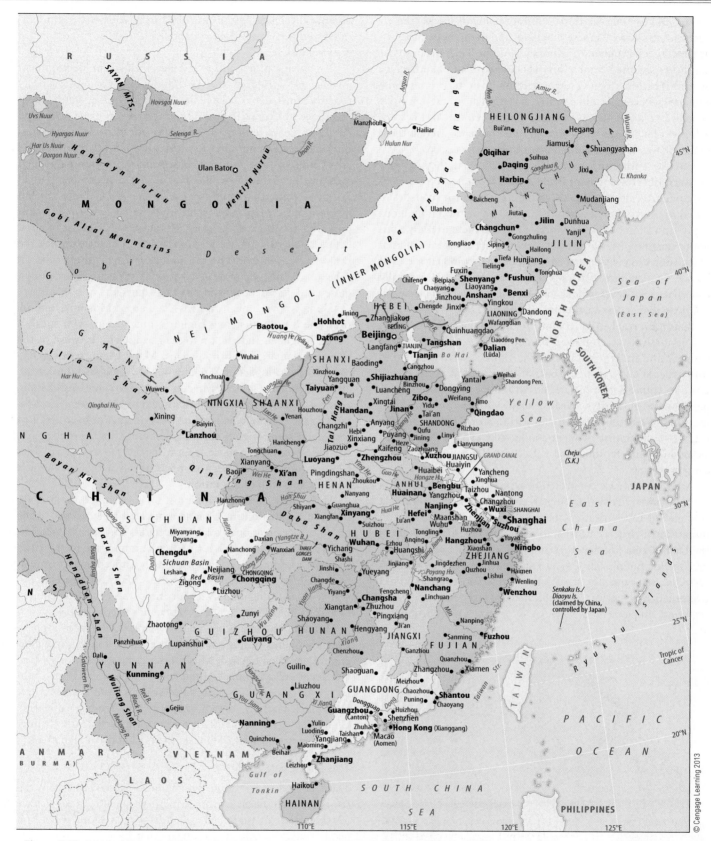

• **Figure 7.50** Principal features of China and Mongolia. On the inset map, the line represents the division between Humid China, or China Proper, to the east and Arid China, or Frontier China, to the west. Humid China has the great majority of the Chinese population and economic activity, and Arid China is home to many minority peoples.

The Hui of Ningxia are Muslims of many different ethnicities who absorbed Chinese language and many Han customs. In contrast, the Uighurs of Xinjiang and the Tibetans have defiantly resisted becoming Chinese—and so Han colonization, or **Hanification**, focuses especially on their territories. The Chinese government aims either to assimilate these minority peoples into a broader, Han-based Chinese culture or to establish a strong enough Han demographic presence among them to dispel any hopes of autonomy or independence.

Chinese authorities have also initiated what they call the **Western Big Development Project** with the stated objective of improving locals' livelihoods enough to diminish their desire for ethnic and political separatism. Focused on the autonomous regions and four other provinces in which there are large minority populations, this project seeks to improve infrastructure like airports and roads and to boost output from factories and farms. The Uighurs, Tibetans, and other minorities generally fear that this "development" is just a cover for demographic, political, and economic domination by the Han.

Buddhism reached Tibet in the middle of the eighth century, and its arrival was followed by decades of conflict between Chinese and Indian Buddhists. The Chinese were defeated in this struggle and expelled from Tibet at the end of the eighth century. China sometimes reexerted loose control over Tibet, but Chinese authority vanished again with the overthrow of the Qing (Manchu) Dynasty in 1911. From 1912 until the Chinese Communists' violent conquest of 1951, Tibet existed as an independent state, with its capital and main religious center at Lhasa.

After the Chinese completed roads to Tibet in late 1954, an increase in restrictive measures by the Communist government contributed to the rise of Tibetan guerrilla warfare. This culminated in a large-scale Tibetan revolt in 1959 and the flight of the **Dalai Lama** (the spiritual and political leader of Lamaism, or Tibetan Buddhism; •Figure 7.51) and many other refugees to India. Subsequently, the Chinese drove most of the monks from their monasteries, expropriated the large monastic landholdings, implemented socialist programs, and prohibited

• **Figure 7.51** The Dalai Lama is the enduring symbol of the nonviolent Tibetan quest for autonomy from China.

organized religion in the rest of China. Large-scale immigration by ethnic Han picked up; about half of Lhasa's population is now Chinese.

Beijing insists it is helping Tibet with major development investment, including the $4 billion Qinghai-Tibet railway recently constructed between Beijing and Lhasa. But Tibetan resistance to Chinese rule continues. From his place of exile in Dharamsala, India, and on frequent international road trips, the Dalai Lama continues to appeal for Tibetan freedom. He asks China to grant what he calls **genuine autonomy** to Tibet, meaning that Tibetans would have authority over most matters except foreign affairs and defense, which the Chinese government would control. His peaceful cause has widespread support. But some Tibetans want stronger, even violent, action. China's government has so far refused to negotiate with the Dalai Lama. For many years, India had promoted the cause of Tibetan autonomy, but as its trade ties with China have grown more vital, its advocacy for Tibet has quieted.

China's government regards Xinjiang's Muslims, mainly from the 8-million-strong Uighur ethnic group, who speak a Turkic language and are kin to the Kazakhs to the west, as the most problematic minority groups. Many Uighurs would like to pull at least a part of Xinjiang out of China's orbit to create an independent Chinese Muslim state. Two short-lived independent Uighur republics, both known as the Eastern Turkestan Islamic Republic, existed in 1933 and again from 1944 to 1949, and some Uighur separatists are fighting for one again. Beyond execution of those separatists deemed to be terrorists, China's response has been to press ahead with development projects and Han immigration.

At the outset of the Communist takeover of China in 1949, the Han population in Xinjiang amounted to 6 percent of the province's population. Today, it accounts for over 40 percent, and some 250,000 more Han Chinese immigrate to the cities (especially Urumqi and Yining) annually, enticed in part by relatively high salaries and other incentives. Many of these Han migrants are in Xinjiang only a short time; they stay long enough to make some money and then return to eastern China.

Human rights monitors report systematic Han abuses of the Uighurs. Scores of mosques have been torn down; Uighur literature has been burned; Muslim clerics have been forced into "reeducation" training; Islamic schools have been closed; public prayer is now forbidden; head scarves have been banned; and prodemocracy demonstrations are brutally quashed. China's struggle to maintain a grip on its ethnically distinct regions is reminiscent of Soviet efforts in non-Russian areas. [134]

The Three Gorges Dam

The Chang Jiang ("Long River"), also known as the Yangtze River, has been central to China's identity and welfare for thousands of years. The river delivers precious water and fertile soils, creating environments that have been intensively settled and farmed, especially for rice and wheat. But this great river has also delivered tragedy in the form of devastating floods. Flooding in August 1998 affected 300 million people along the river and did an estimated $24 billion in damage, for example. Inevitably, questions arose about the advantages to be gained—especially in flood control, drought relief, and hydroelectricity production—if the river were to be dammed. The concept of a giant dam on the river dates to 1919, when it was proposed by Sun Yat-sen. His vision has been realized on a scale far grander than he could have imagined in the massive Three Gorges (Sanzxia) Dam, begun in 1994 and completed in 2009 (•Figures 7.52 and 7.53; see also location on Figure 7.50).

As with any large dam project, like the Aswan Dam seen earlier (page 181), the Three Gorges may best be evaluated by considering its pros and cons. On the plus side, this largest dam ever built—1.3 miles (2.1 km) wide and 610 feet (186 m) high—is creating a reservoir 385 miles (620 km) long that will dramatically reduce the threat of downstream flooding. Stored waters will also help alleviate drought. The Three Gorges Dam will also improve navigation and therefore enhance

• **Figure 7.52** Changed landscape of the Three Gorges Dam region; NASA visualization.

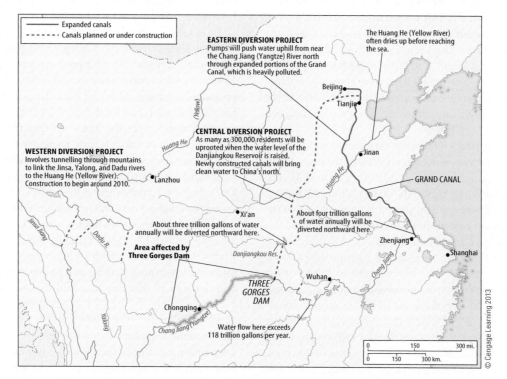

• **Figure 7.53** The Three Gorges Dam and the Chang Jiang Water Transfer Project.

trade, especially by connecting the burgeoning inland city of Chongqing to the world abroad. Historically, the Chang Jiang was able to accommodate freighters of 10-foot (3 m) draft all the way from the East China Sea to Wuhan. Above Wuhan, the shallower river was traditionally plied by smaller freighters to Yichang, situated just below the river's spectacular gorges. Yichang was a classic **break-of-bulk point**: the point at which cargo is unloaded and broken up into smaller units for delivery, thus minimizing transport costs. In this case, freight and passengers were offloaded and put on flat-bottomed junks and sampans drawn upstream through the gorges to Wanxian by human trackers who pulled and rowed the vessels (you can see all these locales in Figure 7.50. With the completion of the Three Gorges Dam project, vessels with much deeper draft are able to sail 1,500 miles (2,400 km) upstream from Shanghai all the way to Chongqing in the Sichuan Basin. Small vessels are able to pass over the dam in a ship elevator, a vast steel box that lifts the ship the height of the dam in 42 minutes. Most ship traffic uses a system of locks with five levels, taking three hours in transit. With these means, shipping to Chongqing has increased fivefold. Some observers say the increased commerce is in the process of transforming Chongqing into a "new Hong Kong." Chinese authorities view Chongqing as a new model for urban growth, as discussed on page 269.

Also on the plus side, the world's largest hydropower plant that is built into the dam provides about 85 billion kilowatt-hours of energy annually, enough to satisfy 15 percent of China's electricity needs. Some of the power is sold to private users, offsetting some of the dam's huge construction costs.

Most of the negatives are related to the vast reservoir forming behind the dam. The new 500 feet (150 m) deep lake has inundated 4,000 villages, 140 towns, 13 cities, numerous archeological sites, and nearly 160 square miles (over 400 sq km) of farmland. Because their homes have been drowned, as many as 1.4 million people have been resettled. Many of the vacated settlements are quite ancient, and their inhabitants had deep cultural roots in them. The old villages were typically situated close to fertile farmlands that were also being inundated. In the new, higher communities into which people were located, soil conditions were generally inferior.

The reservoir has had aesthetic consequences too, erasing the world-class wondrous wild river running through scenic canyons. The natural marvels of the three gorges drew 17 million Chinese tourists and hundreds of thousands of foreign visitors every year. There are also practical concerns about the silt that is building up behind the dam. It is possible that this increased soil load will have the ironic effect of causing flooding upstream from the dam, even as the dam prevents flooding below. There are fears of the almost unthinkable: that the dam could burst, causing unimaginable devastation and suffering downstream. Finally, there may also be seismic consequences of the dam. Central China suffered massive economic and human losses (about 70,000 dead and 5 million homeless) in the magnitude 8.0 Great Sichuan Earthquake of May 12, 2008. There is growing scientific evidence to suggest that this quake was triggered by the weight of a reservoir close to the fault on which the quake occurred. The Chinese government admits that the Three Gorges Dam has created seismic and landslide risks, and that its reservoir has been plagued by algae carpets and waste dumped by communities on its shores.

Although recently admitting some environmental consequences, the government of China pursued the mighty Three Gorges Dam project without remorse and without tolerance for dissent within China. The enormous cost of the project, nearly $30 billion, has been born solely by China. Like megadams elsewhere, the Three Gorges Dam is a statement that its builder is a great power to be reckoned with on the world stage. [181]

Chang Jiang Water Transfer Project

China has one of the world's lowest per capita water supplies and most uneven distributions of water. More than 40 percent of China's people live in the north, but less than 15 percent of the country's water is there. Chinese engineers see water transfer on a massive scale as the way to redress this imbalance.

One project being considered is the transfer of seawater through a pipeline all the way to the dried up salt lakes and desert basins of Xingiang. The hope is that the water will evaporate to induce rainfall over dry northern and northwestern China.

Another project is underway: as part of the massive **Chang Jiang Water Transfer Project**, the Three Gorges Dam will provide water to arid North China (Figure 7.53). This spectacular engineering project—more ambitious even than the Three Gorges Dam—will move water from south to north through three canals. One of its most difficult challenges is to blast through some of the giant mountains of the Himalayas to divert water from the upper reaches of the Yangtze River to the headwaters of the Yellow River. This feat will be the first time in history that people have merged two river basins.

In contrast with their approach to the Three Gorges Dam, Chinese leaders view the Chang Jiang Water Transfer Project with some trepidation. The project's concept originated with Chairman Mao and was approved before China's present leaders came to power. Many of these leaders feel the project will have unfortunate, unintended consequences, both social and environmental. Like the Three Gorges Dam, this project will require people to relocate, and there are fears of social unrest. And the environmental consequences of changing the course of rivers and reallocating their waters are very difficult to foresee. Dai Qing, an environmental and political activist, had this to say of the Chang Jiang Water Transfer Project: "[It] will be considered a disaster within twenty years. Under Chairman Mao, we often heard the slogan 'Humans must conquer nature,' and that is still the mentality of so many people today—but gradually we are learning that human beings cannot conquer nature and that this philosophy inevitably leads to disaster." [17]

What's Next for Industrial China?

As described earlier (pages 233–234), China's economy has been booming. But as the global economic crisis spread beginning in 2008, it was clear that even this economic dragon had weaknesses.

China's extraordinary recent successes are tempered by several concerns. One is freedom of expression. In 1989, the Chinese government cracked down brutally on prodemocracy demonstrations centered on Tiananmen Square in Beijing. Years later, there continue to be strictures on academic, media, artistic, and other freedoms that some observers believe are conducive to long-term economic well-being. China's government is one of few in the world (including those of Myanmar, Malaysia, Thailand, Singapore, Turkmenistan, Uzbekistan, Saudi Arabia, Syria, the United Arab Emirates, Tunisia, and Iran, Vietnam, Cuba, Sudan, Pakistan, Yemen, and North Korea) to censor the Internet, which is arguably one of the most important tools of economic development today. With what has become known as the "Great Firewall of China," the government requires all computers sold in China to be preinstalled with software blocking political and religious websites and pornography, thus enabling the monitoring of personal communications and the Web sites visited by computer users. The government also monitors mobile phones, the use of which is growing 30 percent annually in China, to ensure that they too do not become instruments of protest (as they did 196 so profoundly in the Arab Spring of 2011). Believe it or not, fearing the potency of a revolutionary symbol, China's government began destroying jasmine plants in the wake of Tunisia's Jasmine Revolution (see page 192).

Another potential problem is that China's leaders have been breathlessly trying to feed what they call the "socialist market economy" with gigantic projects, thereby hoping to create jobs and stimulate economic growth. They reckon that the country needs to register at least 7 percent economic growth annually just to prevent massive unemployment and social unrest. Continuing the infrastructure investment for the 2008 Olympic Games, one of the world's most impressive building booms ever is continuing: new subway systems, railroads, highways, bridges, sewage systems, dams, water diversion projects, a natural gas pipeline, a national electricity project, urban improvements, airports, industrial parks, and more. These efforts cost a lot of money, and the government is drawing down its national treasury to fund them. State banks are pouring billions of dollars into federal projects whose returns are uncertain. (Incidentally, China is the United States' largest foreign lender. Buying up U.S. Treasury Bonds, it helps finance America's national debt.)

A further hazard is that with its galloping growth, China is developing a classic economic bubble that will eventually burst, probably with costly results. Historically, few economies have been able to maintain torrid growth for long, especially when that growth is fueled by speculative cash flows and questionable bank loans, as has been the case recently in China, and as had been witnessed with the 2008 global credit crisis.

There are also environmental costs for such spectacular growth. Due to generally lax environmental quality standards, China's skies and waters are often badly polluted. China has more entries on the World Health Organization's list of most polluted cities than any other nation. China's government acknowledges the problems. The State Environmental Protection Administration (SEPA) reported that one-quarter of the water in China's largest river systems was so polluted that it was dangerous for people and other living things to come into contact with or had "lost the capacity for basic ecological function." Polluted water was killing tens of thousands of people every year and reducing crop yields. SEPA warned that China's approach of "growth through industrialization" was pushing its environment "close to the breaking point."[18]

Having overtaken the United States as the world's largest producer of carbon dioxide, China (which gets 80 percent of its electricity by burning coal) presumably has the largest role in climate change. When the Kyoto Protocol on climate change was negotiated, China was exempted because it was an LDC. Technically this is still true, but China has surprised [35] many by taking some responsibility for climate change. China has also resolved to become the global leader in green technology. China sees this as an opportunity both to confront its own problems with climate change, apparently including both droughts and floods, and to create new jobs. To this end, China has quickly become the world's largest manufacturer of both wind turbines and solar panels, and aims to become the world's biggest builder of nuclear power plants. China has recently passed the United States to become the world's largest market for wind turbines, which are sprouting up on the Gobi Desert and elsewhere.

There is concern that the benefits of China's economic growth have not been distributed evenly throughout the population. Prosperity and consumption have increased enormously, but in a remarkably unequal manner. In 1980, China had one of the world's most even distributions of wealth, but that equation has changed dramatically, with the flamboyant super-rich showing off their wealth to the very poor. These inequalities are comparable to those that existed when the revolution brought Chairman Mao's Communists to power on the promise of redistributing wealth.

Remember that, despite its phenomenal growth, China is still relatively poor (with a per capita GNI PPP of $6,890 (comparable to Turkmenistan) and agricultural (with a 50 percent rural population; see Table 7.1). The agricultural face of China is changing rapidly, though, as the cities boom (see the discussion of megacities on page 226). Most of the economic activity and growth is in the east, and especially in the southeast in the five **Special Economic Zones (SEZs)** designated to attract investment and boost production through tax breaks and other incentives (see •**Figures 7.54 and 7.55**).

One of the results of the clustering of these cities, and the economic power they represented, has been a serious problem of regional imbalance. Most of the country's industrial development and urban growth occurred along its east coast, near the traditional port and river cities that were strongly influenced by Western economic interests from the mid-nineteenth century on. Even the rural areas of eastern China experienced much more growth relative to those in the west, especially because they were well linked to expanding markets and prospering cities. Meanwhile, people in western regions of China, mountain villages all through Humid China, and landscapes

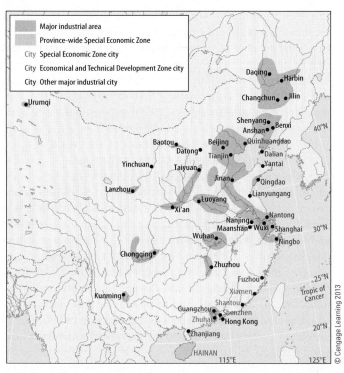

• **Figure 7.54** Major industrial resources, industries, and industrial cities of China.

• **Figure 7.55** The futuristic Pudong section of Shanghai is China's modern face to the world, and is one of the country's SEZs.

lacking in surface transportation in the southwest and parts of upland China were frustrated. They saw relatively little evidence of China's much touted economic prosperity.

Predictably, large numbers of people have migrated from the less prosperous regions of the country to the promising SEZs and other urban centers of the east and south. Millions of migrant workers have left their villages—then populated primarily by children and the elderly—to work in construction or menial labors in the coastal cities, sending much of their meager monthly earnings of $50 to $100 home as remittances. People in the cities often discriminate against these migrants, known derisively as "outsiders," in education, housing, and jobs, treating them like illegal aliens. Many technically are illegal because they fail to sign work contracts that would both register them and require factories to pay fees to local cities.

Chinese cities are characterized by a two-tiered society of legal residents and nonresidents, created by a registration system known as *hukou*. Residency allows people access to better education and health care, and it is financially prohibitive for a nonresident to obtain legal residency. The nonresident migrants provide cheap labor not only to factories but to affluent urbanites who hire them as nannies and servants. Police routinely "shake down" workers without papers, seeking bribes from the frightened migrants. Some of the factories—like the Foxconn complex in Special Economic Zone of Shenzhen known as iPod City because these Apple devices and the iPhone and iPad are assembled here—are world-class, but others are **sweatshops**, where working conditions border on the nightmarish. But even at Foxconn, conditions are so stressful that suicide has become

a problem (employees there are now required to sign a contract promising they will not commit suicide!). Management typically suppresses strikes, bans independent trade unions, and fails to enforce minimum-wage laws for workers who toiled through 18-hour workdays and caught brief rest in prisonlike dormitories. It is little wonder that many of the eastern boomtowns see as much as 10 percent annual turnover in the workforce and soaring crime rates.

The Chinese government has a social contract with its people: it pledges that it will create and maintain a "harmonious environment" for them. The basic formula of China's government and economic system is that employment and economic growth needs to be kept high in order to avert social unrest and, in the worst case scenario, revolution. China's leaders watched the Arab Spring with great concern. Although they are often not reported by media, there are growing numbers of labor strikes by workers demanding double-digit salary increases. Recently, the government raised the minimum wage by 20 percent, and employers like Foxconn have granted even higher raises.

Chinese authorities remain convinced that urbanization is the ticket to greater prosperity for China's huge, poor, rural population. But they do not want the magnet cities of the east to absorb all of these people; many would end up unemployed and living in slums. The strategy is therefore that the larger cities of the interior, such as Chongqing, should absorb much of the migrant traffic that has been flowing farther east and south. Already home to 8 million people, Chongqing, for example, is poised to grow much more through its "one-hour economy circle plan." In this plan, by 2017, 4 million rural residents will be moved into newly urbanized urban areas within an hour's drive of the city center. Millions of new jobs are thus being created in the long-neglected urban centers of China's interior. At the same time, wages have been rising sharply all over the country. Once pushed and pulled toward the coast, workers are now finding competitive jobs closer to home. The result is a sudden and surprising shortage of wage laborers in the coastal factories. Once-confident sweatshop managers now worry that the global marketplace will shift jobs to lower-wage countries like nearby Vietnam.

Other trends also point toward the end of the Chinese model. Not only are wages rising, but so too are land prices. A serious "property bubble," partly built up through the same kind of questionable loans that plagued the United States before its credit crisis, is part of the economic bubble building in China. As the population ages, the labor force is shrinking. Cheap capital, cheap labor, cheap energy, and cheap land were at the heart of the Chinese economic miracle, but each of these is becoming pricier. Inflation and unemployment are growing. The danger is, of course, social unrest.

If the bubble does not pop, and if growth continues at about 9–10 percent per year, as many as 70 percent of the Chinese will be middle class or above by 2020. That would represent a huge market of potential consumers. The government wants the Chinese economy to be less reliant on manufacturing and exports, and to become more dependent on domestic consumer spending.

Taiwan and the Two Chinas Problem

Known by Westerners for centuries as Formosa (Portuguese for "beautiful"), Taiwan is separated from South China by the Taiwan, or Formosa, Strait, 100 miles (160 km) wide (see Figure 7.1). The island is nearly 14,000 square miles (c. 36,200 sq km) in area and is home to 23 million people, nearly all of Chinese origin. Driven from the mainland in 1949, the Chinese Nationalist government fled to Taiwan with remnants of its armed forces and many civilian followers. Nearly 2 million Chinese were in this migration. Here, protected by American sea power, the government reestablished itself as the Republic of China with its capital at Taipei. The Nationalists, originally an authoritarian regime but with increasing elements of political democracy, were very successful in fostering industrial development. They united inexpensive Taiwanese labor with foreign capital to build one of Asia's first urban-industrial countries and one of its Asian Tigers. Its major exports are machinery, electronics (notably computers, cell phones, modems, routers,

233

and global positioning systems), metals, textiles, plastics, and chemicals, especially to mainland China and the United States. The population is 78 percent urban; its infant mortality rate is very low; and its birth rate is one of the lowest in Asia (see Table 7.1). The average Taiwanese citizen is four times wealthier than the average mainland Chinese citizen. In almost all respects, Taiwan is clearly recognizable as a more developed country. A major hurdle against even stronger growth has been a lack of energy resources. Some coal and natural gas exist, and some hydropower has been harnessed, but the island depends heavily on imported oil.

Taiwan's Republic of China continues its claim as the legitimate government of China. Meanwhile, the People's Republic of China claims Taiwan as part of its own territory. The United States backed the Nationalist claim for a time, but during the 1970s the United States and the mainland People's Republic developed closer relations. After years of opposition, in 1971 the United States supported the revocation of Taiwan's seat in the United Nations and its replacement by the People's Republic. In 1979, bowing to the wishes of the People's Republic, the United States withdrew its official recognition of Taiwan, instead recognizing China's claim of sovereignty over Taiwan. This **One China Policy** prevents the United States from having formal diplomatic relations with Taiwan. However, the United States opposes the annexation of Taiwan to China by force, vowing to defend Taiwan from attack, just as it opposed Taiwan's call for a Nationalist effort to retake the Chinese mainland during periods of internal weakness there. In the meantime, the United States is a strong backer of Taiwan, supplying it with generous packages of weapons and economic aid.

Taiwan continues to resist mainland China's overtures for reunification, which include promises of broad autonomy, and is moving increasingly away from any type of reunion. That trend reflects Taiwan's ethnic makeup. Less than 15 percent of the island's people descend from the Nationalist refugees, who with their recent ties to the mainland are known as "mainlanders." Native non-Chinese people make up about 2 percent of the population. The vast majority of the island's inhabitants are descendents of Han peoples who emigrated from China as many as four centuries ago, notably the Fujianese, who came from the southeastern coastal province of Fujian, and the Hakka, from Guangdong province. This majority is much less mainland-oriented in political outlook. Despite discrimination by the Nationalists, these "indigenous" Chinese of Taiwan have enjoyed increasing political power.

The two countries must be cautious about their economic and political relations. Taiwan's economy has become very dependent on mainland China, where Taiwanese companies take advantage of low-cost mainland labor (about one-fifth the cost in Taiwan for the manufacture and assembly of products) and where they have a huge market for their products. Any efforts to sever most ties with China and assert Taiwan's independence more forcefully could have large economic consequences and invoke military intervention from Beijing—and an American military response.

Japan and the Koreas

Japan's Postwar and Its Costs

On the Way Up . . . Without colonies or empire, Japan (•Figure 7.56) became an economic superpower after World War II. The nation's explosive economic growth after its defeat was one of the most remarkable developments of the late twentieth century and, echoing the broader region's experience, is known widely as the Japanese "miracle." Japan watchers cite different reasons for the country's economic success. Proponents of dependency theory argue that Japan, having never been colonized, escaped many of the debilitating relationships with Western powers that hampered many potentially wealthy countries. Some analysts believe that the country's postwar economic miracle grew from an intense spirit of achievement and enterprise among the Japanese. Notably, many Japanese

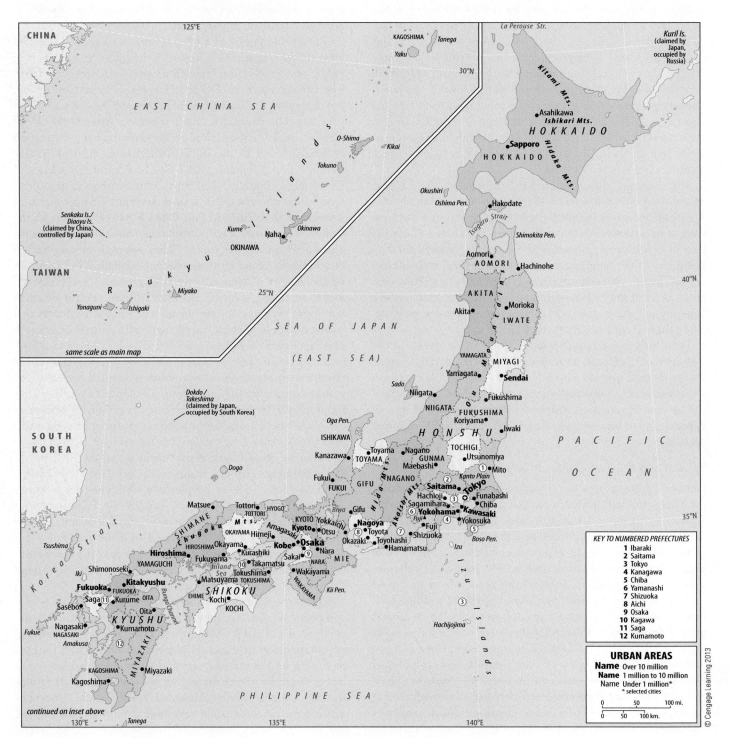

• **Figure 7.56** Principal features of Japan.

attribute this industrious spirit to Japan's geography as a resource-poor island nation. To overcome the constraints nature has placed on them, the Japanese people feel they must work harder. Still others attribute Japan's postwar achievements to its association with the United States following World War II. Some of the postwar U.S.-imposed reforms worked well economically, and relations between the two countries since the war have also generally stimulated Japan's industrial productivity.

Economists also point to several unique features of Japanese management and employment to explain the country's meteoric postwar gains. One has been recruitment through an extremely challenging (some say brutal) educational system that emphasizes technical training. Although recent years have seen a move to a more lenient and more comprehensive education, a rigorous and stressful testing system still determines access to higher education. Japanese management strategies emphasize benevolence toward employees, encouragement of employee loyalty, and participation of workers in decision making. About 20 percent of Japanese workers enjoy guarantees of lifetime employment in their firms, and many large Japanese companies help provide housing and recreational facilities to their employees.

One essential factor in Japan's postwar economic growth was a high level of investment in new and efficient industrial plants. Investment capital was made available when government policies cut expenditures on amenities and services such as roads, antipollution measures, parks, housing, and even higher education. Japanese industrial cities grew explosively and became highly polluted areas of dense and inadequate housing, with few public amenities and snarled transportation. The money "saved" by not being spent on amenities and services was made available for investment in industrial growth, but the costs to Japanese society were high.

Some analysts cite elements of Japan's political culture to explain the country's economic successes. One political party, the **Liberal Democratic Party (LDP)**, has with few exceptions been elected to power since Japan regained its sovereignty in 1952. It is a conservative and strongly business-oriented and business-connected organization. The party has promoted Japanese exports with policies to keep the yen (the Japanese unit of currency)—and thus Japanese goods—inexpensive. The party has also cooperated closely with Japanese business in the development of new products and new industries. However, that very coziness between government and business—the "crony capitalism" seen in several other East and Southeast Asian countries—was partly to blame for the economic malaise that spread over Japan in the 1990s, and support for the LDP began to erode.

... and on the Way Down Despite all of these factors in Japan's economic favor, the Japanese miracle did not last. The peak of Japan's postwar success came in the 1980s. A powerful economic boom led to speculative rises in stock prices and land prices. At the end of the 1980s, the total value of all land in Japan was four times greater than that in the United States.

The Tokyo Stock Exchange was the world's largest, based on the market value of Japanese shares. Then the **bubble economy** burst, and real estate and stock prices fell by more than 50 percent. Japan had near-zero economic growth through most of the 1990s and into the early 2000s. Japanese industries whose growth had seemed unstoppable experienced an increasing loss of market share to U.S. and European producers. Part of the reason the economy could not regain its footing was official Japanese commitment to prop up faltering industries and farms with huge subsidies. Because of the disproportionate political influence of small towns and rural areas, the Japanese tried to modernize remote villages and islands with expensive and underutilized public works projects. Due to such projects, Japan is now the world's most indebted country. Overall, the system succeeds in subsidizing its past at the expense of its future.

Attitudes toward ethnic groups play a significant role in Japan's economic development, or more precisely the lack of development. Japan has one of the world's most homogeneous populations; it is 99.5 percent ethnic Japanese. This lack of racial and ethnic diversity has had mixed results for Japan. Many observers believe that it has helped the country achieve a sense of unity of purpose, allowing the Japanese to persist through periods of adversity, especially the postwar years of reconstruction. However, the Japanese have also earned a reputation for intolerance of ethnic minorities. This attitude may be costly for Japan's future economic development. One way to increase productivity in the face of Japan's declining population would be to import skilled workers from abroad, but most people in the country remain averse to immigration. Inward-looking tendencies have also left Japan short of people capable of and interested in learning English, a language routinely sought by aspiring businesspeople in nearby South Korea and China.

There is definitely another upside to Japan's homogeneity. Capitalist Japan has had a remarkably egalitarian society, with about 80 percent of the population solidly middle-class. The richest third of the population has a total income just three times as great as that of the poorest third (compared with a fivefold differential in the United States). The Japanese have historically embraced the concept of *wa*, or harmony, based in part on the principle of economic equality. However, the recessions of the 1990s and early 2000s, followed by an economic rebound until the global financial crisis of 2008, resulted in growing joblessness and homelessness in Japan and a widening gap between haves and have-nots. The country's official unemployment rate stood at 4.7 percent in 2011, compared with 2.1 percent in 1990. There are growing ranks of "permanent temporary workers," mainly young people who live with their parents to keep their expenses low.

The government has balked at many suggestions for economic reform because of the threat they pose to a socially harmonious Japan. There have been loud international appeals for Japan to reform its economic system by allowing inefficient businesses to collapse (rather than prop them up with expensive subsidies) and to resist the temptation to satisfy marginal

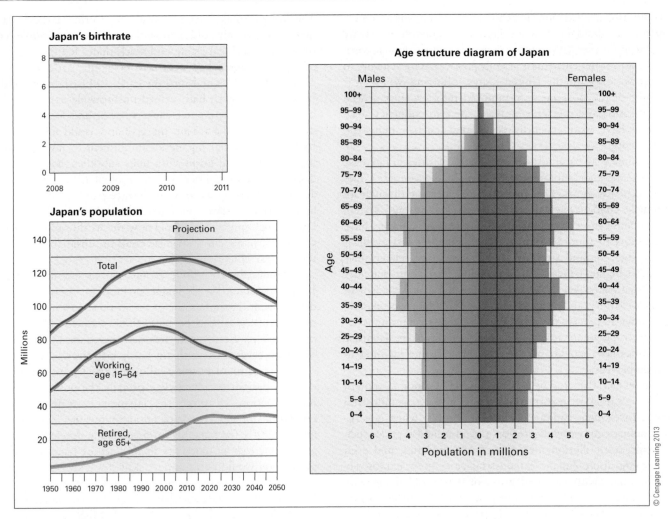

Japan's birthrate

Japan's population

Age structure diagram of Japan

• **Figure 7.57** Many countries face problems of overpopulation, but Japan is concerned about not having enough people to keep its economic engine running in the future.

rural populations with costly education and public works projects.

The legendary Japanese work ethic has had its advantages and drawbacks. It has helped the Japanese create a prosperous country, but it has also created a nation of workaholics beset with the same problems—stress, suicide, depression, and alcoholism—experienced by workers in the world's other MDCs. There is a growing incidence of what Japanese call *karoshi*, or death by overwork. The educational system encourages children to be highly successful but also to conform; critics say this stifles creativity and innovation. College graduates who are not hired during the annual recruiting season face difficult obstacles to entering the job market. And for those it affects, the practice of lifetime employment makes it difficult to change jobs. The country's welfare system, pensions, and social security are inadequate, compelling workers to work harder for savings. In their long hours at work, men are accustomed to spending little time with their spouses and children.

Women have not achieved parity with men in the workplace, and they complain increasingly of discrimination and sexism. In struggling to increase their footing in Japanese society, growing numbers of Japanese women are both working longer hours and marrying later, contributing to Japan's remarkably low birth rate (7 per 1,000 annually, one of the lowest in the world). This trend and the others make Japan's declining population stand out clearly in •Figure 7.57; note especially Japan's chimney-shaped age structure diagram, such a remarkable contrast with the diagrams of LDCs. Ironically, [53] Japan's falling birth rates will only increase the burdens of its workers. The population is both shrinking (it is expected to go from 128 million in 2011 to 100 million in 2050 and 67 million in 2100) and aging (already one in five people were over age 65 in 2011, and 1 million people a year join those ranks). The Japanese of working age will face an increasing load of taxes and family obligations to meet the needs of older citizens.

The Law of the Sea

Some 90 miles (145 km) between the shores of Japan and South Korea in the Sea of Japan (East Sea) lie 34 small, inhospitable islands known to outsiders as the Liancourt Rocks, as Takeshima to the Japanese, and as Dokdo (or Tokdo) to the Koreans (see Figure 7.56). South Korea has controlled the islands since 1956, but with only a handful of Korean civilians and some Korean coast guard personnel living on the islands; Japan has periodically asserted its right to them. Since the 1990s, these remote rocks have been the focus of a dispute between Japan and Korea, not because of any riches they contain but because of a 1970s United Nations treaty known as the **Convention on the Law of the Sea**. This treaty would permit their sovereign power to have greater access to surrounding marine resources.

237, 427

The Law of the Sea began as an effort to apportion ocean resources as equitably as possible and to avoid precisely the kind of dispute that developed between Japan and South Korea. The treaty gives a coastal nation mineral rights to its own continental shelf, a territorial water limit of 12 miles (19.3 km) offshore, and the right to establish an **exclusive economic zone (EEZ)** of up to 200 miles (320 km) offshore (in which, for example, only fishing boats of that country may fish). The power that controls offshore islands such as Takeshima and Dokdo can extend its EEZ even further.

By early 1996, some 85 nations had ratified the treaty. South Korea ratified it late in 1995. Japan was preparing to ratify it in 1996, when news reached Tokyo that South Korea had plans to build a wharf on the islands. To avoid provoking Japan, North Korea, and China, South Korea avoided declaring an exclusive economic zone off its waters that would include the islands. Japan ratified the treaty. But fears that South Korea may build facilities and place more people on the islands as a step toward establishing an EEZ, and thereby excluding Japanese fishermen from the area, caused Japan to restate its claim to the islands in 1996. South Korean officials answered with military exercises near the islands, and South Korean civilians staged loud demonstrations outside Japan's embassy in Seoul. South Korea declared the islands a national monument in 2002. Long simmering, the issue boiled up again in 2006 when Japan announced it would send ships to survey the area and South Korea answered that it would form a naval blockade against the Japanese vessels. The problem may have to be resolved in the international legal arena.

Even more important than Japan's competition with South Korea over territorial waters and islands is its contest with China. Both lay claims to the islands in the East China Sea that the Japanese know as the Senkakus and the Chinese as the Diaoyu Islands (see the inset in Figure 7.56). Again, both are interested in staking out an EEZ in which oil reserves might be found. China does not accept the UN definition of the EEZ beginning from a country's coastline; it insists on a wider zone that starts at the continental shelf.

The Senkaku or Diaoyu Islands are uninhabited, but their control is championed by nationalists on both sides and is therefore a potentially explosive issue. In 2004, Chinese activists landed on one of the islands to stake China's claim there, but Japanese authorities promptly arrested them, and Japan took formal possessions of the islands the following year. Since then, there have been several encounters between Japanese and Chinese ships near the islands. The worst was the 2010 incident known as the "Senkaku Shock," which began when a Chinese fishing vessel rammed a Japanese patrol boat near

the islands. Japan detained and refused to release the Chinese captain. China responded by halting exports of "rare earth" minerals needed in Japanese high-tech industrial products. By geological accident, and good fortune for China, a very large proportion of the world's supply of 17 rare earth elements are found exclusively or mainly in China. These are indispensable in cell phones, laptops, and many "green" technologies, like wind turbines and hybrid vehicles. They are also important in military technology, and the United States regards access to them as a matter of national security.

Japan's formal claim to the islands dates to 1895, when Japan declared that the islands had previously been *terra nullius*, or no one's property. But China insists that there is Chinese documentation about control of the islands dating to the Ming Dynasty in the fifteenth century. The islands are close to Taiwan, which also lays claim to the islands, so any move by mainland China to secure access to them could also precipitate conflict between the two Chinas.

North and South Korea: Night and Day

Just 110 miles (177 km) from westernmost Japan is the Korean Peninsula, roughly the size of Minnesota or Portugal (•Figure 7.58). The two Koreas have occupied an unfortunate

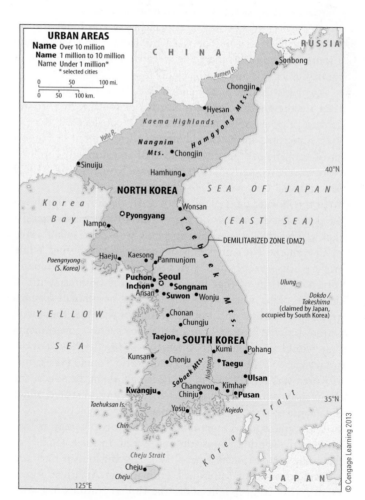

• **Figure 7.58** Principal features of the Korean Peninsula.

location in historic geopolitical terms. North Korea adjoins China along a frontier that follows the Yalu and Tumen Rivers; it faces Japan across the Korea Strait; and in the extreme northeast, it borders Russia for a short distance. These small countries are thus located near larger and more powerful neighbors—China, Russia, and Japan—that have frequently been at odds with one another and with the Koreans. China and Russia have traditionally feared Japanese rule in Korea because it might serve as a springboard for invasion of their countries. Meanwhile, to Japan, Korea has been seen as a Chinese or Russian dagger aiming at its heart. Japan therefore wanted to occupy or block Korea. For many centuries, the Korean Peninsula has served as a bridge between Japan and the Asian mainland. From an early time, both China and Japan have been interested in controlling this bridge, and Korea was often a subject or vassal state of one or the other.

Despite external ambitions of conquest and division, from the late seventh century to the mid-twentieth century, Korea was a unified state, sometimes invaded and forced to pay tribute but never destroyed as a political entity. The decline of Chinese power in the nineteenth century was accompanied by the rise of modern Japan, whose influence grew in Korea, and from 1905 until 1945, Korea was firmly under Japanese control. In 1910, it was formally annexed to the Japanese empire. A legacy of hostility and occupation continues to cast a shadow over relations between the Koreas and Japan, and disputes periodically emerge between them over issues such as control of islands and fishing rights (see page 272).

Japan lost Korea at the end of World War II, setting the stage for the division between North and South Korea. To understand Korea's split, it is important to consider the Korean War, sometimes in the United States called "America's Forgotten War." The conflict cost the lives of an estimated 2 million North Koreans, perhaps 300,000 South Koreans, and 33,629 Americans.

In the closing days of World War II, the Soviet Union entered the Pacific war as an ally of the United States against Japan. The two sides drew up plans to accept Japan's surrender on the Korean Peninsula, arbitrarily drawing a line at the 38th parallel. The Soviet Union would accept Japan's surrender north of that line and the United States south of that line. This was not meant to be a permanent boundary, but it became one. On either side of the line, the Soviet Union and the United States moved to set up governments that would be friendly to them. By 1948, the Soviet-style, Soviet-backed, Soviet-armed Democratic People's Republic of Korea was established in the north, while in the south, the Western-oriented Republic of South Korea was created under the auspices of the United Nations.

War came soon. On June 25, 1950, North Korean troops invaded the south and quickly took Seoul. In those Cold War days, the United States feared Soviet expansionism into and beyond South Korea. President Truman sent American forces in, and they soon fought under the United Nations flag with support from 16 other countries. Truman selected General Douglas MacArthur to serve as supreme commander of the effort. By September, MacArthur's forces had pushed the North Korean advance all the way back to the Yalu River on the Chinese border. The general proposed using nuclear weapons to press forward from there, but the president refused, and his protests over the issue cost MacArthur his job.

China, seeing enemy forces on its doorstep, reacted by sending huge numbers of forces across the border, driving the allies south across the 38th parallel and taking Seoul on January 4, 1951. Seoul was recaptured on March 15, and the battlefront stabilized generally along the 38th parallel. An armistice was signed at the border site of Panmunjom on July 27, 1953, by the Chinese, the North Koreans, and the United Nations command. It was only a cease-fire, and to this day, there has not been a full treaty or reconciliation between the Koreas.

The border between the two Koreas, often cited as the tensest boundary on earth, still follows this armistice line. The **demilitarized zone (DMZ)** between them, 150 miles (240 km) long and 2.5 miles (4 km) wide, is a virtual no-man's-land of mines, barbed wire, tank traps, and underground tunnels. The world's largest concentration of hostile troops faces off on either side of it. Remarkably, with the near absence of human activity, rare birds like the Manchurian crane and other endangered animal species have taken refuge in this narrow strip. The DMZ thus has the potential to become an international "peace park" like that taking shape along the old Iron Curtain in Europe. [89]

The Korean armistice line divides one people, united by their ethnicity and language, into two very different countries (•Figure 7.59; also see Table 7.1). South Korea is a republic that has fluctuated between attempts at democracy and a repressive military dictatorship. It has a capitalist economy heavily dependent on relationships with the United States and Japan. North Korea is a rigid and very tightly controlled Communist

• **Figure 7.59** The two Koreas are like night and day. A satellite image taken at night shows the profound economic distinctions, with prosperous South Korea awash with urban and industrial lights while North Korea is nearly dark.

state that promotes what it calls *juche* (pronounced "*djoo'-cheh*"), or self-reliance. North Korea's leadership has used this political philosophy's three principles of political independence, economic self-sustenance, and national self-defense to limit economic and other exchanges with the outside world. North Korea's principal Cold War ties shifted between China and the Soviet Union. With the collapse of the USSR, China became North Korea's main ally and benefactor, and the Russian Federation is emerging as one of North Korea's few trading partners.

From World War II until his death in 1994, the dictator Kim Il Sung, whom his citizens called "The Great Leader," governed North Korea. His son and successor, Kim Jong Il (dubbed "The Dear Leader"), perpetuated the country's militaristic character. The same is widely expected of Kim Jong-Un, "The Great Successor" and son of Kim Jong Il, who died in 2011. North Korea spends a staggering 23 percent of its gross domestic product on its military, compared with South Korea's expenditure of only 2.7 percent, China's 4.3 percent, and Japan's 0.8 percent. North Korea's lavish expenditures on defense have been one cause of its severe economic decline. North Korea's economy is only 3 percent the size of South Korea's. No data are available for North Korea, but South Korea's GNI PPP is $27,240, similar to that of Israel.

Beginning in 1995, successive waves of flood and drought brought famine to North Korea. The natural calamities only contributed to the already plummeting economic health of the country that came in the wake of the collapse of the Soviet Union, formerly North Korea's main benefactor and trading partner. North Korea's almost impenetrable veil of secrecy made it difficult to calculate the losses, but estimates of the number of people who died in the famine range from 900,000 to 2.4 million, or up to 10 percent of the country's prefamine population. As one crop after another failed, North Korea gradually and reluctantly sought food aid from abroad. It coaxed emergency supplies of wheat from the United States in part by threatening to withdraw from its 1994 agreement with the United States to halt its nuclear weapons program (see the discussion on pages 239–240). Although food aid from the United States, China, and South Korea helped alleviate the immediate crisis, the poor conditions of North Korea's health system, drinking water supplies, and electrical generation have perpetuated malnutrition and disease.

Another physical contrast between the two Koreas is that most of the nonagricultural natural resources are in North Korea. All of the peninsula's major resources—coal, iron ore, some less important metallic ores, hydropower potential, and forests—are more abundant in the North than in the South. North Korea was thus originally the more industrialized state, featuring a typical Communist emphasis on mining and heavy industry, together with hydroelectric production and timber products. Viewed strictly in terms of resource potential for development, North Korea should have become the more prosperous power. Its decision to close its doors to the outside world and insist on strict state control of industry and agriculture provides a fascinating opportunity to appreciate how different political systems produce different economic results.

While the North focused on defense and strict socialism, the South took over industrial leadership in the 1970s and 1980s by developing a dynamic and diversified capitalist industrial economy.[19] South Korea's economy first began to boom between 1961 and 1980 under a state capitalism approach rather than free-market and free-trade capitalism. This was a time during which the government favored and strongly funded a few small companies that it nurtured into giant conglomerates called *chaebols*. Hyundai is a good example. Originally a small rice milk company, it diversified into trucks and buses. The government believed in its potential and supported Hyundai financially, supervising its growth into a world-class corporation renowned for the manufacture of automobiles and ships. Samsung, now one of the best electronics companies in the world, followed a similar path. Altogether about 15 of these gigantic, interlocking, family-controlled conglomerates came to dominate the economy. Large investments from Japan and the United States, as well as access to markets in those countries, aided in the explosive development of *chaebol* industries. Also important was the availability of inexpensive and increasingly skilled Korean labor. Finally, the factor South Koreans most frequently point to in their success has been an emphasis on education. They sell valued possessions if need be to afford their children's educations at home or abroad, and their culture prizes teachers and professors. South Korea has one of the world's highest numbers of Ph.D.s per capita.

With the world's 13th largest economy, South Korea is now among the top 10 countries worldwide in the production of automobiles, ships, steel, computers, and electronics. The development of its high-tech industries has concerned competitors in Japan and the United States; for example, Samsung has established an increasing market share in an industry dominated by Japanese firms. South Korea is the seventh largest trading partner of the United States, which has long subscribed to a "domino theory" that if South Korea's economy fell, so would Japan's and perhaps even its own. In 2007, the United States and South Korea signed a free-trade pact that will even more strongly enmesh their economies. The United States ratified this agreement in October 2011, and in November Korea did the same.

In sum, South Korea is one of Asia's success stories, an Asian Tiger that made an extraordinary postwar recovery, especially by using brainpower to overcome its resource limitations, and has enjoyed periods of rocketing economic growth. Although South Korea is an MDC by the strict technical GNI PPP definition, it is still something of an "in-between" country, not entirely an MDC or an LDC, not all urban but not all rural either, and both industrial and agricultural. The government still lavishes huge subsidies to protect its rice growers, for example, while also promoting the virtues of free trade in industry.

The transition from LDC to MDC has brought some growing pains, seen especially in the social changes of more women in the workforce, higher divorce rates, and falling birth rates. Like Japan, South Korea wants to maintain its cohesive society and does not want to invite foreigners to fill its impending shortage of productive young people. The government once encouraged family planning but is now asking its people to have more babies and to stop giving up their own for adoption abroad. Due to a traditional preference for boys, there are fewer women than men in the Korea, and due to young women's migration to the cities for work, there is a pronounced shortage of eligible Korean women in the countryside. The result is a full-blown industry of bride-shopping outside Korea, especially in China, Vietnam, and other parts of Asia.

The bride-seeking Korean farmer typically pays a broker for the full suite of services, from selecting his bride to the wedding ceremony. Prospective brides are lined up for his consideration—in Vietnam's Ho Chi Minh City, for example. He picks one and with no courtship marries her and takes her home to Korea. The Vietnamese and other brides are aliens in the homogeneous Korean society and are commonly ostracized by the grooms' extended families and by society at large. Most husbands and wives do not even speak one another's languages. There are many accounts of the wife's loneliness, depression, physical abuse, and even suicide. Mixed-race babies often suffer domestic abuse and bullying. Happily, there are also reports of successful unions and harmonious families, against the odds.

Despite having some experimental free-trade zones, North Korea is an economically isolated, poor country almost frozen in time, intent on pursuing the command economy model that has failed in every other nation that employed it. Despite its motto of self-reliance, it depends on huge imports of food and energy, especially from Russia. Its principal exports are apparently missiles and other military hardware; heroin, amphetamines, and other synthetic and semisynthetic drugs; and counterfeit money. Its people, whose sentiments and hopes are unknown because they are not allowed contact with the outside world, periodically hover close to famine. Only a select few are allowed access to the Internet, although the infrastructure is there: much of the country is wired with a fiber-optic intranet system. The North Korean leadership has a dilemma: If it gives up on *juche* and opens its doors, the economy might improve, but the people of this country often called the **Hermit Kingdom** will discover the bounties just across the border and might revolt.

Could the Koreans simply reunite as the two Germanys did? The prospect for Korean reunification actually appears to be very remote, even as relations between the two countries improve. North Korea continues to argue that the two countries should unite peacefully and without foreign influence. Outside observers generally believe, however, that North Korea is not really interested in reunification because it would probably doom the country's regime. For its part, South Korea is fearful of the enormous economic cost it would have to pay for absorbing its much poorer neighbor, just as West Germany did in absorbing East Germany. Reunification would probably begin with a huge stream of

poor northerners moving south. The nearby great powers are also quietly pleased with Korea's lingering division. The two Koreas provide dual buffer zones between the historical adversaries China and Japan. And if the two Koreas united, finally taking advantage of North Korea's strong natural resource base, China and Japan could suddenly have a rival great power to contend with.

For now, there is little thought of Korean unification. Even with the progress of the Six Party Talks and North Korea's pledge to disable its nuclear weapons facilities (see page 240), there is much more concern with simply avoiding war. Some incidents between the Koreas have been so serious that they threaten to spill over into war. For example, in 2011 North Korean forces sank a South Korean ship claimed to have crossed the Northern Limit Line (drawn hastily by the UN at the end of the Korean War) into North Korean waters. Forty-six South Koreans died. Soon after, North Korean artillery shelled the South Korean island of Yeonpeong, killing four. Because of treaty obligations, incidents in these places or in the DMZ could escalate tension between the United States and North Korea, and even be a tripwire for a great conflict (•**Figure 7.60**). U.S. troops could not avoid being involved, and Chinese troops might be brought in. Some analysts fear that Korea may yet become a nuclear battlefield. Japan worries that it would also be drawn into any conflict between the Koreas. The economic and political stability of these three small countries on the western Pacific Rim is critical to the well-being of the global system.

The book now takes us from that rim out to sea, into the vast Pacific realm.

• **Figure 7.60** Two South Korean soldiers keep a wary eye on their North Korean counterpart, just steps away in the Joint Security Area of the Panmunjom "truce village" in the DMZ. The low concrete slab running left-to-right on the left of the photo marks the border between the two countries. From time to time, a North Korean defector runs across the border, sometimes pursued by North Korean soldiers. Firefights involving both U.S. and South Korean troops against North Korean soldiers have accompanied some of these incidents.

summary

- South and East Asia include the countries of Japan, North and South Korea, China, Taiwan, Pakistan, India, Sri Lanka, Bangladesh, Bhutan, Nepal, Maldives, Myanmar (Burma), Laos, Thailand, Cambodia, Vietnam, Malaysia, Singapore, Indonesia, the Philippines, Brunei, Timor-Leste, and numerous islands scattered along the edges of this major continental bloc.

- This is the most populous world region, with 54 percent of the world's people. It includes the world's most populous countries, China and India. Population growth rates in the region vary widely, from zero growth in Japan to over 3 percent per year in Timor-Leste.

- Three concentric arcs make up the broad physiography of the region: an inner arc of the high mountain ranges, the Himalaya, Karakoram, and Hindu Kush; a middle arc of major floodplains, deltas, and low mountains; and an outer arc of thousands of islands, including the archipelagoes of the East Indies, the Philippines, and Japan.

- Major rivers include the Indus, Ganges, and Brahmaputra in South Asia; the Irrawaddy, Chao Praya (Menam), Mekong, and Red in Southeast Asia; and the Chiang Jiang (Yangtze) and Huang He (Yellow) in East Asia.

- The region's climate types and biomes include tropical rain forest, savanna, humid subtropical, humid continental, steppe, desert, and undifferentiated highland. Shifting cultivation and wet rice cultivation are important forms of agriculture. Wet rice cultivation produces very high yields and is associated with dense human populations.

- Although South and East Asia have some of the world's largest cities, about 60 percent of the region's population is rural. Many uniquely Asian traditions helped shape village settlement planning and home design.

- South and East Asia's ethnic and linguistic compositions are diverse. Major language families are Indo-European, Dravidian, Sino-Tibetan, Altaic, Austric, and Papuan.

- Major religions and sociopolitical philosophies of Asia are Hinduism, Islam, Buddhism, Confucianism, Daoism, and Christianity.

- Great Britain, the Netherlands, France, and Portugal were the most important colonial powers in this region. Most of these domains were relinquished by the middle of the twentieth century, and the later British return of Hong Kong and the Portuguese return of Macao (both to China) closed the colonial period.

- China has the region's strongest economy, and globally is second only to that of the United States. Recent decades have seen strong economic growth among the Tigers and Tiger Cubs of the region, including South Korea, Taiwan, Singapore, Thailand, and Malaysia. The most important emerging economic power is China, whose large and inexpensive labor force has recently been attracting investment away from other parts of the region. India's well-educated and inexpensive labor force is contributing to its strong economic growth. There are fears in the United States about the outsourcing of American jobs to India.

- The Green Revolution is a broad effort to increase agricultural productivity in dominant crops. Biotechnology has produced crops that are more drought- and pest-resistant and capable of creating much higher yields, but genetic engineering of crops has perceived risks. Profits and other benefits from the Green Revolution are not uniformly spread.

- There are several major geopolitical issues in South and East Asia. The traditional enemies Pakistan and India (both pivotal countries from the U.S. point of view) now possess nuclear weapons. There are fears that destabilization in Pakistan might allow the weapons to fall into the wrong hands. Pakistan's support of the United States in its war on terrorism is very risky because a popular backlash could bring down the government. North Korea has nuclear weapons and is apparently trying to use them as bargaining chips to get more food aid and other assistance from the West. Anti-government and anti-Western Islamists are active in Indonesia, another country viewed as pivotal by the United States. Strife and fragmentation in Indonesia could threaten vital oceanic shipping lanes.

- Afghanistan is one of the world's poorest nations. It is landlocked and has limited resources, poor internal transportation, and little foreign trade. Opium is its main export. Its location along the ancient caravan routes and adjacent to large oil reserves has made it the target of stronger powers. Backed by the United States, rebels succeeded in driving out a Soviet occupation force in the 1980s. That costly conflict was followed by a period of civil war during which outside countries neglected Afghanistan. The militant Islamist Taliban came to power and protected the presence and training of al-Qa'ida militants. Osama bin Laden and others planned anti-Western attacks, including 9/11, from Afghanistan. Al-Qa'ida was driven into hiding from there in U.S.-led attacks following 9/11, and the Taliban were deposed from power. Reconstruction of the country is progressing, but a Taliban insurgency continues.

- The subcontinent is home to many different faiths, including Hinduism, Sikhism, Christianity, Islam, Buddhism, Zoroastrianism, and Jainism. Religious, ethnic, and other differences underlie serious and often violent conflicts. The main ones have been Hindus versus Muslims in India, Sikhs versus the government of India, and Tamils versus Sinhalese in Sri Lanka.

- Until 1947, what are now Pakistan, Bangladesh, and India formed the single country of India. For over a century, it was the most important unit in the British colonial empire—the "jewel in the crown."

- With independence, India was partitioned between avowedly Muslim Pakistan (which later divided into two countries, Pakistan and Bangladesh) and secular, primarily Hindu India. The partition had many consequences, including violence between

Hindus and Muslims and their large-scale migrations, disputes over the sharing of Indus River waters, and resources and industries stranded on either side of the new partition lines.

- The most severe product of the partition has been lasting conflict in Kashmir. Several wars have been sparked or fought in the still volatile frontier province, which, despite its Muslim majority, was joined with India in 1947.

- Although it is an overwhelmingly poor and rural country, India has many modern industries that, along with privatization of former state-run firms, have contributed to its substantial rate of economic growth in recent years. These include the software and computer industries of Bengaluru and the film industry of Mumbai. Pakistan and Bangladesh are much less industrialized and rely heavily on textile exports. India and Bangladesh have been successful in lowering birth rates, but population growth is a serious issue for both countries. India has managed to feed its huge population.

- South Asia provides large numbers of workers to other countries, ranging from professional to semiskilled and unskilled. Remittances sent home by these workers are important to the South Asian economies. Many of the professionals do not return, contributing to the region's brain drain.

- Some of India's large minorities have had serious difficulties with the majority Hindu population. Hindu nationalism is a problem in India, fanning the flames of communal violence between Hindus and Muslims. In the 1980s, some of the Punjab's Sikhs agitated for an independent country.

- Pakistan's government faces challenges from its Pashtun population in the country's west and northwest and from strong Islamist sentiment throughout the country.

- Bangladesh and the Maldives are threatened by serious natural hazards from the sea. Strong typhoons (hurricanes) regularly devastate large parts of low-lying Bangladesh, and most of the Maldives would be inundated if sea levels were to rise by 3 feet (90 cm).

- Sri Lanka, the world's largest tea exporter, has a Buddhist Sinhalese majority and a Hindu Tamil minority. For more than 20 years, Tamil factions have led a violent struggle to establish an autonomous homeland on the island's east, north, and west periphery. The violence has been punctuated by peace talks that have not yet arrived at a permanent solution.

- A megathrust earthquake off the northeastern coast of Sumatra caused the great tsunami of 2004, which killed more than 200,000 people in 14 countries. A similar earthquake in 2011 triggered the tsunami that caused massive devastation on the coast of northeast Japan.

- These are more Southeast Asian villagers than urbanites.

- Southeast Asia's tropical forests are being destroyed at a faster rate than others of the world, mostly for commercial logging for Japanese markets. In biodiversity terms, Indonesia is a megadiversity country whose forests are under intense pressure.

- Southeast Asia as a whole is among the world's poorer regions. The poorest countries—Myanmar, Cambodia, Laos, and Vietnam—have suffered from warfare and, in the case of Myanmar, from unwise political management.

- Drug production, drug abuse, and HIV/AIDS are serious problems in Myanmar. Neighboring Thailand has some of these problems but has dealt more effectively with them.

- Prodemocracy efforts in Myanmar have been thwarted by official repression.

- U.S. forces succeeded the French withdrawal from Vietnam in the 1950s. American efforts to win a war against communism in Vietnam failed. Normal relations now exist between the United States and Vietnam, and although Communist, Vietnam encourages private enterprise and has seen recent strong economic growth.

- Numerous dams are being constructed on the Mekong River. These will produce hydroelectricity but will damage the productive fishery of this basin.

- Much unrest in Southeast Asia is related to tensions between ethnic groups. Indonesia is in danger of fragmentation. Catholic Timor-Leste has gained impendence from Indonesia, and the provinces of Aceh and Papua are seeking self-rule or independence. Muslim insurgents in the southern Philippines and Communist rebels throughout the country have stepped up their campaigns against the government.

- China may be physically divided into the arid west and the humid east. An arc drawn from China's border with Myanmar northeast to Harbin is the rough dividing line between the arid west and the humid east. The majority ethnic Han are associated with the more densely populated humid east, and non-Han realms are primarily in the arid west. The Huang He (Yellow River) and the Chang Jiang (Yangtze River) descend from the Tibetan highlands.

- The Three Gorges Dam, under construction on the Chang Jiang, is designed to control floods, provide hydropower, and allow transfer of water to the more arid north. The project has been both celebrated and lamented because of the scale of environmental change and social dislocation associated with it.

- China has modeled much of its most dynamic economic growth after capitalist rather than Communist economic and political models.

- Inexpensive labor helped fuel the booming industries of the coastal cities. There is still poverty in China, with the cities generally wealthier than rural areas. The most prosperous areas are in the southeast, along the east coast (including the Special Administrative Regions of Hong Kong and Macau, and several Special Economic Zones), and in farmlands near large cities, whereas regions in the west, southwest, uplands, and Arid China have had less economic development. Jobs and wages are now growing in more areas, and labor shortages have developed in the coastal boomtowns.

- Taiwan (the Republic of China) continues to claim to represent the true government of China and has a tense relationship with the mainland People's Republic of China. Taiwan is heavily invested in mainland China's development. The United States does not have diplomatic relations with Taiwan but supports it militarily and economically.

- Japan experienced phenomenal economic growth after its crushing defeat in World War II, but it has recently suffered economic recession.

- Due to its relative poverty in natural resources, Japan must import most of its raw materials, most energy supplies, and a large share of its food.

- The quest for possession of offshore resources near islands is a source of tension between Japan and Korea and between Japan and China.

- The hardworking Japanese are experiencing many symptoms of workaholism, and many complain about their low level of amenities relative to that of other prosperous countries.

- The Korean Peninsula has an unfortunate location in geopolitical terms, sandwiched between the greater powers of Russia, China, and Japan. Korea has had a turbulent recent political history, first as a colony of Japan and then as a land divided between the Communist North and the more dynamic and democratic South.

- There are marked physical and economic contrasts between North and South Korea, with North Korea being much more rugged and having more natural resources for industry. Capitalist South Korea is far more prosperous than Communist North Korea.

- North Korea has suffered famine recently but has been reluctant to open up to trade and assistance because North Koreans might revolt.

- There are prospects for reconciliation and, more remotely, reunification, but also for war between North and South Korea. Reunification is not favored by the South because it would be too costly. China and Japan would apparently be happier to have a weaker divided Korea than a stronger unified Korea.

Key Terms + Concepts

Ainu (p. 226)
ancestor veneration (p. 228)
animism (p. 228)
archipelago (p. 214)
Aryans (p. 225)
Asian Century (p. 210)
Asian Tigers (p. 233)
Association of Southeast Asian Nations (ASEAN) (p. 256)
Austric language family (p. 226)
  Austro-Asiatic subfamily (p. 226)
  Austronesian subfamily (p. 226)
  Hmong-Mien subfamily (p. 226)
  Tai-Kadai subfamily (p. 226)
autonomous regions (p. 262)
blowback (p. 242)
break-of-bulk point (p. 266)
bubble economy (p. 271)
Buddhism (p. 277)
"Burmese Way to Socialism" (p. 255)
Cao Dai (p. 228)
caste (p. 248)
chaebols (p. 275)
Chang Jiang Water Transfer Project (p. 266)
communal violence (p. 247)
Comprehensive Test Ban Treaty (p. 236)

Confucianism (p. 231)
contingent sovereignty (p. 255)
Convention on the Law of the Sea (p. 273)
Dalai Lama (p. 264)
Dalits (p. 248)
Daoism (Taoism) (p. 231)
demilitarized zone (DMZ) (p. 274)
"denying the countryside to the enemy" (p. 258)
domino effect (p. 240)
Dravidian language family (p. 225)
endemic species (p. 254)
exclusive economic zone (EEZ) (p. 273)
fallow (p. 219)
feng shui (p. 224)
Free Aceh Movement (GAM) (p. 261)
genuine autonomy (p. 264)
geomancy (p. 224)
Golden Crescent (p. 242)
Golden Triangle (p. 256)
Great Game (p. 241)
Green Revolution (p. 235)
Han Chinese (p. 226)
Hanification (p. 264)
Hermit Kingdom (p. 276)
Hinduism (p. 228)
Indo-Iranian subfamily (p. 223)

Indus Waters Treaty (p. 247)
insolation (p. 217)
insourcing (p. 235)
intensive subsistence agriculture (p. 219)
Jainism (p. 231)
Jains (p. 231)
Jemaah Islamiah (JI, or the Islamic Group) (p. 240)
juche (p. 275)
karma (p. 230)
karoshi (p. 272)
Khalistan (p. 274)
Laskar Jihad (p. 240)
laterite (p. 219)
lateritic soils (p. 219)
levee (p. 222)
liberal autonomy (p. 261)
Liberal Democratic Party (LDP) (p. 271)
Mahayana Buddhism (p. 230)
megacity (p. 226)
megadiversity country (p. 254)
Mekong River Commission (p. 259)
Mohajirs (p. 244)
Mon-Khmer (p. 226)
monsoon (p. 217)
  summer monsoon (p. 217)
  winter monsoon (p. 217)
mutually assured destruction (MAD) (p. 237)

Naxalites (p. 247)
New Asian Tigers (p. 233)
nirvana (p. 230)
Nuclear Nonproliferation Treaty (NPT) (p. 236)
offshore outsourcing (offshoring) (p. 235)
One China Policy (p. 269)
Operation Enduring Freedom (p. 243)
overseas foreign workers (OFW) (p. 212)
paddy (padi) (p. 220)
Pancasila (p. 261)
pandemic (p. 213)
Papuan languages (p. 227)
Papuans (p. 262)
Parsis (p. 231)
pivotal countries (p. 237)
plinthite (p. 219)
sectarian violence (p. 247)
shifting cultivation (p. 219)
Shintoism (p. 227)
Sinhalese (p. 247)
Sino-Tibetan language family (p. 226)
  Sinitic subfamily (p. 226)
  Tibeto-Burman subfamily (p. 226)
slash-and-burn cultivation (p. 219)
Special Economic Zones (SEZs) (p. 267)

State Law and Order
  Restoration Council
  (SLORC) (p. 255)
strategic hamlet (p. 223)
"String of Pearls" (p. 248)
sweatshops (p. 268)
swidden cultivation (p. 219)
syncretism (p. 228)

Taliban (p. 243)
Tamils (p. 247)
Tamil Tigers (p. 248)
Taoism (p. 231)
terraces (p. 220)
theory of Himalayan
  environmental degradation
  (p. 222)

Theravada Buddhism (p. 230)
Tiger Cubs (p. 233)
untouchables (p. 248)
value-added manufacturing
  (p. 233)
Viet Cong (p. 258)
Wallace's Line (p. 254)
"Water Towers of Asia" (p. 250)

Western Big Development
  Project (p. 264)
wet rice cultivation (p. 220)
yang (p. 224)
yin (p. 224)
Zomia (p. 221)
Zoroastrianism (p. 231)

Review Questions

1. What countries make up South and East Asia?

2. Use the concept of the three arcs to trace the outlines of South and East Asia's geography. What are the largest plains and river valleys of the region? What are the most important mountain ranges? What are the major island groups?

3. What are the region's most populous countries? Where are the highest and lowest population densities? Which countries have the highest and lowest population growth rates?

4. What are the monsoons? What roles do they play in climates and human activities?

5. What types of environments are shifting cultivation and wet rice cultivation practiced in? What are some of the basic methods and land use considerations of both types of agriculture?

6. What is the geographical concept of Zomia, and what is James Scott's argument about the people who live in Zomia?

7. What roles have *feng shui* and Confucianism played in the site selection and design of homes, schools, and villages in parts of Asia?

8. What are some of the innovations and products that originated in South and East Asia and diffused from there?

9. What are the major linguistic and ethnic groups of South and East Asia? What are the principal religions and sociopolitical philosophies?

10. What were the major colonial powers in South and East Asia? What regions did they colonize?

11. What are the region's most significant economic trends?

12. Why are India, Pakistan, and Indonesia considered pivotal countries for U.S. interests? Has the United States been forced to take sides in the traditional enmity between India and Pakistan?

13. Why are there far fewer women than men in some Indian provinces? What problems are associated with this phenomenon?

14. What is the caste system? How does it affect employment and social relations in India? What evidence of change in this ancient system is there?

15. Where is Ayodhya? What site there is critical in relations between Hindus and Muslims in India?

16. What ethnic and political differences have given rise to strife in Sri Lanka?

17. What particular problems do Bangladesh and the Maldives have in terms of climate and climate change?

18. What caused the great tsunami of 2004, and what were some of its effects?

19. What countries are the largest producers and consumers of Southeast Asian tropical hardwoods?

20. What major changes are coming to the Mekong River basin?

21. What political events hampered development in Vietnam?

22. What are the major political and resource-related issues faced by Indonesia?

23. What factors may threaten the continuation of China's economic boom?

24. How can China's economy be characterized? What is its "average" standard of living based on per capita GNI PPP? How effectively is the wealth distributed? Where is it concentrated geographically, and how is the government hoping to change that pattern?

25. Where are China's Special Economic Zones? What does their distribution suggest about China's economic geography?

26. What are the major political attributes of Taiwan?

27. What is Japan's ethnic makeup? What relations have the Japanese majority had with internal minorities and foreign groups at various times?

28. How has Japan been able to overcome its poverty in natural resource assets by commercial, military, or other means?

29. What are some of the features of Japanese society today, especially those related to the country's economy?

30. What precipitated the division of the Korean Peninsula into North and South Korea?

31. What are the major differences between North and South Korea in natural resources, political systems, and economic development?

32. What are the prospects for reunification of the Koreas? How do some outside powers and the Koreans themselves view reunification?

Notes

1. John Reader, British anthropologist and traveler, cited in Fred Pearce, "Terraces: The Other Wonders of the World," *Eurozine,* March 2001, http://www.eurozine.com/articles/2001-03-01-pearce-en.html. Accessed July 11, 2007.

2. Jack D. Ives and Bruno Messerli, *The Himalayan Dilemma: Reconciling Development and Conservation* (New York: Routledge, 1989).

3. Quoted in Pankaj Mishra, "Caste Adrift." *Financial Times,* January 23, 2011, p. 13.

4. This discussion is based largely on two unpublished manuscripts, "Living Spaces of Korean Architecture" and "In Tune with Nature: Rural Villages and Houses in Korea," by the Korean geographer Sang-Hae Lee, of Sungkyunkwan University, and personal talks with him. I am grateful for his contributions.

5. Quoted in David Pilling, "Megacities." *Financial Times,* November 6, 2011, p. HH-1.

6. Descriptions of Buddhism and Daoism are based on the *Microsoft Encarta Reference Library,* 2004.

7. Michael Schuman. 2009. *The Miracle: The Epic Story of Asia's Quest for Wealth.* New York, NY: HarperBusiness.

8. Patrick Barta, July 27, 2009, "Asian Nations Revisit Safety Net in Effort to Bolster Spending." *The Wall Street Journal,* p. A2.

9. Quoted in Bob Davis, "Massive Population Lifts Nation's Growth." *The Wall Street Journal,* p. A12.

10. Alan S. Blinder, quoted in Anand Giridharadas, "India's Edge Goes beyond Outsourcing." *New York Times,* April 4, 2007, p. C4.

11. Quoted in Edward Wong, "China Navy Reaches Far, Unsettling the Region." *New York Times,* June 15, 2011, p. A11.

12. Robert Chase, Emily Hill, and Paul Kennedy, "Pivotal States and U.S. Strategy." *Foreign Affairs,* January–February 1996, pp. 33–51.

13. Fukuyama, quoted in Martin Wolf, "The End of History Man." *Financial Times,* May 29, 2011, p. 3.

14. James Hookway, "Tensions Flare over Disputed Asian Sea." *The Wall Street Journal,* June 10, 2011.

15. Quoted in "Unquenchable Thirst." *The Economist,* November 19, 2011, p.. 27.

16. Samsar Chand Kour, *Beautiful Valleys of Kashmir and Ladakh* (Mysore, India: Wesley Press, 1942), p. 2.

17. See UNESCO, *Assessment of Capacity Building Requirements for an Effective and Durable Tsunami Warning and Mitigation System in the Indian Ocean: Consolidated Report for the Countries Affected by the 26 December 2004 Tsunami* (Paris: UNESCO, 2005), http://unesdoc.unesco.org/images/0014/001445/144508e.pdf. Accessed July 15, 2007.

18. Quoted in Jamil Anderlini, "A Blast from the Past." *Financial Times,* December 15, 2009, p. 11.

19. Jamil Anderlini and Mure Dickie, "Taking the Waters: China's Gung-Ho Economy Exacts a Heavy Price." *Financial Times,* July 24, 2007, p. 11.

20. I thank the Korea Society for sponsoring my field study of Korea, on which much of this discussion is based. Professor Byong Man Ahn of the Hankuk University of Foreign Studies and Professor Taeho Bark of the Graduate School of International Studies at Seoul National University graciously provided much of the information on Korean economies and politics.

Online Resources

CourseMate: Make the most of your study time by accessing everything you need to succeed in one place. Read your textbook, take notes, review flashcards, watch videos, complete activities, take practice quizzes, and more—online with CourseMate. Log in at **www.cengagebrain.com**.

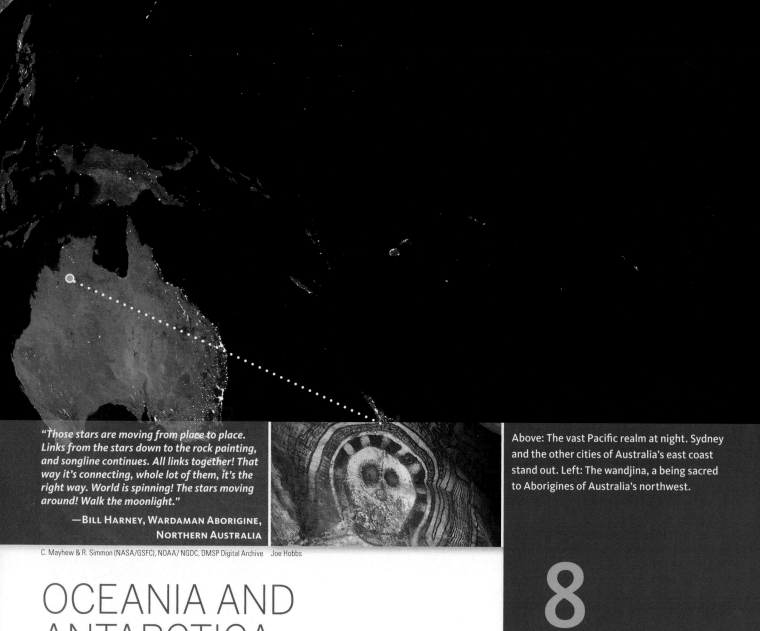

"*Those stars are moving from place to place. Links from the stars down to the rock painting, and songline continues. All links together! That way it's connecting, whole lot of them, it's the right way. World is spinning! The stars moving around! Walk the moonlight.*"

—BILL HARNEY, WARDAMAN ABORIGINE,
NORTHERN AUSTRALIA

Above: The vast Pacific realm at night. Sydney and the other cities of Australia's east coast stand out. Left: The wandjina, a being sacred to Aborigines of Australia's northwest.

C. Mayhew & R. Simmon (NASA/GSFC), NOAA/ NGDC, DMSP Digital Archive   Joe Hobbs

OCEANIA AND ANTARCTICA

8

Covering fully one-third of the earth's surface, **Oceania** (pronounced "oh-shee-*an*-ee-uh") is mostly water. The world's largest ocean, the Pacific, defines and dominates this region. It is bigger than all the earth's continents and islands combined. Before World War II, the Western world spawned legends about this ocean and its islands as a kind of utopia. There has long been trouble in this paradise, however. On many islands, foreign traders, whaling crews, labor recruiters, and other opportunists exploited the indigenous peoples, reducing their numbers and disrupting their cultures. The military battles of World War II further shattered the idyllic qualities of many islands. Yet today, the Pacific mystique, which has been perpetuated in books and films, forms part of the allure for booming tourism development across the region. This chapter presents the cultural and natural geographies of Oceania, summarizes the impacts of colonial enterprises in the region, and discusses some of the obstacles to development confronting the countries of this watery realm.

chapter objectives

This chapter should enable you to

- Appreciate the economic prominence of larger, Europeanized Australia and New Zealand in a vast sea of small, mainly indigenous political units.

- Recognize the associations between the physical geographies of islands, their typical social and political organizations, and their economic characteristics.

- Consider the impacts of human agency, especially deforestation and the introduction of exotic species, on island ecosystems.

- Hear the concern expressed by low-lying island countries about the production of greenhouse gases in faraway industrialized nations.

- Recognize the peculiar dependence of some Pacific populations on the military interests of distant nations.

- Understand the obstacles faced by indigenous people in winning legal recognition of ancestral claims.

- Evaluate the impacts of exotic species on island ecosystems.

- Appreciate the process by which Australia and New Zealand are loosening ties with their ancestral European homeland and strengthening their regional orientation.

- Learn about "unclaimed claims" to the continent of Antarctica.

Also covered in this chapter is Antarctica, an icy continent with no permanent human inhabitants, but where much important scientific research is being conducted.

8.1 Area and Population

Oceania, also known as the Pacific World, includes Australia, New Zealand, and the islands of the mid-Pacific lying mostly between the tropics (•Table 8.1). The Pacific islands nearer the mainlands of East Asia, Russia, and the Americas are excluded here on the basis of their close ties with the nearby continents. Hawaii belongs to this region and is dealt with both here and in Chapter 11's discussion of the United States. Large areas of the eastern and northern Pacific that contain few islands are also not considered here. Australia and New Zealand have strong political and economic interests in the smaller tropical islands and a similar insular character in many ways, and they share ethnic affiliations with the original inhabitants of the smaller islands. But they are also uniquely European in character and dwarf the other countries in economic and political clout. Therefore, these two countries, along with the continent of Antarctica, are examined in more detail later in the chapter.

Major Divisions of the Region

Geographers have long divided the Pacific islands into three principal regions: Melanesia, Micronesia, and Polynesia (•Figure 8.1). The islands of **Melanesia** (Greek for "black islands"), lying northeast of Australia, are relatively large. New Guinea, the largest, is about 1,500 miles (c. 2,400 km) long and 400 miles (c. 650 km) across at the broadest point. It is divided between two countries. The western section, called Papua (formerly Irian Jaya), is part of Indonesia and is discussed in Chapter 7. The eastern section is an independent country, Papua New Guinea. The other Melanesian countries are the Solomon Islands, Vanuatu, and Fiji. New Caledonia, also in Melanesia, is controlled by France.

Micronesia (Greek for "tiny islands") includes thousands of scattered small islands in the central and western Pacific, mostly north of the equator. The Micronesian countries are Palau, the Federated States of Micronesia, the Republic of the Marshall Islands, Nauru, and the western part of the nation of Kiribati (pronounced "keer-uh-bahss"). Micronesia also includes two possessions of the United States: Guam and the Northern Mariana Islands.

Polynesia (Greek for "many islands") occupies a greater expanse of ocean than Melanesia or Micronesia. It is shaped like a rough triangle with corners at New Zealand, the Hawaiian Islands, and remote Easter Island. Excluding Hawaii and New Zealand, Polynesia's independent countries are the eastern part of Kiribati, Tuvalu, Samoa, and Tonga. Possessions of other countries in Polynesia include French Polynesia and France's Wallis Islands, American Samoa, New Zealand's Tokelau and Cook Islands, Britain's Pitcairn Island, and Chile's Easter Island.

The typical Pacific island country (excluding Australia and New Zealand) has about 100,000 to 150,000 people in an area of 250 to 1,000 square miles (c. 650 to 2,600 sq km); consists of a number of islands; is poor economically; is an ex-colony of Britain, New Zealand, or Australia; and depends heavily on foreign economic aid. The total land area, including Australia and New Zealand, is 3.3 million square miles (8.5 million sq km), or about 90 percent of the size of the entire United States (•Figure 8.2).

The People and Where They Live

The region's total population is 37 million (•Figure 8.3). Aside from Australia's 23 million people, populations range from 7 million in Papua New Guinea, which is exceptionally large in population and area for this region but has a low population density, to 9,300 on tiny Nauru. The highest population densities are in the smallest island groups, notably the Tuamotu Islands in French Polynesia and the Ellice Islands of Tuvalu. Population growth rates vary widely, from the predictably low 0.7 and 0.8 in prosperous Australia and New Zealand, respectively, to a high 2.6 percent in much poorer Papua New Guinea.

There are some apparent problems of overpopulation, especially in Polynesia. Considerable emigration occurs, especially

Table 8.1   Oceania: Basic Data

Political Unit	Area (thousands; sq mi)	Area (thousands; sq km)	Estimated Population (millions)	Estimated Population Density (sq mi)	Estimated Population Density (sq km)	Annual Rate of Natural Increase (%)	Human Development Index	Urban Population (%)	Per Capita GNI PPP ($US)
Australia and New Zealand	**3,093.4**	**8,011.906**	**27.1**	**9**	**3**	**0.7**	**0.925**	**83**	**36,770**
Australia	2,988.9	7,741.3	22.7	8	3	0.7	0.929	82	38,510
New Zealand	104.5	270.7	4.4	42	16	0.8	0.908	86	27,790
Others	**213.4**	**552.6**	**10.2**	**48**	**18**	**2.3**	**0.550**	**24**	**2,620**
American Samoa (U.S.)	0.2	0.5	0.05	250	97	0.0	N/A	90	N/A
Cook Is. (N.Z.)	0.09	0.2	0.02	222	86	N/A	N/A	70	N/A
Fiji	7.1	18.3	0.9	127	49	1.4	0.688	51	4,530
French Polynesia (Fr.)	1.5	3.8	0.3	200	77	1.3	N/A	51	N/A
Guam (U.S.)	0.2	0.5	0.2	1,000	386	1.4	N/A	93	N/A
Kiribati	0.3	0.7	0.1	333	129	2.0	0.624	44	3,310
Marshall Islands	0.1	0.2	0.1	1,000	386	2.5	N/A	68	N/A
Micronesia, Federated States of	0.3	0.7	0.1	333	129	1.9	0.636	22	3,240
Nauru	0.009	0.02	0.01	1,111	429	1.9	N/A	100	N/A
New Caledonia (Fr.)	7.2	18.6	0.3	42	16	1.2	N/A	58	N/A
Northern Mariana Islands (U.S.)	0.2	0.5	0.07	350	135	−4.0	N/A	94	N/A
Palau	0.2	0.5	0.02	100	39	0.6	0.782	77	N/A
Papua New Guinea	178.7	462.8	6.9	39	15	2.6	0.523	13	2,260
Samoa	1.1	2.8	0.2	182	70	2.3	0.688	22	4,270
Solomon Islands	11.2	29	0.5	45	17	2.7	0.510	20	1,860
Tonga	0.3	0.7	0.1	333	129	2.0	0.704	23	4,570
Tuvalu	0.01	0.02	0.01	1,000	386	1.4	N/A	47	N/A
Vanuatu	4.7	12.1	0.3	64	25	2.6	0.617	24	4,290
Summary Total	**3,306.8**	**8,564.5**	**37.28**	**11**	**4**	**1.1**	**0.822**	**67**	**27,440**

Sources: World Population Data Sheet, Population Reference Bureau, 2011; Human Development Report, United Nations, 2011; World Factbook, CIA, 2011.

from the Cook Islands, Samoa, Tonga, and Fiji to New Zealand and North America. Many educated young Australians are also lured abroad, especially to the United States. Because birth rates are low, people in both Australia and New Zealand share the typical postindustrial fear of not having enough people to support the countries' economies and their aging populations. Both countries, however, are averse to liberal immigration policies that would boost their populations.

70, 271, 393

8.2 Physical Geography and Human Adaptations

The climates, vegetation, landscapes, and cultural geographies of the islands scattered across the vast Pacific vary, but a few notable patterns emerge (•Figures 8.4 and 8.5; note that only the islands of the southwestern Pacific are large enough to have meaningful data depicted at this map scale).

Climates and Biomes

Climatically, most of the region is tropical; all of Micronesia and Melanesia and most of Polynesia (except for northern Hawaii and New Zealand and its outlying islands) lie in the tropics. Most of New Guinea, which lies entirely in the tropics, has tropical rain forest climate and biome types, but its high mountains have undifferentiated climates and vegetation types, varying with elevation all the way to tundra above 11,000 feet (3,300 m). The great majority of the smaller Pacific islands lying in the tropics are influenced by the easterly trade winds that bring abundant precipitation for tropical rain forest climate and biome types.

The Tropic of Capricorn bisects Australia. Cooling midlatitude westerly winds bring much of New Zealand and parts of coastal southern Australia a marine west coast climate, with associated temperate mixed forests. Some coastal areas of southern Australia also have Mediterranean and semiarid steppe climates, with associated Mediterranean scrub

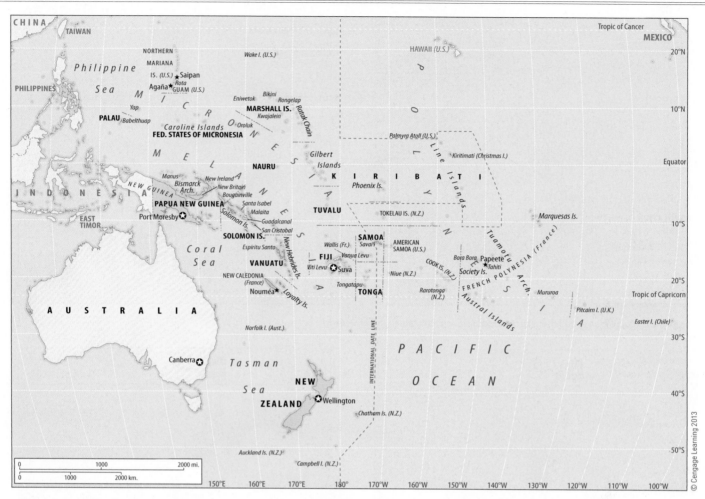

• **Figure 8.1** Principal features of Oceania. Note the peculiar jog in the International Date Line near Kiribati (along the equator), discussed on page 11.

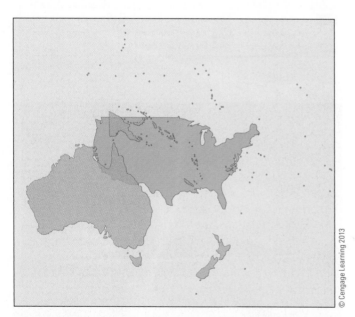

• **Figure 8.2** Oceania compared in area (but not latitude) with the conterminous United States.

vegetation and prairie and steppe grasses. As in California and in the Mediterranean region itself, Australia's Mediterranean climate is ideal for viticulture and production of the wines available in your local markets. Although much of the interior of Australia has a desert climate and vegetation, coastal regions of northern Australia generally have climates and vegetation of tropical savanna, with small patches of tropical rain forest climate and vegetation. More details on the climates, biomes, and environmental problems of Australia come later in the chapter.

Island Types

Topographically, Oceania features three general types of islands. **Continental islands** are continents or were attached to continents before sea level changes and tectonic activities isolated them. These include New Guinea, New Britain, and New Ireland in the Bismarcks; New Caledonia, Bougainville, and smaller islands in the Solomons; the two main islands of Fiji; Australia; and the North and South Islands of New Zealand.

Some of these have lofty peaks, including several over 16,000 feet (c. 4,900 m) in New Guinea, whereas Australia is

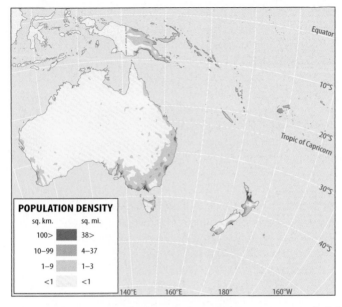

• **Figure 8.3** Population distribution (left) and population cartogram (right) of Oceania.

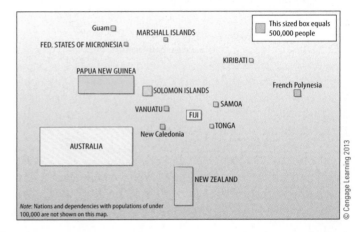

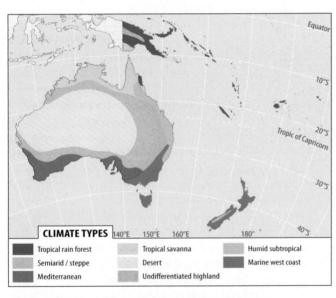

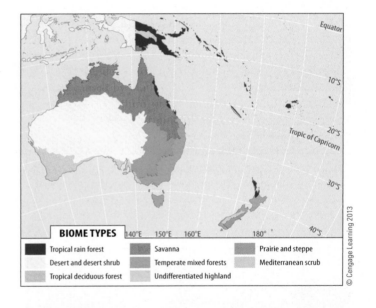

• **Figure 8.4** Climates (left) and biomes (right) of Oceania.

• **Figure 8.5** (a) Moorea Island, in Tahiti, is a classic and much romanticized high island. (b) Low islands, such as the one seen here, are most associated with the "coconut civilizations" of the Pacific. Low islands are extremely vulnerable to the dangers that would be posed by global warming.

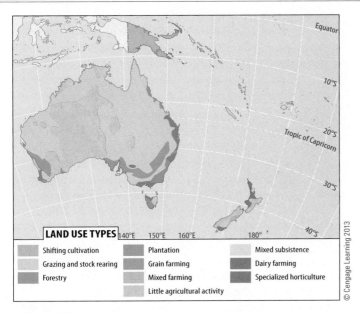

• **Figure 8.6** Land use in Oceania.

LAND USE TYPES

- Shifting cultivation
- Grazing and stock rearing
- Forestry
- Plantation
- Grain farming
- Mixed farming
- Little agricultural activity
- Mixed subsistence
- Dairy farming
- Specialized horticulture

Table 8.2	Oceania: Metropolitan Populations Data	
1	Sydney, Australia	3.7
2	Melbourne, Australia	3.5
3	Brisbane, Australia	1.8
4	Auckland, N.Z.	1.3
5	Perth, Australia	1.3
6	Adelaide, Australia	1

Population in millions.

overall rather low. The region's other islands are categorized as either **high islands** or **low islands** (Figure 8.5). Most high islands are the result of volcanic eruptions. The low islands are made of coral, a material composed of the skeletons and living bodies of small marine organisms that inhabit tropical seas.

These very different settings offer sharply different opportunities for human livelihoods (•**Figure 8.6**). Overall, however, except in urban, industrialized Australia and New Zealand, the Pacific peoples are rural; statistically, the region excluding those two countries is 76 percent rural. People live in small farming or fishing villages, where they have long depended on root crops (such as taro and yams), tree crops (such as breadfruit and coconuts), ocean fish, and pigs (•**Figure 8.7**). Outside

Australia and New Zealand, there are only four cities with a population greater than 100,000: Port Moresby in Papua New Guinea, Suva in Fiji, Nouméa in New Caledonia, and Papeete in Tahiti (•**Table 8.2**).

Most of the region's high islands are not continental but volcanic in origin and have a familiar pattern. There is a steep central peak, with ridges and valleys radiating outward to the coastline. Permanent streams run through the valleys. A coral reef surrounds the island, and between the shore and the reef is a shallow lagoon. Many of the high volcanic islands are spectacularly scenic (see Figure 8.5a). Classic high islands in the region include the Polynesian groups of Hawaii, Samoa, and the Society Islands.

Some of the volcanic high islands of the Pacific comprise island chains. These are formed when the oceanic crust slides over a stationary **geologic hot spot** in the earth's mantle, where molten magma is relatively close to the crust (•**Figure 8.8**). [21] As the crust slides over the geologic hot spot, magma rises through the crust to form new volcanic islands. The Hawaiian Island chain is an excellent example. The big island of Hawaii is now situated over the hot spot and is still active; it is one of the chain's youngest members. The Pacific Plate of the earth's crust is moving northwestward here; older volcanic islands that were born over the hot spot have moved to the northwest and have become inactive. Still older islands farther northwest in the chain have submerged to become **seamounts**, or underwater volcanic mountains. Other seamounts have yet to break the surface, including one in the southeast that will emerge in about 10,000 years to become Hawaii's newest island.

The geographer Tom McKnight explained that the high island's topography has historically influenced its cultural and economic patterns.[1] Each drainage basin formed a relatively distinct unit that was governed by a different chief. Within each unit, there were several subchiefs, each of them controlling a bit of each type of available habitat: coastal land, stream land, forested land, and gardening land. Because available resources were distributed relatively equally among chiefdoms, there were few power struggles or monopolies over resources. It is important to note that these are general rather than universal patterns of association between physical geography and human adaptations in the Pacific. Geographers reject the outdated notion of **environmental determinism**, which insisted that certain environmental conditions always correlate with specific culture traits.

Today, the main cities and seaports of Oceania, and the largest populations, are on the volcanic high islands and the

• **Figure 8.7** A market in Vanuatu, where root crops are a dietary staple.

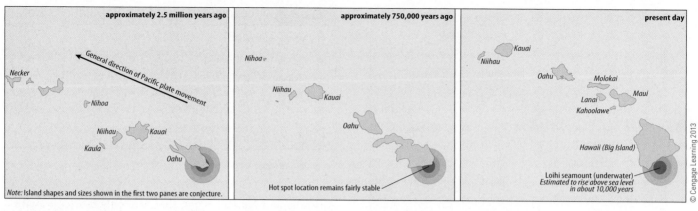

approximately 2.5 million years ago

Necker

General direction of Pacific plate movement

Nihoa

Niihau

Kaula

Kauai

Oahu

Note: Island shapes and sizes shown in the first two panes are conjecture.

approximately 750,000 years ago

Nihoa

Niihau

Kauai

Oahu

Hot spot location remains fairly stable

present day

Kauai

Niihau

Oahu

Molokai

Lanai

Maui

Kahoolawe

Hawaii (Big Island)

Loihi seamount (underwater)
*Estimated to rise above sea level
in about 10,000 years*

© Cengage Learning 2013

• **Figure 8.8** The Hawaiian Islands were formed (and continue to form today) as a piece of the earth's crust has slid over a geologic hot spot.

continental islands. Their rich soils, typically of volcanic origin, support a diversity of tropical crops. These islands are generally more prosperous than the low islands.

The low islands are formed of coral. Most take the shape of an irregular ring surrounding a lagoon; such an island is called an **atoll**. Generally, the coral ring is broken into many pieces, separated by channels leading into the lagoon, but the whole circular group is commonly considered one island. Charles Darwin devised a still widely accepted explanation for the three-stage formation of atolls (•**Figure 8.9**). First, coral builds what is known as a "fringing reef" around the edge of a volcanic island. Then, as the island slowly sinks, the coral reef builds upward and forms a barrier reef separated from the shore by a lagoon. Finally, the volcanic island sinks out of sight, and a lagoon occupies the former land area, whose outline is reflected in the roughly oval form of the atoll.

The low islands are generally smaller than the volcanic high islands and lack the resources to support dense populations. Their shorelines are typically dotted with the waving coconut palms that are a mainstay of life and the trademark of the South Sea isles. The Gilbert Islands of Kiribati in Micronesia are typical of the picturesque low island atolls idealized by Hollywood and travel brochures (see Figure 8.5b). Other major atoll groups in Micronesia are the Caroline and Marshall Islands. The Tuamotu Archipelago is a Polynesian atoll group, and atolls are also scattered across Melanesia.

Despite their idyllic appearance, these islands pose many natural hazards to human habitation. The lime-rich soils are often so dry and infertile that trees will not grow, and shortages of drinking water limit permanent settlement. Their low elevation above sea level provides little defense against storm waves, tsunamis, and sea level rise due to climate change. Coastal regions of higher islands also suffer from these events; a 1998 tsunami killed 2,100 people in northern Papua New Guinea, for example.

Before the intrusion of outsiders, the low-island economies were based heavily on subsistence agriculture (with strong reliance on coconuts), gathering, and fishing. Tom McKnight explained that sparse resources were distributed with relative uniformity among the low islands' populations.[2] The political and social units were smaller and less structured than those of the high islands. These resource-poor low islands have a long history of population-limiting factors such as war, infanticide, and abortion.

Many low-island countries now rely heavily on modern commercialized versions of the coconut and fishing economy. The coconut is so important in their subsistence that many of these islands have developed what has been characterized as a

(a) Volcanic island with fringing reef

(b) Slight subsidence, barrier reef

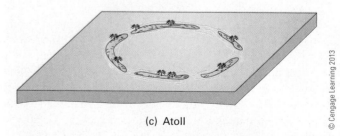

(c) Atoll

© Cengage Learning 2013

• **Figure 8.9** Charles Darwin's explanation of the development of an atoll. (a) First, a fringing reef of coral is attached to the volcanic island's shore. (b) As the island subsides, a barrier reef forms. (c) With continued subsidence, the coral builds upward, and the volcanic center of the island finally becomes completely submerged, forming an atoll.

• **Figure 8.10** Pacific islanders know how to make many "appliances" and wares from coconut tree parts. This Fijian man is weaving a mat. Note the husked coconuts in the foreground.

"coconut civilization." Coconuts provide food and drink for the islanders, and the dried meat, known as copra—which is processed for coconut oil used in foods, cosmetics, and soaps—is the only significant export from many islands. The husks and shells of the nuts have many uses, as do the trunks and leaves of the coconut palms. People make baskets and thatching from the leaves, for example, and use timber from the trunk for construction and furniture (•**Figure 8.10**).

Why Are Oceania's Ecosystems So Vulnerable?

Island ecosystems in the Pacific region, like those across the globe, are typically inhabited by endemic species of plants and animals—those found nowhere else in the world. They result from a process in which ancestral species colonize the islands from distant continents and, over a long period of isolation and successful adaptation to the new environments, evolve to become new species. Gigantic and flightless animals, like the moa bird of New Zealand and the giant tortoises of the Galápagos (in the Pacific) and the Seychelles (in the Indian Ocean), are typical of endemic island species. More than 70 percent of the native plant species of New Caledonia and Hawaii are endemic.

Because they have developed in the absence of natural predators and inhabit relatively small areas, island species are especially

• **Figure 8.11** The luxuriant forests of New Guinea contain an exceptional wealth of plant and animal species.

vulnerable to the activities of humankind. People have intentionally or accidentally introduced to the island alien plant and animal species, known as **exotic species** and including rats, goats, sheep, pigs, cattle, and fast-growing plants, that often end up preying on or overtaking the endemic species. Habitat destruction and deliberate hunting also lead to extinction. For example, New Zealand's moas and the dodo birds of the Indian Ocean island of Mauritius are now, well, "dead as dodos." Indigenous inhabitants of New Zealand set fires to hunt the giant ostrich-like moa and had already killed off most of the birds by the time the Maori people colonized New Zealand around 1350 C.E. By 1800, the moa was extinct.

From one end of Oceania to the other, concern about such human-induced environmental changes has been mounting. The Hawaiian Islands have become known to ecologists as the **"extinction capital of the world"** because of the irreversible impacts that exotic species, population growth, and development have had on indigenous wildlife there. Environmentalists are also concerned about the impact of commercial logging on the island of New Guinea (•**Figure 8.11**). About 22,000 plant species grow there, fully 90 percent of them endemic. And ecologists consider nearby New Caledonia one of the world's biodiversity hot spots (not to be confused with the geologic hot spots like Hawaii) because of the ongoing human impact on its unique flora and fauna.

Human-induced extinctions are not the only threats to island species. Volcanic eruptions, typhoons (hurricanes), and rising sea levels affect island plant and animal populations, sometimes in surprising ways (see page 298). When Hurricane Iniki struck the Hawaiian Islands in 1992, for example, it destroyed numerous chicken coops on the island of Kauai. Today, as a result, feral chickens are thick on Kauai's grounds. Most animals find it difficult or impossible to evacuate in the event of a natural catastrophe, and an island's distance from other lands may hinder or prevent recolonization. So although island habitats have a higher

| INSIGHTS | Deforestation and the Decline of Easter Island: A Parable? |

"Easter Island is Earth writ small," wrote the naturalist and geographer Jared Diamond.[a] Recent archaeological and paleobotanical studies suggest that the civilization that built the island's famous monolithic stone statues destroyed itself through overpopulation and abuse of natural resources (•Figure 8.A). Deforestation has serious repercussions wherever it occurs, but perhaps nowhere are these impacts so apparent as on Easter Island, a remote outpost of Chile lying 2,200 miles off Chile's coast (see Figure 8.1).

When the first colonists from eastern Polynesia reached remote Easter Island (known as Rapa Nui to locals) in about 400 c.e., they found the island cloaked in subtropical forest. Plant foods, especially from the Easter Island palm, and animal foods, notably porpoises and seabirds, were abundant. The human population grew rapidly in this prolific habitat. The complex, stratified society that emerged on the island grew from an estimated 7,000 to 20,000 people between 1200 and 1500 c.e., when most of the famous statues were built. Apparently in association with their religious beliefs, Easter Islanders erected more than 200 statues, some weighing up to 82 tons and reaching 33 feet (10 m) in height, on gigantic stone platforms. At least 700 more statues were abandoned in their quarry sites and along roads leading to their would-be destinations, "as if the carvers and moving crews had thrown down their tools and walked off the job," Diamond observed.[b]

"Its wasted appearance could give no other impression than of a singular poverty and barrenness," the Dutch explorer Jacob Roggeveen wrote of the island on the day he discovered it—Easter Sunday 1722.[c] Not a single tree stood on the island. The depauperate landscape bears testimony to the fate of the energetic culture that built the great statues. By 800 c.e., people were already exerting considerable pressure on the island's forests for fuel, construction, and ceremonial needs. By 1400, people and the rats they introduced to the island caused the local extinction of the valuable Easter Island palm. Continued deforestation to make room for garden plots and to supply wood to build canoes and to transport and erect the giant statues probably eliminated all of the island's forests by the fifteenth century. By then, people had hunted to extinction many terrestrial animal species and could no longer hunt porpoises because they lacked the wood needed to build seagoing canoes. Crop yields declined because deforestation led to widespread soil erosion. There is evidence that in the ensuing shortages, people turned on each other as a source of food. By about 1700, the population began a precipitous decline to only 10 to 25 percent of the number who once lived on this isolated Eden. Today, about 4,000 people live there, welcoming more than 45,000 tourists every year.

THOMAS J. ABERCROMBIE/National Geographic Stock

• **Figure 8.A** The Polynesian people who erected at least 900 moai statues between 1200 and 1500 C.E. deforested most of Easter Island—in part to supply levers and rollers for the statues—and thus precipitated the demise of their civilization.

[a-c] All quotes are from Jared Diamond, "Easter's End." *Discover, 16*(8), 1995, p. 64.

percentage of endemic species than would be found on mainlands, environmental difficulties sometimes cause the number of different kinds of species (that is, species diversity) to be lower. Far more than natural hazards, however, it is the human agency that altered the Pacific island ecosystems, in some cases completely transforming the natural landscape (see Insights, this page).

8.3 Cultural and Historical Geographies

Their indigenous cultures having been greatly diminished in numbers and influence, Australia and New Zealand today are mainly European in culture and ethnicity. The region's other cultures, however, are overwhelmingly indigenous and quite diverse. Aside from Australia and New Zealand, with their majority European populations, and Fiji, New Caledonia, and Guam, each of which has populations about half indigenous and half foreign, approximately 80 percent of the Pacific's people are indigenous. The balance is 13 percent Asian and 7 percent European. Of the indigenous populations, Melanesians make up about 80 percent, with 14 percent Polynesian and 6 percent Micronesian.

The Indigenous Peoples of Oceania

Recent DNA analysis suggests that **Aborigines** arrived in Australia in a single major migration about 50,000 years ago (but archaeology may well push this date deeper into history). Some ethnographers refer to the Aborigines as a distinct Australoid race. However, they share racial characteristics with other indigenous groups in Asia, including the Mundas of central India, the Veddas of Sri Lanka, and even the Ainu of Japan. All of these groups probably descended from a common ancestral race in Asia, and the Aborigines developed their distinctive features over tens of thousands of years of living in Australia. The **Torres Strait Islanders**, a distinct indigenous group of Australians living mainly in Queensland, came to Australia from nearby Papua New Guinea.

During the human migration to Australia about 50,000 years ago, sea levels were much lower and straits were narrower and easier to navigate. The settlers probably came to Australia across the land bridge that linked New Guinea and Australia until about 10,000 B.C.E. (the historical continent made up of these two landmasses is known as Sahul). Once established in Australia, they developed a unique and enduring culture. Their **Aboriginal languages**, classified into two major groups—**Pama-Nyungan** in the southern 90 percent of Australia and **non-Pama-Nyungan** in the north—are not clearly related to any languages outside the continent.

New waves of migrants known as **Austronesians** began a long process of diffusion across the western Pacific and eastern Asia after 5000 B.C.E. Their ancestral stock probably originated in Taiwan and southern China. Their descendants migrated to the mainland and islands (initially the Philippines and Indonesia) of Southeast Asia. Their livelihoods were based on fishing and simple farming, especially of taro, yams, sugarcane, breadfruit, coconuts, and perhaps rice. Their domesticated animals included pigs and probably dogs and chickens, but they had no herd animals like cattle, sheep, or goats. From their Southeast Asian bases, over a period of several thousand years, the Austronesians embarked on voyages to Madagascar, New Zealand, Easter Island, Hawaii, and perhaps Chile (around 1400 C.E.), "island-hopping" all the way and accomplishing extraordinary feats of navigation in their outrigger canoes. Their settlement of Polynesia came rather late in these adventures, within the last 1,500 years. Today's Micronesians and Polynesians are mainly their descendants, but considerable mixing has occurred over the centuries.

These seafaring peoples spread their characteristic agriculture and a language called **Proto-Austronesian**—the ancestor of all modern Austronesian languages—across the Pacific. The Pacific region's linguistic diversity today is extraordinary (•Figure 8.12). Most of the indigenous peoples of the region speak languages in the **Austronesian language family**, which

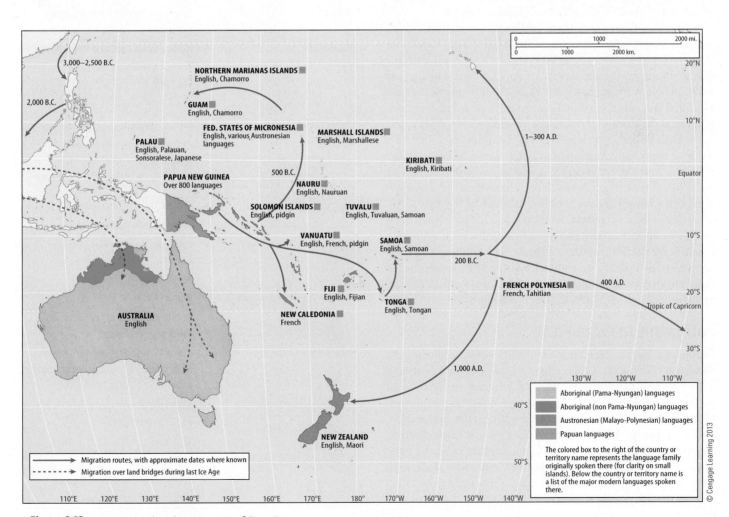

• **Figure 8.12** Languages and settlement routes of Oceania.

Joe Hobbs

• **Figure 8.13** Can you read the language of Vanuatu? This advertisement for a cell phone carrier is in Pidgin. Try the first two lines.[a]

is divided between the **Formosan** and the **Malayo-Polynesian language subfamilies.** The Melanesian peoples, however, speak Papuan languages, which are not a distinctive linguistic family but an amalgam of often unrelated and mutually unintelligible tongues. There are about 700 Papuan languages. Some Torres Strait Islanders speak a Papuan language, and others speak an Aboriginal Pama-Nyungan tongue.

Papua New Guinea is home to 860 languages (mainly Papuan but also Austronesian), or 20 percent of the world's total number of languages, making it by far the world's most linguistically diverse country, with an average of roughly one language per 5,000 people. Vanuatu has fewer than 200,000 people and 105 identified languages, earning that country's population the distinction as the world's most linguistically diverse on a per capita basis, with roughly one language per 2,000 people. There are fewer of the mainly Austronesian languages in Micronesia and Polynesia, and more of them are mutually intelligible. Reflecting the colonial past, English and French are official languages in some of the islands and are spoken widely across the region. Another lingua franca is **pidgin**, which consists of English and other foreign words mixed with indigenous vocabulary and grammar. Pidgin is the official language of Papua New Guinea, where it is known as Tok Pisin. Its kinship to English may be appreciated in these phrases: *kanu i kapsait* means "the canoe capsized," and *mi lukim dok* means "I saw the dog"[3] (•**Figure 8.13**).

Europeans in Oceania

Europeans began exploring the Pacific islands during the sixteenth century, and their impact was especially profound after 1800. Spanish and Portuguese voyagers were followed by Dutch, English, French, American, and German explorers. Many famous names are connected with Pacific exploration, including Ferdinand Magellan, Abel Tasman, Louis-Antoine de Bougainville, Jean-François de La Pérouse, and the most famous of all, British Captain James Cook, who undertook three great voyages in the 1760s and 1770s, only to be killed

[a]They read "Talk talk plenty. No more worry."

by Hawaiian Islanders in a cultural misunderstanding in 1779. Europeans were interested mainly in the high islands because of their relative resource wealth. They sought sandalwood, oyster shells, whales, and local people to indenture as whaling crews and as manual laborers. On island after island, European penetration decimated the islanders and disrupted their cultures. The intruders introduced sexually transmitted and other infectious diseases, alcohol, opium, forced labor, and firearms, which worsened the bloodshed in local wars.

The Europeans also came as Christian missionaries. Some ancient indigenous belief systems thrive in parts of the region, often nestled within Christianity and other recently founded introduced faiths (•**Table 8.3**; see also **Perspectives from the Field**, next page). 228, 322, 365

Overwhelmingly, the peoples of the Pacific adopted and continue to practice Christianity, with Methodists, Mormons,

Table 8.3	Religions of Oceania
Country or Territory	**Major Religions**
American Samoa (U.S.)	Christian Congregationalist, Roman Catholic, Protestant
Australia	Anglican, Roman Catholic, other Christian
Cook Is. (N.Z.)	Cook Islands Christian Church
Fiji	Hindu, Methodist, Roman Catholic, Muslim
French Polynesia (Fr.)	Protestant, Roman Catholic
Guam (U.S.)	Roman Catholic
Kiribati	Roman Catholic, Protestant
Marshall Islands	Protestant
Micronesia, Federated States of	Roman Catholic, Protestant
Nauru	Protestant, Roman Catholic
New Caledonia (Fr.)	Roman Catholic, Protestant
New Zealand	Anglican, Presbyterian, Roman Catholic, Methodist, other Christian
Northern Mariana Islands (U.S.)	Roman Catholic
Palau	Roman Catholic, other Christian, Modekngei religion
Papua New Guinea	Roman Catholic, Lutheran, Presbyterian, other Christian, indigenous beliefs
Solomon Islands	Anglican, Roman Catholic, United, Baptist, other Christian
Tonga	Christian
Tuvalu	Church of Tuvalu
Vanuatu	Presbyterian, Anglican, Roman Catholic, indigenous beliefs

Source: CIA, *World Factbook*, 2008.

In twentieth-century Melanesia, there was an extraordinary development in the history and geography of religion. A number of so-called **cargo cults** emerged in New Guinea and elsewhere, especially during World War II, as "Stone Age" people were exposed for the first time to chocolates, cigarettes, soft drinks, bottles, metals, manufactured cloth, and other material trappings brought by American soldiers and other Westerners. Many Melanesians attributed such "cargo" to spiritual forces. There were many variations among the cargo cults, but most were **millenarian movements**, maintaining that a supernatural event would trigger a new and more prosperous age for their followers. Intervention by white foreigners would end, and the locals would be mystically delivered the massive "cargo" of material possessions long denied them. Some of the believers built jetties, runways, and storehouses in anticipation of those events. Most withered with time, but one—the John Frum movement—endured with little change on Tanna, the southernmost island of Vanuatu (for location of Vanuatu, see Figure 8.1).[4]

As an undergraduate student of anthropology at the University of California–Santa Cruz, I was fortunate to be able to take a class on cargo cults taught by a visiting professor from New Zealand. I was captivated by his account of the John Frum "cult." It was not only a religion but a resistance movement against the colonial and missionary laws that would deprive locals of their culture (which they call *kastom*), including the vital ceremonial drinking of kava (•**Figure 8.B**). I wanted so much to meet these people who believed that one day a messianic

• **Figure 8.B** The ritual and recreational drinking of kava (*Piper methysticum*) are important features of most Melanesian and some Micronesian and Polynesian cultures. This is a kava market in Fiji. Visible are the roots, which contain the plant's active compounds; powder of the root, from which the drink is prepared; and the kava beverage, ready to consume.

• **Figure 8.C** Mt. Yasur is a perpetually active volcano at the southern end of the Vanuatu island chain. It is sacred to the followers of the John Frum faith. It is also a secular destination for "volcano tourism," a niche of ecotourism. The author and others stood on the rim peering down into the boiling caldera. Some unfortunate tourists take a step too far and fall to their deaths.

American G.I. named John Frum would return to their village bearing the coveted cargo.

My wish came true in 2009. My wife, Cindy, and teenage daughters, Katie and Lily, and I made the long pilgrimage from Missouri to Tanna. On the island, we hired a local guide, who took us first by four-wheel drive to the foot of Mount Yasur, the active volcano whose nighttime glow led Captain Cook to discover the Vanuatu islands (for some time known as the New Hebrides) in 1774 (•**Figure 8.C**). We hiked up to the cold, windswept rim of the volcano and peered nervously down into its fiery mists. Every few minutes, Yasur would deliver a "lava bomb," a fiery red-hot projectile that shot far overhead and fell back into the crater. As it blasted off, each lava bomb issued a loud thud and shook the earth underfoot.

From the foot of Mount Yasur, we drove across a moonscape of recently made black earth, pocked with gurgling hot springs and steam vents. Just inside the edge of the forest, and still in the shadow of Mount Yasur, we arrived at Lamakara, one of Tanna's two villages where people honor John Frum. As in so many other religions, there had been a schism that split the approximate 6,000 believers.

The headman of this village is Chief Isaak Wan Nikiau, and I was practically trembling with delight when he received us (•**Figure 8.D**). As we spoke of John Frum, I learned many new things about the faith, most importantly that John Frum actually lived *inside* fiery Mount Yasur,

• **Figure 8.D** Chief Isaak Wan Nikiau of Lamakara, Tanna Island.

from where he would emerge on that great day to meet his believers. Every February 15 since 1947, Chief Isaak said, all the villagers donned their homemade G.I. uniforms and with wooden models of rifles marched in formation to call upon John Frum to return. There were special martial ceremonies on each John Frum Day, but even on this ordinary day at dusk we saw men bring "the colors" down the flagpoles and carefully fold them (•**Figure 8.E**). These were flags of the United States, the U.S. Marines, and Vanuatu.

I was very surprised to learn that the soft-spoken, humble Chief Isaak had visited the United States in 1995. He had been invited to speak about John Frum at a symposium on cargo cults. He travelled to New York, Washington, and Los Angeles. I wanted to know what this "man from the Stone Age" thought about flying high over the Pacific and seeing firsthand the great cities of the United States. As night fell I asked more questions about him, his fellow villagers, and John Frum.

• **Figure 8.E** The ritual parade ground of Lamakara, with the flags representing the allegiances of John Frum's followers. The Stars and Stripes are at the far right.

Joe Hobbs

• **Figure 8.14** The Hindu Temple of Sri Siva Subramaniya in Nadi, Fiji. The towering structure at left is known as a *goparum*. These structures are unique to South India's form of Hinduism. Their presence in Fiji, so far from India, is an element on the cultural landscape that tells you something (in this case, British colonialism) brought South Indians to Fiji.

Joe Hobbs

• **Figure 8.15** A train carries freshly cut sugar cane from the fields to a sugar processing plant on the northern side of Fiji's island of Viti Levu.

and Catholics the largest denominations. Where Asians have settled, there are pockets of Hinduism, Buddhism, and Islam; Fiji, for example, is a distant outpost of Hindu India (•**Figure 8.14**).

Europeans created new settlement patterns, disrupted old political systems, and rearranged the demographic and natural landscapes. In Polynesia, for example, arriving Europeans typically sought shelter for their vessels on the lee side of an island. That safe place became the island's European port. According to Professor Knight, whatever chief happened to be in control of that spot typically, because of his association with the Europeans, became more powerful and wealthier than other island leaders. Eventually, he became high chief and often expanded his control to other islands; thus the first Polynesian kings emerged.

Europeans, North Americans, and East Asians changed the natural history of the islands by introducing exotic crops and animals, including arrowroot and cassava (manioc); bananas; tropical fruits such as mangoes, pineapples, papayas, and citrus; coffee and cacao; sugarcane; and cattle, goats, and poultry. They established extensive sugar plantations on some islands, notably in the Hawaiian Islands, Fiji Islands (where sugar is still the major export; see •**Figure 8.15**), and Saipan in the Marianas. The valuable Honduran mahogany trees planted by British colonists in Fiji early in the twentieth century are only now beginning to mature, and competition for export profits from this resource has become intense.

It is remarkable that as late as 1840, almost all of the Pacific islands had yet to be formally claimed by outside powers; only Spain had taken the Marianas. Then, rapidly, Britain, France, Germany, and the United States vied for power. By 1900, the entire Pacific Basin was in European and American hands. Germany held Western Samoa (today's independent Samoa), part of New Guinea, the Bismarck Archipelago, Nauru, and Micronesia's most important islands. Britain controlled part of New Guinea, the Solomon Islands, Fiji, and smaller island groups, and held the kingdom of Tonga as a protectorate. France

controlled New Caledonia, French Polynesia, and with Britain, the condominium (jointly ruled territory) of the New Hebrides. The United States held Hawaii, Guam, and eastern Samoa.

When Germany lost its Pacific territories at the end of World War I, Japan took control of German Micronesia, New Zealand took Western Samoa, and Britain and Australia took the German section of New Guinea and the Bismarcks. Nauru came to be administered jointly by Britain, New Zealand, and Australia.

With the outbreak of World War II, Japan quickly overran much of Melanesia. The ensuing ferocious battles between Japan and the Allies for control of New Guinea, the Solomon Islands, and much of Micronesia brought enormous changes to the region. Many of the cultures were in the Stone Age as the war began and by its end had seen the Atomic Age; Micronesia had become a nuclear proving ground. Every-298 where, traditional peoples were drawn into cash economies and affected by other global forces originating far beyond their watery horizons.

Most of the colonists are gone, but some of the seeds of strife they planted are still germinating. Many conflicts in the region today reflect the demographic changes Europeans wrought to meet their economic interests. In many cases, the Europeans introduced laborers from outside the region because local peoples were too few in number or refused to work in the cane fields. Successive infusions of Chinese, Portuguese, Japanese, and Filipinos brought a polyglot character to Hawaii, with little apparent ethnic discord. But ethnic strife is ongoing in Fiji and the Solomon Islands, which became independent from Britain in 1970 and 1978, respectively.

The Solomon Islands' troubles date to 1942, when U.S. Marines fought to push Japanese troops off the island of Guadalcanal. The American forces enlisted the support of thousands of people from the neighboring island of Malaita. After the successful offensive, many of these Malaitans elected to stay on Guadalcanal, where under British colonial rule they came to dominate economic and political life. The indigenous people of

Guadalcanal resented their dominance and in particular what they saw as the loss of their lands to the Malaitans (who purchased the lands). Conflict between the groups finally erupted in 1999, with the majority Guadalcanalans murdering scores of Malaitans and driving tens of thousands off the island; most returned to Malaita. An Australian-led peacekeeping mission arrived in 2003 to quell the unrest and put civil servants to work rebuilding the country's shattered economy.

The violence in the Solomon Islands was inspired by ethnic unrest in Fiji. In the early 1900s, colonial British plantation owners imported mainly Hindu Indians as indentured laborers to work Fiji's sugarcane plantations. Today, indigenous Fijians now only slightly outnumber ethnic Indians (**Indo-Fijians**). Here too, the indigenous inhabitants resent the dominance of the foreign ethnic group; ethnic Indians control most of Fiji's economic life. A coup led by ethnic Fijians in 2000 succeeded in ousting Fiji's ethnic Indian prime minister and throwing out a 1987 constitution that granted wide-ranging rights to ethnic Indians. Accused of lawlessness, the coup leaders were threatened with economic sanctions by the European Union (which pays higher than market prices for most of Fiji's sugar crop) and by the United States, Australia, and New Zealand.

The prospect of economic crisis led the army to take control of Fiji's affairs, imprison the coup leader, and promise a general election and return to civilian government in 2002. The unrest caused a crash in tourism, Fiji's most valuable industry. Tourist numbers rebounded after a peaceful election that brought the indigenous Fijian political party back to power, but a military-led coup in 2006 temporarily diminished the vital industry again. Ever resilient, tourism bounded back yet again, with more than half a million visitors arriving yearly. Lack of progress toward democracy led to Fiji's ejection from the British Commonwealth of Nations—a rare action that Fijians feel quite ashamed of.

8.4 Economic Geography

A lack of industrial development characterizes the Pacific island economies (other than those of Australia and New Zealand), and the poverty typical of LDCs prevails in the region. Contributing to this underdevelopment are what could be described as tyrannies of size and distance. Many of the countries consist of little more than small islands of limestone sprinkled with palm trees and situated hundreds and often thousands of miles from potential markets and trading partners. Most of the countries suffer large trade deficits and must import far more than they can export.

Making a Living in Oceania

There are seven major economic enterprises in the Pacific region, only one of them (textiles) involving value-added processing of raw materials: exports of plantation crops (especially palm products), fish, and minerals; services for Western military interests; information technology; textile production; and tourism. Significant manufacturing outside Australia and New Zealand is limited mainly to the Cook Islands and Fiji. Until worldwide quotas on textile trading were lifted in 2005, giving an immediate advantage to China, Saipan in the Northern Marianas had a flourishing garment industry owned by South Korean companies and operated by Chinese laborers. Polynesia's Cook Islands, a self-governing territory in free association with New Zealand, has an industrial economy based on exports of labor-intensive clothing manufacture. Fiji also began exporting manufactured clothes after the mid-1990s, when Australia granted preferential tariff treatment to Fijian-made garments.

A few Pacific countries are beginning to participate, in rather unlikely and sometimes unscrupulous ways, in the information technology revolution that has generated so much global wealth since about 1990. One example is tiny Tuvalu in Polynesia. In 2000, an American entrepreneur agreed to pay the Tuvalu government $50 million to acquire Tuvalu's two-letter Internet suffix, .tv. Although the new owner of the .tv domain has been busily profiting by selling Internet addresses, Tuvalu is delighted with its windfall. It used part of the money to finance the process of joining the United Nations, build new schools, add electricity and other infrastructure to outlying islands, and expand its airport runway. Vanuatu enjoys significant hard-currency revenue by permitting its country's phone code to be used for 1-900 sex calls. And Nauru and other island nations use telephone and digital technology to practice money laundering (more on this shortly).

Mining takes place on only a few islands. The major reserves are on the continental islands, notably petroleum, gold, and copper on the main island of Papua New Guinea; copper on Bougainville (an island belonging to Papua New Guinea); and about one-fourth of the world's nickel on the French-held island of New Caledonia. In recent decades, issues of control over mineral wealth have brought unrest to New Caledonia and Bougainville. French Polynesia exports some of the world's finest and largest sea pearls, which are classified as gems.

The 9,300 people of the tiny oceanic island of Nauru, which gained independence from Australia in 1968, once enjoyed a high average income from the mining of phosphate, the product of thousands of years of accumulation of seabird excrement, or **guano**, which is used as fertilizer. Until recently, Nauru's economy depended totally on annual exports of 2 million tons of phosphate and on revenues earned from overseas investments of profits from phosphate sales. This resource is approaching total depletion, however, and because the mining of phosphate has stripped the island of 80 percent of its soil and vegetation, turning it into a virtual lunar landscape, many people in Nauru are now looking for a new island home (•**Figure 8.16**).

Many Pacific island countries, formerly including Nauru, have earned revenue through **money laundering**. These island entrepreneurs have undertaken offshore banking, in which they can operate loosely controlled banks anywhere but in their home islands. Many of these are **shell banks**, existing only on paper and specializing in turning money earned in illegal ways into money that looks legal. Although Nauru no

TORSTEN BLACKWOOD/AFP/Getty Images

• **Figure 8.16** The 10 square miles (26 sq km) of land that is Nauru have been devastated by the phosphate mining that once made Nauruans among the wealthiest people per capita on earth.

longer launders money, like many Pacific islands it has been sought out by businesses and individuals as a tax haven (levying little or no taxes on enterprises).

After 2001, hard-up Nauru began earning revenue by serving yet another unlikely function: detention center for refugees from the war in Afghanistan. By 2004, more than 1,000 Afghan refugees who were apprehended at sea in transit to Australia, where they hoped to gain asylum, had been diverted to Nauru. Australia paid the government of Nauru about $15 million yearly to hold the detainees while their cases were reviewed in the Australian courts. Nauruans were dismayed when the newly elected Australian prime minister brought an end to this practice in 2007, diverting the boat people to Australia's own Christmas Island. The next Australian prime minister eyed Timor-Leste as the next best place to locate Australia's "regional processing center."

The most dynamic element of the Pacific economies is the rapid growth of tourism, with most visitors coming from the United States, Australia, and Japan and a rapidly growing number from China. Tourism is the largest industry in the Pacific region, providing substantial income and employment and stimulating overall economic growth by encouraging foreign investment and facilitating trade.

The Pacific islands continue to exert an irresistible appeal to people awash in the conveniences and congestion of the industrialized world. One online entrepreneur, hoping to capitalize on disenchantment with the modern world, is building an e-community of people who will contribute funds to, and sometimes visit, his idyllic Fijian island (see http://tribewanted.com). The most popular tourist destinations are Hawaii, Australia, New Zealand, and French Polynesia, but there has been

strong growth in ecotourism, archaeological tourism, and other tourism niches in destinations including Palau, Vanuatu, and Easter Island.

Some of the countries have developed unique appeals for international tourists. In a bid to attract visitors by being the first country in the world to greet the new millennium, Kiribati decided unilaterally in 1997 to shift the International Date Line eastward by more than 2,000 miles (3,200 km). The impact of this decision stands out on any map depicting the date line (see Figure 8.1). Samoa in 2011 also moved the line eastward, in this case for the perceived economic advantages of being closer to the workdays of Australia and East Asia than to the United States. Palau advertises itself as a paradise for scuba divers, including those wanting close encounters with sharks. Palau has used its shark tourism to make an interesting pitch about the tradeoffs between conservation and exploitation. Diver tourism contributes about 39 percent of Palau's $218 million annual gross domestic product. About 21 percent of the divers come specifically to see the sharks; therefore sharks contribute about 8 percent of GDP. There are roughly 100 sharks, so each is worth $179,000 annually to the country's tourism industry. Each shark has a lifetime value of $1.9 million. Killed and sold only for the fins and meat, those sharks would be worth only $10,800.[5] A similar study has been done of the Maldives' sharks; each is worth $33,500 annually. These studies argue strongly for the principle of sustainable development that conservation and development *are* compatible (see page 61).

Finally, the military interests of outside powers, particularly the United States and France, generate much revenue for some of the Pacific economies, notably Guam, American Samoa, and French Polynesia. This military presence is a mixed blessing, however, because of its environmental impacts and because it perpetuates dependence on foreign powers. These are among the geopolitical concerns discussed next.

8.5 Geopolitical Issues

Once entirely colonial, Oceania today is a mix of units still affiliated politically with outside countries as well as others that have become independent. Since the end of World War II, the United States, Britain, Australia, and New Zealand have abandoned most of their colonies in the region. Only France has insisted on holding on to all of its colonies.

Why Are Foreign Powers Interested in the Pacific?

In some cases, colonial powers have delayed independence to would-be island nations. They explain that such territories are too small and isolated or are not economically viable enough for independence. Some islands remain dependent because they confer unique military or economic advantages on the governing power. For example, French Polynesia was until recently the venue for French atomic testing (see Regional Perspective, next page), and Guam and American Samoa are still useful to the United States for military purposes. Guam, which has been a U.S. possession since 1898 and a U.S. territory since 1950 (its residents are American citizens), is especially vital in U.S. strategic thinking. Pentagon planners classify it as a **power projection hub** that can be used as a forward base that is five full days' sailing time closer to Asia than Hawaii is. It has been used as a jumping-off point for American B-52 jets on bombing missions to Iraq and would be used in any U.S. military action in Korea or elsewhere in East Asia. In the event of such hostilities, the United States would need Japan's permission to operate from bases in Japan, but no such approval would be necessary to act from Guam.

A generally unspoken but widely acknowledged U.S. military objective in building up readiness in Guam is that the island would be critical to any confrontation with China, should that giant country ever become a belligerent power. China is well established in its "string of pearls" footholds across the Pacific and Indian Oceans (see page 248), and in a contest reminiscent of the Cold War, the U.S. is positioning itself geo-strategically to match China. Chinese officials were upset when, while visiting Australia in 2011, President Barack Obama resolved to station American troops and ships in Australia. "The U.S. is a Pacific power and we're here to stay," the President said.[6] One of the goals of the President's trip was to assert U.S. access to the South China Sea. Geopolitically, both Washington and Beijing perceive the Pacific region as the *Asia-Pacific* region.

Australia and New Zealand have Pacific-oriented defense and security agreements, somewhat precariously balanced with U.S. interests in the region. In 1951, they joined the United States to form the **Australia New Zealand United States (ANZUS) security alliance**. Australian troops supported American forces in the Vietnam War. Since 1987, however,

when its liberal government declared that New Zealand would henceforth be a nuclear-free country, New Zealand's relations with the United States have been strained. Legislation banned ships carrying nuclear weapons or powered by nuclear energy from New Zealand's ports. The United States refused to confirm or deny whether its ships violated either restriction, so New Zealand denied port access to them. This action removed New Zealand from ANZUS. New Zealand paid some heavy political and economic costs for its nonnuclear position; for example, it was pointedly left out of negotiations that led to the Australia–United States Free Trade Agreement (AUSFTA) that came into force in 2005.

Despite its small army (only 50,000 strong), Australia has begun to provide a selective security blanket over its interests in the region and elsewhere. In 1999, Australian troops were deployed to head up the United Nations peacekeeping force to oversee the referendum that gave independence to East Timor and its subsequent transition to full independence as Timor-Leste. Australia (and New Zealand) also lent troops and other support in the U.S. buildup to and execution of the war against Iraq in 2003. Al-Qa'ida vowed revenge and apparently delivered it with a bombing in Bali, Indonesia, that killed many young Australian tourists in 2002. Australia's apparent concern about a long-term threat from North Korea also prompted it to join the United States in the development of the Strategic Defense Initiative missile shield program. Australia is sometimes snubbed by its Southeast Asian neighbors in political and economic affairs because it is seen as something of a regional "police officer" or as a United States "deputy."

Oceania's Environmental Future

Another set of geopolitical concerns in the Pacific region is environmental. The remote islands of Oceania, so far from the industrial cores of North America, Europe, China, and Japan, are actually the places most likely to suffer first and most from the most feared impacts of industrial carbon dioxide emissions. Along with the Maldives described earlier (page 250), the islands would be among the earliest and hardest hit victims of any rise in sea level due to climate change. There have already been reports of unprecedented tidal surges on Pacific island shores. If the trend continues, the first Pacific islands to be totally submerged would be Kiribati, the Marshall Islands, and Tuvalu (where the highest point is 15 feet above sea level), while Tonga, Palau, Nauru, Niue, and the Federated States of Micronesia would lose much of their territories to the sea. There are already discussions about how "**deterritorialized states**"—those that have been drowned by the sea—might continue to have sovereignty and rights to underwater and other resources.

These island nations are among the 39 countries comprising the **Alliance of Small Island States**, which argued forcefully but unsuccessfully at the 1997 Kyoto Conference on Climate Change that by 2005 global greenhouse gas emissions should be reduced to 20 percent below their 1990 levels. What they did get was a pledge by most industrialized nations (not including the United States), in the form of the Kyoto Protocol, to cut

The remote locales and nonexistent or small human populations of some Pacific islands have made them irresistible sites for weapons tests by the great powers controlling them (•Figure 8.F). In the 1940s and 1950s, U.S. authorities displaced hundreds of islanders from Kwajalein atoll in the Marshall Islands to make way for intercontinental ballistic missile (ICBM) tests. During those years, the United States also used nearby Bikini atoll as one of its chief nuclear proving grounds (the island's fame at the time led French designers to name a new two-piece bathing suit after it). In 1946, U.S. government authorities relocated the indigenous inhabitants of Bikini to another island, promising they could return when the tests were complete.

Over the next 12 years, there were 23 nuclear blasts on Bikini, out of a total of 67 conducted as **Operation Crossroads** in the Marshall Islands. A 15-megaton hydrogen bomb detonation on Bikini in 1954 had the power of 1,000 Hiroshima bombs, blowing a mile-wide crater in the island's reef and producing radioactive dust that fell downwind on the Rongelap atoll, where children played in the dust "as though it were snow," one observer wrote.[a] For decades, the U.S. Atomic Energy Commission claimed that a sudden shift in winds took the fallout to inhabited Rongelap, but recently revealed records proved that authorities knew of

• **Figure 8.F** The nuclear explosion, codenamed *Seminole* in *Operation Redwing*, at Eniwetok atoll in the Pacific Ocean on June 6, 1956. This atomic bomb was detonated at ground level and had the same explosive force as 13,700 tons (13.7 kilotons) of TNT. *(U.S. Navy/SPL/Photo Researchers, Inc.)*

Photo courtesy of National Nuclear Security Administration / Nevada Site Office

the wind shift three days prior to the test. The Marshallese today suffer a legacy of cancers and deformities linked to the weapons tests on Bikini and Eniwetok atolls. So far, the United States has provided more than $60 million to compensate Marshall Islanders made ill by the blasts.

The Bikinians have never accepted their exile. But they have been concerned about returning to their radioactive native soils. A repatriation of the Bikinians in the 1960s was found to be premature, as radiation persisted, and the nuclear nomads moved again. More recent tests have determined that Bikini has little ambient radioactivity and is essentially habitable, except for the cesium-137 that permeates the soil and becomes concentrated in fruits and coconuts. Bikinians do not want to go home to stay until they have assurances from the United States that the problem of poisoned produce can be resolved and that the island is completely safe. Along with the people of Eniwetok, they are seeking massive funding from the United States for complete environmental decontamination of their homeland and for further compensation for their travails.

Now that Bikini is at least safe to visit, some Bikinians are capitalizing on its formerly off-limits status. The island's lack of human inhabitants for more than 50 years has had the remarkable effect of rendering Bikini a true marine wilderness. In the late 1990s, it opened as a tourist destination, with naval ships sunk by the atomic blasts a highlight for many divers. The dive masters and tour guides are Bikinians.

Like the United States, France has had a nuclear stake in the Pacific. In 1995, after a three-year hiatus, France resumed underground testing of nuclear weapons on the Mururoa ("Place of Deep Secrets") atoll in French Polynesia. There was an immediate backlash from some national governments and from environmental and other organizations. France soon bowed to international pressure, ceasing its nuclear tests in the Pacific and signing the Comprehensive Test Ban Treaty in 1996.

Ironically perhaps, many islanders fear the retreat of French military interests in the region. Despite the nuclear tests, most inhabitants of French Polynesia have long favored continued French rule. Tourism and other local businesses contribute only about 25 percent of the territory's revenue; the remainder comes from French economic assistance, largely in the

military sector. Many locals, whose parents and grandparents gave up subsistence fishing and farming for jobs in the military and its service establishments, are worried about their future.

Similar fears stalk Palau, the Marshall Islands, and the Federated States of Micronesia (FSM), three independent nations carved from the former U.S. Trust Territory of the Pacific. The United Nations awarded that territory, which includes the Bikini and Eniwetok atolls, to the United States in 1946 as the world's first and only political-military zone. In 1986, the United States granted independence to the Marshall Islands and the FSM in an agreement known as the **Compact of Free Association**. In this agreement, the countries agreed to rent their military sovereignty to the United States in exchange for tens of millions of dollars in annual payments and access to many federal programs. The United States uses its leases to conduct military tests—for example, on Strategic Defense Initiative (SDI, also known as "Star Wars") technology. The United States also maintains a military presence in Palau and provides large cash grants and development aid to the nation, which became independent from the United States in 1994 but remains in free association with it.

In these cases, the American government handed over the monies with no strings attached. The funds have been embezzled and misspent, and there is little development to show for them. The United States has begun to reassess the payments, with an eye to their eventual curtailment. Many islanders fear that their economies will subsequently plummet. The people of the Marshall Islands and the Federated States of Micronesia are especially worried because more than half of the gross domestic product of each country is comprised of economic aid from the United States. The Marshall Islands' leaders, however, are taking a more defiant stand, saying that the United States has no right to question what becomes of the aid money and pointing out that the U.S. gets much in return in the form of extensive military rights. The United States is hard-pressed here because the Kwajalein Missile Range in the Marshalls is the only reasonable place in the world where, Pentagon officials say, they can test many of the SDI program components.

[a]Nicholas D. Kristof, "An Atomic Age Eden (but Don't Eat the Coconuts)." *New York Times,* March 5, 1997, p. A4.

these emissions to at least 5 percent below their 1990 levels by 2012. Already claiming to be a victim of lost coastline, higher storm surges, and more storms as a result of global warming, Tuvalu in the early 2000s lobbied other countries to join it in a lawsuit against the United States and Australia. Tuvalu wanted to make the case that these large countries' failure to ratify the Kyoto Protocol is a principal cause of global warming. As so often in the past, the peoples of the Pacific today feel vulnerable to the actions and policies of more powerful nations far from their tranquil shores.

8.6 Regional Issues and Landscapes

Australia and New Zealand

Becoming Less British, More Asian-Pacific

These two unusual countries (**see •Figures 8.17 and 8.18**) are akin in population, cultural heritage, political problems and orientation, type of economy, and location. Australia and New Zealand are among the world's minority of prosperous countries. Australia's per capita GNI PPP of $38,510 is higher than that of France and Germany, although below that of the United States and the most prosperous European countries. Australia has been dubbed the "Lucky Country," and a recent World

Bank study ranked Australia as the world's wealthiest country based on its natural resource assets divided by the total population. With China's seemingly insatiable appetite for minerals and other raw materials, resource-rich Australia has been particularly lucky lately, with a gold rush and a coal boom to name just two developments. New Zealand is less affluent but is still prosperous compared to most nations of the world. In both countries, there are relatively few people among whom to spread the wealth; Australia's 23 million people and New Zealand's 4 million together amount to only about 73 percent of the people living in California.

Both countries owe their prosperity to the wholesale transplantation of business culture and technology from the industrializing United Kingdom to the remote Pacific beginning in the late eighteenth century. Australia (established originally as a penal colony for British convicts) and New Zealand are products of British colonization and strongly reflect the British heritage in the ethnic composition and culture of their majority populations. They speak English, live under British-style parliamentary forms of government, acknowledge the British sovereign as their own, and attend schools patterned after those of Britain. Despite their independence (for Australia in 1901 and New Zealand in 1907), both countries maintain loyalty to Britain. They still belong to the British Commonwealth of

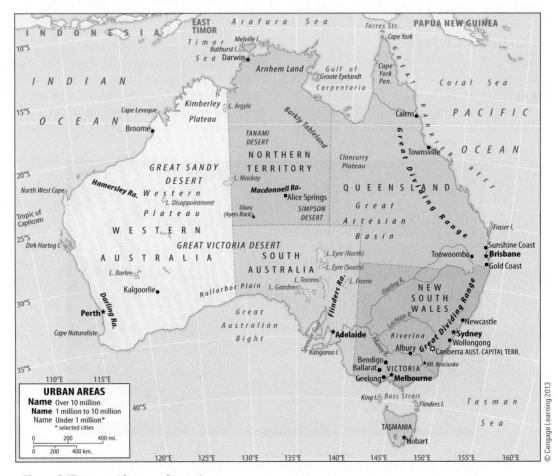

• **Figure 8.17** Principal features of Australia.

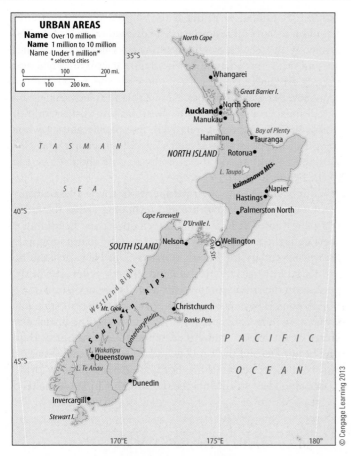

URBAN AREAS
Name Over 10 million
Name 1 million to 10 million
Name Under 1 million*
* selected cities

© Cengage Learning 2013

• **Figure 8.18** Principal features of New Zealand.

Nations. In both world wars, the two countries immediately came to the support of Britain and lost large numbers of troops on battlefields far from home (•Figure 8.19). In World War I, Australia lost 60,000 of the 330,000 soldiers it sent to fight in Europe, by far the highest proportion of dead among any of the Allied countries. In World War II, U.S. forces helped prevent a threatened Japanese assault on Australia. Since that time, Australia and New Zealand have sought closer relations with the United States, and British influence has waned.

Australia and New Zealand are seeking stronger roles in the economic growth projected for the Pacific Basin. Britain is now a rather insignificant trading partner of both Pacific nations. Seven of Australia's 10 largest markets are in the Asia-Pacific region, with China, Japan, and the United States its leading trading partners (in that order). Australia, China, the United States, and Japan have become New Zealand's leading trade partners.

Perhaps the best symbol of Australia's new Pacific perspective is the debate over whether Australians should convert the country into a republic, ending more than 200 years of formal ties with Britain. Even now, according to Australia's constitution (which was written in 1901 by the British), Australia's head of state cannot be an Australian; it has to be England's king or queen (this continues to be the case in 16 of the 55 countries that make up the Commonwealth). Republic status would change that and allow Australians to have their own elected or appointed president. The monarch's portrait would be removed from Australian currency. The British Union Jack would be removed from its position in the upper left corner of the Australian flag.

In 1998, an Australian constitutional convention voted to adopt this republican model, and Australia's avowedly monarchist prime minister, John Howard, reluctantly agreed to put the issue to a referendum in 1999. In essence, the referendum asked Australians where they preferred their economic and political future to rest—either with the other nations of Asia and the Pacific or with Britain and the Commonwealth. A narrow majority (55 percent) voted to keep Australia's lot with Britain and the Commonwealth. The debate continues, with

Joe Hobbs

• **Figure 8.19** The ANZAC War Memorial in Sydney, Australia, commemorates the sacrifice of thousands of young Australian and New Zealander lives in combat in World War I, including in the terrible Battle of Gallipoli at the Turkish Straits chokepoint.

the most recent polls showing growing support for Australia's traditional model.

The monarchy-versus-republic debate has led to something of an identity crisis for Australia and has focused more scrutiny on the sensitive issue of nonwhite immigration. Between 1945 and 1972, some two million people, mainly British and continental Europeans, emigrated to Australia. They were allowed in by a **"white Australia" policy** that excluded Asian and black immigrants because of fears of invasion from Asia and a desire to increase the country's population with white, skilled, English-speaking immigrants. That policy was eventually dropped, and doors opened considerably to skilled nonwhites. About half of the annual quota of 90,000 legal immigrants are Asian, who now make up about 7 percent of Australia's population (about 92 percent of Australians are Caucasians). But Asians will form up to one-fourth of Australia's population by 2025, according to some projections. Australia has recently passed legislation requiring new immigrants to settle in cities other than Sydney (•**Figure 8.20**), where services threaten to be overwhelmed by growth.

Recent years have seen a marked surge in racism, targeted mainly at Asians but also at Middle Easterners; a wave of white violence targeted at ethnic Lebanese swept over the country in 2005, for example. There has been a growing tide of illegal immigration from new sources, mainly Afghanistan and Iraq. These "boat people" are generally trafficked by Indonesian racketeers, who charge a refugee as much as $7,000 for the service. Prime Minister John Howard's vow to tighten restrictions against refugees and asylum seekers helped win him a rare third term in office in 2001. Howard won a fourth term in 2004 but went down to defeat in 2007.

Ethnic tensions aside, Australia and New Zealand continue to forge important ties with their neighbors around the vast Pacific and with each other. Both countries adopted a free-trade pact in 1990 called the **Closer Economic Relations Agreement**, which eliminated almost all barriers between them to trade in farm and industrial goods and services. Both countries belong to the 21-member **Asia-Pacific Economic Cooperation (APEC)** group, which has agreed to establish "free and open" trade and investment between member states by the year 2020. Australia

and the United States are the most aggressive member states in arguing for the abolition of trade quotas and the reduction of tariffs imposed on imported goods. Along with New Zealand, both are members in the growing **Trans-Pacific Partnership**, a free-trade initiative that as of 2012 had seven more members from South America and Asia. China is conspicuous by its absence. To join, China would have to reduce state ownership of economic enterprises and encourage more private enterprises, as well as work harder to protect intellectual property rights. 234

Australians refer to their Asian neighbors not as the Far East but as the **Near North**. Australia's Northern Territory city of Darwin is well positioned geographically to take advantage of new trading relations with the Near North. Darwin is closer to Jakarta, Indonesia's capital, than it is to Sydney, and Darwin is promoting itself as Australia's "gateway to Asia." The city has established a trade development zone to provide manufacturing facilities and incentives for overseas companies and for Australian companies wanting to do business overseas. A deep-water port was built in the 1990s, and a recently completed transcontinental Darwin-Adelaide railway, called the Ghan is expected to help fuel trade between Australia and its neighbors to the northeast. Australia's Asia-oriented perspective has already changed the nature of one of the country's export staples—beef—as ranchers have been shifting to breeds of cattle better suited for live shipment by sea to Indonesia, the Philippines, and Thailand.

Australia's Original Inhabitants Reclaim Rights to the Land

Both Australia and New Zealand have minorities of indigenous inhabitants. The native Australian Aborigines and Torres Strait Islanders have an extraordinarily detailed knowledge of the Australian landscape and its natural history, and a complex belief system about how the world came about. According to the Aborigines, the earth and all things on it were created by the "dreams" of humanity's ancestors during the period of the **Dreamtime**. These ancestral beings sang out the names of things, literally "singing the world into existence." The Aborigines call the paths that the beings followed the **Footprints of the Ancestors** (or the **Way of the Law**; whites know them as **Songlines**). In a ritual journey called **Walkabout**, the Aboriginal boy on the verge of adulthood follows the pathway of his creator-ancestors. "The Aborigine clings to his native soil with every fiber of his being," wrote the ethnographer Carl Strehlow. "Mountains and creeks and springs and water holes are to the Aborigine not merely interesting or beautiful scenic features. They are the handiwork of ancestors from whom he himself has descended. The whole country is his living, age-old family tree."[7]

An estimated 300,000 to 1 million Aborigines inhabited Australia when the first Europeans arrived in the seventeenth century. Colonizing whites slaughtered many of the natives and drove the majority into marginal areas of the continent. The Aborigines today number about 510,000, living mainly in the tropical north of the country. Some carry on a traditional way of life in the wilderness, but most live in generally squalid conditions on the fringes of majority-white settlements.

• **Figure 8.20** Sunrise over Australia's beautiful Sydney. Note the iconic Opera House and Harbour Bridge just right of center.

Joe Hobbs

As in the United States, newcomers to Australia forged a nation by dispossessing the ancient inhabitants of the land. And as in the United States, Australia has a long history of white racism, discrimination, and abuses against people of color, who in Australia's case were the original inhabitants. Aborigines were not mentioned in the 1901 constitution, not allowed to vote until 1962, and not counted in the national census until 1967. The most disadvantaged group in Australian society, Aborigines suffer from high infant mortality (four times that of the country's whites), low life expectancy (15 to 20 years less than that of white Australians), and high unemployment (officially 20 percent, although Aboriginal sources claim 50 percent, versus 4 percent for whites). Like Native Americans, a disproportionately large share of Aborigines fall prey to the economic and social costs of alcoholism. Statistics indicate they are 13 times more likely than non-Aborigines to end up in jail. Once in jail, they are more likely than non-Aborigines to commit suicide or be killed by guards. The Australian government now has a program to improve the justice system for native Australians.

One of the largest issues of contention between Australia's indigenous people and the white majority (known to the Aborigines as "whitefellas") is land rights. As in the United States, European newcomers pushed the native people off potentially productive ranching, farming, and mining lands into special reservations on inferior lands. The Europeans used a legal doctrine called *terra nullius*, meaning "no one's land," to lay claim to the continent.

Under the 1976 Land Rights Law, Aborigines were permitted to seek title to vacant state land ("Crown land") to which they could prove a historical relationship, but few succeeded. Following new legislation in 1992, the government began to return titles on a parcel-by-parcel basis—generally, in very marginal lands—to Aborigines who had argued their rights successfully. However, most Aborigines were unhappy with the terms of the returned titles, which allowed the native title to coexist with, but not supersede, the established Crown title. The government retained mineral rights to the land and could therefore lease native land to mining companies.

In 1993, after the longest senate debate in the country's history, Australia's foreign minister pushed the **Native Title Bill** through Australia's parliament. The new legislation addressed the Aborigines' major objections, giving them the right to claim land leased to mining concerns once leases expire, and allowing them to buy land to which they have proved native title. The Native Title Bill is of great consequence in the states of South Australia and Western Australia, where there is much vacant Crown land, and many Aborigines are able to file claims on it. Those states are also heavily dependent on mining. Members of Australia's Mining Industry Council are unhappy with this legislation and worry that their mine leases will run out before the minerals do, which will require them to negotiate with the Aboriginal titleholders.

Potential Aboriginal claims to land expanded vastly in 1996 with another Australian High Court ruling on what is known as the **Wik Case** (named after the indigenous people of northern Queensland who initiated it). The court concluded that Aborigines could claim title to public lands held by farmers and ranchers under long-term "pastoral leases" granted by state governments. These lands comprise about 42 percent of Australia's territory.

What Aboriginal claims to these lands would mean is bitterly debated. White farmers and ranchers protest that they would be ruined economically if they had to share profits with Aborigines claiming land rights. Generally, however, Aborigines have insisted that their title rights would not mean running whites out of business but would simply confer nominal recognition of their title and access to the land, especially for visitation to sacred sites.

Within the past two decades, about 100 Aboriginal communities have won official land rights, typically in areas of little economic value. However, there are important exceptions. On the northwest coast near Broome, the best site to build a liquefied natural gas plant is on James Price Point, on traditional Aboriginal land. The opportunity to build the plant has split Aboriginal opinions between those anxious to earn lucrative royalties from the development ($1.5 billion over 30 years) and those who wish to preserve the sacred landscape. An Aboriginal supporter of the development said, "These resources booms come along maybe every 50 to 100 years, big ones like this. So if we don't get positioned during this next stage, the chances are we're going to be locked out of the opportunities that are going to help build the economic basis for our families in the future. We can't sit back."[8] An Aboriginal opponent of the project, noting that James Price Point is the nexus of a sacred songline, said he was defending his "sense of country." Aboriginal supporters of the project, he said, were just "out for the money. They don't know how to connect to country anymore. They walk around with a dead feeling, these people, inside them."[9]

In the Northern Territory, which has its own land rights legislation, vast tracts of land have already been given back to the Aborigines. Aborigines there make up one-quarter of the population and now control more than a third of the land. Aborigines also hold title to Australia's two greatest national parks, Uluru (Ayers Rock) and Kakadu, both in the Northern Territory. In a **comanagement** arrangement that has become a model worldwide for management of national parks where indigenous people reside, the Aboriginal owners of the reserves have leased them back to the Australian government park system, which manages them jointly with the Aborigines.

Tourism programs in Uluru and Kakadu now highlight the cultural resources of the parks in addition to their natural wonders. At both parks, visitors can enjoy interpretive natural history walks that also emphasize Aboriginal culture and that are conducted by the land's oldest and most knowledgeable inhabitants (•Figure 8.21). At Uluru, the Anangu Aborigines who own the rock hope that cultural awareness will cut down the number of tourists who climb the rock—estimated at 70 percent of the 400,000 annual total—because it is sacred to the Anangu.

The Aborigines' struggle for legal recognition of traditional land tenure is representative of a growing trend among indigenous peoples in many parts of the world. These cultures are increasingly enlisting the aid of geographers, anthropologists, and other social scientists to document, measure, and analyze traditional land claims. The objective is to produce harder evidence of traditional land ownership that can be successfully presented in national and international court cases to win land rights. The methods often include **counter-mapping**, in which the accepted

Joe Hobbs

• **Figure 8.21** A woman of the Anangu Aboriginal clan takes visitors on walking tour around the base of Uluru (Ayers Rock), a site sacred to these people. In this national park program, visitors have the rare chance to learn about the meanings and uses of Aboriginal land. She is carrying a tool (not visible in this photo) that she uses to dig for grubs and other "bush tucker" (wild foods) as well as to plant crops.

"truths" of colonial and other cartographies are challenged by mapping indigenous and local perceptions of place (•**Figure 8.22**). Issues such as these are discussed at international meetings like the International Forum on Indigenous Mapping. Among the contributors at these meetings are geographers fresh from the field with global positioning system (GPS) data and maps testifying to traditional land claims. Many belong to the active Indigenous Peoples Specialty Group of the Association of American Geographers.

Exotic Species on the Island Continent

Exotic species are nonnative plant and animal species introduced through natural or human-induced means into an ecosystem. Their impact tends to be pronounced and often catastrophic to native species, particularly on island ecosystems where resources and territories are limited. In Ecuador's Galápagos Islands National Park (in the eastern Pacific), for example, cats, rats, pigs, goats, wasps, and other animals introduced by people have disrupted nests and food supplies and thereby reduced populations of endemic tortoises and other species.

The severe impacts of exotic species demonstrate that Australia truly is an island; islands tend to be very sensitive to ecological disruption. Both inadvertently and purposely, people have introduced foreign plants and animals that have multiplied and affected the island, sometimes in biblical plague proportions. English settlers imported the rabbit for sport hunting, confident in the belief that as in England, there would be natural checks on the animal's population. They were mistaken (•**Figure 8.23**). The rabbits bred like rabbits, eating everything green they could find, with ruinous effect on the vegetation. Observers described the unbelievable concentrations of the lagomorphs as "seething carpets of brown fur." In the mid-twentieth century, the government mounted a huge eradication effort, employing fences, snares, dogs, guns, fires, and numerous other means, and exhorting combatants with the slogan "The Rabbit War Must Be Won!" An introduced virus called "white blindness" eventually helped reduce the vast numbers, but the animal is still prolific throughout large parts of the country.

Other remedies have been applied, with varying success, to correct numerous other problem species, including foxes, mice, water buffalo, cane toads, and the prickly pear cactus. Many of the problem populations, notably cats and water buffalo, are **feral animals**—domesticated animals that have abandoned their dependence on people to resume life in the wild. An estimated 12 million feral cats infest the country and are held responsible for causing the extinction or endangerment of 39 native animal species. One Australian politician promoted legislation that would kill all cats on the continent by 2020.

Exotic livestock such as sheep and cattle are also a problem in the Australian environment. These hard-hoofed imports cut into the soil, promoting erosion and desertification. Some range management scientists and conservationists believe that the future of Australian ranching lies with the country's 25 million

Joe Hobbs

• **Figure 8.22** Australia's Aborigines say that spirits "dreamed" the world and all its features into existence. This is a rock painting of one of those creator beings, the *wandjina*, a deity of wind and storms.

Hulton-Deutsch Collection/CORBIS

• **Figure 8.23** Rabbits were one of Australia's many human-caused scourges and are still a problem, but not nearly as much as in the 1920s, when people went on "rabbit drives" like these to chase, corral, and kill the declared vermin.

kangaroos. These soft-hoofed marsupials are adapted to the Australian landscape and do not have such a damaging impact on the soil. Farmers and ranchers have traditionally eradicated about two million of them each year as pests. However, if more consumers at home and abroad could learn to appreciate kangaroo meat, there would be a strong incentive for ranchers to reduce their cattle and sheep herds and allow kangaroos to proliferate and be harvested. South Africa already imports about 20 tons of kangaroo meat yearly from Australia, and nearly 100 more tons go each year to countries including France, the Netherlands, and Germany. Russia was a big consumer until an *E. coli* outbreak led to a ban. Kangaroo exporters are now eyeing China, where there is a huge market for wild animal meats. Kangaroo skin is also used to make athletic shoes.

Australia is an unlikely exporter of another quadruped, the dromedary camel. Australia's nearly 1 million camels are feral descendants of those brought from South Asia and Iran early in the twentieth century to provide transport across the arid continent. Australian camels are exported even to Saudi Arabia, where their wild temperaments suit them to racing. Oddly, camels are part of Australia's "carbon farming initiative," which allows the country to cull (kill) the feral camels and claim carbon credits in the global carbon emissions trading scheme to mitigate global warming (Australia ratified the Kyoto Protocol). Every year a flatulent camel emits the equivalent of 99 pounds (45 kilograms) of methane, a greenhouse gas that is even more potent than carbon dioxide. The camel also eats

about a ton of carbon dioxide–absorbing vegetation each year. On the carbon ledger, each camel's death earns an "emissions avoidance benefit" equal to about 15 tons of carbon dioxide.

From Australia, we move to the bottom of the world, to a continent twice its size, the white wilderness of Antarctica.

Antarctica: The White Continent

Antarctica is the world's fifth largest continent, with an area of 5.5 million square miles (14,245,000 sq km), lying south of the tip of South America and virtually filling the Antarctic Circle (•Figure 8.24). It is the setting of enormous human dramas in exploration, bravery, and foolhardiness as people have crossed the continent with dogsleds, on skis, on foot, and in airplanes and helicopters. Some, including all members of Britain's Scott Expedition of 1910–1912—which had hoped to beat the Norwegians to the South Pole but arrived five weeks too late—did not return alive.

Antarctica's distinctiveness comes in part from its winter of darkness, alternating with summer "whiteouts" caused by light refraction on snow and ice surface covering about 95 percent of the continent. Average temperatures during the summer months barely reach 0°F (−18°C). Winter averages are the coldest in the world, with winter mean temperatures averaging −70°F (−57°C). Winter's months of darkness of the polar night are an astronomer's dream, and many scientific teams, including one from China, have set up observatories here. Antarctica is also the world's windiest and driest continent and is very sunny during its half year of

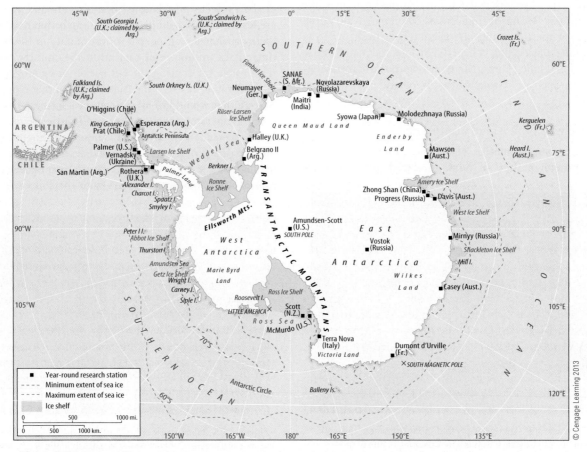

• **Figure 8.24** Principal features of Antarctica.

• **Figure 8.25** Land and water on Antarctica's coast. The blue ice is part of a solitary iceberg that was "calved" from a nearby glacier. You are literally seeing the tip of the iceberg here. About 90 percent of the iceberg lies underwater, presenting a threat to navigation. The underwater portion of an iceberg fatally damaged the *Titanic*.

midnight sun; the South Pole on average receives more sunlight each year than the world's equatorial regions. Although the continent is capped with glacial ice up to 10,000 feet (c. 3,000 m) thick, only 2 to 10 inches (5 to 25 cm) of precipitation fall each year.

Antarctica has a prominent position in global concerns about climate change. The "ozone hole" in the earth's stratosphere is concentrated seasonally over Antarctica, and scientists keep constant watch as it dwindles, thanks to the Montreal Protocol treaty to phase out CFCs. Records show that the Antarctic Peninsula has warmed by 4.5°F (2.5°C) since 1944, causing both the peninsula's fringing ice sheet of sea ice and its inland glacial ice to begin to melt. Further melting of the land ice would lead inevitably to rising sea levels worldwide.

In 2011, the U.S. National Academy of Sciences published a report entitled *Future Science Opportunities in Antarctica and the Southern Ocean*. It identified key questions that should drive research in the region over the next two decades. Most of these questions focus on Antarctica's role in global climate change. Scientists see the southern polar region, like the northern polar region, as critical in understanding the global climate system and for creating models to predict future changes. Of particular concern is the Pine Island Ice Shelf, which is melting unusually rapidly. Scientists are using remote sensing to understand how relatively warm water is undercutting the ice and causing its melting. The fear is that accelerated melting will cause sea levels to rise more rapidly than anticipated.

Antarctica is a vast international scientific laboratory. There is virtually no human settlement beyond the research teams that have built facilities on small areas of exposed land that lie near the outer edges, on the island of Little America in the Ross Sea, and at the South Pole itself. About 1,000 researchers are on the continent in the winter and about 4,000 in the summer. The

United States has blazed a 1,000-mile (1,600 km) "ice highway" from its McMurdo Research Station to the South Pole, angering Antarctic environmentalists.

Public, economic, and geopolitical interests in this distant world are increasing. Expensive voyages by sea bring about 40,000 visitors each year (up from 5,000 in 1990) to mingle with penguins and enjoy some of the wildest scenery on earth (•Figure 8.25). Seven countries have staked claims (some of which overlap) to portions of the continent as national territories: nearby Chile and Argentina, the Southern Hemisphere powers of Australia and New Zealand, the United Kingdom, France, and Norway. No other countries recognize these claims. Both the United States and Russia have avoided laying claim to parts of Antarctica but maintain that they have the right to do so. Such claims, which have been technically "frozen" since 1961, may be moot. The **Antarctic Treaty**, negotiated in 1959 and signed by 45 countries, including the seven laying claims, forbids any exploitation of the continent's natural resources until 2048. There has long been speculation about the potential for considerable resource wealth lying beneath the ice, but to date no mineral deposits of significant economic value have been found.

Antarctica is the world's largest nature preserve. But even at this end of the Earth there are pressures on resources. Some of Antarctica's waters are overfished. Japanese whaling vessels take about 1,000 whales from Antarctica waters every year. Australia has taken on Japan in the International Court of Justice to put an end to Antarctic whaling. In the meantime, conservationists argue that many more marine reserves like one created under the Antarctic Treaty in 2009 are needed.

From Antarctica, we move north (the only way to go) to Africa, another continent that boasts much wilderness and that is the cradle of humankind.

summary

- Oceania encompasses Australia, New Zealand, and the islands of the mid-Pacific lying mostly between the tropics. Tropical rain forest climates and biomes are most common, but Australia and New Zealand have several temperate climate and biome types.

- The Pacific islands are commonly divided into three principal regions: Melanesia, Micronesia, and Polynesia.

- Oceania is ethnically complex, having been settled by peoples of various Asian origins. Polynesia was the last to be populated. Papua New Guinea is the world's most linguistically diverse country. Christianity is the majority faith in this region.

- Although countless islands are scattered across the Pacific Ocean, there are three generally recognized types: continental islands, high islands, and low islands. Continental islands are either continents themselves (such as Australia) or were connected to continents when sea levels were lower (such as New Guinea). Most high islands are volcanic. Low islands are typically made of coral, a material composed of the skeletons and living bodies of small marine organisms that inhabit tropical seas.

- The island ecosystems of the Pacific region are typically inhabited by endemic plant and animal species—species found nowhere else in the world. Island species are especially vulnerable to the activities of humans, such as habitat destruction, deliberate hunting, or the introduction of exotic plant and animal species.

- Europeans began to visit and colonize the Pacific islands early in the European Age of Exploration and brought mainly negative impacts to island societies. However, a steady process of decolonization has accompanied a recent surge of Western interest and investment in the region.

- There are ethnic conflicts, related mainly to maldistribution of income, between Malaitans and indigenous Guadalcanalans on the Melanesian island of Guadalcanal and between indigenous Fijians and Indo-Fijians in Fiji. Interest in securing more income from minerals has pitted the people of New Caledonia against the ruling power, France.

- Aside from a few notable exceptions, the poverty typical of less developed countries prevails throughout most of the Pacific region.

- In general, the Pacific islands' economic picture is one of nonindustrial economies. Typical economic activities include tourism, plantation agriculture, mining, and income derived from activities connected with the military needs of occupying powers. Several countries are profiting from offshore banking and telemarketing.

- During the 1940s and 1950s, the United States used the Bikini atoll in the Marshall Islands as one of its chief testing grounds for nuclear weapons. Strong negative reaction arose throughout the region in the 1990s as the French resumed underground testing of nuclear weapons on the Mururoa atoll in French Polynesia. That testing has since ceased. The United States relies on the region for testing of its Strategic Defense Initiative missile technology. Many inhabitants of French and American military zones are fearful of the economic repercussions of the withdrawal of foreign military presence.

- Some of the low-island countries are fearful that global warming might raise sea levels and inundate them, and Tuvalu considered legal recourse against the United States and Australia for failing to ratify the Kyoto Protocol. *international treaty reduce g gases*

- Australia and New Zealand are products of British colonization. The British heritage is strongly reflected in the ethnic composition and culture of the two nations' majority populations.

- Both Australia and New Zealand have minorities and indigenous inhabitants. Native Australians are known as Aborigines, and the dominant indigenous group in New Zealand is the Maori.

- In Australia, one of the largest issues of contention between the Aborigines and the white majority is land rights. In 1993, the Native Title Bill addressed the Aborigines' major concerns and made several land rights concessions.

- Exotic species and human pressures have had catastrophic effects on native species and natural environments in Australia and New Zealand.

- Australia and New Zealand have been reorienting their focus toward the Pacific Rim and away from the United Kingdom. An important expression of this change in focus is the ongoing debate about whether Australia should become a republic, thus ending more than 200 years of formal ties with Britain.

- Antarctica is the world's coldest, driest, and least populated continent. There is no permanent population, only scientific researchers. The region's melting sea and land ice may reflect global warming, and the "hole" in the earth's stratospheric ozone layer concentrates over Antarctica. Seven countries stake claims to the continent, but no other countries recognize these claims, and so far they have not had any economic significance.

Review Questions

1. What three principal subregions comprise Oceania? What are some of the countries and colonial possessions of each?

2. Describe the major differences between high and low islands. According to the geographer Tom McKnight, how did the physical characteristics of the islands relate to social and political developments? What are the typical demographic and economic qualities of each?

3. What is a geologic hot spot, and how does it relate to Pacific islands?

4. What unique ecological attributes do islands generally have? What are the major threats to them?

5. What is an atoll, and how does it form?

6. What are the major ethnic groups, languages, and religions of Oceania?

7. How was Nauru able for a time to avoid the poverty typical of most Pacific islands? How is Nauru's economy diversifying now that its principal asset is dwindling?

8. What are the major economic activities in the Pacific islands, excluding Australia and New Zealand? Why and where have military interests been prominent?

9. What explains the peculiar jog in the International Date Line?

10. What fears do some Pacific countries have about potential global warming?

11. What features do Australia and New Zealand have in common? How are their identities changing?

12. What is the significance of the Dreamtime to Australia's Aborigines? What issues of land ownership concern them?

13. What advantages might kangaroo husbandry have over that of conventional livestock?

14. What are the major physical characteristics of Antarctica? What is the continent's political status?

15. What are "cargo cults?" Are they still active in the Pacific region?

Key Terms + Concepts

Aboriginal languages (p. 291)
  non-Pama-Nyungan (p. 291)
  Pama-Nyungan (p. 291)
Aborigines (p. 290)
Alliance of Small Island States (p. 297)
Antarctic Treaty (p. 305)
Asia-Pacific Economic Cooperation (APEC) group (p. 301)
atoll (p. 288)
Australia New Zealand United States (ANZUS) security alliance (p. 297)
Austronesians (p. 291)
Austronesian language family (p. 291)

Formosan language subfamily (p. 292)
Malayo-Polynesian subfamily (p. 292)
Proto-Austronesian (p. 291)
cargo cults (p. 293)
Closer Economic Relations Agreement (p. 301)
"coconut civilization" (p. 289)
Compact of Free Association (p. 298)
continental islands (p. 285)
comanagement (p. 302)
counter-mapping (p. 302)
deterritorialized states (p. 297)
Dreamtime (p. 301)

environmental determinism (p. 287)
exotic species (p. 289)
"extinction capital of the world" (p. 289)
feral animals (p. 303)
Footprints of the Ancestors (p. 301)
geologic hot spot (p. 287)
guano (p. 295)
high islands (p. 287)
Indo-Fijians (p. 295)
low islands (p. 287)
Melanesia (p. 283)
Micronesia (p. 283)
millenarian movements (p. 293)
money laundering (p. 295)

Native Title Bill (p. 302)
Near North (p. 301)
Oceania (p. 282)
Operation Crossroads (p. 298)
pidgin (p. 292)
Polynesia (p. 283)
power projection hub (p. 297)
seamount (p. 287)
shell banks (p. 295)
Songlines (p. 301)
terra nullius (p. 302)
Torres Strait Islanders (p. 291)
Trans-Pacific Partnership (p. 301)
Walkabout (p. 301)
Way of the Law (p. 301)
"white Australia" policy (p. 301)
Wik case (p. 302)

Notes

1. Tom L. McKnight, *Oceania: The Geography of Australia, New Zealand and the Pacific Islands* (Englewood Cliffs, N.J.: Prentice Hall, 1995), p. 181.

2. Ibid.

3. Jeff Siegel, "Tok Pisin," http://www.une.edu.au/langnet/definitions/tokpisin.html.

4. An excellent article on Tanna's John Frum Cargo Cult is Paul Raffaele's "In John They Trust," *Smithsonian*, February 2006.

5. David Jolly, "Priced Off the Menu? Palau's Sharks Are Worth $1.9 Million Dollars Each, a Study Says." *New York Times*, May 2, 2011, p. A12.

6. Quoted in David Pilling, "How America Should Adjust to the Pacific Century." *Financial Times*, November 17, 2011, p. 9.

7. Quoted in Geoffrey Blainey, *Triumph of the Nomads* (Melbourne: Sun Books, 1987), p. 181.

8. Norimitsu Onishi, "Rich in Land, Aborigines Split on How to Use It." *New York Times*, February 13, 2011, p. 5.

9. Ibid.

Online Resources

CourseMate: Make the most of your study time by accessing everything you need to succeed in one place. Read your textbook, take notes, review flashcards, watch videos, complete activities, take practice quizzes, and more—online with CourseMate. Log in at **www.cengagebrain.com**.

"I dream of the realization of the unity of Africa, whereby its leaders combine in their efforts to solve the problems of this continent. I dream of our vast deserts, of our forests, of all our great wildernesses."

—NELSON MANDELA

Africa at night. Note the lights of relatively prosperous South Africa on the world's poorest continent. Left: Doum palms rise from the banks of the Uaso Nyiro River in Kenya's Samburu National Park

Sub-Saharan Africa

9

Geographers typically recognize the continent of Africa as having two major divisions: North Africa (the predominantly Arab and Berber realm of the continent, described in Chapter 6), and ethnically diverse Sub-Saharan Africa (•Figure 9.1). This chapter sketches the broad outlines of the geography of Sub-Saharan Africa and examines issues in the major African subregions.

The region of Sub-Saharan Africa is culturally complex, physically beautiful, and problem ridden. This introduction to Africa surveys the region's diverse environments, peoples and ways of living, population distributions, European colonial legacies, and major current issues.

chapter objectives

This chapter should enable you to

✗ Understand what caused Sub-Saharan Africa to become and remain the world's poorest region.

• Know how the region came to have the highest HIV/AIDS infection rates in the world and how the epidemic could be reversed there.

• Appreciate the pressures on African wildlife and some unique approaches taken to protect the animals.

• Know what is uniquely African about African cultures.

• See how Africans have leap-frogged communications challenges with mobile phones, and use phones to overcome the "digital divide."

• Recognize why, after more than a decade at the sidelines, Sub-Saharan Africa is considered important again in geopolitical affairs.

✗ Recognize how European colonial favoritism of some ethnic groups over others sowed seeds of modern strife and warfare.

• Appreciate what corrupt leadership has done to impoverish people in countries with enormous oil and other natural resource wealth.

• Understand what causes the "resource curse."

• Understand how drought can trigger a process of environmental degradation.

• Consider the causes of the genocide and "ethnic cleansing" that marred Africa in the 1990s, particularly in East and East-Central Africa.

For many people, Sub-Saharan Africa is the most poorly understood or misunderstood of the world's regions. It seems to have so many countries, so much violence and disease, and so much poverty that it is simply too difficult to think about. Hollywood still perpetuates stereotyped images of an Africa rich in wildlife but with dangerous or backward peoples. This chapter will show that getting to know the real Africa is manageable, rewarding, and interesting. This region's diverse cultures and natural environments are sources of endless wonder. Conditions for many Africans are improving in a number of ways. There are many reasons for hope and much to celebrate in the study of this fascinating region. Just ask the people: a Gallup survey of 50,000 citizens around the world found that Africans are the most optimistic.[2]

9.1 Area and Population

Sub-Saharan Africa (including Madagascar and other nearby Indian Ocean islands, and the newest country of South Sudan) has the second largest land area of all the major world regions described in this book. It covers 8.7 million square miles (22.4 million sq km) and so is more than twice the size of the United States (•**Figure 9.2**).

Sub-Saharan Africa's population was 847 million in 2011. Even with the loss of population due to AIDS, the rate of natural population increase in Sub-Saharan Africa is 2.5 percent per year, or about five times that of the United States. The region's population is expected to double by 2050. As in most LDCs, African parents generally want large families for several reasons: to have extra hands to perform work; to be looked after when they are old or sick; and in the case of girls, to receive the "bride wealth" a groom pays in a marriage settlement. Large families also convey status. However, there are signs of significant change 50 in this pattern of population growth: birth rates have been dropping in every country in this region over the past two decades.

In spite of the large population, much of the region is sparsely populated. The region's average population density is 312 slightly more than that of the United States. Most of the people in this region live in a few small, densely populated areas (•**Figure 9.3**). The main areas are the coastal belt bordering the Gulf of Guinea in West Africa, from the southern part of Africa's most populous country, Nigeria, westward to southern Ghana; the savanna lands of northern Nigeria; the highlands of Ethiopia; the highland region surrounding Lake Victoria in Kenya, Tanzania, Uganda, Rwanda, and Burundi; and the eastern coast and parts of the high interior plateau of South Africa.

Sub-Saharan Africa is the world's most rural region (68%), with rural populations of most countries between 40 and 80 percent. The most rural are two East African nations with very fertile soils: Burundi (89% rural) and Malawi (86%). The most urbanized are Gabon (86%), where people are flocking to partake in an oil boom that is facilitating urbanization, and Djibouti (76%), where there is virtually no arable land and a small number of people are clustered in a port city. The region's major cities and large towns are magnets that attract many poor, rural people. The rate of urban growth is high; in 2011 there were twice as many cities with more than a million residents as there were in 1990 (•**Tables 9.1 and 9.2**). It is projected that 50 percent of Africans will live in cities in 2025. But for now, life in villages is the rule for the majority. A typical rural home is a small hut made of sticks and mud, with a dirt floor, thatched roof, and no electricity or plumbing (•**Figure 9.4**).

Africa has the world's youngest population, with 41 percent of the people under 15 years old (compared with about a third of the populations of Latin America and Asia). The median age is now 20 (compared with 30 in Asia and 40 in Europe). There are different ways of viewing this young, growing population. Some economists see the youthful median age as a "demographic dividend," as large numbers of workers will be able to contribute to the region's economic growth. Others fear that the Malthusian scenario will come about because of the gulf between population growth and agricultural output in these mainly agrarian societies. Since the 1960s, the population of Sub-Saharan Africa has grown at a rate of about 5 percent annually, whereas food production in the region has grown at only about 2 percent annually. This is the only world region where per capita food production declined in that period.

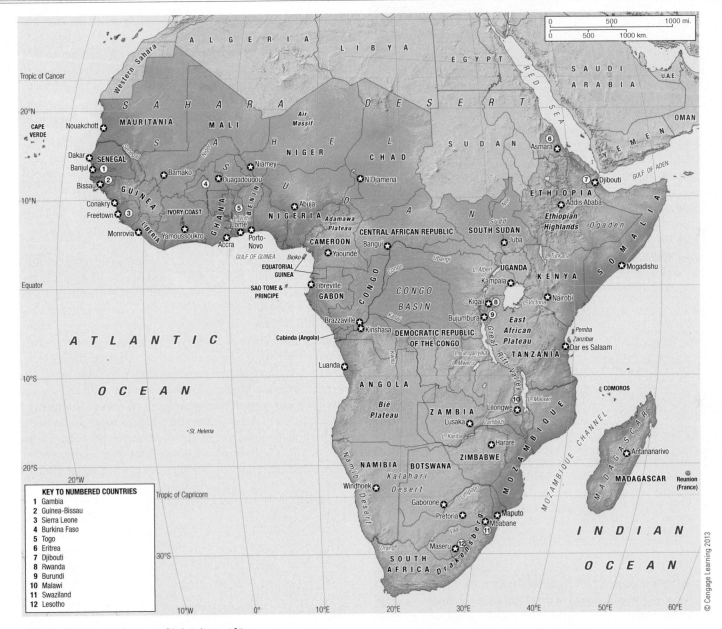

• **Figure 9.1** Principal features of Sub-Saharan Africa.

The wild card in Africa's population deck is the **human immunodeficiency virus (HIV)**. The disease it causes—**AIDS (acquired immunodeficiency syndrome)**—has inevitably been identified by some as the Malthusian "check" to the region's population growth (see Medical Geography, page 316).

9.2 Physical Geography and Human Adaptations

Sub-Saharan Africa is both rich in natural resources and beset with environmental challenges to economic development. It is home to some of the world's greatest concentrations of wildlife and some of the most degraded habitats.

The Landscapes of Africa

Most of Africa is a vast plateau, actually a series of plateaus, with a typical elevation of more than 1,000 feet (c. 300 m). Near the Great Rift Valley in the Horn of Africa and in southern and eastern Africa, the general elevation rises 2,000 to 3,000 feet (c. 600 to 900 m), with many areas at 5,000 feet (1,520 m) and higher (see Regional Perspective, page 317). The highest peaks and largest lakes of the continent are located in this belt. The loftiest summits lie within 250 miles (c. 400 km) of Lake Victoria. They include Kilimanjaro (19,340 ft/5,895 m) and Kirinyaga (Mount Kenya, 17,058 ft/5,200 m; •Figure 9.5), which are volcanic cones, and the Ruwenzori range (up to 16,763 ft/5,109 m), a nonvolcanic massif produced by faulting. Lake Victoria, the largest lake in Africa, is surpassed in area

Africa's Biomes and Climates

The equator bisects Africa, so about two-thirds of the region lies in the low latitudes, having tropical climates and biomes; Africa is the most tropical of the world's continents (•Figure 9.6). Areas of tropical rain forest climate center on the great rain forest of the Congo Basin in central and western Africa. The forest merges gradually into a tropical savanna climate on the north, south, and east. This is the climatic and biotic zone supporting most of the magnificent large mammals of Africa. The savanna areas in turn trend into steppe and desert on the north and southwest. A broad belt of drought-prone tropical steppe and savanna bordering the Sahara on the south is known as the Sahel. There is desert on the coasts of Eritrea, Djibouti, and Somalia in the Horn of Africa. In South Africa and Namibia, a coastal desert, the Namib, borders the Atlantic. The Kalahari Desert, which lies inland from the Namib, is better described as steppe or semidesert than as true desert. Along the northwestern and southwestern fringes of the continent are small but productive areas of Mediterranean climate, while eastern coastal sections and adjoining interior areas of South Africa have a humid subtropical climate. Bordering the subtropical climate region is an area of marine west coast climate.

Total precipitation in the region is high but unevenly distributed; some areas are typically saturated, whereas others are perennially bone dry. Even in many of the rainier parts of the continent, there is a long dry season, and wide fluctuations occur from year to year in the total amount of precipitation. One of the major needs in Africa is better control over water. In the typical village household, women carry water from a stream or lake or a shallow (and often polluted) well. Use of more small dams would help provide water storage throughout the year.

Drought is a persistent problem in most of the countries. Although all droughts create problems, some last for years, with devastating effects in this heavily agricultural region. Droughts have been particularly severe in recent decades in the Sahel and in the Horn of Africa. The number of food emergencies, most of them tied to drought, has tripled since the mid-1980s in Sub-Saharan Africa.

The consensus among scientists is that this region will pay the highest price for global climate change, even though it contributes the fewest greenhouse gases to the atmosphere. United Nations studies predict even more severe droughts and crop losses, along with the destruction of coastal infrastructure that will accompany rising sea levels. Such dangerous environmental changes may be contributing to conflict in the region as well. Pastoralists and farmers already compete for scarce land and water resources in many places—this was one of the problems in Sudan's Darfur conflict, for example—and droughts intensify those struggles.

The patterns of Africa's land use (•Figure 9.7) reveal that the most productive lands are on river plains, in volcanic regions (especially the East African and Ethiopian highlands), and in some grassland areas of tropical steppes (notably the High

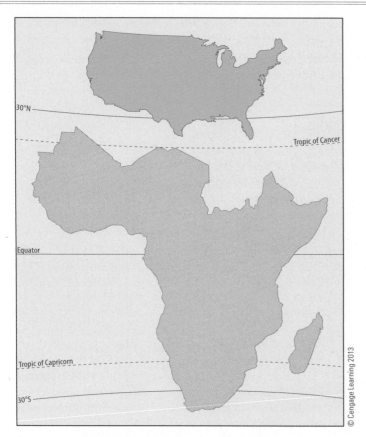

• **Figure 9.2** Sub-Saharan Africa compared in area and latitude with the conterminous United States.

among inland waters of the world only by the Caspian Sea and Lake Superior. It is relatively shallow, however, and the large numbers of people living on its shores are taxing its resources. Other very large lakes in East Africa include Lake Tanganyika and Lake Malawi.

The physical structure of Africa has influenced the character of African rivers. The main rivers, including the Nile, Niger, Congo, Zambezi, and Orange, rise in interior uplands and descend by stages to the sea. At some points, they descend abruptly, particularly at plateau escarpments, with rapids and waterfalls interrupting their courses. These often block navigation a short distance inland. The many waterfalls and rapids do have a positive side: they represent a great potential source of hydroelectric energy. There are major power stations on the Zambezi, Congo, Niger, Volta, and Blue Nile Rivers. But only about 5 percent of Africa's hydropower potential has been realized (compared to about 60% in North America). Many of the best sites are remote from large markets for power. In some cases, geopolitical considerations pose obstacles to dam construction. Downstream Egypt, for example, has expressed concern and even hostile rhetoric about dams and water diversions of the Nile and its tributaries by upstream Sudan, Ethiopia, Uganda, Kenya, and Tanzania (see page 180 in Chapter 6).

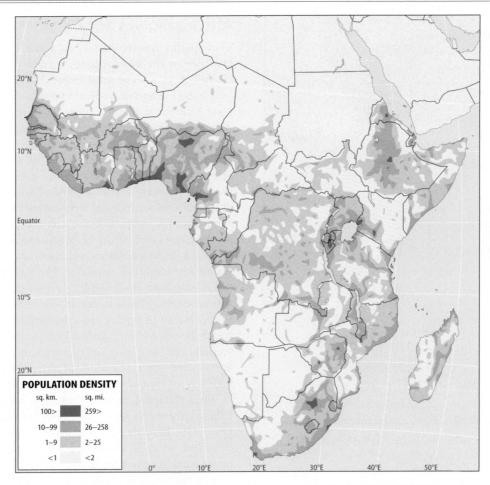

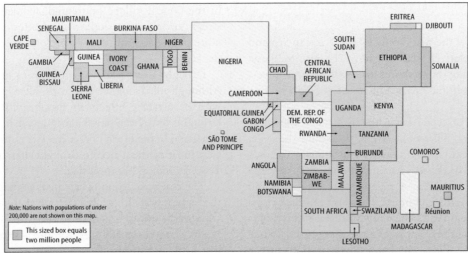

• **Figure 9.3** Population distribution (top) and population cartogram (bottom) of Sub-Saharan Africa.
Source: The Oxford School Atlas edited by Patrick Wiegand (OUP, 1997), copyright © Oxford University Press. Used by permission of Oxford University Press.

Veld in South Africa). Soils of the deserts and regions of Mediterranean climate are often poor. In the tropical rain forests and savannas, there are reddish, lateritic tropical soils that are infertile once the natural vegetation is removed and can support only shifting cultivation. Agriculture is discussed in more detail beginning on page 328.

219,
345,
384

Africa's Wildlife

Although cattle raising is widespread throughout the savannas, cattle are largely excluded from extensive sections both north and south of the equator by the disease called **nagana**, which is carried by the tsetse flies that also transmit sleeping sickness

Table 9.1   Sub-Saharan Africa: Basic Data

Political Unit	Area (thousands; sq mi)	Area (thousands; sq km)	Estimated Population (millions)	Estimated Population Density (sq mi)	Estimated Population Density (sq km)	Annual Rate of Natural Increase (%)	Human Development Index	Urban Population (%)	Per Capita GNI PPP ($US)
The Sahel	**2,296.3**	**5,947.6**	**86.8**	**38**	**15**	**2.8**	**0.324**	**27**	**1,110**
Burkina Faso	105.8	274.0	17	161	62	3.1	0.331	24	1,170
Cape Verde	1.6	4.1	0.5	313	121	1.7	0.568	62	3,530
Chad	495.8	1,284.1	11.5	23	9	2.9	0.328	28	1,160
Gambia	4.4	11.4	1.8	409	158	3.0	0.420	59	1,330
Mali	478.8	1,240.1	15.4	32	12	3.1	0.359	33	1,190
Mauritania	396	1,025.6	3.5	9	3	2.4	0.453	42	1,940
Niger	489.2	1,267.0	16.1	33	13	3.6	0.295	17	680
Senegal	76	196.8	12.8	168	65	2.8	0.459	43	1,810
South Sudan	248.7	644.3	8.2	33	13	N/A	N/A	N/A	N/A
West Africa	**812.4**	**2,104.1**	**246.1**	**255**	**98**	**2.5**	**0.450**	**49**	**1,820**
Benin	43.5	112.7	9.1	209	81	2.9	0.427	43	1,510
Ghana	92.1	238.5	25	271	105	2.3	0.541	52	1,530
Guinea	94.9	245.8	10.2	107	41	2.7	0.344	28	940
Guinea-Bissau	13.9	36.0	1.6	115	44	2.3	0.353	30	1,060
Ivory Coast	124.5	322.5	22.6	182	70	2.3	0.400	51	1,640
Liberia	37.2	96.3	4.1	110	43	3.1	0.329	47	290
Nigeria	356.7	923.9	162.3	455	176	2.5	0.459	51	2,070
Sierra Leone	27.7	71.7	5.4	195	75	2.2	0.336	39	790
Togo	21.9	56.7	5.8	265	102	2.8	0.435	37	850
West Central Africa	**1,576**	**4,081.8**	**99.4**	**53**	**20**	**2.6**	**0.347**	**43**	**1,140**
Cameroon	183.6	475.5	20.1	109	42	2.2	0.482	59	2,190
Central African Republic	240.5	622.9	5	21	8	2.1	0.343	39	750
Congo Republic	132	341.9	4.1	31	12	2.5	0.533	63	3,040
Congo, Democratic Republic of	905.4	2,345.0	67.8	75	29	2.8	0.286	36	300
Equatorial Guinea	10.8	28.0	0.7	65	25	2.3	0.537	40	19,330
Gabon	103.3	267.5	1.5	15	6	1.8	0.674	86	12,450
São Tomé and Principe	0.4	1.0	0.2	500	193	2.7	0.509	63	1,850
East Africa	**703**	**1,820.8**	**143.4**	**155**	**60**	**2.9**	**0.460**	**20**	**1,300**
Burundi	10.7	27.7	10.2	953	368	3.2	0.316	11	390
Kenya	224.1	580.4	41.6	186	72	2.7	0.509	18	1,570
Rwanda	10.2	26.4	10.9	1,069	413	2.1	0.429	19	1,130
Tanzania	364.9	945.1	46.2	127	49	2.9	0.466	27	1,360
Uganda	93.1	241.1	34.5	371	143	3.4	0.446	15	1,190
Horn of Africa	**727**	**1,882.9**	**103.8**	**152**	**59**	**2.7**	**0.363**	**20**	**920**
Djibouti	9	23.3	0.9	100	39	1.9	0.430	76	2,480
Eritrea	45.4	117.6	5.9	130	50	2.6	0.349	22	580
Ethiopia	426.4	1,104.4	87.1	204	79	2.7	0.363	17	930
Somalia	246.2	637.7	9.9	40	16	2.8	N/A	38	N/A

(continued)

Table 9.1	Sub-Saharan Africa: Basic Data (*continued*)

Political Unit	Area (thousands; sq mi)	Area (thousands; sq km)	Estimated Population (millions)	Estimated Population Density (sq mi)	Estimated Population Density (sq km)	Annual Rate of Natural Increase (%)	Human Development Index	Urban Population (%)	Per Capita GNI PPP ($US)
Southern Africa	**2,310.8**	**5,985.0**	**142.4**	**52**	**20**	**1.9**	**0.486**	**45**	**5,440**
Angola	481.4	1,246.8	19.6	41	16	2.8	0.486	59	5,190
Botswana	224.6	581.7	2	9	3	1.2	0.633	59	12,840
Lesotho	11.7	30.3	2.2	188	73	1.2	0.450	23	1,800
Malawi	45.7	118.4	15.9	348	134	2.7	0.400	14	780
Mozambique	309.5	801.6	23.1	75	29	2.8	0.322	31	880
Namibia	318.3	824.4	2.3	7	3	1.7	0.625	39	6,350
South Africa	471.4	1,220.9	50.5	107	41	0.6	0.619	62	10,050
Swaziland	6.7	17.4	1.2	179	69	1.5	0.522	22	4,790
Zambia	290.6	752.7	13.5	46	18	3.1	0.430	36	1,280
Zimbabwe	150.9	390.8	12.1	80	31	1.9	0.376	29	N/A
Indian Ocean Islands	**229.7**	**595**	**24.6**	**107**	**41**	**2.6**	**0.493**	**34**	**1,740**
British Indian Ocean Territory (U.K.)	0.02	0.05	0.003	150	58	N/A	N/A	100	N/A
Comoros	0.9	2.3	0.8	889	343	2.8	0.433	28	1,180
Madagascar	226.7	587.2	21.3	94	36	2.8	0.480	31	990
Mauritius	0.8	2.1	1.3	1,625	627	0.5	0.728	42	13,270
Mayotte (Fr.)	0.1	0.3	0.2	2,000	772	3.0	N/A	50	N/A
Réunion (Fr.)	1	2.6	0.9	900	347	1.1	N/A	94	N/A
Seychelles	0.2	0.5	0.1	500	193	1.0	0.773	56	16,790
Summary Total	**8,655.2**	**22,417.2**	**846.5**	**98**	**38**	**2.5**	**0.423**	**32**	**2,080**

Sources: World Population Data Sheet, Population Reference Bureau, 2011; Human Development Report, United Nations, 2011; World Factbook, CIA, 2011.

to humans (see Insights, page 321). In tropical rain forests, tsetse flies are even more prevalent and few cattle are raised, but goats and poultry are common (as they are in tsetse-frequented savanna areas).

Africa has the planet's most spectacular and numerous populations of large mammals. The tropical grasslands and open forests of Africa are the habitats of large herbivorous animals, including the elephant, buffalo, zebra, giraffe, and many species of antelope, as well as carnivorous and scavenging animals, such as the lion, leopard, and hyena. The tropical rain forests have fewer of these "game" animals (as Africans call them); the most abundant species here are insects, birds, and monkeys, with the hippopotamus, the crocodile, and a great variety of fish in the streams and rivers draining the forests and wetter savannas.

Although documentary films promote a perception outside Africa that the continent is a vast animal Eden, the reality is less positive. Human population growth, urbanization, and agricultural expansion are taking place in Africa, as elsewhere in the world, at the expense of wildlife. Hunting and competition with domesticated livestock also take their toll. Many species are safeguarded by law, especially in the protected areas that make up about 15 percent of the region. Such laws are difficult to enforce, however, and poaching has devastated some species. Still, Africa remains home to some of the world's most extraordinary and successfully managed national parks, including South Africa's Kruger National Park, Tanzania's Ngorongoro Crater National Park, and Kenya's Amboseli National Park (at the foot of Kilimanjaro; see Figure 9.5). International tourism to these parks is a major source of revenue for some countries, particularly South Africa, Tanzania, Kenya, Namibia, and Botswana.

Elephants and rhinoceroses are Africa's largest herbivores and the most vulnerable to global market demand for wildlife products. In 1970, there were about 2.5 million African

Table 9.2	Sub-Saharan Africa: Metropolitan Populations Data	
1	Lagos, Nigeria	14.9
2	Kinshasa, D.R.C.	10.9
3	Johannesburg, South Africa	8.7
4	Abidjan, Ivory Coast	6.7
5	Ibadan, Nigeria	5.8
6	Cape Town, South Africa	5.4
7	Luanda, Angola	5.3
8	Kano, Nigeria	4.9
9	Accra, Ghana	4.6
10	Nairobi, Kenya	4.5
11	Durban, South Africa	3.7
12	Harare, Zimbabwe	3.5
13	Addis Ababa, Ethiopia	3.4
14	Dar es Salaam, Tanzania	3.4
15	Lusaka, Zambia	3.3
16	Dakar, Senegal	2.6
17	Pretoria, South Africa	2.6
18	Benin City, Benin	2.5
19	Bamako, Mali	2.5
20	Yaoundé, Cameroon	2.4

Population in millions.

© Cengage Learning 2013

Joe Hobbs

• **Figure 9.4** Homes in African villages are typically made of local materials including thatch, which if replaced regularly do a fine job of keeping rain out. They are usually elevated slightly above to stave off the threat of flooding. Village roads are seldom paved. People live in close proximity with their livestock, in part to protect the animals from predators. This village is in northwestern Madagascar.

elephants living on the continent. Poaching and habitat destruction had reduced their numbers to less than 500,000 in 2012. Poaching for ivory was having a catastrophic impact on African elephants when delegates of the 112 signatory nations of the Convention on International Trade in Endangered Species (CITES, now with 175 members) met in 1989. The organization succeeded in passing a worldwide ban on the ivory trade, and since then the precipitous decline has eased.

MARK C. ROSS/National Geographic Stock

Robert Mackinlay/Getty Images

• **Figure 9.5** Africa's highest mountains, both volcanic, are Tanzania's Kilimanjaro (left; photographed from Kenya's Amboseli National Park) and Kenya's Kirinyaga (right), also known as Mt. Kenya.

AIDS is a global problem. The earth has not seen a **pandemic** (very widespread epidemic) like this since the bubonic plague devastated fourteenth-century Europe and smallpox struck the Aztecs of sixteenth-century Mexico. AIDS had killed over 25 million people worldwide by 2012, and there are almost 3 million new infections around the world each year. Authorities say that worldwide 7,400 people are infected every day. Sub-Saharan Africa is where the virus originated (apparently among chimpanzees; the first human case was reported in the Belgian Congo—today's Democratic Republic of Congo—in 1959) and is today the epicenter of this health crisis. The statistics are startling. In 2012, two-thirds of the world's estimated 33 million people infected with HIV lived in Sub-Saharan Africa.

The epidemic is most severe in southern Africa. In some countries, including Botswana, Lesotho, and Swaziland, as much as one-quarter of the adult population is HIV positive (•Figure 9.A). South Africa has more HIV cases (about 5.7 million) than any other country in the world. This scourge has been causing sharp reductions in life expectancy in Africa and, unless contained, will dramatically alter the region's population and development trends.

The good news is that HIV/AIDS is starting to be contained. The rate of new infections is down about 25 percent in the worst affected countries, mainly in southern Africa. Death rates, which were rising precipitously, are falling. This is brightening the prospects for development in this least developed of the world's regions. The incidence of HIV infection is particularly high among the region's most educated, skilled, and ambitious young urban professionals, including teachers, white-collar workers, and government employees. These are the people, in addition to truckers and merchants, who travel the most, have higher incomes, and are therefore the most likely to have many sexual liaisons.

What can be done to fight the AIDS epidemic? Two countries that have had remarkable success in stemming the AIDS tide provide some answers. Uganda and Senegal have used relatively inexpensive public health tools, especially education, in the fight against the disease. To reduce the huge stigma associated with AIDS and to encourage counseling, testing, and condom use, the countries' political and religious leaders have been outspoken about AIDS. The results have been impressive, with infection rates dropping in Uganda (from 18% to 6%) and staying low in Senegal.

In addition, wealthier countries outside Africa could invest more in combating and preventing the disease. The United Nations in 2001 established the Global Fund to Fight AIDS, Malaria, and Tuberculosis. Bono and other celebrity activists continue to pressure governments to pledge more monies to the Global Fund. But their efforts are flagging due to the global Great Recession and "donor fatigue."

The United States has its own effort to combat AIDS in 12 countries in Africa and 2 (Haiti and Guyana) in the Americas. In a program called **PEPFAR**, the President's Emergency Plan for AIDS Relief, the George W. Bush administration pledged over $45 billion to the effort. The United States gives more money to combat AIDS than all other donors combined. This is an enormous amount of money, and yet it is dwarfed by the scale of the problem: it is enough to treat only about one in four people infected with HIV. The Bush and Obama administrations spent most of those funds to put patients on life-lengthening **antiretroviral (ARV) drugs** and other drugs. About 5 million people worldwide were on these drugs in 2012, and another 10 million needed but could not get them.

The Obama Administration could not afford the escalating costs of treatment, and shifted the focus to prevention and mother/

CITES member states won the ivory ban over the strong objections of several southern African states, led by Zimbabwe, that were experiencing what they saw as an elephant overpopulation problem. At the 1989 CITES meeting, these countries appealed for an exemption from the ivory ban so that they could earn foreign export revenue from a sustainable yield of their elephant populations. The majority of CITES members rejected this appeal, arguing that any loophole in a complete ban would subject elephants everywhere to illegal poaching. Since 1989, CITES has periodically allowed Zimbabwe, Namibia, Botswana, and South Africa one-time sales of ivory under rigorous monitoring as a reward for their positive wildlife policies. All continue to cull (kill) their "excess" elephants, with the meat going to needy villagers and crocodile farms.

If elephant ivory is like gold, rhinoceros horns are like diamonds. Men in the Arabian Peninsula nation of Yemen prize daggers with rhino horn handles (•Figure 9.10). Although Western scientists deny the medicinal efficacy of powdered rhino horn, traditional medicine in East Asia makes wide use of it, including as an aphrodisiac and a cure for cancer. These demands, and the current black-market value of tens of thousands of dollars per horn ($45,000 per pound, more than gold, cocaine, or heroin), have led to a precipitous decline in population of black rhinoceroses in Africa. There were an estimated 65,000 black rhinos in Africa in 1982; in 2012, there were an estimated 3,600. Despite aggressive anti-poaching efforts, the Western Black Rhino became extinct in the wild in 2011, leaving three other subspecies with critically low numbers. Even rhinos mounted on display in European museums have had their horns poached!

9.3 Cultural and Historical Geographies

Many non-Africans are unaware of the achievements and contributions of the cultures of Sub-Saharan Africa. The African continent was the original home of humankind. Recent DNA studies suggest that the first modern people (*Homo sapiens*)

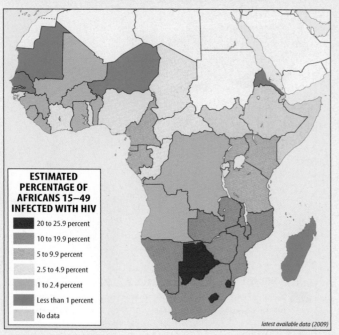

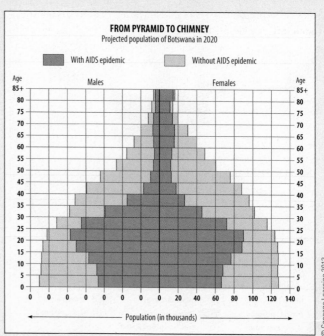

• **Figure 9.A** HIV/AIDS has had a devastating impact in Sub-Saharan Africa and threatens to redraw the demographic profiles of some countries in unprecedented ways.

child health. For sexual transmissions, the main weapons are male circumcision, a vaginal microbicidal gel, and of course, condoms. Needle-exchange programs are most important in preventing transmission among drug users. So far though, the math is heartbreaking. In the best-case scenario, the Global Fund and PEPFAR jointly will allow 10 million people to be put on treatment by 2014. An estimated thirty-six million will be infected by then. For every 100 people put on treatment, 250 more are newly infected. The overwhelming majority of the newly infected will die untreated. AIDS activists are angry about the spending focus of existing programs, saying that children will be saved now only to die later of AIDS.

to inhabit Asia, Europe, and the Americas were descendants of a small group that left Africa via the Isthmus of Suez about 100,000 years ago. After about 5000 B.C.E., indigenous people were responsible for agricultural innovations in four culture hearths: the Ethiopian Plateau, the West African savanna, the West African forest, and the forest-savanna boundary of West Central Africa. Africans in these areas domesticated important crops such as millet, sorghum, yams, cowpeas, okra, watermelons, coffee, and cotton. From Africa, these diffused to populations in other world regions.

49

82

REGIONAL PERSPECTIVE | **The Great Rift Valley**

One of the most spectacular features of Africa's physical geography is the **Great Rift Valley**, a broad, steep-walled trough extending from the Zambezi Valley (on the border between Zimbabwe and Zambia) northward to the Red Sea and the valley of the Jordan River in southwestern Asia (see Figure 9.1 and also Figure 6.9, page 163 and Figure 2.1, page 22). Its relationship to the tectonic movement of crustal plates is still poorly understood. However, most earth scientists believe it marks the boundary of two crustal plates that are rifting, or tearing apart, causing a central block between two parallel fault lines to be displaced downward, creating a linear valley. This movement will eventually cut much of southern and eastern Africa away from the rest of the continent and allow seawater to fill the valley.

The Great Rift Valley has several branches. Lakes, rivers, seas, and gulfs already occupy much of it. It contains most of the larger lakes of Africa, although Lake Victoria, situated in a depression between two of its principal arms, is an exception. Some, like Lake Tanganyika—at 4,823 feet (1,470 m), the world's second-deepest lake, after Russia's Lake Baikal—are extremely deep. Most have no surface outlet. Volcanic activity associated with the Great Rift Valley has created Kilimanjaro, Kirinyaga, and some of the other great African peaks, along with lava flows, hot springs, and other thermal features. Faulting along the Great Rift Valley in Ethiopia, Kenya, and Tanzania has also exposed remains of the earliest known ancestors of *Homo sapiens*.

39

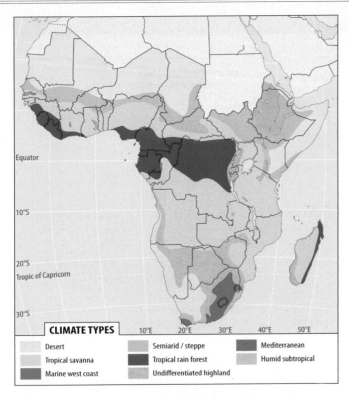

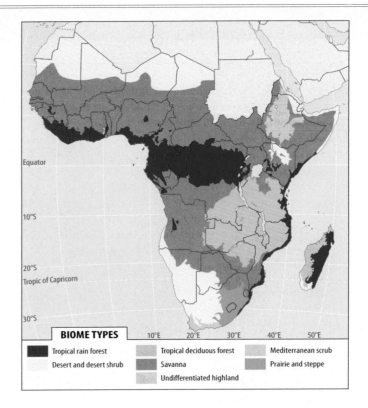

•**Figure 9.6** Climates (left) and biomes (right) of Sub-Saharan Africa.

Sources: (left) Based on Rand McNally's Classroom Atlas, 2003. (right) Based on World Wildlife Fund ecoregions data, 1999.

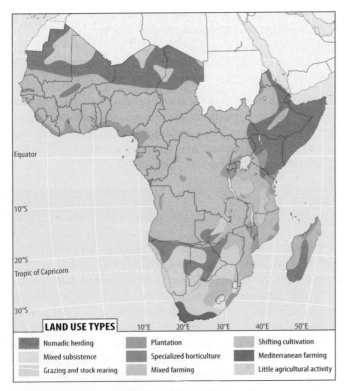

•**Figure 9.7** Land use in Sub-Saharan Africa.

Source: The Oxford School Atlas edited by Patrick Wiegand (OUP, 1997), copyright © Oxford University Press. Used by permission of Oxford University Press.

•**Figure 9.8** Mother and child in Zimbabwe. Women plant and harvest most of Africa's food and care for most of its children.

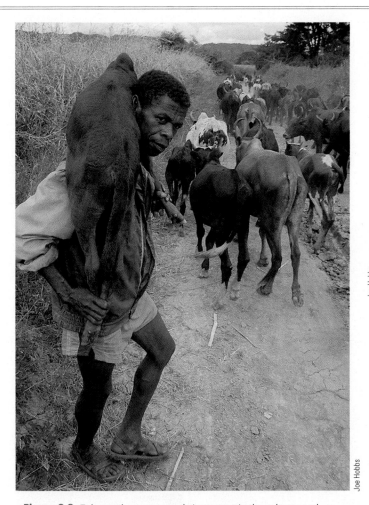

Joe Hobbs

• **Figure 9.9** Zebu cattle are extremely important in the cultures and household economies of rural Madagascar. There are about 11 million cattle in this country of 21 million people.

Joe Hobbs

• **Figure 9.10** Daggers are a nearly universal dress accessory for men in the Arabian Peninsula nation of Yemen. The most prized dagger handles are of rhino horn, a custom that has had a devastating impact on African rhinos thousands of miles away.

Civilizations and empires emerged in Ethiopia, West Africa, West Central Africa, and South Africa. In the first century c.e., a Christian empire based in the Ethiopian city of Axum controlled the ivory trade from Africa to Arabia. Ethiopian tradition holds that a shrine in Axum still contains the biblical Ark of the Covenant and the tablets of the Ten Commandments, which
170 disappeared from the Temple in Jerusalem in 586 B.C.E. Several Islamic empires, including the Ghana, Mali, and Hausa states, emerged in West Africa between the ninth and nineteenth centuries. All of these agriculturally based civilizations controlled major trade routes across the Sahara. They profited from the exchange of slaves, gold, and ostrich feathers for weapons, coins, and cloth from North Africa. Three kingdoms arose between the fourteenth and eighteenth centuries in what are now the southern Democratic Republic of Congo and northern Angola. These included the Kongo kingdom, which had productive agriculture and was the hub of an interregional trade network for food, metals, and salt. In what is now Zimbabwe, the Karanga kingdom of the thirteenth to fifteenth centuries built its capital city at the site known as Great Zimbabwe. Its skilled metalworkers mined and crafted gold, copper, and iron, and merchants traded these metals with faraway India and China.

The Languages of Africa

Recent scientific studies have concluded that humans first began using language in southwestern Africa between 50 and 100 thousand years ago. Today, there is great linguistic diversity in Sub-Saharan Africa (•**Figure 9.11**). Africans speak about 2,000 of the world's 6,000 languages, belonging to four broad language groups.

The **Niger-Congo language family** (sometimes considered a subfamily of the Niger-Khordofanian family) is the largest. It includes the many West African languages and the roughly 400 Bantu subfamily languages that fall into seven branches: **Benue-Congo, Kwa, Atlantic, Mandé, Gur, Adamawan,** and **Khordofanian**. Most of these are spoken south of the equator. The *Bantu* language of the Benue-Congo branch is the most widespread.

The Afro-Asiatic language family includes Semitic languages (such as the *Amharic* language of Ethiopia and *Arabic*) and tongues of the **Cushitic** (*Oromo* and *Somali* of the Horn of Africa, for example) and **Chadic** (especially the *Hausa* of northern Nigeria) branches. People living in the area adjoining the Sahara, from West Africa to the Horn of Africa, speak these languages. Even some of the Niger-Congo languages originating south of the Sahara, such as the *Swahili (Kiswahili)* tongue spoken widely in East Africa, have borrowed much from Arabic and other languages with roots elsewhere. The prominence of Arabic words in Swahili reflects a long history of Arab seafaring along the Indian Ocean coast of Africa; in fact, Swahili means "coastal" in Arabic.

The **Nilo-Saharan language family** of the central Sahel region, the northern region of West Central Africa, and parts of East Africa includes about 100 languages in three branches: *Songhai* (a single language, spoken mainly in Mali) and the **Saharan** and **Chari-Nile** language groups.

The **Khoisan languages** of the **San** and related peoples in the western portion of southern Africa are nearly extinct. The

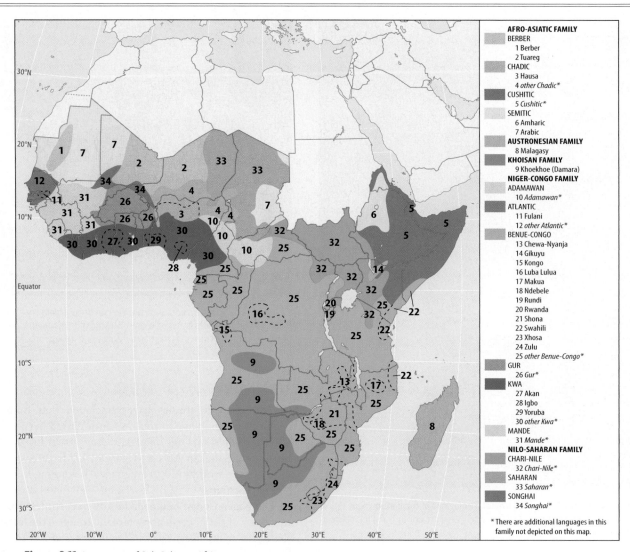

AFRO-ASIATIC FAMILY
BERBER
  1 Berber
  2 Tuareg
CHADIC
  3 Hausa
  4 *other Chadic**
CUSHITIC
  5 *Cushitic**
SEMITIC
  6 Amharic
  7 Arabic
AUSTRONESIAN FAMILY
  8 Malagasy
KHOISAN FAMILY
  9 Khoekhoe (Damara)
NIGER-CONGO FAMILY
ADAMAWAN
  10 *Adamawan**
ATLANTIC
  11 Fulani
  12 *other Atlantic**
BENUE-CONGO
  13 Chewa-Nyanja
  14 Gikuyu
  15 Kongo
  16 Luba Lulua
  17 Makua
  18 Ndebele
  19 Rundi
  20 Rwanda
  21 Shona
  22 Swahili
  23 Xhosa
  24 Zulu
  25 *other Benue-Congo**
GUR
  26 *Gur**
KWA
  27 Akan
  28 Igbo
  29 Yoruba
  30 *other Kwa**
MANDE
  31 *Mande**
NILO-SAHARAN FAMILY
CHARI-NILE
  32 *Chari-Nile**
SAHARAN
  33 *Saharan**
SONGHAI
  34 *Songhai**

* There are additional languages in this
family not depicted on this map.

• **Figure 9.11** Languages of Sub-Saharan Africa.
Source: Adapted with permission from Bernard Comrie, et al., The Atlas of Languages, Revised Edition (New York: Facts On File, 2003).
Reprinted with permission of the Publisher.

largest, *Khoekhoe (Damara)*, is spoken by fewer than 200,000 people in Namibia. Recognizable by their prominent "clicking" sounds, these may be humankind's oldest languages.

Quite distinct from those four groups, the people of Madagascar speak a language without African roots. This is *Malagasy*, an Austronesian tongue that originated in Southeast Asia.

Africa's list of lingua francas—the languages most likely to be recognized on the continent and those that Africans can use to speak to the wider world—is provided by the six official languages of the African Union, the continent's chief supranational organization. They are English, French, Portuguese, Spanish, Swahili, and Arabic. Use of English is growing most rapidly, even in the formerly French-dominated "Francophone" countries. Countries that formerly were British colonies have a distinct advantage in the world of outsourced services: South Africa, Kenya, and Ghana, for example, have already opened call centers that they hope will rival those of India. Some Francophone countries also have call centers catering to French industries and clients.

The dominance of foreign languages in computers and other digital technologies could endanger African languages, many

of which are already in peril because they are spoken by so few people. As elsewhere, many languages in Africa are threatened by language death. Many linguists, however, see the computer as a tool for *saving* the roughly 3,000 endangered languages of the world. They argue that the computer will make it much easier for people to learn and preserve their traditional tongues. Australian linguist James McElvenny uses software to revitalize vanishing languages (look up the Enduring Voices Project on the web). He gives young language learners portable electronic reference tools that give the definitions and sounds of words that are otherwise no longer spoken. Dr. Tucker Childs, a linguist working in the West African nation of Sierra Leone, is using a geographic information system (GIS) to plot locational variations of Kim and other endangered languages. With these findings, linguists can decide where to focus their efforts.

Africa's Belief Systems

The religious landscape of Africa is complex and fluid. Spiritualism is extremely strong, but spiritual affiliations and practices are more interwoven and flexible than in most other

INSIGHTS | **Africa's Greatest Conservationist**

Not much larger than the common housefly, the tsetse fly of Sub-Saharan Africa packs a wallop. This insect carries two diseases, both known as **trypanosomiasis**, which are extremely debilitating to people and their domesticated animals. People contract **sleeping sickness** from the fly's bite, and cattle contract nagana. Since the 1950s, widespread efforts have been made to eradicate tsetse flies so that people can grow crops and herd animals in currently fly-infested wilderness areas (•**Figure 9.B**). Where the efforts have been successful, people have cleared, cultivated, and put livestock on the land. The results are mixed. Although people have been able to feed growing populations in the process of opening up these lands, they have also eliminated important wild resources and in many cases caused erosion, desertification, and salinization of the land. Where the tsetse fly has been eliminated, so has the wilderness. The diminutive tsetse fly thus may be characterized as a **keystone species**—one that affects many other organisms in an ecosystem. The loss of a

keystone species—in this case, a fly that keeps out humans and cattle—can have a series of destructive impacts throughout the ecosystem.

For its role in maintaining wilderness in Sub-Saharan Africa, some wildlife experts call the tsetse fly "Africa's greatest conservationist."

Joe Hobbs

• **Figure 9.B** Strenuous and successful efforts to eradicate tsetse flies have opened vast new areas of Africa to human use. This is a tsetse fly trap with two components: a liquid chemical called "simulated cow's breath" that attracts the fly and a pesticide-soaked tarp that kills it.

---

world regions. It is not uncommon for family members to follow different faiths or for an individual to change his or her religious beliefs and practices in the course of a lifetime.

Broadly, however, some dominant patterns of religious geography can be recognized (•**Figure 9.12**). Islam is the

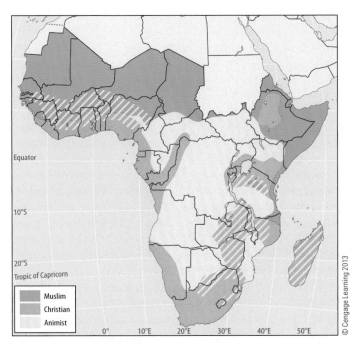

• **Figure 9.12** Religions of Sub-Saharan Africa.

dominant religion in North Africa and the countries of the Sahel on the southern fringe of the Sahara. The **Ethiopian Orthodox Church**, closely related to the Coptic Christian faith of Egypt, makes Ethiopia an exception to the otherwise Islamic Horn of Africa region (•**Figure 9.13**). Islam is also the prevailing religion of the East African coast, where Arab traders introduced the faith. Muslims are a majority or strong minority in rural northern Nigeria and Tanzania, and there are minority Muslim populations in cities and towns across the continent. Christians are a majority in southern Nigeria, Uganda, Lesotho, and parts of South Africa. Both Christianity and Islam are strongest in the cities, whereas traditional religions prevail or overlap with these monotheistic faiths in rural areas.

Christians and Muslims have mounted efforts to dispose of "heretical" notions in African traditional belief systems, but these concepts are very resilient. Even where the large monotheistic faiths nominally prevail and their followers are devout, a strong substrate of indigenous beliefs has persisted and usually comfortably merged with the "official" faith—a syncretism also present in other cultures of the world. One example is the mingling of indigenous *vodun* ("voodoo") beliefs with Christianity and Islam in West Africa. 228, 365

Indigenous African religions have many animistic elements, emphasizing that spiritual forces are manifested everywhere in the environment. Many spirits are tied to particular places in the landscape. Such beliefs contrast with the introduced Christian and Muslim faiths, which tend to see

Joe Hobbs

• **Figure 9.13** The Ethiopian Church has been the dominant religion in Ethiopia since the 4th Century. This priest's flock worships at the 12th Century Church of St. George in Lalibela. When my Mom and I visited Ethiopia in 1983 we were apparently the only tourists in the country! Ethiopia had a Marxist government that discouraged tourism and outlawed religion. The country's devout Christians and Muslims defied the ban.

nature as separate from God and people as apart from and superior to nature.[3] Across the spectrum of African cultures, the use of mediums to contact spirits of deceased ancestors, creator beings, or earth genies that can intercede on one's behalf is widespread (see Perspectives from the Field, next page). Reverence for ancestors is strong, reflected also in great respect for the elderly who are living.

Outsiders have tried, seldom successfully, to give accurate names to indigenous African belief systems, practices, and personnel: ancestor worship, animism, *force vitale* ("life force"), living dead, and witch doctor among them. Although it would be gratifying to have precise names and generalizations for African beliefs, they are unique and can best be understood by careful reading, observation, and discussion on a case-by-case basis.

The Origins and Impacts of Slavery

Until about 1,000 years ago, the cultures of Africa south of the Saharan desert barrier remained largely unknown to the peoples north of the desert. Egyptians, Romans, and Arabs had contact with the northern fringes of this region, and some trade filtered across the Sahara, but to most outsiders, Africa was the "Dark Continent," a self-contained, tribalized land of mystery. Even at the opening of the twentieth century, vast areas of interior tropical Africa were still little known to Westerners.

The tragic impetus for growing contact between Africa and the wider world was slavery (•Figure 9.14). Over a period of 12 centuries, as many as 25 million people from Sub-Saharan Africa were forced to become slaves, exported as merchandise

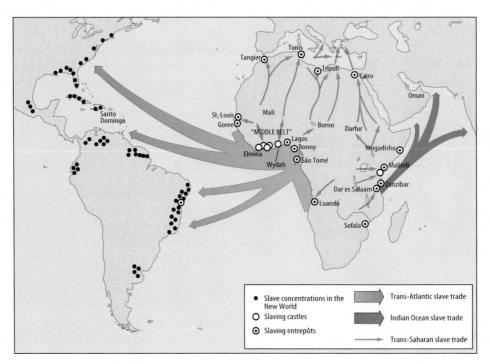

• **Figure 9.14** Slave export trade routes from Sub-Saharan Africa.
Source: Based on Stock, Robert, Africa South of the Sahara.

In the austral autumn of 2000, I went to Madagascar to investigate the relationship between people and caves on the island. It was a long journey: 42 hours from the time I left home in Columbia, Missouri, until I arrived in Antananarivo (Tana), Madagascar's capital. During a three-hour delay of the Air Madagascar flight from Paris, excess weight, including my baggage, was unloaded from the aircraft. Twelve hours after takeoff, I arrived exhausted and without provisions (except for 30 pounds of camera and video gear in my hand luggage) but very excited about being in this exceptional country.

Fortunately, once in Tana, I was in the capable hands of a dear friend and colleague, Steve Goodman. A zoologist based at Chicago's Field Museum, Steve is a pioneer in the natural history study of this extraordinary island, where nearly 90 percent of all plants and animals are endemic. Steve has lived here since 1989 and has virtually become Malagasy. His wife, Asmina, is of the island's coastal Sakalava culture. Steve, Asmina, and their son, Hisham, reside with an extended family network, and as soon as I arrived from the airport, I was immersed in special festivities surrounding the impending wedding of a family member. That afternoon, these included the *fumba fumba,* a ceremony in which the bride's and the groom's families seek approvals from their respective ancestors (*razana*) for the wedding. It is impossible to overstate the significance of the spirits of one's ancestors in this culture. The Malagasy peoples believe them to influence or control virtually every aspect of daily life.

When my baggage arrived two days later, Steve and I set out across the High Plateau, bound for the caves of northwestern Madagascar's region of Majunga. Not 3 hours into what should have been a 15-hour journey, Steve's vehicle developed a fuel line problem and began to hemorrhage gasoline. There was no choice but to return to Tana. At least this setback allowed me the opportunity to observe in more detail the environmental holocaust wrought by 2,000 years of human activity on the island (see Figure 9.31). Relentless clearing of forest for rice cultivation and cattle pasture has created a virtual moonscape. The rivers run blood-red with what is left of the island's precious topsoil.

Steve was drawn away by other obligations, but I had the good fortune to be joined by his sister-in-law, Patty Vavizara, who would be my French and Malagasy interpreter and caving companion for the next two weeks. We flew to Majunga, hired a four-wheel-drive vehicle, and made our way toward the vast and beautifully decorated caves of Anjohibe and Anjohikely.

My favorite cave is Andoboara, which we reached late at night. We had not set out until dusk, and our two local guides had difficulty getting oriented in the darkness. There was no road to follow, and their views were obscured by grasses nearly twice the height of the vehicle. But they finally located Andoboara Cave, small and uninteresting but for its human and spiritual associations and therefore of huge interest to me.

The people of the nearby village of Ambalakedi consider this cave sacred because on three separate occasions, most recently in 1998, grief-stricken parents whose children had wandered into the forest had recovered them alive here. The children had been abducted by a sometimes-malevolent spirit called *kalanoro*, who sought to rebuke the parents for not taking proper care of their children. The parents sought guidance from another spirit called *tromba*, who possessed a fellow villager and recited a laundry list of demands made by the *kalanoro* that would have to be met for the child to be found. Once the parents fulfilled their obligations by, for example, leaving honey and sacrificing zebu cattle in the place designated by the *tromba*, the *kalanoro* revealed via the *tromba* the lost child's location, and in each case the joyous reunion took place in the Andoboara Cave. As I was to learn in the coming weeks, there are countless caves in Madagascar bearing rich accounts of this kind.

Near the end of my stay, Patty and I took a five-day walk through the Ankarana Special Reserve near the island's northern tip. The park is dominated by a long wall of Jurassic-era limestone, eroded by wind and water into razor-sharp formations locally called *tsingy-tsingy* (meaning "ouch ouch," as in how your feet feel when you tread the stuff). The massif is perforated by scores of caves, some so enormous that even a powerful Petzl headlamp like mine could not illuminate their ceilings. Others are less grand in scale but graced with lovely calcite formations.

The Ankarana's sinkholes and the perimeter of the limestone massif contain protected remnants of dry tropical forest that are home to abundant wildlife. Each night, members of the primitive primate family of lemurs invaded our camp in search of food, calling so loudly that sleep came only through sheer exhaustion from the day's walking and caving. As a reptile lover, I was in paradise, indulging in the capture, photography, and release of boas and other snakes, riverine turtles, and—my favorites—the chameleons.

But the caves and their human connotations were of greatest interest. At one night's campfire, a man from a village just outside the park told me how his ancestors, members of the royal family of the Antakarana tribe, sought refuge in the caves for nearly three years during the 1800s. They were escaping would-be slavery at the hands of the Merina tribe of highland Madagascar, whose goal was to unite (or pacify and enslave) the island's 17 ethnic groups. The Antakarana had taken advantage of the Merina's hopeless disorientation in the underground labyrinth of the Ankarana massif and survived in its caves while hunting, foraging, and growing some crops in the hidden karst canyons and sinkholes. The first four Antakarana kings were buried in caves here and are still visited on a ceremonial pilgrimage once every five years by the current king, Issa, and his subjects. I met with him to seek his permission to enter the royal cave tombs, but he said it was *fady* (taboo) even for him to do so, except on the specified occasion; he would welcome me to take part in the next event.

One particular day in the Ankarana was among the most extraordinary in my life. With an excellent guide named Aurelian Toly, we walked through the massive Andrafiabe Cave for several hundred meters before reaching a giant sunlit chasm containing what botanists call a "sunken forest," essentially a tropical paradise surrounded by karst desert wilderness. Lemurs howled and parrots cried in this lost world. I thought, "Spielberg couldn't top this!" Crossing the forested ravine, we entered Cathedral Cave, whose entrance is littered with the hearths and pots used by the Antakarana in their time of refuge here. Well into the cave, we crossed a 600-foot-long (200 m) hill of guano built by a massive roost of insectivorous *Hipposideros* bats. Fresh deposits fell like raindrops on our helmets. Our noses were choked by the stench of ammonia and the dust of the

(continued)

guano trodden underfoot. Our headlights shined upon seething masses of cockroaches and centipedes feasting on the treasures dropped by the bats. We couldn't help but crunch this fauna as we walked, and with the shrieking of the annoyed bats, I thought again of a Spielberg movie set.

My last excursion into underground Madagascar was in Crocodile Cave. It is appropriately named, for running through it is a river in which crocodiles seek habitat while the world outside is parched in the long dry season of April–October. On the cave's sand banks, we saw abundant crocodile tracks, some fresh, and we crossed the cave stream with trepidation. Instead of finding the great reptiles (some up to 19.7 ft [6 m] long have been seen here), about 3,300 feet (c. 1,000 m) into the cave we were astonished to come upon a dugout canoe beached on the bank. Next to the pirogue were three poles, each bearing on either end a strand of eight to ten cormorants and other waterfowl tied by their feet. Some had already died and others struggled for life. In the next stretch of cave passage, our headlamps revealed the three poachers responsible, and a confrontation ensued. The two park guides in my party of seven cried out, "We are from

the national park service. Come forward. Drop your weapons because the warden has a gun and will use it against you!" I was surprised to learn that I was the warden, and the GPS pouch on my belt bore the gun. The ruse was enough to convince the poachers, who now pleaded for leniency. They said they had known hunting was forbidden in the park, but the need to feed their families had driven them to take this long and strange canoe voyage into the cave.

I still did not understand. Why were there waterfowl in a cave? The park rangers demanded that the hunters return immediately to release the birds where they had taken them, and marching behind the poachers 1.25 miles (2 km) further into the cave, I had my answer (•**Figure 9.C**). The cave opened up into another lost world, a great circular cavity in the Ankarana massif, this one filled not with forest but with marshes and lakes teeming with wildfowl. By this time, few of the birds were living, but the hunters did release them. The guides asked me (still feigning as warden) what to do with the criminals and their dead birds. "Would they do this again?" I asked. There was some discussion. Then the three poachers made what seemed a solemn vow that they would

never again take wildlife from the park, lest the spirits of the ancestors of the Antakarana seek vengeance upon them. It was a particularly appropriate Malagasy way of dealing with a very understandable problem: the need of poor men to feed their families. For the geographer, the entire incident provided an especially fascinating insight into the relationship between people and caves in an extraordinary country.

• **Figure 9.C** The poachers (with birds on poles) after being apprehended by the Ankarana park staff, here escorting them out of the Crocodile Cave.

from their homelands. The trade began in the seventh century, with Arab merchants using trans-Saharan camel caravan routes to exchange guns, books, textiles, and beads from North Africa for slaves, gold, and ivory from Sub-Saharan Africa. As many as two-thirds of the estimated 9.5 million slaves exported between the years 650 and 1900 along this route were young women who became concubines and household servants in North Africa and Turkey. Male slaves usually became soldiers or court attendants (some of whom eventually assumed important political offices). From the eighth to nineteenth centuries, about 5 million more slaves were exported from East Africa to Arabia, Oman, Persia (modern Iran), India, and China. Again, most were women who became concubines and servants.

The cruel and lucrative traffic in slaves provided the initial motivation for European commerce along the African coasts, inaugurating the long era of European exploitation of Africa for profit and political advantage. The European-controlled slave trade was the largest by far, particularly with the development of plantations and mines in the New World. Between

the sixteenth and nineteenth centuries, the capture, transport, and sale of slaves was the exclusive preoccupation of trade between the European world and West Africa. Portuguese and Spaniards began the trade in the fifteenth century, and within 100 years English, Danish, Dutch, Swedish, and French slavers were participating. The peak of the transatlantic slave trade was between 1700 and 1870, when about 80 percent of an estimated total of 10 million slaves made the crossing. In escape attempts, in transit, and in the famines and epidemics that followed slave raids in Africa, probably more than 10 million others died.

Slaves were a prized commodity in the **triangular trade** linking West Africa with Europe and the Americas. European ships carried guns, ammunition, rum, and manufactured goods to West Africa and exchanged them there for slaves. They then transported the slaves to the Americas, exchanging them for gold, silver, tobacco, sugar, cotton, rum, and tropical hardwoods to be carried back to Europe. As "raw material" and as the labor working the mines and plantations of Latin America, the West Indies, and Anglo America, slaves generated much of

the wealth that made Europe prosperous and helped spark the
Industrial Revolution.

45,
82

Although Europeans carried out the triangular trade, their physical presence was limited to coastal shipping points. Africans were the intermediaries who actually raided inland communities to capture the slaves and assemble them at the coast for transit shipment. West African kingdoms initially acquired their own slaves in the course of waging local wars. As the demand for slaves grew, these kingdoms increasingly went to war for the sole purpose of capturing people for the trade. As the exports grew, so did the practice of Africans keeping African slaves. Even after Britain (in 1807) and the other European countries abolished slavery (1870), it continued to flourish in Africa. By the end of the century, slaves made up half the populations of many African states.

Slavery has not yet died out in the region (and indeed, a modern-day form of slavery exists in the form of human trafficking elsewhere in the world; see pages 145 and 395). In Mauritania, some Moors still enslave blacks, although the national government has outlawed this practice several times. Enslavement of children persists in West Africa. The typical pattern is that impoverished parents in one of the region's poorer countries, such as Benin, are approached by an intermediary who promises to take a child from their care and see that the boy or girl is properly educated and employed. This involves a fee (as little as $14) that, unbeknown to the parents, is a sale into slavery. The intermediary sells the child (for $250 to $400, on average) to a trafficker who sees that the child is transported to one of the region's richer countries, such as Gabon. There, the child ends up working without wages and under the threat of violence as a domestic servant, plantation worker (especially on cacao and cotton plantations), or prostitute. Benin is the leading slave supplier and trafficking center; Gabon, Ivory Coast, Cameroon, and Nigeria are the main buyers of slaves. Slavery also exists in Sudan. In the 1990s, well-intentioned church groups and other organizations in the West began paying hard cash to buy freedom for slaves in Sudan. Although many real slaves were freed, Sudanese profiteers also moved in with "counterfeit slaves," ordinary people hired to act as slaves until their "rescue" had been paid for.

The Impacts of Colonialism

In the sixteenth century, European colonialism began to overshadow and inhibit the growth of indigenous African civilizations. This is a diverse region, but one thing its peoples have in common is a sense that in the past they were humiliated and oppressed by outsiders.[4]

Portugal was the earliest colonial power to build an African empire. The epic voyage of Vasco da Gama to India in 1497–1499 via the Cape of Good Hope was the culmination of several decades of Portuguese exploration along Africa's western coasts. During the sixteenth century, Portugal controlled an extensive series of strong points and trading stations along both the Atlantic and Indian Ocean coasts of the continent. European penetration of the African interior began in 1850 with a series of journeys of exploration. Missionaries like David Livingstone, as well as traders, government officials, and adventurers and scientific explorers such as James Bruce, Richard Burton, and John Speke, undertook these expeditions. By 1881, when Africans still ruled about 90 percent of the region, foreign exploits had revealed the main outlines of inner African geography, and the European powers began to grab colonial territory in the interior.

Much of the "scramble for Africa" took place at the Conference of Berlin in 1884 and 1885, when the French, British, Germans, Belgians, Portuguese, Italians, and Spanish established their respective spheres of influence in the region. By 1900, only Ethiopia and Liberia had not been colonized. For more than half a century, Sub-Saharan Africa was a patchwork of European colonies (•Figure 9.15), and Europeans in these possessions were a privileged social and economic class.

At the outbreak of World War II in 1939, only three countries—South Africa, Egypt, and Liberia—were independent. The United Kingdom, France, Belgium, Italy, Portugal, and Spain controlled the rest. But after the war, mainly in the 1960s and 1970s, a sustained drive for independence was mounted. Ceasing to be a colonial region, Africa emerged with more than a quarter of the world's independent countries. This was a peaceful process in most instances, but bloodshed accompanied or followed independence in several countries.

Dependency theorists often point to Africa as a prime example of how colonialism created lasting disadvantages for the colonized. European colonization produced or perpetuated many negative attributes of underdevelopment. These included the marginalization of subsistence farmers, notably those who colonial authorities, intent on cash crop production, displaced from quality soils to inferior land. In addition, European use of indigenous labor to build railways and roads often took a high toll in human lives and disrupted countless families. Furthermore, the colonizers often corrupted traditional systems of political organization to suit their needs, sowing seeds of dissent and interethnic conflict.

44

47

338

The European colonial enterprise did have some positive impacts. The colonies, and the independent nations that succeeded them, were the beneficiaries of new cities and the transport links built with forced or cheap African labor; new medical and educational facilities (often developed through Christian missions); new crops and better agricultural techniques; employment and income provided by new mines and modern industries; new governmental institutions; and government-made maps useful for administration and planning. Such innovations were very helpful, but they were distributed unequally from one colony to another and were inadequate for the needs of modern societies when independence came.

One of the persistent problems in Africa's political geography is that many modern national boundaries do not correspond

• **Figure 9.15** Colonial rule in 1914. Germany lost its colonies to the British, French, and Belgians after World War I.

Source: Based on John Haywood, ed., Atlas of World History (Barnes and Noble Books, 1997).

with indigenous political or ethnic boundaries. In most cases, this is another legacy of colonialism: British, French, and other occupying powers created arbitrary administrative units that were transformed into countries as the colonial powers withdrew. Nigeria is a good example of an ethnically complex, unnaturally assembled nation. On the political map, it appears as an integral unit, but its boundaries have no logical basis in physical or cultural geography. Some Nigerians still refer to the **"mistake of 1914,"** when British colonial cartographers created the country despite its ethnic rifts (see Figure 9.22). The greatest divide is between Muslim north and Christian south, but there are at least 250 ethnic groups within the country. The British colonizers invested more in education and economic development in the south and built army ranks among northerners, whom they thought made better fighters; so the subsequent pattern is that the northern military leaders have tried to blunt the economic clout of southerners. Colonial administration and boundary drawing thus sowed seeds of modern conflict in Nigeria, a problem described in more detail in Section 9.6.

Although formal political colonialism has vanished, most countries still have important links with the colonial powers that formerly controlled them, and many foreign corporations that operated in colonial days still maintain a significant presence. France has a long history of post-independence intervention in the political and military affairs of its former African colonies. France is the only ex-colonial power to keep troops in Africa (with the highest numbers in Djibouti, Ivory Coast, Gabon, and Chad). France took steps to ensure that most of its former colonies trade almost exclusively with France and in turn supported national currencies with the French treasury. But France found this paternalistic approach to be extremely expensive and has been reducing its military presence and other costly assistance to its African clients.

9.4 Economic Geography

Sub-Saharan Africa ranks at the very bottom of every statistical indicator of global quality of life. Great poverty is characteristic of the region; all but two of the world's 20 poorest countries are there (see Table 9.1). About half of this region's people subsist on less than $1.25 per day, according to the World Bank. This is the only world region that has grown poorer in the last 25 years. It is the most difficult world region in which to conduct business, in part because of corruption and bureaucratic obstacles. A recent study of the world's civil wars since 1960 determined that there were three important risk factors for conflict: poverty, low economic growth, and high dependence on natural resources. Sub-Saharan Africa is richly endowed with these ingredients of war.

These are familiar statistics and connotations of Africa. But there is no denying that the region is changing for the better. Despite the global economic downturn, the region's economic growth of 5 percent annually between 2000 and 2010 was twice its pace in the previous two decades. Private companies are creating jobs. There is an emerging middle class; in other words, the number of people able to spend money on things other than life's bare necessities is growing. Roads, bridges, dams, and other infrastructure are improving, in many cases through Chinese investment. Governance is improving too, especially as corruption is being tackled. The number of significant conflicts was cut in half between the 1990s and 2000s. This attractive mix of factors has led to a surge in foreign direct investment. Some financial authorities believe that Sub-Saharan Africa as a group should be just as attractive for investment as the BRIC nations (Brazil, Russia, India, and China). Some even insist on expanding the acronym to BRICS, with the "S" standing for South Africa.

Let us take a closer look at the region's challenges and the efforts and opportunities that may overcome them.

Commodities: Boom!

Africa's place in the commercial world is mainly that of a producer of primary products, especially cash crops and raw materials (particularly minerals), for sale outside the region. In most nations, one or two products supply more than 40 percent of all exports—for example, in Angola oil and diamonds make up 99 percent of the country's exports, and in Kenya coffee and tea are overwhelmingly important (•**Figure 9.16**). Such countries are vulnerable to international oversupply of an export on which it is vitally dependent; oversupplies will cause prices to crash. But when demand for these exports is high, the money pours in.   45

The recent upward prices for commodities, driven especially by demand in China, India, and the older industrialized economies, have generally been a blessing for this resource-rich region. The challenge is to keep the blessing from becoming a mixed blessing, or worse yet a curse (see A Closer Look, next page).

Subsistence Agriculture

Famine is an all-too frequent visitor to Sub-Saharan Africa. Per capita food output in most of Sub-Saharan Africa has declined or remained flat since independence. The average   331, 332

• **Figure 9.16** Kenya's economy is highly dependent on the export of coffee. Cash crops and other raw materials are typical exports of Sub-Saharan Africa.

A **closer** LOOK

The Resource Curse

Many countries in Africa have been described as suffering from the **resource curse**, also known as the "paradox of plenty." This is a paradox in that a country with a great abundance of a valuable natural resource often experiences lower economic growth than countries without such abundance. Nigeria provides a good example. Three hundred billion dollars in oil revenues have poured into the country in the past three decades, yet Nigerians have become poorer, with 70 percent of Nigerians living below the poverty line (see page 336 for more discussion of Nigeria's oil problems). Venezuela's oil minister in the 1960s was quoted famously as saying that oil was not black gold; it was "the devil's excrement."*

What causes the resource curse? There are economic reasons: it is difficult to plan when commodity prices (like Nigeria's oil) swing wildly up and down; oil prices, for example, can reach lofty heights and then plummet. These boom and bust cycles lead to risk-taking, debt, and overinvestment. With busts, there are budget crises that hurt the poor. Oil-fueled growth does not always create many jobs; although oil can account for as much as 80 percent of a country's revenues, it often employs fewer than 10 percent of the workforce. Against this backdrop of economic inequality, the revenue earned from oil crowds out other sectors, such as agriculture and manufacturing.

The resource curse is sometimes called the "Dutch Disease," because in the 1950s natural gas was discovered in the Netherlands' waters. As large foreign exchange inflows poured in, local costs went up, so imports undercut domestic manufacturing and agriculture. All domestic industries except the fossil fuel industry went into a tailspin.

There are also political reasons for and consequences of the resource curse. A glut of money from the resource provides incentives for corruption and repression of dissent. Officials and company executives get rich, and the poor get poorer. Oil-producing LDCs often spend as much as 10 times what poorer LDCs do on their militaries, and are more likely to get involved in conflicts.

*Quoted in Moises Naim, "Oil Can Be a Curse on Poor Nations." *Financial Times*, August 19, 2009, p. 7.

African consumes 10 percent fewer calories than two decades ago, and malnutrition afflicts almost half the region's children. Rapid population growth and drought are partly responsible for this trend. To support growing populations, farmers across Sub-Saharan Africa have shortened fallow periods and pressed their lands to yield more crops. The result has been an unprecedented degradation of the resource. Fully 75 percent of the region's farmland is severely low in the nutrients needed to grow crops, up from 40 percent a decade earlier. Studies suggest that at current rates, crop yields will fall as much as 30 percent by 2022. Most African farmers cannot afford fertilizers and are not familiar with the soil conservation techniques that would help reverse this ominous trend.

Over half of Africa's peoples practice subsistence agriculture (farming their own food but producing little surplus for sale) and pastoralism. Women do a large share of the farmwork—they produce 80 to 90 percent of Africa's food—in addition to household chores and the bearing and nurturing of children (**Figure 9.8**). Mechanization is rare, fertilizers are expensive, and so crop yields are low. In the steppe of the northern Sahel, both rainfall and cultivation are scarce. The more dependable rainfall of the southern Sahel creates a major area of rain-fed cropping, with unirrigated millet, sorghum, corn (maize), and peanuts the major subsistence crops. In the tropical savannas south of the equator, corn is a major subsistence crop in most areas, with manioc (cassava) and millet also widely grown. Corn, manioc, bananas, and yams are the major food crops of the rain forest areas.

Many peoples, particularly in the vast tropical grasslands both north and south of the equator, are pastoral. Herding of sheep and hardy breeds of cattle is especially important in the Sahel. As happened in Sudan's tragedy of Darfur, an increasing problem is that farmers often drive pastoralists from traditional grazing lands. Confined to smaller areas in which to browse and graze, the nomads' cattle, sheep, and goats often overgraze vegetation and compact the soil. Access to drinking water for livestock and people is also a major problem.

Most Africans who live by tilling the soil also keep some animals, even if only goats and poultry. Among African peoples such as the Maasai of Kenya and Tanzania and the Tutsi (Watusi) of Rwanda and Burundi, livestock not only contribute to the daily diet but are also an indispensable part of customary social, cultural, and economic arrangements. Cattle are particularly important, with sheep and goats playing a smaller role. Traditional Maasai pastoralists are probably the most famous African example of close dependence on cattle. They milk and carefully bleed the animals for each day's food and tend them with great care. The Maasai give a name to each animal, and herds play a central role in the main Maasai social and economic events through the year.

Madagascar is a good example of an African country in which cattle represent status, wealth, and cultural identity. In order to enhance their social standing, Malagasy livestock owners tend to want larger numbers of animals even more than they want better quality animals (**Figure 9.9**). A Malagasy family practicing the ritual commemoration of deceased ancestors sacrifices a large number of zebu cows for fellow villagers, and similar feasts accompany other important ritual dates. Due to such demands, the population of zebu cattle on the island is about 11 million. Their forage needs have grave consequences for Madagascar's rain forests and other wild habitats. People clear the forests and repeatedly set fire to the cleared lands to provide a flush of green pasture for their livestock, causing a rapid retreat of the island's natural vegetation.

Commercial Agriculture and Marginalization

There is enormous potential to improve African agriculture. The region is a net food importer, yet 60 percent of the world's uncultivated arable land is in sub-Saharan Africa. Much of the land is inherently difficult to farm, but cultivation methods, roads, food storage, and other infrastructure can be improved.

Who will benefit from these improvements? The poor farmer, the wealthy landowners, or both?

Food shortages often result when governments promote the cultivation of cash crop commodities for export, instead of subsistence food crops for local people. Coffee, cacao, cotton, peanuts, and oil palm products in particular provide a means of gaining foreign exchange with which to buy foreign technology, industrial equipment, military armaments, and consumption items for the elite. Secondary cash crops include sisal (grown for its fibers; see Figure 10.22), pyrethrum (used in insecticides), tea, tobacco, rubber, pineapples, bananas, cloves, vanilla, cane sugar, and cashew nuts.

The expansion of commercial agriculture requires more arable land, and it is fascinating and disturbing to see where this land is coming from. There is a new and distressing trend for small farmers whose land is owned by the government. In a classic example of marginalization, small farmers are being driven off their land as it is sold to investors from China, South Korea, Libya, Saudi Arabia, and other countries. In some cases, there are well-established villages in what was sold to the investors as "vacant" land. The process is creating a new source of landless, unemployed poor in African cities. None of the income from rice and sugar cane and other products grown on these lands goes to local people. Sometimes the people push back.

When Madagascar's government was about to sell about half of the country's arable land to a South Korean firm, opposition to Madagascar's president swelled and he was voted out of office.

Although cash crops are important sources of revenue in these poor countries, excessive dependence on them, as with other commodities, can be harmful to a country's economy. The income they bring often goes almost exclusively to already prosperous farmers and corporations. The prices they fetch are vulnerable to sudden losses amid changing world market conditions. The plants are susceptible to drought and disease. Finally, they are often grown instead of food crops, and cash crops do not feed hungry people.

Mineral Resources

Mining and mineral exports have had a strong impact on the physical and social geographies of Sub-Saharan Africa. This is a well-endowed region with, for example, at least 10 percent of the world's petroleum reserves, 60 percent of its diamonds, and a third of its cobalt (used in the manufacture of glass and ceramics). The three primary mineral source areas are South Africa and Namibia; the Democratic Republic of Congo–Zambia–Zimbabwe region; and West Africa, especially the areas near the Atlantic Ocean (•Figure 9.17). Notable

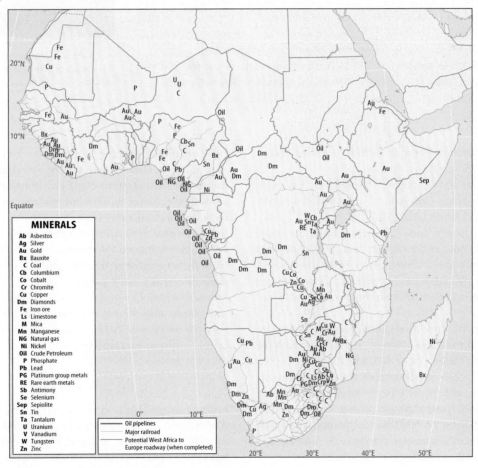

• **Figure 9.17** Minerals, oil pipelines, and transportation links in Sub-Saharan Africa.
Source: Adapted from country mineral maps from USGS.

"Diamonds are a girl's best friend," sang Marilyn Monroe. *Diamonds Are Forever*, proclaimed the title of a James Bond Film. Another film did not have such an endearing name: *Blood Diamond*. This film tells part of a sordid story about diamonds: as recently as 2004, as much as 15 percent of the world's annual production of rough diamonds was made up of **dirty diamonds** (also called **conflict diamonds** or **blood diamonds**), defined by the United Nations as "rough diamond used by rebel movements or their allies to finance conflict aimed at undermining legitimate governments." Gems of such questionable origin financed at least three African wars. But diplomatic, nongovernmental, and business efforts have largely stemmed the tide of Africa's dirty diamonds.

In 2002, following four years of negotiations, 45 countries endorsed a U.N.-backed certification plan, called the **Kimberley Process**, designed to ensure that only legally mined rough diamonds, untainted by violence, reach the market. Rough diamonds must be sent in tamper-proof containers with a certificate guaranteeing their origin and contents. The importing countries (the biggest of which are China and India, which have replaced Belgium and Israel as the world's leading diamond

cutters and polishers) must certify that the shipments have arrived unopened and reject any shipments that do not meet the requirements. Only countries that subscribe to the Kimberley Process are allowed to trade in rough diamonds.

Corporate interests and consumer ethics have cleaned up the diamond trade. Diamond certification has been spearheaded by the world's leading diamond company (with about two-thirds of the market), the South African-based multinational named De Beers. De Beers was embarrassed by a report that it had bought $14 million worth of diamonds from Angolan rebels in a single year. Perceiving that a public relations debacle could lead to a business disaster, as happened with an organized boycott against fur products in the 1980s, De Beers seized the initiative. In the name of Africa's welfare, but certainly also as a means of increasing demand and profit, De Beers introduced **branded diamonds**, certified as coming from nonconflict areas. Other diamond companies followed suit, especially in an effort to please Americans—who buy half the world's finished diamonds (•**Figure 9.D**).

Diamonds are not the only potentially "dirty" African mining products. Recently passed

American legislation requires that products made of tin, tungsten, tantalum, and gold, sold in the United States, be certified as free of conflict in the eastern Democratic Republic of the Congo (the labels read "DRC Conflict Free"). This requirement affects sales of airplanes, computers, cell phones, television sets, and other goods.

• **Figure 9.D** Many consumers are unaware that this symbol of love and beauty may have had a sinister origin in Africa.

mineral exports from these regions include precious metals and precious stones (particularly diamonds; see Problem Landscape, this page), iron alloys, copper, phosphate, uranium, petroleum, and high-grade iron ore, all destined principally for Europe, the United States, and China.

Some of Africa's recent and current conflicts, in many cases fought by child soldiers, have not been about ideology or ethnicity but simply about control over resources such as diamonds, oil, and other minerals. It is therefore of growing interest in the consuming countries that these commodities not be tainted by violence.

Large multinational corporations, financed initially by investors in Europe and the United States and most recently by China, do most of the mining in Africa. Mining has attracted far more investment capital to Africa than any other economic activity. Money is invested directly in the mines, and many of the transportation lines, port facilities, power stations, housing and commercial areas, manufacturing plants, and other elements in the continent's infrastructure have been developed primarily to serve the needs of the mining industry.

Great numbers of workers in the mines are temporary migrants from rural areas, often hundreds of miles away. The

recruitment of migrant workers has had important cultural and public health effects. Millions of Africans who work for mining companies have come into contact with Western ideas as well as those of other ethnic groups and have carried these back to their villages. Unfortunately, migrant labor has also helped spread HIV, the virus that causes AIDS. Most miners are single or married men who spend extended periods away from their wives or girlfriends, and it is common that they contract the virus from prostitutes and then return home to their villages, where they spread it still further. The HIV infection rate among migrant miners working in South Africa was more than 25 percent in 2011—a high rate, but down from 40 percent just three years earlier.

Africa's Fragile Infrastructure

Poor transportation hinders development in Sub-Saharan Africa. Few countries can afford to build extensive new road or railroad networks, and much of the colonial infrastructure has deteriorated (•**Figures 9.17, 9.18**). The region critically needs a good international transportation network, together with a lowering of trade barriers, to enlarge market opportunities.

Poor transportation is also often a contributing factor to famine. In Ethiopia, for example, the road network is so poor that it is extremely difficult to get food from the western part of the country, which often has crop surpluses, to the eastern part, which has chronic food shortages. On more than one occasion, it has proved cheaper to ship food from the United States to eastern Ethiopia than to truck it across this African country! Similarly, a World Bank study showed that it cost $50 to ship a metric ton of corn from Iowa 8,500 miles (13,600 km) to the Kenyan port of Mombasa but $100 to move it from Mombasa 550 miles (880 km) inland to Kampala, Uganda. Transportation problems contribute to the high costs of agricultural inputs like fertilizers, which in this region cost two to three times what they do in Asia. Poor communication, especially in digital technology applications, also hinders development. But that is changing quickly, and for the better (**see A Closer Look**).

Cell phones are at the heart of a technological revolution in Africa's rural regions. Every phone owner needs to recharge the phone, but with most homes lacking electricity, that requires a trip to town and a recharge fee. The increasingly popular alterative is the purchase of an $80 Chinese-made photovoltaic power system that can recharge a phone and power a lamp. The investment pays for itself quickly. The solar panel is "green" in many ways, not only producing no greenhouse gases but also averting the need to buy candles, charcoal, batteries, wood, and kerosene. Perhaps the most impressive result of the purchase is a marked improvement in children's grades at school: they now have light for studying.

Whether by mobile phone or computer, in beginning to overcome the digital divide, African peoples are ramping up their development prospects. The World Bank reported that for every 10 percentage points of increase in high-speed Internet access,

A **closer** LOOK

The Mobile Phone Revolution in Africa

There are more telephone land lines in the city of Tokyo than in all of Africa. So what, who needs land lines? Not Africans. Africa has the fastest growing mobile phone market worldwide—and of all the world regions, has been the most underserved by both mobile phone and Internet access. Mobile phone network coverage was available to just 10 percent of Sub-Saharan Africa's people in 1999, but that grew to 60 percent by 2008 and 85 percent in 2010, by which time there were more than 100 million users. There has also been explosive growth in broadband Internet coverage, made possible by recent projects laying tens of thousand of miles of undersea fiber optic cables.

Mobile phones are penetrating Africa far more quickly than the Internet, and African peoples are using them in many ways: to transfer money, buy and sell goods, and even turn on water wells. Without doubt, cell phones are a critical tool in rural development. In Uganda, for example, farmers in remote areas are communicating with plant scientists in the nation's capital to report the incidence and spread of a disease infecting bananas. The farmers take photos, mark Global Positioning System (GPS) locations, write remarks, and then transfer all these data by mobile phone to the scientists. The scientists then prepare an attack plan against the disease, and communicate disease treatment procedures back to the farmers. Using Google text messaging and an operator service, villagers are also able to learn more about farming practices and other critical issues like health care. Text messaging is improving farmers' access to markets. In time, web-enabled cell phones will make it much easier for millions of Africans to jump to the Internet without having to use computers.

• **Figure 9.18** Many crucial African river crossings have no bridges; people must rely on ferries instead. Here, vehicles and people board a ferry to cross an estuary north of Mombasa on the Kenyan coast. Backup of traffic at these bottlenecks may take hours or even days to clear.

economic growth rises 1.3 percentage points. Internet connections are critical for the establishment of new businesses and the expansion of existing ones. Voice-over-Internet technology (VOIP, such as Skype) is making telephone calling more affordable for hundreds of millions of Africans and is also making African call centers more competitive with the established Indian market. Don't be surprised if one of your next customer service agents speaks with an African rather than Indian or Filipino accent.

In the meantime, please have a look at a valuable Internet tool created in Africa and used there mainly on cell phones: **Ushahidi** (ushahidi.com). Ushahidi (meaning "witness" or "testimony" in Swahili) is at the intersection of critical current events and their geospatial context. It originated as a means of tracking the violence accompanying the disputed 2008 presidential election in Kenya. Eyewitnesses sent text messages or wrote emails about what they saw, and each of these communications was "georeferenced" (placed on a known coordinate mapping system, in this case Google maps). It was therefore possible to see in real time where the hot spots of violence were. Authorities, supporters, opponents, and others were able to act or make plans with this information. This kind of information can be thought of as "crowdsourcing," or obtaining eyewitness 13 accounts from a large group of people.

There are no limits to the possible applications of Ushahidi. It is particularly valuable as a means of reporting the fallout from human and natural disasters such as the tsunami in Japan and the violence in eastern Congo described later, and as a means of calling people together for social action, as it was in 192, Occupy Wall Street and Arab Spring protests. Have a look at 414 the site to see what recent events Ushahidi has covered, and reflect on how inherently geographic this tool is. Having begun as "activist mapping," bringing together individual testimony with activist causes in their geospatial context, Ushahidi—like Twitter and Facebook—now has no bounds, with one caveat: Whatever Ushahidi is covering, it is doing so geospatially.

There is a widespread perception that Africa is always behind. Certainly much of the discussion in this chapter suggests that Africa has a lot of catching up to do. But Ushahidi is just one example of the surging innovation and creativity in African hands. It is as Pliny the Elder (23–79 C.E.) wrote: "There is always something new out of Africa."

Trade and Aid

One of the major obstacles to African economic development is that many countries outside the region have effectively closed their doors to African imports. These restrictions typically take the form of subsidies, high tariffs, or low quotas imposed on agricultural products (like cotton) or manufactured goods (like textiles). Potential importing countries, such as the United States, impose these restrictions to protect their own industries. For instance, the United States and African countries are the world's leading exporters of raw cotton. The U.S. government spends $2 billion each year to subsidize its 25,000 cotton farmers. This means higher U.S. production and exports, and consequently lower cotton prices and incomes for African farmers.

"Trade Not Aid!" is a growing mantra of African economics. An example of what reduced trade barriers might do for Africa is provided by the **African Growth and Opportunity Act (AGOA)**, passed by the U.S. Congress in 2000 and due to expire in 2017. AGOA reduced or ended tariffs and quotas on more than 1,800 manufactured, mineral, and food items that could be imported from Africa. African exports to the United States have surged, especially in the clothing industry. Apparel made in Lesotho, Kenya, Mauritius, South Africa, Madagascar, and elsewhere in Africa is now sold in Target and Walmart stores. The United States is considering extending AGOA's benefits with the group of countries comprising the **Southern African Customs Union (SACU)**: South Africa, Lesotho, Swaziland, Botswana, and Namibia. SACU is one of several trade pacts and organizations within Africa. These entities make it easier for international trade organizations to do business with Africa. In many cases, jobs and incomes for Africans have grown because of new opportunities created by trade pacts. AGOA, for example, helped tiny Lesotho establish a market niche in "sweatshop-free" garment production.

For decades, almost all of the countries of Sub-Saharan Africa were heavily in debt to foreign lenders. Young nations undertook costly development projects with borrowed money, particularly after the mid-1970s. Western financiers, planners, and contractors gave optimistic assessments of the benefits to be expected from such projects, and African leaders were ready to accept loans as a way to reap quick benefits from newly won independence. Countries soon found they had great difficulty in meeting even the interest payments on their debts, and their efforts to do so often resulted in further economic woes and destructive environmental effects. 46

Western donors began attaching more strings to loans as means of promoting democracy. African leaders were told that if they did not institute democratic reforms, hold fair elections, or improve bad human rights records, their countries would no longer be eligible for development aid or lending programs. Often the leaders made just enough concessions to win the aid without instituting real reform—a phenomenon known as **donor democracy** (•**Figure 9.19**).

In the West, increased foreign aid in the form of gifts or grants to Africa became controversial. Critics argued that foreign aid only increases dependence and deflects national attention away from problems underlying famine. International aid monies often free up funds from national treasuries to be spent on unfortunate uses. In the Horn of Africa, for example, when foreign money became available for famine relief, it allowed Ethiopia and Eritrea to spend more on weapons used in their war. Aid monies are often simply stolen, and monitoring by donors of how they are spent has been inadequate.

Since about 1990, donors have proceeded more cautiously, more often funneling money through private and religious groups working in health care, education, and small-business development using microcredit practices (see Try It, next page). Foreign aid has been very effective in advancing primary and secondary education in the region. However, many African governments view universities as potential hotbeds of dissent and discourage investment in higher education. Africa's universities are crumbling, and the classic brain drain of African talent to Western institutions means there is that much less intellectual capital left at home to invest in the region's development.

Africa's untenable debt situation, and the slow flow of aid to the region, began to change around 2005. Two major factors were at work. First, in the West, awareness of the humanitarian costs of not directing more resources to Africa was growing. Media, political, and corporate stars like the singer Bono, former U.S. President Bill Clinton, and Microsoft founder Bill Gates kept up a steady appeal for governments and citizens of the MDCs to be more generous to impoverished Africa. Secondly, just as it was becoming clear that the West was falling behind on its promised course to double aid to Africa, an Eastern giant stepped into the gap.

China Steps In

The second and more powerful shift in Africa's debt, trade, and aid picture came after 2006 with the meteoric rise of China's engagement with the region. China's trade with Sub-Saharan Africa increased 10-fold between 1999 and 2006, when China displaced Britain as the region's third most important trading partner (after the United States and France). By 2010, China was the region's leading trade partner. In its 2007

Microcredit

A promising alternative to conventional development based on large, top-down projects funded with nationwide borrowing and debt is **microcredit** (or microlending, microfinance), the lending of small sums to poor people to set up or expand small businesses. Typically, poor people have no collateral—assets to pledge against loans they seek—and therefore cannot borrow from commercial banks. This drives them into the arms of local loan sharks, who may charge huge interest rates of up to 10 to 20 percent *per day*. But in the interest of fostering development, microlending organizations have begun to encourage poor borrowers to form a small group that will cross-guarantee all members' loans. One member of the group may take out a loan of $25, for example, to help start up a small business such as a roadside restaurant. Only when that individual has paid it back may the next person in the group receive a loan. Peer pressure minimizes the default rate, and the accumulated history of timely paybacks increases the entire group's creditworthiness.

Microlenders generally prefer to lend to women because women are more likely to use

---

the money in ways that will feed, clothe, and otherwise benefit their children, whereas men may be more likely to spend the money on alcohol or other frivolous pursuits. One fear in Africa is that microloans will not be repaid if the borrower or someone in the borrower's family contracts HIV or some other debilitating illness. Therefore, the loans are often extended only after the borrower receives some amount of public health training, for example, in the use of birth control and in general food preparation hygiene.

One microlending project that has received widespread and generally positive attention is Kiva. Many people find it attractive because they can lend small amounts and track the progress of the individuals they are helping (see kiva.org). Want to try it? There is a chance you would lose some or all of your investment of $25 or more, but the Web site states that the repayment rate is 98.91 percent. You may want to at least explore the site to get a sense of how individuals and small groups around the world are seeking to improve their well-being through microborrowing.

---

dialogue with the African Development Bank (AfDB), China pledged $20 billion in infrastructure and trade financing in the region.

Those funds, to be expended over a three-year period, represent a doubling of China's previous inputs and rival the value

• **Figure 9.19** In Scotland, an aid worker raises funds to fight poverty in Africa.

of those pledged by Africa's traditional financiers in the West. There is an important difference, however. Western institutions tend to have strings attached to loans and aid, insisting that African countries adopt democratic and other reforms to be eligible for the monies. China, in contrast, has a "no strings attached" policy. It is not insisting on political reforms, is not expressing concern about environmental or social responsibilities, and much to the chagrin of the West is even engaging with repressive governments that the West is trying to isolate. At the same time, Western governments and financial institutions are increasingly alarmed that China is making rapid inroads into African markets at the expense of Western economies and geopolitical interests.

237, 335, 367, 373

Even Africans look with some trepidation of China's advance. Some complain about China's record on not hiring locals, instead bringing in Chinese nationals for virtually all labor and management positions. In other cases, Chinese firms have reportedly mistreated African workers and paid them what have been described as "slave wages." Some Chinese projects have been documented as having negative environmental impacts.

China is unabashed about its interests in Sub-Saharan Africa: it needs and wants the region's raw materials—especially oil, iron ore, copper, and cotton—to feed its insatiable industrial appetite. To many Africans, China is building the same kind of exploitive mercantile relationship that characterized the colonial West: it takes Africa's raw materials, transforms them into value-added products, and sells them mainly to its own advantage. South Africa's president warned in 2007 that Africa risked becoming an economic colony of China. But desperate for assistance, even the wary are inclined to accept China's terms. The most typical arrangement is the "mines for infrastructure" deal, in which the African country grants mining concessions to the Chinese, and the Chinese build railways, roads, schools, and hospitals. There is especially great appreciation for the schools—there are simply not enough of them in many African countries.

367

China's model of state-led capitalism has proven attractive to many African governments. South African government officials go to Beijing to learn how to run state-owned companies more profitably. China is helping several African countries to build special economic zones, just as China did to spur its own economic growth.

268

China and the West do not call all the shots. There are a growing number of African institutions designed to encourage economic growth and political stability. In 2002, the 53-member **African Union (AU)** was formed to replace the old and often inconsequential Organization of African Unity (OAU). One of the African Union's main articulated goals is to implement self-inspection or "peer review" to help promote democracy, good governance, human rights, gender equity, and development. It has formed an agency responsible for this monitoring, known as the **New Partnership for Africa's Development (NEPAD)**. NEPAD's premise is that improvements in such human affairs will improve the climate for international investment and assistance to Africa. The African Union has also created a 265-member Pan-African Parliament to make regional laws.

9.5 Geopolitical Issues

Sub-Saharan Africa has waxed and waned as a theater of geopolitical interest since the end of World War II. During much of the twentieth century, the great powers were interested in the region. To boost their competing aims during the Cold War years, the Soviet Union and the United States played African countries against one another, arming them with weapons with which to wage **proxy wars**. The superpowers also extended aid generously to many African nations, but not simply for altruistic reasons.

178, 242

Great power concerns about Africa changed with the end of the Cold War around 1990, and the nature of conflict in Africa changed. The great powers withdrew support, but their weapons remained to prolong smoldering conflicts. The United States helped build the arsenals of eight of the nine countries involved in the recent Democratic Republic of Congo conflict, for example. Cold War-era weapons like AK-47 assault rifles were "dumped"—sold at low prices—in Africa, where they were not considered obsolete and where they fanned the flames of conflict. Increasingly, fighting began to spread across international borders. Previously, the United States and the Soviet Union had maintained a kind of security balance that kept warfare from becoming internationalized, but that restraint no longer exists.

339

The scenario of regional or even Africa-wide wars was realized with Africa's first "world war," centered on the Democratic Republic of Congo (see page 340). In the absence of regional or international powers to keep the peace, some countries (for example, Sierra Leone) simply imploded, fragmenting into fiefdoms run by factions that claimed to be revolutionaries but were essentially profiteers. Other national governments (as in Ethiopia) spent vast sums on expensive military aircraft while their people suffered from malnutrition.

The end of the Cold War constricted (and in the case of Russia, ruptured) aid pipelines. The United States cut its economic assistance to Sub-Saharan Africa by 30 percent between 1985 and 1992 and reduced it another 10 percent between 1992 and 2001. Why the waning concern? Former U.S. President George W. Bush was quoted as saying, "While Africa may be important, it doesn't fit into the national strategic interests, as far as I can see them."[5]

Then came 9/11. Al-Qa'ida's attack on the United States and the U.S. counterstrike against al-Qa'ida in Afghanistan raised fears that the Islamist organization might find new recruits and training grounds in Africa. Al-Qa'ida had already bombed U.S. embassies in Kenya and Tanzania in 1998 and in 2002 struck again in the Kenyan city of Mombasa. The United States maintains a focus on **terrorism hot spots** in the region, including Kenya, Somalia, Djibouti, Niger, Chad, and Mali, all of which are regarded as real or potential training grounds for al-Qa'ida-affiliated groups. As related on the next page, U.S. forces have made the tiny Horn of African country of Djibouti a major military center for potential action against nearby Somalia, Yemen, and other countries where Islamist militants are active. So far, U.S. interest in the hot spots has translated into military expenditures and partnerships with African governments, but not into wider development assistance.

In addition to the terrorism hot spots, and the Somali piracy that occasionally threatens commercial traffic in the region's vital sealanes, the United States has strong geopolitical interests in the region's oil-producing countries. These too became more critical after 9/11 and the U.S. war in Iraq, which heightened the sense that Middle Eastern oil supplies were vulnerable. The United States is seeking more stable oil supplies and has pinned many hopes on Africa. Petroleum-rich Nigeria, Cameroon, Gabon, Congo, Angola, and Equatorial Guinea together in 2011 supplied 14 percent of the oil used by the United States, a figure projected by the U.S. Central Intelligence Agency to rise considerably in the coming years.

Increasingly, the United States will be competing with China for these dwindling finite fossil reserves, and given current trends, the geopolitical landscape of Sub-Saharan Africa will see Asia on the rise and American clout in decline. Countries that are not pleased with Washington's wishes or demands will find partners elsewhere, particularly in Asia. Some military analysts describe the emerging U.S.-China rivalry in Africa as a modern "Great Game", comparable to the British-Russian contest over central Asia in the nineteenth century.

For many years following the deaths of U.S. service personnel in Somalia in 1993 (the "Blackhawk Down" incident), there was great reluctance to project U.S. military power in African trouble spots. Even Al-Qa'ida's simultaneous attacks on U.S. embassies in Kenya and Tanzania in 1998 did not change American military commitments in the region. The 9/11 attacks did, however. Since 2001, the United States has maintained a military base at Camp Lemonnier in Djibouti. Along with bases in Ethiopia and the Seychelles, it has been very useful in projecting U.S. power into Somalia and Yemen. Unmanned drones equipped with laser-guided weapons have been the weapons of choice against suspected al-Qa'ida (and in Somalia, al-Qa'ida's affiliate al-Shabab) targets.

The United States views South Africa as the region's economic powerhouse and as the key to regional stability. South Africa's military strength, which included nuclear weapons until they were decommissioned in 1990, contributes to its global significance. It is strategically important for its frontages on both the Atlantic and Indian Oceans, its possession of Africa's finest transport network (vitally important to many African countries that trade with and through South Africa), and its diversified mineral wealth.

Finally, HIV/AIDS in the region is also a geopolitical concern. Despite Africa's apparent remoteness from the United States, the two are linked by hundreds of air traffic routes traveled by thousands of people each day. Hundreds of HIV-positive Africans, perhaps most of them unaware of their virus, pass through U.S. airport gateways daily. Some African strains (subtypes) of the virus are more easily transmitted by heterosexual contact than the strains prevalent in the United States, so the risk to the population at large is greater. There are other dimensions of the potential burden on the United States. As long as the epidemic rages in Africa, there may be large flights of "medical refugees" seeking asylum in the United States. There could also be AIDS-related political instability or civil wars that invite U.S. military intervention in the region. How the world should respond to poverty and recurrent crises is one of the key questions facing the huge, resource-rich, problem-ridden but promising region of Sub-Saharan Africa. The following section examines some of these challenges in greater detail.

9.6 Regional Issues and Landscapes

The Sahel

Drought and Desertification in the Sahel

The Sahel region extends eastward from the Cape Verde Islands to the Atlantic shore nations of Mauritania, Senegal, and the Gambia, and inland to Mali, Burkina Faso ("Land of the Upright Men," formerly Upper Volta), Niger, Chad, and the new nation of South Sudan (•**Figure 9.20**).

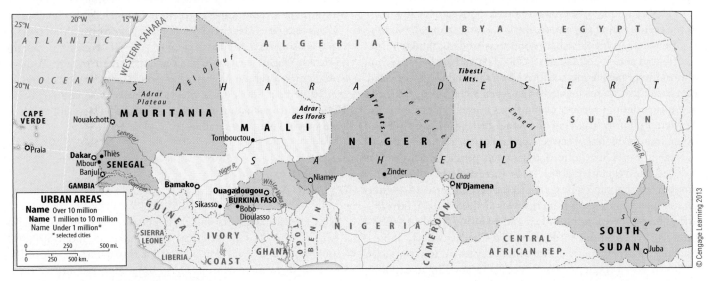

• **Figure 9.20** Principal features of the Sahel.

The Sahel region, like much of Sub-Saharan Africa, experiences periodic drought. Drought is a naturally occurring climatic event in which rain fails to fall over an area for an extended period, often years. When rain does return, the arid and semiarid ecosystems of the Sahel come to life with a profusion of flowering plants, insects, and herbivores like gazelles, whose populations climb when foods are abundant.

These ecosystems have high **resilience**, meaning they are able to recover from the stress of drought and have mechanisms to cope with a natural cycle that includes periods of dryness and rain.

Desertification is the destruction of that resilience and the biological potential of arid and semiarid ecosystems. It is an unnatural, human-induced condition that has afflicted the Sahel periodically and severely since the late 1960s, and thus it is different from the natural phenomenon of drought. However, the recent history of the Sahel suggests that drought can be the catalyst that initiates the process of desertification. During decades of good rains prior to the late 1960s, the Fulani (Fulbe), Tuareg, and other Sahelian pastoralists allowed their herds of cattle and goats to grow more numerous. The animals represent wealth, and people naturally wanted their numbers to increase beyond the immediate subsistence needs of their families. However, due to declining death rates, the numbers and sizes of families keeping livestock were also very large. Thus, the unprecedented numbers of livestock that built up during the rainy years made the Sahelian ecosystem vulnerable to the impact of drought on an unprecedented scale.

Drought struck the region in 1968, persisted through 1973, and returned over many years through 1985. With the drought, annual plants failed to grow, so the large herds of cattle and goats turned to acacia trees and other perennial sources of fodder. They ate all of the edible vegetation. This destruction was a critical problem because plants play an important role in maintaining soil integrity by helping intercept moisture and funnel it downward through the root system. Plant litter and decomposer organisms working around the plant contribute to soil fertility and help stabilize the soil. With hungry livestock in the region eating the plants, the landscape changed. No longer anchored and replenished, good soils eroded. "Junk plants" like Sodom apple (*Calotropis*) replaced palatable plants, and an almost impermeable surface formed on the land. This degraded ecosystem lost its resilience so that even when rains returned, vegetation did not recover promptly as it usually does.

Meteorologists studying the Sahel drought discovered another important link between precipitation and the removal of vegetation. They noticed that when people and livestock reduced the vegetative cover, they increased the **albedo**, the amount of the sun's energy reflected by the ground. Fewer plants meant more solar energy was deflected back into the atmosphere, lowering humidity and reducing local precipitation. This connection is known as the **Charney effect**, named for Jules Charney, the scientist who documented it in the early 1970s. It was a tragic sequence: People responding to drought actually perpetuated further drought conditions.

The 1968–1973 drought devastated the great herds of the Sahel. An estimated 3.5 million cattle died. Two million pastoral nomads lost at least 50 percent of their herds, and many lost as much as 90 percent. The Niger and Senegal rivers dried up completely. The cost in human lives was estimated at 100,000 to 250,000. Environmental refugees poured into towns unprepared to deal with such a large and sudden influx. Many pastoral nomads gave up herding and settled down to become wage earners and, where possible, farmers.

This crisis resulted in an international effort to combat desertification, using recommendations developed by the United Nations in 1977. The U.N. report, however, soon became a case study in how *not* to deal with an environmental crisis in the developing world. Most of the recommendations were universal, high-technology solutions that proved impossible to implement in the villages and degraded pastures of the Sahel. For example, many international agencies attempted to relieve the suffering of Sahelian pastoralists by using techniques such as digging deep wells to water livestock. But these agencies failed to anticipate that these wells would act like magnets for great numbers of people and animals; although there was plenty to drink, many animals starved to death when they decimated what vegetation remained around the new water supplies.

Since the disappointment of the 1977 U.N. plan, and with the lessons relief organizations learned as a result, there has been greater focus on sustainable development, with its local rather than universal solutions. In Niger, for example, hit by a devastating drought as recently as 2005, farmers have ceased the traditional practice of clearing trees before planting crops and are now protecting trees and growing crops under and between them. They are also harvesting and selling many useful tree products. A key factor for the farmers' new concern for trees is that Niger's government released its long-held claim that all trees belonged to the state; now farmers recognize the trees as private property. New attitudes and practices like these, and a sustained period of good rains (which unfortunately have prompted periodic locust swarms), have combined to produce a remarkable regreening of much of the Sahel since 1985. Satellite imagery shows significant regrowth of vegetation in southern Mauritania, northern Burkina Faso, northwestern Niger, central Chad, and southern Sudan. Annual millet and sorghum harvests have risen in some cases as much as 70 percent above their pre-1985 levels. This change in the Sahel's fortunes is an important lesson that desertification is not inevitable or irreversible.

West Africa

The Poor, Oil-Rich Delta of Nigeria

West Africa extends from Guinea-Bissau eastward to Nigeria (•Figure 9.21). Its nine political units make up about 800,000 square miles (c. 2 million sq km), or nearly one-fourth of the area of the United States. These nine countries are Guinea-Bissau, Guinea, Sierra Leone, Liberia, Ivory Coast, Ghana,

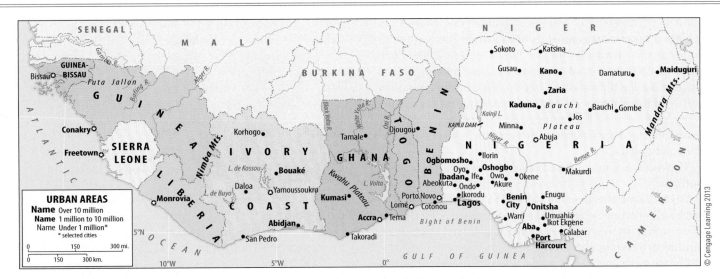

● **Figure 9.21** Principal features of West Africa.

Togo, Benin, and Nigeria. The spatial, demographic, political, and economic giant of the region is Nigeria.

Oil is the economic lifeblood of populous Nigeria. Africa's largest oil producer, the country is well-endowed, ranking 10th in the world's proven reserves. Most of the production is concentrated in the Niger River Delta, home to about 12 million mostly Christian people (●**Figure 9.22**). They belong to several different ethnic groups who have one thing in common: they say they have derived few benefits and have suffered greatly from oil development in their homeland. They complain that for more than 30 years, thousands of oil spills have tainted their croplands and water, destroying their crops and fisheries, while the flaring off of natural gas has polluted their air and caused acid rain. They note that despite the enormous

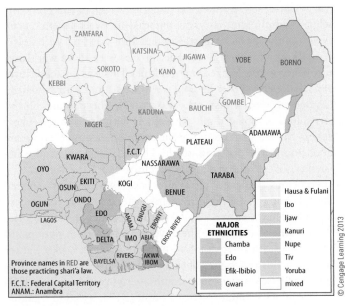

● **Figure 9.22** Nigeria's ethnic map.

revenue generated from oil drilled on their land, little money has returned to the area; most goes to foreign oil companies like Shell and Chevron and to the Muslim-dominated government in the north. Meanwhile, most of them live in palm-roofed mud huts; half of the delta region lacks adequate roads, water supplies, and electricity. Schools have few books, and clinics have few medical supplies.

Where does Nigeria's oil money go? There is not enough transparency to answer this question. Incredibly, until June of 2011, it was illegal in Nigeria to publish official government data and statistics, including government accounts. Recently, $22 billion from the government's Excess Crude Account simply "went missing." An estimated $400 billion has been pilfered from Nigeria's treasury since the country became independent in 1960.

Since 1990, several ethnically based organizations have been agitating for change. Ethnic **Ogoni** activists founded the **Movement for the Survival of the Ogoni People (MOSOP)**, issuing a bill of rights in which they declared the right to a safe environment and more federal support of their people. One of the movement's leaders, a popular author, playwright, and television producer named Ken Saro-Wiwa, also called for self-determination for the Ogoni. Despite international condemnation, Nigeria's government executed him after a questionable murder trial.

Despite the return to civilian rule in Nigeria that came after Saro-Wiwa's death, dissent in the delta continued to grow. Ogoni defiance and contempt of government and the oil companies spread to the region's other major ethnic groups: the **Ibo (Igbo)**, **Yoruba**, and **Ijaw**. The Ijaw (who make up about 5% of Nigeria's population) have been especially strident in demanding that some of the oil money be used to fund roads, electricity, running water, and medical clinics. Ijaw activists have spoken of secession and self-determination if their demands are not met. They have periodically registered dissatisfaction by seizing onshore stations and offshore rigs belonging to foreign oil companies, temporarily disrupting Nigeria's oil exports on both occasions. In 2006, a mainly Ijaw group calling itself the **Movement for the Emancipation of the Niger Delta (MEND)** escalated these tactics into frequent kidnappings of foreign oil workers. Opportunistic profiteers without well-established political views also joined in the action, and kidnapping became big business. An oil worker is typically freed unharmed upon payment of a ransom of several hundred thousand dollars.

This militancy in the delta has sent shockwaves through the world economy. Nigeria had touted itself as the safe alternative to Middle Eastern oil but has had to cut production by as much as 25 percent when tensions have been particularly high. The resulting shortages on world markets helped send oil and gasoline prices to record levels.

Questions of equity in a potentially rich country thus pose a threat to national, regional, and international stability. The peoples of the eight oil-producing states in the delta have been promised 13 percent of all petroleum revenues (compared with 5% previously), and have been told that the region's foul gas flares will be shut off. But locals say the promised revenues have been embezzled, while the flaring off of gas continues.

One of Nigeria's ironies is that in this oil-rich country, there are continuous shortages of gasoline and electricity. Neglect, mismanagement, corruption, and theft are all responsible. One of the cruelest and most repetitive Nigerian news stories is the immolation of scores and even hundreds of poor delta villagers who, in an activity known as "scooping" or "bunkering," illegally puncture gasoline or oil pipelines in an effort to collect and sell the fuels, only to inadvertently ignite the fires that consume them.

East Africa

No More Divisionism?

Five countries make up East Africa: Kenya and Tanzania, which front the Indian Ocean, and landlocked Uganda, Rwanda, and Burundi (•**Figure 9.23**). Their total area is 703,000 square miles (1.8 million sq km), roughly the size of the U.S. state of Texas, with a total population in 2011 of 143 million (nearly half the population of the United States).

Episodes of bloodshed have punctuated East Africa's history. This section and the next section on west-central Africa tell a complex and tragic tale of genocide. These details are related here because the potential for renewed conflict is high and because the world community vowed not to forget what happened here.

Rwanda and Burundi have had tragic disputes between their majority and minority populations. About 85 percent of the population in Burundi and 90 percent in Rwanda are composed of the **Hutu (Bahutu),** and most of the remainder of the two populations is **Tutsi (Watusi).** These were not originally separate tribal or ethnic groups; the peoples speak the same language, share a common culture, and often intermarry. The distinction between Hutu and Tutsi was instead based on socioeconomic classification. Historically, the Tutsi were a ruling

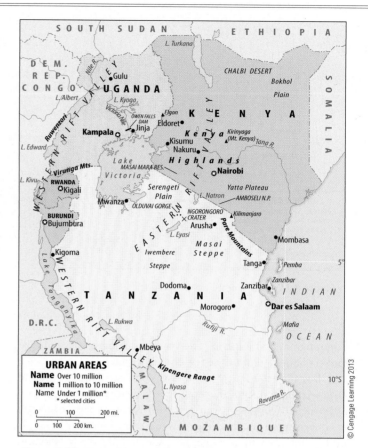

• **Figure 9.23** Principal features of East Africa.

class that dominated the Hutu majority. Tutsi power and influence were measured especially by the vast numbers of cattle they owned. Tutsi who lost cattle and became poor came to be identified as Hutu, and Hutu who acquired cattle and wealth often became Tutsi.

Colonial rule in East Africa attempted to polarize these groups, increasing antagonism between them. German and Belgian administrators differentiated between them exclusively on the basis of cattle ownership: Anyone with fewer than 10 animals was Hutu, and anyone with more was Tutsi. Europeans mythologized the wealthier Tutsis as "black Caucasian" conquerors from Ethiopia who were a naturally superior, aristocratic race whose role was to rule the peasant Hutus. The colonists replaced all Hutu chiefs with Tutsi chiefs. These leaders carried out colonial policies that often imposed forced labor and heavy taxation on the Hutus. Education and other privileges were reserved almost exclusively for the Tutsis.

Ferocious violence between these two peoples marred Rwanda and Burundi intermittently after the states became independent in the early 1960s. After independence, the majority Hutus came to dominate the governments of both nations. In 1994, the death of Rwanda's Hutu president in a plane crash (in which Burundi's president also died) sparked civil war in Rwanda between the Hutu-dominated government and Tutsi rebels of the Rwandan Patriotic Front (RPF), based in Uganda. Hutu government paramilitary troops, still vengeful about decades of domination by Tutsis, systematically massacred Tutsi civilians with automatic weapons, grenades, machetes, and nail-studded clubs. Women, children, orphans, and hospital patients were not spared. An estimated 800,000 Tutsis and "moderate" Hutus (those who did not support the genocide) died. International media tracked these horrors, but outside nations did nothing to stop the fighting.

Well-organized and motivated Tutsi rebels seized control of most of the country and took power in the capital, Kigali, in 1994. That precipitated a flood of 2 million Hutu refugees mainly into neighboring Zaire, now the Democratic Republic of Congo (see page 340). Although the Tutsi-dominated government promised to work with the Hutus to build a new multiethnic democracy, Hutu insurgents, based mainly in eastern Zaire, continued to carry out attacks against Tutsi targets in Rwanda.

Rwanda now has a national unity policy aimed at reconciling Hutus and Tutsis; both occupy important government posts, and although the Tutsis still hold the political upper hand, they have encouraged the resettlement of Hutu refugees who fled the country during its troubles. In 2004, marking the 10-year anniversary of the genocide with a new policy meant to prevent its recurrence, Rwanda's government outlawed ethnicity. Now any Rwandan who speaks or writes of Hutus and Tutsis may be fined or imprisoned for practicing "divisionism." The country's past nevertheless continues to shape its future; for example, there is an ongoing defection of the faithful from Rwanda's majority Catholic Church to other religions, especially Islam but also smaller Christian denominations. Most of the country's worst massacres took place in Catholic churches, with the apparent knowledge and even participation of church leaders.

In Burundi, the Tutsi minority traditionally dominated politics, the army, and business. In 1993, there was a brief shift of power to the Hutu majority when Burundians elected the country's first Hutu president. Tutsis murdered him in October 1993, and a large-scale Hutu retaliatory slaughter of Tutsis ensued. An estimated 300,000 died in a decade of violence. Burundi's Tutsis and Hutus now have a power-sharing agreement in which the presidency rotates between them.

The peace has been maintained by the first-ever deployment of the African Union's **African Standby Force**. To help ensure the stability needed for economic and political progress throughout the region, the African Union established this continental military force. It takes orders from the AU's Peace and Security Council. The force will, in theory, deploy peacekeepers or peacemakers to ensure that the map of Africa, so filled with deadly conflicts in the past, has far fewer trouble spots in the future. The African Union has plans to create an African Court on Human and People's Rights and, like the European Union after which it is modeled, may also eventually seek a common currency.

Collective shame and embarrassment in the international community about the failure to stop the killing in Rwanda and Burundi led in subsequent years to numerous new anti-genocide measures. These include a United Nations "early warning" system to prevent genocide and a new forum for crimes against humanity in the **International Criminal Court**, established in 2002 in The Hague, Netherlands.

West Central Africa

The subregion of West Central Africa is flanked by Cameroon and the Central African Republic to the north and the Democratic Republic of Congo to the south (•Figure 9.24). It contains seven countries: Cameroon, Gabon, Central African Republic, Congo Republic (often known by the shorthand Congo-Brazzaville), Democratic Republic of Congo (known by the shorthand Congo-Kinshasa or DRC), São Tomé and Príncipe, and Equatorial Guinea.

Colonialism and Modern Struggles

During the last quarter of the nineteenth century, the Congo Basin was virtually a personal possession of Belgium's King Léopold II, whose agents ransacked it ruthlessly for wild rubber, ivory, and other tropical products gathered by Africans. This lawless era inspired Joseph Conrad's famous novel *Heart of Darkness* (1902), in which the trader Kurtz, at the point of death, evokes the ravaged Congo region with the cry, "The horror! The horror!"

In 1908, the Belgian government formally annexed the greater part of the Congo Basin, creating the colony of Belgian Congo. In 1960, the Congo colony became the independent Republic of the Congo. In 1971, it took the name Zaire, meaning "river." Following the overthrow of Zaire's government in 1997, the country was renamed Democratic Republic of Congo.

In recent decades, the region's troubles have focused on its major power, the Democratic Republic of Congo, and have been inseparable from events in neighboring East Africa.

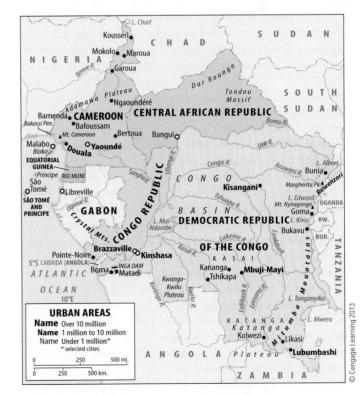

• **Figure 9.24** Principal features of West Central Africa.

During his tenure in office from 1965 to 1997, then-Zaire's autocratic ruler, Mobutu Sese Seko, had neglected the country's development but acquired a massive fortune. Political opposition and public discontent with unemployment and inflation led to unrest in the early 1990s.

All-out war began after 1997, when Zairean army troops attempted to expel a force of indigenous Tutsis living in the country's far eastern region (for discussion of the region's Tutsis and Hutus, see page 338). The assault backfired when these Tutsis, supported by neighboring Tutsi-dominated governments in Rwanda and Burundi and by other sympathetic countries in the region, took up yet more arms and initiated fierce attacks on government forces.

The Tutsi counterattack quickly widened into an antigovernment revolt comprised of many ethnic factions and led by a non-Tutsi guerrilla named Laurent Kabila. Kabila's forces pushed westward virtually unopposed through the huge country and in May 1997 occupied Kinshasa with little loss of life. President Mobutu fled Zaire on the eve of Kabila's triumph. Kabila then declared himself president of the country he now called the Democratic Republic of Congo.

Like his predecessor Mobutu, he funneled much of the country's wealth and power into the hands of family and friends. He also obstructed international human rights investigations into the deaths and disappearances of thousands of Hutu refugees during the heady days of rebellion against Mobutu. And he failed to bring stability to the far east of his huge country, where an environment of chaos had given birth to his own bid for power.

In 1998, Hutu forces based in the eastern Congo-Kinshasa launched successive devastating attacks on Tutsi interests in Rwanda. Although the Tutsis of Rwanda had helped bring Kabila to power in Congo-Kinshasa, they now turned against him, forming the backbone of a rebel alliance bent on bringing down Kabila's young regime. The unrest led to something rather rare in the region, a broader fight dubbed **"Africa's First World War"** that involved nine countries and 20 rebel movements, and resulted in the deaths of more than 5 million people.

Although the fighting was sometime ferocious, it was not warfare that directly claimed most of the lives lost in this struggle between 1996 and 2001. Most of the casualties were civilians who died from starvation and disease or in widespread massacres directed against ethnic groups, usually Hutus killing Tutsis, but sometimes the reverse. The country appeared to be on the verge of dissolution when Laurent Kabila was assassinated by a bodyguard in 2001. He was succeeded by his son Joseph.

Conflict in Congo-Kinshasa has been enormously profitable for the warring factions, some foreign companies, and foreign governments (Uganda and Rwanda, for example, looted eastern Congo-Kinshasa of its gems, minerals, timber, agricultural produce, and wildlife, including elephant ivory from some of the country's national parks). Since 1998, most of the fighting has in fact been unabashedly about control over the Congo's vast mineral resources. Warlords and rebel proxies have been sustained by exploitation of minerals in the areas they controlled. Most of Congo's vast reserves of gold, diamonds, and copper are in the south center and west of the country, where

the government is negotiating with the Chinese a vast minerals (coltan, tin, copper, diamonds, gold)-for-infrastructure deal (see page 334). In the east, where fighting to overthrow the government was centered, there is more gold. In this region, armed groups, now in concert with the Congo's army, control mines. There is a good chance that your cell phone contains coltan mined in this region and that it was not mined and exported "cleanly" as most diamonds are nowadays.

Although peace has been negotiated in this area, war could easily break out again. Now integrated into the Congolese army, ethnic Tutsi have been able to extend the area and the mineral wealth they control. Rival Rwandan Hutu militias, who control their own mineral wealth, have been carrying out reprisal attacks against Congolese civilians. Strife between Rwanda and the Congo is also enabled by expatriate sympathizers of the groups in Europe and the United States.

The Horn of Africa

The Galápagos Islands of Religion

In the extreme northeastern section of Sub-Saharan Africa, a great volcanic plateau rises steeply from the desert. This highland and adjacent areas occupy the greater part of the Horn of Africa, named for its projection from the continent into the Indian Ocean. This subregion includes the countries of Ethiopia, Eritrea, Somalia, and Djibouti (•Figure 9.25).

Of all these countries' peoples, Ethiopia's are the most ethnically and culturally diverse. About 45 percent, including the politically dominant **Amhara** peoples, practice **Ethiopian Orthodox Christianity**, an ancient branch of Coptic Christianity that came to Ethiopia in the fourth century from Egypt.

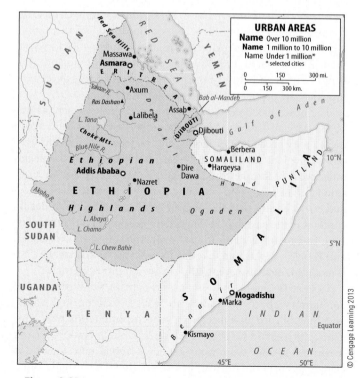

• **Figure 9.25** Principal features of the Horn of Africa.

• **Figure 9.26** The churches of Lalibela, carved from volcanic rock in the twelfth and thirteenth centuries, are among Ethiopia's many Christian cultural treasures. For scale, note the men at left and right.

The entire area has had important cultural and historical links with Egypt, the Fertile Crescent, and Arabia. The Ethiopian monarchy based its origins and legitimacy on the union of the biblical King Solomon and the Queen of Sheba, who, tradition holds, gave birth to the first Ethiopian emperor, Menelik. Until a Marxist coup brought an end to the emperorship in the 1970s, Ethiopia's rulers were always Christian. Ethiopia has many outstanding Christian artistic and architectural treasures, including the 11 churches of Lalibela, carved from solid rock in the twelfth and thirteenth centuries (•Figure 9.26). Most of the rest of Ethiopia's people are either Muslims (who make up about 40% of the population) or members of Protestant, Evangelical, and Roman Catholic churches. There are still small numbers of **Falashas**, or **Ethiopian Jews**, in Ethiopia, a remnant of a very ancient and isolated Jewish population. The majority, about 100,000 people, have fled to Israel since the mid-1980s. Because this mountainous country has long served as an isolated refuge for such unique groups, it has been nicknamed the **Galápagos Islands of Religion**.

Perhaps the most unusual religion with roots in Ethiopia is that of Rasta, or the Rastafari movement of the Jah People (called **Rastafarianism** by non-Rasta). This faith, which originated in the 1930s in the West Indies island of Jamaica, has the central doctrine that the Ethiopian Emperor Haile Selaisse (1892–1975) was the earthly incarnation of Jah or Jehova (God). Ras Tafari was the Emperor's name before his coronation, and he was crowned on November 2, 1930, as the "King of Kings, Elect of God, and Conquering Lion of the Tribe of Judah (•Figure 9.27)."[6]

• **Figure 9.27** The Lion of Judah.

Many poor Jamaicans gravitated to him for leadership and spiritual inspiration; he was the only black leader at the time to be recognized as legitimate in international circles. Their deep respect for him evolved quickly to reverence and worship, and the Rasta movement was born. The Rastafari faithful believe that blacks are the true children of Israel, and that on Judgment Day, Haile Selaisse will call them to come home to Zion, which they identify with Africa rather than Jerusalem. During his reign, Emperor Selaisse indulged the wishes of some Rastas to "repatriate" to Africa, allowing them to settle on his land in Ethiopia.

On April 21, 1966, Emperor Haile Selaisse's plane touched down on the airport tarmac in Kingston, Jamaica. For the 200,000 Rastafaris who had gathered, God Himself had come for a visit. He told the assembled that they should not return to Ethiopia until they first liberated Jamaica. Haile Selassie's short stay bestowed new legitimacy to the Rasta faith, and Rasta culture flowered and diffused rapidly after that. One of its principal carriers was Bob Marley, whose emotive cries for freedom thrust reggae onto the international music stage.

Southern Africa

In the southern part of Sub-Saharan Africa, five countries—Angola, Mozambique, Zimbabwe, Zambia, and Malawi—share the basin of the Zambezi River and have long had important relations with each other (•Figure 9.28). The first two were former Portuguese colonies, and the latter three were formerly British. Still farther south are South Africa and four countries whose geographic problems are closely linked to South Africa: the three former British dependencies of Botswana, Swaziland, and Lesotho, and the former German colony of Namibia.

Ethnicity, Colonialism, Strife, and Reconciliation in South Africa

Visitors from Western Europe and North America will find in South Africa (•Figure 9.29) most of the institutions and facilities to which they are accustomed. However, the Europeanized cultural landscape does not reflect the majority culture of this unusual country. Whites represent only about 10 percent of the total population, and blacks make up 79 percent. There are two other large racial groups: the so-called **coloreds**, of mixed origin (9%), and the Asians (2%), most of whom are Indians. Many peoples make up the black African majority. The **Zulu** and **Xhosa** (pronounced "*khoh*-suh") are both concentrated in hilly sections of eastern South Africa near the Indian Ocean and are the most populous of the nine officially recognized tribal groups.

A huge economic gulf separates South Africa's impoverished black Africans from about 5 million whites of European descent who dominate the economy and enjoy a lifestyle comparable to that of people in Western Europe and North America.

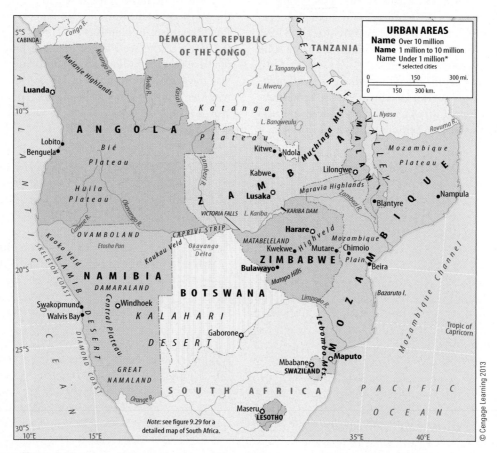

• **Figure 9.28** Principal features of southern Africa.

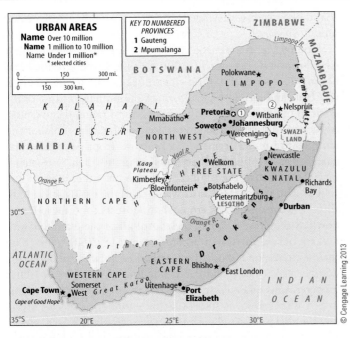

URBAN AREAS
Name Over 10 million
Name 1 million to 10 million
Name Under 1 million*
   * selected cities

KEY TO NUMBERED PROVINCES
1 Gauteng
2 Mpumalanga

© Cengage Learning 2013

• **Figure 9.29** Principal features of South Africa.

South Africa's overall per capita GNI PPP of $10,050 is comparable to that of Brazil. For the unskilled and semiskilled labor needed to operate their mines, factories, farms, and services, the whites of South Africa have always relied mainly on low-paid black workers in most of the country and colored workers in Western Cape province. The white-dominated economy could not operate without them, and the Africans are in turn extremely dependent on white payrolls. Although the races are interlocked economically, relations between them have historically been very poor and have been structured to benefit the whites. A country whose people were legally segregated by race, South Africa was one of the world's most controversial nations until its new beginning in 1994. Following is a summary of how the country's racial structures—a critical issue in its modern geography—evolved.

Racial segregation characterized South African life from 1652 onward, but the **Afrikaners** (descendents of Dutch, German and French settlers) after 1948 systematized it with laws supporting the official policy of "separate development of the races," subsequently called "multinational development" and commonly known as **apartheid** (pronounced "uh-*pahr-*tayt"). The new laws imposed racially based restrictions and prohibitions on the entire population, but they weighed most heavily on black Africans and denied them political power. The laws fragmented and displaced families and ethnic groups, rearranging the country's cultural landscape, and culminated in the establishment of makeshift tribal states, or **homelands** (also known as **native reserves, Bantustans**, and **national states**), which were nominally intended to achieve "independence" as black nations. Whites ejected several million Africans not wanted in "white" South Africa from their homes and transferred them to the homelands, which were located on the most

undesirable and least productive lands. Thus, great numbers of blacks were removed from cities and white farms and dumped in poverty-stricken, crowded, resource-poor areas, often far from where they had ever lived before.

The apartheid system drew furious criticism from many other nations. Most of the world community ostracized South Africa, and many countries, including the United States, imposed economic sanctions against it. Black unrest directed against apartheid and the general underdevelopment of the African majority became so widespread and violent between 1984 and 1986 that the government declared a state of emergency. But even after the state of emergency was lifted in 1990, violence continued. Much fighting took place between two large, tribally based rival factions: the Xhosa-dominated **African National Congress (ANC)**, led by Nelson Mandela, and the Zulu-dominated **Inkatha Freedom Party (IFP)**, led by Chief Mangosuthu Buthelezi.

With the country on the brink of anarchy, a complete official turnabout on the issue of apartheid resolved South Africa's ongoing racial crisis. It began in 1989 after Frederik W. de Klerk, an insider in the Afrikaner political establishment, became president. Urged on by white South African business leaders, antiapartheid activists, the powerful force of South African and international political opinion, the impact of international economic boycotts against South Africa, and his own stated convictions concerning the injustices and unworkability of apartheid, de Klerk launched a broad program to repeal the apartheid laws and put South Africa on the road to revolutionary governmental changes under a new constitution. Years of negotiations led to an all-race election in 1994, in which voters turned out in massive numbers to elect a new national parliament. The parliament then chose Nelson Mandela as president. (The white government had freed Mandela only four years earlier after a 27-year period of political imprisonment.) This historic election ended white European political control in the last bastion of European colonialism on the African continent. The apartheid laws became null and void, and the homelands were abolished.

Mandela, "South Africa's George Washington," retired in 1999 and was succeeded by a second black president, Thabo Mbeki, who resigned from office in 2008. The country's political landscape is remarkably stable, but challenges remain. The races are continuing on their long path of reconciliation. Some of the healing process has been formalized in the national **Truth and Reconciliation Commission**, which allows those who were party to racial violence during the apartheid era to confess their misdeeds and, in a sense, be absolved of them.

There is still a huge economic gulf between the haves and have-nots, but there are now more blacks—over 10 million—in the ranks of the middle and upper classes. Growing black wealth has boosted racial integration in neighborhoods. The black underclass continues to grow, however, and an overall unemployment rate of 25 percent fuels an ongoing epidemic of petty and violent crime, especially in the cities. Crime, the world's worst HIV/AIDS epidemic, a low level of education among workers, and relatively high labor costs have dampened foreign investment and economic growth in South Africa. No longer restrained by apartheid-era economic sanctions, South Africa has at the same time become a major investor in other African economies, buying banks, railways, cell phone networks, power plants, and breweries across the region.

There is a real need for land reform in South Africa because the white minority owns more than 70 percent of the productive land. A plan is in place to help reduce the imbalance by redistributing 30 percent of the country's farmland from white to black hands by 2014 on a willing-seller, willing-buyer basis. Progress is slow, however, and some landless blacks vow a takeover of white farms as occurred in neighboring Zimbabwe. A program is also under way to compensate the estimated 3.5 million blacks forcibly displaced by the government to the homelands between 1960 and 1982. More broadly, South Africa has an aggressive affirmative action program to help redress the economic imbalance between the races. The **Employment Equity Act** does not impose quotas but requires employers to move toward "demographic proportionality" based on the national proportions of race and gender. In another effort to secure more income for the black majority, the South African government assumed control of all the country's mineral resources in 2002. Mining companies now can exploit these resources only under state license and, presumably, with more supervision to ensure that profits are not drained excessively abroad or to whites.

The Indian Ocean Islands

Madagascar and the Theory of Island Biogeography

The islands and island groups off the Indian Ocean coast of Africa are unique in their cultures and natural histories. Madagascar, the Comoro Islands, Reunion, Mauritius, and the Seychelles have had African, Asian, Arab, European, and even Polynesian ethnic and cultural influences (•**Figure 9.30**). As

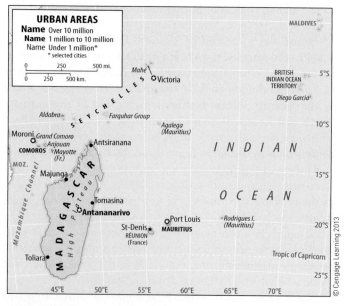

• **Figure 9.30** Principal features of the Indian Ocean islands.

island ecosystems, they are home to many endemic plant and animal species.

Madagascar (in French, *La Grande Ile*) is the fourth largest island in the world, nearly 1,000 miles (c.1,600 km) long and about 350 miles (c.560 km) wide. It lies off the southeast coast of Africa and has geological formations similar to those of the African mainland. Its distinctive flora and fauna include most of the world's lemur and chameleon species. These plants and animals are under tremendous pressure from people; Madagascar has 21 million inhabitants, many of them subsistence farmers who clear the island's forests to meet their needs.

The entomologist Edward Wilson and the biologist Paul Ehrlich have forecast that if tropical rain forests continue to be cut down at the present rate, a quarter of all of the plant and animal species on earth will become extinct by 2040.[7] They base their estimate on a model that correlates habitat area with the number of species living in the habitat. This **theory of island biogeography** emerged from observations of island ecosystems in the West Indies. It states that the number of species found on an individual island correlates with the island's area, with a 10-fold increase in area normally resulting in a doubling of the number of plant and animal [species]. If island A, for example, is 10 square miles in area and has 50 species, 100-square-mile island B may be expected to support 100 species.

What makes the theory useful in projecting species losses is the inverse of this equation: a 10-fold reduction in area will result in a halving of the number of species; therefore, 1-square-mile island C can be expected to hold only 25 species. In applying the model, ecologists treat habitat areas as if they were islands. Thus, if people cut down 90 percent of the tropical rain forest of the Amazon Basin, for example, the theory of island biogeography suggests that they would eliminate half of the species of that ecosystem. Scientists caution that the theory is only a tool meant to help in making rough estimates; the actual number of species lost with habitat removal may be higher or lower.

As a rough guideline, the theory of island biogeography is useful in projecting and attempting to slow the rate of extinction in the world's biodiversity hot spots, such as Madagascar. More than 90 percent of Madagascar's plant and animal species are endemic, occurring nowhere else on earth. Extinction of species was well under way soon after people arrived on the island; the giant flightless elephant bird (*Aepyornis*) was among the early casualties. But human activities, particularly the clearing of forests to grow rice and provide pasture for zebu cattle, are eliminating habitat areas on the island at a faster rate than ever before. Meanwhile, scientists are anxious to learn whether some of Madagascar's remaining plants might be useful in fighting diseases such as AIDS and cancer. Already Madagascar's rosy periwinkle has yielded compounds effective against Hodgkin's disease and lymphocytic leukemia. Other species could become extinct before their useful properties ever become known.

How urgent is the task to study and attempt to protect plant and animal species in Madagascar? Scientists turn to the theory of island biogeography for an answer. Although people

Joe Hobbs

Joe Hobbs

• **Figure 9.31** The subsistence needs of a growing human population have had ruinous effects on Madagascar's landscapes and wildlife. Before people came some 2,000 years ago, most of the island was forested, as in the Montagne d'Ambre National Park (left). Today, less than 10 percent of the island is forest, and a characteristic feature of its High Plateau is a barren gully, an erosion scar known locally as *lavaka* (right). The reddish color comes from the island's lateritic soil (see page 219)

have lived on Madagascar for less than 2,000 years, they have succeeded in removing 90 percent of the island's forest, setting the stage for some of the most ruinous erosion seen anywhere on earth (•**Figure 9.31**). The theory of island biogeography suggests that in the process, they have caused the extinction of roughly half of the island's species. With Madagascar's human population on track to double in 25 years, and with pressure on the island's remaining wild habitats expected to increase accordingly, the task of conservation is extremely urgent.

Africa fits like a piece of a jigsaw puzzle into neighboring South America. From Africa, we cross the Atlantic to explore that neighboring piece and the greater region of Latin America.

290

summary

- Africa is the cradle of humankind, where hominids originated and from where they diffused. The region has seen many indigenous civilizations and empires and is ethnically and linguistically diverse.

- The majority of the region's people are rural. There are several clusters of dense population.

- HIV/AIDS is taking a large toll, dramatically lowering life expectancy and projections for population growth, especially in southern Africa. But progress is being made in the fight against the epidemic.

- Sub-Saharan Africa has a relatively low population density overall, but a majority of this region's people live in a small number of densely populated areas.

- Most of Africa consists of a series of plateau surfaces dissected by prominent river systems.

- One of the most spectacular features of Africa's physical geography is the Great Rift Valley, a broad, steep-walled trough. The feature marks the boundary of two crustal plates that are rifting, or tearing apart.

- Although about two-thirds of the region lies within the low latitudes and has tropical climates and vegetation, Sub-Saharan Africa contains a great diversity of climate patterns and

biomes, some resulting from elevation rather than latitudinal position.

- Africa's diverse wildlife is often threatened by human population growth, urbanization, and agricultural expansion. Management of elephants and rhinoceroses involves international market demand and legal restrictions.

- In the four culture hearths of Sub-Saharan Africa, early indigenous people were responsible for several agricultural innovations, including the domestication of millet, sorghum, yams, cowpeas, okra, watermelons, coffee, and cotton.

- The four major language families of the region are Niger-Congo, Afro-Asiatic, Nilo-Saharan, and Khoisan. Malagasy, spoken only on Madagascar, is an Austronesian language. Islam and Christianity are major faiths, but indigenous belief systems are often mixed with them or exist on their own in some locales.

- The tragic impetus for growing contact between Africa and the wider world was slavery. Pockets of slavery still exist in the region.

- Most of Sub-Saharan Africa fell under European colonialism after the Conference of Berlin in 1884 and 1885.

- The majority of the people of Sub-Saharan Africa are poor, live in rural areas, and practice subsistence agriculture.

In most countries, per capita food output has declined or has not increased since independence. Export crops include coffee, cacao, cotton, peanuts, and oil palm products.

- Frequent droughts, lack of education, poor transportation, and serious public health issues have hindered development in Sub-Saharan Africa. Most countries are under-industrialized and overly dependent on the export of a few primary products.

- The export of minerals has had a particularly strong impact on the physical and social geography of Sub-Saharan Africa. The notorious trade in dirty diamonds has been largely cleansed.

- Most countries are heavily in debt to foreign lenders. Economic and humanitarian assistance to the region slowed considerably after the end of the Cold War, but it has picked up again since 9/11. China is the latest major economic power to engage in Africa and has a "no strings attached" policy when it comes to aid and trade with the region.

- Although many countries have been under authoritarian governments since independence, there has been some progress toward democracy. Serious political instability is characteristic of many African countries. Important links with the colonial powers that formerly controlled them remain strong in many countries of the region.

- The recently established African Union is a supranational organization dedicated to solving African problems without the intervention of outside powers.

- So often judged to be marginal in world affairs, Sub-Saharan Africa deserves and is receiving increased international attention because of its humanitarian problems, the global implications of its public health and environmental situations, problems in the management of its natural resource wealth, its oil reserves, and concerns about terrorism.

- Sub-Saharan Africa can be divided into seven subregions: the Sahel, West Africa, West Central Africa, East Africa, the Horn of Africa, southern Africa, and the Indian Ocean islands.

- The Sahel region extends eastward from the Cape Verde Islands to the Atlantic shore nations of Mauritania, Senegal, and the Gambia and inland to Mali, Burkina Faso, Niger, and Chad. The name Sahel in Arabic means "coast" or "shore," referring to the region as a front on the great desert "sea" of the Sahara.

- Between the late 1960s and 1985, the Sahel was subjected to severe droughts, which, in combination with increased human pressure on resources, prompted a process of desertification. Human and natural factors have helped restore the region recently.

- The most significant mining development in West Africa has been Nigeria's emergence as a producer and major exporter of oil. There are serious conflicts among ethnic, religious, and political groups in Nigeria, some resulting from the maldistribution of income from the country's oil wealth.

- The subregion of West Central Africa is flanked by Cameroon and the Central African Republic to the north and the Democratic Republic of Congo (DRC) to the south. The DRC was the epicenter of "Africa's First World War."

- East Africa includes the countries of Kenya, Tanzania, Uganda, Rwanda, and Burundi. Ethnic rivalries and conflicts have beset all of the East African countries. Warfare between the Hutus and Tutsis of Rwanda and Burundi in the 1990s was the most serious and was related to the violence in the Congo.

- Comprised of a great volcanic plateau and adjacent areas, the Horn of Africa includes the countries of Ethiopia, Eritrea, Somalia, and Djibouti. Ethiopia has been called the Galápagos Islands of religion.

- Southern Africa includes the countries of Angola, Mozambique, Zimbabwe, Zambia, Malawi, South Africa, Botswana, Swaziland, Lesotho, and Namibia.

- Racial segregation characterized South African life from 1652 onward, but it was systematized after 1948 under a body of laws known as apartheid.

- In 1994, Nelson Mandela was elected president of South Africa after an all-race election was held in the country. After years of conflict, this historic election ended apartheid and white European political control. Reconciliation between the races is continuing. The ranks of middle- and upper-class blacks are growing, but the gap between the haves and have-nots in South Africa is huge.

- The Indian Ocean islands of Madagascar, the Comoros, Réunion, Mauritius, and the Seychelles exhibit African, Asian, Arab, European, and even Polynesian ethnic and cultural influences. Madagascar has a wealth of endemic species under tremendous pressure by humans, who have cleared 90 percent of the original vegetation cover in the past 2,000 years.

Key Terms + Concepts

"Africa's First World War" (p. 340)
African Growth and Opportunity Act (AGOA) (p. 332)
African National Congress (ANC) (p. 343)
African Standby Force (p. 339)
African Union (AU) (p. 334)
Afrikaners (p. 343)
AIDS (acquired immunodeficiency syndrome) (p. 310)
albedo (p. 336)
Amhara (p. 340)
antiretroviral (ARV) drugs (p. 316)
apartheid (p. 343)
Bantustans (p. 343)
blood diamonds (p. 330)
branded diamonds (p. 330)
Chadic language subfamily (p. 319)
Charney effect (p. 336)
coloreds (p. 342)
conflict diamonds (p. 330)
Cushitic language subfamily (p. 319)

desertification (p. 336)

dirty diamonds (p. 330)

donor democracy (p. 332)

Employment Equity Act (p. 344)

Ethiopian Orthodox Christianity (p. 340)

Ethiopian Orthodox Church (p. 340)

Falashas (Ethiopian Jews) (p. 341)

"Galápagos Islands of Religion" (p. 341)

Great Rift Valley (p. 317)

homelands (p. 343)

human immunodeficiency virus (HIV) (p. 310)

Hutu (Bahutu) (p. 338)

Ibo (Igbo) (p. 337)

Ijaw (p. 337)

Inkatha Freedom Party (IFP) (p. 343)

International Criminal Court (p. 339)

keystone species (p. 321)

Khoisan language family (p. 319)

Kimberley Process (p. 330)

microcredit (p. 333)

"mistake of 1914" (p. 326)

Movement for the Emancipation of the Niger Delta (MEND) (p. 338)

Movement for the Survival of the Ogoni People (MOSOP) (p. 337)

nagana (p. 312)

national states (p. 343)

native reserves (p. 343)

New Partnership for Africa's Development (NEPAD) (p. 334)

Niger-Congo language family (p. 319)

Adamawan subfamily (p. 319)

Atlantic subfamily (p. 319)

Benue-Congo subfamily (p. 319)

Gur subfamily (p. 319)

Khordofanian subfamily (p. 319)

Kwa subfamily (p. 319)

Mandé subfamily (p. 319)

Nilo-Saharan language family (p. 319)

Chari-Nile subfamily (p. 319)

Saharan subfamily (p. 319)

Songhai (p. 319)

Ogoni (p. 337)

pandemic (p. 316)

PEPFAR (p. 316)

proxy war (p. 334)

Rastafarianism (p. 341)

resilience (p. 336)

resource curse (p. 328)

San (p. 319)

sleeping sickness (p. 321)

Southern African Customs Union (SACU) (p. 332)

terrorism hot spot (p. 334)

theory of island biogeography (p. 344)

triangular trade (p. 324)

Truth and Reconciliation Commission (p. 343)

trypanosomiasis (p. 321)

Tutsi (Watusi) (p. 338)

Ushahidi (p. 332)

Xhosa (p. 342)

Yoruba (p. 337)

Zulu (p. 342)

Review Questions

1. Where are the five principal areas of population concentration in Sub-Saharan Africa?

2. What factors account for the high HIV infection rates in Sub-Saharan Africa? What can be done to fight the epidemic? What is HIV/AIDS doing to life expectancy and the age structure profiles in countries where it is epidemic?

3. What are the principal climatic zones and biomes of Sub-Saharan Africa?

4. Where were the major slave trade routes in and from Sub-Saharan Africa?

5. Where were the major colonial possessions of various European powers in Sub-Saharan Africa?

6. What is the significance of cattle in many cultures of Sub-Saharan Africa?

7. What are the region's major export crops? Why does dependence on them make a country vulnerable?

8. What is a keystone species? Why is the tsetse fly considered a keystone species in Sub-Saharan Africa?

9. How have cellphones revolutionized African communications, and what can their potential benefits include?

10. What do the terms *donor democracy* and *donor fatigue* mean, and what do they reveal about Africa's political geography?

11. What impacts did the end of the Cold War have on many African countries?

12. How is China's ascendancy affecting Africa's economies and geopolitical status?

13. What are the seven subregions of Sub-Saharan Africa? In the case studies of section 9.6, what are some of the most significant countries in each of these subregions?

14. In what ways is the Sahel both a physical and cultural boundary zone in Africa?

15. What country has the greatest demographic, political, and economic clout in West Africa? What unique difficulties does this country face with respect to its oil wealth?

16. What was the original distinction between the Hutus and the Tutsis? What issues brought them into conflict in the 1990s?

17. Why is Ethiopia known as the "Galápagos Islands of Religion"?

18. What tensions emerged between blacks and whites over the issue of land tenure in Zimbabwe? What effects has government policy had on land reform?

19. What are the main ethnic groups in South Africa? What were their traditional social and economic roles? How have these changed since the end of apartheid?

20. How are deforestation and species loss measured in Madagascar?

Notes

1. Quoted in "Rebranding Africa," *New York Times*, July 9, 2009, p. A21.

2. Lydia Polgreen, "Misery Loves Optimism in Africa," *New York Times*, March 5, 2006, p. WK-1.

3. Robert Stock, *Africa South of the Sahara*, 2nd ed. (New York: Guilford Press, 2004), p. 41.

4. Ibid.

5. Quoted in Ian Fischer, "Africans Ask If Washington's Sun Will Shine on Them," *New York Times*, February 8, 2001, p. A3.

6. This information comes from an excellent online description of the Rasta, http://en.wikipedia.org/wiki/Rastafarian#Haile_Selassie_and_the_Bible (accessed September 9, 2005). "The Galapagos Islands of Religion" was the title of a Marjorie Coeyman's *Christian Science Monitor* article on Ethiopia's religions, on March 30, 2000.

7. Edward O. Wilson, "Threats to Biodiversity," *Scientific American*, 261(3), 1989, pp. 60–66.

Online Resources

CourseMate: Make the most of your study time by accessing everything you need to succeed in one place. Read your textbook, take notes, review flashcards, watch videos, complete activities, take practice quizzes, and more—online with CourseMate. Log in at **www.cengagebrain.com.**

"At first I thought I was fighting to save rubber trees, then I thought I was fighting to save the Amazon rainforest. Now I realize I am fighting for humanity."

—CHICO MENDES

Latin America at night, from space. Note the dark vastness of the Amazon rain forest. Manaus, at the confluence of the Amazon and Rio Negro, shines brightly. Left: This cathedral in highland Oaxaca is characteristic of Latin America's cultural landscape

C. Mayhew & R. Simmon (NASA/GSFC), NOAA/ NGDC, DMSP Digital Archive

Joe Hobbs

10

Latin America

The land portion of the Western Hemisphere south and southeast of the United States is commonly known as Latin America. This name reflects the importance of cultural traits inherited mainly from the European colonizing nations of Spain and Portugal, whose languages evolved from Latin. Native American cultures were flourishing when Columbus arrived in the Americas in 1492, shortly followed by an onslaught of other Europeans. The native cultures still persist to varying degrees and are especially strong in Mexico and Guatemala and in several South American countries. But "Latin" influences predominate today in the region as a whole.

This chapter will acquaint you with the varied indigenous and European cultures that have shaped the region. The exceptional topographic and environmental variations in Latin America are introduced here too. The economic prospects for this developing region are very good and are tied particularly closely to relationships with the United States and China. These issues are described here, along with broader involvement of the United States in affairs "south of the border."

chapter objectives

This chapter should enable you to

- Appreciate how topographic variety creates a predictable range of environmental conditions and livelihood opportunities.

- Recognize how the uneven distribution of resources, particularly of quality farmland, has contributed to dissent and war.

- Understand how the economic boom in China, where labor costs are very low, has taken a toll on Latin American industries.

- Know how diverse and accomplished the indigenous cultures of the region were, and how European conquest and colonization decimated and changed these cultures.

- Recognize the predominant ethnic patterns of the region and how ethnicity correlates with livelihood, wealth, and political power.

- Evaluate the region's efforts to shift from raw-material exports to manufacturing through participation in free-trade agreements.

- Understand how U.S. interests have shaped the region's political and economic systems.

- See why fertile volcanic uplands came to be the core regions of most of the mainland countries.

- Appreciate the vital role played by tourism in many of the nations' economies.

- Recognize the paradox of how a blessing of oil wealth can paralyze national economies and political systems, and be rejected by the poor.

- Appreciate the challenges that rebel forces, empowered by resource wealth, have posed to national governments.

- Understand why "saving the rain forest" is not a simple problem and how various countries and national groups see the problem differently.

- Appreciate how excessive dependence on raw materials can cause an uncontrollable cycle of economic boom and bust.

10.1 Area and Population

With its 38 countries, Latin America consists of a mainland region that extends from Mexico south to Argentina and Chile, together with the islands of the Caribbean Sea (•**Figure 10.1**; •**Table 10.1**). Its maximum latitudinal extent of more than 85 degrees, or nearly 5,900 miles (c. 9,500 km), is greater than that of any other major world region. Its maximum east–west measurement, amounting to more than 82 degrees of longitude, is also impressive.

Latin America's two main subregions are neatly offset from one another geographically. The northern part, known as Middle America, includes Mexico; the countries of Central America (Guatemala, Belize, El Salvador, Honduras, Nicaragua, Costa Rica, and Panama); Haiti and the Dominican Republic on the island of Hispaniola; Cuba; and the smaller islands of the Caribbean. It trends sharply northwest to southeast before intersecting with the north–south orientation of the continent of South America. South America protrudes much farther into the Atlantic Ocean than the Caribbean realm or Latin America's northern neighbor, North America. In fact, the meridian of 80 degrees west longitude (80°W), which touches the west coast of South America in Ecuador and Peru, passes through Pittsburgh, Pennsylvania. Brazil also lies less than 2,000 miles (3,200 km) west of Africa, to which it was joined in the supercontinent of Pangaea until around 110 million years ago. ²¹

Latin America has a land area of slightly more than 7.9 million square miles (20.5 million sq km), with South America accounting for 6,893,000 square miles (17,869,000 sq km) and Middle America, 1,047,000 square miles (2,713,000 sq km). The total is about 2.5 times the size of the conterminous United States (•**Figure 10.2**).

Latin America's 2011 population of 596 million represented 9 percent of the world total. The population has uneven distributions and densities (•**Figure 10.3**). Most of Latin America's people are packed into two major geographic alignments. The larger of these two areas, known as "the rim" or "the Rimland," is a discontinuous ring around the margins of South America. The second, generally a highland, extends along a volcanic belt from central Mexico southward into Central America.

About two-thirds of Latin America's people live along the South American Rimland. There are two major segments of the rim. The much larger segment, in terms of both population and area, extends along the eastern margin of the continent from the mouth of the Amazon River in Brazil southward to the humid **pampa** (subtropical grassland) around Buenos Aires, Argentina. The second segment is located partly on the coast and partly in the high valleys and plateaus of the adjacent Andes Mountains. It stretches around the north end and down the west side of South America. This crescent begins in the vicinity of Caracas, Venezuela, on the Atlantic coast and arches all the way around to the vicinity of Santiago, Chile, on the Pacific side of the continent.

This second segment is more fragmented, broken in many places by steep Andean slopes and coastal desert. A strip of hot, rainy, and thinly populated coast lies between Caracas and the mouth of the Amazon River. In the far south on the Atlantic side of the continent, the population rim is again broken in the rugged, relatively inaccessible, rain-swept southern Andes and the dry lands of Argentina's Patagonia in the Andean rain shadow. These territories have only a sparse human population, far exceeded by the millions of sheep that graze in Patagonia and adjacent Tierra del Fuego.

The second major alignment of populated areas in Latin America lies on the mainland of Middle America. It runs

KEY TO NUMBERED COUNTRIES
1 Guatemala
2 El Salvador
3 Puerto Rico (U.S.)
4 Virgin Islands (U.S. & U.K.)
5 Anguilla (U.K.)
6 St. Martin (Fr. & Neth.)
7 St. Kitts & Nevis
8 Antigua & Barbuda
9 Guadeloupe (Fr.)
10 Dominica
11 Martinique (Fr.)
12 St. Lucia
13 St. Vincent & the Grenadines
14 Grenada
15 Aruba (Neth.)
16 Netherlands Antilles (Neth.)

• **Figure 10.1** Principal features of Latin America.

Table 10.1	Latin America: Basic Data

Political Unit	Area (thousands; sq mi)	Area (thousands; sq km)	Estimated Population (millions)	Estimated Population Density (sq mi)	Estimated Population Density (sq km)	Annual Rate of Natural Increase (%)	Human Development Index	Urban Population (%)	Per Capita GNI PPP ($US)
Middle America	**957.5**	**2,479.9**	**158**	**165**	**64**	**1.5**	**0.733**	**72**	**11,760**
Belize	8.9	23.1	0.3	34	13	2.1	0.699	44	5,990
Costa Rica	19.7	51.0	4.7	239	92	1.2	0.744	65	10,930
El Salvador	8.1	21.0	6.2	765	296	1.4	0.674	65	6,420
Guatemala	42	108.8	14.7	350	135	2.4	0.574	50	4,570
Honduras	43.3	112.1	7.8	180	70	2.1	0.625	52	3,710
Mexico	756.1	1,958.3	114.8	152	59	1.4	0.770	78	14,020
Nicaragua	50.2	130.0	5.9	118	45	1.9	0.589	58	2,540
Panama	29.2	75.6	3.6	123	48	1.4	0.768	75	12,180
Caribbean	**90.3**	**233.9**	**41.4**	**459**	**177**	**1.1**	**0.653**	**65**	**9,400**
Anguilla (U.K.)	0.04	0.1	0.01	250	97	N/A	N/A	100	N/A
Antigua and Barbuda	0.2	0.5	0.1	500	193	0.9	0.764	30	17,670
Aruba (Neth.)	0.07	0.2	0.07	1,000	386	N/A	N/A	51	N/A
Bahamas	5.4	14.0	0.4	74	29	0.8	0.771	84	N/A
Barbados	0.2	0.5	0.3	1,500	579	0.5	0.793	45	N/A
Cayman Islands (U.K.)	0.1	0.3	0.04	400	154	N/A	N/A	N/A	N/A
Cuba	42.8	110.9	11.2	262	101	0.4	0.776	75	N/A
Dominica	0.3	0.8	0.1	333	129	0.7	0.724	67	8,460
Dominican Republic	18.8	48.7	10	532	205	1.6	0.689	66	8,110
Grenada	0.1	0.3	0.1	1,000	386	1.3	0.748	40	7,710
Guadeloupe (Fr.)	0.7	1.8	0.4	571	221	0.7	N/A	98	N/A
Haiti	10.7	27.7	10.1	944	364	1.8	0.454	53	N/A
Jamaica	4.2	10.9	2.7	643	248	1.0	0.727	52	7,230
Martinique (Fr.)	0.4	1.0	0.4	1,000	386	0.6	N/A	89	N/A
Puerto Rico (U.S.)	3.5	9.1	3.7	1,057	408	0.4	N/A	99	N/A
St. Kitts and Nevis	0.1	0.3	0.1	1,000	386	0.6	0.735	33	13,640
St. Lucia	0.2	0.5	0.2	1,000	386	0.7	0.723	28	8,860
St. Vincent and the Grenadines	0.2	0.5	0.1	500	193	0.9	0.717	50	8,830
Trinidad and Tobago	2	5.2	1.3	650	251	0.6	0.760	14	24,970
Turks and Caicos Islands (U.K.)	0.1	0.3	0.02	200	77	N/A	N/A	46	N/A
Virgin Islands (U.S.)	0.1	0.3	0.1	1,000	386	N/A	N/A	47	N/A
South America	**6,898.4**	**17,866.9**	**396.4**	**38**	**15**	**1.1**	**0.728**	**84**	**10,200**
Argentina	1,073.5	2,780.4	40.5	38	15	1.1	0.797	93	14,090
Bolivia	424.2	1,098.7	10.1	24	9	1.9	0.663	67	4,250
Brazil	3,300.2	8,547.5	196.7	60	23	0.9	0.718	87	10,160
Chile	292.1	756.5	17.3	59	23	0.9	0.805	87	13,420
Colombia	439.7	1,138.8	46.9	107	41	1.2	0.710	75	8,600
Ecuador	109.5	283.6	14.7	134	52	1.6	0.720	68	8,100
Falkland Islands (U.K.)	4.7	12.2	0.003	1	0	N/A	N/A	85	N/A
French Guiana (Fr.)	34.7	89.9	0.2	6	2	2.5	N/A	77	N/A

Table 10.1 Latin America: Basic Data *(continued)*

Political Unit	Area (thousands; sq mi)	Area (thousands; sq km)	Estimated Population (millions)	Estimated Population Density (sq mi)	Estimated Population Density (sq km)	Annual Rate of Natural Increase (%)	Human Development Index	Urban Population (%)	Per Capita GNI PPP ($US)
South America *(continued)*									
Guyana	83	215.0	0.8	10	4	1.6	0.633	29	3,270
Paraguay	157	406.6	6.6	42	16	1.8	0.665	58	4,430
Peru	496.2	1,285.2	29.4	59	23	1.5	0.725	77	8,120
Suriname	63	163.2	0.5	8	3	1.1	0.680	70	6,730
Uruguay	68.5	177.4	3.4	50	19	0.5	0.783	93	12,900
Venezuela	352.1	911.9	29.3	83	32	1.6	0.735	94	12,220
Summary Total	**7,946.2**	**20,580.7**	**595.8**	**75**	**29**	**1.2**	**0.724**	**79**	**10,560**

Sources: World Population Data Sheet, Population Reference Bureau, 2011; Human Development Report, United Nations, 2011; World Factbook, CIA, 2011.

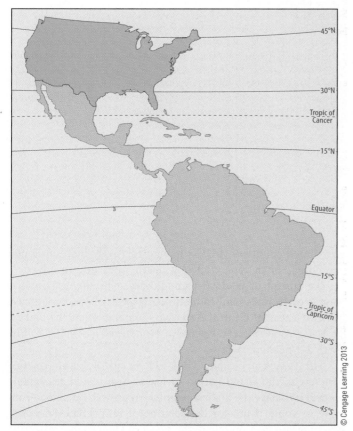

• **Figure 10.2** Latin America's size and latitude relative to the conterminous United States.

along an axis of volcanic land dominating central Mexico and from there reaches southeastward through southern Mexico and along the Pacific side of Central America to Costa Rica. This belt is characterized by good soils, enough rainfall for crops, and elevations high enough in most places to moderate the tropical heat. In Pacific Central America, more people live in the highlands than the lowlands.

Average population densities disguise the fact that nearly every Latin American mainland country reflects a basic spatial configuration: a well-defined population core (or multiple cores) with one or more outlying, sparsely populated **hinterlands** (peripheries). Figures on population density for the greater part of Latin America are extraordinarily low, averaging fewer than two people per square mile over approximately half of the entire region.

The pattern of core and hinterland is especially noticeable in Brazil and Argentina (see Figure 10.3). Most Brazilians live along or near the eastern seaboard south of Belém, a port city in northeastern Brazil, whereas large areas in the interior are still thinly populated. In Argentina, about three-fourths of the population is clustered in Buenos Aires and the adjacent humid pampas. Mexico and the Andean countries also have this kind of spatial imbalance in their populations.

Not every country, however, displays this pattern of core and hinterland. El Salvador and Costa Rica (both with densely settled volcanic lands) and most Caribbean islands are different. The combined population of the Caribbean is much less than that of central Mexico. But these islands have heavy population densities, grown over a long period of expanding population on restricted territories. The crowding is often aggravated by high rates of natural population increase or by steep slopes limiting opportunities for farming and settlement.

Very high rates of urban growth have characterized Latin America in recent times. Overall, the region is 79 percent urban, compared to the world average of 51 percent. A major element in the rapid growth of Latin American cities is rural-to-urban migration caused by both push and pull factors. The region's largest cities are listed in •Table 10.2.

Most Latin American countries are in the second stage of the demographic transition, clearly recognizable as less

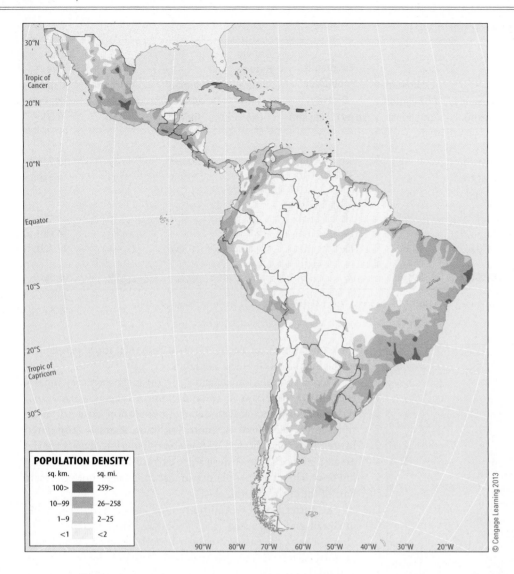

POPULATION DENSITY

sq. km.	sq. mi.
100>	259>
10–99	26–258
1–9	2–25
<1	<2

© Cengage Learning 2013

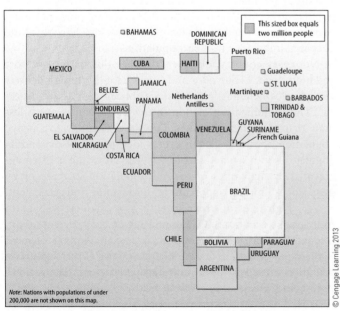

This sized box equals two million people

Note: Nations with populations of under 200,000 are not shown on this map.

© Cengage Learning 2013

• **Figure 10.3** Population distribution (top) and population cartogram (bottom) of Latin America.

developed countries (LDCs) having relatively high (although declining) birth rates and low death rates due to advances in the spread of medical technologies. The highest population growth rates tend to be in the least developed countries, notably those of Central America; Guatemala has the highest in Latin America at 2.4 percent; Honduras and Belize follow with 2.1 percent each.

The lowest per capita GNI PPP in all of Latin America belongs to the island nation of Haiti. You would reasonably expect it to have the highest population growth rate, and probably it would if not for the scourge of HIV/AIDS. Haiti has the highest infection rate in the Western Hemisphere, affecting some 2.6 percent of adults. Haiti's annual population growth rate is just 1.8 percent. The lowest annual rate of population change is 0.4 percent in Cuba, due more to steadfast authoritarian promotion of family planning rather than general prosperity. The overall rate of increase for the population of Latin America is 1.2 percent, which is lower than the average for the world's less developed countries. This regional rate of increase is down considerably from the high of 3 percent in 1960, largely because the region's economies are generally improving.

Table 10.2	Latin America: Metropolitan Populations Data	
1	Mexico City, Mexico	27.7
2	São Paulo, Brazil	20.5
3	Buenos Aires, Argentina	14.7
4	Rio de Janeiro, Brazil	12.3
5	Bogotá, Colombia	8.7
6	Lima, Peru	8.1
7	Belo Horizonte, Brazil	6.1
8	Santiago, Chile	5.2
9	Caracas, Venezuela	4.7
10	Guadalajara, Mexico	4.5
11	Monterrey, Mexico	4.2
12	Salvador, Brazil	4.1
13	Guatemala City, Guatemala	4
14	Porto Alegre, Brazil	4
15	Recife, Brazil	3.7
16	Santo Domingo, D.R.	3.6
17	Curitiba, Brazil	3.6
18	Fortaleza, Brazil	3.5
19	Medellín, Colombia	3.5
20	Puebla, Mexico	3.1

Population in millions.

© Cengage Learning 2013

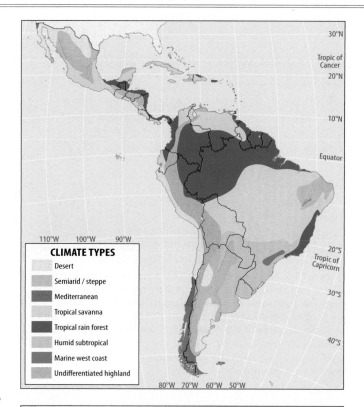

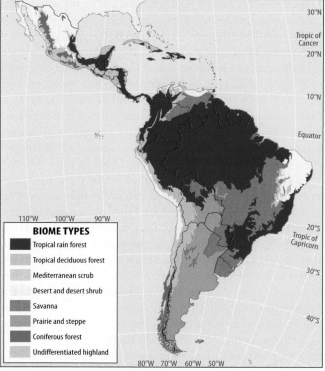

• **Figure 10.4** Climates (top) and biomes (bottom) of Latin America.

Sources: (top) Based on Rand McNally's Classroom Atlas, 2003. (bottom) Based on World Wildlife Fund ecoregions data, 1999.

10.2 Physical Geography and Human Adaptations

Dramatic differences in elevation, topography, biomes, and climates characterize Latin America (•Figure 10.4). Low-lying plains drained by the Orinoco, Amazon, and Paraná-Paraguay River systems dominate the north and central part of South America and separate geologically older, lower highlands in the east from the rugged Andes of the west. In Mexico, a high interior plateau broken into many basins lies between north-to-south-trending arms of the Sierra Madre. High mountains, largely within the Sierra Madre and the Andes, form a nearly continuous landscape feature from northern Mexico to Tierra del Fuego at the southern tip of South America. Most of the smaller islands of the Caribbean Sea's West Indies are volcanic mountains, although some islands of limestone or coral are lower and flatter. The largest islands of the Caribbean have a more diverse topography, including low mountains.

Climates and Vegetation

The climatic and biotic diversity of Latin America is extraordinary (•Figure 10.5). Even within some single countries (Ecuador, for example), conditions range over a very short horizontal distance from sea-level tropical rain forests to alpine tundra. The

tropical rain forest climate and biome, with heavy year–round rainfall, continuous heat and humidity, and superabundant vegetation dominated by large broadleaf trees, is generally a lowland type lying mainly along and near the equator, with segments extending to the tropical margins of the Northern and Southern

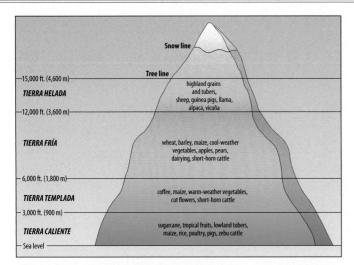

• **Figure 10.5** Altitudinal zonation in Latin America. Specific patterns of land use have evolved to take advantage of the region's diverse altitudinal zones. This graph shows elevations and general land use patterns, which may also be compared with land uses in Figure 10.6. Elevations shown here are for the equatorial region and are somewhat lower toward the poles for the respective land uses.

Source: Based on data from Clawson, 1997. Latin America and the Caribbean. Dubuque, Iowa: William. C. Brown Publishers.

Hemispheres. The largest segment, representing the world's largest continuous expanse of tropical rain forest, lies in the basin of the Amazon River system. There is more tropical rain forest in southeastern Brazil, eastern Panama, the western coastal plain of Colombia, on the Caribbean side of Central America and southern Mexico, and along the eastern (windward) shores of some Caribbean islands, particularly Hispaniola and Puerto Rico. Much of the original tropical rain forest vegetation of the islands has been lost to human activity, and even the vast Amazon forest is under considerable pressure (this problem, which is largely a reflection of Brazil's development, is described on pages 383–386).

On either side of the principal region of tropical rain forest climate, the tropical savanna climate extends to the vicinity of the Tropic of Capricorn in the Southern Hemisphere and, with more fragmentation, to the Tropic of Cancer in the Northern Hemisphere. Still farther poleward, in the eastern portion of South America, lies a large area of humid subtropical climate, with cool winters unknown in the tropical zones. Its Northern Hemisphere counterpart is north of the Mexican border in the southeastern United States. In South America, this climate is associated mainly with prairie grasses in the humid pampas of Argentina, Uruguay, and extreme southern Brazil. On the Pacific side of South America, a small strip of Mediterranean or dry-summer subtropical climate in central Chile is similar to that in southern California. This is an ideal climate for wine production, and Chilean wines are now well established on world markets.

The humid climates of Latin America have a fairly predictable spatial arrangement. But the region's dry climates and biomes—desert and steppe—are the product of local circumstances such as rain shadows or nearby cool ocean water. In northern Mexico, aridity is due partly to the rain shadow effect of high mountain ranges on either side of the Mexican plateau (a rain shadow is an area that is dry because it is sheltered from rain-bearing winds

by a mountainous barrier). These dry climates are also partly associated with the global pattern of semipermanent zones of high pressure that create arid conditions in many areas of the world along the Tropics of Cancer and Capricorn.

In Argentina, the extremely high and continuous Andean mountain wall accounts for the aridity of large areas, especially the southern region called Patagonia. Here, the Andes block the path of the prevailing westerly winds, creating heavy orographic (mountain-induced) precipitation on the Chilean side of the border but leaving Patagonia in rain shadow. However, in the west coast tropics and subtropics of South America, the Atacama and Peruvian deserts cannot be explained so simply. Here, shifting winds that parallel the coast, cold offshore currents, and other complexities, as well as the Andes Mountains, combine to create the world's driest area. The Atacama Desert is drier even than the Sahara! The mountains restrict this area of desert to the coastal strip. Arid and semiarid conditions also prevail along the northernmost coastal regions of Colombia and Venezuela and in the region called the Sertão in northeastern Brazil.

Elevation and Land Use

With respect to human adaptations, one of the most significant features of Latin America's physical geography is a series of highland climates arranged into zones by elevation. This zonation results from the fact that air temperature decreases with elevation at a normal "lapse rate" of approximately 3.6°F (1.7°C) per 1,000 feet (c. 300 m). At least four major zones are commonly recognized in Latin America (Figures 10.5 and •10.6): the *tierra*

• **Figure 10.6** Land use in Latin America reflects a remarkable range of opportunities posed by varying latitudes and elevations.

Source: The Oxford School Atlas edited by Patrick Wiegand (OUP, 1997), copyright © Oxford University Press. Used by permission of Oxford University Press.

• **Figure 10.7** Tending the coffee crop on a *finca* (plantation) in the *tierra templada* of southern Mexico. Note the coffee beans in the boy's right hand.

caliente (**hot country**), the *tierra templada* (**cool country**), the *tierra fría* (**cold country**), and the *tierra helada* (**frost country**). At the foot of the highlands, the *tierra caliente* is a zone embracing the tropical rain forest and tropical savanna climates. The zone reaches upward to approximately 3,000 feet (c. 900 m) above sea level at or near the equator and to slightly lower elevations in parts of Mexico and other areas near the margins of the tropics. In this hot, wet environment, the favored crops are rice, sugarcane, bananas, and cacao. Latin America's blacks are concentrated in many of the *tierra caliente* zones, a legacy of the slave trade, when they were forced to work the region's plantations.

The *tierra caliente* merges almost imperceptibly into the *tierra templada*, which flanks the rugged western mountain ranges and is the uppermost climate in the lower uplands and highlands to the east. Sugarcane, cacao, bananas, oranges, and other lowland products reach their uppermost limits in the *tierra templada*, but this zone is most famous as coffee habitat (•**Figure 10.7**). The upper limits of this zone—approximately

6,000 feet (1,800 m) above sea level—tend to be the upper limit of European-introduced plantation agriculture and modern commercial crops in Latin America. Densely inhabited sections occupy large areas in southeastern Brazil, Colombia, Central America, and Mexico. Although broadleaf evergreen (tropical rain forest) trees characterize the moister, hotter parts of this zone, coniferous evergreens replace them in parts of the zone's poleward margins. In places such as the highlands of Brazil and Venezuela, where there is less moisture, scrub forest or savanna grasses prevail.

Seven metropolises exceeding 2 million in population—São Paulo, Belo Horizonte, Brasília, Caracas, Medellín, Cali, and Guadalajara—are in the *tierra templada,* and two others—Mexico City and Bogotá—lie slightly above it. Four smaller cities that are national capitals and the largest cities in their respective countries are located in the *tierra templada*: Guatemala City, San Salvador, Tegucigalpa, and San José. Others, like Rio de Janeiro, which are situated at lower elevations, have close ties with predominantly residential or resort towns in these cooler temperature areas.

The *tierra fría* is comprised of high plateaus, basins, valleys, and mountain slopes within the great mountain chain that extends from northern Mexico to Cape Horn. By far the largest areas of this habitat are in the Andes, with significant areas also in Mexico (•**Figure 10.8**). Around the equator, the *tierra fría* begins at about 6,000 feet (c. 1,800 m) and extends upward to the **upper limit of agriculture** (represented by such hardy crops as potatoes and barley) and the **tree line** (the upper limit of natural tree growth) at about 12,000 feet (c. 3,600 m).

The *tierra fría* experiences frost and is often the habitat of a Native American economy with a strong subsistence component. In a classic process of marginalization, European colonization of Latin America drove some Native American settlements upslope and into the *tierra fría* zone, although some major populations (notably the Inca of Peru) had already

• **Figure 10.8** Characteristic *tierra fría* landscape and agriculture in Andean South America.

• **Figure 10.9** The *páramo* at about 12,000 feet (c. 3,600 m) in the Andes of Colombia.

selected upland locations for their settlement before the arrival of Columbus. These upland settlements are most extensive in Ecuador, Peru, and Bolivia and are also very evident in Colombia, Guatemala, and southern Mexico.

Another zone lies above the other three and consists of the alpine meadows, known as *páramos*, along with still higher barren rocks and permanent fields of snow and ice (•**Figure 10.9**). This is the *tierra helada*. It generally lies between 12,000 feet (c. 3,600 m) and the lower edge of the snow line. It supports some grains and livestock (llama, alpaca, sheep) but is largely above the mountain flanks that are central to upland native settlement and agriculture. In the highlands of Latin America, the *tierra fría* tends to be a last retreat and a major home of the indigenous peoples, except in Guatemala and Mexico. Most of their settlements are rural villages based on subsistence agriculture. In some cases, notably in Bolivia and Peru, valuable minerals like tin and copper are located here, attracting large-scale mining enterprises into the *tierra fría* and the *tierra helada*.

The roof of South America is Mount Aconcagua in Argentina, rising to 22,841 feet (6962 m) in the Andes. Its slopes, like many of this higher Andean reaches, are covered in glacial ice. Andean glaciers are following the worldwide trend of melting. As the glaciers melt, they take consistent, dependable water supplies with them. Some scientists predict that the Bolivian city of La Paz, with its million residents, will be the first large urban casualty of climate change. There are other natural threats as well.

Natural Hazards in Latin America

Adjoining a large section of the Pacific Ring of Fire and fronting two seasonal hurricane regions, Latin America is beset by natural hazards. A belt of land stretching from northern Mexico all the way to Tierra del Fuego, the "Land of Fire" [21] at the southern tip of South America, has a violent history of earthquakes and volcanic eruptions. On the map of global tectonic plates (Figure 2.1), you can see the Nazca Plate and South American Plate confronting one another, with tremendous geological consequences including the rise of the Andes Mountain chain and those volcanoes and earthquakes. The largest earthquake ever recorded on earth was the titanic 9.5 quake in 1960, just off the coast of southern Chile.

Figure 2.1 also depicts an active tectonic area on the eastern side of the Americas: the Caribbean Plate is also in motion, causing earthquakes and a variety of associated tectonic features. You can see that one of this plate's boundaries passes through the island of Hispaniola, where the countries of Haiti and the Dominican Republic are located (see Figure 10.1 and Geography of Natural Hazards, next page).

Hurricanes, giant low-pressure systems that feed on warm waters of the Atlantic and Pacific Oceans, have sometimes devastating impacts. A single storm can wreck the economy of an island nation, as Hurricane Ivan did by wiping out Grenada's nutmeg crop in 2004. The monstrous Hurricane Mitch (category 5, highest on the Saffir-Simpson scale) was an unusually late storm that made landfall in Honduras on October 28, 1998, and roamed for five days over that country and Nicaragua, El Salvador, Guatemala, and Mexico. Its ferocious winds and rainfalls of up to 2 feet (60 cm) brought floods, landslides, and storm damage—intensified by the region's widespread deforestation—that killed an estimated 15,000 people; inflicted enormous damage on property, crops, roads, power, and other infrastructure; and left disease epidemics in its wake. As if Mitch were not enough, the region had also suffered from drought and fires caused by El Niño in 1997 and 1998.

What is El Niño? It is best understood as a complete reversal of a normal climatic pattern. Every few years, things seem to go wrong with the winds and waters of the Pacific Ocean. Normally, the winds blow east to west, helping to promote the upwelling of cold, nutrient-rich water from the depths to the surface of the tropical eastern Pacific (•Figure 10.10). But sometimes, usually beginning in December, the winds reverse direction, blowing from west to east, suppressing the upwelling water and thus raising the surface temperature of the water by as much as 20°F (11°C). Because of its common occurrence in December, that event has been dubbed **El Niño**, meaning "the Baby" in reference to the Christ child, whose birth is celebrated in December. Less prosaically, meteorologists know it as **El Niño Southern Oscillation (ENSO)**.

El Niño conditions sometimes last a year or more, causing global climatic disruptions. These vary between events, but the typical El Niño pattern is illustrated in •**Figure 10.11**. Closest to the source of the condition, Peru experiences torrential rains. Unusually high rainfall also occurs in southern Brazil, the southeastern United States, western Polynesia, and East Africa. Drought or unusually high temperatures settle

GEOGRAPHY OF NATURAL HAZARDS | Heartbreak in Haiti

On January 12, 2010, an earthquake with a magnitude of 7.0 struck Hispaniola on a strike-slip fault zone just 16 miles (25 km) from Haiti's capital, Port-au-Prince. In a more developed country better prepared with earthquake-resistant infrastructure, such a quake might not have produced extensive damage. But in Haiti, by far the poorest country in the Western Hemisphere and one of the poorest on earth, this temblor was catastrophic. It killed an estimated 315,000 people and made nearly a million homeless (•Figure 10.A).

A unique combination of geologic, historic, and political factors made Port-au-Prince especially vulnerable to seismic catastrophe. Not only is the city close to the fault zone on which the earthquake struck, but much of Port-au-Prince lies atop soft sedimentary rock that amplified the seismic waves. Then there were the population issues. United States forces occupied Haiti for about two decades after World War I, building roads and other infrastructure that centralized urbanization in Haiti: Port-au-Prince became a magnet for settlement and development. Far surpassing the metric of a primate city, the capital was home to one in three Haitians when the earthquake hit. Most of the city's housing stock was made up of **informal settlements**, a term often used interchangeably with *slums* but meaning the occupants lack legal claims to the land and that the housing is not in compliance with planning and building regulations. Much of the housing is fashioned from recycled materials. Meeting the requirements of earthquake resistance is far beyond the financial means of the people of Port-au-Prince, where two-thirds of the labor force lacks formal employment. Much of Port-au-Prince is a sprawling slum, described this way by journalist Amy Wiletnz:

"a heart-tugging city that functioned on a complicated, hypercharged fuel of chaos, exposed wiring, pig slop, smog, gingerbread turrets, hot cooking oil, rum, cockfights and bougainvillea."[1]

The utter devastation of the 2010 earthquake gave Haitians the opportunity to re-fashion their urban geography. Assisted by geographers and other scientists from the University of Miami, a group of Haitian urban planners has come up with a plan to decentralize Haiti, redistributing large numbers of people to smaller Haitian cities at safer distances from fault zones. The plan will completely transform Haiti from a country dominated by the primacy of Port-au-Prince to one with a network of smaller urban "growth poles." These would be more suited to the economic assets Haiti could best develop: agriculture and tourism. Planners will locate schools, markets, and hospitals and health care centers in smaller towns and villages so that people will not need to leave these settlements and gravitate again to Port-au-Prince. Homes, stores, and offices will be constructed with earthquakes in mind and grow over time in modular fashion from basic shelters into permanent fixtures.

Unfortunately, as a metropolis ripe for seismic ruin, Port-au-Prince is probably not alone. Other vulnerable cities, which seismologists call "rubble in waiting," include Istanbul, Turkey; Karachi, Pakistan; Katmandu, Nepal; Tehran, Iran; and Lima, Peru. All of these cities continue to swell from rural to urban migration, and all share a surplus of informal settlements built from such dangerous materials as brick and mud-brick. University of Colorado seismologist Roger Bilham discussed the seismic hazards of "an unrecognized weapon of mass destruction: houses."[2]

Xinhua /Landov

• **Figure 10.A** A magnitude 7.0 quake devastated Port-au-Prince, the Haitian capital, in December 2010.

in over northern Brazil, southeastern Canada and the northeastern United States, southern Alaska and western Canada, the Korean Peninsula and northeastern China, Indonesia and northern Australia, and southeastern Africa. When warmed, Pacific waters are far less rich in nutrients, so typically fish catches plummet and food chains are disrupted, leading to massive wildlife losses. The most severe recent El Niño events occurred in 1982–1983 and in 1997–1998, but it is not unusual for three or four moderate events to occur every decade.

10.3 Cultural and Historical Geographies

It is perhaps unfortunate that this region came to be known as Latin America because there were no "Latins" among its inhabitants before the end of the fifteenth century. In 1492, when the first Europeans arrived (perhaps after seafaring Polynesians reached the west coast of South America; see page 291), the region was home to an estimated 50 to 100 million **Native**

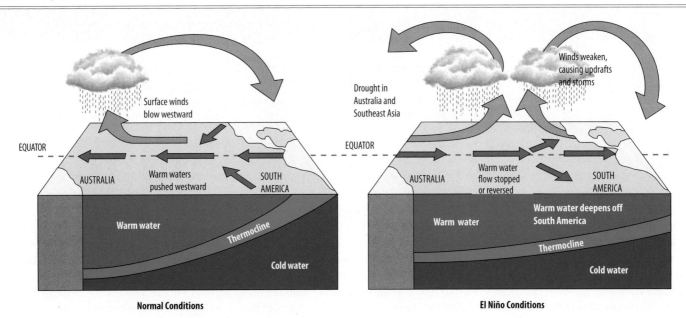

• **Figure 10.10** Normal and El Niño conditions in the Pacific Ocean.
Source: From MILLER/SPOOLMAN, Environmental Science, 13E. © 2011 Cengage Learning.

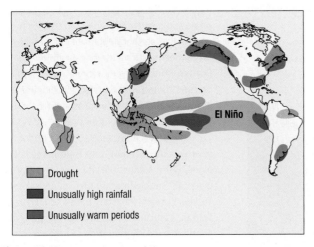

• **Figure 10.11** Climatic impacts of El Niño.
Source: From MILLER/SPOOLMAN, Environmental Science, 13E. © 2011 Cengage Learning.

Americans, also known as **Amerindians** or, as Columbus (who thought he had landed in India) misidentified them, **Indians**. After their early migrations starting around 10,000 B.C.E. from the Asian continent, apparently across a land bridge now drowned by the Bering Strait, they developed many distinctive livelihoods and cultures. Some Native Americans practiced hunting and gathering, while others developed agriculture and took the same kind of pathway to urban life and civilization followed by several Old World cultures. Culture hearths associated with civilization emerged in the Andes region of South America and in southern Mexico and adjacent Central America (•Figure 10.12).

Civilizations Predating the Europeans' Arrival

In 1492, as they do today, the **Maya** inhabited southern Mexico, Belize, and Guatemala. They practiced agriculture based on maize (corn), squash, beans, and chili peppers and had by 200 C.E. developed a highly complex civilization in both lowland tropical rain forests and highland volcanic regions. They created monumental religious and residential structures, including pyramids, temples, and astronomical observatories. They built stone roadways through dense forest areas but notably had no wheeled vehicles, nor did they have the sailing vessels, the plow, beasts of burden, and some other tools and trappings associated with civilizations elsewhere. The Maya had highly developed systems of mathematics, astronomy, and engineering and an extremely precise calendar (with an ending of December 12, 2012—a fact that sparked a rash of apocalyptic prose and films leading up to the date).

The Maya also had a hieroglyphic writing system that survives on stone monuments and in a handful of books. Regarding them as heretical, Spanish priests put almost all of the written volumes to the torch. On July 12, 1562, Bishop Diego de Landa destroyed at least 30 Mayan books in a bonfire outside a church in Mexico's Yucatán Peninsula. "They contained nothing in which there were not to be seen superstitions and falsehoods of the devil," de Landa later wrote. "We burned them all." Many Maya watched this event, which, de Landa recalled, "they regretted to an amazing degree and which caused them much affliction."[3]

The Maya civilization peaked around 900 C.E. and then began a slow decline. It is still unknown why this advanced civilization apparently abandoned its massive religious and urban centers and melted back into the forests of the Yucatán Peninsula and points south. Theories include natural disasters, agricultural failures, and revolts against ruling elites. When the Spaniards arrived, the great Maya cities were overgrown. Many can be seen in their restored glory today. Chichén Itzá in Mexico's Yucatán Peninsula and Tikal in northern Guatemala are very significant sources of international tourism revenue for the respective countries.

• **Figure 10.12** Major Native American groups and civilizations in Latin America on the eve of the Spanish conquest.

Source: Based on The History Atlas of South America, Edwin Early, et al., eds., Macmillan Reference USA, 1998, The Gale Group.

North of the Maya realm, in the Valley of Mexico about 30 miles (40 km) from present-day Mexico City, arose Teotihuacán, the first true urban center in the Western Hemisphere. Its construction began about 2,000 years ago, and by 500 C.E. it covered about 8 square miles (21 sq km) and had 200,000 residents, comparable to London a millennium later. The city was laid out on a grid aligned with stars and constellations that were central to the ceremonial timekeeping of the **Teotihuacános**. There were some 2,000 apartment compounds, a 20-story Pyramid of the Sun, a Pyramid of the Moon, and a 39-acre (16,000 sq m) civic and religious complex called the Great Compound or Citadel (•**Figure 10.13**). The origins, language, and culture of Teotihuacános are still not fully understood, and it is hoped that future archaeology will shed more light on these extraordinarily creative people.

The **Aztecs**, who called themselves the Mexica (pronounced "meh-*shee*-kuh") and came later than the Teotihuacános, honored the earlier city as sacred space and a religious center when they took over the Valley of Mexico in the 1400s. They founded their capital, Tenochtitlán (now overlain by Mexico City), in 1325. With a population of about 250,000, Tenochtitlán had

a sacred center surrounded by active marketplaces, all linked by stone-surfaced roads to other parts of the urban settlement. There were some outlying districts that provided goods for trade, and craftspeople and merchants came to the city to work. In the middle of the fifteenth century, the Aztecs joined

• **Figure 10.13** Teotihuacán's Pyramid of the Sun.

Joe Hobbs

Greg Hobbs

• **Figure 10.14** Machu Picchu, the legendary "lost city of the Inca," has a dramatic site above Peru's Urubamba Valley.

with rulers of two neighboring cities to form the "Triple Alliance". From this base, the Aztecs and their allies developed an empire stretching across central Mexico. They tried but failed to conquer the kingdom of an adjacent civilization, the Tarascan. The Aztecs left a legacy of much merit in metallurgy (which diffused to Mexico from the Andean region), woodworking, weaving, pottery, and especially urban design.

In South America, the **Inca** built an empire about 2,000 miles (3,200 km) long from northern Ecuador to central Chile. They controlled it for only about a century before 1532, when the Spanish conquistadors, under Francisco Pizarro, began their subjugation. The Inca's origins, dating to about 1200 C.E., are not fully known. From their governing center in Cuzco, Peru, the Inca engineered a system of roads, suspension bridges, and settlements that connected and supported their empire. The road system ran more than 2,500 miles (c. 4,100 km) and was busy with movement of trade and tribute. The Inca also achieved great skill in stonework, terrace construction, and irrigation networks (•Figure 10.14). Remarkably, they had neither paper nor a writing system, but kept mathematical records by knotted ropes called *quipus* (pronounced "*kee*-pooz").

Still another group, the little-known **Nazca** culture, left a unique legacy on the landscape of South America about 2,000 years ago. Their "Nazca lines" in the desert of southern Peru are a complex of nearly 200 square miles (c. 520 sq km) of massive carvings in the sandy surface, depicting birds, insects, and fish, the smallest of which is an animal figure some 80 feet (c. 23 m) in length. These designs can only be appreciated from the air, giving rise to the improbable theory that they were built as landmarks for "ancient astronauts." No written records of the Nazca culture survive, so the purpose of their lines may never become known. Another of the less celebrated civilizations of South America is the **Chibcha** of what is now Colombia. The Chibcha lived in small agricultural villages rather than large cities, and their exquisite gold work is a major legacy of their culture. Speakers of Chibchan

languages now live in the Andes of Colombia and Ecuador and as far north as Costa Rica. Beneath the dominant imported languages of Spanish and Portuguese, the linguistic substrate of Latin America is still rich.

Languages in Latin America

With the exception of the Nazcans (whose record ends about 600 C.E.), the groups just discussed were only the largest major civilizations at the time of the Spanish conquest. There were smaller civilizations and others that predated the conquest. Far more numerous were the groups that practiced hunting and gathering or agriculture but did not build cities and civilizations. Their legacy survives today in the cultures and languages of a rich variety of Native American ethnic groups. Linguistic geographers disagree sharply over Native American language distributions, so only very general patterns are depicted in •Figure 10.15.

Six Native American language families are represented in Mexico and Central America. The **Hokan-Siouan language family** has speakers in Mexico's Baha Peninsula and in El Salvador and adjacent Honduras. The **Aztec-Tanoan language family** includes the *Nahuatl* that the Aztecs spoke and that is still spoken in northern Mexico today. The **Oto-Manguean language family** includes seven languages spoken mainly in northern Mexico. There is a small **Totonac language family** region on the Gulf of Mexico. Southern Mexico is home to

• **Figure 10.15** Indigenous and European languages of Latin America.
Source: Based on Bernard Comrie, et al., The Atlas of Languages, Revised Edition (New York: Facts on File, 2003).

the **Penutian** and **Mayan language families**. The indigenous cultures these languages represent form a significant minority overall in Mexico at 30 percent, but they comprise over 50 percent in some southern states of the country. The Mayan language family has five major subfamilies, representing more than 60 million speakers living all across the realm of the ancient Maya, especially in Guatemala. Some Maya, particularly in highland Guatemala, never did mix with outsiders and are ethnically little changed from their forebears. The Maya of highland Guatemala tend to speak Mayan as their first language, but many are bilingual and also speak Spanish. With its 23 Maya groups, Guatemala's population is more than 40 percent Native American overall.

South America—which had a Native American population as early as 12,500 years ago—has an even richer language palette, with about 600 languages belonging to 16 families. They include the **Quechua-Aymaran language family**, to which the language of the ancient Inca belonged. More than 7 million people today, mainly in Peru but also in Colombia, Ecuador, Bolivia, and Argentina, speak one of the five languages in this family. Most still speak *Quechua* as a first language, but many are bilingual with Spanish.

Indigenous languages are used far more widely in South America than elsewhere in Latin America, and there are more speakers there that do not speak a second language. Very high proportions of the populations in highland South America are Native Americans. The largest populations of Native Americans are found in Bolivia (55% indigenous) and Peru (45%), with smaller but still significant numbers in the basin of the Amazon River and in Panama, where scattered groups of lowland peoples maintain their traditional cultures. Remarkably, in 2011 remote sensing imagery of the upper Amazon region revealed 18 "uncontacted tribes," meaning that these isolated peoples have had little and perhaps no interaction with the wider world. Anthropologists do not know what ethnic groups they belong to or what languages they speak. These scientists are debating the ethical issues of whether they should initiate contact with the groups. The fear is that interaction even with well-meaning researchers would begin a process of cultural change and loss.

European languages joined the linguistic mix in Latin America at the end of the fifteenth century (see Figure 10.15). At first, only the colonizers spoke them, but they became more widespread as colonial administration progressed and expanded. The language distributions today reflect the pattern of colonial rule. Spanish, now one of the world's most spoken languages, or **megalanguages** (along with Chinese, English, Hindi, and Arabic), is the most widespread and is the prevalent European language throughout the region, with these exceptions: Portuguese in Brazil; French in French Guiana, Haiti, Guadeloupe, and Martinique; Dutch in Suriname; and English in Guyana and Belize. These languages are also spoken in the dependencies and former colonies of several other small Caribbean islands. The European languages generally became the sole official languages of both the colonial administrations and the independent countries that succeeded them. They serve as welcome lingua francas in countries like Mexico and Peru that have a bewildering diversity of indigenous languages. Bolivia, Peru, Paraguay, and Guatemala have official languages that are both European and indigenous.

Another linguistic product of European colonialism in Latin America was the evolution of **creole languages**. These are tongues that developed among the black slaves and indentured servants brought to work the plantations of the Caribbean islands and the Atlantic and Caribbean coasts of South and Central America. Their vocabularies come mainly from the respective colonizers' languages of Spanish, Portuguese, French, English, and Dutch. Sometimes a speaker of the parent language can recognize the creole; for example, in Costa Rican Creole English, "*Mi did have a kozin im was a boxer, kom from Panama*" is easily enough understood as "I had a cousin from Panama who was a boxer."[4] But it is not always that simple. Like Pidgin in the South Pacific (see page 292), Creoles are distinct languages and have their own grammars.

None of the indigenous African languages that the slaves brought to the New World survived there to the present day. Some vocabulary and other elements of the ancestral African languages do survive in some of the creoles, such as in the *Garifuna* spoken on the coast of Belize.

The European Conquest

The arrival of Christopher Columbus and his three Spanish ships in 1492 brought more than a new language to the New World. It marked the beginning of profound changes in almost every aspect of life in what would become Latin America.

First came death, both deliberate and unintended. A series of expeditions to the New World convinced the Spaniards that there were riches worth pursuing in that far-off land. The Spanish conquistador Hernando (Hernán) Cortés landed on Mexico's coast near present-day Veracruz in 1519 and led his cohort of invaders inland, building alliances with ethnic groups opposed to the Aztecs and killing those who would not join him. Unfortunately for the Aztecs, Cortés's arrival seemed to fulfill the prophecy that Quetzalcoatl, a fair-skinned, bearded Aztec god, would return one day, approaching from the east. In Tenochtitlán, the Aztec leader, Moctezuma, received Cortés and his army as emissaries of Quetzalcoatl. Cortés and just 30 men took Moctezuma prisoner and were unopposed, apparently because of the widespread belief in Quetzalcoatl's return. Using a divide-and-conquer strategy that enlisted local forces, Cortés achieved a military victory over the Aztecs in 1521, destroying Tenochtitlán and laying the foundation for Mexico City in its ashes.

Spain soon dispatched new expeditions to conquer other peoples of Mexico and Central America. When Francisco Pizarro landed in Peru in 1531 with only 168 soldiers and 102 horses, he was warmly greeted by an Incan leader, Atahualpa. The Spanish conquistador imprisoned Atahualpa, freed him for a huge ransom, imprisoned him again, and then had him hanged (he was to have been burned at the stake, but Pizarro modified the execution because Atahualpa agreed to be baptized). Two years later, Pizarro's forces held the Incan capital Cuzco, and from that point the colonization of all of South America by the Spanish—and in Brazil, the Portuguese—proceeded swiftly.

Disease brought death to the indigenous people on a vast scale. The Native Americans had never been exposed to, and hence had no natural immunities against, the host of diseases that the Spanish and the Portuguese introduced: bubonic plague, measles, influenza, diphtheria, whooping cough, typhus, chickenpox, tuberculosis, and smallpox, joined later by the malaria and yellow fever that came with African slaves. The worst was smallpox, which killed between one-third and one-half of all Native Americans in affected areas. The diseases, famine, and enslavement that came with Spanish and Portuguese colonization brought what was probably one of humankind's largest population declines ever. The numbers are elusive but were certainly huge. The Spanish chronicler Bartolomé de Las Casas wrote that in Spanish-controlled Latin America, 40 million Native Americans had died by 1560. The geographer David Clawson estimates that by 1650, 90 to 95 percent of the Native Americans who had come into contact with Europeans had died.

European settlement patterns emerged as the conquest proceeded. The development of ports as bases for subsequent penetration of the interior was the first major settlement task. Large coastal cities had not existed prior to European colonization. Many early ports eventually grew into large cities, and a few became major metropolitan centers. In some cases—notably Lima, Caracas, and Santiago—the main city developed a bit inland but retained a close connection with a smaller coastal city that was the actual port.

Around the ports, agricultural districts developed, spread, and shipped an increasing volume of trade products overseas. Plants and animals domesticated by Native Americans would revolutionize diets and habits in Europe and elsewhere in the Old World: corn, tobacco, potatoes, cacao, and turkeys were among them (•Figure 10.16; Table 4.3, page 82). In time, on the other side of the Colombian Exchange, the European-introduced horses, cattle, sheep, donkeys, wheat, sugarcane, coffee, and bananas would revolutionize New World agriculture. In the second half of the 1500s, slave ships began to bring the Africans who provided the principal labor on the plantations created and owned by Europeans, primarily in the *tierra caliente* tropical lowlands with the generally fertile alluvial soils of the coastal plains.

The powerful allure of gold and silver stimulated deeper European penetration of the Andes and the Brazilian Highlands in South America. After capturing the stores of precious metals that had been accumulated by the indigenous peoples for ceremonial use, tribute, and trade, the foreigners opened or reopened mines and established market centers to service them. Some highland settlements became centers of new ranching and plantation enterprises. A few highland cities, of which the largest is Bogotá in Colombia, grew into large metropolises. Separated from the seaports by difficult terrain, these cities were eventually connected to the coasts

Joe Hobbs

• **Figure 10.16** Corn (maize), one of the world's great staple crops, was domesticated by Native Americans. It continues to be the most important food in much of Latin America. This is a street scene in Oaxaca, Mexico.

by feats of great engineering skill. Some of the main seaports and highland cities became important centers of colonial government as well as economic nodes, and several are national capitals today, including La Paz in Bolivia and Quito in Ecuador.

At the heart of each Spanish and Portuguese city established in Latin America, the newcomers placed their familiar combination of a plaza dominated by a massive Roman Catholic cathedral (•Figure 10.17). Roman Catholicism was introduced—imposed, actually—as the one and only acceptable faith in the New World, and 80 percent of the region's people are Catholic today. The British and Dutch brought their Protestant faiths to their possessions. Today, Protestant offshoots, notably Evangelical and Pentecostal religions—characterized by high-energy, participatory forms of worship and their appeal to disrupted families and the poor—are making inroads in traditionally Catholic communities throughout the region. Jamaica has embraced Rastafari, a faith championing black empowerment

• **Figure 10.17** A plaza and cathedral are at the center of most Latin American cities. This is the *Zócalo* of Mexico City. From lower right to center are three male traditional healers, or shamans, practicing their trade.

and holistic living dating to the early twentieth century, with unlikely links to Ethiopia.

341

Many indigenous beliefs have become syncretized with Catholicism, but some of them dating to **pre-Columbian times** (before 1492) remain alive among Native Americans in Latin America. The Catholic Maya of Guatemala, for example, still make offerings to the "earth lords" who dwell in caves (•Figure 10.18). Innumerable icons testify to the blending of cultures and beliefs. One of the most enigmatic is Maximón, also known as San Simón, a Guatemalan folk saint with pre-Columbian and Catholic attributes. Supplicants donate cigars, beer, and money to his effigy, hoping to fulfill votive prayers for such matters as good health, crops, and marriages (•Figure 10.19). Other syncretic faiths include that of **María Lionza,**

"Saint Mary of the Jaguar," whose devotees in Venezuela have a blend of Catholic, West African, and indigenous beliefs; and the Caribbean region's **Santería**, or the "Way of the Saints," similarly imbued with that mix of beliefs and, in particular, a strong thread of Yoruba culture brought by slaves from West Africa.

Ethnicity in Latin America

Despite the dominance in Latin America of culture traits derived from Europe, a large majority of the original European settlers or their descendants intermarried with Native Americans or blacks. Only Argentina, Uruguay, and Costa

• **Figure 10.18** Cave offering of the Q'eqchi' Maya, Alta Verapaz, Guatemala.

• **Figure 10.19** The effigy of Maximón (San Simón) in a private home in Santiago de Atitlan, Guatemala.

(a) (b)

(c) (d)

All photos courtesy of Joe Hobbs

• **Figure 10.20** In addition to the Native Americans of Latin America, the leading racial types are **(a)** Europeans, **(b)** blacks, **(c)** mestizos, and **(d)** mulattoes.

Rica have significant white European ethnic groups today (•Figure 10.20a). These are mainly families of immigrants from Europe who arrived relatively recently, during the late nineteenth and early twentieth centuries. Scattered districts in other countries are also predominantly European.

Black Latin Americans of relatively unmixed African descent live in the greatest numbers on the Caribbean islands and along the Atlantic coastal lowlands in Middle and South America (Figure 10.20b). These are the areas to which African slaves were brought during the colonial period, mainly to work on sugar plantations. An estimated 3 to 4 million were sold in Brazil alone. Slavery was gradually abolished during the nineteenth century, although not until the 1880s in Brazil and Cuba. By that time, slavery had generated large fortunes for many owners of plantations and slave ships. The African peoples it introduced to the region have made cultural contributions that today provide a varied, vibrant, and important part of Latin American civilization.

Latin America has escaped many of the racial tensions and violence that have battered much of the world. This may be because a majority of Latin Americans are of mixed heritage,

the most common being a mixture of Spanish and Native American that has resulted in a heterogeneous group known as **mestizos** (Figure 10.20c). A person of mixed African (black) and European (white) ancestry, sometimes with Native American blood as well, is known as **mulatto** (Figure 10.20d); in some locales, the term used is, like the linguistic term, **Creole**. It must be noted, however, that in most places, socioeconomic discrepancies are strongly associated with ethnicities in the region. A thin stratum of people of mainly European origin often dominate government and business sectors, with mestizos and mulattoes making up the bulk of the middle and lower classes. Native Americans almost always have the lowest standing, and in some countries such as Guatemala, systematic exclusion and violence has been directed against them from time to time.

Mestizos have not lost all traits of their Amerindian forebears. Although it can be hazardous to generalize about ethnicities, the mestizos' inheritance from their Native American roots includes the collective exploitation of land, the accumulation of lifelong personal bonds, the devotion to a whole series of supernatural beings (who have Christian names but non-Western origins), the ritualized remembrance of deceased relatives, and reverence for political leaders.[5] As will be seen later in the chapter, reverence for political leaders cannot be taken for granted in this region today.

Despite enormous impacts, some Native American cultures have proven remarkably resilient, as we will see throughout the rest of the chapter. Some, like the indigenous peoples of Nicaragua's Miskito coast, are actively pushing for independence (they call themselves the Communitarian Nation of Mosquitia). Many are technologically savvy, using the Internet and social media to elicit understanding and political and economic support in their struggles. Some are championed by Western celebrities and activists, but others, including some of the groups pushing back against oil development in the Ecuadorian Amazon, have managed to gain a footing without such assistance.

Some governments actively protect the lives and ways of life of Amerindians. New constitutions expanded indigenous rights in Colombia in 1991, Peru in 1993, Bolivia in 1994, and Ecuador in 1997. In these countries, Native Americans now have more rights over land, and the governments recognize communal rights over some resources. Brazilian law forbids mineral extraction by outsiders on aboriginal lands, but illegal diamond and gold mining has led to many bloody skirmishes between Native Americans and prospectors in several Indian reserves in the country. On the negative side of this protection, as the Native Americans see it, many resources of Brazilian and other national Indian reserves are placed under government control, and often indigenous peoples are forbidden from setting up their own businesses.

10.4 Economic Geography

Latin America is generally a region of less developed countries (LDCs). An estimated one-third of its residents live in poverty. However, both poverty and unemployment have diminished in recent years for the region as a whole, and the middle class is growing. In most countries, this is thanks to an abundance of raw materials; Mexico is an exception in that it relies more on exports of manufactured goods rather than commodities. The commodities boom is tied mainly to China's dynamic growth. As in Africa and elsewhere, Chinese interests are scouting out every possible prospect in Latin America that could help feed 333 that nation's voracious appetite for minerals, energy, and food. Latin America's trade with China was 16 times bigger in 2009 than it was in 1999, and China now stands as Latin America's leading trade partner.

In some Latin American countries, this raises the same fears as in Africa: that China is seeking to develop a classic mercantile relationship with Latin America, taking raw materials (both agricultural and industrial) and profitably selling its own manu45 factured goods. But for now, the influx of Chinese capital is welcomed. It helped Latin America to weather the Great Recession and emerge with economic growth rates generally well above those of industrialized countries like the United States.

Although Latin America's overall situation is improving, the region suffers from notoriously large discrepancies between the haves and the have-nots. The glitter of the great metropolises like Mexico City and Rio de Janeiro, with their forests of new skyscrapers, may distract attention from poverty, but portions of these cities are massive, squalid slums, known as *favelas* in Brazil and *barrios* in Spanish-speaking countries (•Figure 10.21). Such **shantytowns**, which are practically universal in the world's LDCs, are full of unemployed, underemployed, and sometimes undernourished people. "*Ni ni*" these people are called throughout much of the region: neither educated nor employed. Their sites and situations are noteworthy: remarkably, many *favelas* immediately border some of the most costly real estate in Rio de Janeiro and other cities; others are in the least desirable, most hazard-prone areas of the rural landscape, making headlines when mudslides slam through them, taking scores of lives. Many people prefer their plight in the *favelas* to the conditions of the depressed rural landscapes from which they came. Glimpses of life in these slums may be seen in the Brazilian film drama *City of God*. Brazil's preparations to host the 2014 Soccer World Cup and the 2016 Summer Olympics included major works to improve conditions in the *favelas*.

Latin American economies have born many of the traditional hallmarks of LDCs. Trying to overcome their economic problems and constrain popular discontent, for example, many Latin American governments have borrowed heavily from the international banking community. In the early 2000s, unpaid loans had reached staggering proportions, and the Latin American debtor nations were having great difficulty in mustering even the annual interest payments, let alone generating surplus capital to pay toward the principal of these troubling loans—a dilemma all too characteristic of the world's LDCs. Like other 47 LDCs, Latin America also has historically been over-reliant on non-value-added goods like cash crops and minerals. As we will see, free-trade agreements form the foundations of a recent push to move away from commercial agriculture and raw materials and toward manufactured exports.

Commercial Agriculture

Although the total number of Latin Americans employed in farming has not declined much in recent decades, the value of

•Figure 10.21 Many of Latin America's poorest people live in ramshackle housing. This is the shantytown of Belen in Iquitos, Peru. The great Amazon River is just out of view to the left.

Joe Hobbs

Bananas are the most popular fruit in the United States, with more than 8 billion sold annually—or about 26 bananas per person per year. Now it is possible to be a discriminating banana buyer. At many North American and European grocery stores, one may find bananas, pineapples, other fruits, chocolates, and coffees bearing "Fair Trade Certified" stickers (•Figure 10.B). These are higher in price than the standard products, but consumers who buy them are generally motivated by a sense of social responsibility. Consumers keen to reduce their environmental footprint are also willing to pay a premium for fair trade products that are organically produced.

In standard trade transactions, only a tiny fraction—2 cents per pound of bananas, for example—of the final price paid in the supermarket actually goes to the grower. Intermediaries, including importers, distributors, and "ripeners" (who spray ethylene gas to turn the bananas yellow), all get a cut. Fair trade buyers deal directly with the farmer cooperatives they helped organize, avoiding brokers and intermediaries and guaranteeing higher prices—typically, 18 cents per pound for bananas—for the farmers.

The fair trade organizations also help the farmers establish schools and health clinics.

Before socially conscious consumerism became popular in the United States, the **fair trade movement** developed in Europe as people

• **Figure 10.B** Fair trade certification of coffee sold in a U.S. health food store reflects changes in consumers' awareness and an improvement in the producers' livelihoods.

began taking notice of the impacts of collapsing coffee prices. Whereas farmers could earn $3.18 per pound for their coffee in 1999, a glut in the world coffee supply, caused largely by surging production in Vietnam, sent prices down to 47 cents per pound by 2001—the lowest in a century, adjusted for inflation. Europeans learned that African and Latin American coffee farmers, never wealthy to begin with, were now destitute, so poor that their young children were compelled to work in the fields to help make ends meet. Green Party politicians and others stepped forward with the fair trade solution, seeking to guarantee what they considered a living wage for the producers of these commodities.

Coffee is the world's second largest traded commodity (after oil), so fair trade coffee could affect many people. Now more than a million coffee growers in Tanzania, Rwanda, Ecuador, Nicaragua, and other countries belong to the cooperatives that sell their products through fair trade channels instead of to commercial producers. Fair trade products are becoming more mainstream in the United States, especially as even large retailers like Walmart realize that social and environmental justice can be good for their bottom line.

---

agriculture in national economies has experienced a marked percentage decline as economies have become more diversified. But in many countries, more than half of all export revenue is still derived from farm products, and some countries rely mainly on one or two agricultural commodities. As in other LDCs, overreliance on a narrow range of exports makes the Latin American countries economically vulnerable to changes in market conditions, competition from other sources, and changing consumer appetites. Reliance on coffee and bananas has whipsawed the economic fortunes of many countries, especially those in Central America such as Honduras and Guatemala that came to be known disparagingly as "banana republics." In colonial fashion, the American-owned United Fruit Company owned vast tracts of land for the production of bananas and other agricultural commodities in these countries and had powerful influence on local politics. In Guatemala, President Árbenz developed land reform policies that were seen as a threat to the United Fruit Company. He was subsequently ousted in 1954 in a CIA-led coup (it has been noted that CIA Director Dulles was a significant United Fruit Company shareholder). Today's fair trade movements represent a reversal of the exploitative relationships that marked the United Fruit Company era (see Insights, this page).

Types of Farms

Farms in Latin America can be divided into two major classes by size and system of production. Large estates with a strong commercial orientation are known as *latifundia* (sing., *latifundio*). These estates, whether called *haciendas*, plantations, or some other name, are owned by families or corporations (•Figure 10.22). Some have been in the hands of the same family for centuries. The desire to own land as a form of wealth and a symbol of prestige and power has always been a strong characteristic of Latin American societies. Spanish and Portuguese sovereigns granted huge tracts to members of the military nobles who led the way in exploration and conquest. In sporadic land reform programs, some of this land has been reallocated to small farmers by government action. But in many countries, a very large share of the land is still in the hands of a small, wealthy, landowning class.

Minifundia (sing., *minifundio*) are smaller holdings with a strong subsistence component. The people who farm them generally lack the money to purchase large and fertile properties; they farm marginal plots, often on a sharecropping basis. Individual landowners are often burdened by indebtedness, the threat of foreclosure of farm loans, and the fragmented

minerals. These countries have been able to develop significant infrastructure, including new highways, power stations, water systems, schools, and hospitals, in the process boosting employment.

Most of the extracted ores from Latin America are shipped to China and other overseas consumers in raw or concentrated form. The region has scattered iron and steel plants, as well as smelters of nonferrous ores, which process metals for use in Latin American industries or for export. Deposits of high-grade iron ore in Venezuela and central Brazil are the largest known in the Western Hemisphere and are among the largest in the world; the two countries are Latin America's main producers and exporters of ore. In Latin America, Brazil is the main producer of iron and steel by far, with Mexico second. Most of the region's production of bauxite—the major source for aluminum—comes from Jamaica, Brazil, Suriname, and Guyana. The deposits in these countries are generally located near the sea and are of critical importance to the industrial economies of the United States, Canada, and China. Venezuela has huge bauxite deposits and the sources of energy needed to process the ore; the country is a major aluminum exporter.

Chile is the largest copper producer in Latin America and also ranks number one in the world. Most of Latin America's known reserves of tin are in Peru, Bolivia, and Brazil, and these countries produce most of the region's output. The silver of Mexico, Peru, and Bolivia—sought since the very beginning of Spanish colonization—is found mainly in mountains or rough plateau country. Mexico and Peru are the largest Latin American producers of silver, lead, and zinc. Latin America's petroleum has long been extracted in the Caribbean Sea and Gulf of Mexico area, particularly the central and southern Gulf coast of Mexico and northern Venezuela. Other oilfields in Latin America are widely scattered, with the main ones located along and near the Atlantic coast in Brazil, Argentina (in Patagonia), and Colombia, and in lowlands along the eastern flanks of the Andes in every country from Trinidad and Tobago to Chile. Natural gas is extracted in many areas that produce oil, but Latin American production is not yet of major world consequence. Mexico, Venezuela, Argentina, and Trinidad and Tobago are the largest gas producers in Latin America. Venezuela is also the only Latin American nation in the Organization of Petroleum Exporting Countries (OPEC) and was one of its founding members in 1960.

Strong growth characterized Latin American economies after 2003, particularly because of China's superheated economic growth. Brazilian iron ore, Chilean and Peruvian copper, and Venezuelan oil have been in ever-increasing demand to feed China's appetite for commodities. The flip side is that if China's boom were to go bust, Latin American economies would fall too. These countries need to diversify to protect themselves. ²³⁴

Free-Trade Agreements

Latin America's heavy reliance on unprocessed minerals has periodically dealt crushing economic blows. Many Latin American countries are trying to reduce their dependence on

• **Figure 10.22** Henequen (sisal) plants on a plantation in Mexico's Yucatán region brought great wealth to the ancestors of the hacienda's current owners.

nature of farms, which become smaller as they are subdivided through inheritance and the reversal of promised governmental programs of land distribution. Such farms produce food for family use and also for the local market (•**Figure 10.23**). The crops most commonly raised, especially in Middle America, are maize (corn), beans, and squash, although many other crops are locally important, especially in the different climatic zones of the highland regions.

Minerals and Mining

Latin America is a large-scale producer of a small number of key minerals, notably petroleum, iron ore, bauxite, copper, tin, silver, lead, zinc, and sulfur. Only a handful of Latin American nations gain large revenues from exporting these

• **Figure 10.23** Highland Maya market, Guatemala.

• **Figure 10.24** Latin American countries belong to a variety of economic associations.

Source: Based on The Economist, April 11, 1998, p. 125.

raw materials and boost their exports of value-added manufactured products, thus improving their financial situations. To do so, they have formed or joined free-trade agreements (FTAs), particularly with their giant neighbor to the north (•**Figure 10.24**) (see Regional Perspective, page 371).

Four nations in South America—Brazil, Argentina, Uruguay, and Paraguay—belong to the **Southern Cone Common Market**, known as **Mercosur**, which was established in 1991. In value of products traded, this is the third largest trade group in the world, after NAFTA and the European Union. Mercosur reduced tariff and other long-standing barriers that had existed between the countries, promoting a huge increase in trade. Its members have worked to coordinate economic policies and a better negotiating position with the United States and other powers. The more than 250 million people of this union can live and work in any of the member countries and have the same rights as citizens of those countries. This stipulation brought immediate relief to an estimated 3 million illegal workers.

Patterned after NAFTA, another U.S.-brokered trade organization was ratified by the U.S. Congress in 2005. This was the **Dominican Republic–Central America Free Trade Agreement (DR-CAFTA)**, composed of the United States, Costa Rica, El Salvador, Guatemala, Honduras, Nicaragua, and the Dominican Republic. Using NAFTA as its foundation and DR-CAFTA as a building block, the United States is taking the lead in establishing a hemisphere-wide trade organization called the **Free Trade Area of the Americas (FTAA)**. Like the political

organization known as the **Organization of American States (OAS)**, it would include all of the countries of North, Middle, and South America, with the exception of Cuba. Originally envisioned for establishment in 2005, this agreement is proving very difficult to achieve because some of its loudest critics in Latin America—Brazil in particular—insist that the United States is including too many restrictions in its vision of "free" trade. Mercosur members, including Brazil, want to see the United States drop the huge subsidies of many agricultural products and steel that keep American farmers and factory workers in business but that shut out Brazilian products. In return, Mercosur would endorse patent, copyright, and other **intellectual property rights (IPR)** protection of U.S. industries. With the FTAA stalled, the United States has chosen to negotiate one-on-one trade agreements with members of the proposed FTAA, beginning with Chile in 2004 and Peru in 2007.

There are other regional trade associations, including the Central American Common Market, the Andean Community, and the Caribbean Community (CARICOM) (see Figure 10.24).

Sending Money Home

More and more Latin American families, particularly from Mexico, Central America, and the Caribbean, have come to rely on having at least one member work abroad to help out the family economy. Among the largest sources of income for Latin American economies are the **remittances**, or earned savings, sent home by people working abroad, especially in the United States. In 2011, Latin Americans working in the United States sent home more than $61 billion, about $3 billion less than in 2008 before the Great Recession began, but significantly more than in 2009 and 2010. The workers typically send these payments home in periodic electronic transfers of $200 to $300 each. Mexico is the world's third largest remittance market (after China and India), with $24 billion infused yearly into the country from Mexican laborers in the United States.

The greatest numbers of these estimated 35 million immigrant workers, of whom perhaps a third are illegal or undocumented, are in the U.S. states of California, New York, Texas, Florida, and Illinois, but all 50 states have at least some. An estimated 75 percent of them send monies home, in many cases to grandparents caring for the working parents' children; families often remain divided for many years while the father or both parents work abroad. Collectively, these remittances are so valuable—far surpassing the amount supplied to Latin America in foreign aid and foreign investment—that home countries are searching for ways to tax the income or otherwise find ways to funnel more of it into national development. Some critics say that Mexico and other Latin American countries have become so dependent on remittances that they neglect to invest their own resources in national development. Honduras is one such country: remittances amount to 20 percent of the country's GDP.

One recent and important twist in the traditional one-way flow of remittances is that some experienced workers are

394

REGIONAL PERSPECTIVE | The North American Free Trade Agreement (NAFTA)

The most economically significant FTA is the **North American Free Trade Agreement (NAFTA)**, of which Canada, the United States, and Mexico are members (•Figure 10.C). In terms of both economic value and geographic area, it is the largest trading bloc in the world. Citizens of these countries make up about half of the 942 million people of the hemisphere. The essential goal in establishing this free-trade agreement, like most, was to reduce duties, tariffs, and other barriers to trade among the member countries, theoretically strengthening all the economies. The United States had a strong if perhaps understated NAFTA goal of reducing Mexican immigration into the United States by boosting Mexico's prosperity.

NAFTA was already controversial before it went into effect in 1994, and it remains so today. Among U.S. citizens, there were fears that factories would close down and jobs would flow south to Mexico. American environmentalists feared that the U.S. environment would suffer because Mexican factories and other polluters, unable to afford the cleanup of their industries, would set new and lower standards that U.S. companies could then match. Mexican officials feared that a surge in their consumption of American-made products would so increase their trade deficit with the United States that Mexico's monetary reserves would be exhausted.

Some U.S. factories did close, in part because production was cheaper south of the border—notably in the *maquiladora* factories

(discussed on page 377)—but NAFTA was not the exclusive cause of the shutdowns. Lower—cost manufacturing for American companies shifted to a host of countries outside NAFTA and the Western Hemisphere. Trying to maximize benefits for its three members in the global marketplace, NAFTA does require all its member countries to abide by **rules of origin** specifying that half or more of the components of any manufactured good must originate in Canada, the United States, or Mexico. This is designed to prevent a fourth country from using Mexican labor in the assembly of its own component parts. Critics of NAFTA say the agreement should have demanded that an even larger proportion of the component parts originate in the member countries to boost their economies.

NAFTA did create more manufacturing jobs in Mexico but in the process contributed to a growing disparity between the wealthier north and the poorer south. The higher-paying jobs became concentrated in a belt in the north. In the more agricultural south of the country, where few of these factories were established, Mexican farmers lost incomes or livelihoods because of falling prices for corn, rice, beans, and pork. Ironically, this pushed a new stream of poor immigrants into the United States. NAFTA gradually eliminated most Mexican tariffs on agricultural imports from the United States. The large, heavily subsidized American agricultural businesses can raise grain or livestock at prices below production cost. A pig, for example, can be raised and sold in Iowa for one-fifth of what

it would cost to raise and sell it in Mexico. Cheap American agricultural products flooded the Mexican market, making it impossible for Mexican farmers to compete and driving many out of business; one-third of Mexico's 18,000 swine producers went out of business soon after NAFTA went into effect. On the positive side, Mexican companies have been able to turn the cheaper U.S. pork into sandwich meats sold back at a profit to American consumers.

Corn occupies a cherished place in Mexico's culture and nutrition, and increasing integration of the Mexican and U.S. economies led to a corn crisis there in 2006. The United States began diverting more of its corn crop into the production of ethanol, a biofuel. Subsequent shortages of corn on the food market drove prices higher in the United States. Those increases were passed on in corn exported to Mexico and matched by domestic corn producers in Mexico. The young government of President Calderon found itself besieged by protesters demanding that Mexico's corn producers be shielded from U.S. imports (then accounting for one-fourth of Mexico's corn) and that the government intervene to keep this most basic of all foodstuffs affordable to the poor. At the same time, genetically modified corns began appearing in southern Mexican fields, and environmentalists and other anti-GM activists chimed in their demands for "food sovereignty" with the protection of Mexico's corn heritage. These events swirling around corn came to be known as Mexico's **"tortilla wars."**

In sum, in the United States and Canada, NAFTA has been beneficial to farmers but not to factory workers, while the reverse is true in Mexico. Both the United States and Canada have trade deficits with Mexico, importing more from that country than exporting to it. The hoped-for reduction of emigration to the United States did materialize, but for the "wrong" reason: the economic downturn in the United States made emigration from Mexico less attractive. Meanwhile, NAFTA appears to be losing some steam, with trade among its three members declining about 10 percent between 2001 and 2011.

• **Figure 10.C** The trinational NAFTA flag symbolizes the integrated economic ambitions of its members.
Source: NAFTA

86

Joe Hobbs

• **Figure 10.25** Warm, turquoise waters of Caribbean and other Latin American beaches are an irresistible draw for tourists from El Norte. This is the "Maya Riviera" of the Yucatán Peninsula's east coast.

returning to their home countries to establish companies that invest in the United States. These so called **multi-Latina companies** are aiding the revival of certain downtrodden sectors of the U.S. economy, including cement and steel mills.

Tourism in Latin America

Tourism has become a major regional economic asset to Latin America, generating critical foreign exchange; only oil exports are more valuable. Countries in which tourist dollars amount to more than 50 percent of the nation's foreign exchange include Antigua and Barbuda, the Bahamas, and Barbados. During what many North Americans see as their too-long winters, they are bombarded by television and print advertisements of vacation in paradise: Jamaica and hosts of other destinations of the Caribbean basin and western Mexico. A wide variety of vacation experiences are available, from the hedonistic resort complexes that tourists seldom leave (one resort on Jamaica actually is called Hedonism), to the insular sea voyage that offers day trips ashore in exotic destinations, to whitewater and rain forest adventures of the ecotourist variety (•**Figure 10.25**).

The region benefits from the proximity of its year-round warmth to the wealthy and mobile and seasonally chilled U.S. and Canadian populations. Tourism revenues reflect that **distance-decay relationship**: the highest tourism receipts in Latin America flow to Mexico, the nearest neighbor to the wealthy countries, but tend to fall off for more far-flung destinations.

10.5 Geopolitical Issues

The central geopolitical reality of Latin America is that it is in America's backyard. For better or worse, the United States has staked its geostrategic claim to the region. Much of the

United States itself is comprised of land wrested from Mexico in conflicts of the early nineteenth century. In 1823, U.S. intentions for Latin America were formulated in a policy known as the **Monroe Doctrine**, after U.S. President James Monroe announced that the United States would prevent European countries from undertaking any new colonizing activities in the hemisphere. Many subsequent interventions, wars, investments, and other U.S. activities may be seen in light of this seminal policy. Even more consequential was the 1904 **Roosevelt Corollary** to the Monroe Doctrine, whereby the United States declared that it had the right to supervise the internal affairs of Latin American countries to ensure U.S. national security.

Ever since the Monroe Doctrine was proclaimed, the United States has had a particularly firm hand in Central America. Washington has dealt unapologetically and sometimes harshly with perceived threats to the United States from this region. Examples include U.S. actions to support governments against nationalist insurgencies, or insurgents against leftist governments: El Salvador and Nicaragua in the 1980s, Cuba in the 1960s, and Nicaragua in the 1950s were the most significant instances. Perhaps the most-cited intervention took place in 1954, when the U.S. Central Intelligence Agency ousted the Guatemalan leader in a coup. Among other things that earned American denunciation as communist, President Árbenz pushed for the redistribution of land in a country where 2 percent of the population owned 70 percent of the arable land. ⁶¹

The Panama Canal

Built between 1904 and 1914 by American contractors, the Panama Canal is a legacy of American hegemony in Central America. The Canal was intended mainly to serve U.S. commercial and strategic interests. For decades, it was sovereign U.S. territory, carefully monitored by an American military presence, and the United States kept more than half of the transit fees charged to ships using it.

The Panama Canal today is one of the world's most critical chokepoints: 14,000 ships a year, traveling 80 shipping routes and representing about 5 percent of the world's cargo volume (and 16% of the United States'), use the 50-mile (80 km) shortcut from the Atlantic Ocean to the Pacific. A 14-day sea ¹⁷⁷ voyage between New York and San Francisco via the canal cuts 7,800 miles (12,530 km) and 20 days from the trip (•**Figure 10.26**; see also Figure 10.31). The United States is a major user; about one-seventh of all U.S. foreign trade is carried through the Canal.

In the late 1970s, the administration of U.S. President Jimmy Carter accepted the Panamanians' complaint that U.S. control of the canal was an outmoded vestige of colonialism. Carter negotiated the transfer of authority over the canal to Panama in a treaty that went into effect in 1999. Now the revenues from ships transiting the canal are exclusively Panama's (more information on Panama's ownership of the canal begins on page 378).

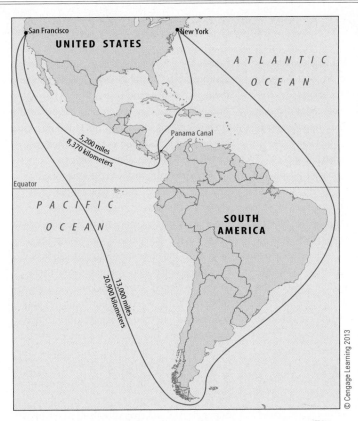

• **Figure 10.26** Distance, time, and money are saved by use of the Panama Canal.

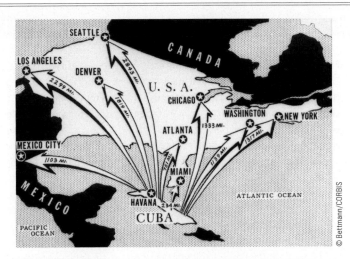

• **Figure 10.27** The 1962 Cuban Missile Crisis brought the United States and the Union of Soviet Socialist Republics close to the brink of nuclear holocaust. This 1962 map shows the range that Soviet missiles would have if stationed in Cuba.

From a strategic and economic defense standpoint, however, the United States is still deeply involved with the canal. The U.S. Navy is concerned that its two access points (Balboa in the Pacific and Cristóbal in the Atlantic) are now controlled by a company with strong backing from China, the canal's second largest user. The United States has already effectively invoked the Roosevelt Corollary since returning control of the canal; in 1989, U.S. troops invaded Panama and ousted its ruler, Manuel Noriega, whom the Americans accused of facilitating the illegal trade of drugs into the United States. The U.S. war on drugs has been another central geopolitical issue of the region since the early 1970s (see Geography of Drug Trafficking, next page).

Cuba

The most consequential test of U.S. commitment to its interests in Latin America came with the **Cuban Missile Crisis** of 1962. Cuba is as close as 90 miles (144 km) from the United States. The frosty relations between the United States and Cuba's Marxist regime, led after 1959 by a firebrand revolutionary named Fidel Castro, were emblematic of the Cold War. Those tensions came dangerously close to war and even nuclear holocaust in 1962, when the Soviet Union began to build facilities in Cuba that would have accommodated nuclear missiles capable of striking Washington, D.C.,

New York, and other cities in the eastern United States (•Figure 10.27). As it so often does today, remote sensing played a vital intelligence role then. Photographs taken from U-2 aircraft provided undeniable evidence of missile launch infrastructure on the ground in Cuba.

Although the missile threat was defused under intense pressure from the U.S. government, led at the time by President John F. Kennedy, Cuba has remained a storm center of political affairs in the Western Hemisphere. After several attempts to assassinate Castro—even with the unlikely ploy of an exploding cigar—and a disastrous attempt by Kennedy's administration to overthrow him by military force (in an incident known as the **Bay of Pigs invasion**), the U.S. administration settled on a course of economic and political isolation of Cuba.

Despite the resumption of diplomatic and economic relations with most of the remaining Communist countries in the world—China most notable among them—the United States has remained steadfast in imposing sanctions on Cuba. This anomaly to some extent reflects domestic political realities in the United States, where a large population of **Cuban Americans** (including 800,000 in Florida alone) who fled Castro's Cuba tend to vote for the politicians who keep up the most pressure against Castro. The Cuban Americans are against any relaxation of the virtual ban on American tourism to Cuba, which is in place because the **Trading with the Enemies Act** forbids Americans from spending more than $300 in Cuba. The failing health of Fidel Castro and the discovery of significant oil reserves off Cuba's shores led to unprecedented discussions among U.S. lawmakers about reopening American doors to Cuba. One fear was that U.S. businesses would lose important opportunities if the doors remained closed. Recent legislation allows U.S. states to sell food and medicine to Cuba on a cash-only basis, and about 40 states do so.

GEOGRAPHY OF DRUG TRAFFICKING | The War on Drugs

From the Nixon administration of the early 1970s to the present, the United States has led a determined, expensive effort known as **the war on drugs**. The Nixon administration was the only one to focus significant efforts on combating demand for heroin, cocaine, and other drugs within the United States. All successive administrations have concentrated their efforts on the supply side, hoping to eradicate drugs at their sources—particularly in South America. Although it is far from over, the war on drugs has produced some definitive and sometimes predictable patterns (•Figure 10.D).

One of the most consistent geographic patterns is the so-called **balloon effect**, referring to the fact that when a balloon is pressed in one place, the air shifts elsewhere. In the case of drugs, concentrated eradication efforts in one place almost always shift production elsewhere, sometimes even from one continent to another. During the Nixon years, for example, U.S.-funded efforts succeeded in all but eliminating production of opium poppies (the plant from which heroin is derived) in Turkey. With demand still high in the United States, Europe, and elsewhere, however, poppy production surged in a region where it had been only modest: the Golden Triangle of Myanmar, Laos, and Thailand. Increasing pressure on poppy producers in the Golden Triangle, the Bekaa Valley of Lebanon, and briefly in Afghanistan, led to the eruption of the poppy industry in Mexico, Guatemala, and Colombia, oceans away from its Old World hearth.

The balloon effect is clearly visible in the geography of coca production in South America. From 1998 to 2008, Colombia's efforts with the U.S.-funded **Plan Colombia** succeeded in reducing coca production in Colombia by about 60 percent. In the same period, coca growing increased by one-third in Peru and doubled in Bolivia. In Colombia, those growers who remained adapted in innovative ways. They dispersed their farms into smaller and more widely scattered plots, often in new areas. They also practiced selective breeding to produce plants that contain more active alkaloids, are more resistant to defoliants, and can grow with more shade cover, thereby escaping detection from the air. Most of the growers who lost their crops to aerial spraying come right back: an estimated 70 percent of the affected fields are replanted in coca. Despite the billions of dollars spent, the amount of land planted in coca was roughly the same in 2011 as it was in 1981.

Much of the force behind the balloon effect is economic: when eradication is successful in one area, supply falls and even more value is added to the already high value of cocaine and heroin. Growers and traffickers are easily tempted by the money to be made under such market conditions. Much more money is made in the distribution than in the production of drugs: a kilogram (2.2 lbs.) of cocaine is worth $1,000 when it leaves the Caribbean coast of Colombia, and $100,000 as it passes through Central America.

Mexican drug cartels have taken the lead in drug trafficking. Due to the government's aggressive war on drugs in Mexico, the drug cartels have moved their distribution centers southward to Central America, where governments are weaker and systems more corrupt. About 90 percent of the cocaine bound for the United States is exported from Honduras, Guatemala, and El Salvador, where the Mexican cartels work with local youth gangs.

Alternative proposals to the long-fought, conventional war on drugs typically have two elements. First, advocates of these other approaches argue that much more attention must be paid to the demand side, using education, treatment, and other means to wean addicts from drugs and discourage others from using them. The United States now spends about two-thirds of its drug war budget on foreign supply sources and one-third on domestic prevention and treatment. The second element focuses on the supply side. Peasant farmers who grow drugs are generally not trying to get rich but to secure a livelihood where jobs are scarce. Few crops offer the reliable market and certain returns that coca or poppies do, and **crop substitution** schemes to get farmers to switch to flowers or coffee almost always fail. Successful substitutions have occurred only in the context of so-called **alternative development** schemes, in which new crops are just one part of an overall effort to enhance the education, health, and livelihoods of rural communities. Alternative development is not a component of Plan Colombia, which may account for the high replanting rate for coca.

A successful war on drugs, won through efforts on the supply side, the demand side, or both, would have large environmental benefits, among others. Clearing of forest for new cultivation of coca has done enormous damage to South American forests, particularly on the steep eastern slopes of the Andes, where probably no other cultivation would ever be attempted. The processing of coca leaves to coca paste and finally powdered cocaine involves the use of highly toxic chemicals that make their way into water supplies, harming ecosystems and human health. A successful war on drugs could also help restructure U.S. interactions with the countries of Latin America. At present, countries deemed to be uncooperative in the war on drugs are subject to **decertification**, meaning that U.S. aid and investment in those countries are curtailed. This punishment is sometimes arbitrary; a country like Mexico often fails all the tests for certification but is too important to U.S. interests to be decertified.

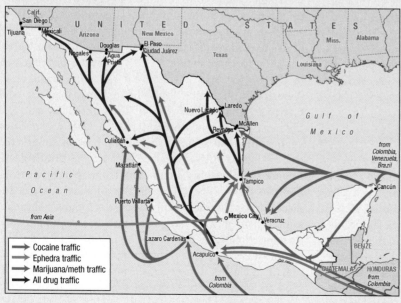

• **Figure 10.D** Drug trafficking routes in the Americas.
Source: Adapted with permission from Stratfor Enterprises, LLC.

The government of Cuba would like to see the country opened to American tourists. It would also like to see the United States withdraw from Guantánamo Bay, an enclave the United States secured in the 1903 treaty that addressed Cuba's fate following the Spanish-American War. The United States dutifully goes through the motions of paying Cuba the $4,000 annual rent for the lease on Guantánamo, which it has used as a military base, and Cuba dutifully refuses to accept it. After 9/11, the United States established a prison camp for suspected al-Qa'ida members and other terrorists at the base. Its location outside the borders of the United States allowed the U.S. government to suspend the usual legal and constitutional rights awarded to criminal suspects on U.S. soil. Presidential candidate Obama vowed to shut down the prison at "Gitmo," but as president he failed to find an alternative to it.

Colombia

Several U.S. interests in Latin America converge on the nation of Colombia. The war on drugs was long a dominant concern. U.S. priorities also have focused increasingly on securing access to Colombia's considerable petroleum reserves. The United States already obtains more of its oil from Latin America than from the Middle East, and it views Colombia, Venezuela, and Mexico in particular as long-term counterweights to the volatile Middle East oil region.

Washington's thinking has been that the best way both to obtain Colombia's oil and to stem the drug tide is to take on Colombia's main rebel army, the Revolutionary Armed Forces of Colombia (FARC; see page 380). FARC became the quintessential **narcoterrorist organization**, deriving most of its funding from the profitable cocaine industry based mainly in the areas it controlled. With U.S. funding and training as part of the Plan Colombia program (see page 374), which began in 2000, Colombian forces undertook conventional drug war tactics, such as stepped-up herbicide spraying of coca crops, confiscation of traffickers' assets, and shooting down of suspected drug-carrying planes. In a related operation dubbed **Plan Patriot**, U.S. military personnel civilian contractors targeted FARC to both reduce drugs and secure oil. To date, these efforts have succeeded in wresting much control from FARC. The government-backed "paramilitary" forces that both fought these rebels and had their own stake in the drugs trade have also been weakened, and the security situation in Colombia is much improved.

The Washington Consensus ...

Between about 1980 and 2000, U.S. administrations exercised a hands-on policy in Latin America. Through a combination of blunt U.S. military interventions and the countries' own efforts, Latin America witnessed an overall transformation of political systems from authoritarian to multiparty democratic. In the thinking of successive U.S. administrations, those greater political freedoms should have ushered in more prosperity. This was one component of the so-called **Washington Consensus**. In this political-economic philosophy, the United States would press the democratic governments of Latin America to open markets to international trade, reduce tariffs, expand quotas, sell off state companies to private investors, allow more foreign investment, cut government spending, and reduce bloated bureaucracies. These reforms would bring the prosperity that would solve the region's problems. If serious problems emerged, the United States could help; it did act, for example, to prevent a financial crisis that shook Mexico in 1994 from spreading. But generally, the United States neglected Latin America, and did so at its peril: into this vacuum stepped China, with a familiar pattern of economic aid and infrastructure.

... And Latin America's Pushback

After 2000 and especially after 9/11, aside from its boots on the ground in Colombia, the U.S. government appeared to adopt a more hands-off policy toward Latin America, perhaps distracted by regions that seemed more threatening in the war on terrorism. Washington watched from the sidelines while a succession of elections and polls showed a drift away from American wishes for the region. Widespread disillusionment with economic conditions led to successive elections of reformist, left-leaning and generally anti-Washington Consensus or anti-U.S. leaders. First came Brazilians' choice in the 2002 and 2006 presidential elections: leftist labor union leader Luiz Inacio da Silva, known as "Lula," who vowed to fight the tide of globalization. Similar changes in administration took place in Argentina, Venezuela, Bolivia, Ecuador, and Nicaragua. In Venezuela, President Hugo Chávez was elected in 1999 and again in 2006, promising to form a regional trade association that would act against U.S. interests. He vowed to use the country's oil wealth to fund social programs. Hoping to do the same with his country's natural gas and declaring that "the worst enemy of humanity is U.S. capitalism," a former coca grower, Evo Morales, became president of Bolivia in 2006. That same year, Ecuadorian President Rafael Correa was elected on promises to close a base used by U.S. aircraft to bomb drug fields, reject free-trade deals, and spend more oil revenue on the poor. All of these leaders, Chávez most vocal among them, had in common a stated intention to reverse the U.S.-backed push toward free trade and free enterprise, returning to state-owned companies and protected markets that would assist the poorest segments of their populations.

These leaders were generally supported by the indigenous Americans, who felt that the descendants of European colonizers of the New World, along with U.S. and other multinational firms, were robbing them of both their natural resources and their political clout. The Native American Aymara majority of Bolivia actually succeeded in driving that country's president from office in 2003, setting the stage for the election of Morales, the first indigenous head of state in Bolivia since the European conquest.

Bolivia had been a strong advocate of a proposed development to export Bolivia's natural gas, mainly to the

United States, through a pipeline that would cross Chile to a Pacific port. Native American Bolivian peasants, backed by labor unions, student groups, and opposition politicians, complained that royalties for Bolivia from the gas sales would be so low that the common people would never realize any benefit from them. They cited Bolivia's history as a colony stripped of its silver and tin to enrich the Spaniards and local aristocrats, and insisted that again only foreigners and Bolivia's elite would gain from the gas project. These opponents wanted the gas to remain in Bolivia, where it could form the basis for new local industries. (They also resented the pipeline's passing through Chile, a focus of Bolivian anger ever since Chile, following an eighteenth-century war, seized lands that cut Bolivia off from the sea.) Their votes were instrumental in bringing Evo Morales, himself an Aymara, to office. Morales championed their cause by announcing that his government would nationalize (impose government control over) all of Bolivia's critical energy and other resources and use them for the benefit of the people. The so-called "ideology of fury" expressed by the Aymara in Bolivia rattled nerves among the elites in other countries of Latin America with large indigenous populations, including Ecuador, Peru, Paraguay, Guatemala, and even Mexico.

The following section explores more issues among the peoples and lands of the region, first in Middle America and then in South America.

10.6 Regional Issues and Landscapes

Middle America

The region most geographers prefer to call Middle America, also known as Mesoamerica, includes Mexico (its largest country in area and population); the much smaller Central American countries of Guatemala, El Salvador, Belize, Honduras, Nicaragua, Costa Rica, and Panama; and the numerous island countries in and near the Caribbean Sea, the largest being Cuba, Haiti, the Dominican Republic, and Jamaica. The region is a vibrant patchwork of physical features, ethnic groups, cultures, political systems, population densities, and livelihoods. Most of the political units are independent, but a few are still dependencies of outside nations.

Mexico: Higher and Further

The federal republic of Mexico—officially, *Los Estados Unidos Mexicanos* (United Mexican States)—is the largest, most complex, and most influential country in Middle America (•**Figure 10.28**). The country's great population of 115 million in 2011 makes Mexico by far the world's largest nation in which Spanish is the main language. Compared to most of the world's countries, Mexico is a spatial giant, but its large size tends to

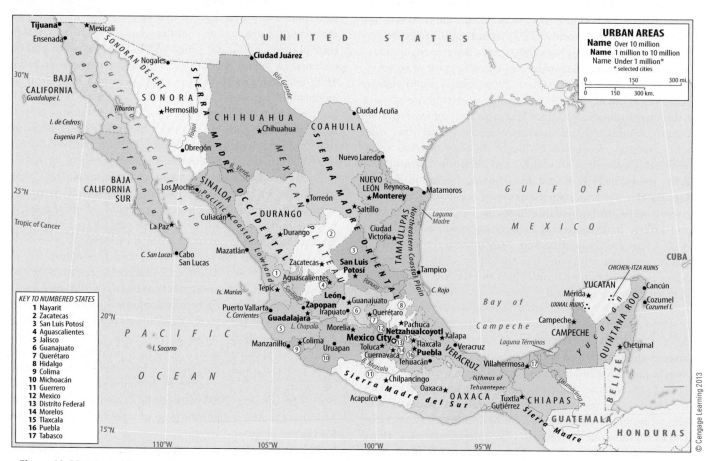

• **Figure 10.28** Principal features of Mexico.

go unrecognized because it lies in the shadow of the United States. Triangular in shape, Mexico's area of 756,000 square miles (c. 2 million sq km) is nearly eight times that of the United Kingdom, and its elongated territory would stretch from the state of Washington to Florida if superimposed on the United States (see Figure 10.2). Its capital, Mexico City, stands now as the world's second largest city in population (after Tokyo-Yokohama), with 27 million inhabitants. Mexico is one of the most urbanized countries of Middle America, with about 78 percent of its people living in cities. In these and many other attributes, Mexico in the Latin American context seems to live up to its motto: "Higher and Further."

Mexico has been described as one of the globe's nine "pivotal countries," a country whose economic or political collapse could cause international refugee migration, war, pollution, disease epidemics, or other international security consequences. The country's stability is enormously important, particularly to its giant neighbor to the north. Mexico's natural and human capital is much greater than many outsiders realize. Mainly because of the wealth of its tropical forests in the south, Mexico ranks as the world's third most important megadiversity country, after Brazil and Indonesia. Mexico ranks highly among the world's producers of such commodities as metals, oil, gas, and sulfur, with oil the country's most important export. The Mexican oil industry is a government monopoly (carried on by a government corporation, Petróleos Mexicanos, or PEMEX), and the United States is its largest foreign customer, taking in 90 percent of the country's oil exports. Mexican oil represented 11 percent of oil imported to the United States in 2011, making Mexico the second largest foreign supplier after Canada.

237, 240

Mexico's oil will run out—some analysts say by 2030. The country is trying to diversify and grow its economy, the second largest in Latin America (after Brazil), away from reliance on commodities like oil. Mexico has made substantial gains in its development of manufacturing and component assembly 371 industries, especially since the initiation of NAFTA. The manufacturing sector is now Mexico's leading source of revenue, followed by oil and remittances from abroad. High-technology exports grew from 8 percent of total exports in 1990 to 22 percent in 2011. About one-third of all manufacturing employees in Mexico work in the thousands of factories located in metropolitan Mexico City.

The large *maquiladoras* of Mexico, churning out automobiles, textiles, electronics, and computers, are the symbols of the new Latin American free-trade-oriented economies. These have become an exception to the previous spatial pattern of Latin American manufacturing enterprises. Larger operations were almost always located in the larger cities, including the greatest producing metropolises of Mexico City, São Paulo, and Buenos Aires, representing the region's three greatest manufacturing countries, Mexico, Brazil, and Argentina, respectively. Consider the geographical concept of *situation*: these *maquiladoras* are placed where location serves them best: as close as possible to the great market of the United States—the principal destination for most of their output—or deep in areas like the country's southern Yucatán region where labor costs are lowest.

Exports from Mexico to the United States have more than tripled since NAFTA went into effect in 1994. This bilateral trade is in Mexico's favor, with imports from Mexico exceeding exports to Mexico by 20 percent. Mexico is the third largest provider of imports to the United States, after China and Canada. For its part, the United States exports more goods to Mexico than to any other country but Canada (13 percent of U.S. exports go to Mexico).

The *maquiladoras* have surged in growth and diversification since NAFTA went into effect; about 2,700 of these factories, employing nearly 1 million Mexicans, are now operating in the U.S. border region alone. The *maquiladora* is designed to boost exports of manufactured goods by using tax breaks, cheap labor, and other incentives to attract foreign industries. The factories produce jeans, cars, computers, and other goods, mainly from components that are imported duty-free because they are assembled for export only (•Figure 10.29). There has also been growth in industries supporting the *maquiladoras*, such as small-scale retail and transportation services. Among

• **Figure 10.29** The *maquiladora* is globalization in a nutshell. This woman is assembling electronic devices for Samsung, a Korean firm, in a *maquiladora* in the Mexican border town of Tijuana for sale in the United States.

AP Photo/Damian Dovarganes

the most recent developments are **energy** *maquiladoras*. These are power generation plants—built by American firms just inside Mexico, where pollution standards are much lower than in the United States—that provide electricity to the U.S. grid. Their power source, natural gas, comes from Texas, and Mexico supplies their coolant waters.

In the global marketplace, Mexico's *maquiladoras* are evolving. After 2000, Mexico lost some of the competitive edge represented by its inexpensive labor. Hundreds of *maquiladoras* closed, and hundreds of thousands of their workers lost jobs. Why? Mexico's clothing, textile, and other low-tech industries could not match Chinese rock-bottom labor costs. Mexico itself became flooded with Chinese imports, even of pottery imitating Mexico's traditional wares.

While conceding the loss of low-tech, low-skill industrial advantage to China (and also nearby El Salvador, where labor is also cheaper), Mexican manufacturers hung on to Mexico's edge in exports of flat-screen televisions, computers, aircraft parts, and other high-tech and heavier goods that are more expensive to ship from China. The *maquiladoras* began a slow and steady recovery and became more attractive as labor wages climbed in China. They have also begun to merge into specialized clusters, for example, with Tijuana concentrating on flat-screen televisions and Ciudad Juárez on car parts. It is remarkable that, at least so far, the *maquiladoras* in these border towns have not been affected by the drug-related violence that has claimed more than 40,000 lives in Mexico.

There are both inseparable ties and divisive issues between Mexico and the United States. Huge disparities in wealth and power come between them. Mexicans have not forgotten that Mexico lost more than half national territory to the *gringos* in the **Mexican War** (known to Mexicans as the **American Invasion**) in 1847. Mexicans have enormous national and cultural pride and often feel misunderstood and underappreciated by their northern neighbors. In the next chapter, we explore immigration and other issues in relations between the United States and Mexico.

Central America

The Panama Canal

Between Mexico and South America, the North American continent tapers southward through an isthmus containing seven less developed countries known collectively as Central America. Five of these countries—Guatemala, Honduras, Nicaragua, Costa Rica, and Panama—have seacoasts on both the Pacific and the Atlantic Oceans (via the Caribbean Sea; •Figure 10.30). The other two have coasts on one body of water only: El Salvador on the Pacific and Belize on the Caribbean. If the seven countries were one, they would have an area only about one-fourth that of Mexico (see Table 10.1), with a total population about 40 percent that of Mexico.

At the southern end of Central America, the narrow Isthmus of Panama separates the Pacific and the Atlantic oceans by a mere 50 miles (80 km). Even looking at a map or globe from a distance, you can easily appreciate that this is the best place to

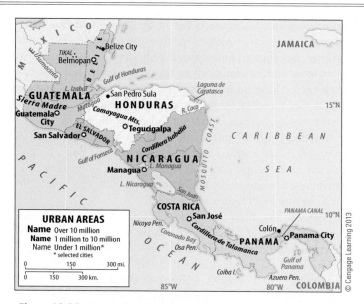

• **Figure 10.30** Principal features of Central America.

locate transcontinental shipment routes. In 1855, the Panama Canal Railway became the first transcontinental railway, and it remains in use today. Colombia is considering whether it should build a second, Chinese-funded railway across its portion of the Isthmus.

The Panama Canal, as discussed earlier (page 372), is of great importance to both Panama and the world (•Figure 10.31). The idea of a shortcut between the Atlantic and Pacific Oceans

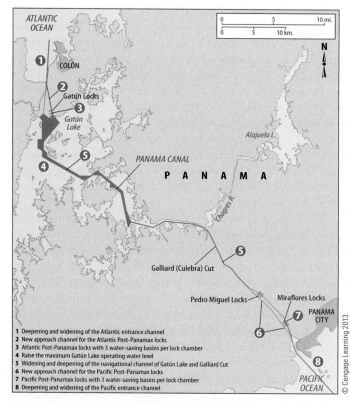

1 Deepening and widening of the Atlantic entrance channel
2 New approach channel for the Atlantic Post-Panamax locks
3 Atlantic Post-Panamax locks with 3 water-saving basins per lock chamber
4 Raise the maximum Gatún Lake operating water level
5 Widening and deepening of the navigational channel of Gatún Lake and Galliard Cut
6 New approach channel for the Pacific Post-Panamax locks
7 Pacific Post-Panamax locks with 3 water-saving basins per lock chamber
8 Deepening and widening of the Pacific entrance channel

• **Figure 10.31** Profile and plan of the Panama Canal.

had been around for a long time before the Panamanian site was chosen. In the early 1800s, Nicaragua was the first choice because of natural landscape features that appeared better able to accommodate a crossing without massive construction (Nicaragua is still considering building this canal, which, unlike even the upgraded Panama Canal, would accommodate the world's megaships). In 1846, however, Colombia (which then controlled Panama) and the United States signed a treaty supporting a project to cut a canal across the much narrower Isthmus of Panama.

Interest in the project picked up substantially in 1849 with the discovery of gold in California. At that time, prospectors and speculators had to sail to the port of Colón on the east side of Panama, make the 50 miles of land crossing in any way possible, and then reboard a ship in Balboa and sail to San Francisco. After 1855, the Panama Canal Railway (funded by New York business interests) provided a reliable mode of transit from port to port. This train, however, required a break of bulk of goods and passengers a canal 266 would not.

In 1878, Colombia granted permission to a French firm to build a canal across the isthmus near the railroad line. Ferdinand de Lesseps, who had seen the Suez Canal to completion in 1869, designed the project. The French quickly bought control of the railroad line and began digging in 1882, with a plan to build a sea-level transit requiring no **locks** (the devices that allow a ship to move from one water level to another). This venture went bankrupt in 1889 and was followed by another French attempt, which also failed. Construction difficulties and a high death toll due to yellow fever and other diseases were to blame. By the beginning of the twentieth century, the French had left. The United States had just participated in the 1898 Spanish-American War and had more geopolitical interest than ever before in this prospective waterway.

In 1903, Panama broke away from Colombia, and within weeks, the United States recognized the new government. Shortly thereafter, the two governments struck an agreement to build the canal, which ultimately required three sets of locks. Construction began in 1904 and, at a cost of $380 million and many lives lost to yellow fever, bubonic plague, and malaria, was completed in 1914. The first crossing was made under a banner that read: "THE LAND DIVIDED; THE WORLD UNITED."

Work on the canal has never ended (Figure 10.31). To meet anticipated demand, especially from surging exports of China's manufactured goods, the canal recently underwent the largest expansion ever. Panama's citizens approved the expensive project in 2006 with the expectation that it would quadruple income from transit fees. The construction project aimed to overcome problems delaying and prohibiting the passage of many ships. To double capacity and accommodate larger cargo vessels up to the size known in the shipping trade as "post-Panamax," new channels with longer and deeper locks were added at each end of the canal. These and other measures doubled the canal's capacity. Thanks to growing trade, expanding ports, and residential and commercial development, Panama's

economy has boomed in recent years; it has invited comparison with that of Dubai, another place that has capitalized on its strategic crossroads location. 168

South America

Venezuela's Petroleum Politics

Oil resources in the coastal area around and under the large Caribbean inlet called Lake Maracaibo (•Figure 10.32) have been a huge blessing for Venezuela, the world's 13th largest oil producer in 2011. Oil makes up about 95 percent of the country's exports by value. The country was one of the founding members of the Organization of Petroleum Exporting Countries (OPEC) in 1960 and is still a member. Oil revenues 176 have fueled urbanization and industrialization and generated wealth. Yet Venezuela is rather like Nigeria or Angola, an oil-rich but oil-cursed country where the vast majority of people remain poor. 328

The most prominent member in the lineup of left-leaning Latin American leaders elected in recent years was Venezuelan President Hugo Chávez. Invoking the revered name of Simón Bolívar (1783–1830), the Venezuelan-born independence leader who spearheaded a regional quest for liberation from Spain, he proclaimed to be undertaking a **"Bolivarian Revolution."** (Chávez had his hero exhumed and examined in 2010, ostensibly to determine how Bolívar died.) Chávez had his own ideas about liberation: he rewrote the constitution, put cronies on the supreme court, cracked down on the media, and jailed political foes. The Chávez agenda appealed to the poorest classes by promising a redistribution of wealth like that championed by Cuba's Fidel Castro (Chávez's strongest ally in the region). Chávez's political philosophy was not communism, but what he called "twenty-first-century socialism" (and others dubbed "Chavismo"), with the armed forces and

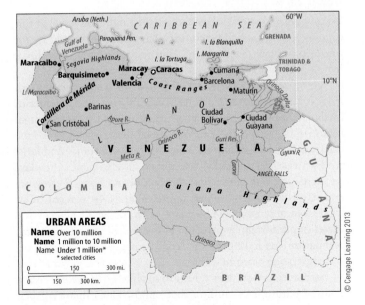

• **Figure 10.32** Principal features of Venezuela.

the president controlling everything but a small and mainly foreign-owned private sector.

Venezuela's professionals—scientists, engineers, doctors, and businesspeople, for example—left Chávez's Venezuela in unknown numbers, probably hundreds of thousands, settling mainly in the United States and Canada. Meanwhile, there was a large influx of immigrants, including Chinese, Lebanese, Jordanians, Syrians, Haitians, and Colombians, drawn by high wages and new opportunities in Venezuela.

His power consolidated, Chávez used Venezuela's oil wealth to advance his political agenda across the Western Hemisphere. He built alliances with Latin America's leftist leaders and courted the center and right governments of the region to move his way. His main goal was to build a coalition of countries to counter U.S.-led free-trade efforts in the region. He cast himself as an advocate of the poor by offering Venezuela's oil to Cuba and to select populations in other countries at little or no profit. He embarrassed U.S. politicians by supplying cheap heating oil to hard-pressed American northeasterners in the cold winter of 2005.

While shunning Chávez and his politics, Washington needs Venezuela's nearby non–Middle Eastern oil and has taken no action that would jeopardize its fourth largest source of imported oil. Chávez, for his part, periodically threatened to cut off the supply of Venezuelan oil to the United States. "Ships filled with Venezuelan oil, instead of going to the U.S., could go somewhere else," he said. "The U.S. market is not indispensible for Venezuela."[6]

Meanwhile, Venezuela stepped up its oil exports to Washington's regional old-fashioned, Cold War–style nemesis, Cuba. Old-fashioned, Cold War–style rhetoric between the United States and Latin America's leftist governments made a comeback. A U.S. Defense Department official said of Venezuela's Chavez, "A guy who seemed like a comic figure a year ago is turning into a real strategic menace."[7]

Colombia's Potential and Peril

Colombia (•Figure 10.33) has had an outstanding reversal of fortune in recent years. The country has abundant resources for future economic development, and they are under less threat from opposition forces than they were a few years ago.

Energy resources include a large hydropower capacity; coalfields in the Andes, the Caribbean coastal area, and the interior lowlands; oil in the Magdalena Valley and in the interior; and natural gas in the area near Venezuela's Lake Maracaibo. Iron ore from the Andes supplies a steel mill near Bogotá, and large Andean reserves of nickel are under development. Oil plays an increasing role in the country's export economy. Petroleum and its products represented approximately 40 percent of all legal Colombian exports by value, with coal at 12 percent and coffee representing 7 percent. However, illegal cocaine and marijuana may have been the largest category of exports, possibly up to one-third by value.

Both anti-government rebel groups like FARC (see page 375) and the paramilitary militia that oppose them financed their

• **Figure 10.33** Principal features of Colombia.

operations through taxes on drugs. Widespread insecurity related to their conflicts and to the firepower and violence of the drugs trade prompted Colombians in 2002 to elect Alvaro Uribe as president. Uribe had promised to crack down on rebellion and restore order to the country. After assuming office, he pleased both his citizens and the U.S. Congress in Washington, which wanted to see some return for its huge investment in Colombia. He gained considerable confidence both there and among a majority of Colombians for his efforts to improve security.

Brazil, Stirring Giant

Brazil is the giant of Latin America (•Figure 10.34). It has an area of about 3.3 million square miles (8.5 million sq km)—about 5 percent larger than the conterminous United States—and a population of 197 million as of 2011, second only to the United States in the Western Hemisphere (see Figure 10.3).

Brazil has an increasingly important role in hemispheric and world affairs—so much so that it is a candidate for a permanent seat on the United Nations Security Council. Its growth is especially notable in its overall population, in the explosive urbanism that has made São Paulo and Rio de Janeiro two of the world's largest cities, in the rise of manufacturing, and in the diversification of export agriculture. Brazil is the "B" in the dynamic BRIC economies, with the largest economy in South

• **Figure 10.34** Principal features of Brazil.

America and the tenth largest in the world. It is widely projected to have the world's fourth largest economy by 2050. Its expanding domestic market and labor force have accelerated economic development in Brazil.

Brazil has a small ultra-rich population, a substantial and growing middle class, and large numbers of technically skilled workers, but the benefits from development have been distributed so unevenly that 20 percent of Brazilians live in poverty. Brazil has the unfortunate distinction of having Latin America's most unequal distribution of national wealth (it ranks fourth in the world, behind three African countries). Just 1 percent of the population owns 45 percent of the country's farmlands. According to the World Bank, the poorest 10 percent of the population receive just 1 percent of the total income, and the richest 10 percent receive almost half. (For comparison, in the United States, the poorest 10 percent receive about 2 percent of the total income—making them the poorest population of all the MDCs—while the richest 10 percent gets 30 percent.) Brazil's income inequity is showing signs of shrinking, however, due in part to more generous welfare payments from the government.

Brazil is also burdened with a large foreign debt owed to the International Monetary Fund and other international lenders for its ambitious and costly development projects launched after the mid-1970s. However, Brazil has many things going for its future, including abundant and varied natural resources, many allies among the world's advanced economic powers, and remarkable internal harmony among its diverse ethnic and cultural populations.

The pace of development in Brazil accelerated after 1945, with periodic fluctuations. Economic growth took place under elected governments between 1945 and 1964 and occurred most rapidly in the so called "miracle years" between 1964 and 1985 under a military dictatorship with technocratic leanings and great ambitions. Unfortunately, Brazil's hoped-for economic takeoff as an industrialized power coincided with the devastating impacts of a quadrupling of oil prices in the mid-1970s due to events in the volatile Middle East. In 1985, the dictatorship relinquished power to a new elected government, which then had to face the accumulated problems of that stalled takeoff. The crushing national debt that had accumulated to finance the country's development had to be paid down, and widespread economic misery came with that effort.

Much financing went into building an effective transportation system with long-distance paved roadways, including the Trans-Amazon Highway system, with an emphasis on trucks as movers of goods. The new roads serve as ribbons of settlement, moving the flow of pioneer settlers to remote areas and connecting communities to markets for rubber, minerals, timber, fish, agricultural products, and tourism. Still, only 10 percent of the country's roads are paved, and upgrading transport infrastructure is a critical challenge in Brazil's development. Some of the huge advantages Brazil enjoys in agriculture are diminished by the high costs associated with getting produce to market and overseas shipping ports.

Major investments in agriculture have contributed to a recent rebound in the country's economy and in the wake of the Great Recession established Brazil as one of the world's major breadbaskets. Brazil has the extraordinary ambition of overtaking the United States to become the world's largest food exporter by 2025. Brazil leads the world in exports of sugar, beef, poultry, coffee, orange juice, and tobacco. There is particular interest in expansion of sugar and soybeans. Brazil wants to double sugarcane production from 2008 levels by 2017, mainly to produce the biofuel **ethanol** for domestic consumption and exports. Often neck-and-neck with the U.S. in ethanol production, Brazil has an advantage; it is much more efficient to produce biofuel from sugar than from corn, 416 as the United States does. Together producing 70 percent of the world's ethanol, the two countries are exploring ways to maintain and increase that share through cooperative research and development. Brazil may be able to help the United States get the kinds of secondary uses it does from its cane processed for ethanol: the cane waste, called *bagasse*, is burned in steam boilers to produce electricity used in ethanol production and for surplus sale back to the electricity grid.

According to critics, major drawbacks to biofuel production are that it does nothing to enhance the welfare of Brazil's landless and rural poor. It puts more land in the hands of agribusiness and the elite—a classic illustration of the marginalization 47 process. It is also grown as a monoculture, cutting down on both natural biodiversity and crop diversity. Critics levy the same accusations at Brazil's soy industry. In the 1960s, Brazil became a major soybean producer and exporter, with production centered in the subtropical southeast and in a new district in the tropical Brazilian Highlands. Soybean production is now pushing into the Amazon Basin and is a major source of deforestation there. With the application of chalk and other additives, the infertile soils of the Amazon can be made more productive.

This expansion is in large part a response to growing demand for soy products in China, the world's largest soy consumer.

Much of Brazil's soy is also bought by American-based companies, including Cargill or Archer Daniels Midland, and made into cattle feed consumed in China and Europe. Brazil is neck and neck with the United States as the world's leader in soybean exports and is second only to the United States in soybean production. To gain an even larger market share, Brazil has given up its long-standing opposition to genetically modified (GM) foods and is producing GM soybeans and other crops. The country is 86 trying to maintain a delicate balance between its North American markets, which have no problem with GM foods, and its European markets, which shun them. Brazil claims it can produce GM and non-GM foods to satisfy both. American farmers look nervously at Brazil's surging agricultural exports. Having won the case it made to the World Trade Organization that U.S. farmers were too heavily subsidized, making Brazilian cotton and other exports uncompetitive, Brazil could undercut American agricultural exports on the world market if the United States abides by the WTO ruling.

Brazil arguably has a "neocolonial" relationship with China, as many other developing countries do. Raw material and ag- 45 ricultural commodities—especially iron ore and soybeans—make up about 85 percent of Brazil's exports to China, and 98 percent of its imports from China are manufactured goods, including cheap cars.

In its relationship with China, Brazil's manufacturing sector is overshadowed. But this sector has experienced rapid expansion. Early-stage industrial products like textiles and shoes are present, along with steel, machinery, chemicals, plastics, automobiles and trucks (Brazil is Latin America's leading automobile producer), ships, aircraft, and weapons. São Paulo is the industrial core, producing nearly two-thirds of Brazil's manufactured goods. Following a global trend, Brazil is outsourcing an increasing share of some of its lower-tech manufacturing, for example, of shoes, to China.

Brazil's metals make an important contribution to the country's industry. The Brazilian Highlands are very mineralized, with ores of iron, bauxite, manganese, tin, and tungsten. Brazil has about an eighth of the world's proven iron ore reserves and is the world's third largest producer and exporter of iron ore. Because of these reserves, Brazil is now in the top 10 of world steel exporters, thanks to China's voracious appetite for this important product.

For decades, Brazil's greatest resource deficiency was in energy. Coal and petroleum reserves were inadequate, threatening the country's continuing industrial development. Tropical hardwoods were widely used in place of fossil fuels in some manufacturing industries. Today, however, the country is in an enviable position with respect to energy: it is virtually self-sufficient, even meeting 90 percent of its needs for oil from domestic supplies. Large natural gas and petroleum reserves have been found offshore and in the remote central Amazon region. The proven offshore oil reserves are so large and growing that some analysts suggest they will rival Saudi Arabia's. But drilling them will be a daunting task: they are very far offshore, and are far deeper even than the Gulf of Mexico wells, and any accident would be difficult to contain. 417

At high environmental cost, and largely with Chinese financial assistance, more than 20 dams have been built or are scheduled for construction in the Amazon Basin to feed Brazil's appetite for electricity, almost 90 percent of which is produced by hydropower. The proposed Belo Monte Dam on the Xingu River is opposed by a coalition of indigenous and environmental groups who argue that it will destroy pristine rain forest environment and that the 100,000 laborers and other outsiders it will attract will be detrimental to Indian cultures. The Indians themselves have taken action, as they are constitutionally allowed to do: in Brazil, Bolivia, and Peru, indigenous groups have the right of consultation on large infrastructure projects. In 2010, leaders of 13 tribes created what they called a new tribe of 2,500 people to occupy the Belo Monte Dam site for as long as needed to stave off its development. Their right of consultation also held up the dam in Brazil's court system. At one point, the Native Americans were joined by one of the dam's most vocal critics, James Cameron of the film *Avatar* fame. The director was even able to show his film to tribal elders and later was overcome with emotion listening to their passionate concerns about the dam.

The Amazon, Its Forest, and Its People

The Amazon is the world's great river (•Figure 10.35). It is not the longest; its 3,915 miles (6,264 km) are surpassed in length by the Nile. But by all other standards, the Amazon rules. It handles more volume than any other river. As much as one-fifth to one-fourth of all the world's available freshwater (excluding what is locked up in ice) is in the Amazon and its tributaries at any given moment. So much water flows from the mouth of the Amazon that it forces back the salty waters of the Atlantic as far as 200 miles (320 km) offshore. The river is so impressively large that early Portuguese explorers called it the **"River Sea."** Many creatures normally found only at sea have adapted themselves to its freshwater vastness, including dolphins, sharks, and rays.

The river drains the Amazon Basin, covering some 2.7 million square miles (4.4 million sq km), or about 10 times the size of Texas. About 60 percent of this basin is in Brazil, with the remainder sprawling into eight other countries. One of the most pristine sections is in Venezuela, where the hard rock of the Guyana Shield makes agriculture virtually impossible and where in the 1980s the government passed legislation protecting the forest and ensuring rights of indigenous peoples. The basin is home to the world's largest remaining expanse of tropical rain forest and some of the world's most remote populations of indigenous peoples. It is also a region that promises, or seems to promise, prospects for economic development. The issues of pristine forest, Native American populations, and development are intertwined, often problematically. Here is a brief look at these problems.

The Amazon Basin rain forest is the world's largest storehouse of plant and animal species, the great majority of which have not yet been described by science. Ecologists argue that preservation of the rain forest's biodiversity is essential for ensuring the genetic variability that nature requires for change and evolution, and for helping ensure valuable future supplies of food, medicines, and other resources for people. The forest also acts as a carbon sink to mitigate possible global warming due to excess greenhouse gas emissions. In economic terms, the amount of carbon that Amazon vegetation captures every year is worth $13 billion. Halting deforestation may be seen as the least expensive way to cut global carbon emissions; deforestation accounts for

• **Figure 10.35** A view from the *middle* of the Amazon River, more than 2,000 miles upstream from its mouth, near Iquitos, Peru. Although so far upstream, this stretch of the river is navigable by oceangoing vessels.

20 percent of global carbon emissions. The challenge is to use mechanisms like the United Nations' REDD (Reducing Emissions from Deforestation and Forest Degradation) Program to persuasively make the case that conservation pays; the ecological services provided by the forests are enormously valuable. The governments of Brazil and other Amazon Basin countries generally acknowledge the importance of the area's natural state but also argue that the area is so vast that portions of it can be reasonably developed to help advance their economies.

Since the 1970s, Brazil in particular has aggressively pursued development in five major areas of the Amazon: exploitation of iron ore and other minerals, resettlement of "excess" populations from Brazil's crowded southeast, farming and ranching, timber exploitation, and hydroelectric development. Each of these serves as an agent of deforestation. All are ultimately linked to Brazil's expanded road network in Amazonia. To begin its development of the Amazon, in the 1970s Brazil began construction of the **Trans-Amazon Highway** and its feeder system (•Figure 10.36). The main line of the interoceanic highway runs roughly east to west from the Atlantic toward its ultimate, long-awaited destination: the Peruvian coast, finally reached in 2011. This road construction was part of an international project called the **Regional Initiative for the Infrastructure Integration of South America**. The final link crossing the Andes is intended to give Brazilian exporters an easier outlet to Asian markets. Principal feeder lines include the controversial BR-364 through the Brazilian states of Acre and Rondônia; BR-163, running north–south through the heart of the region; and BR-319, now under construction between the free-trade zone of Manaus and Rondônia. BR-319 is of particular concern to environmentalists because it runs through pristine forest to the "arc of deforestation" that was deforested precisely because roads like BR-364 gave access to the forest. This arc

of deforestation, on the southern edge of the Amazon Basin, is visible as the mustard color in Figure 10.36.

Here is how that arc was created: the first wave of settlers into Amazonia consisted mainly of these hardscrabble farmers and ranchers. Brazilian authorities advertised the Amazon as a "land without people for people without land." Brazil began transferring its landless poor from the overcrowded southeast of the country, up along the feeder roads into the wilderness of the Amazon where, with government cash subsidies in hand, they began a destructive cycle of land use. The process is a classic example of marginalization as described on page 47, with poverty fueling environmental destruction.

Road construction in Amazonia initiates a sequence of events, illustrated well by Highway BR-364 (seen in •Figure 10.37). First, the main road is cut as a swath through the forest. Landless peasants, lured by cheap or free land, cut and burn the rain forest along the main road and the smaller lanes linked to it. Typical of farmers' experiences with tropical slash-and-burn agriculture elsewhere, they may be able to wrest only three to five years of corn, rice, or other crops from the soil before it is exhausted. They then move on to clear and cultivate new lands—the typical pattern of shifting cultivation.

Tropical hardwoods are harvested and sold as well. About 40 percent of the wood cut from the Amazon is shipped overseas. The largest single destination is the United States (consuming about a third of the total), followed by China, but collectively, the European Union nations take in 40 percent. In 2007, Brazil announced that it would open vast new tracts of the Amazon for large-scale logging but promised there would be strict monitoring to ensure that the resource was not overharvested and that none of the 70 percent of the forest owned by the government would end up in private hands. Brazil also shifted some of the burden back to the consuming countries, insisting that they should buy timber only from these zones.

Brazil and other Amazon Basin countries have passed legislation to slow or halt illegal logging and other unlawful uses that contribute to the rain forest's destruction. Authorities in Brazil have slowed the illegal logging of valuable mahogany trees, but the problem has boomed in the Peruvian Amazon. Nearly 80 percent of the estimated 1.6 million cubic feet (45,000 cu m) of mahogany exported each year from Peru to the United States—where it becomes furniture, acoustic guitars, home decks, and coffins—is illegal. Despite U.S. support for laws prohibiting the illegal logging of mahogany, enforcement has been lax, and legal and illegal mahogany is often imported to the United States in the same shipments.

If Amazonian lands used for farming and timber were abandoned and left to regrow, they would return to mature tropical rain forest in a matter of decades. In Amazonia, however, the typical pattern is that large cattle operations move in to use the areas conveniently cleared for them by the farmers and loggers. As about 60 million cattle (one-third of Brazil's total cattle population) feed and tread over these Amazonian lands, the soils are exposed to excessive cycles of wetting and drying and to relentless bombardment by ultraviolet radiation. This turns the soil into the bricklike substance known as laterite,

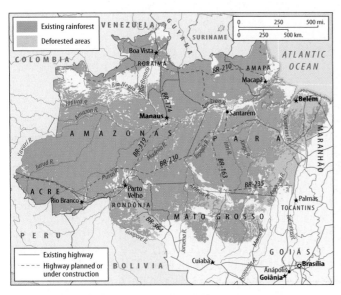

• **Figure 10.36** Deforestation and road development in the Brazilian Amazon.
Source: Adapted with permission from the April 22, 2004 issue of The Christian Science Monitor (www.CSMonitor.com). © 2004 The Christian Science Monitor.

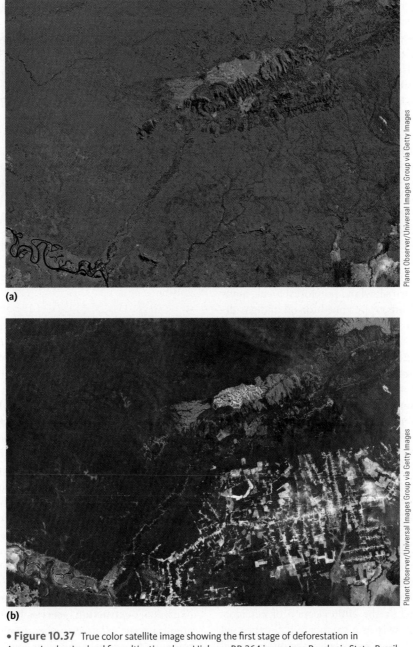

Planet Observer/Universal Images Group via Getty Images

(a)

Planet Observer/Universal Images Group via Getty Images

(b)

• **Figure 10.37** True color satellite image showing the first stage of deforestation in Amazonia: clearing land for cultivation along Highway BR-364 in western Rondonia State, Brazil.

219, 345 on which it is all but impossible for forest regrowth to occur. When ranchers have exhausted the land in this fashion, they move their cattle on to the next plots cleared by farmers. This **shifting ranching** follows the shifting cultivation; it is part of a cycle that is exceptionally destructive of the rain forest.

The Amazon is gradually becoming urbanized in the process. Between 1990 and 2010, the population of Brazil's Amazon region grew by 50 percent to 20 million, the majority of these living in cities and towns. There are now five cities in Amazonia with more than 300,000 residents—a threshold that is attracting Brazil's national retail chains. American-style shopping malls have begun sprouting in Amazonia.

These processes and the other uses of the forest, especially the boom in commercial soybean cultivation, have resulted since the 1960s in the removal of about 20 percent, or 1.6 million square miles, of the tropical rain forest cover of the Amazon Basin. In recent years, the annual rate of deforestation in Brazil has slowed somewhat to about 4,600 square miles (12,000 sq km), or about the size of Connecticut. National law in Brazil stipulates that 80 percent of every tract in the upper Amazon, and 50 percent in more developed regions, must remain forested, but there is little enforcement. About 20 percent of the Amazon Basin's forest, or 1.6 million square miles, has been razed. The World Wildlife Fund projects that at current rates of deforestation, 55 percent of the Amazon forest will be gone by 2030. Brazilian authorities have announced a plan to reduce deforestation by 72 percent by 2017. But there is concern among ecologists that the Brazilian forest has reached a "tipping point," from where the vegetation will shift into a savanna biome with less biodiversity. Forest dieback has been accelerated recently by unprecedented droughts. These unusual dry spells are related to El Niño and to another weather phenomenon called the Atlantic dipole, which warms Atlantic waters. There were widespread forest fires in 2005 and again in 2010. A positive feedback loop is at work: with less foliage there is less moisture in the water cycle, making the local climate hotter and drier.

Climate change threatens not only flora and fauna but also indigenous cultures relying on hunting, gathering, and simple agriculture to survive. Previously, their adaptation to environmental stress was to move, but that has become nearly impossible as other Amazonian interests stake claims to the forest. Development in the Amazon has also increasingly provoked conflicts with the region's indigenous populations. There is particular concern now about how fossil fuel production will affect the Kichwa, Achuar, Schuar, and other Native Americans. After 2004, 70 percent of Peru's lowland rain forest areas, many inhabited by Native Americans, were zoned for oil and gas development, and indigenous advocates worry that exploration will introduce new diseases to the native populations. There are plans to build 800 miles (1,280 km) of pipelines that would take natural gas from lowland rain forest areas to coastal refineries and ports. Another big natural gas project is planned for a part of the Brazilian Amazon inhabited by indigenous people. In Ecuador, where oil already makes up half the country's exports by value, there are plans to double production, again mainly in areas where there are significant Native American populations.

In almost all cases, the Native Americans in the areas affected believe that little, if any, benefit will come to them from these developments. They are becoming activists, using a combination of threats against the oil interests and pleas for more benefits from the projects. Some of the native groups have powerful advocate allies abroad and have come to rely on the Internet to make their cause known; an example is the Kichwa people of Sarayaku (http://www.sarayaku.com), who live in one of the Ecuadorian oil blocks slated for development.

The Kichwa were buoyed by the 2006 election of Ecuadorian President Rafael Correa, who as a candidate had expressed his support for their cause. Correa appealed to the international community to use debt-for-nature swaps or other means to finance the protection of sensitive cultural and environmental sites in the rain forest. In exchange for the funding, Ecuador would prohibit oil drilling in these areas. Correa set forth particular challenges; for example, if the international community will give

• **Figure 10.38** Are ecotourism and cultural tourism helping to protect indigenous people and natural environments in the Amazon? This community of Bora Native Americans in Peru's Amazon region promotes itself as a "genuine Indian village." Its residents wear native costume for visitors.

Ecuador $3.6 billion, exactly half the projected earnings from extracting oil in the Yasuni National Park, Ecuador will leave the ground there untouched. European powers were interested in this cause until their sovereign debt crisis made such generosity difficult. Some Native American communities are developing their own ecotourist and cultural tourist enterprises, hoping both to demonstrate to governments that there are alternative economic uses for the rain forest areas and to gain valuable support and publicity from international visitors (•**Figure 10.38**).

Amazon rain forest destruction has become a *cause célèbre* for environmentalists around the world, who denounce the development efforts of Brazil and the other countries. Most of the region's governments and people see it differently, denouncing foreigners as hypocrites who have cut down their own forests and who are intent on stunting development of the rain forest countries. There are alternative development schemes, such as the commercial extraction of valuable exports like rubber and brazil nuts from intact tropical forests. These take place in areas like Brazil's **extraction reserves**, but they are very limited in economic impact relative to the conventional development enterprises in Amazonia, and their long-term prospects for growth are uncertain. "Sustainable development" of the Amazon is elusive but, if achieved, would have enormous benefits for people and environments within and far beyond the forest biome.

The book continues, and ends, with the United States and Canada.

summary

- Latin America lies south of the United States and includes Mexico, the Central American countries, the island nations of the Caribbean Sea, and all the countries of South America. It is called Latin America because of the post-fifteenth-century dominance of Latin-based languages and the Roman Catholic Church imposed by European colonizers. The term

Middle America refers to Mexico, Central America, and the Caribbean.

- Latin American landscapes and livelihoods are correlated in altitudinal zones, which include *tierra caliente, tierra templada, tierra fría,* and *tierra helada.*

- Large areas of humid climates and biome types are in the lowlands of Latin America, with tropical rain forests occupying major river basins, especially in Brazil and Venezuela. Grasslands are found poleward of the tropics, and there is a zone of Mediterranean climate on the western coast of Chile.

- Dry climates and biomes, including desert and steppe, exist because of rain shadow and other factors. There are major desert regions on the west coast of South America and northern Mexico.

- Latin America's population is about 8.5 percent of the world's total, and its growth rate is above the world average. Most of the people live on the rim of the landmasses, with major population bands in the mountain valleys and, in some cases, on high plateaus or on the flanks of the Andes Mountains in western South America.

- Before the arrival of the Europeans in the late 1400s, Latin America had been home to highly advanced indigenous civilizations, notably the Aztecs, Maya, and Inca. There were many other Native American groups that did not develop urban livelihoods. All these cultures declined precipitously after Europeans arrived. Although some of the losses are associated with war and with mining and other land uses, the biggest killer was European diseases, especially smallpox.

- The region's major ethnic groups today are Native Americans, mestizos, mulattoes, blacks, and Europeans. Political and economic systems still tend to favor people of European extraction, but non-Europeans have found increasing power through democratic political systems. Many reject the free-trade and capitalist systems that they believe favor the ethnic Europeans.

- Latin America has been urbanizing rapidly, with now more than two-thirds of its population living in cities. There continue to be steady rural-to-urban migration and a steady stream of legal and illegal immigrants northward into the United States. The remittances of those workers are an important resource for Latin American economies.

- *Latifundia* and *minifundia* are two major agricultural systems. There is a strong plantation economy, especially in Central America, the Caribbean, and coastal Brazil, with sugar and fruits dominant. The fair trade movement raises prices for consumers of coffee, bananas, and other produce but helps ensure better wages for Latin American farmers. Ownership of farmland is concentrated in the hands of relatively few people, creating imbalances that land reform efforts have so far not redressed.

- Minerals have played a major role in the history, economic development, and trade networks of Latin America. Petroleum is of growing importance, especially in Mexico and Venezuela. The global commodities boom related especially to China's growth has helped Latin American economies in recent years.

- The Latin American countries have formed or joined several free-trade agreements. The largest in volume and value is the 1994 North American Free Trade Agreement (NAFTA). One of its goals is to promote economic growth in Latin America so that migration to the United States becomes less attractive to unemployed Latin Americans.

- Tropical Latin America's proximity to temperate and affluent North America has helped make tourism one of the region's most important industries.

- U.S. interests in Latin America dominate the region's geopolitical themes and issues. The Monroe Doctrine and its Roosevelt Corollary were designed to justify U.S. activities and interventions in the region. Historical efforts have been directed at suppressing Communist or leftist interests in the region and at safeguarding passage through the Panama Canal. Modern interests focus on promoting trade, fighting drug trafficking, and guaranteeing secure access to oil in Venezuela and Colombia.

- Mexico is the largest, most populated, and economically most developed of the Middle American nations. It is the largest Spanish-speaking country in the world.

- The North American Free Trade Agreement (NAFTA) has played a major role in economic change in Mexico. Much of Mexico's economy is based on production in the *maquiladora* factories of the U.S. border region, where NAFTA has benefited the country. NAFTA has been less beneficial to the mainly agricultural south, where farmers have a difficult time competing with subsidized U.S. farmers.

- Central America is made up of Guatemala, Belize, El Salvador, Honduras, Nicaragua, Costa Rica, and Panama. Panama's history over the past century has been shaped by U.S. interests in the Panama Canal, which is now being enlarged.

- In Venezuela, major settlement is concentrated in the uplands. Discovery of oil a century ago in the coastal regions around Lake Maracaibo prompted rapid industrial growth. Venezuelan President Hugo Chávez has led a self-proclaimed "Bolivarian Revolution," ostensibly aimed at using the country's oil wealth to enhance the prospects for the poor.

- Until recently, much of Colombia's territory was controlled by two rebel groups. With U.S. economic and military assistance, the government is fighting back and has wrested much control from the rebels.

- Brazil is the giant of South America. It is larger than the 48 lower United States but has more than 100 million fewer people. Settlement clusters along the coast, but Brazil is promoting settlement in the interior savanna and rain forest. Brazil's economy is the largest in South America, and Brazil has the region's widest disparities between wealthy and poor people.

- Brazil has a large realm of tropical rain forest centered in the Amazon River system, which drains almost 3 million square miles (7.8 million sq km).

- Periods of economic change in Brazil have been marked by booms and busts in sugarcane, gold, diamonds, rubber, and coffee. Soybeans, especially for the Chinese market, are increasingly important in Brazil's agricultural export economy. Beef and orange juice concentrate are also prominent exports. Sugarcane is grown mainly for Brazil's ethanol fuels, which are in increasing demand as fossil fuel costs rise. Iron ore and steel are vital nonagricultural exports.

- The Amazon has a greater volume of water than any other river in the world. The Amazon Basin extends into nine countries, and about half of it lies in Brazil. Development in the Amazon region is controversial, in part because of the rich biodiversity in its rain forest. The Trans-Amazon Highway and its feeder roads are opening Amazonian wilderness to development. Shifting cultivation is often followed by shifting ranching in a very destructive land use cycle.

- Oil in both Ecuador and Peru has begun to be extracted in the tropical lowlands draining into the Amazon River. Its development is controversial because many indigenous peoples live on lands slated for oil development, and they are generally opposed to the industry.

Key Terms + Concepts

alternative development (p. 374)
American Invasion (p. 378)
Amerindians (p. 360)
Aztec-Tanoan language family (p. 362)
Aztecs (p. 361)
balloon effect (p. 374)
barrio (p. 367)
Bay of Pigs invasion (p. 373)
"Bolivarian Revolution" (p. 379)
Chibcha (p. 362)
Creole (p. 366)
creole languages (p. 363)
crop substitution (p. 374)
Cuban Americans (p. 373)
Cuban Missile Crisis (p. 373)
decertification (p. 373)
distance-decay relationship (p. 372)
Dominican Republic–Central America Free Trade Agreement (DR-CAFTA) (p. 370)
El Niño (p. 358)
El Niño Southern Oscillation (ENSO; El Niño) (p. 358)

energy *maquiladoras* (p. 378)
ethanol (p. 382)
extraction reserve (p. 386)
fair trade movement (p. 368)
favela (p. 367)
Free Trade Area of the Americas (FTAA) (p. 370)
hacienda (p. 368)
hinterland (p. 353)
Hokan-Siouan language family (p. 362)
hurricane (p. 358)
Inca (p. 362)
Indians (p. 360)
intellectual property rights (IPR) (p. 370)
informal settlements (p. 359)
latifundia (p. 368)
lock (p. 379)
maquiladora (p. 371)
María Lionza (p. 365)
Maya (p. 360)
Mayan language family (p. 363)
megalanguage (p. 363)
Mercosur (p. 370)
mestizo (p. 366)

Mexican War (p. 378)
minifundia (p. 368)
Monroe Doctrine (p. 372)
mulatto (p. 366)
multi-Latina companies (p. 372)
narcoterrorist organization (p. 375)
Native Americans (p. 359)
Nazca (p. 362)
North American Free Trade Agreement (NAFTA) (p. 371)
Organization of American States (OAS) (p. 370)
Oto-Manguean language family (p. 362)
pampa (p. 350)
páramos (p. 358)
Penutian language family (p. 363)
Plan Colombia (p. 374)
Plan Patriot (p. 375)
pre-Columbian times (p. 365)
Quechua-Aymaran language family (p. 363)
Regional Initiative for the Infrastructure Integration of South America (p. 384)

remittances (p. 370)
"River Sea" (p. 383)
Roosevelt Corollary (p. 372)
rules of origin (p. 371)
Santería (p. 365)
shantytown (p. 367)
shifting ranching (p. 385)
Southern Cone Common Market (Mercosur) (p. 370)
Teotihuacános (p. 361)
tierra caliente (hot country) (p. 356)
tierra fría (cold country) (p. 357)
tierra helada (frost country) (p. 357)
tierra templada (cool country) (p. 357)
tortilla wars (p. 371)
Totonac language family (p. 362)
Trading with the Enemies Act (p. 373)
Trans-Amazon Highway (p. 384)
tree line (p. 357)
upper limit of agriculture (p. 357)
war on drugs (p. 374)
Washington Consensus (p. 375)

Review Questions

1. Where are the main areas of population concentration in Latin America?

2. What and where are the principal climate and biome types of Latin America?

3. What are the four elevation zones? What human uses are generally associated with them?

4. Why was Port-au-Prince especially vulnerable to earthquake damage, and what plans are there to remake Haiti's urban geography in the wake of the 2010 earthquake?

5. What causes El Niño? What effects are usually associated with it in particular regions?

6. What and where were the major Native American civilizations prior to the arrival of the Europeans?

7. What are the major pre-Columbian and modern languages of Latin America, and where are they spoken?

8. What are the principal faiths in Latin America?

9. What are the principal cash crops and minerals exported from Latin America?

10. What are the region's most important existing and planned free-trade agreements?

11. What are remittances? What role do they play in Latin American economies?

12. What country outside the region is most significant in the geopolitics of Latin America, and how has it exercised its influence in the region?

13. What drives the economies of towns on the Mexico–U.S. border?

14. What are energy *maquiladoras*? What resources do they use, where is their product exported, and why are they located where they are?

15. What considerations went into the construction of the Panama Canal? What steps are being taken to improve the canal?

16. What is the balloon effect, and what is its relationship to the "war on drugs"?

17. In what ways is Brazil the giant of South America?

18. What is the Trans-Amazon Highway system? What role does it play in the region's development and deforestation?

19. What cycle of land use is particularly destructive to the Amazon rain forest?

20. Why are Native Americans of the Amazon Basin generally opposed to fossil fuel development there?

21. What were the hallmarks of Brazil's economic history? What exports are particularly strong today, and why?

22. How committed is Brazil to ethanol, and what interests do other countries have in this fuel source?

Notes

1. Amy Wilentz, "The Dechoukaj This Time," *New York Times,* February 7, 2010, p. WK-12.

2. Quoted in Andrew C. Revkin, "In Quake-Threatened Cities, Quick Growth Invites Disaster," *New York Times,* February 25, 2010, p. A-1.

3. Quoted in Alfred M. Tozzer, "Landa's *Relación de las Cosas de Yucatán*: A Translation," in *Papers of the Peabody Museum of American Archaeology and Ethnology,* vol. 18 (Cambridge, MA: Harvard University Press, 1941), p. 78.

4. Bernard Comrie, Stephen Matthews, and Maria Polinsky, *The Atlas of Languages* (New York: Facts on File, 2003), p. 151.

5. Simon Collier, Harold Blakemore, and Thomas E. Skidmore, eds., *The Cambridge Encyclopedia of Latin America and the Caribbean* (Cambridge: Cambridge University Press, 1985), p. 159.

6. Mike Ceaser, "An Oil for Food Policy, or Buying Off U.S. Influence?" *The Christian Science Monitor,* August 25, 2005, p. 7.

7. David S. Cloud, "Like Old Times: U.S. Warns Latin Americans Against Leftists," *New York Times,* August 19, 2005, p. A3.

Online Resources

CourseMate: Make the most of your study time by accessing everything you need to succeed in one place. Read your textbook, take notes, review flashcards, watch videos, complete activities, take practice quizzes, and more—online with CourseMate. Log in at **www.cengagebrain.com**.

C. Mayhew & R. Simmon (NASA/GSFC), NOAA/ NGDC,
DMSP Digital Archive

Joe Hobbs

"Whatever you are, be a good one."

—ABRAHAM LINCOLN

North America from space, at night.
Look for some of the cities you know. What
about where you live — can you see it? Left:
Monument Valley, in the Arizona-Utah
borderlands, is one of America's iconic
landscapes.

11

The United States
and Canada

The United States and Canada, together with Mexico and Central
America, make up the continent of North America. Mexico and Central America are so
linked with South America in culture, language, and tradition that they are best classified
within the separate region of Latin America. The United States and Canada as a culture
region are sometimes called "Anglo America," emphasizing their British colonial origins
and their contrasts with Latin America. But that term is increasingly irrelevant today, as
multiculturalism has become a hallmark of both societies. This chapter depicts a complex
cultural mosaic overlaying an essentially British political, cultural, and economic foundation.

The island of Greenland belongs politically to Denmark but rises from the same
continental crust on which Canada sits, so it is also described briefly as part of this region
(see the region's reference map in •**Figure 11.1**).

chapter outline

chapter objectives

This chapter should enable you to

- Appreciate the wide range of Native American adaptations and ways of life in this region's diverse environments.

- Recognize Canada and the United States as countries shaped mainly by British influences but also by a wide range of foreign immigrant cultures.

- Understand how the United States acquired its vast land empire.

- View the prosperity of the United States and Canada in part as products of their vast natural resource wealth.

- Trace the rise of the United States to a position of global economic and political supremacy and to recent rivalries from other giants.

- Appreciate how depletion of fish stocks has contributed to the economic decline of the Atlantic region.

- Witness the scramble for resources in Arctic lands and waters that has accompanied warming temperatures in the region.

- Understand the economic rationale for Greenlanders' wanting to retain their ties with distant Denmark.

- Understand the environmental and political issues that complicate development of coal, oil, and natural gas in the United States but promote the use of biofuels.

- Follow the decline of traditional heavy industries in the process of deindustrialization and their replacement by high-technology and service industries.

- Evaluate the depopulation of the Great Plains, the decline of small towns, and the potential for communication technology and foreign immigration to reverse these trends.

- Gain more insight into the United States' ethnic geography and immigration issues.

11.1 Area and Population

Canada is the world's second largest country. Its 3.85 million square miles (9.97 million sq km) makes it slightly larger in area than the United States' 3.72 million square miles (9.63 million sq km). The region's map comparison with Europe is on page 72, and the continental United States is compared with Latin America on page 353. Canada had a population of 35 million in 2011, compared with 312 million in the United States (•Table 11.1 and •Figure 11.2). Together, the two countries have 5 percent of the world's population, on 13 percent of its land surface.

The United States reached the milestone of 300 million people in 2006 and kept growing. A snapshot of the population on a given day in 2011 would show 11,000 babies born and 2,800 immigrants arriving—both much larger figures than the number of deaths and emigrants. If the 2011 growth rate were to be sustained, it would put the population at 400 million—double its 1967 population—around 2040.

The majority of this region's 346 million people live in the eastern half of the region, south and east of the Saint Lawrence Lowlands and Great Lakes. The average population density of the two countries is only 84 per square mile (32/sq km) for the United States and 9 per square mile (3/sq km) for Canada, which has a tremendous expanse of sparsely settled northern lands. With its vast icecap, Greenland is even more sparsely populated, with just 57,000 people living in an area of 836,000 square miles (2.17 million sq km), or 0.1 people per square mile (0.02/sq km).

Canada and the United States share the longest international border in the world—5,527 miles (8,895 km). About 90 percent of the Canadian population lives within 100 miles (161 km) of this border, on only 12 percent of Canada's territory. In contrast, only a small percentage of the U.S. population lives within 100 miles of the Canadian border. Americans (in this chapter referring to inhabitants of the United States) think of their northern border region as a winter icebox, but to Canadians that area is almost balmy.

Canadians and Americans are overwhelmingly urban, with city dwellers accounting for about 80 percent of both Canadians and Americans (•Table 11.2). Canada's population is concentrated in the cities of four main regions: the Atlantic region of peninsulas and islands at Canada's eastern edge; the culturally divided core region of maximum population and development along the lower Great Lakes and Saint Lawrence River; the Prairie region in the interior plains between the Canadian Shield on the east and the Rocky Mountains on the west; and the Vancouver region on and near the Pacific coast at Canada's southwestern corner (see Figures 11.1 and 11.2).

In the United States, people are concentrated in urban areas of the Northeast and portions of the West, with more than half that population living within 50 miles of the coast. About 55 million people, or 18 percent of the national population, live in the Northeast on 5 percent of the nation's area. The Northeast's population density is several times that of the South, Midwest, or West. The density is much more extreme within the narrow urban belt stretching about 500 miles (c. 800 km) along the Atlantic coast from metropolitan Boston (in Massachusetts) through metropolitan Washington, D.C. The belt is known as the Northeastern Seaboard, Northeast Corridor, Boston-to-Washington Axis, **megalopolis** (a chain of metropolitan areas with a combined population over 10 million), or **"Boswash."** Here, seven main metropolitan areas, including the country's largest—New York—have about 47 million people combined, or 1 in every 7 Americans. In the West are another 73 million people, nearly one-quarter of the nation. More than half live in California.

From the beginning of European settlement until the 1970s, this region had a rapidly growing population. By 1800, about

435

• **Figure 11.1** Principal features of the United States, Canada, and Greenland.

200 years after the first European settlements, there were more than 5 million people in the United States and several hundred thousand in Canada, and both immigration and natural increase continued to swell their ranks. Now population growth rates (excluding migration) of the United States and Canada are low (0.5% and 0.4%, respectively), testifying to the passing of these countries into the final phase of the demographic transition.

Migration into North America

The United States is the only MDC in the world that is experiencing significant population growth (this is an outstanding characteristic). This is due mainly to immigration rather than natural increase, and both the United States and Canada are best appreciated as nations of immigrants and of continuing immigration. America's ethnic diversity has made it truly a rainbow nation. There were 63 ethnic/racial groups to choose from on the 2010 U.S. census form! It was not unusual for an individual to check the boxes Asian, Hispanic, Native American, and white to describe herself or himself. In 2012, more than 8 percent of marriages were interracial. A half-century ago, such relationships were stigmatized, and in some states were even illegal.

Themes of migration are carried into the discussion of cultures later in this chapter, but an overview of modern migration

Table 11.1   North America: Basic Data

Political Unit	Area (thousands; sq mi)	Area (thousands; sq km)	Estimated Population (millions)	Estimated Population Density (sq mi)	Estimated Population Density (sq km)	Annual Rate of Natural Increase (%)	Human Development Index	Urban Population (%)	Per Capita GNI PPP ($US)
North America	**8,403.8**	**21,765.8**	**346.3**	**41**	**16**	**0.5**	**0.909**	**79**	**44,800**
Canada	3,849.7	9,970.7	34.5	9	3	0.4	0.908	80	37,280
Greenland (Den.)	8,36.3	2,166.0	0.05	0.1	0.02	0.0	N/A	80	N/A
United States	3,717.8	9,629.1	311.7	84	32	0.5	0.910	79	45,640

Sources: World Population Data Sheet, Population Reference Bureau, 2011; Human Development Report, United Nations, 2011; World Factbook, CIA, 2011.

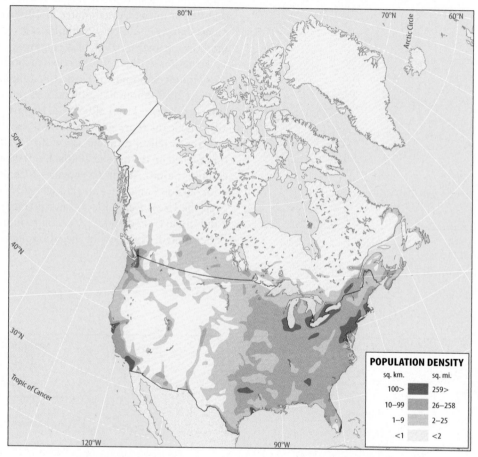

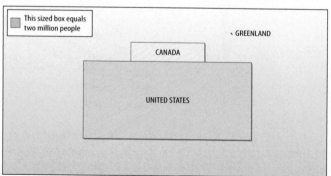

• **Figure 11.2** Population distribution (top) and population cartogram (bottom) of the United States, Canada, and Greenland.

Source: The Oxford School Atlas edited by Patrick Wiegand (OUP, 1997), copyright © Oxford University Press. Used by permission of Oxford University Press.

Table 11.2	North America: Metropolitan Populations	
1	New York, U.S.	22
2	Los Angeles, U.S.	17.8
3	Chicago, U.S.	9.8
4	Washington, U.S.	8.5
5	Boston, U.S.	7.5
6	San Francisco, U.S.	7.4
7	Dallas, U.S.	6.7
8	Philadelphia, U.S.	6.5
9	Houston, U.S.	6
10	Atlanta, U.S.	5.6
11	Toronto, Canada	5.5
12	Miami, U.S.	5.5
13	Detroit, U.S.	5.2
14	Seattle, U.S.	4.2
15	Phoenix, U.S.	4.2
16	Montreal, Canada	3.8
17	Minneapolis, U.S.	3.6
18	Denver, U.S.	3.1
19	San Diego, U.S.	3
20	Cleveland, U.S.	2.8

Population in millions.

© Cengage Learning 2013

flows is provided briefly here (•Figure 11.3; see also Figure 3.15, page 56).

The Great Recession that began in 2008 hammered birth rates in the United States; there was a 7.3 percent decline from 2007 to 2010. Economic distress has a predictably negative impact on birth rates. As discussed below, immigration to the United States also declined during this period. Together, these factors accounted in 2011 for the lowest increase in population since 1945.

Each year, 65,000 people are granted permission to enter the United States as "guest workers." Far greater numbers do not obtain permission and enter as **illegal aliens**; most undocumented workers manage to obtain temporary or longer-term employment despite their illegal status. The Great Recession [70, 370] that began in the United States in 2008 reduced job opportunities sharply. Numbers of immigrants collapsed along with the housing market that had employed so many of them as construction workers. In the construction sector, for example, busts followed booms in Florida, Virginia, and Nevada. From 2000 to 2005, about 850,000 immigrants came into the United States each year. From 2005 to 2010, just 300,000 came annually. Mexicans represented about 60 percent of that flow.

Canada tells a different story. Birth rates there are low, but they inched up during the 2000s. But the bigger story in Canada is immigration. About 20 percent of Canada's population was born outside Canada (compared with 13% in the United States). Even the Great Recession did not detract migrants recently, as it did in the United States. Aversion to immigrants in Canada is rather muted. Numbers of illegal immigrants are smaller than

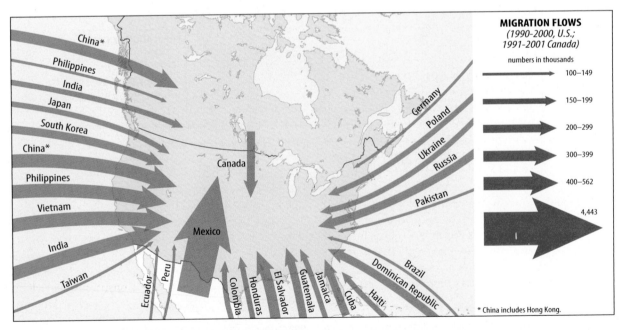

• **Figure 11.3** Migration flows into the United States and Canada.

Source: Data from the U.S. Census Bureau and Statistics Canada.

in the United States—from 2000 to 2005 they were 500,000 and from 2005 to 2010, 150,000, annually. About 280,000 immigrants came to Canada legally in 2011, the most since the 1950s. Provinces in Canada compete for these immigrants, with the more rural communities trying to draw them away from the popular "MTV" cities (Montreal, Toronto, Vancouver).

An estimated 11 million illegal immigrants lived in the United States in 2011, down from 12 million in 2007. They make up 5 percent of the workforce. About 39 million people in the nation, or 13 percent of the population, are foreign-born (compared to 9.5 million, or about 5 percent of the population, in 1970). Latin America is the largest source of both legal and illegal migration into the United States. Other principal source countries of immigrants, in descending order, are China, India, and the Philippines; the same three countries are the largest sources of immigration to Canada.

Not all immigrants come to the United States and Canada of their own will or work at the jobs they had hoped to find; as victims of human trafficking, some are modern-day slaves. Struggling to escape poverty in countries such as Thailand, India, and Russia, tens of thousands of people in the United States and Canada end up working as sweatshop laborers, strippers, and prostitutes. Their circumstances are very similar to those described of modern-day slaves coming out of the former Soviet countries (see pages 144–145).

What to Do about Illegal Immigration?

The status and future of illegal immigrants is a particularly contentious issue in the United States, where an estimated 8 million, nearly two-thirds of the total, are employed. Many American citizens fear that an unstoppable tide of poor immigrants will take their jobs and bleed social and other services. Others, particularly in the business community, argue that the low-wage immigrants are vital for the American economy, generally taking jobs shunned by most Americans (but also displacing some of the mainly uneducated Americans who would take those jobs), while also contributing to the economy through their purchases. The satirical film *A Day without a Mexican* depicts a U.S. economy in disarray when deprived of its undocumented workers. Their low wages also help keep some jobs in the country that might otherwise be outsourced or offshored. Although they do burden schools and other social services, most undocumented workers also pay federal taxes (a price they pay for using fake Social Security and taxpayer ID cards).

Contrary to stereotypes, foreign workers are not confined to backbreaking, less-than-minimum-wage jobs. Most earn at least minimum wage, and many work outside construction and other labor-intensive industries. Legal immigrants add much to America's pool of professional talent and expertise. One in five U.S. doctors is foreign-born, as are one in five computer specialists and one in six engineers. Forty percent of the country's Ph.D.s are foreign-born. A quietly celebrated trend in immigration is that increasing numbers of foreign nationals—more than 60 percent—who have earned their Ph.D.s in science and engineering in the United States are overstaying their visas.

How do poor illegal immigrants get into the United States? The typical Mexican or other Latin American migrant manages to muster the funds to be smuggled by a "coyote" or "chicken rancher" across weak points in the U.S. border defenses. Organ Pipe Cactus National Monument in southern Arizona, along with other wild sections of the Sonoran Desert frontier, is a favored crossing point. Some immigrants traveling such desert routes in the extreme conditions of summer and winter succumb to the elements. Anticipating their travails, "Good Samaritans" on the U.S. side put out water and other provisions for the migrants. Other Americans who are opposed to illegal immigration operate as vigilantes, patrolling the border to turn the travelers back.

The United States does not have a firm or effective policy for managing illegal immigration. The great majority of illegals caught within the United States are returned to the Mexican side of the border without prosecution, and an estimated half of them cross back into the United States. Many are caught yet again, perpetuating a "revolving door" of illegal immigration.

Measures are being taken to extend physical and other barriers that would deter new arrivals, including terrorists seeking easy entry into the United States. In the Secure Fence Act of 2006, the U.S. Congress passed legislation calling for lengthening the existing 15-foot (4.5 m) border fences (to 700 miles or 1,120 km). Critics feel that determined immigrants will find their way over and around physical barriers, which, in any case, they said, will probably end up being built by illegal immigrants! There was an alternative for a while: through its Secure Border Initiative, known as SBInet, the government began building a so-called **virtual fence**, a multibillion-dollar, mainly high-tech surveillance system for the border region. The project was judged to be too expensive, and funding for it was frozen in 2011. Environmentalists, joined in Arizona by the Tohono O'odham Indians (who view wildlife as kindred spirits), worry about the impacts of long fences on wildlife, particularly migrating animals. The virtual fence would have presented no such obstacles, but its effectiveness in reducing or preventing border crossings is much debated.

How effective were the fences and virtual fences along the U.S.-Mexican border? Not very. Nine out of 10 crossing attempts succeeded. But stronger border enforcement created greater demands for coyotes. In 2011, it cost the Mexican about $3,000 to cross the border, versus $700 in the early 1990s. Illegal entry from Central and South America is very costly, roughly $10,000, and the migrant usually pays this off in installments after he or she gets work in the United States.

How about flying over the obstacles, or avoiding them altogether? Some very wealthy Chinese families have come up with an extraordinary strategy to give their children an advantage right out of the gate: have their babies in America. In the practice of **birth tourism**, an expectant mother flies to the United States explicitly for the purpose of having her baby. The Fourteenth Amendment to the U.S. Constitution awards citizenship to any child born in the country. The trip can cost as much as much as $30,000, but parents expect the investment to be repaid many times over as the child is more likely to get into a good college, and then to secure a well-paying job. In addition,

Phuong Dung Nguyen

• **Figure 11.4** Did you know that any baby born in the United States, regardless of the circumstances, is automatically a U.S. citizen? There is debate about whether this constitutional right should be amended to discourage immigration. Young Thai Nguyen Khang is American, and his parents Thai Thong and Doan Nguyen are Vietnamese students working on their graduate degrees at the University of Missouri.

at age 21 the citizen can petition for a grant of permanent residence in the United States for her or his parents; here again, the family value of a child caring for parents is prominent. In this role, the child is an **"anchor baby"** who can later facilitate immigration for family members (•Figure 11.4). This term may be pejorative, as it has made its way into the raucous debate on immigration. Birth tourism is most prominent among moms from South Korea and Taiwan as well as China, and the second most favored destination after the United States is Canada.

U.S. lawmakers are struggling over whether and how illegal immigrants should be granted what is effectively **amnesty**; they would not be prosecuted for being in the country illegally if they met certain conditions and would be granted permanent resident status (which would eventually make them eligible for citizenship). Detractors say that amnesty programs will simply attract more illegal immigrants.

These apparently conflicting signals about immigration have led some observers to describe U.S. policy on immigration as "schizophrenic"; as an exasperated U.S. Border Patrol agent put it, "America loves illegal immigrants but hates illegal immigration."[1]

America's 2010 Census revealed some notable population trends that will be discussed in more detail in following sections. There was much faster regional growth in the South and West than in the Midwest and Northeast. Overall, the South and West contributed 84 percent of the population growth during the decade 2000 to 2010. Nevada was the fastest-growing state, with a 35 percent increase during that decade. It has been the fastest-growing state for five decades. Only one state—Michigan—lost population during between 2000 and 2010. Cities with the largest population losses were Detroit (where the population dropped 25 percent from 2000 to 2010), Cleveland, Chicago, Pittsburgh, Buffalo, Baltimore, St. Louis, and New Orleans. Many of these are cities of the Rust Belt, described on page 419.

11.2 Physical Geography and Human Adaptations

The natural environments of the United States and Canada are remarkably diverse and include some of the most spectacular wild landscapes on the planet. They have presented people with a vast array of opportunities for land use and settlement. It is useful to consider how landforms (and what lies beneath and above them) have promoted or hindered human uses and how climates have also done the same. This wide range of environmental settings is associated with a great variety of natural hazards (**see Geography of Natural Hazards box, next page**).

Landforms and Land Uses

•Figures **11.1**, **11.5**, and **11.6** depict the landforms, climates and biomes, and land uses of North America. Let's have a closer look at conditions in these mapped areas.

In the far northeast lies Greenland, the world's largest island. About 80 percent of Greenland is covered in what is classified as "permanent" ice up to 10,500 ft (3200 m) thick. As we have seen, there is great concern about this ice as well as Antarctic ice melting and causing sea levels to rise. Permanent ice also covers small areas of nearby islands of the Canadian Arctic. 33, 305, 428

The ancient geological core of North America, with rocks up to 3 billion years old, is called the **Canadian Shield**, covering roughly the area from Nunavut south to Minnesota and northeast to Labrador. Human settlement is sparse in this vast region, and agriculture is limited by poor soils and a harsh climate similar to that of northeastern Siberia. The Canadian Shield was 119 scoured by glaciers until about 10,000 years ago, and today the area is dotted with many large lakes and thousands of smaller ones. The surface is mostly rolling, but there are areas of hills and low mountains such as the Superior Upland of Minnesota and Wisconsin. Hydropower, wood, iron, nickel, and uranium are the major resources of the Canadian Shield.

Southeast of the Canadian Shield lie the Appalachian Mountains and associated highlands. The Appalachians were possibly the highest mountains in the world when they were formed 400 million years ago, but erosion has reduced the elevations to between 2,000 feet (600 m) and 6,684 feet (2,037 m). Rising in northern Alabama and Georgia and running northeast toward the Gaspé Peninsula of Quebec (the island of Newfoundland is also Appalachian in origin), the Appalachians are a complex system where mountain ranges, ridges, and rugged dissected plateaus are interspersed with narrow valleys, isolated lowlands, and rolling uplands. The western Appalachians are known for their large coalfields. Coal mining has long been a hazardous and heartbreaking occupation for the people here.

About 80 percent of the Appalachians and adjacent lowlands of New England are forested, a remarkable turnaround from the nineteenth century, when over two-thirds of the original forest had been removed. After 1850, farmers gravitated toward the much richer soils of the Midwest, and hydropower and fossil fuels replaced timber as fuel. These shifts eased

GEOGRAPHY OF NATURAL HAZARDS | **Nature's Wrath in the United States**

The North American continent has more natural hazards than any other, and the United States has more natural hazards than any other country in the world (•**Figure 11.A**). The western edge of the continent lies on a zone where two major plates of the earth's crust collide (see Figure 2.1, page 22), producing both earthquakes and volcanoes. The most consequential quake in U.S. history was the 1906 event of magnitude 8.0 on the Richter scale that produced a displacement of up to 20 feet (6 m) on the San Andreas Fault near San Francisco, California. The earthquake itself and, even more devastatingly, the fires fed by the ruptured gas lines in the city destroyed roughly 30,000 of San Francisco's buildings and killed about 3,000 people. The San Andreas is just one of the largest and best known among the numerous fault lines that trend mainly north–south through California. The pressure built up by tectonic movement is enormous and must eventually be released, so many more earthquakes are in California's future. Some of them will be catastrophic, although the effects will be mitigated somewhat by strict building codes and emergency preparedness.

A network of faults at the edge of the continent threatens Anchorage and other cities in Alaska. The coasts of Alaska, western Canada, and the western United States must also be on guard for tsunamis, the waves that result when offshore earthquakes displace huge volumes of water.

In the western U.S. states of Oregon and Washington are the many volcanoes of the Cascade Range. Several of them, notably Mount Rainier and Mount Hood, still have the potential for eruption and, depending on wind flow and other variables, could do serious damage to Seattle and other populated areas. The destructive potential of these mountains was demonstrated on May 18, 1980, when Washington's Mount St. Helens blew up with an explosive force equivalent to 400 million tons of TNT. Luckily, few casualties resulted in this sparsely populated area.

Surprisingly, one of the most potentially destructive forces in the continental United States is a very ancient fault zone roughly paralleling the course of the southern Mississippi River. In 1811 and 1812, on the New Madrid fault in Missouri, three earthquakes of magnitude 7 to 8 on the Richter scale reportedly reversed the flow of the Mississippi River for some moments and rang church bells 1,000 miles (1,600 km) away in Boston, Massachusetts. The mid-Mississippi region had a small population at the time, and human losses were minimal, but a quake of that magnitude in the same region today would be very destructive to Saint Louis, Memphis, and other cities.

Many Americans of the eastern United States think Californians are crazy to live in earthquake zones, but they regularly contend with their own menacing natural hazards. Storms often pound this part of the country. It is a humid region where rainfall potential is high and where different agents trigger precipitation. The convection associated with daytime heating and the collision of air masses with different temperatures give rise to destructive thunderstorms, with heavy rain, lightning, hail, and one of the most powerful natural forces on earth, tornadoes. Across a swath of the Midwest known as **"Tornado Alley"** (the "high risk of tornadoes" region of Figure 11.A), these intense vortices of very low

16, 254

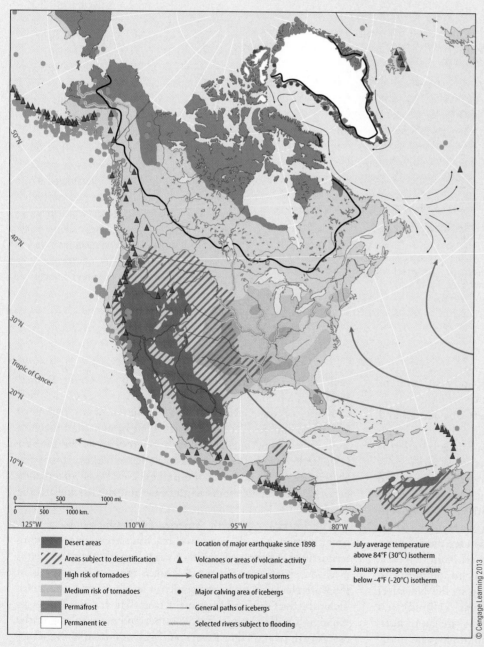

Desert areas

//// Areas subject to desertification

High risk of tornadoes

Medium risk of tornadoes

Permafrost

Permanent ice

● Location of major earthquake since 1898

▲ Volcanoes or areas of volcanic activity

→ General paths of tropical storms

● Major calving area of icebergs

→ General paths of icebergs

Selected rivers subject to flooding

July average temperature above 84°F (30°C) isotherm

January average temperature below -4°F (-20°C) isotherm

© Cengage Learning 2013

•**Figure 11.A** Natural hazards of North America, Middle America, and northern South America.

(continued)

GEOGRAPHY OF NATURAL HAZARDS *(continued)*

atmospheric pressure cut paths of destruction, sometimes causing large losses of life and property. Storms in May 2011 caused many deaths in Mississippi and Missouri. A tornado ranking EF5 (the strongest possible) on the Enhanced Fujita scale simply leveled a long swath of Joplin, Missouri (•Figure 11.B).

Successive days or even weeks of thunderstorm activity sometimes produce localized or widespread flooding. In the summer of 1993, weeks of heavy rains over portions of

the Missouri and Mississippi River watersheds produced a flood of historic proportions. Flooding and wind damage are typical results of the hurricanes that batter the East Coast and Gulf of Mexico regions between June and October. The most catastrophic in U.S. history in terms of lives lost was a hurricane in 1900 that killed more than 8,000 people, mainly in storm surges, in Galveston, Texas. In storm damages, the most costly prior to 2005 was Hurricane Andrew, which came ashore south

of Miami, Florida, in 1992, causing $27 billion in property losses. Those losses were dwarfed by the impacts of Hurricane Katrina, which roared onto the coasts of Louisiana, Mississippi, and Alabama in August 2005 (see discussion on pages 437–439). Yet another precipitation-related natural hazard is the **blizzard**, a combination punch of heavy snowfall and high winds that strikes the United States—typically in the Midwest and Northeast—an average of 11 times each year.

A more pernicious natural hazard, a kind of disaster in slow motion, is the drought that sometimes rakes the country, especially in the West. The worst was the eight-year drought that produced the **Dust Bowl** of the 1930s, ruining crops and livelihoods and causing mass migrations of broken farming families to leave Oklahoma and adjacent areas, mainly to settle in the West. This story is told superbly in John Steinbeck's novel *The Grapes of Wrath*. Recently, a drought lasting longer than the Dust Bowl years afflicted the western and especially southwestern United States. Climatological records suggest that such lower precipitation may be the norm for this region, which may previously have enjoyed decades of abnormally high precipitation. Human populations in the region boomed in those years, perhaps far beyond levels that can be sustained by local water supplies.

Joe Hobbs

• **Figure 11.B** The monstrous EF 5 tornado that struck Joplin, Missouri, on May 22, 2011, killed 124 people and destroyed this high school, a hospital, and many homes. This was among the first of a string of deadly twisters that struck "Tornado Alley" that year.

pressure on the forests, and with their rebound also came increasing numbers of moose, beaver, and other wildlife.

Immediately east of the Appalachians, between Georgia and the vicinity of New York City, lies the rolling to hilly Piedmont ("foothills"), an area with a large population and good soils for farming. The Piedmont is fairly narrow, and its eastern edge is marked by the **fall line**, a natural boundary where rapids and waterfalls commonly mark the head of navigation on many rivers and have long supplied hydropower. Cities such as Columbia, South Carolina; Richmond, Virginia; Washington, D.C.; and Philadelphia, Pennsylvania, are located along the fall line.

East of the fall line is the Atlantic Coastal Plain, which is continuous with the Gulf Coastal Plain to the south. Forming a large crescent from eastern Texas up to the New Jersey shore, the coastal plains are characterized by sandy and largely infertile soils (although river valleys are more agriculturally productive), a generally flat topography, and many large river mouths and bays indenting the coast. There are many important wetland areas along the coast and inland,

such as the Everglades in Florida and the Great Dismal Swamp in Virginia.

North of the Gulf Coastal Plain lies a small region of highlands composed of the Ozark Plateau and the Ouachita Mountains, sometimes collectively referred to as the Interior Highlands. Divided by the Arkansas River Valley, this is the largest area (but not with the highest elevations) of uplands between the Appalachians to the east and the Rockies to the west.

The interior of North America is dominated by a vast, roughly triangular-shaped plain, reaching from Texas to Ohio, to northern Alberta. Millions of years ago, this land was submerged beneath a shallow sea. Today, most of this area is in the watershed of the Mississippi River and its many tributaries, including the Ohio and Missouri Rivers. The Interior Plains, as the area is collectively called, is hillier and more rolling in the east, north of the Ohio River, but becomes more level farther west, even as average elevation increases markedly from east to west. The eastern sections (the prairie), once dominated by long

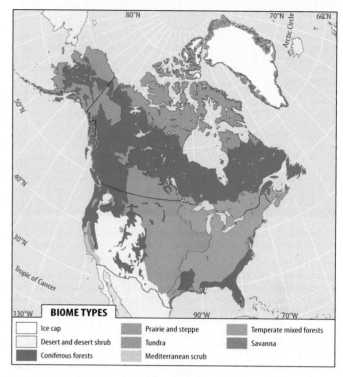

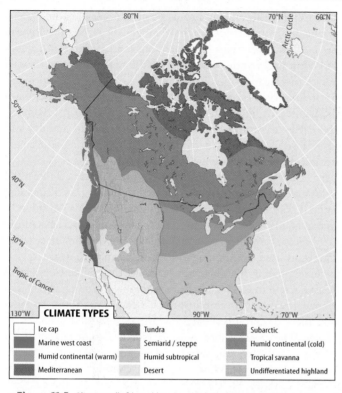

• **Figure 11.5** Climates (left) and biomes (right) of the United States, Canada, and Greenland.

Sources: (left) Based on Rand McNally's Classroom Atlas, 2003. (right) Based on World Wildlife Fund ecoregions data, 1999.

grasses with scattered wooded areas, have mostly been turned into built-up and agricultural land; only about 10 percent of the original ecosystem is intact. The western third of the region is called the Great Plains, a semiarid steppe dominated by short

• **Figure 11.6** Land use in Canada, the United States, and Greenland.

Source: The Oxford School Atlas edited by Patrick Wiegand (OUP, 1997), copyright © Oxford University Press. Used by permission of Oxford University Press.

grasses that is mostly flat but higher in elevation than many of the peaks of the Appalachian Mountains (see •**Figure 11.7**). The Great Plains, once referred to as the Great American Desert before its agricultural potential was realized, has some of the world's finest soils and most productive farmlands but is only lightly populated. A more rugged area that includes the Black Hills (site of Mount Rushmore; •**Figure 11.8**) and the colorful terrain known as the Badlands disrupts the otherwise generally level terrain of the western plains in the western Dakotas and eastern Montana.

The Great Plains end abruptly to the west alongside the Rocky Mountains, a long chain of mountain ranges stretching

• **Figure 11.7** The Appalachian Mountains in western New York state. This highly scenic portion of the Appalachians is called the Allegheny Mountains.

• **Figure 11.8** The Black Hills are sacred to many Native American groups. While nearby Mt. Rushmore was being sculpted, Indian chiefs asked the sculptor Korczak Ziolkowski to carve an image of the Oglala Lakota warrior Crazy Horse in the Black Hills. Work began in 1948. When completed, this will be the largest sculpture in the world, exceeding the Mother Russia statue depicted on **page 129**.

from British Columbia to New Mexico (•**Figure 11.9**). The tallest mountain in the Rockies is Mount Elbert in Colorado at 14,433 feet (4,399 m). High and rugged, with few easy passes, the Rockies are the source of many major rivers, including the Rio Grande, Colorado, Missouri, Fraser, Saskatchewan, and Columbia. Recreation, especially skiing and visiting parks such as Yellowstone National Park in Wyoming and Banff National Park in Alberta, is a major draw to the Rockies.

West of the Rockies is a geologically complex region of alternating basins and lower mountain ranges. The Columbia Plateau lies in the northern part of this region; south of there, it includes the Great Basin to the west and the Colorado Plateau to the east and is often called the "basin and range country." Arid in the north and true desert in the south, as tall mountains block rainfall to the east (in the Rockies) and west (in the Sierra Nevada), this is the driest area of North America. Canyons and dry valleys carve many plateaus, in many cases presenting stunningly beautiful landscapes. The Grand Canyon (5,000 ft [1,500 m] deep) and the lowest point on the continent, Death Valley (282 ft [86 m] below sea level), are located in this region. Ranching, mining, and energy extraction (oil, gas, and coal) are the economic mainstays of this region. Despite the dryness, agriculture is also important here, with major dams impounding river water that is diverted toward irrigation projects and urban areas.

A series of mountain ranges parallel the Pacific Coast from Alaska south to California. These ranges—including the Alaska Range, the Coast Mountains, the Cascades, the Sierra Nevada, and the Coast Ranges—are all tall, rugged, and glaciated in places. The tallest mountains in North America (Denali, or Mount McKinley, 20,320 ft [6,194 m]), in Canada (Mount Logan, 19,551 ft [5,959 m]), and in the contiguous United States (Mount Whitney, 14,505 ft [4,421 m]) are located in these ranges. The Pacific Coast is prone to earthquake activity, and several active volcanoes are found in Alaska and the Pacific Northwest. Fishing and logging are important to the economy, as is agriculture in two very productive lowlands, the Willamette Valley of Oregon and the Central Valley of California (•**Figure 11.10**).

Stretching across northern Alaska and into the Canadian territories lies the Arctic Coastal Plain, a flat, marshy area that has few resources except for vast quantities of oil lying offshore in the Beaufort Sea.

• **Figure 11.9** The Maroon Bells Wilderness in the Colorado Rockies.

• **Figure 11.10** California's Central Valley, one of the breadbaskets of the United States. The rolling hills bordering the western side of the valley (at left) belong to California's Coastal Range. The two aquaducts carry water from Northern California to Southern California.

Climates and Land Uses

The United States has more climatic types than any other country in the world, and even Canada is quite varied. The wide range of economic opportunities afforded by this climatic variety helps to make this world region so wealthy.

Canada's and Alaska's tundra climate, with its long, cold winters and brief, cool summers, promotes vegetation of mosses, lichens, sedges, hardy grasses, and low bushes. The subarctic climate zone, also in Canada and Alaska, has long, cold winters and short, mild summers, with a natural vegetation of conifer-
121 ous (boreal) forest resembling the Russian taiga. In 2003, an unusual coalition of energy and forestry companies, environmental groups, and Native American communities joined an agreement to preserve at least 50 percent of Canada's subarctic forests and to develop the other half in sustainable ways. Population is very sparse in the subarctic and tundra climate areas. Scattered peoples engage in trapping, hunting, fishing, mining, logging, and military activities; many live largely on welfare. The severe winter in vast sections of the tundra and subarctic zones requires unusual adaptations—like having to wait until the spring thaw to bury loved ones who died during the winter.

The humid continental climate with short summers ("humid continental cold" in **Figure 11.5**) is characterized by long, cold winters and short, warm summers. Agriculturally, emphasis is on dairy farming except in the Far West, where spring wheat production dominates. The humid continental long-summer climate ("humid continental warm") has cold winters and long, hot summers; agriculturally, this belt, which includes the agricultural riches of the Midwest, emphasizes dairy farming in the east and corn, soybeans, cattle, and hogs in the midwestern (interior) portion.

In the humid subtropical climate zone, winters are short and cool—with some cold snaps—and summers are long and hot. Agriculture varies, with local production of cattle, poultry, soybeans, tobacco, cotton, rice, peanuts, fruits, and vegetables.

Within each of these three humid climate regions, the pattern of natural vegetation is complex, with areas of coniferous evergreen softwoods, broadleaf deciduous hardwoods, mixed hardwoods and softwoods, and prairie grasses. Far southern Florida has a small area of tropical savanna climate, which is a major climate in adjacent Latin America. The state of Hawaii has a tropical rain forest climate, but most of the original as-
289 sociated vegetation has been cleared (•**Figure 11.11**). Along the Pacific shore of the United States and Canada, the narrow strip of marine west coast climate is associated with the barrier effect of high mountains near the sea, ocean waters that are warm in winter and cool in summer, and winds prevailing from the west. The mild, moist conditions have favored a magnificent growth of giant conifers (most notably the redwood and the Douglas fir) that provides the basis for the lumber industry, and the now untouched giants, the sequoias (see Figure 2.6i). Lush pastures support dairy farming, as they do in the corresponding climatic region in northwestern Europe.

The Mediterranean, or dry-summer subtropical, climate region of central and southern California, where almost all

• **Figure 11.11** Na Pali coast, island of Kauai, Hawaii.

rain comes in winter, is associated with irrigated production of cattle feeds, vineyards, vegetables, fruits, cotton, and numerous other crops (see Figure 11.10). These products, together with associated livestock, dairy, and poultry, make California the leading U.S. state in total agricultural output.

In the semiarid steppe climate region, occupying an immense area between the Pacific coast of the United States and the landward margins of the humid East and extending north into Canada, temperatures range from continental in the north to subtropical in the south. The natural vegetation of short grass, bunch grass, shrubs, and stunted trees supplies forage for cattle ranching, which is the predominant form of agriculture. More moist areas are used for wheat (both winter wheat and spring wheat). Other crops are grown in scattered irrigated areas, often associated with major rivers such as the Columbia, Snake, Arkansas, and Rio Grande.

The desert climate of the U.S. Southwest is associated with scattered irrigated areas and settlements emphasizing mining, recreation, and retirement. The high, rugged mountains of the Rockies and the Sierra Nevada have undifferentiated highland climates varying with latitude, elevation, and exposure to moisture-bearing winds and to the sun (•**Figure 11.12**).

• **Figure 11.12** The Sierra Nevada range and Mono Lake (top, just left of center).

You cannot assume that the long-standing distribution of biomes as mapped in Figure 11.5b will be accurate in the coming years. There is growing evidence that as global temperatures rise, agricultural and natural vegetation zones will shift poleward in the United States, as they are doing elsewhere. Large-scale maps that you may be familiar with are beginning to show the changes. For example, the 2012 United States Department of Agriculture map of hardiness zones for typical garden plants in the United States showed that many bands are a full zone warmer, and in some places two zones warmer, than when the last map was published in 1990. The NASA Goddard Institute for Space Studies has predicted that by 2050 extreme droughts will occur every other year in the United States. As discussed later, despite the growing body of science to the contrary, 4 in 10 Americans do not accept that the climate is changing.

• **Figure 11.13** Anasazi pueblo dwelling in Canyon de Chelly, Arizona.

11.3 Cultural and Historical Geographies

As mentioned earlier, many geographers call the region of the United States and Canada "Anglo America," but the decision not to do so here is an acknowledgment of the rich ethnic roots and branches of this region. Native Americans began their migrations into the region as Asians, crossing what was then a land bridge between Alaska and Siberia at least 12,500 years ago. However, there is some scientific evidence, still being scrutinized, that the migrations may have begun as early as 33,000 years ago. Other controversial science holds that Polynesian sea voyagers also reached the shores of the Americas in very ancient times. Migrations across the Asian land bridge persisted until about 3,000 years ago, and a diverse pattern of indigenous ethnic groups, languages, and lifeways emerged.

Native American Civilizations

The indigenous cultures of Middle and South America have been described on pages 359–363 in Chapter 10. The indigenous cultures of what are now Canada and the United States, especially those of what is now the U.S. Southwest, were related to them in many ways. Some of them developed civilizations, the rather complex, agriculture-based ways of life associated with permanent or semipermanent settlements and stratified societies. In what are now the southwestern states of Arizona, Utah, Colorado, and New Mexico, three dominant Native American civilizations emerged: the **Mogollon, Hohokam,** and **Anasazi**. Despite their arid environment, these peoples developed productive agricultural systems that borrowed from and interacted with the Aztec and other cultures of Middle America. The same basic crop assemblages were found across all these cultures: corn, beans, squash, and chili peppers were the staples.

The Hohokam of what is now southern New Mexico had a very productive system of irrigated agriculture. Their culture flourished between 100 B.C.E. and 1500 C.E. The Anasazi developed a dwelling pattern based on the **pueblo**, in which interconnected mud-brick (adobe) residences and ceremonial centers, with beamed roofs of mud, sticks, and grass, were built into cliff sides or on the flat-topped mesas of the region (•Figure 11.13). The Mogollon culture that thrived between 300 B.C.E. and 1400 C.E. inherited these building traditions. Their culture was effectively absorbed by the Anasazi, whose roots extend to about 1200 B.C.E. The Anasazi heartland was in the Four Corners region where Colorado, New Mexico, Arizona, and Utah meet.

Around 1300 C.E., the Anasazi began to abandon their pueblos and the productive agriculture associated with them. The reasons for the Anasazi decline remain unknown, but drought, invasion by hostile neighbors, or simply political decisions to relocate may have been responsible. Some Native American groups of the region today, including the Hopi, Zuni, and Tiwa peoples, carry on the architectural legacy of the Anasazi by continuing to dwell in pueblos.

Farther to the east and north, in much of the eastern half of what is now the United States, the so-called **Mound Builder civilizations** developed. There were four main culture groups among them: the **Poverty Point** and the **Mississippian** cultures of the Southeast, the **Hopewell** in the Midwest, and the **Adena** in the Northeast. The earlier Poverty Point and Adena cultures (c. 2000 B.C.E. to 200 C.E.) grew some crops but were mainly hunters and gatherers who thrived in the productive humid environments of the East. The Hopewell (c. 200 B.C.E. to 700 C.E.) grew more crops, including corn, beans, and squash. The Mississippian (700–1700 C.E.), with an urban and suburban settlement pattern in many sites, was the greatest farming culture and had an extensive trading network. Of all the mound-building cultures, the Mississippian peoples were the greatest builders of earthen mounds. In all the cultures, these structures served as ceremonial sites, and in some they were also tombs.

Indigenous Culture Groups and Lifeways

The Native American groups that were not associated with complex societies, ceremonial centers, and civilizations were

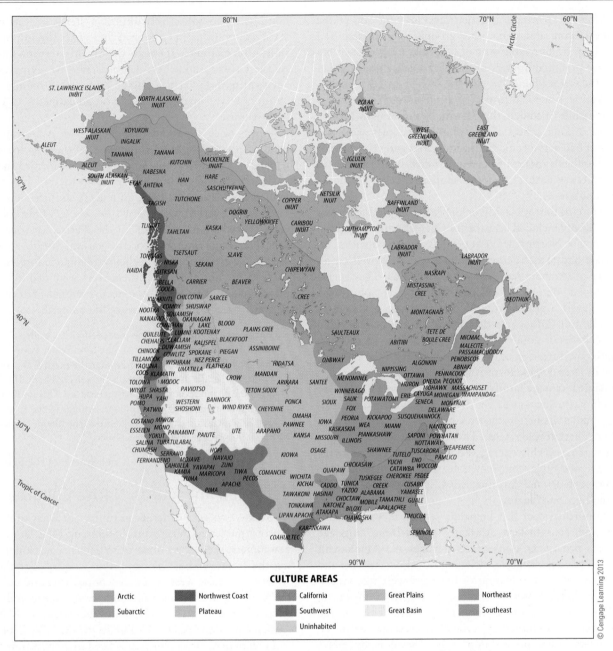

• **Figure 11.14** Native American culture areas and selected tribes of Canada, the United States, and Greenland.

varied and numerous. As many as 18 different culture regions for these groups are recognized; this text uses 11 (•Figure 11.14). The best way to categorize their essential features is to acknowledge some of the major groups within each region, and their original languages. Although the past tense is used in this discussion (because these cultures flourished in pre-Columbian times), none of the languages mentioned is extinct, and speakers of all of them may still be found—although for most, English is their first or second language.

Seven Native American language families are represented in the United States, Canada, and Greenland by more than 250 languages. The Aztec-Tanoan language family that sprawled across most of northern Mexico was represented in the U.S.

Southwest by languages including Hopi, Comanche, Shoshone, and Papago. Some of the groups that spoke these languages carried on the agricultural and pueblo-dwelling lifestyles pursued by the earlier Mogollon, Hohokam, and Anasazi civilizations, whereas others were nomadic hunter-gatherers who sometimes supplemented their diet by raiding the pueblo-dwelling peoples.

The larger part of what is now the contiguous United States was the hearth of the Hokan-Siouan language family, which includes such languages as Iroquoian, Mohawk, and Lakhota. Among its more famous culture groups were the **Sioux**, including the **Lakota** and **Dakota**, who developed their trademark subsistence pattern after the Europeans arrived in the New

World. After the Spaniards and their successors introduced the horse, the Lakota and some other groups gave up what had been a village and farming way of life, migrating west to become horse-mounted hunters of the bison (buffalo) and other game on the Great Plains.

In what is now California and Oregon, the Penutian language family included the Miwok and Klamath-Modoc languages, among others. Most of these peoples were not farmers but had very productive hunting and gathering economies that took advantage of the region's diverse and abundant resources. These natural bounties supported relatively dense human populations.

Farther to the north, but still on the Pacific coast, the **Mosan language family** included branches of Chemakuan, Salish, and Wakashan. The Northwest Coast peoples who spoke these languages also had rather dense populations that thrived on an abundance of sea mammals, fish, and land game. They generally developed complex societies in which the custom of the potlatch played a prominent role. **Potlatch** was a system in which rank and status were determined by the quantity and quality of material goods one could give away—an interesting variant and inversion of the "keeping up with the Joneses" tradition of later Anglo North America.

The **Algic language family** members were spread across all 10 Canadian provinces and 20 U.S. states, mainly in the northern tier. Some of its better-known languages include Arapaho, Blackfoot, Cheyenne, and Cree. The tribes that spoke the Algic languages relied on the region's vast forests for game animals, fuel, shelter, and tools, complementing their hunting and gathering with fishing and farming; some groups were seminomadic and lived in villages.

To the northwest of this linguistic region, in west central Canada and southeastern Alaska, the **Na-Dene language family** was prominent. This group includes the **Athabascan subfamily** with its languages, including Koyukon; Apache; and in a remote outlier, the Navajo of the Southwest (the largest tribe in the United States, with more than 270,000 members; see Figure 1.5, page 7). Most of these peoples lived either in the subarctic region of coniferous forests and wetlands or in the tundra, where agriculture was impossible. They hunted caribou and other game, and their populations were limited.

Across the rest of Alaska, throughout northern Canada, and in Greenland, the **Eskimo-Aleut language family** dominated. The peoples who carried these tongues into North America were among the last to migrate from Asia, probably between 2500 and 1000 B.C.E., and they did so in boats, as the former land bridge was inundated by then. The so-called **Eskimos**—an Algonquian term meaning "raw meat eaters"—prefer to call themselves **Inuit**, meaning "the people." The Inuit and the **Aleuts** were indeed characterized by a carnivorous diet, as their livelihood was based on hunting sea mammals, caribou (known as reindeer in the Old World), and fish.

What all these diverse groups had in common was an exceptionally well-developed set of adaptations for living in the local environment. Native American economies spanned a remarkably wide continuum, from simple foraging and

hunting to complex agricultural systems. These were not static adaptations but changed and generally improved over time—for example, through greater mastery of finer tools. As the Lakota and others demonstrated by giving up village agriculture in favor of a nomadic life based on horses, these cultures were capable of dramatic adjustments to changing opportunities.

It was once supposed that the Native Americans always lived within the limitations imposed on them by local ecosystems, but it is now known that they transformed landscapes in significant ways by fire and other means and may even have hunted populations of large animals to extinction; this is the Pleistocene overkill hypothesis (see page 40). One trait apparently shared by most if not all the Native American groups was their deep reverence for the natural world, as is clearly seen in Chief Seattle's words below. Animals, forests, geological features, and the people themselves were tied together intimately in animistic belief systems that emphasized the kinship between these diverse elements of life on earth.

228, 302, 322

So that they will respect the land, tell your children that the Earth is rich with the lives of our kin. Teach your children that we have taught our children that the earth is our mother. Whatever befalls the Earth befalls the sons of Earth. If men spit upon the ground, they spit upon themselves. This we know: the Earth does not belong to man; man belongs to the Earth. This we know. All things are connected like the blood which unites one family. All things are connected.

—Chief Seattle of the Suquamis

European Impacts on Native Cultures

There are two very different narratives of what took place in North America in the centuries following 1492. For European newcomers, these were times of settlement, development, taming the frontier, and "civilizing the savages." For the Native Americans, these were times of depopulation and cultural demolition. Popular T-shirts and bumper stickers on Indian reservations in the western United States these days read "NATIVE AMERICANS: FIGHTING TERRORISM SINCE 1492."

How many Native Americans lived in what is now Canada, the United States, and Greenland when Columbus made landfall in 1492 is uncertain. The most common estimate is 1 million, with about three-quarters of that population in what is now the United States and almost all the rest in Canada. As in Latin America and Australia, the European contact initiated years of population losses among Native Americans through disease, famine, and warfare. In the United States, Native American populations crashed to an all-time low of fewer than 250,000 between 1890 and 1910, but they are now back up to 1.8 million. Their growth has been particularly strong in the Great Plains, where there are now more Native Americans (and bison) than there were in the late 1870s.

292, 364

In Canada, the current population of Native Americans is about 1 million. Collectively, these peoples refer to themselves as the **First Nations** in acknowledgment of their pre-Columbian cultures and claims to the land. In Canada, this

• **Figure 11.15** Modern Native American reservations and other lands in the United States and Canada.
Source: From Carl Waldman, Atlas of the North American Indian (Facts On File, 1985). Reprinted with permission of the Publisher.

definition excludes the Inuit, who were relative latecomers but who are referred to as among the **First Peoples**. The indigenous Greenlanders, known as the *Kalaallit*, are related linguistically and ethnically to the Inuit of Canada, Alaska, and Siberia. The official name of their island is Kalaallit Nunaat, meaning "Land of the Greenlanders."

In the United States, most of the Native American populations are in the west; in Canada, they are in the north and west. The two countries have dealt in different ways with their indigenous populations and claims to land and other rights, with government–Native American relations historically better in Canada.

In the United States, an estimated one-third of all Native Americans today live on the **reservations** that the government established as remnants of their former tribal territories or sometimes set up thousands of miles from native homelands, forcing the people to relocate there (•**Figure 11.15**). "The Res," as Native Americans call it, is among the poorest communities of the United States, often plagued by high rates of incarceration, alcoholism, drug abuse, depression, broken families, teen suicide, and unemployment. The Lakota's Pine Ridge Reservation in South Dakota (•**Figure 11.16**), for example, has an annual per capita income of just $7,770 (compared to the national U.S. average of $27,330). However, it must also be noted that even these lamentable conditions represent a considerable improvement. Native Americans have growing political and economic clout, with poverty and unemployment rates dropping and education levels rising.

Gambling revenues have contributed to economic and other improvements in many Indian communities. The semiautonomous status of their territories allows casinos to circumvent

• **Figure 11.16** Lakota at the Wounded Knee Massacre Site, Pine Ridge, South Dakota. Their reservation is prominent in the poverty map on **page 413**.

Joe Hobbs

• **Figure 11.17** The Seneca Allegany Casino in Salamanca, western New York State, is in the Allegany Indian Reservation. Most of the people living here are from the Seneca, Iroquois, and Cayuga tribes.

many restrictions that prohibit casinos elsewhere (•Figure 11.17a). About one-third of the roughly 563 federally recognized Native American tribes in the United States have developed casinos since Congress legalized gambling on their lands in 1998. This **gaming industry** has been a mixed blessing for the Native Americans, with a wide variety of successes and failures in different places. In some, gambling revenues have supported community education and student scholarships, health care, and cultural centers, but in others they have lined a few pockets and failed to benefit many members of the community. Some Native Americans complain that several of the groups that have established casinos are not legitimate, historical tribes to begin with. Non-native people often resent casinos because they view them as a windfall of special treatment for individuals who have not earned it. In all cases, casinos have introduced significant changes.

In Hawaii, as in North America, incursions by Europeans —Captain Cook's visits were the first in 1778 and 1779— brought huge changes to the native people. Polynesians may have populated the island group as early as 300 c.e., making a living by growing bananas, coconuts, breadfruit, and
291 taro and by fishing. Hawaiian society was based on a ranked caste system, and taboos governed many aspects of daily life and relationships with the spiritual world. The islands were a united indigenous kingdom in 1893, when a group of American businessmen, backed by the U.S. military, overthrew the monarchy, and took control (for more information on Polynesian geography and culture, see Chapter 8). The United States formally annexed Hawaii five years later. Diseases, including smallpox and influenza, introduced accidentally by outsiders, nearly wiped out the indigenous population: there were an estimated 1 million ethnic Hawaiians in the late eighteenth century but just 22,000 in 1919. Today, the **Kanaka Maoli**, as the native Hawaiians call themselves, number 250,000. There has been much ethnic mixing among this population. The

native *Hawaiian* language a member of the Malayo-Polynesian subfamily of the Austronesian family, is undergoing a major revival following a long decline. Hawaiian became an official state language in 1978—the only indigenous language in the United States to have this distinction. Major education efforts ensued. As recently as 1983, only 50 children spoke the language, but now more than 2,000 do.

The situation for indigenous peoples in Canada is unique, even on a worldwide scale. In 1999, Canada ceded nearly one- 301 quarter of its total land area to the Inuit peoples, who called their land **Nunavut** ("our land" in Inuktitut; see Figure 11.15). These 733,600 square miles (1.9 million sq km) were carved from Canada's Northwest Territories and generally lie north of 60°N, surrounding the northwest part of Hudson Bay and extending well within the Arctic Circle. The total population is approximately 30,000, scattered through about 30 communities. About nine-tenths of the capital required to keep the territory viable comes from Ottawa, the federal capital of Canada, making the per capita cost of maintaining Nunavut higher than for any other political unit in Canada. The Canadian government has allocated more than $1 billion to be distributed in the form of services, supplies, and subsidies until 2016 to the Inuit population, representing about $38,000 a year per person. Large subsidies also go to Native American populations in the Northwest Territories.

All across Canada, settlements of indigenous claims to ancestral lands have led to federal government financial support of the communities, a measure of self-government for the indigenous peoples, and royalty income to them from mines and fossil fuel industries. Indigenous peoples who have relocated to Canada's cities are showing major improvements in education and employment. Still, the average income of Canada's indigenous peoples is about a third lower than the population as a whole. Canada's record with its indigenous peoples is not unblemished; in the 1950s, for example, the government forcibly relocated a group of Ungava Inuit from Hudson Bay 1,200 miles (1,920 km) north to inhospitable Ellesmere Island. The move, which caused these people enormous suffering, was meant to help confirm Canadian sovereignty over the island.

European Settlers and Settlements

European settlement of what are now Canada and the United States took place in a series of waves propelled by religious persecution in Europe, colonization of new lands by European powers, and then the expansionist efforts of newly independent Canada and the United States. Canada's core region of Québec and Ontario developed with the French entry into the interior of North America. The French founded Québec City in 1608, at the point where the Saint Lawrence estuary leading to the Atlantic narrows sharply. Their fortifications on a bold eminence allowed control of the river there.

From Québec, fur traders, missionaries, and soldier-explorers soon discovered an extensive network of river and lake routes with connecting portages reaching as far as the Great Plains and, via the Mississippi River, to the Gulf of Mexico. Montréal, founded later on an island in the Saint Lawrence River, became the fur trade's forward post for the interior wilderness. Between Québec City and Montréal, a thin line of settlement grew along the Saint Lawrence, forming the agricultural base for the colony. Population grew slowly in this northerly outpost where the winter was harsh and only French Catholics were welcome.

French Canadians numbered only about 60,000 when Britain conquered the region in 1759 (France was expelled from Canada in 1763). They did not join the English-speaking colonists of the Atlantic Seaboard in the American Revolution and by their refusal laid the basis for the division of the continent of North America into two separate countries (see page 423). Until the late twentieth century, French Canadians increased rapidly in numbers. About 20 percent of Canada's people today are French in language and culture. They are concentrated in the lowlands along the Saint Lawrence River, where they form the majority population in the province of Québec. The French Canadians have clung tenaciously to their distinctive language and culture, with some attempting twice since 1980 to gain independence for the province of Québec.

A few British settlers came to Québec after the conquest, and this immigration increased rapidly during and after the American Revolution of 1775–1783. These migrations set the foundations for the British segment of Canada's core. Many of the early English-speaking immigrants were Loyalist refugees from the newly independent United States, whose rebellion French Canada had refused to join. They were soon joined by more English settlers emigrating directly from Europe. These newcomers formed a large minority in French Québec, including a British commercial elite in Montréal that persists to this day, but they also settled west of the French on the Ontario Peninsula. By 1791, the British government decided to separate the two cultural areas into different political divisions, known at that time as Lower Canada (Québec) and Upper Canada (Ontario), taking the directional terms from their positions on the Saint Lawrence River.

Farther west, the prairie region of Canada was settled much later, in the decades after 1890 (see •Figure 11.18). Settlers found some good land in an area with long and harsh winters, short and cool summers, and marginal precipitation. They could grow only hardy crops for subsistence. Later came specialization in spring wheat for export, and the prairie region became the Canadian part of the North American Spring Wheat Belt.

The settlers of this region were mainly English-speaking people from eastern Canada and the United States, but they included notable minorities of French Canadians, Germans, Ukrainians, and others. The ethnic makeup of the area still

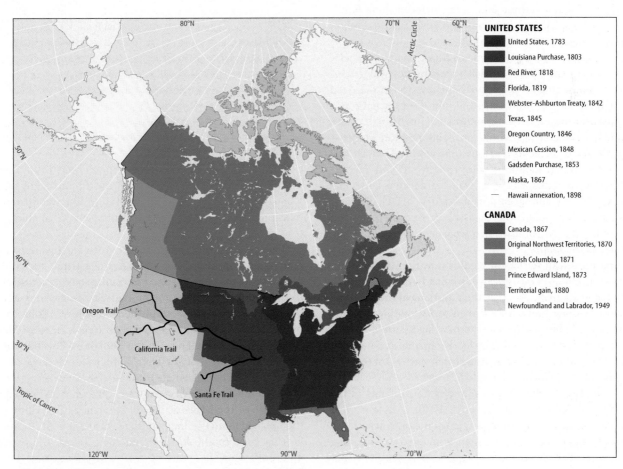

UNITED STATES
- United States, 1783
- Louisiana Purchase, 1803
- Red River, 1818
- Florida, 1819
- Webster-Ashburton Treaty, 1842
- Texas, 1845
- Oregon Country, 1846
- Mexican Cession, 1848
- Gadsden Purchase, 1853
- Alaska, 1867
- — Hawaii annexation, 1898

CANADA
- Canada, 1867
- Original Northwest Territories, 1870
- British Columbia, 1871
- Prince Edward Island, 1873
- Territorial gain, 1880
- Newfoundland and Labrador, 1949

• **Figure 11.18** Territorial acquisitions of the United States and Canada.

Source: Based on John Haywood, ed., Atlas of World History (Barnes and Noble Books, 1997).

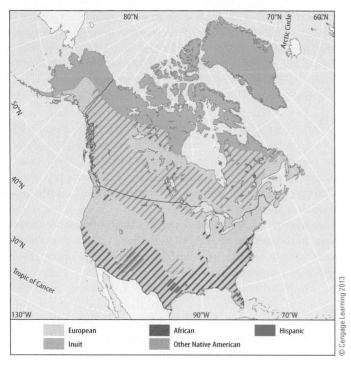

© Cengage Learning 2013

• **Figure 11.19** Major ethnic groups in the United States and Canada.

Table 11.3	Ethnic Groups of Canada and Their Origins

Major Ethnic Groups	Dates
Native peoples	pre-1600
French	1609–1755
Loyalists from the United States	1776–mid-1780s
English and Scots	mid-1600s on
Germans and Scandinavians	1830–50, c. 1900, 1950s
Irish	1840s
African Americans	1850s–70s
Mennonites and Hutterites	1870s–80s
Chinese	1855, 1880s, 1990s
Jews	1890–1914
Japanese	1890–1914
East Indians	1890–1914, 1970s on
Ukrainians	1890–1914, 1940s–50s
Italians	1890–1914, 1950s–60s
Poles	1945–50
Portuguese	1950s–70s
Greeks	1955–75
Hungarians	1956–57
West Indians	1950s, 1967 on
Latin Americans	1970s on
Vietnamese	1970s on
Filipinos	1990s

Sources: E. Herberg, *Ethnic Groups in Canada: Adaptations and Transitions*. Copyright © 1989 Nelson Canada, Scarborough, Ontario; and Census of Canada, 2001.

reflects them. Canada's ethnic mix today includes about 6 million people of English origin, 3 million Irish, and 4 million Scottish (see •**Figure 11.19** and **Table 11.3**).

Other ethnic groups would come to join Canadian society, which by law in the 1988 **Multiculturalism Act** was recognized as a multicultural society. These laws acknowledged that Canada's Native American and non-European groups would probably never become assimilated in Canadian society and that the country would probably be more stable—avoiding the kind of racial tensions that wracked the United States—if traditional identities were recognized and encouraged. Most Canadians embrace this legislation, but some argue that Muslim immigrants in particular are challenging Canada's freedoms; mirroring an issue in Britain, there has been a debate over Muslim women wearing the face-cloaking
71 veil known as the *niqab*.

The first European settlers of what is now the United States were Spanish, Dutch, English, Swedish, and French, with the majority being English. Most of their settlements were along the Eastern Seaboard and were linked to Europe by the long but generally reliable sea passage across the Atlantic. Northeastern urban development began early. In the early 1600s, English settlers founded Boston, and Dutch settlers established New Amsterdam (which the English annexed and renamed New York in 1664). The English founded Philadelphia in the late 1600s and Baltimore in the early 1700s. All four were major urban centers by the time of the American Revolution.

Development of the interior began with exploration along waterways: the British up the Chesapeake Bay and its tributaries, the Dutch up the Hudson in the early 1600s, and the French up the Saint Lawrence and Mississippi rivers. Westward

expansion proceeded very slowly; up to about 1800, there was little European settlement beyond the Appalachian region of what is now the eastern United States. The floodgates of expansion then opened, and wave after wave of immigration took place.

There were several phases of territorial acquisitions in what Americans considered their **Manifest Destiny**—the conviction that Americans were destined to acquire new lands all the way to the Pacific, "from sea to shining sea" as the song goes. (see Figure 11.18). First, they established themselves along the Ohio and eastern Mississippi watersheds following the American Revolution in 1783. Then, in the **Louisiana Purchase** of 1803, the United States bought from France (for 3 cents per acre!) about 23 percent of what are now the lower 48 states. This included much of the western watershed of the Mississippi and the lands of the Missouri River watershed. The United States then picked up major lands in the south and west. It purchased Florida from Spain in 1819 and took Texas, the Southwest, and California as spoils of war with Mexico in 1845 and 1848. It established the Oregon Territory in 1846. In the 1853 378 **Gadsden Purchase**, the United States bought a small strip of present-day Arizona and New Mexico from Mexico. Finally, it bought Alaska from Russia for 2 cents per acre in 1867. The young nation thus acquired a vast land empire much like that built earlier by Russia, although by different means.

127

The **Homestead Act** of 1862 made many of these new regions almost irresistible magnets for settlement, as it allowed a pioneer family to claim up to 160 acres of farming land for a fee of about $10. Two major trails, the ancestors of today's interstate highway system, were developed in the 1840s to channel the newcomers into the frontier zones (see Figure 11.18). From a trailhead in Independence (near Kansas City), Missouri, the **Oregon Trail** took a northerly track across what are now Nebraska, Wyoming, and Idaho to the fertile Willamette Valley of the Oregon Territory, with a southern branch (the **California Trail**) leading to what is now Sacramento, California. The southern route from Independence, known as the **Santa Fe Trail**, snaked across Kansas and New Mexico. Army posts were set up along the way to facilitate and protect the westbound migrants. They also served as military garrisons in the forceful subjugation of Native American peoples, whose stories are told in broken treaties, massacres, and decimation of political leadership.

Accessed in part by the Oregon and Santa Fe trails, the U.S. Great Plains region was settled rapidly between 1860 and 1890. A major impetus for settlement of the far west was the **California Gold Rush** of 1849. No Panama Canal existed at that time, so most of the prospectors, known as **Forty-Niners**, made their way either tortuously along the Oregon and California trails, across the Isthmus of Panama by land, or by sea for a 17,000-mile (27,200 km) voyage around South America, which could take up to seven months.

372

Most of the homesteaders, Forty-Niners, and other settlers of the American frontier were of English, German, Scandinavian, and other western and northern European ethnicities. The United States ultimately developed into a nation in which about 99 percent of the people were either immigrants or of immigrant stock from a vast array of ethnic backgrounds. The larger groups of today include more than 45 million of German origin, over 20 million of Irish origin, and over 20 million of English origin. Although some kept their own ethnic identities—still found in ethnic neighborhoods, especially in the East, or in small rural communities in the Midwest—most were blended into the melting pot that produced what might be called the **mainstream European American culture**.[2] The ethnographer David Levinson describes that culture as one based on institutions such as the English language and British educational principles brought by the British colonists, combined with core American values that developed on U.S. soil. These include English as the national language, religious tolerance, individualism, majority rule, equality, a free-market economy, and the idealization of progress.

Added later to the mix were immigrants from Eastern and Southern Europe, represented today by more than 11 million Americans of Italian origin, over 6 million of Polish origin, and over 5 million Jews. Many of these immigrants settled initially in the larger cities of the U.S. Northeast, where earlier immigrants of mainly northern and western European extraction often

• **Figure 11.20** One poor immigrant group after another – Irish, Italian, Polish – stayed in the tenements of Manhattan's Lower East Side before moving up the economic ladder.

discriminated and directed violence against them (•Figure 11.20). It took several generations for these newer immigrants to become part of the mainstream ethnic fabric of America. Other large groups have had more difficult careers with the white majority—which is projected to become a minority soon after 2050—or are relatively new contributors to the American ethnic mix.

Ethnic Minorities in the United States and Canada

Former U.S. Secretary of State Condoleezza Rice aptly described slavery as "America's birth defect."[3] The plantation economy that supported the American South was built on the exploitation of more than 500,000 black slaves brought to America from Africa between 1619 and 1807. People of mainly African ancestry, known as **blacks** and **African Americans**, occasionally as **Afro-Americans**, and formerly as **Negroes**, number about 39 million, or 13 percent of the U.S. population. The group also includes many people of mixed black and other ethnic extraction and blacks who came from regions other than Africa. Blacks were acknowledged as the largest minority in the United States until 2000, when census figures revealed that the number of people identifying themselves as Hispanic slightly exceeded the number of African Americans.

322

Unlike the French Canadians, African Americans are not largely confined to one part of the country. Some regions, especially in the eastern half of the United States, have a very strong black cultural presence (Figure 11.19), and African Americans are majorities in a number of large cities. In the so-called **Great Migration** between 1915 and 1970, about 6 million blacks moved out of the former Confederacy states of the Old South, settling especially in the inner cities of the Northeast and Midwest. They tended to have more education than the blacks who stayed in the South, and had higher employment numbers and a more stable family life than Northern-born blacks. These are the general characteristics of the "migrant advantage" shared by Irish, Italian, and other immigrants in U.S. historical geography.

Race relations have played an important role in the country's political and social history. Conflicting attitudes toward black slavery were among the issues that nearly split the United States into two countries in the American Civil War. The **Thirteenth Amendment** to the Constitution ended slavery in 1865, but the economically and politically subordinate position of blacks has been slow to change. As recently as the 1960s, there were legal struggles to end official **segregation** between the races in parts of the United States—for example, in the state of Alabama, where there were separate sections for whites and blacks on buses and where the races were expected to live in different neighborhoods, attend different schools, and never intermarry.

Today, although a segment of the African American population has moved into the middle class and above, a large portion resides in impoverished and troubled **ghettos** of central cities, where there are problems in housing, jobs, and crime (•**Figure 11.21a**). About 40 percent of all people incarcerated in U.S. prisons are black. The legal battles to eliminate segregation have ended, but "hearts and minds" have been slower to change, and African Americans continue to struggle for equality. The 2008 election of Barack Obama as the 44th President of the United States marked a milestone in the nation's history. As with the growing trend of interracial marriages, the inauguration of President Obama raised widespread hopes that remaining racial barriers in the United States would weaken and fall.

In the 1840s, when the United States won from Mexico the great swath of the American Southwest, West, and Texas, it also became home to many Spanish speakers of European or mestizo descent. The growth rate of people of Spanish or mixed Spanish background—the Hispanic American (or **Latino**) population—has in recent years been more rapid (about 3% per year) than that of any other group, both domestically and through the major migration stream flowing actively from Latin America (**Figure 11.21b**). About half of the growth is from natural increase and another half from immigration, both legal and illegal. About 50 million people, or 16 percent of the U.S. population, are Hispanic. The majority is of Mexican origin and is concentrated in California, Texas, New Mexico, Arizona, and Colorado. Other important Hispanic groups include people of Puerto Rican origin, most heavily represented in New York City and in the Northeast and increasingly in Florida, and people of Cuban origin, who live mainly in southern Florida. Previously, Hispanic immigrants settled overwhelmingly in the "gateway" states of the southwestern United States, but today, many leapfrog these states and put down roots all across the country.

Minorities make up about one-third of the U.S. population, and they are rapidly gaining momentum toward that date in the 2040s when non-Hispanic whites will drop below 50 percent of the population. Demographically and politically, a gap is emerging in the United States between a young, ethnically diverse population and an older, mainly white one. There are four states with majority populations of ethnic minorities: Hawaii (77%), New Mexico (60%), California (60%),

(a)

(b)

(c)

• **Figure 11.21** Ethnic urban landscapes of the United States: **(a)** a black-run business in Los Angeles; **(b)** a Latino neighborhood in Chicago; **(c)** Chinatown in San Francisco.

and Texas (55%). Twenty-one percent of the nation's minority populations live in California, and 12 percent live in Texas.

Asian Americans now make up about 4 percent of the U.S. population. Immigrants from China were especially important in the construction and service sectors of the West Coast economies of both Canada and the United States between 1849 and 1882. Chinese laundries and the Chinese "coolies" who did the grueling labors of building western railways became the stuff of stereotype and legend, and there were sore points in Chinese-Anglo relations. Chinese immigrants brought an indelible cultural presence in the form of "Chinatown" landscapes in major cities, particularly in the nineteenth and early twentieth centuries, and especially in the western regions of both countries (Figure 11.21c). Chinese food has become even more widespread than Mexican food across the continent.

Asian Americans have long been steadily moving out of the central city and into white and mixed neighborhoods; they include among their ranks not only the Chinese but also large populations of ethnic Japanese, Korean, Vietnamese, and Filipinos, among others. The closest approach to Asian minority dominance of a state is in multiethnic Hawaii, where Asian Americans, especially of Japanese descent, make up more than 40 percent of the population. Race relations between the Anglo majority and Asian groups in both the United States and Canada have generally been good in recent decades. The last major episode of tension was during World War II, when large numbers of civilian Japanese Americans, regarded as potential collaborators with the enemy, were rounded up and interned in prison camps in the West.

Asian Americans in the United States today have the distinction of being the most highly educated of all the ethnic and religious groups except for Mormons and Jews. Among their ranks, the ethnic Japanese, Koreans, Chinese, East Indians, and Filipinos have generally prospered economically, but the ethnic Vietnamese, Lao, and Cambodians are poorer. The largest Asian populations in the United States are the Chinese and Filipino, each over 1 million. The largest in Canada are ethnic Chinese (about 1 million, more than double their numbers in 1984), East Indian (mainly of the Sikh religion), and Filipino. Recent large Asian migrant flows into Canada include Koreans and Vietnamese. Asian minorities have intermarried with other ethnic groups in both countries, and their mixed offspring call themselves **Hapas**, a Hawaiian word meaning "half" that people of mixed Pacific Island and other ethnicities also use to describe themselves.

These brief sketches only scratch the surface of the ethnic mosaics of the United States and Canada. Nearly 200 non–Native American ethnolinguistic groups are represented in these countries, which are rather rare examples of nations that have generally welcomed immigrants into their social fabric. The "melting pots" of Canadian and American societies are truly remarkable. You can travel all around the world and never see anything like their great wealth of ethnic diversity.

Nonindigenous Languages and Faiths

The non–Native American religions and languages prevailing in Canada and the United States reflect both early colonial influences and more recent waves of immigration. English and

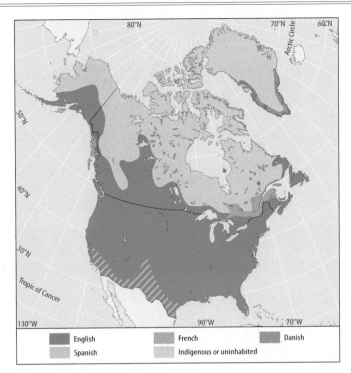

• Figure 11.22 Nonindigenous languages of the United States, Canada, and Greenland.

Source: Based on Bernard Comrie, et al., The Atlas of Languages, Revised Edition (New York: Facts on File, 2003).

French are Canada's official languages, with French usage confined largely to the traditionally French southeast (•Figure 11.22). The United States does not have an official language (despite repeated efforts to pass a constitutional amendment that would establish it as English), but 96 percent of U.S. residents speak English. The language map well illustrates the growth of Hispanic populations in the Southwest and southern Florida. **"Spanglish,"** a hybrid tongue of Spanish and English, is moving from Hispanic neighborhoods of southern California into the mainstream media and culture of the United States. Spanglish uses "code switching," substituting words or phrases from one tongue when speaking in another (for example, *"Vamos a la store para comprar milk,"* meaning "Let's go to the store to buy some milk") and also coining new words, like *perro caliente* for "hot dog."[4] In the larger cities of both the United States and Canada, there is a dizzying array of languages associated with ethnic neighborhoods.

Laws guarantee religious freedoms in both the United States and Canada, and both countries have a rich fabric of faith (•Figure 11.23). Christianity prevails in both countries, with numerous Christian churches represented. The largest single denomination in both countries is Roman Catholic (26% of all Americans and 43% of Canadians), brought to the United States by Irish, Italian, Polish, Filipino, Hispanic, and other immigrants, and to Canada mainly by the French. Hispanics are replenishing the ranks of American Catholics who have left the church in recent decades and are also introducing their own forms of charismatic worship into the country. The next largest group in the United States is Baptist (about 17%), then

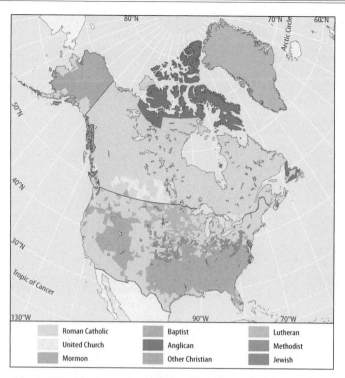

• **Figure 11.23** Religions of the United States, Canada, and Greenland. Source: Based on Edwin Scott Gaustad, Philip L. Barlow, and Richard W. Dishno, New Historical Atlas of Religion in America (Oxford University Press, 2001).

Methodist (about 7%), and Lutheran (about 5%). The other Abrahamic faiths are well represented by about 5 million Jews and 1 million Muslims. Islam is the fastest-growing religion in the United States, as it is in the world.

11.4 Economic Geography

The United States and Canada are very wealthy nations. Per capita GNI PPP figures for the countries are $45,640 and $37,280, respectively, compared to the world's highest, Luxembourg, at $59,590. Both countries belong to clubs representing the world's strongest economic powers: the Group of 8 (G-8) and the Group of 20 (G-20). Together, the United States and Canada generate an impressive 21 percent of the global gross national product.

The United States has the world's largest economy and is by far the world's largest producer and consumer of goods and services. It consumes far more than it produces. It is also the world's leading example of "consumption overpopulation" (see page 58). With just about 5 percent of the world's population, the United States has a third of the world's wealth and each year consumes, for example, more than 18 percent of the world's energy, 27 percent of the world's commercially harvested wood, and 40 percent of the world's illegal drugs. Even with China's surging consumerism, the average American consumes 53 times more goods and services than the average Chinese.

Generally, in the globalized economy, the United States may be seen as having built an economic structure of consumption. By contrast, China and most other major economies have built economies of production and export. The United States

borrows money from other countries (it has the world's largest federal foreign debt—much of it to China—as well as the world's largest national debt), and it sells off assets to finance its consumption. The nation has a trade deficit (for which China alone accounts for about one-third), exporting $1.5 trillion worth of goods while importing $2.3 trillion annually. Once the world's largest exporter of goods, the United States lost that position to Germany in 2003 and became third (after China) in 2007. The United States' share of global gross domestic production fell from 31 percent to 23 percent between 2000 and 2011. In the same period, the four leading emerging economies of Brazil, Russia, India, and China—the so-called BRICs—saw their share of global gross domestic production rise by 3 percent to a combined 11 percent.

133, 233, 380

This discussion of American affluence may lead many of you to wonder who is being written about. In relative terms, the United States is extremely well-to-do on the global scale. But also in relative terms, the "average" American has been going nowhere economically for a number of years. Adjusted for inflation, the median income of the American family (with half making more, half making less) in 2011 was $49,445, roughly where it was in 1996, and is more than 7 percent below its 1999 peak. The gains of the boom years in the 2000s have been wiped out. A typical man's earnings are less than they were in 1978, again adjusted for inflation. Although a typical woman makes 77 percent of what the typical man does, her income has been rising more than his. There was a clear correlation between education and income, leading many people during the Great Recession to go to or go back to college.

When I travel abroad, people often do not believe me when I say there are poor people in America. But there are. They were more than 46 *million*, or 15 percent of the population, in 2011. About the same number of people lack health insurance. Twenty-two percent of American children are poor. In many of the countries I visit—Egypt, Vietnam, or Madagascar, for example—the $22,314 that made up the poverty line for a family of four in the United States in 2011 sounds like a great fortune, but of course it does not go far in many places in America (*where* you live is very important, isn't it?) (•**Figure 11.24**).

Surprisingly perhaps, the majority of the urban poor (55%) lives in the suburbs rather than the inner cities. The biggest asset in Americans' net worth is their homes, which have declined in value in recent years. Many are "underwater," meaning that they are worth less than the money owed on them. In the heady days before the Great Recession, homebuyers were lured by seductive credit terms into buying homes that were far beyond their means.

421

In the wake of the Great Recession, unemployment was highest in the West (with Nevada and California in the number one and number two spots, respectively) and in the South. This trend represents a reversal from the pattern of previous decades, and it is not clear whether it is temporary or will be long-lasting. Nationwide, as much as 9 percent of the eligible workforce was unemployed in 2011, up from 4.4 percent in 2007, on the eve of the Great Recession. It is still not bad when compared to Zimbabwe's 95 percent unemployment rate! That

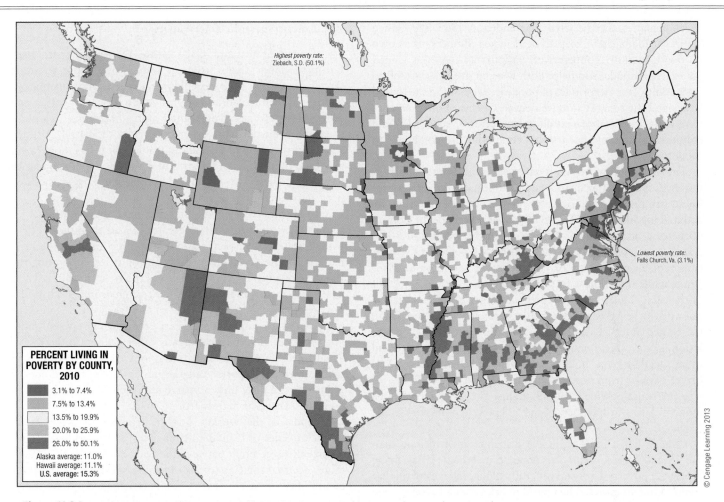

Highest poverty rate:
Ziebach, S.D. (50.1%)

Lowest poverty rate:
Falls Church, Va. (3.1%)

**PERCENT LIVING IN
POVERTY BY COUNTY,
2010**

3.1% to 7.4%
7.5% to 13.4%
13.5% to 19.9%
20.0% to 25.9%
26.0% to 50.1%
Alaska average: 11.0%
Hawaii average: 11.1%
U.S. average: 15.3%

© Cengage Learning 2013

• **Figure 11.24** Poverty in America. If we want to tackle poverty, we can use this map to show us where to work.

is the highest in the world, followed by Turkmenistan with a 70 percent unemployment rate. The European microstate of Monaco registered a zero percent unemployment rate.

There are many ways to measure poverty in the United States. Again, there is a clear correlation between education and affluence. The United States can boast about being on the fortunate side of the digital divide globally, but there is a digital divide within the country that contributes to a cycle of poverty. Only 4 out of 10 households with incomes of less than $25,000 had Internet access at home in 2010. Being cut off from the Net is not trivial: it translates into lower-quality health services, education, and career opportunities. About 30 percent of Americans lack any kind of access to the Internet. Many will come on board initially with smartphones, but it is almost impossible to fill out job applications or type up a résumé on a smartphone.

Technically, the United States made its way out of the Great Recession but recovery was slow. Falling tax revenues meant huge cuts in city, state, and federal budgets. Canada weathered the recession much better than the United States did, by 2011 recouping the number of jobs lost in 2008–09. In 2011, Canada had a jobless rate of 7 percent.

Sources of the Region's Affluence

This region, particularly the United States, is blessed with very large endowments of some of the world's most important natural assets. A number of geographic, political, and other circumstances have complemented this resource wealth in the development of these economically powerful countries. Both countries are among the five largest in the world. A wide range of environmental settings has both allowed and stimulated full use of human ingenuity in economic development. The combined population of both countries is large, yet neither country is "people overpopulated." A large population represents both a big pool of prospective labor and talent to promote economic growth, and a vast market for the goods and services produced. Both countries developed mechanized economies early and under favorable conditions. Innovations continue to replace human labor with machine labor, generally increasing efficiency and productivity (while also leading to unemployment in traditional manufacturing sectors).

Peace and stability within and between these countries have provided a good climate for investment and economic development. Since the Civil War ended in 1865, the United States has managed to keep all of its major conflicts off its own shores. Generally, cooperative relations between the United States and Canada also help. Both have an overall sense of internal unity despite episodes of poor race relations in the United States and the separatist movement in Québec. Both countries have a track record of continuity in political, economic, and cultural institutions. Such stable conditions are conducive to economic growth.

The distribution of national wealth in both countries, however, is imbalanced. In America especially, there is a profound gap between the rich and poor. The top 20 percent accounted for 50 percent of the nation's wealth, while the bottom 40 percent accounted for 12 percent of the wealth, in 2010. Another measure was that the wealthiest 1 percent of Americans took in more than 20 percent of the country's total income, while the poorest 10 percent took in less than 2 percent of the total. This represents the greatest gap between rich and poor of all the MDCs and has led to growing social discord in the United States. Large segments of the United States' population lack a "safety net" in the form of health insurance and other social services. In 2011, a movement called **Occupy Wall Street** swelled from protests directed against that epicenter of wealth to nationwide protests against the "1 percent"—the monied class—by members of the "99 percent"—everyone else (•**Figure 11.25**).

Canada has a more equitable distribution of income than the United States, a smaller population (about 10%) living below the poverty line, and generally higher indexes of quality of life. Long dominated by liberal politics, Canada has a more socialistic approach to its population than the United States does (for example, with taxpayer-funded universal healthcare) and tends to be more in line with Europe than with the United States on most social and political issues. Taken as a whole, however, Canada and the United States make up a region where most people enjoy the "good life." 90

An Abundance of Resources

The United States and Canada are the epitome of "neo-Europes," lands that European immigrants chose to settle largely because of their resemblance to the European environments and their potential for production of wheat, cattle, and other vital products. The United States has more high-quality arable land than any other country in the world. A much smaller proportion of Canada is arable, but it has more farmable land than all but a handful of countries. Such abundance has helped these two countries become the largest food-exporting region of the world. 45

Canada and the United States also have abundant forest resources. Canada's forests are greater in area than those of the United States, but not as varied. Wood has been an essential material in the development of both countries, prompting the less developed countries of the world today, such as Brazil and Malaysia, to cry foul when American environmentalists denounce tropical deforestation. Even now, the United States cuts more wood for lumber each year, produces more wood pulp, and produces more paper than any other country in the world. The United States both exports and imports large quantities of wood, with the exports going largely to Japan. U.S. domestic industries can meet only 60 percent of their lumber needs, and their imports come mainly from Canada. 35, 386

Canada is the world's largest exporter of wood (•**Figure 11.26**). Sales of Canadian timber to the United States have long been a sore point between the two countries as the United States accused the Canadians of timber subsidies (pricing this commodity artificially low). Canada is delighted to have found a new customer in China, having convinced China that wood is superior to concrete in construction of homes and schools in earthquake-prone areas. The United States had bought 60 percent of Canada's timber exports, but by 2011 that figure fell to 40 percent as China bought up 20 percent of Canada's exports. 420

The United States and Canada are also very rich in mineral resources. These played a major role in the earlier development of the region's economy, and mineral output is still huge.

• **Figure 11.25** The Occupy Wall Street movement that erupted in 2011 expressed widespread dismay about the growing gap between the haves and have-nots in the United States.

Reuters /JESSICA RINALDI/LANDOV

• **Figure 11.26** Canadian timber en route to East Asia.

Richard Kolar/Animals Animals

Canada is a major exporter of oil and natural gas, mainly to the United States, which in 2012 obtained 21 percent of its imported oil from Canada—more than from any other country. Although still a major mineral producer, the United States is also a major importer, especially of oil and industrial raw materials. This situation reflects the partial depletion of some American resources and the enormous demands and buying capacity of the U.S. economy. Recent years have seen impressive growth in proven reserves of fossil fuels in Canada and the United States. Exploitation of new reserves of oil in North Dakota and elsewhere within its borders made the United States at least temporarily reverse course in its trend of dependence on foreign oil imports. In 2005, the United States imported 60 percent of its oil; in 2011 that figure was 47 percent.

A spectacular development in fossil fuel resources in recent years has been the exploitation of natural gas (and, to a lesser extent, oil) lying in shale deposits. When energy prices were low, it was uneconomical to coax this energy from the ground, but as prices rose, so did the incentive to tap these hydrocarbons. The resources are obtained through an energy and water-intensive process known as **hydraulic fracturing**, or **fracking**. Water, sand, and chemicals are pumped underground at high pressure through a well into shale deposits thousands of feet deep. This causes fissures and layers in the rock to crack. The sand particles keep the fissures open, allowing natural gas to flow up the well. At the surface, the gases are separated into dry gases and natural gas liquids. One of the most important products is ethylene, used in making many plastics. Fracking is a special challenge and even a dilemma in drought-prone areas like the U.S. West. In using enormous amounts of water, the process draws down on aquifer waters that are also needed for irrigation and other human uses.

Some Canadian provinces and American states are particularly well endowed with these resources, Alberta, British Columbia, North Dakota, Texas, Arkansas, Oklahoma, Virginia, West Virginia, and Pennsylvania among them (•Figure 11.27). Some of these states are in the Rust Belt, and the shale gas industry has the potential to revitalize their long-depressed economies. There are tens of thousands of new jobs, not only for those who work directly to get the energy out of the ground,

419

but also in manufacturing the equipment and parts for the rigs and supplying services like food and lodging.

Is all of this news too good to be true? Yes: There are known and suspected downsides to fracking. Gas can leak into water reservoirs. You may have seen YouTube videos of someone turning on the kitchen water tap and igniting their "flammable water."

A number of other countries, including China, Argentina, Mexico, and South Africa, have deposits of shale gas, but the United States is far ahead in the technical capacity to exploit them.

Abundant energy resources have been essential to the region's economies. Coal was the largest source of energy during the nineteenth-century industrialization of the United States. Today, the country leads the world in estimated coal reserves (about 25% of the total). In coal production, China and the United States are first and second, with China producing 35 percent and the United States 18 percent. Canada is also an important coal exporter, with reserves and production that are small on a world scale but large in terms of Canada's needs.

Environmental concerns about the impacts of this most polluting of fossil fuels raise long-term questions about what the United States can do with its generous coal endowment. As it stands, the United States gets more than half of its electricity from coal-fired plants, with natural gas supplying 22 percent and nuclear power 20 percent. Drawing on large uranium resources in each country, nuclear power is an important secondary source of electricity for both (see Geography of Energy, page 416).

Petroleum fueled much of the development of the United States in the twentieth century. For decades, the country was the world's greatest producer, having such large reserves that the question of long-term supplies received little thought. The United States is still exploiting its 21 billion barrels, an estimated 1.6 percent of the world's proven oil reserves (a figure that may increase following the recent find of a new oilfield deep under the Gulf of Mexico). But its huge economy demands 7 billion barrels a year—as much as the four next largest consumers, China, Japan, India, and Brazil, combined—far more oil than the country can produce. The United States does possess reserves of petroleum that can be produced from oil shale, but only at great economic and environmental cost. This is why the United States imports so much oil. In 2011 it imported 11 percent more than it did during the 1973 energy crisis, when President Carter became the first American leader to insist that the U.S. must reduce its dependence on foreign oil. The top five suppliers of imports are, in descending order, Canada, Mexico, Saudi Arabia, Nigeria and Venezuela. Note that the top two suppliers are America's nearest neighbors, and that only one Middle Eastern country (Saudi Arabia) is among these five.

With demand and prices so high, the United States has been pressed to develop its own domestic petroleum supplies. It is going after increasingly technically challenging and environmentally risky reserves, including the Macondo Prospect in the United States' territorial waters of the Gulf of Mexico. 417

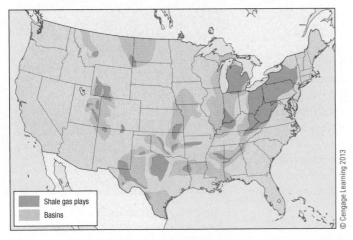

• **Figure 11.27** Principal U.S. shale gas reserves.

Shale gas plays
Basins

© Cengage Learning 2013

GEOGRAPHY OF ENERGY | **Energy Alternatives in the United States**

A notable trend in American energy policy recently has been to take an "all of the above" approach, exploiting conventional and alternative sources to feed an enormous energy appetite.

The United States is the world's largest producer of nuclear power and derives 20 percent of its electricity from that source. This energy source has an uncertain future in the United States, however. In 1979, an accident occurred in the Three Mile Island nuclear power plant in Pennsylvania, releasing little radiation to the environment but spurring enormous public fear of nuclear power. The truly disastrous Chernobyl accident in 1986 in Ukraine heightened the perceived peril of nuclear energy. Japan shut down its vital nuclear power industry after the gigantic 2011 earthquake crippled its Fukushima reactors. This added to perceptions of a troubled and dangerous industry. Due to such fears and to cost overruns and other problems, no nuclear power plants have gone online in the United States since 1978. The same concerns have troubled Canada, which gets 15 percent of its electricity from nuclear power. It has not built a new nuclear power plant since the 1970s. Nevertheless, rising fossil fuel costs have forced a serious reevaluation of nuclear energy in both countries. United States and Canadian authorities anticipate dozens of applications to build new nuclear plants.

The more benign sources of alternative energy—including windmills, photovoltaic panels, hydrogen fuel cells, and hydropower—meet just 14 percent of U.S. energy needs. The United States did not endorse the Kyoto Protocol or other international legislation that would compel it to embrace alternative energies. Renewable energy sources have received only modest federal funding for research and development since the energy crisis of the 1970s. Until 2009, few tax breaks or other incentives existed for the development of these alternatives by private companies. Official policy was that alternatives should develop in the free market rather than through federal funding and legislation. Finally in 2009, as part of a large stimulus bill to help bring the United States out of the Great Recession, a hefty infusion of federal subsidies helped produce a boom in solar and wind power.

Becoming accustomed to not having help from Washington, many states struggled to take affairs into their own hands. They wanted to cut carbon dioxide emissions through strictly mandated standards for vehicles and factories and the adoption of "greener" technologies. California has led these efforts. The state's **Global Warming Solutions Act of 2006** requires that greenhouse gas emissions be cut to 1990 levels by 2020, by which time utilities will have to generate one-third of their energy from renewable resources. Proving the maxim that "there's no such thing as a free lunch," some environmentalists in California are up in arms over some of the carbon-reducing steps proposed. For example, they oppose Los Angeles' proposition to build a "green path" corridor for transmitting solar and geothermal power across 80 miles (c. 130 km) of southeast California's desert wildlands to the city.

Federal legislation has improved energy efficiency, especially in the vital transportation sector that accounts for about 28 percent of U.S. energy needs. In 1975, in the wake of the energy crisis, Congress enacted the **Corporate Average Fuel Economy (CAFE) standard**, requiring that the fleets of passenger cars produced by all the auto companies achieve an average of 27.5 miles per gallon. The standard for light trucks was set lower, at 20.7 miles per gallon. Sport utility vehicles (SUVs) are classified as light trucks, allowing them to meet this lower standard—and allowing American consumers to pursue their love affair with these "gas guzzlers." Rising fuel costs may bring an end to this affair, however, and new legislation will require the CAFE standard to rise for entire fleets (including light trucks) to 35 miles per gallon by 2020.

Remarkable changes in the U.S. auto and farming sectors have accompanied the rapid embrace of ethanol fuels (•Figure 11.C). Ethanol, made primarily from corn in the United States and typically mixed with gasoline to be sold as "super unleaded" fuel, can help to reduce gasoline consumption. The corn-producing states of the Midwest cheered ethanol development, which represented rare good news in the farming sector. Boosted by federal ethanol subsidies of 45 cents per gallon, more and more land went into corn production (one-third of the total farmed in 2011) and more corn (40% of the total in 2011) was turned into ethanol. With 190 ethanol plants operating or under construction, the United States surpassed Brazil in ethanol production in 2011. That lead may be eroded, as the American ethanol subsidies came to an end in 2012. The next generation of ethanol in the United States may well focus on cellulosic ethanol produced from plant waste and from nonfood crops like switchgrass and other plants having fewer conflicting uses and requiring lower energy inputs than corn.

The "no free lunch" principle certainly applies to the apparently benign fuel source of ethanol. Much fossil fuel-based energy and fertilizer goes into its production, raising questions about net energy gains: more energy may be used to create ethanol than the ethanol itself yields. Diversion of the corn crop into ethanol also raises food prices. Not only does corn as a human food source become more expensive, but so does the price of feed corn for hogs and cattle, leading to higher costs of pork, milk, and other produce.

Scott Olson/Getty Images

• **Figure 11.C** Pictured here is a byproduct of corn ethanol production, which can be combined with other ingredients to make high protein livestock feed.

One of the U.S.-owned and British Petroleum (BP)–operated offshore rigs there, the *Deepwater Horizon*, was drilling at a depth of 5,000 feet below the Gulf of Mexico seafloor (at a combined depth of 10,000 feet) when an explosion took place on April 20, 2010. Over the next three months, oil gushed virtually unabated into the Gulf of Mexico (•Figure 11.28). It caused still undetermined damage to marine, saltwater marsh, and shoreline wildlife and to the fish, shrimp, oyster, and tourism industries of the Gulf.

Louisiana has long presented the public relations case that these enterprises and ecosystems get along well with the oil industry. For more than 75 years, the state has hosted the seemingly incongruous annual Shrimp and Petroleum Festival, for example. But the *Deepwater Horizon* disaster was ruinous. With about 40 percent of the waters in the Gulf closed to commercial fishing, the shrimp catch fell by almost 60 percent in the year following the accident, for example. BP paid over $11 billion to clean up the spill and cover fines and claims. The impact on tourism was estimated up to $23 billion. BP waged a public relations campaign to bring back tourism and reassure the public of seafood safety. But the public remained skeptical, and business was brisk for imports of Asian farmed shrimp.

Drilling in the Gulf was suspended briefly but within a year had returned to normal levels. This disaster prompted a reevaluation of the risks of conventional fossil fuel development but did not prompt any serious action on energy alternatives. One effect of it was to make Canadian energy supplies particularly attractive.

Canada's conventional oil reserves, situated mainly in Alberta and offshore of Newfoundland, are small in comparison to those of the United States, representing just 0.4 percent of the world total. However, in Alberta and neighboring Saskatchewan, Canada has the world's largest reserves of **tar sands** (also known as *oil sands*). The potential output of this unconventional oil makes Canada second only to Saudi Arabia in proven oil reserves. There is a catch, however. Tar sands are mineral deposits in which a thick liquid petroleum known as bitumen is mixed with sand and clay from an ancient ocean (•Figure 11.29). Through a water-guzzling, energy-consuming process costing about $25 per barrel, the bitumen can be converted to petroleum. When conventional oil prices are high, as they have been, tar sands are economically attractive to exploit. The price surge in oil beginning in 2003 led to strong growth in production from Canada's Athabascan tar sands, with most of the output going to the United States.

The tar sands account for fully half of Canada's oil production. One way to ease the process is to ship the coarse mixture of bitumen and other materials through a pipeline from their Canadian point of origin to the refineries of the U.S. Gulf of Mexico, near New Orleans. This would be done through the roughly 1,700-mile-long **Keystone XL Pipeline**. Environmentalists opposed the project on the grounds that the pipeline would cross portions of the critical Ogalala Aquifer in Nebraska. An oil spill could taint these precious waters. There was an option to reroute the pipeline across less sensitive lands. Proponents liked the prospect of 20,000 U.S. jobs added by the

NASA/GSFC, MODIS Rapid Response

• **Figure 11.28** Oil from the *Deepwater Horizon* rig pours into the Gulf of Mexico in the summer of 2010. The amphibious landscape of the Mississippi Delta region is apparent in this satellite image.

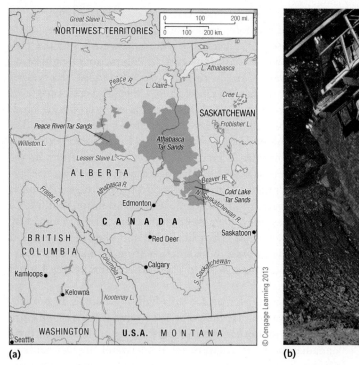

(a)

(b)

• **Figure 11.29** **(a)** Tar sands locations in Canada. **(b)** Oil production at the Athabasca tar sands facility.

construction effort. Turned down, the opportunity would mean opening the prospect of Canada exporting more of this energy to China.

One big obstacle to tar sands development for domestic consumption was Canada's 2004 ratification of the Kyoto Protocol, which would have required the country to reduce its carbon dioxide emissions to 6 percent below their 1990 levels. This commitment would have required a massive cutback in the use of fossil fuels, including the natural gas that must be burned to extract oil from the tar sands. Alberta's provincial government was furious at the federal government's support of the Kyoto treaty, arguing that it would cost the province billions of dollars in lost revenue and thousands of jobs. Alberta's politicians even threatened that the province might secede from Canada if the treaty were ratified. Canada's domestic political and economic circumstances trumped its international commitments: Canada withdrew from the Kyoto Protocol in 2011.

Natural gas has become an increasingly important fossil fuel to these two economies. The United States is the world's largest consumer of gas and is a major importer, especially from Canada. It ranks first in the world in output, producing 18 percent of the world's gas from its 3 percent of proven reserves worldwide. That equation means that the present high output cannot be sustained for long. With recent high energy prices, states like Montana and Colorado that still have reserves of natural gas are weighing whether to exploit them or to keep the lands above them relatively pristine for a "new economy" based on hunting, recreation, and tourism. Canada, with smaller but rapidly increasing proven reserves, produces about 5 percent

of world natural gas output and is a major exporter. Shale gas will significantly add to Canada's natural gas totals.

The potential for future hydropower development in the United States and Canada is small. Development has been so intense in the past that the two countries rank with China and Brazil as the world leaders in hydropower capacity and output. Canada gets about 60 percent of its electricity from hydropower, and the United States, 6 percent. The countries' dams are controversial because of the environmental changes they produce and also because they often result in the inundation of indigenous peoples' lands. Some dams in the United States, including several in Washington and Oregon, are scheduled for **decommissioning** (major modification or removal) primarily for environmental reasons, especially to allow a dozen endangered salmon species to reach their traditional upstream spawning grounds. Fish ladders and weirs have been constructed to aid salmon migrations, but salmon populations have nevertheless plummeted from 16 million before the dams were built to about 1 million today. A coalition of Native American groups, environmentalists, and sport fishing enthusiasts has long been demanding that dams on Washington's Columbia and Snake Rivers and the Klamath River of Oregon and California be torn down. Recently, even long-standing supporters of the dams, including farmers, have begun discussing a future without the four Snake River dams. Wind-generated electricity would replace the hydropower lost when the dams are decommissioned.

Other mineral resources in North America have presented a similar picture of abundance. Many of the ores easiest to mine and richest in content are now mostly depleted, and even large mineral outputs in the United States cannot meet the huge demands of the domestic economy. In many such cases, Canada has been able to assume the role of leading foreign supplier. A major example is iron ore. The United States originally had huge deposits of high-grade ore, but by the end of World War II in 1945, this ore had been seriously depleted. New sources of high-grade ore were then developed in Québec and Labrador, mainly for the U.S. market.

Mechanization, Services, and Information Technology

Although raw materials contribute much to their wealth, the United States and Canada have become prosperous mainly because of machines and mechanical energy, complemented in recent decades by a boom in information technology (IT). From an early reliance on waterwheels powering simple machines, especially along the fall line in the Northeast (see page 398), the United States increasingly exploited the power of coal-fired steam engines. Later came oil, internal-combustion engines, and electricity, all harnessed for greater and more efficient energy use.

In this energy-abundant and mechanized economy, the United States was able to take advantage of unique circumstances. Its abundant resources attracted foreign capital and stimulated domestic accumulation of capital through large-scale and often wasteful exploitation. There was also a labor shortage that attracted millions of immigrants as temporary low-wage workers but also promoted higher wages and labor-saving mechanization in the long run. A culture of striving for advancement emerged. This emphasis on productivity, dubbed the **Protestant work ethic** but shared by people of many faiths and ethnicities, characterized the struggle for economic success and the **American dream** of a comfortable life. Only

occasionally did national energies have to be diverted excessively into war and defense.

Most of these American conditions were duplicated in Canada. However, Canada lacked the large, unified, and growing internal market that fostered growth in its neighbor to the south. In recent decades, the advantage of the giant American domestic marketplace has been eroded by the expansion of the global market. Transportation and communication have become so relatively cheap and rapid, and so many artificial trade barriers have fallen, that the world market, rather than the domestic market, has become the essential one for major industries. This globalization of the economy has presented huge challenges to the United States, Canada, and the other MDCs, which are trying to prevent their traditional industries from being undercut by cheaper output from abroad. The United States and its industrialized allies thus often find themselves in the awkward position of advocating free trade but erecting trade barriers (see Insights, next page).

Mechanization in these economies has been increasingly replaced in recent decades with a new labor sector: services and information technology. The service sector, consisting of insurance, finance, medical care, education, entertainment, retail sales, and other enterprises, has become a more significant employer than the manufacturing sector in the United States and Canada, as shoes, clothes, toys, cars, and other traditional manufactures have migrated to other countries. As telecommunications technologies have interconnected North America with the rest of the world, the United States has acquired global leadership in information technology even while many of its old heavy industries have expired in the country's **Rust Belt** states, including Illinois, Indiana, Michigan, Ohio, and Pennsylvania (•Figure 11.30).

• **Figure 11.30** The United States gained economic supremacy in the twentieth century in large part because of its production of steel and other manufactured goods in the part of the country now called the "Rust Belt." It lost much of its production to other countries where labor and other costs were lower. South Korea's Posco steel mill, seen here, puts out a product that once epitomized the economy of Pittsburgh, Pennsylvania.

INSIGHTS | **Trade Barriers and Subsidies**

Many of the world's LDCs are convinced that the best way to develop is not to borrow money from the richer countries but to trade with them. After all, with their larger populations and lower labor costs, the LDCs should easily be able to churn out products at much more favorable prices than those sold by the MDCs. Buyers around the world should be snapping up those affordable products, pumping desperately needed money back into the developing countries. This exchange does not take place as freely as it could, however, because of two tools the MDCs wield as much as possible to protect their own producers: subsidies and tariffs.

Subsidies are expenses paid by a government to an economic enterprise to keep that enterprise profitable, even if the market would not. Tariffs are essentially taxes that a government imposes on foreign goods, basically for the same reason, to ensure that domestic products make it to market, even though they are not produced at competitive costs. Alone or in combination, subsidies and tariffs can make it all but impossible for a less developed country to join the world economy as a trading partner.

There are often struggles between the MDCs and LDCs over the issues of subsidies and tariffs, especially in meetings of the World Trade Organization (WTO). At these meetings, the LDCs are represented by a bloc that calls itself the **G-21 (Group of 21)** as a "play on numbers" against the more affluent G-20.

Often bitter debate between these groups focuses on subsidies that the United States, the European Union, Japan, and other MDCs have used for several decades to prop up certain sectors of their economies, especially agriculture. The subsidies were created to help farmers and the countries get through especially difficult times, including the **Great Depression** of the 1930s in the United States, when the livelihoods of so many farmers were threatened, and post-World War II Europe and Japan, when the specter of malnutrition loomed. Over the years, the critical need for subsidies passed, but the sectors that received them were loath to part with them. Federal supporters of subsidies found political capital in maintaining them. Today, farming subsidies are something of a sacred cow in many of the MDCs, and any effort to reduce or remove them would have domestic consequences—especially at the ballot box. Altogether the MDCs pay out $1 billion a day in subsidies that support the farmers who make up less than 1 percent of their population. In the United States, subsidies are concentrated on growers of the big commodity crops such as wheat, rice, corn, soybeans, and cotton (these tend to be large, corporate-owned farms). The government views farmers of the "specialty crops" such as fruits and vegetables as self-sustainable (these tend to be smaller, family-owned farms). Under market pressure from cheaper imports, such as Chinese garlic, these farmers are now clamoring for subsidies too.

Cotton is a fine example of how the subsidies and related tariffs work domestically and what their impact is in the LDCs. The United States spends over $4 billion each year to subsidize its cotton farmers. The country produces far more cotton than it consumes, and most of the prodigious output is exported. The high level of production made possible by the subsidies drives the global price of cotton down to around 50 cents per pound. This is less than the African production cost of around 65 cents per pound. The African cotton producer not only cannot compete in the world market but also loses money doing so. Cotton is, or should be, big business for Africa. If the U.S. subsidies were reduced or removed and world cotton prices rose, the potential economic gains for several African countries would be enormous. The United States, however, has been slow to reform cotton policies except in ways that clearly serve its own interest. In recent legislation, tariffs were reduced on some imported African textiles—but only those made from American cotton. The European Union, Japan, South Korea, and other countries are just as strident in protecting some of their products.

The G-21 countries, speaking for two-thirds of the world's farmers, demand that the MDCs reduce or eliminate such subsidies and tariffs or that the LDCs be compensated for the revenue lost to them. They threaten to begin erecting tariffs of their own (a proposal the MDCs say is bad for the developing economies) if the MDCs do not yield. In a familiar negotiating sequence, the MDCs give up little or no ground in such impasses, causing the G-21 delegates to walk away from trade talks in frustration.

Would removal of subsidies hurt the small American family farmer, whose cause is championed by entertainers Willie Nelson and Neil Young? That would depend on how subsidy cuts were targeted. Sixty-five percent of the subsidies go the top 10 percent of America's agricultural producers, the large corporate farming industries. It is possible that the unkindest cuts could be avoided with sacrifices imposed instead on the bigger businesses, thereby perhaps helping some of the world's poorest countries escape the poverty trap on their own.

Manufacturing now only accounts for about 10 percent of the U.S. economy (down from as much as 30% in the 1950s), having fallen to the forces of globalization that moved traditional industries like textiles to countries with lower manufacturing costs, such as China and, more recently, Vietnam. Manufactured goods make up two-thirds of U.S. exports by value, however, and the nation still has the global lead in many high-tech products. It is also important to keep in mind that manufacturing has not declined as much as the figures might suggest; instead, its relative contribution has been displaced by growth in other sectors of the economy, like finance, insurance, and real estate.

Ironically, perhaps the best prospect for more growth in U.S. manufacturing is investment from China as that country looks to produce goods for the American market that are cheaper to produce in the United States than to export from Asia. Wages for Chinese workers are rising, already beginning to push manufacturing jobs back to the United States; this process is known as insourcing (or re-shoring, that is, the reverse of outsourcing or offshoring; see page 235). For now, though, the United States hopes to maintain its global leadership in design and innovation, profiting from a knowledge economy even while other countries actually make products. As one American economist put it, "We want people who can design iPods, not make them."[5] But many worry that design and innovation will also move offshore, leaving the United States with no competitive edge.

Cities often serve as economic barometers. In the first part of the twentieth century, wealth and jobs clustered in what are now Rust Belt cities like Cleveland and Detroit, where there were large concentrations of heavy industry and capital. Now the economic growth is concentrated in "smart cities" where the knowledge economy is driven by higher education levels and a larger supply of highly skilled workers.

70, 72, 88, 94, 269, 271, 434

The United States is the world leader in the production of software. In the 1990s, there was such a speculative frenzy about IT that a classic **economic bubble** developed. Tens of millions of Americans started investing in stocks for the first time, bidding up the values of companies—even companies that produced no tangible products—to lofty heights. There was talk of a "new economy" in which the old paradigms of corporate investment and growth no longer applied. However, the **tech bubble** burst in 2000. Clever or lucky investors withdrew their investments in time, but others lost fortunes that had existed on paper. An economic recession set in. The economy gradually recovered. The tech sector was approached more cautiously. High-tech industries increasingly built what came to be known as a **platform economy**, with companies like Dell operating headquarters in the United States but outsourcing virtually all work on its products to other countries.

235

394

A far larger bubble burst in 2008. The implosion began in the United States housing sector, where through so-called "subprime loans" millions of under-qualified American borrowers bought homes they could not afford. Banks foreclosed on the unpaid properties, housing prices slumped, and a full-fledged crisis developed in the banking industry. This **credit crisis** in the United States spread quickly worldwide, sparking a sell-off in global stock exchanges, a widespread economic recession, and plummeting prices for oil and the other commodities that had supported booming economies. The Great Recession had begun.

U.S.-Canadian Economic Relations

Each of these countries is a vital trading partner of the other, although Canada is much more dependent on the United States than vice versa. In 2011, Canada supplied 14 percent of all U.S. imports by value and took in 19 percent of all U.S. exports, making Canada the leading country in total trade with the United States. That same year, the United States supplied 50 percent of Canada's imports and took in 75 percent of its exports. Except for Canadian exports of automobiles and auto parts to the United States, the main pattern of trade between the two countries is the exchange of Canadian raw and intermediate-state materials—primarily ores and metals, timber and newsprint, oil, and natural gas—for American manufactured goods.

Free-trade agreements designed to open markets and increase revenues for both countries have facilitated this exchange. Nevertheless, the countries have periodically tried to protect their industries from one another, often with poor results. In the 1990s, there was a so-called **wheat war** between the two countries as they struggled for greater market share in an environment of high yields and low prices. With agricultural prices for grains and pork dropping to the lowest levels in more than 50 years, many farmers both south and north of the U.S.-Canada border went bankrupt. U.S. producers claimed that Canada was **dumping** (selling the product for less than it cost to produce it) its durum wheat on the U.S. market.

There was also a **salmon war** between the two countries, marked by periodic skirmishes until a resolution was reached in 1999. Salmon hatch in Canadian and American rivers, spend most of their lives in the open seas, and return to their river birthplaces to spawn. The dispute centered on the issue of "interceptions," meaning that American fishers often intercepted fish that hatched in Canadian upstream waters but migrated into U.S. waters.

In 2003, a new dispute emerged when the United States, accusing Canada of dumping low-cost lumber in the United States, imposed heavy new tariffs of 27 percent on softwood lumber imported from Canada. An agreement reached in 2006 lowered the tariffs and required Canada to price its lumber more competitively. Other issues between the countries await resolution. There are dozens of small water disputes across the border, including one over North Dakota's plan to divert excess water into the Red River, which flows north into Manitoba's Lake Winnipeg.

Despite such volleys occasionally fired on both sides of the border, the trend has been toward more cooperation and free trade. In 1965, the two countries negotiated a free-trade agreement for the production, import, and export of automobiles. This sparked rapid expansion of the auto industry into southern Ontario (Canada's most industrialized province), on the border with the U.S. automobile-producing state of Michigan. When the auto industry suffered in Detroit, it did so in neighboring Windsor, Ontario. Ontario contributed to the "bailout" of the automakers General Motors and Chrysler. In this bailout, which was widely considered to be successful, the American government infused huge amounts of capital into the ailing auto industries.

The two countries entered into the comprehensive **Canada-U.S. Free Trade Agreement** in 1988, and in 1994 Canada, the United States, and Mexico enacted the North American Free Trade Agreement (NAFTA). Its goals and impacts are described in detail, especially as they affect Mexico, on page 371 in Chapter 10. NAFTA has brought a boom in trade between the three countries but also caused major industrial readjustments, generally characterized by the more labor-intensive firms (beginning with clothing) moving to Mexican locales from Canada's southeastern core region and from the industrial northeast of the United States.

Transportation Infrastructure

Both governments have sought to enhance national unity and economic strength with transportation networks spanning

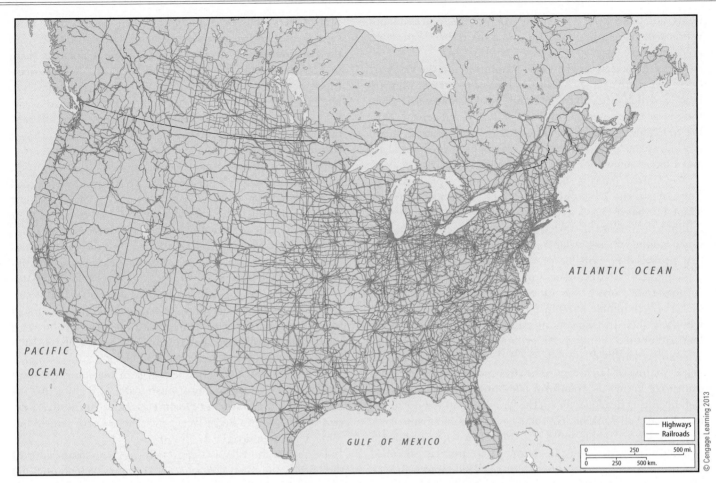

• **Figure 11.31** There is a dense network of highways and railways in the United States and southern Canada.

their vast landscapes (•**Figure 11.31**). In both countries, east–west networks have been built over long distances, against the "grain" of the land, especially the great north–south-trending ranges of the Rocky Mountains. The coasts of both countries were first tied together effectively by heavily subsidized transcontinental railroads completed during the second half of the nineteenth century. Later, national highway and air networks improved transportation.

A landmark development in the United States was the growth of the **interstate highway system** beginning in the late 1950s. Today, this system serves as the primary network for the trucking of cargo across the country. Railways meet these needs in most countries. Although railways still carry the bulk of goods being transported across the country, few new lines are being built. Passenger rail traffic is particularly neglected, and once-popular routes are no longer used. In many cases, the old rails have been torn up and converted productively as "rails to trails," becoming bicycle and walking paths. There is discussion of building high-speed (90–200 mph) passenger rail service, using existing rails rather than building new ones as the Europeans do. A major problem with this plan is that more than 90 percent of the U.S.

rail network is owned by freight companies, and in part for safety reasons they do not want to share their tracks with passenger traffic.

The stagnation or decline of U.S. passenger railways is a direct reflection of the American love affair with the car. Personal prosperity (or at least the ability to take out a loan) and the freedoms associated with being on the road in one's own automobile diminished appetites to travel by rail and bus. Public transportation remained popular only in the cities, where gridlock made it an attractive alternative, but rising energy prices could lead quickly to a reevaluation of public mass transport.

11.5 Geopolitical Issues

Even with the meteoric economic rise of China and the collective strength of the European Union, the United States is the most important economic, military, and diplomatic power in the world. This section first examines the country's strategic relations with its northern neighbor and then with the rest of the world.

Historical Relations between the United States and Canada

For many years after the American Revolution, which secured U.S. independence from Britain, the political division of Canada and the United States—between a group of British colonial possessions to the north and the independent United States to the south—was accompanied by serious friction between the peoples and governments.

There were several sources of antagonism. The northern colonies failed to join the Revolution, and the British used those colonies as bases during the war. A large segment of the northern population had come from **Tory** stock driven from U.S. homes during the Revolution (Tories were the "friends of the king," colonists who wanted to maintain political connections with the British government). And tensions were high over the issue of who would have ultimate control of the central and western reaches of the continent.

The **War of 1812** was fought largely as a U.S. effort to conquer Canada. Even after its failure, a series of border disputes occurred, and the United States openly expressed ambitions to possess this British territory in North America. Officials in the United States threatened annexation throughout the nineteenth century and even into the twentieth.

Canada's emergence as a unified nation came partly as a result of U. S. pressure. After the U.S. Civil War (1861–1865), during which more than 600,000 Americans died, the military power of the reunited nation seemed threatening to Canadian and British eyes. Recalling that the British had been hostile to the ultimately victorious Union (North) during the Civil War, some Americans argued that taking Canadian territory would be a just recompense. In 1867, the British Parliament passed the **British–North America Act**, bringing an independent Canada into existence by combining the separate colonies of Nova Scotia, New Brunswick, Quebec, and Ontario into one dominion named Canada. With that act, Great Britain sought to establish Canada as an independent nation capable of deterring U.S. conquest of the whole North American continent.

Although hostility between the United States and its northern neighbor did not immediately end with the establishment of an independent Canada, relations improved gradually. Canada and the United States today are strong allies with a connection only occasionally tested. But just as the United States overshadows Latin America in geopolitical affairs, it overshadows Canada as well. Canadians, like Mexicans, often feel overlooked and underappreciated by the neighbors across the border. Canada maintains a rather low profile in international affairs, sometimes disagreeing with the United States on such critical issues as the Iraq invasion of 2003. Canada exerts a policy of relative independence and neutrality despite membership with the United States in the North Atlantic Treaty Organization (NATO) and other international agreements. Canadian soldiers have died supporting the NATO mission in Afghanistan. Some American travelers in lands where Americans are often disliked have learned to exploit Canada's relative neutrality for their own protection by sewing badges of Canada's national symbol, the maple leaf, on backpacks and jackets.

The United States' Place in the World

Historically, the United States was not the same kind of empire-building power as many European countries were. The United States did have its overseas possessions—the Philippines, Puerto Rico, and Guam, for example—and still holds a few, but its power has been and continues to be projected more through military action and trade (•Figure 11.32).

The United States long enjoyed the geographic advantage of being something of an island situated far from the world's hot spots—especially Europe and the Middle East. In both world wars that ravaged Europe in the twentieth century, the United States initially clung to **isolationism**—the view that those conflicts were someone else's and that the United States would do best to stay out of them. Only late in World War I, at the unpopular insistence of President Woodrow Wilson, did the United States enter the stalemated war on the side of Britain and France. The United States lost 116,516 soldiers in that war, but in helping its allies secure victory, it also gained unprecedented influence and importance on the world stage.

Again in World War II, the United States managed to stay out of the conflict for more than two years after Hitler's troops invaded Poland and ignited conflagrations across Europe. But on the infamous day of December 7, 1941, Japan forced the United States into the war, with its surprise attack on Hawaii. And again, U.S. leadership and victory on both fronts of that war, which cost 405,399 American lives, helped the country achieve unprecedented strength in global affairs.

That war was succeeded by the Cold War, in which the two leading powers, the United States and the Soviet Union, faced off against one another, flanked by a host of often strong but

• **Figure 11.32** Hanging in the Oval Office of the White House, this is probably one of the most consequential maps in the world.

Pete Souza - White House via CNP/Newscom

88, always less powerful allies. U.S. concerns about the spread of
114, communism from the Soviet Union and China into the newly
139 independent countries of the postwar world—and the so-
called domino theory that one after another of these countries
might fall to communism—led to numerous and sometimes
very costly American military interventions. The Vietnam War
258 was the most significant of these.

The peaceful conclusion of the Cold War around 1990 was
followed by a decade in which the United States sought a new
role for itself on the world stage. There were military interventions to quell the conflict in disintegrating Yugoslavia and to
try to deliver lawless Somalia from famine, but there was a new
sense of security—and, in hindsight, complacency. There was
no longer any defining framework, such as east–west relations,
in international affairs. Terrorism was thought to be a problem
that existed far from home.

Not unlike Pearl Harbor, the attacks of September 11,
2001, brought an end to the notion that geographic distance
protected the United States. The events of that day represented
a great watershed in U.S. geopolitical history. The United
States under President George W. Bush developed a policy
of **preemptive engagement** (also known as the "Bush Doctrine"): whenever and wherever the country perceived a threat
to its security, it would take military action if necessary to
201 defuse that threat. That action would be unilateral, if necessary; in its first term, the Bush administration downplayed the
need for diplomacy and partnership. The preemptive engagement policy was founded mainly on the premise that such actions would prevent potentially devastating terrorist attacks
on the U.S. homeland, perhaps with chemical, biological, or
nuclear (CBN) weapons of mass destruction (WMD). Such
threats could emerge not only from transnational groups like
al-Qa'ida but also from "rogue states," most of whom are on
the United States' list of **official state sponsors of terrorism**,
which includes Iran.

As discussed earlier (page 178), the Bush administration justified the invasion of Iraq in 2003 on the grounds that Iraq
was a state sponsor of terrorism and had WMD that it might
use against the United States or its allies, especially Israel. Detractors wrote of a **new American imperialism**, proposing that
U.S. post-9/11 actions marked the beginning of a new era of
aggressive global involvement for the country. While continuing to prosecute America's wars, President Obama drew distinctions from his predecessor. In ousting Qaddafi from Libya,
for example, his strategy was to "lead from behind," allowing
America's European allies to do much of the heavy military
"lifting."

The United States remains the world's sole **superpower**,
with the strongest economy, military expenditures larger
than those of the next 14 countries combined, dominance of
global popular culture (from films to fast foods), the world's
best universities for graduate studies, and headquarters to
many of the world's leading international organizations
(including the United Nations, the International Monetary
Fund, and the World Bank). Its singular stature suggested a
"unipolar" world in which its dominance is unchallenged.

A shift appears to be in the works, however, as we have
seen repeatedly in this book. China's surging economy is the
principal challenge; it is expected to be the world's largest
by 2027.

The following section takes a closer look at Canada and the
United States.

11.6 Regional Issues and Landscapes

Canada

The Québec Separatist Movement

The core region of Canada (•**Figure 11.33**), with about two-
thirds of Canada's total population, has developed in a narrow strip along Lakes Erie and Ontario and seaward along the
Saint Lawrence River.

Ontario and Québec account for about 40 percent of products marketed from Canadian farms. Three-fourths of Canada's total manufacturing is in this core region, but in the
pattern of deindustrialization also seen recently in Europe and
the United States, China's output has caused factories here to
shut down; household appliances, electrical equipment, plastics, rubber, and textiles have all been hit hard.

The two sections of this vital region differ in their agricultural, industrial, and cultural features. The Ontario Peninsula,
which has strong British roots, mirrors the U.S. Midwest's crop
belts, with corn and livestock production, dairy farming, and
specialty crops like tobacco. Ontario, where about 85 percent
of the people speak English as a first language and only 5 percent speak French as a first language, is the most populous of
the nine mainly English-speaking provinces.

In Québec, 82 percent of the people speak French as a preferred language and 8 percent speak English (•**Figure 11.34**).
This is the only province in which French speakers predominate and control the provincial government, and Québec's potential devolution or even independence from Canada casts a
long shadow over Canadian affairs.

Citizens of more developed countries are generally unaccustomed to the prospect that a major region of their own country might break off to form an independent nation. This is a
real prospect for Canadians, however, if the Québec separatist
movement has its way. The modern separatist movement has
its roots in early Canadian history: the French were early settlers in these lands, but after a major British military victory in
1763, the broad sweep of language, government, and authority
became decisively British. A cultural dichotomy developed that
gave the French dominant influence in Québec but established
only modest influence for them beyond the Saint Lawrence
Lowlands.

After the end of World War II, French Canadian dissatisfaction spread, partly as attention to the plight of minorities
grew worldwide. In Québec, this led to the formation of a
separatist political party, the *Parti Québécois (PQ)*, dedicated to the full independence of Québec from Canada. In

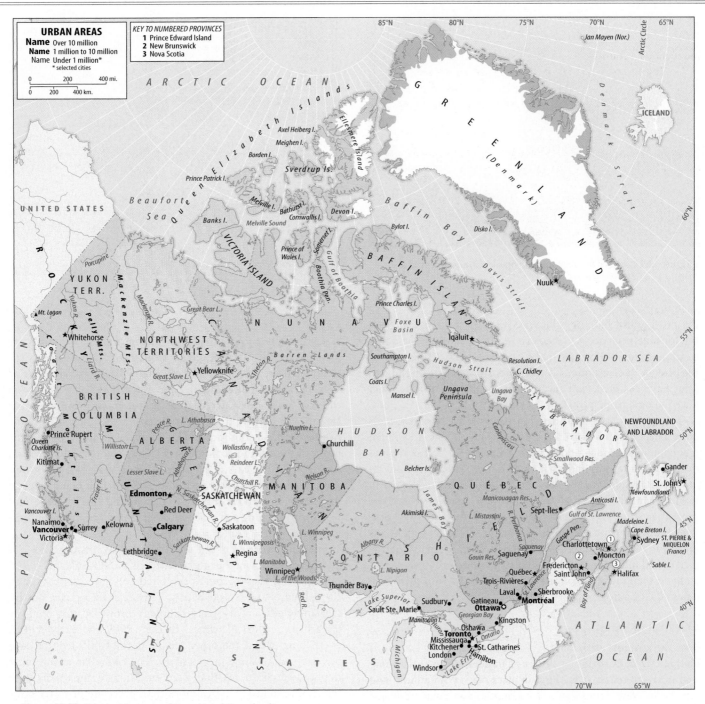

•**Figure 11.33** Principal features of Canada and Greenland.

Source: Based on Getis, Arthur and Judith Getis. 1995. The United States and Canada: The Land and Its People. Dubuque, Iowa: William.
C. Brown Publishers.

1976, the party gained control of Québec's provincial government and made French the language of commerce, requiring that all immigrants to the province be educated in French. Québec refused to ratify the country's 1982 constitution. In 1980 and again in 1985, the separatists forced the country to hold referenda on Québec's independence; both failed. In 1995, the issue was taken to provincial voters again, and those who favored staying a part of Canada again won, but by a very small margin (51%). In the referendum's wake, the federal government officially recognized Québec as a "distinct society" within Canada.

Although the PQ suffered a political setback in 2007, losing ground to a party favoring autonomy rather than independence for Québec, the issue of separatism is still very much alive and will likely stay on the national agenda. Separatists are frustrated by the steady influx of non-French-speaking immigrants seeking jobs and settlement in Québec City and other cities along the Saint Lawrence Seaway, who tend to vote with

• **Figure 11.34** French and English multilingualism is characteristic of Canada, especially in the east.

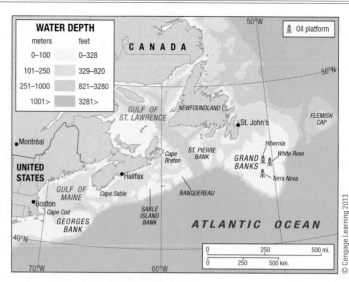

• **Figure 11.35** The Grand Banks.

the pro-English bloc or otherwise have no stake in Québec's bid for independence. Québec can ill afford shutting the doors to immigrants, who now make up more than 10 percent of the province's electorate. The province's historically high birth rates have declined, and future growth in the workforce will have to come from outside.

The effects of the separatist controversy have been considerable. The Canadian government has made French an official language of Canada—along with English—all across the nation. Laws now stipulate that the children of French Canadians born outside Québec may have their children educated in French. Meanwhile, other Canadian provinces have used the separatist strategies to gain more control of their own provincial governments. As Ottawa has given French Canadians more authority in Québec in the hope that they might find the idea of full independence less attractive, other provinces, especially in the Prairie region, have come forward with their own demands for autonomy.

Overfished Waters

The easternmost of Canada's four main populated areas is characterized by strips or nodes of population, mainly along and near the seacoast in the provinces of Newfoundland and Labrador, New Brunswick, Nova Scotia, and Prince Edward Island. The four provinces are commonly called the Atlantic Provinces, and three, excluding Newfoundland and Labrador, have long been known as the Maritime Provinces. Their main populated areas are separated from the far more populous and more prosperous Canadian core region in Québec and Ontario by areas of mountainous wilderness.

The fishing industry was long the economic backbone of Canada's Atlantic region. Before the beginning of European settlement in the early seventeenth century, and probably before Columbus arrived in the Americas in 1492, European fishing fleets worked these waters. This area lies relatively close to Europe, and its waters were always exceptionally rich in fish.

Elevated portions of the sea bottom known as **banks** are located off the Atlantic coast from near Cape Cod to the Grand Banks, which lie just off southeastern Newfoundland (•**Figure 11.35**). Shallow depths and the mixing of waters from the cold Labrador Current and the warm Gulf Stream favor the growth of plankton, the tiny organisms on which many fish feed. These conditions helped put the Grand Banks at the heart of the fishing economy in Canada's northeastern region for nearly 500 years. Early fleets fished mainly for bottom-dwelling cod, which were cured on land before being shipped to European markets. Both the British and the French established early fishing settlements in Newfoundland, but the British drove out the French in 1763.

In 1977, worried about the effects of overfishing, the Canadian government began to enforce a 200-mile (322 km) offshore jurisdiction that prohibited foreign competition in the fishing of the Grand Banks. This delayed a crisis, but increasingly efficient fishing technology propelled this industry toward more difficult times. In the 1990s and early 2000s, overfishing of cod and other species in the banks led to a serious decline in fishing productivity. The Canadian government[29] imposed catch quotas that were enormously unpopular with the fishing industry. After more than a decade of restrictions on fishing, the government in 2003 completely closed several cod-fishing grounds in Newfoundland waters in an effort to stop the slide. Tens of thousands of fishers and fish processors lost their jobs. The fishing tradition that was for so long a major component of the economic base for Atlantic Canada appears to have suffered a great setback and may be restored only if fish stocks can themselves be replenished. Thus far, the ban on fishing has failed to restore the fish populations, for reasons that scientists have not yet determined.

Race for the Arctic

Canada has a vast Arctic territory and wants to exploit it—and areas even farther north, if possible—to maximum advantage.

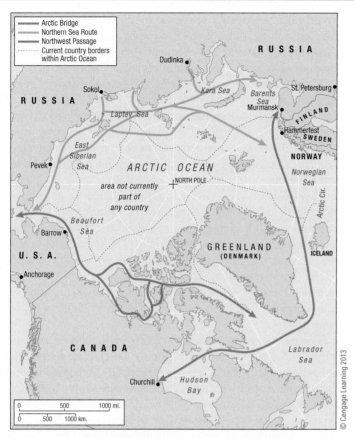

• **Figure 11.36** Recent dramatic reductions in sea ice cover, and hopes of abundant resources, have Arctic Ocean countries staking territorial claims and drawing future trade routes.

If global warming continues on its present course—according to recent studies, the Arctic Ocean will be free of summer sea ice as soon as 2020—more resources will become readily available to Canada and other countries of the Arctic Ocean rim. The race is already on (see •**Figure 11.36**).

Explorers have long sought a **Northwest Passage** for ship navigation through the maze of Canada's Arctic islands. Until recently, sea ice has made that passage impossible except under extraordinary conditions. But if Arctic warming should open up a reliable route, this would be an enormous boon to shipping interests and also to Canada because it could charge shipping fees. This would be the counterpart to Russia's Northern Sea Route (see pages 144 and 148), allowing great time and cost savings: the distance traveled by ships between northeast Asia and Europe would be cut by 40 percent. The United States has rankled Canada by suggesting that the Northwest Passage should be regarded as international waters, not Canadian. Canada and the United States are unlikely to have a serious showdown over this issue. But in 2007, the Canadian prime minister stated that his country's attitude toward its arctic territory was "use it or lose it" and announced plans for construction of both more icebreakers and a deepwater Arctic Ocean port.

Canada and Russia are also talking about opening the **Arctic Bridge**, a shipping route from Canada's port of Churchill, on Hudson Bay, to the northwestern Russian port of Murmansk. That would be an eight-day voyage. Goods arriving in Churchill could be shipped quickly by rail south to ports on the Saint Lawrence Seaway that could otherwise be reached only after a 17-day sea voyage from Murmansk.

Then there are the Arctic Ocean's sea and seabed resources. As Arctic Ocean waters warm, new fishing grounds will open. One-fourth of the world's undiscovered oil and gas reserves lie in the Arctic, according to U.S. Geological Survey estimates. Five Arctic nations—Russia, Canada, Norway, Denmark, and the United States—are scrambling to stake their turf. In some cases, they are taking advantage of loopholes and legal gray areas in the Convention on the Law of the Sea (see page 272). One portion of the treaty allows a country to extend its 200-mile (320 km) exclusive economic zone (EEZ) from the edge of its undersea continental shelf rather than from its shoreline. On this basis and with the use of recent marine cartographic surveys, Russia has already laid claim to the North Pole, even using a submarine to plant a flag at the seabed spot 13,000 feet (c. 4 km) below the surface.

Russia's rhetoric about its claim to seabed resources is heating up. Without naming the country he was accusing, Russia's president said in 2010 that "Regrettably we have seen attempts to limit Russia's access to the exploration and development of the Arctic's mineral resources. That's absolutely inadmissible from the legal viewpoint and unfair given our nation's geographical location and history."[6] President Medvedev signed a paper saying that the polar region must become "Russia's top strategic resource base" by 2020. The document included plans for beefing up military assets in the region in order "to ensure military security under various military political circumstances."[7] Canada was not far behind in lodging its commitment to the Arctic, with a foreign ministry spokeswoman insisting "Canada's sovereignty over lands, island and waters of the Canadian Arctic is long-standing, well-established and based on historical title. We take our responsibility for the future of the region seriously."[8]

Denmark uses Greenland's continental shelf as the basis for marking its EEZ, which also gives Denmark the North Pole. The United States is hampered by its refusal, dating to the Reagan administration of the 1980s, to ratify the Convention on the Law of the Sea. The reason? Conservative lawmakers have regarded ratification as a subordination of U.S. sovereignty to a supranational entity, the United Nations. With so much at stake, negotiations and disputes among the Arctic Ocean countries will be ongoing for years to come.

Greenland: A White Land

An unlikely cultural outlier of Europe, Greenland is geologically part of North America, and because it is closest to Canada, it is discussed briefly here (see Figure 11.33). The world's largest island (840,000 sq mi [2,175,600 sq km]), Greenland has been under Danish control since 1814 and was claimed by Norway for several centuries before that. It became a Danish province in 1953, but in 1979 the Danes granted the island self-government while retaining control over its foreign affairs.

It lies more than 2,000 miles (3,200 km) west of the European nation with which it is associated. In 1985, Greenland withdrew from the European Economic Community (EEC, the precursor to the European Union) because of its frustration over EEC fishing policies and the lack of any EEC development aid. Today, however, the European Union gives Greenland almost $50 million each year in exchange for Greenland's granting of fishing quotas to EU member states in its territorial waters. Denmark subsidizes the massive island with an average support of more than $8,000 per person per year. There is increasing talk of independence among the Greenlanders, but severing the aid pipeline from Denmark and the European Union would make that a very costly proposition. There is one way out of this: striking it rich with oil. Greenland may have done just that: there are estimates that Greenland may have oil reserves as great as Libya's, and a petroleum industry has started up in Greenland.

Greenland has nearly one-fourth the area of the United States, but about 80 percent of the island is covered by an icecap up to 10,000 feet (3,000 m) thick (•Figure 11.37). Until recently, Greenland seemed to have bucked the worldwide trend toward melting of glacial ice, but between 2003 and 2011 lost an average of about 50 cubic miles (c. 200 cu km) each year. Scientists are working to determine whether the melting is related to overall climate change and what its impacts on the "conveyer belt" of Atlantic oceanic waters and subsequent climates might be. Some models suggest that the ice sheet could shrivel completely over a period of about 1,000 years, which would raise the world's sea levels by 20 feet (6 m). With the ongoing melting, Greenland's geography is a work in progress. Maps of the coastline are outdated almost as quickly as they are drafted as new islands and inlets emerge from the thaw.

There was recently a cartographic brouhaha about Greenland's ice. The 2011 edition of Britain's revered *Times Atlas of the World* depicted significant portions of Greenland's coastline ice-free, representing a 15 percent loss of ice cover from 1999 to 2011. Critics insisted that the retreat was exaggerated. *The Times* promised a more accurate revision and called for an investigation.

Ninety percent of Greenland's population is distributed along the southern coast. Greenland has a total population of just over 50,000, giving it a population density of less than one-tenth of a person per square mile. The capital and largest city is Nuuk (originally called Godthab; population 15,000) on the southwestern coast. Under the auspices of NATO, joint Danish-American air bases are maintained there, and the United States has a giant radar installation at Thule in the remote northwest.

The great majority of inhabitants are Inuit, with Danes making up about 15 percent of the population. The leading means of livelihood are fishing, hunting, trapping, sheep grazing, and the mining of zinc and lead. As mentioned above, there may be oil in the island's future. This would make fishing seem like small fry. There is risk involved, however: Greenland is a pristine arctic wilderness, and an oil spill here would be infinitely more difficult to contain than the *Deepwater Horizon*'s.

• **Figure 11.37** Greenland's spectacular glacial landscape.

Joe Hobbs

The United States

Reference Map

The book concludes with United States, a country that is home to most of its readers (•**Figure 11.38**).

Adaptive Reuse and Gentrification

A characteristic urban landscape transformation, particularly in the Northeast and Midwest, is the **adaptive reuse** of older structures. In this process, abandoned commercial or factory buildings are being converted to new economic activities, generating jobs and attracting people back to the area. This shift is often associated with **gentrification**, the process by which a low-cost, run-down neighborhood undergoes physical renovation for residential or office space, resulting in increased property values and new homes and offices for more affluent urban professionals (•**Figure 11.39**). In Connecticut, for example, where the manufacturing economy shrank by nearly 25 percent in the past two decades, vacated mills and other factories were reborn as art shops and antique stores. Others were razed and replaced by big-box stores like Best Buy and discount department stores, including Kohl's and Target. There is still twice as much office space in suburbs as there is in central business districts, but the trend is unmistakable: during the Great Recession, suburban-based businesses like mortgage lenders and home lenders were hit hard, whereas the government offices and big banks of the downtown areas suffered less.

Adaptive reuse takes place most often in old central business districts, but the process has also been occurring in old residential neighborhoods. Adaptive reuse and gentrification give value to the past, especially when historic buildings can be modified and used rather than destroyed (•**Figure 11.40**). But although these processes revitalize run-down areas, attracting new wealth and revenue to them, they have a downside. People who have been living in these neighborhoods often can no longer afford to do so and are displaced to distant housing that is more affordable but far less convenient for work and services (•**Figure 11.41**). There is little they can do but protest—and people have been doing that in growing numbers in the United States as "the American Dream" retreats from their grasp.

419

The Changing Geography of American Settlement

Across the United States, midsize cities with 750,000 to 2 million inhabitants have become more attractive communities for manufacturing and service firms. Traditional locational factors like proximity to coal or iron ore or rivers are no longer so important. Smaller cities have the infrastructure and capable workers that employers need. The Internet and other information technology allow many firms to locate almost anywhere they wish. Operating and living costs are often less than in the larger cities. Midsize cities often provide attractive incentives to new businesses, including low-cost land, reduced taxes, and subsidized transportation.

Although the midsize American city has blossomed, the small town has withered. Between 1900 and the most recent census year, 2010, while the American population more than tripled, 25 percent of U.S. counties lost population. There was a marked migration of people out of the rural Great Plains region in particular (•**Figures 11.42** and **11.43**). Another way to think of this is that an increasing share of America's wealth moved to cities, while the rural Great Plains and Mississippi Delta grew poorer.

A flight from small towns to larger cities was mainly responsible for this depopulation. Small towns had their heyday when they were vital clusters of merchants and townsfolk located on some avenue of transportation—highway, railway, or river—and serving a broad agricultural hinterland. The more fortunate towns became home to small industries like food processing, shoe, or textile factories.

Transportation gradually improved, and better roads led to ever-larger settlements that attracted most new businesses. By offering consistently lower prices and a wider selection of merchandise, Walmart and other large retailers drove out diverse businesses in small towns and in the middle-sized and larger towns in which they were opened; this **"Walmart effect"** has had a huge impact around the country. Smaller towns have lost their economic significance and their residents.

The past two decades have seen some reverse migration. Some small towns in rural states of the Great Plains began actively recruiting international immigrants from countries including Vietnam, Sudan, and India to take up residence and help revive local economies. Others are enticing outsiders with offers of tax breaks, low-interest loans, and land giveaways to those who would build homes, often playing up the virtues of small-town living and the distinctive attributes of these settlements.

Another trend changing American settlement patterns has been the process of **exurbanization**. This represents a kind of re-embrace of rural life, but in response to urban conditions. City dwellers who are either priced out of closer-in suburbs or drawn to the wider spaces of more distant ones move to the "exurb," an outer suburb of the city, typically so far out as to stand almost alone. The more affluent may even buy a rural "ranchette"—a plot of often about 35 acres (c. 14 ha)—well away from the big city. American settlement patterns continue to evolve: there are signs that rising gas prices would lure faraway families back to the cities and their closest suburbs.

The Big Apple

The United States' largest city, New York City, is centrally located on the Northeastern Seaboard's urban strip midway between Boston and Washington. An enormous harbor, an active business enterprise, and a central location within the economy of the colonies and the young United States made New York City the country's leading seaport. Superior access to the developing Midwest also gave New York unchallenged leadership among American cities in size, commerce, and economic impact.

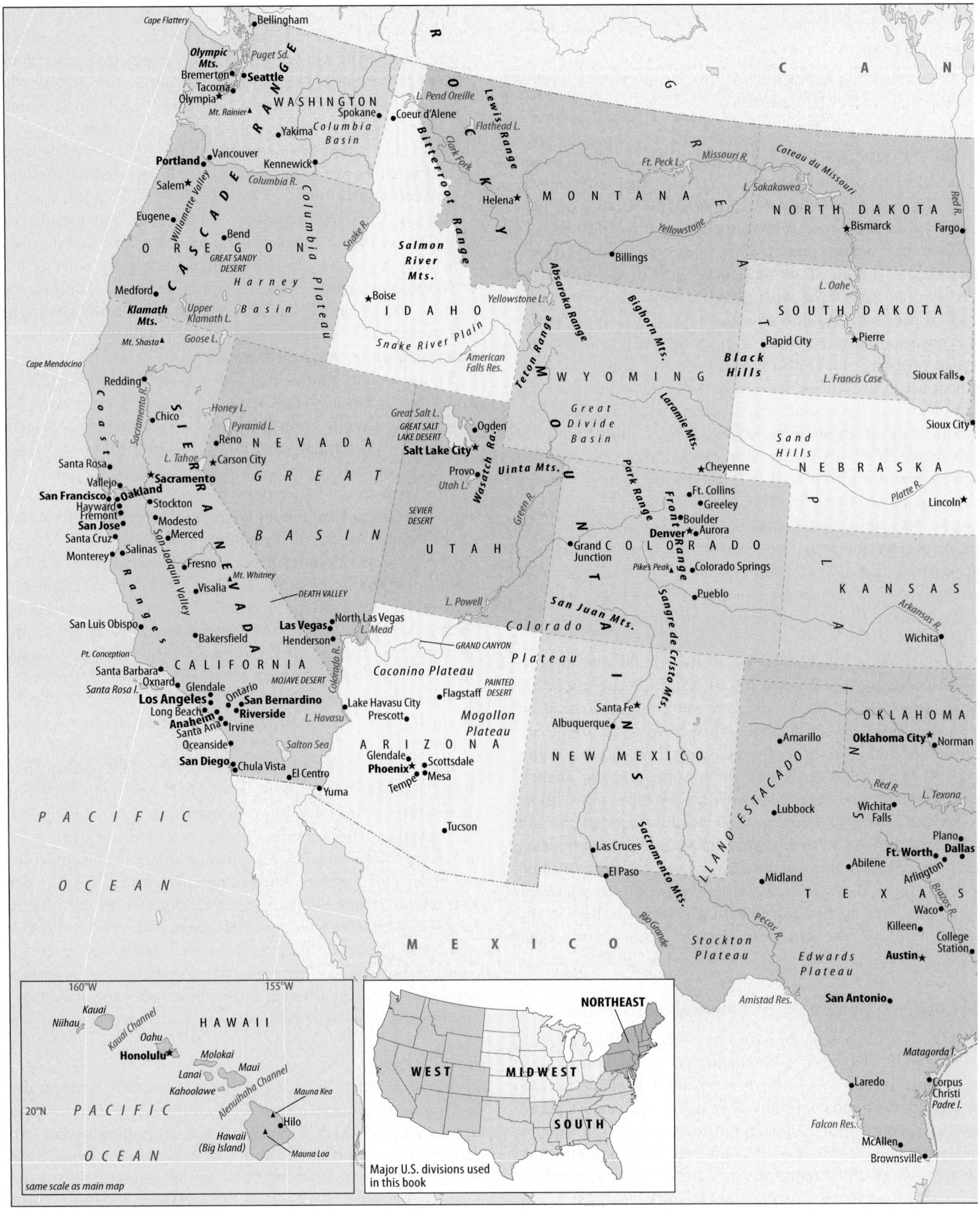

• **Figure 11.38** Principal features of the United States. Alaska is depicted in Figure 11.51.

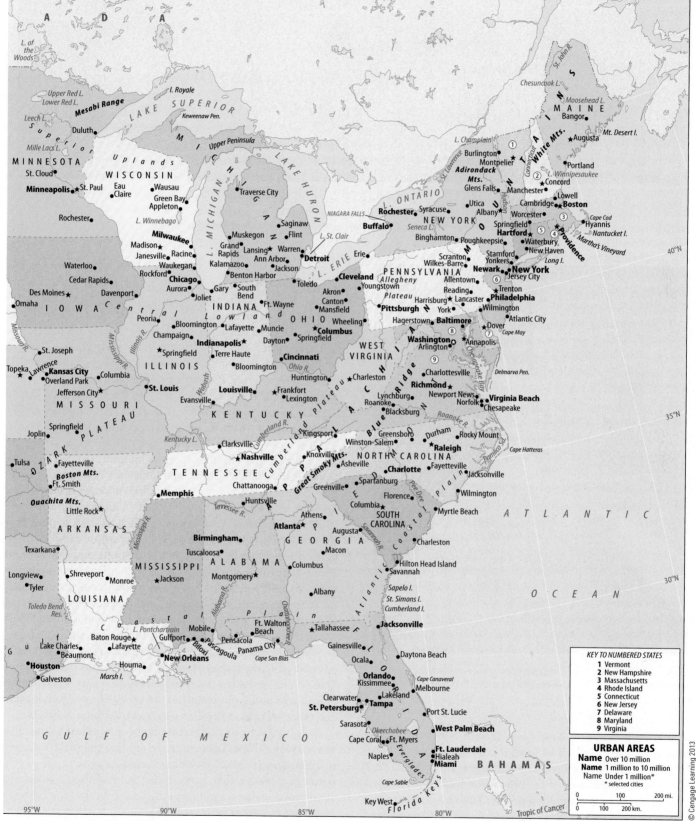

KEY TO NUMBERED STATES
1 Vermont
2 New Hampshire
3 Massachusetts
4 Rhode Island
5 Connecticut
6 New Jersey
7 Delaware
8 Maryland
9 Virginia

URBAN AREAS
Name Over 10 million
Name 1 million to 10 million
Name Under 1 million*
  * selected cities

0   100   200 mi.
0 100 200 km.

• **Figure 11.39** The Baltimore Inner Harbor Development, called Harborplace, is a successful reclamation of an old warehouse and wharf area. It boasts a shopping mall with outdoor restaurants and cafés, an aquarium, power plant buildings converted into elegant townhouses and condominiums, and access to Camden Yards, the home of baseball's Baltimore Orioles.

• **Figure 11.40** Rather than being torn down, an old shoe warehouse in downtown St. Louis has been remade into the interesting, family-friendly and historic City Museum.

New York City itself has a population of 8 million, and the metropolitan area (which includes much of northern and central New Jersey, all of Long Island, southernmost New York State just north and west of the Bronx, and southwestern Connecticut) is home to 22 million people. The city is centered on the less than 24 square miles (62 sq km) of Manhattan Island (•Figure 11.44). The island is narrowly separated from the southern mainland of New York State on the north by the Harlem River, from Long Island on the east by the East River, and from New Jersey on the west by the Hudson River. Manhattan is one of five **boroughs** (administrative divisions) of the city proper. The others are the Bronx, on the mainland to the north; Brooklyn and Queens, on the western end of Long Island; and Staten Island, to the south across Upper New York Bay.

Expressways and mass-transit lines converge on bridges, tunnels, and ferries linking the boroughs to each other and to northern New Jersey. Port activity is concentrated in New Jersey along the Upper Bay and Newark Bay. The whole area is held together in part by an amazing 722 miles (1,155 km) of subway system, parts of which are more than a century old (•Figure 11.45).

Of the city's total population today, more than 35 percent are foreign-born, with the largest contingents coming from the Dominican Republic, China, and Jamaica. Their growing numbers more than offset the out-migration of longer-resident New Yorkers, with the trends of whites and Asians moving elsewhere in the greater New York region, Asians moving to the West Coast, retiring whites to Florida, blacks whose families originated in the South moving back to the South, and some Puerto Ricans returning to their home island. Immigrants make up more than 40 percent of New York City's labor force, including more than half in restaurants and hotels, about 60 percent in construction, and nearly 70 percent in manufacturing. Most live in the outer boroughs, with two-thirds of them in Queens or Brooklyn. Whereas New York historically was an ethnic checkerboard of Irish, Italian, Jewish, Chinese, and other groups, Queens and Brooklyn today are ethnic stewpots where immigrants from numerous countries live cheek by jowl. Restaurants and other services in these areas are extraordinarily diverse.

New York Mayor Michael Bloomberg promised a "sustainable city" for New Yorkers of the future. His plan, unveiled in 2006, had 10 goals, including a huge increase in affordable housing, a pledge that every New Yorker would live within 10 minutes' walk of a public park, an overhaul of the subways and other transport systems, and a target of having the cleanest air of all American cities. Not to be outdone by efforts in California, the mayor insisted in this plan that the Big Apple should reduce greenhouse gas emissions by 30 percent by 2030. 416

Joe Hobbs

• **Figure 11.41** Chicago's Pilsen area has been spiffed up in recent years. As the area becomes more expensive and attractive to urban professionals, its poorer, longtime, mainly Latino residents are driven out. With this mural, the less affluent residents condemn Pilsen's gentrification. Note on the left the sign reading "STOP GENTRIFICATION IN PILSEN!" Eastern European immigrants once lived in Pilsen, and moved out during an earlier cycle of urban change.

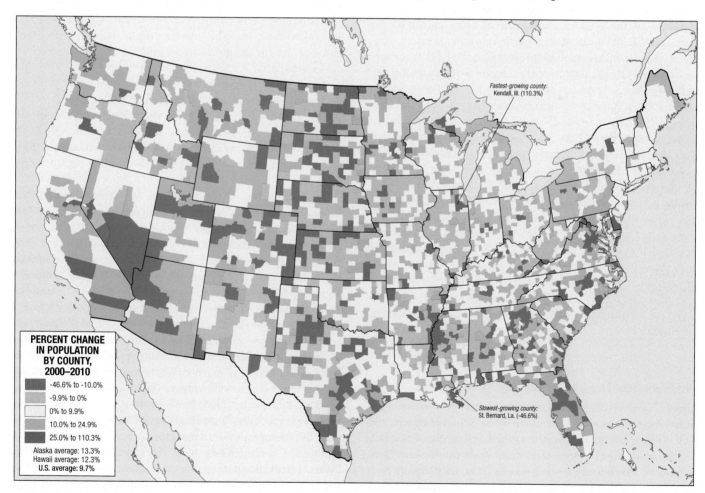

PERCENT CHANGE IN POPULATION BY COUNTY, 2000–2010

- -46.6% to -10.0%
- -9.9% to 0%
- 0% to 9.9%
- 10.0% to 24.9%
- 25.0% to 110.3%

Alaska average: 13.3%
Hawaii average: 12.3%
U.S. average: 9.7%

Fastest-growing county: Kendall, Ill. (110.3%)

Slowest-growing county: St. Bernard, La. (-46.6%)

• **Figure 11.42** This map clearly shows population movement trends in the U.S. Note in particular the depopulation of the Great Plains.

Source: From Timothy Egan, "Vanishing Point: Amid Dying Towns of Rural Plains, One Makes a Stand," From the New York Times, December 1, 2003, p. A18. Copyright © 2003 The New York Times Co. Used by permission and protected by the Copyright Laws of the United States. The printing, copying, redistribution, or retransmission of this Content without express written permission is prohibited.

Joe Hobbs

• **Figure 11.43** Many small towns across the Great Plains are dying—or even dead.

Size, financial importance, cultural complexity and variety, and national history have given New York City an uncontested position of urban prominence in the United States. The heart of this extraordinary city is roughly the lower two-thirds of Manhattan. The island's Midtown hotel, shopping, theater, and office district includes most of the city's best-known hotels, restaurants, department stores, specialty shops, theaters, concert halls, and museums. The Midtown area also has numerous office buildings, including the Empire State and Chrysler buildings. At the southern end of this area are Times Square and Broadway, formerly notorious for seedy nightclubs and strip joints but spiffed up in the 1980s to be a family-friendly, must-see area (•**Figure 11.46**). Here too is the historic Garment District, also called the Fashion District, to present a less industrial image.

The lower Manhattan Financial District includes the imposing cluster of office skyscrapers in the Wall Street area, and it was the home of the twin towers of the World Trade Center until 9/11. This area's critical role in the global economy made it a prime target for al-Qa'ida's September 11 attacks (and an often forgotten, more feeble attempt in 1993), and the prospect of more attacks will always loom over New York. The city's economy has rebounded remarkably well from the direct and indirect costs of the 9/11 attack. Many firms lost in the rubble relocated to central Manhattan and nearby New Jersey. While rebuilding, the Financial District is also being reborn as an affluent bedroom community. In the footprints of the World Trade Center buildings now stands an inspirational, heart-rending memorial.

San Francisco: The City by the Bay

About 7 million Californians—roughly 20 percent of them Asian Americans—live in the San Francisco metropolis, the sixth largest urban area in the United States. San Francisco is our illustration of large- and small-scale maps—see Figure 1.6, page 9. San Francisco—beloved as "The City" by its residents and neighbors—is located on a hilly peninsula between the Pacific Ocean and San Francisco Bay, but the metropolitan area includes counties on all sides of the bay (•**Figure 11.47**). The climate is unusual. Marine conditions moderate San Francisco's

temperatures from an average of 48°F (9°C) in December to only 64°F (18°C) in September. Local marine and topographic circumstances give rise to San Francisco's trademark fog, which can produce very chilly days even in the height of summer. Mark Twain is said to have quipped, "The coldest winter I ever spent was a summer in San Francisco."

The area is in the border zone between the marine west coast and Mediterranean climates. Characteristic of the Mediterranean influence, about 80 percent of the area's rainfall comes in five months from November through March. Mean annual precipitation is 20 inches (50 cm) in San Francisco, enough to produce a naturally forested landscape. There are even large groves of redwood trees within a short distance both north and south of the Golden Gate Strait. East of San Francisco Bay, warmer temperatures and lower precipitation have produced a scrubby forest (chaparral and scattered oak trees) and grassland typical of the Mediterranean biome.

San Francisco began as a Spanish fort (*presidio*) and a mission on the south side of the Golden Gate Strait. It grew as a port for the gold-mining industry and then for the expanding agricultural economy of central California. The **Gold Rush of 1849** centered on the gold-bearing gravels of Sierra Nevada streams pouring into the Central Valley north and south of Sacramento. San Francisco, with its spacious natural harbor, was the nearest port to receive immigrants and supplies for the gold camps, and the city boomed during the race for gold.

Today's San Francisco Bay Area is a major domestic and international tourist destination, famed for its scenery, wines, arts, and other cultural attractions. Along with the greater Los Angeles metropolitan area, it plays a vital role in the California's economy. If considered apart from the United States, California would have had the world's eighth largest economy.

Manufacturing in the Bay Area is varied. In recent decades, computer-related and other electronics businesses have boomed from San Francisco south to San Jose, center of an area known as **Silicon Valley** for the silicon chips that carry microcircuits. The rise of Silicon Valley as a center of high technology is due partly to the allure of the San Francisco area as a place to live and its proximity to two outstanding universities: the University of California at Berkeley (close to Oakland, an industrial center on the east side of the bay), and Stanford University, in Palo Alto south of San Francisco. When the **dot-com bubble** (or tech bubble) popped at the turn of the twenty-first century, much of the software capital fled, and the region suffered a downturn. The local economy had rebounded sharply by 2005, however, in part by reaffirming its place as the country's leading biotechnology hub, with more than 800 companies in that sector. Among the most recently established is the leading center for stem cell research in the United States. But perhaps the best illustration of Silicon Valley's place in global economies and cultures is that it is home to Apple. When the company's CEO Steve Jobs passed away in 2011, many of us paused to reflect on American leadership in innovation and technology.

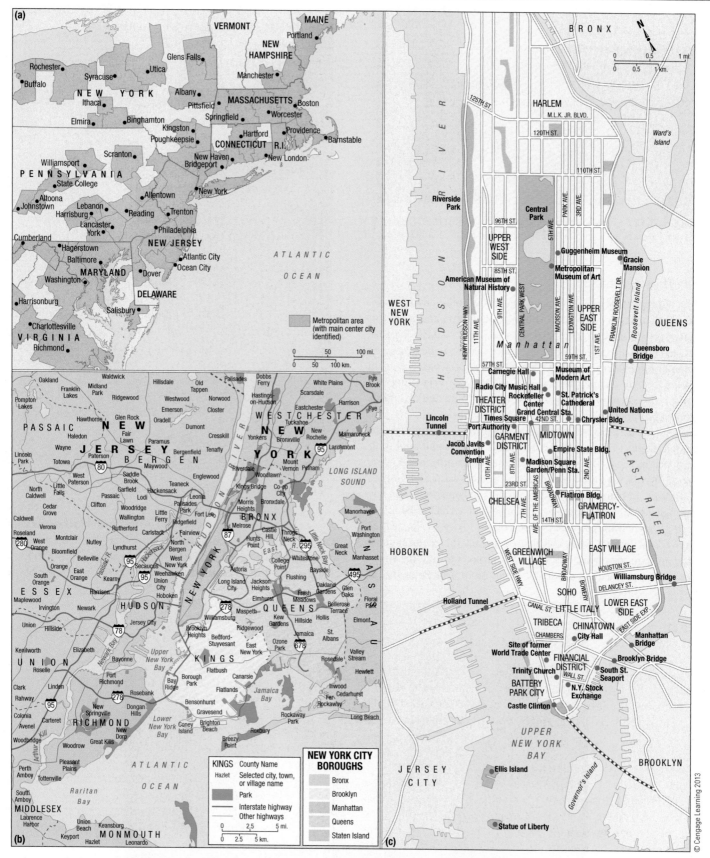

• **Figure 11.44** New York from three perspectives and scales. **(a)** New York is located at the center of a strip of highly urbanized land christened the "megalopolis" by the French geographer Jean Gottmann in 1961. Many waterfronts and other areas shown on the maps of New York City **(b, c)** are being redeveloped for new commercial, residential, and recreational uses. Note the complex of bridges, tunnels, and ferries tying Manhattan Island to the rest of metropolitan New York.

© Cengage Learning 2013

• **Figure 11.45** Lower Manhattan. For anyone familiar with pre-9/11 New York, the twin towers of the World Trade Center are always conspicuously absent.

• **Figure 11.46** Manhattan's Times Square, one of the world's most famed places.

The Thirsty West

The subregion of the American West has some environmental challenges: dry climates, rough topography, and a lack of water transportation except along the Pacific Coast. Because settlement is clustered in places where water is adequate, the population pattern is oasis-like, with oases separated from each other by steppe, desert, and mountains. Prevailing dry conditions often menace the region with wildfire and droughts, ironically punctuated by floods and landslides.

Water is a therefore a critical issue throughout the western United States. Historically, settlements were small and self-sufficient in water. But with explosive suburbanization following World War II, huge numbers of people moved west, especially to California, in search of the American dream. Populations quickly surpassed local water supplies, and new sources had to be found or redirected from elsewhere. The region's older cities, such as Denver and Phoenix, were able to grow because of federal support of giant dam, reservoir, and irrigation projects during the twentieth century. But the realities of living in an arid environment have set in, and newer communities cannot meet their needs with large-scale manipulation of the environment. They must rely far more on water conservation and other ways of adapting to local conditions. Business as usual when it comes to water is out of the

• **Figure 11.47** San Francisco: the top of "The Crookedest Street in the World" (Lombard Street), with Coit Tower and the hills of the East Bay visible in the distance.

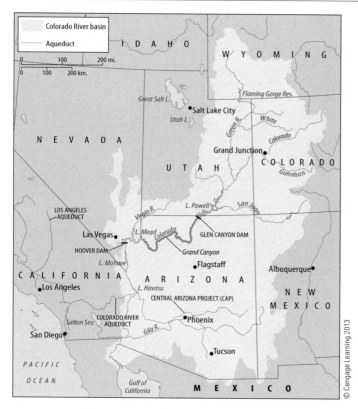

• **Figure 11.48** The Colorado River Basin.

• **Figure 11.49** The Glen Canyon Dam, just upstream from the Grand Canyon, was constructed in the 1950s. Note the low water level in this 2005 photo. A full reservoir would cover the white strip of bank that extends behind the dam.

question. There is a book about Phoenix entitled *Bird on Fire: Lesson's from the World's Least Sustainable City.*

Large-scale twentieth-century waterworks included canals dug to bring water from distant Mono Lake and Owens Lake, in the shadow of the Sierra Nevada, to the Los Angeles metropolitan region (see Figure 11.12). These controversial projects produced decades of discord between northern and southern California. Environmentalists finally succeeded in efforts to restore these debilitated lakes with projects that have resumed water flow to both watersheds.

The region's largest water source is the Colorado River, carrying snowmelt from mountain peaks and flanks in north central Colorado and the Green River in Utah (•Figure 11.48). It has been highly manipulated to suit human needs. Its waters are impounded behind two major dams: the Glen Canyon in Arizona (•Figure 11.49) and the Hoover on the Arizona-Nevada border. Hoover Dam's Lake Mead (with the largest volume capacity of all American reservoirs) supplies a major portion of the water needs for about 25 million people in the three lower basin states of California, Arizona, and Nevada. Almost 20 million people in southern California drink these waters. Lake Powell (behind Glen Canyon Dam) and Lake Mead also generate hydroelectricity for the region. Below the dams, a canal network known as the **Central Arizona Project (CAP)** siphons water from the Colorado River to sustain large, booming, and very thirsty Phoenix and Tucson.

Colorado River water has been a subject of litigation and federal and state disputes ever since the Hoover Dam was completed in the mid-1930s. For decades, California has used more

water from the river than was allocated to it in a 1922 agreement known as the **Colorado River Compact**, whereas the six upstream river basin states (Arizona, Nevada, New Mexico, Utah, Colorado, and Wyoming) have drawn less. Meanwhile, the upstream states, especially Nevada and Arizona, have had explosive urban growth and demand for water. It took threats from the upstream states to actually cut off water to California to bring that powerful state to the negotiating table. In 2003, California agreed to reduce its use of Colorado River water over a period of 14 years, to allow the six upstream basin states to use their share.

There is unhappiness in California; water allocation is a classic zero-sum game in which any benefit to one party equals a loss to another. Meanwhile, booming cities search desperately for new water sources. Las Vegas wants to divert water by pipeline from northern Nevada—a project opposed by people there and in neighboring Utah. Saint George, Utah, a rapidly growing retirement and recreation community, wants a new pipeline from Lake Powell—much to the dismay of Nevada.

Hurricane Katrina and Its Aftermath

Katrina was an unforgettable storm. What do you remember about it? On August 29, 2005, Hurricane Katrina made landfall as a Category 3 storm, with maximum sustained winds of 125 miles per hour (205 kph), along the vulnerable Gulf of Mexico coast from southeastern Louisiana to Alabama. New Orleans was the storm's epicenter of tragedy. Known as the "Big Easy" and the "Crescent City," New Orleans had enjoyed a reputation of carefree soulfulness, but Katrina changed that.

Hurricane Katrina was the worst natural disaster in U.S. history. Yet the city's tragedy was not unforeseeable. Geographically, New Orleans is extremely vulnerable to the hurricane hazard. It sits in a bowl between the waters of Lake Pontchartrain to the north and the Mississippi River to the south, only 15 miles (25 km) from the waters of the Gulf of Mexico (•Figure 11.50). The city already lies an average of 9 feet (2.75 m) below sea level. It continues to slowly sink at a rate of 3 feet (0.9 m) each century as loose sediments are consolidated and cannot

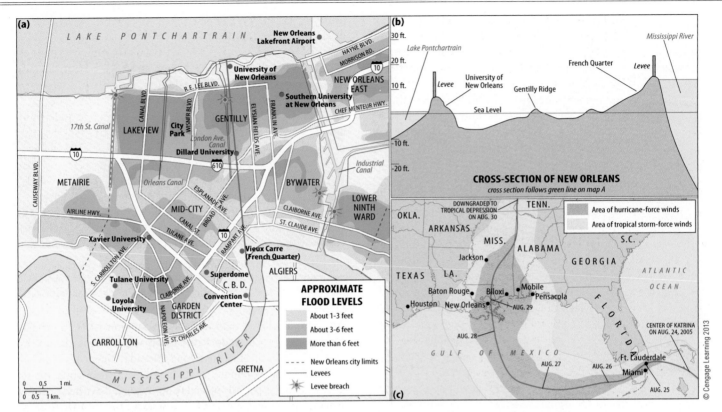

• **Figure 11.50** A monster storm, Hurricane Katrina overwhelmed the inadequate levees defending the vulnerable city of New Orleans.

be replenished: far upstream, dams on Missouri River tributaries hold back 80 percent of the sediment that would naturally reach New Orleans, while the very dikes that protect the city also keep available sediment in the Mississippi at bay.

Hurricane Katrina's **storm surge**—a powerful wave of water—penetrated the city's defenses, a 350-mile (560 km) system of elevated levees and 22 pumping stations. Before Katrina, the U.S. Army Corps of Engineers received less than half of the funds it had asked for to shore up this network, leaving New Orleans buffered by levees that were thought to be able to withstand a Category 3 hurricane. Katrina's storm surge put tremendous pressure on the levees, and they gave way in several places, notably along the 17th Street Canal levee near Lake Pontchartrain. Power supplies failed, and the city's pumps were unable to prevent water from inundating fully 80 percent of the city.

Preparations for the storm were inadequate. The city's mayor issued a mandatory evacuation order well in advance of the storm but did not make transportation available to tens of thousands of people who did not own automobiles. The disaster hit minorities and the disadvantaged especially hard: of the most affected, 4 in 10 were black, 2 in 10 poor, and 1 in 10 elderly.

The emergency response to Katrina was tragically inept. Missteps and poor communication among city, state, and federal officials led to a delay of several days in deployment of security personnel and relief supplies. An estimated 25,000 people sought refuge in the giant Superdome sports stadium and an equal number in the city's Convention Center, but both

sites were short on medical supplies, food, and emergency personnel. Many New Orleans police fled the city, and a few joined in the widespread looting that followed the evacuation. The appointed head of the government body responsible for disaster relief, the Federal Emergency Management Agency (FEMA), had no disaster management experience. Thousands of National Guard troops from Louisiana and neighboring states who might have been rapidly deployed to deal with the crisis were instead serving in Iraq.

New Orleans's population three months after the storm was about 280,000, or just 58 percent of its pre-Katrina level. The poorest and most vulnerable residents of the city had fled to Houston and other distant cities, and it still appears that most will not return to New Orleans. Those who returned as the city rebuilt were more affluent and more predominantly white, but large sections of the city remained wastelands.

The economic costs of repair and cleanup from Katrina were enormous: over $80 billion, making this by far the costliest natural disaster in U.S. history (in lives, Katrina took 1,577, well below the 8,000 lost to a hurricane in Galveston in 1900). Nearly 1 million people along the Gulf Coast were made homeless, and 300,000, jobless. Most of New Orleans's housing stock was made of wood and was irreparably damaged by floodwaters.

As the city continues to be rebuilt, questions persist about whether there should even be a new New Orleans (as we remember from page 70 in Chapter 4, its *situation* makes perfect sense, but its *site* is a nightmare). Not far from the sinking city, the Gulf of Mexico is eating away at the coastline: Louisiana has lost more than 1,560 square miles (4,040 sq km) of coastal land since 1930 and continues to do so at the rate of about one-tenth of a square mile (0.25 sq km) every day. There are concerns that climate change may be providing fuel for stronger hurricanes. Gulf of Mexico waters are about 1°F (0.56°C) warmer now than they were a century ago. The warmer these waters are, the greater their potential to support catastrophic hurricanes. Both Katrina and, later in the 2005 hurricane season, hurricanes Rita and Wilma spent part of their time in the Gulf as Category 5 storms, the strongest possible.

Do you recall the discussion of adaptation and mitigation in dealing with climate change (page 34)? New Orleans will have to adapt to anticipated climate change. How can the city be protected in the future? Work to repair and strengthen existing levees began in Katrina's wake, but engineers agree that these will not prevent another disaster. The U.S. Army Corps has a larger strategy, still in the planning stage, with three legs. One is to rebuild the narrow, steadily eroded barrier islands that form a punctuated chain between Louisiana's mainland and the Gulf of Mexico. Second, restored wetlands along the Gulf of Mexico coast would serve as a natural buffer against storm surge. Finally, federal and other funding would build the Morganza Project, a semipermeable levee system that would allow engineers to control the flow of both freshwater and tidal water across a wide swath of southeastern Louisiana. This **"great wall of Louisiana"** would not protect New Orleans directly but would reduce the losses of much of the state's coastal marshlands. These help buffer the city, serve as a nursery for Louisiana's valuable fishing industry, and are central to Cajun culture.

New Orleans is not the only city calculating and constructing adaptations to climate change. City planners in Chicago expect their city's climate to be similar to that of Baton Rouge, Louisiana, by the end of the century. Conditions would be hotter and wetter, with more frequent snowstorms and extreme events including tornadoes. Workers have already replaced native Illinois species with southern species that thrive in higher temperatures. They are widening sidewalks to allow these trees to produce maximum shade cover. Light-reflecting pavement is being put down to replace the dark surfaces that contribute to the urban "heat island" effect.

Although these cities may be the "canaries in the coal mines" of climate change in the United States, fewer and fewer Americans seem to be concerned. In 2006, 79 percent of Americans believed the Earth was warming. In 2011, that figure was 59 percent. We might assume that Chinese and Indians would put concerns about economic growth ahead of climate change, but 2010 polls showed that 70 percent of both were willing to pay more for energy in order to address climate change. In the United States, just 38 percent were willing.

The Arctic National Wildlife Refuge

Alaska (from *Alyeska,* an Aleut word meaning "great land") (•Figure 11.51) makes up about 17 percent of the United States by area, but it is so rugged, cold, and remote that its total population was only 722,000 in 2011.

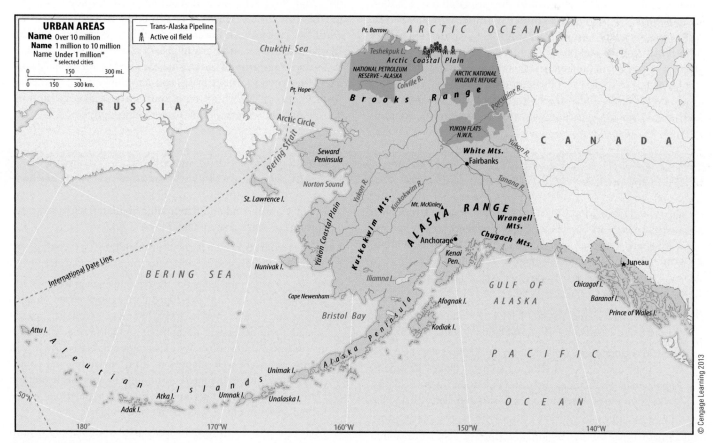

• **Figure 11.51** Alaska.

© Cengage Learning 2013

Alaska's Arctic Coastal Plain is an area of tundra along the Arctic Ocean north of the Brooks Range. Known as the North Slope, it holds the reserves of oil and natural gas that make Alaska the biggest fossil fuel–producing state in the United States. Alaska hopes to pump natural gas from the North Slope to the Lower 48. To date, the focus has been on oil, which supplies more than 80 percent of the state's revenues. Oil income has built a hefty surplus known as the **Permanent Fund**. Interest accrued in this account provides every resident about $1,100 in a typical year just to live in the Union's harsh but beautiful northernmost state.

The 1973 energy crisis delivered a profound shock to U.S. and other industrialized economies. In its wake, there was a push to develop more domestic sources of energy, particularly oil. The most significant reserves were in the North Slope of Alaska, where Prudhoe Bay and nearby oilfields were established. Since 1977, their oil has been transported southward through the Trans-Alaska Pipeline, an extraordinary feat of engineering snaking 800 miles (1,300 km) across some of the most challenging environments on earth, all the way to the port of Valdez on the south coast. The pipeline route may be seen in Figure 11.51.

Originally holding about 10 billion barrels of proven reserves (just under 4% of Saudi Arabia's estimated reserves in 2011), Prudhoe Bay and nearby fields now produce about 11 percent of total U.S. domestic oil production. The fields' output peaked in 1988, when 20 percent of the country's oil was produced there, and has declined steadily since then. Attention has inevitably turned eastward and southward to the estimated 3 to 8 billion barrels beneath the **Arctic National Wildlife Refuge** (**ANWR**, often spoken of as "Anwar"). When the 12,500 square miles (32,000sq km) of the refuge were set aside as wilderness in 1980, the fate of the oil beneath this tundra ecosystem was left for Congress to decide at some later date.

ANWR's oil has been the subject of a bitter and polarizing debate ever since. On the one side are the oil interests, Alaskan politicians, and almost all Republicans in Congress, who have argued that the United States must develop these reserves to give the country more independence from Middle Eastern oil. On the other side is a coalition of environmentalists, many Native Americans, and congressional Democrats who argue that oil production in ANWR will do irreparable damage to the coastal plain's unique environment and the indigenous people who depend on it.

The main Native American group involved is the **Gwich'in**, a 7,000-strong Athabascan tribe that derives much of its livelihood from exploiting caribou (New World reindeer). The Gwich'in live mainly south of ANWR, but the 180,000-strong population of caribou known as the Porcupine Herd spends the winter in Gwich'in territory. The Gwich'in insist that oil drilling and related activity in ANWR will disrupt caribou calving, decimate the herd, and thus endanger the Gwich'in way of life.

The indigenous perspective on ANWR is not united, however; an Inuit group called the **Inupiat** supports oil production there, especially because of the jobs it would bring them. Environmentalists argue that habitats and native cultures would be despoiled to provide just six months of U.S. oil needs, whereas the oil industry says that 20 years of critical U.S. supply is in ANWR. Meanwhile, attention is turning to two other potential oil bonanza sites in Alaska: the North Slope's Teshekpuk Lake, a particularly oil-rich section of the **National Petroleum Reserve**, estimated to have more reserves than ANWR—and one that successive administrations have avoided developing because of its sensitive populations of geese and other wildlife—central Alaska's Yukon Flats National Wildlife Refuge, rich in oil and natural gas.

Our journey around the world has ended in northern Alaska. May your own journeys bring you happiness and knowledge. And when you are thinking about courses, majors, minors, certificates and careers, don't forget Geography!

summary

- The United States and Canada make up a region often known as Anglo America because of British influences. Both had numerous pre-European indigenous cultures and through immigration have been shaped by many others since 1492.

- Greenland is politically a part of Denmark but physically part of this region.

- The United States and Canada have large land areas of about the same size. Their combined population is about 335 million.

- Physical regions of the United States and Canada are the Arctic Coastal Plains, Gulf and Atlantic Coastal Plains, Piedmont, Appalachian Highlands, Interior Highlands, Interior Plains, Rocky Mountains, Intermountain Basins and Plateaus, and Pacific Mountains and Valleys.

- Climate types of the region are tundra, subarctic, humid continental with short and long summers, tropical savanna, Mediterranean, and undifferentiated highland, with associated patterns of vegetation and land use.

- Major resources include agricultural land, forests, coal, petroleum, natural gas, iron ore and other minerals, and waterpower.

- Native American settlement of the region began at least 12,500 years ago. Some groups developed civilizations based on agriculture and permanent settlement, but most were at least partly nomadic and dependent on hunting and gathering. Great linguistic and cultural diversity existed among Native American groups. All were well adapted to local environmental conditions, and all were disrupted profoundly by European contact. Native Americans are among the poorest populations in the United States but have fared better in Canada.

- Early European settlement displaced native populations in both countries, where waves of immigration continually changed demographic and ethnic characteristics. The French are the most significant minority in Canada. Hispanics and African Americans are the largest minority groups in the United States. Asians are prominent minorities in both countries.

- The United States built a land empire through warfare and purchases, including the Louisiana Purchase (from France) of 1803; the acquisition of Florida from Spain in 1819; the addition of Texas, the Southwest, and California as spoils of war with Mexico in 1845 and 1848; the Gadsden Purchase from Mexico of a strip of Arizona and New Mexico in 1853; and Alaska, bought from Russia in 1867.

- Both the United States and Canada industrialized in the nineteenth and twentieth centuries. The economies have changed as manufacturing became less important while employment in service industries and information technology grew.

- Canada and the United States are among the world's richest countries because of many factors. Both have large endowments of natural assets; both are large in area and population; and both industrialized fairly early. Innovations have increased their efficiency and productivity. Peace and stability at home have stimulated investment and economic development. Both have strong internal unity and political stability.

- The United States has the world's largest appetite for fossil fuel. Its energy efficiency has improved since 1973, but energy consumption and the percentage of energy imported have both increased. Nuclear energy production has stalled because of environmental and health considerations. Production of ethanol from corn has surged in the United States, as has production of oil from tar sands in Canada.

- The two countries have had generally good economic and political relations. Each is a major consumer of exports from the other. About 90 percent of the Canadian population lives within 100 miles of the international border.

- The United States is the world's strongest military and economic power. Its influence in world affairs grew in the wake of the two world wars of the twentieth century. The terrorist attacks of September 2001 put U.S. foreign policy on a new course that led to a more aggressively interventionist foreign policy, but in the recent Libyan conflict the United States "led from behind."

- Atlantic Canada consists of the provinces of Newfoundland and Labrador, New Brunswick, Nova Scotia, and Prince Edward Island. Cod fishing was long the mainstay of its economy, but the Grand Banks have become so overfished that government regulation has restricted the harvest, causing much unemployment.

- The Ontario Peninsula is strongly British in origin and character, and Québec is strongly French, with 82 percent of the Québec population of French origin. Québec separatists have been a political presence for the past 40 years, and in 1995 the referendum for an independent Québec was defeated by less than 1 percent. But Québec has gained more provincial autonomy.

- As temperatures warm in the Arctic region, Canada is one of several nations capitalizing on new opportunities for trade and mineral exploitation. Several countries have claimed parts of the Arctic Ocean as sovereign territory, and some have conflicting claims to various areas, including the North Pole.

- Greenland is geologically a part of North America but is a self-governing property of Denmark. Greenland's foreign affairs are controlled by Denmark, which, along with the European Union, spends heavily on subsidies for the 50,000 inhabitants of the world's largest island. Scientists are watchful of the potential melting of the massive icecap that covers much of the island.

- From the fall-line cities, early industrialization, beginning with steel production, spread. The greatest center of major industry, finance, textile manufacturing, publishing, and many other enterprises was New York City.

- Major demographic changes have taken place in recent decades in the United States. Rural areas in the Midwest have been depopulated, whereas southern and western cities have exploded in population.

- Settlement in the West is concentrated in well-watered areas that are, in effect, oases. Topography has a strong role in the region's climates.

- California's cities and farms have huge water requirements that cannot be met from local resources. California has long exceeded its allotment of Colorado River waters, and states upstream in the Colorado Basin have won court actions to increase their share and reduce California's.

- The San Francisco Bay Area is home to Silicon Valley, the dominant software development center in the United States. Like Los Angeles and other Californian cities, the Bay Area is very multicultural.

- Alaska stretches north of the Arctic Circle and has only about 685,000 people. Recent decades have seen a pronounced warming that has affected natural and human environments. There is an ongoing dispute among Congress, the oil industry, environmentalists, and indigenous peoples about whether the oil reserves of northeastern Alaska's Arctic National Wildlife Refuge (ANWR) should be exploited.

Key Terms + Concepts

adaptive reuse (p. 429)
African Americans (p. 409)
Afro-Americans (p. 409)
Aleuts (p. 404)
Algic language family
    (p. 404)

American dream (p. 419)
amnesty (p. 396)
"anchor baby" (p. 396)
Arctic Bridge (p. 427)
Arctic National Wildlife
    Refuge (ANWR) (p. 440)

Asian Americans (p. 410)
bank (p. 426)
birth tourism (p. 395)
blacks (p. 409)
blizzard (p. 398)
borough (p. 432)

"Boswash" (p. 391)
British–North America Act
    (p. 423)
California Gold Rush
    (p. 409)
California Trail (p. 409)

Canada-U.S. Free Trade
 Agreement (p. 421)
Canadian Shield (p. 396)
Colorado River Compact
 (p. 437)
Central Arizona Project (CAP)
 (p. 437)
Corporate Average Fuel
 Economy (CAFE) standard
 (p. 416)
credit crisis (p. 421)
Dakota (p. 403)
decommissioning (of dams)
 (p. 418)
dot-com bubble (p. 434)
dumping (p. 421)
Dust Bowl (p. 398)
economic bubble (p. 421)
Eskimo-Aleut language family
 (p. 404)
Eskimos (p. 404)
exurbanization (p. 429)
fall line (p. 398)
First Nations (p. 404)
First Peoples (p. 405)
Forty-Niners (p. 409)
G-21 (Group of 21) (p. 420)
Gadsden Purchase (p. 408)
gaming industry (p. 406)

gentrification (p. 429)
ghetto (p. 410)
Global Warming Solutions Act
 of 2006 (p. 416)
Gold Rush of 1849 (p. 434)
Great Depression (p. 420)
Great Migration (p. 409)
"great wall of Louisiana"
 (p. 439)
Gwich'in (p. 440)
Hapa (p. 411)
Homestead Act (p. 409)
hydraulic fracturing (fracking)
 (p. 415)
illegal aliens (p. 394)
interstate highway system
 (p. 422)
Inuit (p. 404)
Inupiat (p. 440)
isolationism (p. 423)
Kanaka Maoli (p. 406)
Keystone XL Pipeline
 (p. 417).
Lakota (p. 403)
Latino (p. 410)
Louisiana Purchase (p. 408)
mainstream European
 American culture (p. 409)
Manifest Destiny (p. 408)

megalopolis (p. 391)
Mosan language family
 (p. 404)
Multiculturalism Act (p. 408)
Na-Dene language family
 (p. 404)
 Athabascan subfamily
  (p. 404)
National Petroleum Reserve
 (p. 440)
 Anasazi (p. 402)
 Hohokam (p. 402)
 Mogollon (p. 402)
 Mound Builders (p. 402)
  Adena (p. 402)
  Hopewell (p. 402)
  Mississippian (p. 402)
  Poverty Point (p. 402)
Negroes (p. 409)
new American imperialism
 (p. 424)
Northwest Passage (p. 427)
Nunavut (p. 406)
Occupy Wall Street (p, 414)
official state sponsors of
 terrorism(p. 424)
Oregon Trail (p. 409)
Parti Québécois (PQ)
 (p. 424)

Permanent fund (p. 440)
platform economy (p. 421)
potlatch (p. 404)
preemptive engagement
 (p. 424)
Protestant work
 ethic (p. 419)
pueblo (p. 403)
reservations (p. 405)
Rust Belt (p. 419)
salmon war (p. 421)
Santa Fe Trail (p. 409)
segregation (p. 410)
Silicon Valley (p. 434)
Sioux (p. 403)
"Spanglish" (p. 411)
storm surge (p. 438)
superpower (p. 424)
tar sands (p. 417)
tech bubble (p. 421)
Thirteenth Amendment
 (p. 410)
"Tornado Alley" (p. 397)
Tory (p. 423)
virtual fence (p. 395)
War of 1812 (p. 423)
"Wal-Mart effect" (p. 429)
wheat war (p. 421)

Review Questions

1. Where are populations of the United States and Canada concentrated?

2. What are their main topographic, climatic, and biotic zones and types of the region and the associated land uses?

3. What natural hazards threaten the United States?

4. What are some of the major Native American cultures of the United States and Canada? How did these people live in their diverse environments?

5. How did the French become such an important population in Canada, and how have their aspirations affected the country's politics?

6. What role has immigration had in shaping the populations of the United States and Canada?

7. What are the major nonindigenous faiths and languages of the United States and Canada?

8. What have been the major trends and developments in the economies of the United States and Canada?

9. What are the major characteristics of economic and political relations between the United States and Canada, historically and today?

10. What priorities have preoccupied U.S. foreign policy in recent years?

11. Why did fishing fail to provide the basis for a prosperous economy in Canada?

12. What factors are behind Québec's push for independence? What are the prospects that this province might actually break away from Canada?

13. What new issues and potential conflicts have arisen as temperatures in the Arctic region have increased?

14. With what nation is Greenland politically affiliated? Why do most Greenlanders want to maintain these ties?

15. What is happening to Greenland's extensive ice cover? What explains that phenomenon? What would be the result of a complete melting of that ice sheet?

16. What five regions (as presented in this book) make up the United States?

17. What are the reasons for the importance of New York City?

18. What physical and political circumstances combined to make Hurricane Katrina such a remarkable disaster? What can be done to prevent a recurrence of an event like this?

19. What are the economic and ethnic characteristics of California and its Bay Area?

20. Why have the rural areas and small towns of the Great Plains been losing population? What hopes are there for the revitalization of America's small towns?

21. Why is the term *oasis-like* used to describe urban settlement in the American West? What major problems and controversies have emerged involving the area's water supplies?

22. What are the main issues in the debate about ANWR?

Notes

1. Quoted in Eilene Zimmerman, "Border Agents Feel Betrayed by Bush Guest-Worker Plan." *Christian Science Monitor,* February 24, 2004, p. 3.

2. David Levinson, *Ethnic Groups Worldwide* (Phoenix, AZ: Oryx Press, 1998), p. 396.

3. Quoted in Richard W. Stevenson, "New Threats and Opportunities Redefine U.S. Interests in Africa." *New York Times,* July 7, 2003, pp. A1, A8.

4. Discussion and quotes are from Daniel Hernandez, "A Hybrid Tongue or Slanguage?"

Los Angeles Times, December 27, 2003, pp. A1, A30.

5. Quoted in Louis Uchitelle, "Goodbye, Production (and Maybe Innovation)." *New York Times,* December 24, 2006, p. BU4.

6. From the Associated Press, March 18 2010, "Russian

Leader: Defend Nation's Claim in Arctic." *Columbia Daily Tribune,* p. 11A.

7. Ibid.

8. Ibid.

Online Resources

CourseMate: Make the most of your study time by accessing everything you need to succeed in one place. Read your textbook, take notes, review flashcards, watch videos, complete activities, take practice quizzes, and more—online with CourseMate. Log in at **www.cengagebrain.com**.

Glossary

1948–1949 Arab-Israeli War A war fought between Israel and the Arab countries of Egypt, Iraq, Syria, Lebanon, and Transjordan after Israel declared its independence as a Jewish country. Hundreds of thousands of Arab Palestinians chose or were forced to flee Israel during this war, many settling in the Gaza Strip and the West Bank.

1973 Arab-Israeli War A war launched against Israel by Egypt and Syria in an attempt to reverse the Arab losses of the 1967 war. Egypt succeeded in winning back the eastern side of the Suez Canal.

Abkhaz-Adyghean language family A collection of languages spoken in the Caucasus region.

Aboriginal languages Indigenous languages to Australia that are not clearly related to any languages outside the continent.

Aborigines People indigenous to Australia that likely arrived in a single major migration 50,000 years ago. Europeans colonizing Australia in the eighteenth century slaughtered large numbers of Aborigines and drove them into marginal areas of the continent. Aborigines have long been Australia's most disadvantaged group and were not allowed to vote until 1962.

absolute (mathematical) location Determined by the intersection of lines, such as latitude and longitude, providing an exact point expressed in degrees, minutes, and seconds.

Adamawan A branch of Bantu languages spoken in West Africa.

adaptation Measures designed to cope with and reduce the unavoidable impacts of climate change.

adaptive reuse New uses for older buildings and stores, often accompanied by a shift from decline to steady renewal in an urban neighborhood.

Adena A mound-builder Native American culture that lived in what is now the northeastern United States.

Africa's First World War A widespread war involving nine African countries and 20 rebel movements between 1996 and 2001 that resulted in the deaths of more than 5 million people.

African Americans Americans of African ancestry.

African Development Bank (AfDB) A multi-government-owned bank, based in the Ivory Coast and dedicated to social and economic development (especially in loans for infrastructure) in African countries.

African Growth and Opportunity Act (AGOA) A bill passed by the U.S. Congress in 2000 that reduced or ended tariffs and quotas on many manufactured, mineral, and food items that could be imported from Africa.

African National Congress (ANC) The governing political party in South Africa since 1994. During the apartheid era, the ANC was dedicated to the nonviolent resolution of conflicts between ruling whites and the black majority. The ANC was banned until 1990. ANC leader Nelson Mandela became South Africa's first post-apartheid president in 1994.

African Standby Force A continental military force established by the African Union.

African Union (AU) An organization of most African countries formed to promote democracy, good governance, human rights, gender equity, and development.

Afrikaners Dutch-speaking descendants of Dutch, French Huguenot, and German settlers in South Africa. They are the largest white population in South Africa; also known as Boers.

Afro-Americans See *African Americans.*

Afro-Asiatic language family A language family that includes Semitic languages such as Arabic and Hebrew.

Age of Discovery (Age of Exploration) The three to four centuries of European exploration, colonization, and global resource exploitation and trading led largely by European mercantile powers, beginning with Columbus at the end of the fifteenth century and continuing into the nineteenth century.

Age of Exploration See *Age of Discovery.*

age structure diagram (population pyramid) The graphic representation of a country's population by gender and 5-year age increments.

Agricultural (Neolithic or New Stone Age or Food-Producing) Revolution The domestication of plants and animals that began about 10,000 years ago.

Agricultural Triangle See *Slavic core.*

AIDS (acquired immunodeficiency syndrome) The disease caused by HIV. AIDS is a global pandemic that has killed over 25 million people since the first case was reported in Africa in 1959.

Ainu An indigenous people and language to Japan.

Al-Aqsa *Intifada* A Palestinian uprising, beginning in 2000, triggered by Israeli Prime Minister Ariel Sharon's visit to the Temple Mount in Jerusalem.

Al-Asqa Mosque A sacred Islamic site on the Temple Mount in Jerusalem.

al-Haraam ash-Shariif See "Noble Sanctuary."

Al-Qa'ida A terrorist group formerly led by Osama bin Laden that was behind several attacks against U.S. interests, including the September 11, 2001, attacks.

Al-Qa'ida in Iraq An insurgent group arising in Iraq and affiliated with al-Qa'ida during the Iraq War, led by Abu Mus'ab al-Zarqawi until 2006.

Alawite A minority Shiite Muslim group.

albedo The amount of the sun's energy reflected by the ground. Less vegetation cover correlates with high albedo, and vice versa.

Aleuts An indigenous people of the Aleutian Islands in Alaska.

Algic language family A formerly widespread Native American language family across what is now Canada and the United States.

Allah God in Islam.

Alliance of Small Island States A group of 39 island countries arguing for significant reductions in global greenhouse gas emissions.

Altaic languages A language family that includes Turkish, Korean, and Japanese.

alternative development An overall effort to enhance the education, health, and livelihoods of Latin American villages in an attempt to reduce the amount of cropland devoted to coca and poppies.

American Dream The concept of upward mobility in American society, associated in part with material possessions such as homes and cars.

American Invasion Mexican term for the Mexican War.

Amerindians See *Native Americans.*

Amhara The politically dominant ethnic group of Ethiopia.

amnesty In the U.S. immigration debate, not persecuting illegal immigrants if they meet certain conditions and granting them permanent resident status.

Anasazi A Native American civilization living in what is now the southwestern United States up until about the year 1500. The

Anasazi gradually abandoned the pueblo dwellings they inhabited for unknown reasons.

ancestor veneration A widespread custom in eastern Asia, with shrines and offerings dedicated to preserving the memory of one's ancestors.

anchor baby A baby born in the United States to foreign parents.

animism The belief that natural processes and objects possess souls.

Attila Line See *Green Line* (Cyprus).

Antarctic Circle A line of longitude (65.56° S) commonly used to divide the southern middle latitudes to the north from the southern high latitudes to the south.

Antarctic Treaty An agreement signed by 45 countries forbidding any exploitation of Antarctica's natural resources until 2048.

Anthropocene Epoch A proposed new name for the present geological epoch, currently termed the Holocene.

antiretroviral (ARV) drugs A medication for treatment of HIV infection.

anti-Semitism Anti-Jewish sentiments and activities.

apartheid The Republic of South Africa's former official policy of "separate development of the races," designed to ensure the racial integrity and political supremacy of the white minority.

aquaculture The cultivation of aquatic organisms such as fish for food.

Arab A person of Semitic Arab ethnicity whose ancestral language is Arabic.

Arab-Israeli War of 1956 See *Suez Crisis*.

archipelago A chain or group of islands.

Arctic Bridge A potential shipping route between Churchill, Canada, and Murmansk, Russia, across the Arctic Ocean.

Arctic Circle A line of longitude (65.56° N) commonly used to divide the northern middle latitudes to the south from the northern high latitudes to the north.

Arctic National Wildlife Refuge (ANWR) A wildlife refuge in Alaska that contains 3 to 8 billion barrels of oil.

Armenian genocide The deaths of an estimated 1.5 million Armenians between 1915 and 1918. Turkey and Azerbaijan have not accepted responsibility for their roles in the genocide.

arms race Usually associated with the competition between the United States and the Soviet Union, it refers to rival and potential enemy powers increasing their military arsenals—each in an open-ended effort to stay ahead of the other.

Ashkenazi Jews Jewish immigrants to Israel from Europe after Israel's independence.

Asia-Pacific Economic Cooperation (APEC) group A 21-member organization (including China, Australia, and the United States) that has agreed to establish free and open trade and investment among member countries by 2020.

Asian Americans Americans of Asian ancestry.

Asian Century A term for the twenty-first century, anticipating the large role many Asian countries may play in changing the globe.

Asian Tigers Several Asian regions including South Korea, Taiwan, Hong Kong, and Singapore that had rapidly growing economies in the late twentieth century.

Association of Southeast Asian Nations (ASEAN) A regional economic association of ten Southeast Asian countries.

asylum People given permission to immigrate on the grounds that they would be harmed or persecuted in their country of origin.

atmosphere The layer of gases surrounding the Earth.

atoll Low islands made of coral and usually having an irregular ring shape around a lagoon.

Attila Line See *Green Line* (Cyprus).

Australia New Zealand United States (ANZUS) Security Alliance A defense and security agreement among the three countries, created in 1951. New Zealand has not been an active member since 1987 because of the country's anti-nuclear stance.

Austric language family A language family of Southeast Asia.

Austronesian subfamily The family of languages of most indigenous peoples of the Pacific Islands.

Austronesians Original settlers of many locations in Southeast Asia and the Pacific beginning after 5000 B.C.E. Many of today's Micronesians and Polynesians are their descendants.

autonomous regions A region of a country granted a measure of self-rule without being fully independent of that country.

autonomy Self-rule, generally with reference to Palestinians' rights to run their own civil (and some security) affairs in portions of the West Bank and Gaza Strip allocated to them in the 1993–2000 peace agreements.

ayatollah A revered and powerful cleric at the head of the religious establishment in Iran.

azimuthal A type of map projection in which the developable surface is based on a geometric plane.

Aztec-Tanoan language family A language family that included the Nahuatl language the Aztecs spoke.

Aztecs A Native American people that created an empire across central Mexico in the fifteenth century, which was destroyed by the Spanish in 1521.

baby bounty A financial incentive provided by the Italian government to families to have more than one child, to combat declining population.

Balkanization The fragmentation of a political area into many smaller independent units, as in former Yugoslavia in the Balkan Peninsula.

balloon effect A term referring to the fact that if a balloon is pressed in one place, the air shifts elsewhere. This term is often used to describe areas where drugs are produced and shipped through; when drug supply or shipment is eradicated in one area, it will tend to surge in previously marginal locations.

Baltic languages A language group of Europe that includes Latvian and Lithuanian.

banana war Reference to the incidence in the late 1990s, when the United States wanted to sell bananas grown in Latin America by U.S. corporations to countries of the European Union, but the EU members refused to buy them, as they wanted to protect their investments in banana farming in former European colonies in Africa and the Pacific.

bank In physical geography, an elevated area of the sea bottom.

Bantustans See *homelands*.

barrio A densely settled neighborhood in city space that is characteristically inhabited by migrants of Latino or Hispanic origin.

barter The exchange of goods or services in the absence of cash.

Basques A people, living primarily in Spain and France, who are ethnically and culturally unrelated to those of their host country majorities.

Bay of Pigs Invasion A failed attempt by the John F. Kennedy administration to overthrow Fidel Castro by military force.

bazaar (suq) The central market of the traditional Middle Eastern city, characterized by twisting, close-set lanes and merchant stalls.

Benue-Congo A language family found in central and southern Africa.

Benelux The name used to collectively refer to Belgium, the Netherlands, and Luxembourg.

Berber An ethno-linguistic group of people living in northern Africa.

big bang The addition of ten primarily Eastern European countries to the European Union in 2004.

biodiversity See *biological diversity.*

biodiversity hot spots A ranked list of places scientists believe deserve immediate attention for flora and fauna study and conservation.

biogeography The distribution of living things on Earth.

biological diversity (biodiversity) The number of plant and animal species and the variety of genetic materials these organisms contain.

biomass The collective dried weight of organisms in an ecosystem.

biome A terrestrial ecosystem type categorized by a dominant type of natural vegetation.

biosphere The global ecological system, including all the relationships played out among the lithosphere, hydrosphere, and atmosphere. Also called the ecosphere.

birth dearth The declining population of many European countries from low birth rates.

birth rate The annual number of live births per 1,000 people in a population.

birth rate solution Intentionally lowering birth rates to prevent an overpopulation crisis.

birth tourism A practice in which an expectant mother from an LDC comes to the United States for the purpose of having her baby on U.S. soil, making the baby an automatic U.S. citizen.

black-earth belt An important area of crop and livestock production spanning parts of Russia, Ukraine, Moldova, and Kazakhstan. The main soils of this belt are mollisols.

blacks See *African Americans.*

blizzard A severe winter storm with heavy snowfall and high winds.

blood diamonds See *dirty diamonds.*

blowback Term associated with attacks on U.S. forces by former *mujahadiin* armed and trained by the United States to fight the Soviet occupation in Afghanistan.

Bolivarian Revolution Hugo Chávez's term for his policy agendas after taking control of Venezuela.

Bolshevik Revolution The seizure of the Russian government by Vladimir Lenin's faction of the Communist Party in 1917.

boreal forest A type of forest (also called taiga) primarily composed of coniferous trees that thrive in subarctic climates.

borough In New York City, one of the five administrative divisions: Manhattan, Brooklyn, Queens, Staten Island, or the Bronx.

Boswash A term for the megalopolis stretching from Boston to Washington, D.C.

bourgeoisie In Marxist doctrine, the capitalist class.

brain drain The exodus of educated or skilled persons from a poor to a rich country or from a poor to a rich region within a country.

branded diamonds Diamonds produced by the De Beers company, certified as coming from nonconflict areas.

break-of-bulk point A classic geographic term describing a point in transit when bulk goods must be removed from one mode of transport and installed on another. A trainload of grain carried to a port for transshipment on cargo boats or barges is a common example.

BRIC economy An acronym for Brazil, Russia, India, and China, four countries that may have the world's largest economies by 2050.

British–North America Act An act of the British Parliament in 1867 uniting several British North American colonies to create the independent country of Canada.

bubble economy See *economic bubble.*

Buddhism A religion based on the life and teachings of Siddhartha Gautama, known as the Buddha (or Enlightened One), who lived in the sixth century B.C.E. in India.

Burmese Way to Socialism Rigid state control of the economy in Myanmar.

California Gold Rush The discovery of gold in California, leading to a population boom in that state.

California Trail One of the major westward routes that American pioneers took in the middle of the nineteenth century, leading to what is now Sacramento, California.

Camp David Accords A 1979 agreement between Israel and Egypt that returned the Israeli-occupied Sinai Peninsula to Egypt in exchange for Egypt's recognition of Israel as a sovereign state.

Canada-U.S. Free Trade Agreement A free trade agreement between Canada and the United States in 1988 that was the forerunner to NAFTA.

Canadian Shield The ancient geological core of North America.

Cao Dai A faith originating in Vietnam that blends elements of Buddhism, Daoism, Confucianism, Judaism, Christianity, and Islam.

cap-and-trade system An international system designed to reduce carbon emissions by having richer countries that produce lots of greenhouse gases pay poorer, energy-inefficient countries to cut their emissions. This allows the richer countries to offset their carbon emissions with the cuts in emissions made by the poorer countries. The poorer country is obligated to use the income from cap-and-trade to invest in energy-efficient and nonpolluting technologies.

carbon dioxide (CO₂) A greenhouse gas, the human production of which is very likely the cause of global warming.

carbon sequestration The capture and long-term storage of carbon dioxide. Usually refers to natural processes, but there are also artificial means.

carbon sink An area such as forest, farmland, or ocean that sequesters a large amount of carbon dioxide.

cargo cults Belief systems that appeared in New Guinea and on various Pacific islands during World War II when "Stone Age" peoples were exposed for the first time to Western technology and consumer goods.

carrying capacity The size of a population of any organism that an ecosystem can support.

Carter Doctrine President Jimmy Carter's declaration, following the Soviet Union's invasion of Afghanistan in 1979, that the United States would use any means necessary to defend its vital interests in the Persian Gulf Region. The "vital interests" were interpreted to mean oil, and "any means necessary" interpreted to mean that the United States would go to war with the Soviet Union if oil supplies were threatened.

cartography The craft of designing and making maps, the basic language of geography. In recent years, this traditional manual art has been changed profoundly through the use of computers and geographic information systems (GIS) and through major improvements in machine capacity to produce detailed, colored map products.

cash (commercial) crops Crops produced generally for export.

caste The hierarchy in the Hindu religion that determines a person's social rank. It is established by birth and cannot be changed.

cataracts Areas along the Nile River where its valley narrows and rapids form.

Catholic Republicans One side of "The Troubles" in Northern Ireland, those who wished the British Army to leave the area and for Ireland to be unified.

Celtic Tiger A nickname for Ireland in the 1990s for the country's rapidly growing economy.

Central Arizona Project (CAP) A canal network bringing water from the Colorado River to the populous cities of Arizona.

Chadic A family of languages spoken mainly in northern Nigeria.

chaebols A selection of small companies in South Korea that were heavily funded by the government, eventually growing into giant conglomerates.

Chang Jiang Water Transfer Project A massive engineering project in China that will move water from China's wet south into its populated but drier north.

Chari-Nile A language family spoken in east-central Africa.

Charney Effect Observed by an atmospheric scientist named Charney, this states that the less plant cover there is on the ground, the higher the albedo—solar energy deflected back into the atmosphere—and therefore the lower the humidity and precipitation.

Chernobyl The site in the Ukraine where, in April 1986, the worst nuclear power plant accident in history occurred. It is thought that approximately 5,000 people died, and a zone with a 20-mile (32 km) radius is still virtually uninhabitable; 116,000 people were moved from the area, and cleanup continues to this day.

chernozem A Russian term meaning "black earth." It is a grassland soil that is exceptionally thick, productive, and durable.

Chibcha A small, pre-Columbian civilization in South America renowned for their exquisite gold works.

chokepoint A strategic narrow passageway on land or sea that may be closed off by force or threat of force.

choropleth maps Maps that are drawn to show the differing distribution of goods or geographic characteristics (including population) across a broad area. Such maps are good for generalizations but often mask significant local variations in the presence of the item being mapped.

Christianity One of the world's major faiths, predominant in Europe and many former European colonial areas. Christianity has several major denominations, including Roman Catholicism, Protestantism and Eastern Orthodox.

Chukotko-Kamchatkan A language family of far northeastern Siberia.

Church of the Holy Sepulcher Located in Jerusalem, this is the center of the Christian world, containing the places where tradition says Jesus Christ was crucified and buried.

City of Light A nickname for Paris, France.

civilization The complex culture of urban life.

climate The average weather conditions, including temperature, precipitation, and winds, of an area over an extended period of time.

Closer Economic Relations Agreement A free-trade pact between Australia and New Zealand.

coal A fossil fuel, used for creating heat and electricity, which has many industrial uses. The Industrial Revolution began in European locations situated near large deposits of coal.

coconut civilization A term characterizing the importance of coconuts in the economies and diets of several low-island Pacific countries.

coke Coal with most of its volatile constituents burned off.

Cold War The tense but generally peaceful political and military competition between the United States and the Soviet Union, and their respective allies, from the end of World War II until the collapse of the Soviet Union in 1991.

collective farm (kolkhoz) A large-scale farm in the former Soviet Union that usually incorporated several villages. Workers received shares of the income after the obligations of the collective had been met.

collectivization The process of forming collective farms in Communist countries.

collectivized agriculture See *collective farm.*

colonization The European pattern of establishing dependencies abroad to enhance economic development in the home country.

Colorado River Compact A 1922 agreement between seven western U.S. states that the Colorado River flows through, which determined how much river water each state may take.

Coloreds A South African term for persons of mixed racial ancestry.

Columbian Exchange The transfer of plants and animals between Europe and the Americas after Christopher Columbus's voyages.

comanagement A joint system of national park management in Australia between the Aborigines and the Australian government.

command economy A centrally planned economy typical of the Soviet Union and its Communist allies, in which the government rather than free enterprise determines the production, distribution, and sale of economic goods and services.

commercial crops See *cash crops.*

Common Market An earlier name given to the (current) 25 countries that make up the European Union. In 1957, an initial six countries combined to form the European Economic Community (EEC), and this supranational community has grown to have considerable economic and political importance in Europe. See *European Union.*

Commonwealth of Independent States (CIS) A loose political and economic organization of former Soviet republics (except for Estonia, Latvia, and Lithuania) formed in 1991.

Commonwealth of Nations A voluntary association of 54 countries around the world, most of them former British colonies.

communal violence Term used in India to refer to sectarian violence.

communism A type of national economy characterized by abolition of trade unions, abolition of private ownership in fields such as manufacturing and commerce, varying degrees of state ownership of agriculture, centrally planned and directed economies, and one-party dictatorial government.

Compact of Free Association A 1986 agreement between the United States and the Marshall Islands and the Federated States of Micronesia granting independence and subsidies to the two Pacific countries in exchange for renting their military sovereignty to the United States.

Comprehensive Test Ban Treaty A treaty signed by 149 nations in 1996 that prohibits all nuclear tests.

compromise projections A map projection that does not preserve any one metric but are designed to distort all metrics roughly equally to achieve a map that "looks right."

conflict diamonds See *dirty diamonds.*

conformal projections A map projection that preserves shape.

Confucianism A Chinese sociopolitical philosophy based upon the writings of Confucius, either on its own or blended with other religions.

conic A type of map projection in which the developable surface is based on a cone.

coniferous forest Needleleaf evergreen trees; most bear seed cones.

consumption overpopulation The concept that a few persons, each using a large quantity of natural resources from ecosystems across the world, add up to too many people for the environment to support.

continental drift See *plate tectonics.*

continental islands Once attached to nearby continents, these are the islands north and northeast of Australia, including New Guinea.

contingent sovereignty An agreement in Myanmar between the government and several minority ethnic groups, granting them more civil rights and economic opportunities.

Convention on the Law of the Sea A 1970s United Nations treaty permitting a sovereign power to have greater access to surrounding marine resources.

coordinate system A grid consisting of horizontal and vertical lines used to establish absolute location. On the Earth's surface, latitude and longitude lines form the coordinate system.

Copts A Christian sect in Egypt.

core location An area of the world with significant importance due to a central location relative to others.

Cornucopians People who argue that humans have always been able to conquer food shortages in the past, thanks to technological innovations, and therefore will be able to in the future as well. Also called *technocentrists*.

Corporate Average Fuel Economy (CAFÉ) standard A 1975 U.S. law stating that passenger cars produced by all auto companies must achieve an average of 27.5 miles per gallon.

Cossacks Peasant soldiers of the Russian steppes who were among the first Russian settlers in Siberia in the sixteenth century.

credit crisis A financial crisis originating in the United States in 2008, after the housing bubble burst, that quickly spread worldwide, causing global selloffs in stock exchanges and a widespread economic recession.

Creole Tongues that developed among black slaves and indentured servants in the Americas, consisting mainly of vocabularies from the European colonizers' languages.

Creole languages See *Creole*.

crop substitution Efforts to get Latin American farmers to switch from growing illegal coca or poppies to legal flowers or coffee.

Crusades A series of European Christian military campaigns between the eleventh and fourteenth centuries aimed at recapturing Jerusalem and the rest of the Holy Land from the Muslims.

Cuban Americans Cubans who fled Castro's regime for the United States, and their descendants.

Cuban Missile Crisis The debacle between the United States and the Union of Soviet Socialist Republics in 1962 that put both countries on the brink of nuclear war. Soviet rocket launch facilities were being built in Cuba, presumably for possible nuclear sites on the United States. President Kennedy demanded that the Soviets withdraw the missiles; they did.

cultural landscape The landscape modified by human transformation, thereby reflecting the cultural patterns of the resident culture.

culture The values, beliefs, aspirations, modes of behavior, social institutions, knowledge, and skills that are transmitted and learned within a group of people.

culture hearth An area where innovations develop, with subsequent diffusion to other areas.

Cushitic A family of languages spoken in Ethiopia and Somalia.

cylindrical A type of map projection in which the developable surface is based on a cylinder.

Dakota A Native American people living on the Great Plains.

Dalai Lama The spiritual and political leader of Tibetan Buddhism (Lamaism).

Dalits The bottom of the social hierarchy in India, not part of the caste system. Once known as "untouchables."

Daoism (Taoism) A Chinese school of thought, second only to Confucianism, originating in the *Tao-te Ching* in the sixth century B.C.E. Daoism philosophy is commonly blended with elements of Buddhism.

Dayton Accord A peace agreement creating a joint multiethnic, democratic national government in Bosnia and Herzegovina, responsible for foreign and economic policies, while also recognizing a second tier of government charged with overseeing internal functions composed of the Federation of Bosnia and Herzegovina and the Republika Srpska.

death rate The annual number of deaths per 1,000 people in a population.

death rate solution A catastrophic increase in death rates, seen as "nature's solution" to overpopulation.

debt-for-nature swap An arrangement in which a certain portion of international debt is forgiven in return for the borrower's pledge to invest that amount in nature conservation.

decertification A country's loss of foreign aid from the United States if that country is deemed to be uncooperative in the war on drugs.

decommissioning (of dams) A major modification or removal of a dam primarily for environmental reasons.

deforestation The removal of trees by people or their livestock.

degrees Unit of measurement for longitude and latitude. There are 180 degrees of latitude (90 degrees north of the equator, and 90 degrees south) and 360 degrees of longitude (180 degrees east of the Prime Meridian, 180 degrees west).

deindustrialization An economy shifting away from industries requiring large amounts of energy and labor toward an economy based on services and the production of high-tech goods.

delta The usually triangular-shaped alluvial area at the mouth of a river.

demilitarized zone (DMZ) The border between North and South Korea.

demographic transition A model describing population change within a country. The country initially has a high birth rate, a high death rate, and a low rate of natural increase; then moves through a middle stage of high birth rate, low death rate, and high rate of natural increase; and ultimately reaches a third stage of low birth rate, low or medium death rate, and low or negative rate of population increase.

demography The study of population.

denying the countryside to the enemy The use of explosives and defoliants such as Agent Orange by U.S. troops to create enormous damage to the farms and forests of Vietnam.

dependency theory A theory arguing that the world's more developed countries continue to prosper by dominating their former colonies, the now independent less developed countries.

desert An area too dry to support a continuous cover of trees or grass. A desert generally receives less than 10 inches (25 cm) of precipitation per year.

desert shrub vegetation Scant, bushy plant life occurring in deserts of the middle and low latitudes where there is not enough rain for trees or grasslands. The plants are generally xerophytic.

desertification The expansion of a desert brought about by changing environmental conditions or unwise human use.

development A process of improvement in the material conditions of people often linked to the diffusion of knowledge and technology.

devolution The process by which a sovereign country releases or loses more political and economic control to its constituent elements, such as states and provinces.

Diaspora The scattering of the Jews outside Palestine beginning in the Roman Era.

digital divide The divide between the handful of countries that are the technology innovators and users and the majority of nations that have little ability to create, purchase, or use new technologies.

direct rule The policy of administering Northern Ireland directly by the central government of the United Kingdom.

dirty diamonds Rough diamonds used by rebel movements or their allies to finance conflict aimed at undermining legitimate governments.

distance-decay relationship The geographical concept that the interactions between places (and the people that occupy them) decline as the distance between them grows.

divine rule by clerics Iranian principle of the Guardian Theologian, an infallible supreme leader whose authority is based on Shiite interpretations of Islam.

Dome of the Rock A Muslim shrine built on the Temple Mount in Jerusalem, on the site where two Jewish temples had previously stood.

domestication The controlled breeding and cultivation of plants and animals.

Dominican Republic-Central America Free Trade Agreement (DR-CAFTA) A free-trade organization created in 2005, comprised of the United States, the Dominican Republic, and five Central American countries.

domino effect In political terms, the Cold War concept that if one country (e.g. Vietnam) were to fall to communism, neighboring countries also would.

donor democracy A government that makes just enough concessions in holding elections or human rights to win foreign aid, without introducing any real reforms.

dot-com bubble See *tech bubble.*

dot map Type of map that uses dots to represent a stated amount of some phenomenon within a political unit.

doubling time The number of years required for the human population of a given area to double.

downstream countries A country that lies along a river that has already passed through at least one other country.

Dravidian language family A language family primarily found in southern India.

Dreamtime In Australian Aboriginal myth, the creation of the Earth and all things on it, created by the "dreams" of humanity's ancestors during the Dreamtime.

drought avoidance Adaptations of desert plants and animals to evade dry conditions by migrating (animals) or being active only when wet conditions occur (plants and animals).

drought endurance Adaptations of desert plants and animals to tolerate dry conditions through water storage and heat loss mechanisms.

Druze Members of a small offshoot of Shiite Islam, living mainly in the Golan Heights.

dry farming Planting and harvesting according to the seasonal rainfall cycle.

dumping Selling a product for less than the cost to produce it.

Dust Bowl An eight-year drought in the 1930s that ruined crops and livelihoods in the Great Plains, spurring many people to move west to California.

east longitude See *longitude.*

Eastern hemisphere See *hemisphere.*

ecodevelopment See *sustainable development.*

ecological bankruptcy The exhaustion of environmental capital, leading to potential political and social crises.

ecological footprint The amount of biologically productive land needed to sustain a person's consumption and absorb wastes.

ecologically dominant species A species that competes more successfully than others for nutrition and other essentials of life.

economic bubble The sharply increasing valuations of companies or entire economic sectors to lofty heights that cannot be sustained, often leading to recessions when these bubbles "burst."

economic shock therapy Russia's economic transformation in the early 1990s from a command economy to a free-market economy. Overseen by Boris Yeltsin, this transformation was difficult for a country accustomed to government direction in all economic matters, thus the "shock."

El Nino Southern Oscillation (ENSO; El Nino) A reversal of normal climatic patterns that occurs every few years, when the waters of the eastern Pacific Ocean become much warmer than usual, which can lead to global climatic disruptions.

emigrant A person moving away from a place.

emission-trading system See *cap-and-trade.*

Employment Equity Act A South African law that requires employers to move toward "demographic proportionality" in the workforce, based on national proportions of race and gender.

endemic species A species of plant or animal found exclusively in one area.

energy crisis The petroleum shortages and price surges sparked by the 1973 oil embargo.

energy *maquiladoras* Power generation plants built by American firms just inside Mexico that provide electricity to the U.S. grid but are subject to less stringent pollution controls.

enosis The desire of the Greek majority in Cyprus for political unification with Greece.

environmental determinism The concept that the physical environment has played a sovereign role in the cultural development of a people or landscape. Also known as *environmentalism.*

environmental perception The concept that how people view the world and its landscapes and resources influences their uses of the earth and therefore the condition of the earth.

equal-area projection A type of map projection that preserves area.

equator A line or latitude on the Earth's surface (0 degrees) halfway between the North and South Poles. The dividing line between the northern and southern hemispheres.

equidistant projection A type of map projection that preserves distance from one specific point only to all other points.

Eskimo-Aleut language family A Native American language family found in extreme northern North America.

Eskimos See *Inuit.*

ETA (Basque Homeland and Liberty) A militant Basque group that carried out many terrorist attacks in Spain from the 1960s until 2011.

ethanol A fuel made primarily from corn.

Ethiopian Orthodox Christianity See *Ethiopian Orthodox Church.*

Ethiopian Orthodox Church The Christian church of Ethiopia, related to the Coptic church of Egypt.

ethnic cleansing The relocation or killing of members of one ethnic group by another to achieve some demographic, political, or military objective.

Eurabia A term indicating the growing Muslim population in Europe.

euro Part of the authority of the European Union has been the institution of a new currency that has become "coin of the realm" since early 2002. Not all EU nations accepted the euro.

European core Primarily the countries of Western Europe, including Germany, France and the United Kingdom.

European debt crisis A financial crisis beginning in 2010, when it appeared that Greece might default on its loans. Greece did not want wealthier European countries dictating its spending and taxation policies, whereas wealthier countries such as Germany did not want to spend their own taxes to bail out Greece and other heavy borrowers.

European Economic Community (EEC) An economic organization designed to secure the benefits of large-scale production by pooling resources and markets. The name has been changed to Economic Union. See also *Common Market.*

European Greenbelt A mosaic of national parks and other protected areas stretching across 18 countries.

European periphery countries located in Northern, Southern, and Eastern Europe.

European Union (EU) The current organization, begun in the 1950s as the Common Market. It now is made up of 27 nations (see Chapter 3).

Eurozone The 17 countries that use the Euro as currency.

Evil Empire President Ronald Reagan's term for the Soviet Union in 1983.

exclusive economic zone (EEC) An area up to 200 miles (320 km) offshore of a country, in which only that country can conduct economic activities such as fishing.

Exodus The departure of the Israelites, led by Moses, from Egypt around 1200 B.C.E.

exotic species A nonnative species introduced into a new area.

extensive land use A livelihood, such as hunting and gathering, that requires the use of large land areas.

external costs (externalities) Consequences of goods and services that are not priced into the initial cost of those goods and services.

extinction capital of the world A term used for the Hawaiian Islands because of the irreversible impacts that exotic species, population growth, and development have had on the native wildlife.

extraction reserve Areas of tropical rainforest in Brazil where valuable resources such as rubber are allowed to be commercially extracted.

exurbanization The development of the distant outer suburbs of a city.

facts on the ground The purpose of the Israeli settlements in the West Bank, to make the Israeli presence so entrenched that its withdrawal would be very unlikely.

fair trade movement The selling of commodities such as coffee through cooperatives that provide a living wage for the people in poorer countries growing that product, instead of through commercial producers.

Falashas An ancient Jewish group in Ethiopia, living in isolation for thousands of years.

fall line A zone of transition in the eastern United States where rivers flow from the harder rocks of the Piedmont to the softer rocks of the Atlantic and Gulf Coastal Plain. Falls and/or rapids are characteristic features.

fallow A period of a decade or more in which tropical areas that have been farmed must be rested in order to revert to wild vegetation, thus restoring the soil's fertility.

Falung Gong A quasi-Buddhist movement that is forbidden by the Chinese government.

fault A break in a rock mass along which movement has occurred. A break due to rock masses being pulled apart is a tensional, or "normal," fault, whereas a break due to rocks being pushed together until one mass rides over the other is a compressional fault. The processes of faulting create these breaks.

faulting The processes of rock crowding together or pulling apart along tectonic plate boundaries.

favelas Shantytowns in Brazil.

feng shui A traditional Chinese theory for selecting favorable sites for building homes, cities, and other urban or architectural structures.

feral animals Domesticated animals that have abandoned their dependence on people to resume life in the wild.

Fertile Crescent The arc-shaped area stretching from southern Iraq through northern Iraq, southern Turkey, Syria, Lebanon, Israel, and western Jordan, where plants and animals were domesticated beginning about 10,000 years ago.

Fertile Triangle See "Slavic Coreland."

final status issues Issues deferred to the end of the Oslo Peace Process between Israel and the PLO. Finally dealt with at Camp David in 2000, they included the status of Jerusalem, the fate of Palestinian refugees, Palestinian statehood, and borders between Israel and a new Palestinian state. The negotiations broke down over these final status issues.

Finno-Ugric A subfamily of the Uralic language family, which includes Finnish and Hungarian.

First Nations A term used by Native American groups in Canada to refer to themselves, except for the Inuit.

First Peoples A Canadian term for Native Americans, including the Inuit.

First Temple A Jewish temple built by King Solomon on the Foundation Stone in Jerusalem that was later destroyed by Babylonians.

fish farming See *aquaculture.*

Five Themes of Geography A list created by the National Council for Geographic Education and the Association of American Geographers, summarizing what geography is all about: location, place, human–environment interaction, movement, and region.

flow map A type of map that uses arrows to indicate movement of people or goods from one place to another.

food fights A type of trade war between the United States and the European Union, involving the sales of foods, especially genetically modified foods.

food chain The sequence through which energy, in the form of food, passes through an ecosystem.

Food-Producing Revolution See *Agricultural Revolution.*

Footprints of the Ancestors In Australian Aboriginal creation myth, a term for the paths humanity's ancestors walked.

foraging See *hunting and gathering.*

formal region See *region.*

Formosan A subfamily of the Austronesian language family.

Foundation Stone The great rock in Jerusalem that the First Temple was built upon.

Forty-Niners Gold prospectors making their way to California during the 1849 gold rush.

fracking See *hydraulic fracturing.*

frankenfoods See *genetically modified foods.*

Free Aceh Movement (GAM) A secessionist movement in the Indonesian region of Aceh.

Free Trade Area of the Americas (FTAA) A proposed free-trade area encompassing most countries of the Western Hemisphere.

fuelwood crisis Deforestation in the less developed countries, caused by subsistence needs.

functional region See *region.*

G-21 (Group of 21) An economic group of 21 LDCs strongly involved in discussions of subsidies and tariffs of agricultural goods at World Trade Organization meetings.

Gadsden Purchase A strip of land in southern Arizona that the United States purchased from Mexico in 1853.

Galápagos Islands of Religion A term for the religious diversity of Ethiopia.

gaming industry The development of casinos and an economy dependent upon gambling for many Native American tribes.

Gaza-Jericho Accord See *Oslo I Accord.*

Gazprom A state-owned Russian natural gas company.

genetically modified (GM) foods Products such as rice, corn, and tomatoes, produced in the United States (and banned in Europe), that have been bio-engineered to be more productive and resilient.

genetically modified organism (GMO) foods See *genetically modified foods.*

gentrification The social and physical process of change in an urban neighborhood by the return of young, often professional, populations to the urban core. These people are often attracted by the substantial nature of the original building stock of the place and the proximity to the city center, which generally continues to have major professional opportunities. Although this process brings an urban landscape back into a primary role as a tax base, it does dispossess a considerable number of minority peoples.

genuine autonomy The Dalai Lama's request to China for Tibet to have authority over most internal Tibetan matters except foreign affairs and defense.

geographic information systems (GIS) The growing field of computer-assisted geographic analysis and graphic representation of spatial data. It is based on superimposing various data layers that may include everything from soils to hydrology, to transportation networks, to elevation. Computer software and hardware are steadily improving, enabling GIS to produce ever more detailed and exact output.

geography The study of the spatial order and associations of things. Also defined as the study of places, the study of relationships between people and environment, and the study of spatial organization.

geologic hot spot A small area of Earth's mantle where molten magma is relatively close to the crust. Hot spots are associated with island chains and thermal features.

geomancy The arrangement of homes and villages for maximum harmony with natural and spiritual environments.

geospatial A term pertaining to the geographic location and characteristics of natural or constructed features and boundaries on, above, or below the Earth's surface, especially referring to data that is geographic and spatial in nature.

Germanic languages A language family of northern and central Europe including German, Dutch, Swedish, and English.

ghetto Impoverished areas of central cities where there are many problems, including high crime and lack of jobs.

ghurba The worldwide diaspora of Palestinians.

gigantomania A Soviet preoccupation with fulfilling huge quotas or implementing grandiose schemes.

glacial deposition In the process of continental and valley glaciation, the deposition of moraines that become lateral or terminal in the act of glacial retreat. This same process also leads to glacial scouring as moving ice picks up loose rock and reshapes the landscape as the glacier moves forward or retreats.

glacial scouring See *glacial deposition*.

glaciation A geologic and climatologic process in which great ice sheets form in the Arctic and Antarctic regions and advance toward the equator.

glasnost Policies introduced by Russian leader Mikhail Gorbachev in the 1980s to allow a more democratic political system, more freedom of expression, and a more market-oriented economy.

globalization The spread of free trade, free markets, investments, and ideas across borders and the political and cultural adjustments that accompany this diffusion.

Global Warming Solutions Act of 2006 A California law requiring the state's greenhouse gas emissions to be cut to 1990 levels by 2020.

Gold Rush of 1849 See *California Gold Rush*.

Golden Crescent A term for portions of Iran, Afghanistan, and Pakistan, referring to the large amounts of the world's opium supply being produced there.

Golden Horde Russian term for the Kazan Tatars in the thirteenth century.

Golden Triangle A region in the hinterlands of Myanmar, Laos, and Thailand that is the center of Southeast Asia's illegal drug production.

Good Friday Agreement An agreement between the Unionists and the IRA in 1998 to replace direct rule of Northern Ireland with the power-sharing Northern Ireland Assembly.

Gosplan (Committee for State Planning) The agency in Moscow that formulated the Soviet Union's national economic plans.

GNI PPP Per capita gross national income purchasing power parity (per capita GNI PPP), a method of comparing the real value of output between countries' economies by combining gross national income (GNI) and purchasing power parity (PPP).

graduated symbol map A type of map that uses simple symbols such as circles or bar graphs scaled proportionally to the quantity of the attribute being mapped.

Great Depression The worst economic downturn in U.S. history, beginning in 1929 and lasting through most of the 1930s.

Great Game A rivalry between Russia and the United Kingdom in the nineteenth and early twentieth centuries over strategic influence in Afghanistan and other areas of Central Asia.

Great Migration The migration of about 6 million blacks from the rural South to the cities of the Northeast and Midwest between 1915 and 1970.

Great Rift Valley As the result of tectonic processes, a broad, steep-walled trough extending from the Zambezi Valley in southern Africa northward to the Red Sea and the valley of the Jordan River in southwestern Asia.

Great Volga Scheme The transformation of the Volga River in Russia in the twentieth century to make the river more navigable and to supply hydroelectric power and water for irrigation.

Great Wall of Louisiana Several projects in the planning stage to minimize flooding in Louisiana, including rebuilding the eroded barrier islands in the Gulf of Mexico, restoring the wetlands along the coast, and building a semipermeable levee system.

Greek The ethnolinguistic group inhabiting Greece.

Green Line (Cyprus) A buffer zone dividing Cyprus between the Greek Cypriots and the Turkish Cypriots.

Green Line (Israel) A term referring to Israel's border with the West Bank as of 1967.

Green Revolution The introduction and transfer of high-yielding seeds, mechanization, irrigation, and massive application of chemical fertilizers to areas where traditional agriculture has been practiced.

greenhouse effect The observation that increased concentrations of carbon dioxide and other gases in Earth's atmosphere cause a warmer atmosphere.

greenhouse gases Gases such as carbon dioxide and methane that trap heat from the Sun in the atmosphere, increasing the Earth's temperature.

Greenwich Meridien See *prime meridien*.

gross domestic product (GDP) The value of goods and services produced in a country in a given year. It does not include net income earned outside the country. The value is normally given in current prices for the stated year.

gross national income (GNI) A measure of a country's wealth, counting gross domestic production plus income from abroad such as rents, profits and labor.

gross national product (GNP) The value of goods and services produced internally in a given country during a stated year plus the value resulting from transactions abroad. The value is normally expressed in current prices of the stated year. Such data must be used with caution in regard to developing countries because of the broad variance in patterns of data collection and the fact that many people consume a large share of what they produce.

Group of Eight (G-8) The world's eight most economically powerful and politically influential countries: Canada, France, Germany, Italy, Japan, Russia, the United Kingdom, and the United States.

GUAM A regional grouping of Georgia, Ukraine, Azerbaijan, and Moldova, promoting economic integration, democratic reforms, and orientation toward Europe and away from Russia.

guano Seabird excrement that is also found in phosphate deposits.

guest workers Migrants (mainly young and male) from less developed countries, who are employed, sometimes illegally, in more developed countries.

Gulf Stream The strong ocean current originating in the tropical Atlantic Ocean that skirts the eastern shore of the United States, curves eastward, and reaches Europe as a part of the broader current called the North Atlantic Drift.

Gulf War The 1990–1991 confrontation between a large military coalition, spearheaded by the United States, and the Iraqi forces that had invaded Kuwait. The allies successfully drove Iraqi forces from Kuwait. Earlier, the Gulf War had referred to the 1980–1988 war between Iraq and Iran.

Gwich'in A Native American people in Alaska opposed to oil development in ANWR.

Gypsies See *Roma*.

hacienda A Spanish term for large rural estates owned by the aristocracy in Latin America.

hajj Pilgrimage to Mecca.

Hamas Palestinian Islamist party long dedicated to the destruction of Israel and opposed to the Palestinian-Israeli peace process. A major rival to the moderate al-Fatah party within the Palestine Liberation Organization.

Han Chinese The original Chinese peoples who settled in North China on the margins of the Yellow River (Huang He) and who were central to the development of Chinese culture. Han Chinese make up more than 90 percent of China's population.

Hanification Efforts by the Chinese government to increase ethnic Han presence in China's outlying areas, particularly Tibet.

Hapa The mixed-race offspring between Asian immigrants to the United States and Canada and other ethnic groups.

Hawaiian language The native Hawaiian language, a member of the Malayo-Polynesian subfamily of the Austronesian language family.

Hebrews A term for Jews in Biblical times.

hemisphere Half of the Earth's surface.

Hermit Kingdom A term for North Korea.

hero projects The construction of dams, railways, factories, and the like, encouraged by the Soviet government.

high islands Generally the result of volcanic eruptions, these are the higher and more agriculturally productive and densely populated islands of the Pacific World.

high latitudes See *latitude*.

Hinduism A regionally varied, complex belief system that lacks a definite creed or theology but includes a belief in reincarnation, the reverence of many living things, an unlimited pantheon of deities, and an infinite range of types of permissible worship.

hinterland The region of a country that lies away from the capital and largest cities and is most often rural or even unsettled; it is often seen in the minds of economic planners as an area with development potential.

Hizbullah Iranian-inspired and supported Shiite Lebanese political party holding many seats in Lebanon's Parliament. Broadly regarded as a terrorist organization, and opposed to the existence of Israel.

Hmong-Mien subfamily A language family of Southeast Asia and southern China.

Hokan-Siouan language family A Native American language family in North America.

Holocaust Nazi Germany's attempted extermination of Jews, Roma (Gypsies), homosexuals, and other minorities during World War II.

homelands Ten former territorial units in South Africa reserved for native Africans (blacks). The homelands had elected African governments, and some were designated as "independent" republics, although they were not recognized outside South Africa. Formerly known as *Bantustans*, they were abolished in 1994. Also known as *native reserves* and *national states*.

Homestead Act An 1862 U.S. law that opened up much of the interior of the country for settlement. A pioneer family could claim up to 160 acres of farming land for a $10 fee.

Hopewell A mound-builder Native American civilization that lived in what is now the U.S. Midwest.

homogenous region See *region*.

Human Development Index (HDI) A United Nations–devised ranked index of countries' development that evaluates quality of life issues (such as gender equality, literacy, and human rights) in addition to economic performance.

human–environment interaction Study of the ways human beings use and change the natural environment.

human geography One of the two major branches of geography, primarily studying the world's peoples, their spatial relationships, and their usage of the natural environment.

human immunodeficiency virus (HIV) The virus that causes AIDS.

human trafficking Movement of people against their will or intention, largely for the sex trade and manual labor.

humid continental A climate with cold winters, warm to hot summers, and sufficient rainfall for agriculture, with the greater part of the precipitation in the summer half-year.

humid subtropical A climate that generally occupies the southeastern margins of continents, with hot summers, mild to cool winters, and ample precipitation for agriculture.

humus Decomposed organic soil material. Grasslands characteristically provide more humus than forests do.

hunting and gathering A mode of livelihood, based on collecting wild plants and hunting wild animals, generally practiced by preagricultural peoples. Also known as *foraging*.

hurricane Large tropical storms that form over the warm waters of the Atlantic and Pacific Oceans.

Hutu (Bahutu) The majority population of Rwanda and Burundi. Ethnically, culturally, and linguistically the same as the minority Tutsi; the differences between the groups are socioeconomic, with the Tutsi a ruling class that dominated the Hutu majority.

hydraulic fracturing (fracking) A method of extracting natural gas by pumping water, sand and chemicals underground at high pressure, creating fissures in the rock that allow the natural gas to rise to the surface.

hydrologic cycle The process by which the sun evaporates seawater into water vapor that is later released as freshwater precipitation.

hydropolitics Political leverage and control over water.

hydrosphere All of the world's water features.

Ibo (Igbo) A major ethnic group in southern Nigeria.

Ice Age A period of widespread glaciation from about 2 million to 10,000 years ago.

ice cap A climate and biome type characterized by permanent ice cover on the ground, no vegetation (except where limited melting occurs), and a severely long, cold winter. Summers are short and cool.

Ijaw A major ethnic group in southern Nigeria.

illegal alien A migrant moving to another country illegally.

immigrant A person moving into a place.

Inca An indigenous people of South America who created an empire running from modern-day Ecuador to Chile. The Inca Empire was crushed by the Spanish in 1532.

Indians See *Native Americans*.

Indo-European languages A major language group made up of smaller language groups such as Germanic, Romance, and Slavic.

Indo-Fijians Residents of Fiji who are descended from indentured laborers from India, brought to the islands by the British in the early 1900s.

Indo-Iranian subfamily A language family that contains many of the languages spoken in India, Pakistan, Afghanistan, and Iran.

Indus Waters Treaty A 1960 treaty between India and Pakistan, describing how much water from the Indus River each country can use.

Industrial Revolution A period beginning in mid-eighteenth-century Britain that saw rapid advances in technology and the use of inanimate power.

informal settlements Settlements by people who do not hold legal claims to the land their homes are built upon.

information technology (IT) The Internet, wireless telephones, fiber optics, and other technologies characteristic of more developed countries. Generally seen as beneficial for a country's economic prospects, IT is also spreading in the less developed countries.

Inkatha Freedom Party (IFP) South African political party with mainly Zulu membership. Founded by Zulu chief Mangosuthu Buthelezi, the IFP is a rival of African National Congress (ANC), long led by Nelson Mandela.

insolation Heat from the sun.

insourcing Keeping certain jobs within a country instead of offshoring those jobs to another country.

intensive land use A livelihood such as farming, requiring use of small land areas.

intensive subsistence agriculture Agricultural practices built around the growing of cereals (especially rice) that require a steady output of arduous manual labor.

Intergovernmental Panel on Climate Change (IPCC) A United Nations panel, composed of 2,500 atmospheric scientists from 130 countries, that has concluded that the evidence for global warming is "unequivocal."

internally displaced persons (IDPs) People dislodged and impoverished by strife in their home countries but who have little prospect of emigrating from that country.

International Atomic Energy Agency (IAEA) A United Nations agency that carries out inspections of nuclear facilities.

International Criminal Court A United Nations forum for crimes against humanity, located in the Netherlands.

International Date Line A line roughly concurrent with the 180-degree line of longitude (with several deviations for political and practical reasons) where the beginning of one day and the end of another day meet. The date west of the International Date Line is one day ahead of the date east of the line.

interstate highway system A massive infrastructure development in the United States that started in the 1950s and serves as the primary network for trucking cargo across the country.

Intifada Palestinian uprising against Israeli occupation.

Inuit A people living in Greenland and far northern Canada and Alaska, sometimes known as Eskimos to others.

Inupiat A Native American people in favor of oil production in ANWR.

Iranian A subgroup of the Indo-European language family. Persian and Tajik are Iranian languages.

Irish Republican Army (IRA) A terrorist group based in Northern Ireland that launched many attacks in the United Kingdom, which were designed to drive out the British army from Northern Ireland.

Iron Curtain A term used by Winston Churchill to describe the dividing line in Europe between capitalist democracies to the west and Communist authoritarian regimes to the east.

irredentism Movement by an ethnic group in one country to revive or reinforce kindred ethnicity in another country—often in an effort to promote succession there.

irrigation The artificial placement of water to produce crops, generally in arid locations.

isarithmic map A type of map that uses lines or bands of color to join the points of equal value of some phenomenon across a mapped area.

Islam A monotheistic faith built upon the foundations of Judaism and Christianity. One of the world's major religions, it originated in the seventh century in what is now Saudi Arabia.

Islamic law (sharia) Islamic interpretations of divine intention in legal matters, based on the Qur'an, the Hadith, and Islamic scholars' opinions in matters of jurisprudence.

Islamic Movement of Uzbekistan (IMU) A group dedicated to the overflow of Uzbekistan's president and the creation of an Islamic state in the Fergana Valley of Central Asia.

Islamists Islamic militants.

Islamophobia A fear of Islam and the Muslims that practice it.

isolationism The tendency of a country to stay out of the conflicts of other countries.

isostatic rebound The slow uplift of land that had been pressed down by the great weight of glaciers during the Ice Age.

Israelites A term for Jews in Biblical times.

Jainism A small but influential religion of India renowned for its respect for geographic features and animal life.

Jain A practitioner of Jainism.

Jasmine Revolution The uprising in Tunisia against that country's leader in early 2011.

Jemaah Islamiah (JI, or the Islamic Group) A militant Islamist group in Southeast Asia.

Jewish settlements Israeli citizens settling in the West Bank to strengthen Israel's claims to the area.

Jews Practitioners of Judaism.

jihad A term that for Muslims can have a range of meanings, from expressing one's commitment to a personal conviction to all-out war against nonbelievers.

juche North Korean principle of self-reliance.

Judaism The first significant monotheistic faith, originating at least 4,000 years ago. It does not have a fixed creed or doctrine, and it is not a proselytizing religion.

Judea and Samaria Biblical kingdoms that are now part of the West Bank.

Ka'aba A Muslim shrine located in Mecca.

Kanaka Maoli The term native Hawaiians use for themselves.

karma An important Buddhist concept of a person's acts and the consequences of those acts.

karoshi Japanese term for death by overwork.

Kartvelian (South Caucasian) language A collection of languages spoken in the Caucasus region.

keystone species A species that affects many other organisms in an ecosystem.

Khalistan A desired homeland for India's Sikhs.

Khoisan language family The languages of the San people in southern Africa. These languages, nearly extinct, are among the world's oldest.

Kimberley Process A United Nations–backed certification plan for ensuring that only legally mined rough diamonds, untainted by violence, reach the market.

kleptocracy An economic and political system based on crime.

knowledge economy An economy based on innovation and services.

Kurdistan Workers Party (PKK) The main Kurdish resistance to Turkish rule.

Kurds The largest minority group in Turkey and the largest non-Arab minority in Iraq. Kurds are the world's largest ethnic group without a country.

Kyoto Protocol A treaty on climate change signed by 160 countries in Kyoto, Japan, in 1997 and put into force in 2005. It requires MDCs to reduce their greenhouse gas emissions by more than 5 percent below their 1990 levels by the year 2012.

Labor Party The liberal political party of Israel.

Lakota Native American people living on the Great Plains.

land empire The establishment of colonies in a continental hinterland as opposed to overseas.

Land for Peace Part of the peace process between Israel and Arabs, with Israel swapping Arab lands it occupied in 1967 in exchange for peace with its neighbors.

land hemisphere See *hemisphere*.

land reform In Latin America, the reallocation of land owned by the wealthy to small farmers.

landscape A portion of the earth's land surface. Geographers are interested in the transformation of natural landscapes into cultural landscapes.

landscape perspective The understanding of the changes people have made on the natural environment over time.

language death When the last speaker of a certain language dies

lapse rate The decrease in temperature over a corresponding increase in elevation. The lapse rate is (on average) about 3.6°F (2.0°C) for each increase of 1,000 feet (305 m) in elevation.

large-scale map A map constructed to show considerable detail in a small area.

Laskar Jihad A militant Islamist group in Indonesia.

laterite A material found in tropical regions with highly leached soils; it is composed mostly of iron and aluminum oxides that harden when exposed and make cultivation difficult.

lateritic soils A soil characteristic of many tropical areas. Excessive exposure to the sun and repeated wetting and drying turns the soil into a bricklike substance nearly impossible to cultivate.

latifundia (sing. *latifundio*) Large agricultural Latin American estates with strong commercial orientations.

Latino An American of Hispanic background.

latitude A measurement that denotes position with respect to the equator and the poles. Latitude is measured in degrees, minutes, and seconds, which are described as parallels. Places near the equator are said to be in low latitudes; places near the poles are in high latitudes. The Tropic of Cancer and the Tropic of Capricorn, at 23.5°N and 23.5°S, respectively, and the Arctic Circle and Antarctic Circle, at 66.5°N and 66.5°S, respectively, form the most commonly recognized boundaries of the low and high latitudes. Places occupying an intermediate position with respect to the poles and the equator are said to be in the middle latitudes. There are no universally accepted definitions for the boundaries of the high, middle, and low latitudes.

Law of Return An Israeli law permitting all Jews living in Israel to have Israeli citizenship.

less developed countries (LDCs) The world's poorer countries.

levee A raised riverbank designed to prevent flooding.

liberal autonomy See *autonomous regions*.

Liberal Democratic Party (LDP) The dominant political party of Japan since 1952.

lifeboat ethics Ecologist Garrett Hardin's argument that, for ecological reasons, rich countries should not assist poor countries.

Liga Veneta An Italian organization wishing to create an autonomous region centered on the city of Venice.

Likud Party The conservative political party in Israel.

lithosphere The rocky portion of the Earth's surface.

location A concept central to all geographic analysis. Where something is relates to all manner of influences, from climate to migration

routes, and is a crucial component in trying to understand patterns of historic and economic development.

lock Devices that allow ships to move from one water level to another in a canal.

loess A fine-grained material that has been picked up, transported, and deposited in its present location by wind; it forms an unusually productive soil.

longitude A measurement that denotes a position east or west of the prime meridian (Greenwich, England). Longitude is measured in degrees, minutes, and seconds, and meridians of longitude extend from pole to pole and intersect parallels of latitude.

loose nukes Stolen radioactive materials.

Louisiana Purchase A large purchase of land by the United States from France in 1803 that significantly added to the size of the United States.

low islands Made of coral, these are the generally flat, drier, less agriculturally productive, and less densely populated islands of the Pacific World.

low latitudes See *latitude*.

Ma'adan (Marsh Arabs) People living on marshlands in Iraq, creating a unique culture found nowhere else in the Middle East. Most of their habitat was destroyed by Saddam Hussein after the first Gulf War.

Mahayana Buddhism One of the two branches of Buddhism, originating in India and spreading northward to Tibet and China.

mainstream European American culture A culture based on the English language and British educational principles brought by British colonists combined with core American values that developed on U.S. soil.

major cluster of continuous settlement An area where all habitations lie no more than 3 miles (5 km) from other habitations in at least six different directions, and roads or railroads lie no more than 10–20 miles (16–32 km) away in at least three directions.

Malayo-Polynesian language subfamilies A subfamily of the Austronesian language family.

Mandé A language family of west Africa.

Malthusian scenario The model forecasting that human population growth will outpace growth in food and other resources, with a resulting population die-off.

Manifest Destiny The opening and settlement of new frontiers across the United States from the Atlantic to the Pacific Oceans.

map A representation of various phenomena over all or a part of the Earth, either on a globe or a flat surface.

map projection A way to minimize distortion in one or more properties of a map (direction, distance, shape, or area).

maquiladoras Operations dedicated to the assembly of manufactured goods, generally in Mexico and Central America, from components initially produced in the United States or other places. With the enactment of NAFTA in 1994, there was a massive expansion of *maquiladora* operations.

marginalization A process by which poor subsistence farmers are pushed onto fragile, inferior, or marginal lands that cannot support crops for long and that are degraded by cultivation.

Maria Lionza A syncretic faith in Venezuela that blends Catholic, West African, and indigenous beliefs.

marine west coast A climate occupying the western sides of continents in the higher middle latitudes; it is greatly moderated by the effects of ocean currents that are warm in winter and cool in summer relative to the land.

Maronites A Christian sect in Lebanon.

Marshall Plan The plan designed largely by the United States, after the conclusion of World War II, by which U.S. aid was focused on the rebuilding of the very Germany that had been its enemy in the war just concluded. Secretary of State George Marshall (who had been Chief of Staff of U.S. Army from 1939–1945) was central to the plan's design and implementation.

Masdar A carbon-neutral city under construction in the United Arab Emirates.

mathematical location See *absolute location*.

Maya A Native American ethnic group inhabiting southern Mexico and northern Central America.

Mayan language family A language family of southern Mexico and Guatemala.

medina An urban pattern typical of the Middle Eastern city before the twentieth century.

Mediterranean (dry-summer subtropical) A climate that occupies an intermediate location between a marine west coast climate on the poleward side and a steppe or desert climate on the equatorward side. During the high-sun period, it is rainless; in the low-sun period, it receives precipitation of cyclonic or orographic origin.

Mediterranean scrub forest The xeroyphytic vegetation typical of hot, dry summer, Mediterranean climate regions. Local names for this vegetation type include *maquis* and *chaparral*.

megacity A city with more than 10 million inhabitants.

megadiversity country A country containing a large percentage of the world's animal and plant species.

megalanguage One of the world's most-spoken languages, including Chinese, English, Spanish, Hindi, and Arabic.

megalopolis A chain of metropolitan areas with a population over 10 million.

Mekong River Commission A coalition of Vietnam, Laos, Cambodia, and Thailand, to determine rules to ensure water quality and quantity of the Mekong River.

Melanesia (black islands) A group of relatively large islands in the Pacific Ocean bordering Australia.

mental map A term used to define the geographies encompassed in every individual's mind as a series of locations, access routes, physical and cultural characteristics of places, and often a general sense of the good or bad of locales.

mercantile colonialism The historical pattern by which Europeans extracted primary products from colonies abroad, particularly in the tropics.

Mercator Projection A cylindrical, conformal map projection intended to be used for navigation developed by Gerardus Mercator in 1569.

Mercosur Acronym for Southern Cone Common Market, an economic association among Brazil, Argentina, Paraguay, and Uruguay.

meridian See *longitude*.

mestizo In Latin America, a person of mixed European and Native American ancestry.

Methane Greenhouse gas resulting from human activities (from rice paddies and the guts of ruminating animals like cattle).

Mexican War A war between the United States and Mexico in the 1840s, in which Mexico lost a significant portion of its territory to the United States.

Mezzogiorno A term for southern Italy.

microcredit The lending of small sums to poor people to set up or expand small businesses.

Micronesia (tiny islands) Thousands of small and scattered islands in the central and western Pacific Ocean, mainly north of the equator.

microstate A political entity that is tiny in area and population and is independent or semi-independent.

Middle Eastern ecological trilogy The model of mostly symbiotic relations among villagers, pastoral nomads, and urbanites in the Middle East.

middle latitudes See *latitude*.

migration A temporary, periodic, or permanent move to a new location.

millenarian movements See *cargo cults*.

minifundia (sing. minifundio) Small Latin American agricultural landholdings, usually with a strong subsistence component.

minutes A unit of measurement equal to 1/60th of a degree.

Mississippian A mound-builder Native American culture that lived in what is now the southeastern United States.

misdeveloped country A term sometimes applied to post-Soviet Russia because of the rampant criminality and corruption in its society and its demographic trends of high death rates and low birth rates.

Mistake of 1914 Nigerian term for the creation of the country's modern borders by the British.

mitigation Measures taken to avoid the adverse impacts of climate change in the long term.

Mogollon A Native American civilization that existed in what is now the southwestern United States.

Mohajirs Muslim immigrants to Pakistan from India after the two countries were partitioned in 1947.

mollisol soils Thick, productive, and durable soils, such as the *chernozem*, whose fertility comes from abundant humus in the top layer.

money laundering Ways of turning money earned in illegal ways into money that looks legal.

Mongols Conquerors of most of Russia in the thirteenth century.

Mon-Khmer A branch of the Austro-Asiatic language subfamily that includes the Khmer and Vietnamese languages.

monoculture The single-species cultivation of food or tree crops, usually very economical and productive but threatening to natural diversity and change.

Monroe Doctrine A policy of U.S. President James Monroe stating the United States would prevent European countries from establishing any new colonies in the Western Hemisphere.

monsoon A current of air blowing fairly steadily from a given direction for several weeks or months at a time. Characteristics of a monsoonal climate are a seasonal reversal of wind direction, a strong summer maximum of rainfall, and a long dry season lasting for most or all of the winter months.

Montreal Protocol A 1989 international treaty to ban chlorofluorocarbons.

moor A rainy, deforested upland, covered with grass or heather and often underlain by water-soaked peat. Also, Muslim inhabitants of Spain.

more developed countries (MDCs) The world's wealthier countries.

Mosan language family A Native American language family along the Pacific coast.

Mound Builder civilizations A collection of Native American culture groups named for their creation of large earthen mounds. They created extensive trading networks and urban settlement patterns.

Movement for the Emancipation of the Niger Delta (MEND) A mainly ethnically Ijaw group in Nigeria that has expressed dissatisfaction with the redistribution of Nigeria's oil wealth by kidnapping foreign oil workers.

Movement for the Survival of the Ogoni People (MOSOP) An ethnic Ogoni organization in Nigeria that petitions the government for a safe environment and more federal support for their people.

mujahidin Anti-Soviet rebel bands in Afghanistan during the Soviet invasion of the 1980s.

mulatto A person of mixed European and black ancestry.

mullah Religious leaders in Iran.

Multiculturalism Act A 1988 Canadian law that officially established Canada as a multicultural society, recognizing and encouraging traditional identities.

multi-Latina companies Companies, established in Latin America by Latin American workers who gained experience in the United States, that invest in the United States.

multinational companies Companies that operate, at least in part, outside their home countries.

Muslims Adherents of Islam.

mutually assured destruction (MAD) The concept of nuclear deterrence: if one side in a conflict were to launch nuclear weapons, the other would. Both countries would be obliterated.

Na-Dene language family A Native American language family of western North America, including Apache and Navajo.

nagana See *trypanosomiasis*.

Nakh-Dagestanian A collection of languages spoken in the Caucasus region.

narcoterrorist organization Armed insurgent groups that are funded by the sale of illegal drugs.

National Petroleum Reserve An area along the North Slope of Alaska estimated to have more petroleum resources than ANWR.

nationalism The drive to expand the identity and strength of a political unit that serves as home to a population interested in greater cohesion and often expanded political power.

national minorities A substantial population of ethnic groups originating in other nations but now living in foreign countries.

Native Americans The indigenous peoples of North and South America who originally migrated to the New World from Asia over a land bridge during the Ice Age. Christopher Columbus misidentified them as "Indians" in 1492.

native reserves See *homelands*.

Native Title Bill Legislation passed by the Australian government in 1993, addressing many issues relating to Aborigines' rights to claim and buy land.

natural replacement rate The highest rate at which a renewable resource can be used without decreasing its potential for renewal. Also known as *sustainable yield*.

natural landscape An environment without (or before) transformations caused by humans.

natural resource A product of the natural environment that can be used to benefit people. Resources are human appraisals.

Naxalites Maoist insurgents in India that claim to champion the causes of the rural poor.

Nazca An extinct ethnic group from Peru that left massive carvings in the desert two thousand years ago.

Near Abroad Russia's name for the now-independent former republics, other than Russia, of the Union of Soviet Socialist Republics.

Near North Australian term for neighboring Asian countries.

Negroes An archaic term for African Americans.

neocolonialism The perpetuation of a colonial economic pattern in which developing countries export raw materials to, and buy finished goods from, developed countries. This relationship is more profitable for the developed countries.

neo-Europes Areas colonized by Europeans in lands similar climatically and geographically to Europe, creating "new Europes" with cultural similarities and ties to Europe.

Neolithic Revolution See *Agricultural Revolution*.

Neo-Malthusians Supporters of forecasts that resources will not be able to keep pace with the needs of growing human populations.

new American imperialism The belief that post-9/11 actions by the United States marked the beginning of a new era of aggressive global involvement.

New Asian Tigers A collection of Southeast Asian countries that have had rapid economic growth.

new lands See *virgin and idle lands*.

New Partnership for Africa's Development (NEPAD) An agency of the African Union responsible for monitoring improvements in human rights across Africa.

New Rome A term referring to Constantinople as a center of Christendom in the late Roman period.

newly industrializing countries (NICs) The more prosperous of the world's less developed countries.

Niger-Congo language family A language family of central Africa.

Night Journey In Islam, Muhammad's brief ascension into heaven from the great rock beneath the Dome of the Rock.

Nile Water Agreement A treaty guaranteeing Egypt 75 percent of the Nile River's flow; it forbids any projects downstream that might threaten the volume of water reaching Egypt.

Nilo-Saharan language family A language family spoken in parts of Africa around the Sahara Desert.

NIMBY An acronym for "not in my back yard." Refers to the opposition of local residents to new developments in their area.

nirvana In Buddhism, a transcendent state in which one is able to escape the cycle of birth and rebirth.

Nitrous oxide Greenhouse gas resulting from human activities (from the breakdown of nitrogen fertilizers).

Noble Sanctuary (al-Haraam ash-Shariif) What Muslims call the Temple Mount.

nodal region See *region*.

Non-Pama-Nyungan A branch of the Aboriginal language family spoken in northern Australia.

nonselective migrants People responding to push factors.

nontimber forest products (NTFPs) Goods, other than trees, made from products found in forests, to prevent deforestation.

North American Free Trade Agreement (NAFTA) A free-trade agreement among Canada, the United States, and Mexico.

North Atlantic Drift A warm current, originating in tropical parts of the Atlantic Ocean, that drifts north and east, moderating temperatures of Western Europe. Also known as the *Gulf Stream*.

North Atlantic Treaty Organization (NATO) A military alliance formed in 1949 that included the United States, Canada, many European nations, and Turkey.

North European Plain The level to rolling lowlands that extend from the low countries on the west, through Germany to Poland. These are areas rich in agricultural development, dense human settlements, and a number of major industrial centers. The plain is broken by a number of rivers flowing from the Alps and other mountain systems in central Europe into the North and Baltic Seas.

north latitude See *latitude*.

North Pole A point at 90 degrees north latitude. The highest latitude in the northern hemisphere.

Northern Alliance The main domestic opponent of the Taliban in Afghanistan.

Northern Hemisphere See *hemisphere*.

Northern League An Italian organization calling for the creation of Padania as an autonomous area or independent country.

Northern Sea Route A waterway developed by the Soviets to provide a connection between the Pacific and Arctic oceans.

Northwest Passage A shipping channel through Canada's Arctic islands if global warming decreases sea ice enough to make a reliable route.

Nuclear Nonproliferation Treaty (NPT) See *Comprehensive Test Ban Treaty*.

Nunavut A Canadian territory created in 1999 to be the home of the Inuit.

Oceania A term collectively referring to Australia, New Zealand, and the islands of the Pacific Ocean.

Occupied Territories The territories captured by Israel in the 1967 War: the Gaza Strip and West Bank, captured, respectively, from Egypt and Jordan, and the Golan Heights, captured from Syria.

official state sponsors of terrorism A U.S. list of countries that sponsor terrorism.

offshore outsourcing (offshoring) See *outsourcing*.

Ogoni An ethnic group in Nigeria.

oil embargo The 1973 embargo on oil exports imposed by Arab members of the Organization of Petroleum Exporting Countries (OPEC) against the United States and the Netherlands.

oligarchs Literally "rule of the few," where a small number of wealthy businessmen control much of a state's production.

one-child campaign See *one-child policy*.

one-child policy Official policy of the Chinese government, limiting families to one child.

Operation Barbarossa A German onslaught against the Soviet Union during World War II.

Operation Crossroads A series of 67 nuclear experiments by the United States in the Marshall Islands of the Pacific Ocean after World War II.

Operation Enduring Freedom The U.S. military strike against Afghanistan that began in October 2001.

Operation Iraqi Freedom The U.S.-led ground assault on Iraq that began in March 2003.

Orange Revolution The 2005 Ukrainian presidential election that brought pro-Western Victor Yushchenko to power.

Oregon Trail One of the major routes leading westward that American pioneers took in the middle of the nineteenth century, leading to what is now the Willamette Valley of Oregon.

Organization of American States (OAS) A political organization of every country in North and South America, except Cuba.

Organization of Petroleum Exporting Countries (OPEC) An international organization created to get higher profits from oil sales to its member countries.

orientation The relationship between direction on a map and the corresponding compass direction in reality.

original affluent society A description of hunter-gatherers, who are thought by some scholars to have enjoyed harmony with the natural world and suffered few social and psychological problems.

Oslo I Accord An agreement signed by Israeli prime minister Yitzhak Rabin and Palestinian leader Yassir Arafat, designed to be a pathway to peace between Israel and the Palestinians.

Oslo II Accord Another agreement, not yet fully implemented, between Israel and the Palestinian Authority that would transfer more of the West Bank to Palestinian control in exchange for increased Palestinian efforts to crack down on terrorists and guarantee Israel's security.

Oto-Manguean language family A family of Native American languages spoken in northern Mexico.

outsourcing The flight of jobs and technology from countries with high manufacturing and service costs to countries with lower costs.

overseas foreign workers (OFW) A term referring to the one-quarter of the workforce of the Philippines that works outside that country and sends back remittances that make up more than 10 percent of the Philippines' GDP.

Padania A term for northern Italy.

paddy (padi) The field in which rice is grown.

Palestine Liberation Organization (PLO) The Palestinian military and civilian organization created in the 1960s to resist Israel and recognized in the 1990s as the sole legitimate organization representing official Palestinian interests internationally.

Palestinian Authority (PA) The Palestinian government created as a result of the Oslo I Accord.

Palestinians Arabs who formed the majority of inhabitants of Palestine prior to the creation of the modern country of Israel.

pampa Subtropical grasslands in South America.

Pama-Nyungan branch of the Aboriginal language family spoken over most of Australia.

Pan-African Parliament A lawmaking body of the African Union.

pan-Turkism The goal of uniting all Turkic peoples of Asia.

Pancasila A pan-Indonesian nationalist ideology designed to neutralize all ethnic identities.

pandemic A very widespread epidemic.

Papuan languages A collection of the 750 languages spoken on New Guinea and surrounding islands.

Papuans People of mostly Melanesian origin that live on New Guinea.

parallel A latitude line running parallel to the equator.

páramos Alpine meadows in Latin America.

Parsis An ethnic group in India that practices Zoroastrianism.

Parti Quebecois (PQ) A separatist political party in Quebec.

pastoral nomads Desert peoples of the Middle East and northern Africa who migrate through arid lands with their livestock, following patterns of vegetation and rainfall.

Penutian language family A Native American language family of Mexico and California.

People of the Book Term used by Muslims to refer to Jews and Christians.

people overpopulation The concept that many persons, each using a small quantity of natural resources to sustain life, add up to too many people for the environment to support.

PEPFAR An acronym for President's Emergency Plan for AIDS Relief, in which the George W. Bush administration pledged over $45 billion to combat AIDS in Africa.

per capita GDP A measure of economic well-being found by dividing a country's gross domestic product by its population.

per capita gross national income A measure of economic well-being found by dividing a country's gross domestic product plus any foreign output by domestically-owned producers by its population.

per capita gross domestic product purchasing power parity (GDP PPP) In this figure, annual per capita gross domestic product (GDP)—the total output of goods and services a country produces for home use in a year—is divided by the country's population. For comparative purposes, that figure is adjusted for purchasing power parity (PPP), which involves the use of standardized international dollar price weights that are applied to the quantities of final goods and services produced in a given economy. The resulting measure, per capita GDP PPP, provides the best available starting point for comparisons of economic strength and well-being between countries.

perceptual region See *region*.

perestroika See *glasnost*.

peripheral location An area of the world with less importance due to a distant location relative to others.

permafrost Permanently frozen subsoil.

Permanent Fund A government surplus in Alaska, created from oil revenues.

Persian The language spoken by the majority of Iranians.

physical geography The subdiscipline of geography most concerned with the climate, landforms, soils, and physiography of the earth's surface.

Pidgin The official language of Papua New Guinea, a combination of English and foreign words mixed with indigenous vocabulary and grammar.

PIGS An acronym for Portugal, Ireland, Greece, and Spain, used in reference to the weaker economies of those countries as opposed to those of France and Germany.

Pillars of Islam The five fundamental tenets of the faith of Islam, which are *profession of faith, prayer, almsgiving, fasting,* and the *hajj.*

pivotal countries Those countries whose collapse would cause international refugee migration, war, pollution, disease epidemics, or other international security problems.

place In geographic analysis of a given locale, the nature of place identity becomes a means of understanding people's response to that particular place. Determination of the environmental and cultural characteristics that are most frequently associated with a certain place helps to establish that "sense of place" for a given location.

Plan Colombia A U.S.-funded effort to reduce the amount of coca production in Colombia.

Plan Patriot An operation in Colombia involving U.S. military personnel and civilian contractors targeting FARC to reduce drugs and secure oil reserves.

plates Giant sections of rock the lithosphere is split into, which slowly drift over time.

plate tectonics The dominant force in the creation of the continents, mountain systems, and ocean deeps. The steady, but slow, movement of these massive plates of the earth's mantle and crust has created the positions of the continents and major patterns of volcanic and seismic activity. Areas where plates are being pulled under other plates are called subduction zones.

platform economy Common in high-tech industries, with a corporation headquartered in the United States but most of the work on its products done overseas.

Pleistocene overkill hypothesis A hypothesis stating that hunters and gatherers of the Pleistocene Era hunted many species to extinction.

plinthite The bricklike substance into which lateritic soils can turn.

podzol Soil with a grayish, bleached appearance when plowed, lacking in well-decomposed organic matter, poorly structured, and very low in natural fertility. Podzols are the dominant soils of the taiga.

Polynesia (many islands) A Pacific island region that roughly resembles a triangle with its corners at New Zealand, the Hawaiian Islands, and Easter Island.

population change rate The birth rate minus the death rate in a population.

population explosion The surge in Earth's human population that has occurred since the beginning of the Industrial Revolution.

population implosion A dramatic loss of population, caused by low fertility rates or death rates that are much higher than birth rates.

population pyramid See *age structure diagram.*

population replacement level The number of new births required to keep a population steady. Generally calculated as 2.1 children per woman in MDCs.

postindustrial Description of an economy or society characterized by the transformation from manufacturing to information management, financial services, and the service sector.

potato famine The starvation of 10 percent of Ireland's population in the late 1840s, caused by a crop disease called blight.

potlatch A Native American system in which rank and status were determined by the quantity and quality of material goods one could give away.

Poverty Point A mound-builder Native American culture that lived in what is now the southeastern United States.

power projection hub In geopolitical terms, a strategic outpost from which a country could launch naval, air, and other forces against an enemy.

prairie An area of tall grass in the middle latitudes, composed of rich soils that have been cleared for agriculture. The original lack of trees may have been due to repeated burnings or periodic drought conditions.

Pre-1967 borders The borders of Israel after 1949 and before the Six-Day War of 1967 in which Israel captured and occupied the Gaza Strip, West Bank, and Golan Heights.

precautionary principle The notion that is it appropriate to attempt to reduce or eliminate any risky practice.

Pre-Columbian times North and South America before Christopher Columbus's first voyage in 1492.

precipitation Water falling from the sky to the surface in the form of rain, snow, sleet or hail.

preemptive engagement A U.S. policy developed after 9/11 that allowed the country to take unilateral military action if necessary to defuse threats to its security whenever and wherever those threats occur.

primate city A city that dominates a country's urban scene and is usually defined as being larger than the country's second and third largest cities combined. Primate cities are generally found in developing countries, although some developed countries, such as France, have them.

prime meridian The line of zero degrees longitude, which passes through Greenwich, England. It separates the Western Hemisphere from the Eastern.

privatization The shift of ownership of economic activities from the state to nongovernmental.

proletariat In Marxism, the industrial working class.

Promised Land Canaan, for the Israelites in Biblical times.

Protestant Reformation The schism in Western Christianity that began in 1517 when Martin Luther registered his objections to much of the dogma, rituals, and political organization of the Roman Catholic Church.

Protestant Unionists One side of "The Troubles" in Northern Ireland, those who wished Northern Ireland to remain unified with the United Kingdom and not unify with the rest of Ireland.

Protestant work ethic In the United States, the emphasis on work, productivity, and economic success in an individual's life. Despite its name, this ethic is shared by people of many faiths and ethnicities.

Protestantism One of the denominations of Christianity, which began in the Protestant Reformation of the sixteenth century.

Proto-Asiatic A family of languages spoken largely in northeastern Russia.

proxy war During the Cold War, the situation created when the United States and the Union of Soviet Socialist Republics played African countries against one another, arming them with weapons, to boost their competing aims.

pueblo An interconnected series of residences and ceremonial centers composed of adobe, used by the Hohokam culture of New Mexico.

purchasing power parity (PPP) A method of comparing the real value of output between different countries' economies, considering factors such as differences in relative prices of goods and services.

push and pull forces of migration Emigration caused by so-called push factors, as when hunger or lack of land "pushes" peasants out of rural areas into cities, or by pull factors, as when an educated villager responds to a job opportunity in the city. People responding to push factors are often referred to as nonselective migrants, whereas those reacting to pull factors are called selective migrants. Both push and pull forces are behind the rural-to-urban migration that is characteristic of most countries.

Quechua See *Quechua-Aymaran language family.*

Quechua-Aymaran language family A language family of central South America, to which the language of the Inca belonged. Many residents of Peru and Bolivia still speak Quechua as a first language.

Rastafarianism A religion originating in Jamaica in 1930 that holds that Ethiopian emperor Haile Selaisse was the earthly incarnation of God.

Red-Dead Peace Conduit A joint venture between Israel and Jordan that will carry seawater from the Gulf of Aqaba northward to replenish the shrinking Dead Sea.

Reducing Emissions from Deforestation and Forest Degradation (REDD) A United Nations effort to create a financial value for the carbon stored in forests and offer incentives for developing countries to reduce carbon emissions from forested areas and invest in low-carbon development.

reference map A type of map primarily depicting the locations of various features on the Earth's surface.

refugees Victims of severe push factors such as persecution, political repression, and war.

region A "human construct" that is often of considerable size, that has substantial internal unity or homogeneity, and that differs in significant respects from adjoining areas. Regions can be classed as formal (homogeneous), functional, or vernacular. The formal region, also known as a uniform region, has a unitary quality that derives from a homogeneous characteristic. The United States is an example of a formal region. The functional region, also called the nodal region, is a coherent structure of areal units organized into a functioning system by lines of movement or influence that converge on a central node or trunk. A major example would be the trading territory served by a large city and bound together by the flow of people, goods, and information over an organized network of transportation and communication lines. Vernacular or perceptual regions are areas that possess regional identity, such as the Sun Belt, but share less objective criteria in the use of this regional name. General regions, such as the major world regions in this text, are recognized on the basis of overall distinctiveness.

Regional Initiative for the Infrastructure Integration of South America An international effort to improve the economies of South American countries by linking them together by road networks and other infrastructure projects.

relative location The location of a place defined by relationship to other places.

remittances Money earned by workers in a foreign country sent back to that worker's home country.

remote sensing Through the use of aerial and satellite imagery, geographers and other scientists have been able to get vast amounts of data describing places all over the face of the earth. Remote sensing is the science of acquiring and analyzing data without being in contact with the subject. It is used in the study of patterns of land use, seasonal change, agricultural activity, and even human movement along transport lines. This process relates closely to GIS. See also *geographic information systems (GIS).*

renationalization A process of reversing the trend toward private ownership of economic production in Russia during the presidency of Vladimir Putin.

renewable resource A resource, such as timber, that is grown or renewed so that a continual supply is available. A finite resource is one that, once consumed, cannot be easily used again. Petroleum products are a good example of such a resource and because it takes too much time to go through the process of creation, they are not seen as renewable.

reservations Areas of land reserved for Native American tribes by the U.S. government.

resilience An ecosystem that can recover from the stress of drought and has mechanisms to cope with a natural cycle that includes extended periods of dryness and rain.

resource curse A paradox of plenty, when a country with a great abundance of a valuable natural resource experiences lower economic growth than countries without such abundance.

Richter scale A system for measuring the strength of earthquakes.

right of return The Palestinian Arab principle that refugees (and their descendants) displaced from Israel and the Occupied Territories in the 1948 and 1967 wars be allowed to return to the region.

risk minimization The exploitation of multiple resources so some resources will support a population if others fail.

River Sea A term referring to the enormity of the Amazon River.

road map for peace An effort led by the United States, the European Union, Russia, and the United Nations to reach a peace settlement between Israel and the Palestinians, ultimately leading to an independent Palestinian state.

Roma (Gypsies) Originating in India, they are one of Europe's largest ethnic minorities. Commonly referred to as Gypsies, they often move from place to place in caravans and have been discriminated against for centuries.

Roman Catholic Church The largest denomination of Christianity, led by the Pope in Vatican City.

Romance languages A language family primarily of southern Europe, including Italian, Spanish and French.

Roosevelt Corollary A 1904 policy in which the United States declared it had the right to supervise the internal affairs of Latin American countries to ensure U.S. national security.

Rose Revolution The peaceful transition of power in the 2003 presidential election in Georgia.

rules of origin A NAFTA requirement that half or more of all the components of any manufactured good must originate in Canada, the United States, or Mexico.

rural-to-urban migration See *urbanization.*

Rus Viking adventurers from Scandinavia that settled in Slavic regions during the Middle Ages.

Russian cross The dramatic rise in death rates and fall of birth rates in Russia after the breakup of the Soviet Union.

Russian Revolution The overthrow of Czar Nicholas II of Russia in 1917.

Russification The effort, particularly under the Soviets, to implant Russian culture in non-Russian regions of the former Soviet Union and its Eastern European neighbors.

Rust Belt An area across the Midwest United States that was home to many heavy industries.

sacred space (sacred place) Any locale that people hold in reverence, such as places of worship, cemeteries, and battlefields.

Saharan A subfamily of the Nilo-Saharan language family spoken in north-central Africa.

salinization The deposition of salts on, and subsequent fertility loss in, soils experiencing a combination of overwatering and high evaporation.

salmon war A dispute regarding salmon fishing between the United States and Canada until it was resolved in 1999.

San The proper name for Bushmen, a major group of the Khoisan ethnolinguistic cultures living mainly in Botswana and Namibia. They were among the last people on Earth to practice hunting and gathering.

sand sea A virtual "ocean" of sand characteristic of parts of the Middle East, where people, plants, and animals are all but nonexistent.

Santa Fe Trail One of the major routes leading westward that American pioneers took in the middle of the nineteenth century, leading to what is now New Mexico.

Santeria A syncretic religion originating in the Caribbean, blending Catholicism, indigenous, and Yoruba beliefs.

savanna A low-latitude grassland in an area with marked wet and dry seasons.

scale The size ratio represented by a map; for example, a map with a scale of 1:12,500 is portrayed as 1/12,500 of the actual size.

Schengen Agreement A framework among signatory countries allowing for free movement of people without need for visas, passports, or other border controls.

Schengenland A term for the countries participating in the Schengen Agreement.

scorched earth The wartime practice of destroying one's own assets to prevent them from falling into enemy hands.

seafloor spreading The creation of new crust from molten material rising from the Earth's mantle through ridges on the ocean floor.

seamount An underwater volcanic mountain.

Second Kuwait A term describing the Russian island of Sakhalin and the potential large reserves of oil and natural gas it has offshore.

second law of thermodynamics A natural law stating that high-quality, concentrated energy is increasingly degraded as it passes through the food chain.

Second Russian Revolution The collapse of the Soviet Union in 1991.

Second Temple The second Jewish temple to stand on the Temple Mount, which was destroyed by the Romans.

seconds A unit of measurement equal to 1/60 of a minute, or 1/3,600 of a degree.

sectarian violence Conflict between people of different religions.

security fence A physical barrier constructed by Israel in the West Bank designed to prevent Palestinian suicide bombers from reaching Israel.

sedentarization Voluntary or coerced settling down, particularly by pastoral nomads in the Middle East.

segregation A policy of separation of races.

seismic activity Usually refers to tectonic forces that result in earthquakes.

selective migrants People reacting to pull factors.

semiarid A mainly dry climate, with sparse precipitation.

Semitic language A language family that includes Hebrew and Arabic.

Sephardic Jews Jews who had inhabited Palestine and other parts of the Middle East.

settler colonization The historical pattern by which Europeans sought to create new or "neo-Europes" abroad.

Sephardic Jews Jews who had inhabited Palestine and other parts of the Middle East since early times.

Shanghai Cooperation Organization (SCO) An organization composed of China, Russia, and several Central Asian countries designed to combat terrorism, separatism, and extremism in the region.

shantytown Areas of often squalid slums on the outskirts of large cities throughout the developing world.

shatter belt A large, strategically located region composed of conflicting states caught between the conflicting interests of great powers.

shell banks A bank existing only on paper that specializes in money laundering.

Shia, or Shiite The branch of Islam regarding male descendants of Ali, the cousin and son-in-law of the Prophet Muhammad, as the only rightful successors to the Prophet Muhammad.

shifting cultivation A cycle of land use between crop and fallow years that is needed to work around the infertility of tropical soils. After clearing tropical forest, a farmer may get only a few years of crops before there is no fertility in the soil and must move on to clear more land. In the meantime, forest reclaims the previously farmed plots. Where new lands are not available, fallow periods on old fields are reduced or eliminated, resulting in soil deterioration. Also known as *swidden* or *slash-and-burn cultivation*.

shifting ranching A destructive process in which ranchers raise cattle on areas of former rainforest originally cleared by farmers. When the land is exhausted, ranchers move their cattle on to the next plot of cleared land.

Shintoism The strongly nationalistic religion of Japan, a blend of nature reverence, Japanese folklore, and Chinese ritual.

Silicon Valley A region of the San Francisco Bay area known for its concentration of high-tech companies and workers.

Silk Road A mostly overland trade route that stretched from Venice to Xi'an in China.

Sinhalese The majority ethnic group of Sri Lanka, predominantly Buddhist.

Sinn Fein The political counterpart to the Irish Republican Army.

Sino-Tibetan language family A major language family of Asia, including Chinese and its variants.

Sioux Native American people living on the Great Plains.

site The specific geographic location of a given place.

situation The accessibility of that site and the nature of the economic and population characteristics of that locale.

Six-Day War A 1967 war between Egypt, Syria, and Jordan against Israel, resulting in a dramatic Israeli victory and the Israeli occupation of the Gaza Strip, West Bank, and Golan Heights.

six essential elements of geography 1. The world in spatial terms; 2. Places and regions; 3. Physical systems; 4. Human systems; 5. Environment and society; 6. Uses of geography.

slash-and-burn cultivation See *shifting cultivation*.

Slavic Core (Fertile Triangle or Agricultural Triangle) The large area of the western former Soviet region containing most of the region's cities, industries, and cultivated lands.

Slavic languages A language family in eastern Europe including Russian, Polish and Czech.

sleeping sickness See *trypanosomiasis*.

small-scale map A map constructed to give a highly generalized view of a large area.

socialization State ownership of economic production.

Solidarity An independent trade union in Poland, the creation of which was one of the conditions resulting in the withdrawal of Soviet control in Poland in 1989.

songlines See *Footprints of the Ancestors*.

south latitude See *latitude*.

South Pole A point at 90 degrees south latitude. The highest latitude in the southern hemisphere.

Southeast Anatolia Project (GAP) An agricultural effort in Turkey aiming to double the amount of the country's irrigable farmland.

Southern African Customs Union (SACU) A trade organization including South Africa, Lesotho, Swaziland, Botswana, and Namibia.

Southern Cone Common Market (Mercosur) See *Mercosur*.

Southern Hemisphere See *hemisphere*.

Soviet satellites Communist countries in Eastern Europe between World War II and 1989.

space In geography, the exact placement of locations on the face of the Earth.

Spanglish A hybrid tongue of Spanish and English.

spatial Geographers recognize spatial distributions and patterns in Earth's physical and human characteristics. The term *spatial* comes from the noun *space*, and it relates to the distribution of various phenomena on Earth's surface. Geographers portray spatial data cartographically—that is, with maps.

special economic zones (SEZs) In China, the urban areas designated after the 1970s to attract investment and boost production through tax breaks and other incentives.

spodosols Acidic soils that have a grayish, bleached appearance when plowed, lack well-decomposed organic matter, and are low in natural fertility. Also known as *podzols*.

state farm (sovkhoz) A type of collectivized state-owned agricultural unit in the former Soviet Union; workers receive cash wages in the same manner as industrial workers.

State Law and Order Restoration Council (SLORC) The military government that seized control of Burma in 1988 and changed the country's name to Myanmar.

steppe (temperate grassland) A biome composed mainly of short grasses. It occurs in areas of steppe climate, which is a transitional zone between very arid deserts and humid areas.

storm surge A powerful wave in the ocean that is associated with landfalling hurricanes.

strategic hamlet Fortified communities in Vietnam that the U.S. military forced millions of South Vietnamese farming families into, in a failed attempt to prevent Viet Cong forces from infiltrating South Vietnamese villages.

String of Pearls China's strategy of gaining critical allies along the Indian Ocean shoreline.

subarctic A high-latitude climate characterized by short, mild summers and long, severe winters.

subduction The process of one tectonic plate descending below another.

subduction zone See *plate tectonics*.

subsidies Payments made by governments to domestic producers.

Suez Crisis A British, French, and Israeli invasion of Egypt in 1956 after Egyptian president Nasser nationalized the Suez Canal. International pressure caused the withdrawal of the invading forces, and the canal stayed in Egyptian control.

summer monsoon See "Monsoon."

Sunni The branch of Islam regarding successorship to the Prophet Muhammad as a matter of consensus among religious elders.

superpower A country with a bigger economy, stronger military, and pervasive popular culture than most other countries. Currently the United States is the only global superpower.

supranational organization An international organization in which member countries are united beyond the authority of any single national government and are planned and controlled by a group of nations.

suq A large commercial zone adjoining the ceremonial and administrative heart of a Middle Eastern city. See *bazaar*.

sustainable development (ecodevelopment) Concepts and efforts to improve the quality of human life while living within the carrying capacity of supporting ecosystems.

sustainable yield (natural replacement rate) The highest rate at which a renewable resource can be used without decreasing its potential for renewal.

sweatshops Factories in the LDCs infamous for low wages and long working hours. Part of the dark side of globalization: Chinese and other Asian companies in particular can undercut competition by underpaying and overworking employees.

swidden cultivation See *shifting cultivation*.

syncretism The incorporation of some animistic beliefs into mainstream religions.

taiga A northern coniferous (needleleaf) forest.

Tai-Kadai sufamily

Taliban One of the mujahidiin factions in Afghanistan fighting the Soviet invasion in the 1980s. The Taliban gained control over most of the country in the 1990s and instituted a very strict code of Islamic law. The Taliban were driven from power during the U.S.-led Operation Enduring Freedom but remain a significant insurgent force.

Tamils The minority ethnic group of Sri Lanka, primarily Hindu.

Tamil Tigers Tamil insurgents in Sri Lanka that wanted to establish an independent Tamil homeland. They were defeated by the Sri Lanka government in 2009.

Taoism See *Daoism*.

tar sands Mineral deposits in which liquid petroleum is mixed with sand and clay. Extracting the oil from tar sands is expensive and consumes large amounts of energy and water.

tariffs Tax penalties imposed on imports.

Tatars Nomads of Central Asia that conquered much of Russia in the thirteenth century along with the Mongols.

tech bubble An economic bubble in the United States in the late 1990s, centering on high-tech products and services.

technocentrists (cornucopians) Supporters of forecasts that resources will keep pace with or exceed the needs of growing human populations.

tectonic forces Processes that derive their energy from within the earth's crust and serve to create landforms by elevating, disrupting, and roughening the earth's surface.

temperate grasslands See *steppe*.

temperate mixed forest Dominant forest type in areas with humid subtropical *and* humid continental climates.

Temple Mount A site in Jerusalem sacred to both Jews and Muslims. Two Jewish temples stood on the Temple Mount, and currently the Muslim Dome of the Rock occupies the site.

Teotihuacanos A Native American ethnic group that lived in central Mexico two thousand years ago, who were later replaced by the Aztecs.

terra nullius A legal doctrine declaring a location "no one's land," allowing for claims to that land.

terraces Carefully constructed rice paddies arranged in a stair-step fashion.

terrorism Premeditated, politically motivated violence perpetrated against noncombatant targets by subnational groups or clandestine agents, usually intended to influence an audience.

terrorism hot spot An area regarded by the United States as real or potential training grounds for al-Qa'ida-affiliated groups.

thematic map A type of map showing the spatial distribution of one or more attributes across a given area.

theory of Himalayan environmental degradation The concept that overpopulation on steep slopes in the middle and high elevations of the Himalaya leads to excessive deforestation and erosion through shifting cultivation. This has many knock-on effects downstream, including flooding and siltation of hydroelectric reservoirs.

theory of island biogeography A theoretical calculation of the relationship between habitat loss and natural species loss, in which a 90 percent loss of natural forest cover results in a loss of half of the resident species.

Theravada Buddhism One of the two branches of Buddhism, found largely in Sri Lanka and Southeast Asia. Therevada Buddhists claim to follow the true teachings and practices of the Buddha.

Third Revolution The belief that sustainable development will bring about a marked shift in human ways of interacting with the natural environment, so dramatic that it will be compared with the origins of agriculture and industry.

Third Rome A term used to describe Moscow's importance in Christian affairs (after Rome and Constantinople) in the eleventh century.

Thirteenth Amendment An amendment to the U.S. Constitution in 1865, ending the practice of slavery.

Tibeto-Burman subfamily A language subfamily spoken in Southeast Asia and western China.

tierra caliente A Latin American climatic zone reaching from sea level upward to approximately 3,000 feet (914 m). Crops such as rice, sugarcane, and cacao grow in this hot, wet environment.

tierra fría A Latin American climatic zone found between 6,000 feet (1,800 m) and 12,000 feet (3,600 m) above sea level. Frost occurs and the upper limit of agriculture and tree growth is reached.

tierra helada A Latin American climatic zone above 10,000 feet (3,048 m) that has little vegetation and frequent snow cover.

tierra templada A Latin American climatic zone extending from approximately 3,000 to 6,000 feet (914–1,829 m) above sea level. It is a prominent zone of European-induced settlement and commercial agriculture such as coffee growing.

Tiger Cubs See *New Asian Tigers*.

tipping point In climate change, a time beyond which some changes are irreversible, likely bringing rapid acceleration in temperature and its related effects.

Tornado Alley An area of the Midwestern United States where tornado activity is very common.

Torres Strait Islanders A distinct indigenous people of Australia with origins in Papua New Guinea.

tortilla wars Increases in the cost of corn in Mexico from U.S. imports that threatened to make it too expensive for the poor to buy. Also refers to the controversial appearance of genetically modified corn in Mexican fields at the same time.

Tory In the American Revolution, colonists who wished to remain politically connected with the British government.

Totonac language family A small language family of eastern Mexico.

trade barriers Restrictions on international trade, especially those created by MDCs to protect their industries from competition from cheaper goods produced in LDCs.

trade wars The use of barriers to restrict trade between countries to protect their interests and industries.

Trading with the Enemies Act A U.S. law forbidding Americans from spending more than $300 in Cuba.

Trans-Amazon Highway A highway linking the Atlantic coast of Brazil with the Pacific coast of Peru.

Trans-Pacific Partnership A free-trade initiative between the United States, Australia, New Zealand, and several countries in Asia and South America.

tree line The upper limit (or northernmost limit) of natural tree growth.

Trench Deep linear Earth feature (for example, the Mariana Trench off Japan).

triangular trade The sixteenth- to ninteenth-century trading links between West Africa, Europe, and the Americas, involving guns, alcohol, and manufactured goods from Europe to West Africa exchanged for slaves. Slaves brought to the Americas were exchanged for the gold, silver, tobacco, sugar, and rum carried back to Europe.

Tropic of Cancer A line of latitude located at 23.44°N. The general dividing line in the northern hemisphere between the tropics to the south and the mid-latitudes to the north.

Tropic of Capricorn A line of latitude located at 23.44°S. The general dividing line in the southern hemisphere between the tropics to the north and the mid-latitudes to the south.

tropical deciduous forest The vegetation typical of some tropical areas with a dry season. Here, broadleaf trees lose their leaves and are dormant during the dry season and then add foliage and resume their growth during the wet season.

tropical rain forest A low-latitude broadleaf evergreen forest found where heat and moisture are continuously, or almost continuously, available.

tropical savanna A relatively moist low-latitude climate that has a pronounced dry season.

The Troubles A period of strife in Northern Ireland marked by violence between Catholic Republicans and Protestant Unionists.

Truth and Reconciliation Commission A South African commission that allows those who were party to racial violence during the apartheid era to confess their misdeeds and, in a sense, be absolved of them.

trypanosomiasis A disease spread by the tsetse fly of Africa. Known as sleeping sickness in humans, and nagana in domestic cattle.

tsunami Large waves in the ocean created by undersea earthquakes.

Tuareg An ethnolinguistic group in northern Africa.

tundra A region with a long, cold winter, when moisture is unobtainable because it is frozen, and a very short, cool summer. Vegetation includes mosses, lichens, shrubs, dwarf trees, and some grass.

Turkic A subfamily of the Altaic language family, spoken mainly in western and central Asia.

Turks The predominant ethnic group of Turkey.

Tutsi (Watusi) See *Hutu*.

two-state solution A 1947 United Nations proposal to establish separate Jewish and Arab states in Palestine.

underemployment The underutilization of labor.

underground economy The "black market" typical of the former Soviet region and many LDCs.

undifferentiated highland A climate that varies with latitude, altitude, and exposure to the sun and moisture-bearing winds.

undocumented workers People who are in the United States illegally but have obtained temporary or long-term employment.

uniform region See *region*.

union state The integration of economic, political, and defense systems between two independent countries.

United Nations Resolutions 242 and 338 Resolutions passed by the United Nations after the Middle Eastern wars of 1967 and 1973, respectively, calling on Israel to withdraw from the Occupied Territories.

untouchables See *Dalits*.

upper limit of agriculture The highest elevation at which agricultural activities can be performed, at about 12,000 feet (3,600 m).

upstream countries Countries where a major river originates that have a large amount of control over that river's water, at the expense of countries farther down the river's course.

Uralic languages A language group that includes Finnish, Hungarian, and many Siberian languages.

urbanites In the Middle Eastern ecological trilogy, the educated people who provide manufactured goods and innovations, secular and religious educational instruction, and cultural amenities.

urbanization The population growth of cities, mostly from the movement of people from rural regions to built-up areas.

Ushahidi A nonprofit software firm and platform founded in Kenya. The software uses "crowdsourcing" for activist mapping of political, environmental, and other problems. It is a fine example of local people in an LDC developing a geospatial tool kit to address problems in development.

usufruct Nondestructive mutual use of resources by multiple clans or tribes located within the territory of another.

value-added manufacturing The process of refining and fabricating more valuable goods from raw or semi-processed materials.

value-added products Finished products, worth much more than the raw materials they are created from.

Varangians See *Rus*.

vernacular region See *region*.

Viet Cong An insurrectionist Communist force in South Vietnam, supported by the North Vietnamese government, that aided the North's conquest of the South and reunification of the country during the Vietnam War.

villagers In the Middle Eastern ecological trilogy, the peasant farmers that form the cornerstone of the trilogy.

virgin and idle lands (new lands) Steppe areas of Kazakhstan and Siberia brought into grain production in the 1950s.

virtual fence A high-tech surveillance system for parts of the U.S.-Mexico border.

volcanism Movement of molten material from the Earth's mantle, usually released through volcanoes.

wa A Japanese concept of harmony based upon the principle of economic equality.

walkabout A ritual journey for Australian Aborigines in which a boy on the verge of adulthood follows the pathway of his creator-ancestors.

Wallace's Line A marked biogeographic division located in Indonesia.

Walmart effect The loss of independent or small retail establishments in towns after being unable to compete with lower prices at Walmart and similar large stores.

War of 1812 A war between the United States and the United Kingdom, fought largely as a result of the United States attempting to annex Canada.

war on drugs A long-running U.S. policy of reducing drug usage within the United States, mostly by attempting to eradicate drugs at their sources, especially in South America.

War on Terror An open-ended, unconventional war declared by President George W. Bush in 2001 against the Islamist groups that carried out terrorist attacks against the United States.

Warsaw Pact A military alliance, now dissolved, consisting of the Soviet Union and the European countries of Poland, East Germany, Czechoslovakia, Hungary, Romania, and Bulgaria.

Washington Consensus A political-economic philosophy by which the United States would pressure democratic governments of Latin America to open their markets to international trade, reduce tarriffs, sell off state companies to private owners, and make other economic changes to bring prosperity to the region.

water hemisphere See *hemisphere*.

Water Towers of Asia The glaciers and snowpack of the Himalaya, Karakorum, and other great mountain ranges.

Way of the Law See *Footprints of the Ancestors*.

weapons of mass destruction (WMD) Weapons designed to kill large numbers of people by chemical, biological, or nuclear means.

weather The atmospheric conditions prevailing at one time and place.

welfare state A national government that provides generous social services to citizens, paid for by high taxation rates.

west longitude See *longitude*.

westerly winds An airstream located in the middle latitudes that blows from west to east. Also known as *westerlies*.

Western Big Development Project A Chinese project to improve infrastructure and the economic livelihoods of residents of China's autonomous regions.

Western Hemisphere See *hemisphere*.

Western Wall The most sacred site in the world accessible to Jews, the only remnant of the Second Temple.

wet rice cultivation Rice harvests producing two or three crops per year that can sustain large populations over long periods of time.

wheat war An economic dispute between the United States and Canada in the 1990s, as wheat yields were high and prices were very low.

White Australia policy An official Australian immigration policy until 1972 that excluded Asian and black immigrants.

white gold A term indicating how valuable cotton in Central Asia was to the Soviet economy.

Wik case A 1996 Australian high court ruling that Aborigines can claim title to public lands held by farmers and ranchers under long-term pastoral leases granted by state governments.

winter monsoon See "Monsoon."

Window on the West Russia's city of St. Petersburg, located on the Baltic Sea, giving Russia marine access to Europe.

world city A city that is a center of economic and political power in the global economy.

Xhosa One of the most populous ethnic groups of South Africa.

yang In *feng shui*, the masculine, bright, and positive.

yin In *feng shui*, the feminine, dark, and negative.

Yoruba A major ethnic group in southern Nigeria.

youth bulge A demographic situation in which a large percentage of a country's population is young.

zero population growth (ZPG) The condition of equal birth rates and death rates in a population.

zero-sum game A term referring to a gain by one party resulting in an equivalent loss to another.

Zion A synonym for Jerusalem.

Zionist movement

Zomia A term for the upland areas of Southeast Asia.

Zoroastrianism Religion practiced by the Parsis of India.

Zulu One of the most populous ethnic groups of South Africa.

Index

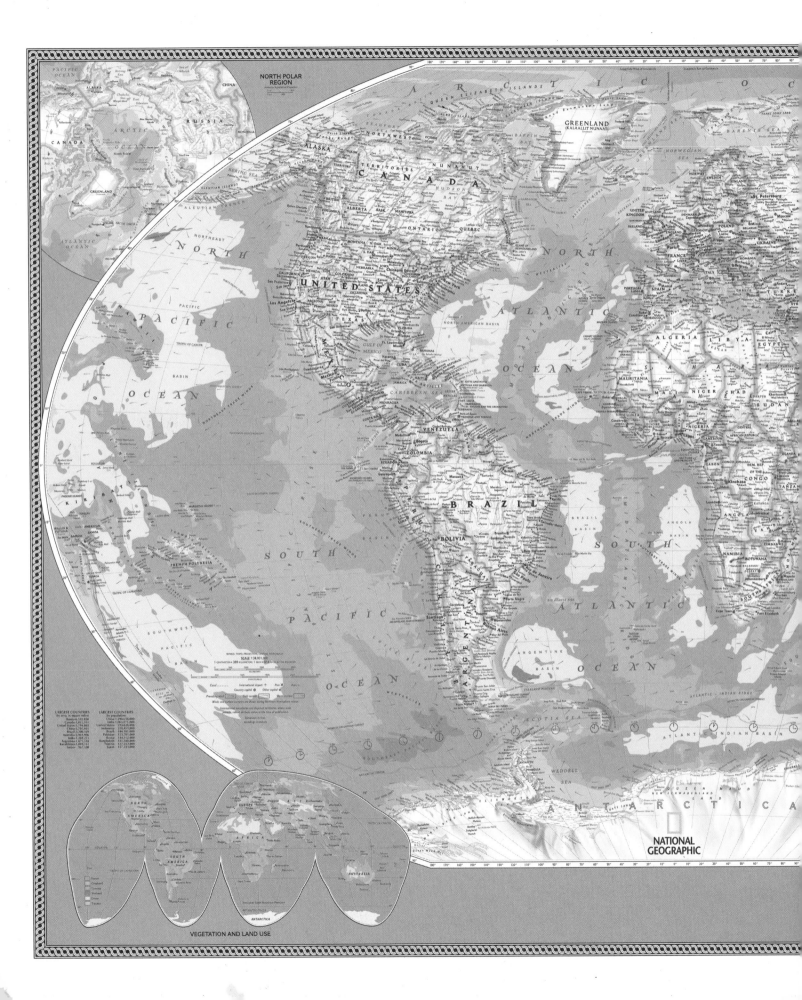

NATIONAL GEOGRAPHIC